D0151208

Modern Motorcycle Technology

Ed Abdo

Australia • Brazil • Japan • Korea • Mexico • Singapore • Spain • United Kingdom • United States

Modern Motorcycle Technology
Ed Abdo

Vice President, Career and Professional Editorial: Dave Garza

Director of Learning Solutions: Sandy Clark

Executive Editor: David Boelio

Managing Editor: Larry Main

Senior Product Manager: Matthew Thouin

Editorial Assistant: Lauren Stone

Vice President, Career and Professional Marketing: Jennifer McAvey

Marketing Director: Deborah S. Yarnell

Marketing Manager: Erin Coffin

Marketing Coordinator: Mark Pierro

Production Director: Carolyn Miller

Production Manager: Andrew Crouth

Content Project Manager: David Plagenza

Art Director: Benj Gleeksman

Technology Project Manager: Chrstopher Catalina

Production Technology Analyst: Thomas Stover

For product information and technology assistance, contact us at
Cengage Learning Customer & Sales Support, 1-800-354-9706

For permission to use material from this text or product, submit all requests online at **cengage.com/permissions** Further permissions questions can be emailed to **permissionrequest@cengage.com**

Library of Congress Control Number: 2008922079

ISBN-13: 978-1-4180-1264-9

ISBN-10: 1-4180-1264-5

Delmar Cengage Learning
5 Maxwell Drive
Clifton Park, NY 12065-2919
USA

Cengage Learning is a leading provider of customized learning solutions with office locations around the globe, including Singapore, the United Kingdom, Australia, Mexico, Brazil, and Japan. Locate your local office at:

international.cengage.com/region

Cengage Learning products are represented in Canada by Nelson Education, Ltd.

For your lifelong learning solutions, visit **delmar.cengage.com**

Visit our corporate website at **cengage.com**

NOTICE TO THE READER

Publisher does not warrant or guarantee any of the products described herein or perform any independent analysis in connection with any of the product information contained herein. Publisher does not assume, and expressly disclaims, any obligation to obtain and include information other than that provided to it by the manufacturer. The reader is expressly warned to consider and adopt all safety precautions that might be indicated by the activities herein and to avoid all potential hazards. By following the instructions contained herein, the reader willingly assumes all risks in connection with such instructions. The publisher makes no representation or warranties of any kind, including but not limited to, the warranties of fitness for particular purpose or merchantability, nor are any such representations implied with respect to the material set forth herein, and the publisher takes no responsibility with respect to such material. The publisher shall not be liable for any special, consequential, or exemplary damages resulting, in whole or part, from the readers' use of, or reliance upon, this material.

Printed in the United States of America
1 2 3 4 5 6 7 11 10 09 08

CONTENTS

CHAPTER 1
Introduction to Modern Motorcycle Technology 1
Learning Objectives 1
Key Terms 1
Introduction 2
A Brief History of the Motorcycle 2
Types of Motorcycles and ATVs 10
Summary 22
Chapter 1 Review Questions 22

CHAPTER 2
Safety First 23
Learning Objectives 23
Key Terms 23
Introduction 24
The Safety Attitude 24
Fire Safety 24
The Fire Triangle 25
The Four Fire Classes 25
Using a Fire Extinguisher 27
Hazardous Chemicals 29
Right-To-Know Laws 30
Electrical Safety 30
Exhaust Gas Safety 32
Safe Operation of Equipment 33
Good Housekeeping Practices 33
Handling Heavy Objects and Materials 34
Using Personal Protective Equipment (PPE) 35
Protecting Your Eyes and Face 36
Protecting Your Lungs 36
Protecting Your Hearing 37
Proper Dress Attire 38
Protecting Your Feet and Legs 38
Protecting Your Hands and Arms 38
Using Tools Safely 38
Safe Riding Practices 39
Chapter Summary 40
Chapter 2 Review Questions 40

CHAPTER 3
Tools 41
Learning Objectives 41
Key Terms 41
Introduction 42
Basic Hand Tools 42
Power Tools 56
Drill Press 57
Bench Grinder 57
Special Tools 59
Purchasing Tools 66
Storing Tools 66
Service Information Library 67
Summary 68
Chapter 3 Review Questions 69

CHAPTER 4
Measuring Systems, Fasteners, and Thread Repair 70
Learning Objectives 70
Key Terms 70
Introduction 71
Measurement Systems 71
Fasteners 72
Inspection, Cleaning, and Repair of Threaded Fasteners 79
Stresses on Threaded Fasteners 80
Tips for Working with Threaded Fasteners 81
Tightening and Torque 82
Repairing and Replacing Broken Fasteners 84
Summary 87
Chapter 4 Review Questions 87

CHAPTER 5
Basic Engine Operation and Configurations 88
Learning Objectives 88
Key Terms 88
Introduction 89
Basic Engine Operation 89
Engine Ratings 91
Basic Four-Stroke Engine Design 96
Basic Two-Stroke Engine Design 99
Engine Cooling 102
Engine Configurations 104
Summary 108
Chapter 5 Review Questions 108

CHAPTER 6
Internal-Combustion Engines 109
Learning Objectives 109

Key Terms 109
Introduction 110
General and Scientific Terms 110
Basic Internal-Combustion Engine Operation 112
Internal-Combustion Engine Operation 114
Basic Four-Stroke Engine Components 115
Four-Stroke Engine Theory of Operation 124
Two-Stroke Engines 127
Two-Stroke Engine Components 127
Two-Stroke Engine Theory of Operation 131
Two-Stroke Engine Induction Systems 135
Comparing Two-Stroke to Four-Stroke Engines 137
Summary 139
Chapter 6 Review Questions 140

CHAPTER 7
Lubrication and Cooling Systems 141
Learning Objectives 141
Key Terms 141
Introduction 142
Lubricants and Lubrication 143
Friction-Reducing Devices 147
Two-Stroke Engine Lubrication 150
Four-Stroke Engine Lubrication 154
Cooling Systems 159
Lubrication System Maintenance 163
Summary 165
Chapter 7 Review Questions 166

CHAPTER 8
Fuel Systems 167
Learning Objectives 167
Key Terms 167
Introduction 168
Fuel 168
Oxygen 169
The Carburetor 169
Fuel Delivery Systems 171
Carburetor Systems and Phases of Operation 176
Types of Carburetors 180
Multiple Carburetors 186
Fuel Injection 187
Summary 198
Chapter 8 Review Questions 198

CHAPTER 9
Drives, Clutches, and Transmissions 199
Learning Objectives 199
Key Terms 199
Introduction 200
Gears 200
Gear Ratios 202
Primary Drives 204
Clutch Systems 205
Transmissions 215
Starting Systems 222
Final Drive Systems 222
Summary 225
Chapter 9 Review Questions 225

CHAPTER 10
Two-Stroke Engine Top-End Inspection 226
Learning Objectives 226
Key Terms 226
Introduction 227
Diagnostics 227
General Tips before Beginning Engine Repairs 227
Repair Procedures 227
Two-Stroke Top-End Disassembly and Inspection 229
Two-Stroke Engine Top-End Inspection 230
Two-Stroke Power Valves 239
Starting the Rebuilt Engine 242
Summary 243
Chapter 10 Review Questions 243

CHAPTER 11
Two-Stroke Engine Lower-End Inspection 244
Learning Objectives 244
Introduction 244
Two-Stroke Engine Removal and Disassembly 246
Two-Stroke Engine Lower-End Inspection 246
Summary 256
Chapter 11 Review Questions 256

CHAPTER 12
Four-Stroke Engine Top-End Inspection 257
Learning Objectives 257
Key Terms 257
Introduction 258

Diagnostics 258
Repair Procedures 258
Four-Stroke Top-End Disassembly and Inspection 259
Four-Stroke Engine Top-End Inspection 261
Four-Stroke Top-End Reassembly 285
Summary 288
Chapter 12 Review Questions 288

CHAPTER 13
Four-Stroke Engine Lower-End Inspection 289
Learning Objectives 289
Key Terms 289
Introduction 290
Repair Procedures 290
Common Lower-End Engine Failures 291
Four-Stroke Engine Lower-End Disassembly 294
Four-Stroke Engine Lower-End Inspection 295
Four-Stroke Lower-End Reassembly 307
Summary 309
Chapter 13 Review Questions 310

CHAPTER 14
Electrical Fundamentals 311
Learning Objectives 311
Key Terms 311
Introduction 312
Safety Precautions with Electricity 313
Basic Principles of Electricity 314
Units of Electricity 320
Electrical Meters and Measurements 324
Magnetism 330
Electronic Devices 334
Electrical Schematics and Symbols 336
Electrical Terms 340
Summary 342
Chapter 10 Review Questions 343

CHAPTER 15
Motorcycle Charging Systems and DC Circuits 345
Learning Objectives 345
Key Terms 345
Introduction 346
Charging Systems 346
Charging System Operation 353
Types of Charging Systems 355
Charging System Inspection 358
DC Electrical Circuits 366
Summary 369
Chapter 15 Review Questions 369

CHAPTER 16
Ignition and Electric Starting Systems 371
Learning Objectives 371
Key Terms 371
Introduction 372
Motorcycle Ignition Systems 372
Basic Ignition System Components 375
Types of Ignition Systems 382
Electric Starter Systems 388
Summary 392
Chapter 16 Review Questions 393

CHAPTER 17
Frames and Suspension 394
Learning Objectives 394
Key Terms 394
Introduction 395
Motorcycle Frames 395
Motorcycle Suspension Systems 401
Rear-Damper Design and Theory of Operation 408
Summary 415
Chapter 17 Review Questions 417

CHAPTER 18
Brakes, Wheels, and Tires 418
Learning Objectives 418
Key Terms 418
Introduction 419
Braking Systems 419
Wheels 430
Motorcycle Tires 434
Summary 437
Chapter 18 Review 438

CHAPTER 19
Motorcycle Maintenance and Emission Controls 439
Learning Objectives 439
Key Terms 439
Introduction 440
Maintenance Intervals 440
Motorcycle Engine Maintenance 443

Motorcycle Chassis Maintenance 461
Motorcycle Storage Procedures 466
Emission Controls, Operation, and Maintenance 467
Summary 470
Chapter 19 Review Questions 471

CHAPTER 20
Motorcycle Troubleshooting 473
Learning Objectives 473
Key Terms 473
Introduction 474
Systematic Approaches to Solving Problems 474
Types of Problems 475
Troubleshooting Engine Problems 476
Fuel System Troubleshooting 480
Related Problems for other Fuel Systems 483
Electrical Problem Troubleshooting 486
Chassis Problem Troubleshooting 491
Abnormal Noise Troubleshooting 494
Summary 496
Chapter 20 Review Questions 496

Glossary 497

Index 511

PREFACE

Modern Motorcycle Technology (MMT) is designed to meet the basic needs of students interested in the subject of motorcycle and all-terrain vehicle (ATV) repair by helping instructors present information that will aid in students' learning experience. The subject matter is intended to help students become more qualified candidates for dealers looking for well-prepared, entry-level technicians.

MMT has been written to make learning enjoyable; the easy-to-read-and-understand chapters and great number of illustrations will assist visual learners with content comprehension. The book comprises twenty chapters starting with the history of the motorcycle and ending with a chapter on troubleshooting various conditions found on any motorcycle. Due to their similarity in usage of technologies, the servicing of ATVs is automatically included in the text.

MMT can be used not only for pre-entry-level technicians, but also as a reference manual for practicing technicians. Motorcycle technicians are currently sought after and will continue to be in demand in the future as technology advances in the manufacturing of modern motorcycle and ATV products. In today's world, an education prior to working in the field is becoming more desirable by hiring dealerships.

I have been in the motorcycle industry on "all sides of the fence." I have been a rider and a racer most of my life, as well as a customer, technician, service manager, motorcycle trade school technical instructor, chief instructor, curriculum developer, manufacturer's service representative, and head of a technical training department tasked to ensure that over 4,000 technicians are up to date with technologies and technical instruction. I have had a passion for motorcycles since my first minibike back in the 1960s, and I have a unique outlook on my job. I love doing what I do! This is something that everyone should strive for, as there is nothing more rewarding.

Ed Abdo

ACKNOWLEDGEMENTS

There were many people that were instrumental in making this book a reality. There were numerous reviews from many in the motorcycle industry; their suggestions were excellent and helped to make this a better textbook. There are also people that I would like to acknowledge who helped make this book possible: Patty Shogren and Nina Hnatov, who worked closely with me editing and proofreading the following pages. Their expertise and assistance have been greatly appreciated. The late George Decker, who many years ago saw something in me that made him think that I could be a technician, hired me for my first job in this industry, and became my mentor. Next, my two children, Anthony and Nick; for their endless love, even though I missed many of their school and after-school sporting events during the writing of this book; and my wife, Carin. It was she who convinced me to take on such a large project and give back to an industry that has done so much for me. Without her support, you would not be reading this.

REVIEWERS

Larry Barrington
Universal Technical Institute
Phoenix, AZ

Michael Baugus
Central Tech
Drumright, OK

Dan Clark
Fullerton, CA

Richard Deuschle
Motorcycle Mechanics Institute
Phoenix, AZ

Roy King
Centennial Colllege
Toronto, Canada

Anthony V. Lambiase
Universal Technical Institute
Phoenix, AZ

Kara Moon
Universal Technical Institute
Phoenix, AZ

Robert Monroig
Lake Washington Technical College
Woodinville, WA

Terry A. Muncy, Jr.
Motorcycle Mechanics Institute
Orlando, FL

John Pfingstag
Universal Technical Institute
Phoenix, AZ

Michael Ross
Phoenix, AZ

Michael Sachs
DeKalb Technical College
Clarkston, GA

CHAPTER 1

Introduction to Modern Motorcycle Technology

Learning Objectives

When students have completed the study of this chapter and its laboratory activities they should be able to:

- Understand a brief history of the motorcycle and the motorcycle industry
- Describe different types of motorcycles
- List some of the many motorcycle industry job opportunities

Key Terms

Advertising and marketing specialists
All-terrain vehicles (ATV)
Custom cruiser
Customer service representatives
Direct drive system
District parts managers
District sales managers
District service managers
Dual-purpose motorcycles
Entry-level motorcycle technicians
Franchised dealerships
General manager
Hot-rod cruisers
Lot attendant
Moto-cross
Motorcycle
Motorcycle repair instructors
Motorcycle technician
Motor scooters
Off-road motorcycles
Parts department
Parts technician
Quality control specialists
Race team support technicians
Research and development engineers
Road racing
Sales department
Service department
Service manager
Service technical training instructors
Service writer
Set-up technician
Sport motorcycles
Sport-touring motorcycles
Standard street motorcycles
Street motorcycles
Technical advisors
Technical illustrators
Technical writer
Three-wheelers
Touring motorcycles
Universal Japanese Motorcycles (UJM)

INTRODUCTION

Motorcycles have a long history that dates over one hundred years and the motorcycle industry has continued to grow at a tremendous rate with yearly sales of new motorcycles reaching over 1 million dollars in 2005 and All Terrain Vehicle (ATVs) sales approaching $800,000. Servicing these machines is different in many ways than in the past. Over the years, constant breakthroughs in engine, chassis, and electronics technology have greatly changed how motorcycles and ATVs are developed and marketed to the consumer. Because of this, trained motorcycle technicians are in high demand throughout the country.

A BRIEF HISTORY OF THE MOTORCYCLE

According to *Merriam-Webster's Collegiate Dictionary*, the word **motorcycle** is defined as "A 2-wheeled automotive vehicle having one or two saddles." Of course, motorcycles are a bit more complex than that simple explanation.

In fact, motorcycles are a direct descendent of the bicycle. The early bicycle (actually called a velocipede, a bicycle with its pedals located on the front wheel) (Figure 1-1) appeared in the early 1860s and was also known as a "*boneshaker*," both for its jarring ride, and its tendency to toss the riders when riding on cobblestone roadways.

Figure 1-1 The motorcycle is a direct descendent of the early bicycle, which was actually called a velocipede because of the placement of the pedals on the front wheel. © National Museum of American History, Smithsonian Institution.

Since its invention, the motorcycle has been considered to be more than just riding a bike with an engine attached. Riding a motorcycle gives people a completely different perspective of the world around them as compared to driving in an automobile. Riding on the open road or along a trail is a feeling like no other. Furthermore, it does not matter what the motorcycle type or brand, all riders share a common bond between them. They are virtually all motorcycle enthusiasts and truly love to ride.

It could be argued that writing a book about the history of the motorcycle could easily take up more space than this entire textbook provides. Throughout the years, there have been literally thousands of different motorcycle manufacturers around the world. Therefore, we will briefly discuss only a few of the many highlights of the motorcycle's vast history.

The Birth of the Motorcycle

There are many different opinions pertaining to who exactly invented the first motorcycle but using the definition by *Merriam-Webster*, it could be considered a steam-powered machine built by Sylvester H. Roper of Roxbury, Massachusetts, in 1868 (Figure 1-2) as the *first* motorcycle.

Figure 1-2 This could be considered the very first motorcycle as it was powered by the use of a steam engine and was created by Sylvester H. Roper of Roxbury, Massachusetts, in 1868. © National Museum of American History, Smithsonian Institution.

Roper's machine was considered to be quite remarkable as it looked very similar to the bicycle of that era but utilized a small vertical steam boiler under the seat. The boiler supplied two pistons that powered a crank drive system to the rear wheel. The throttle was controlled by twisting the handlebar forward and back. The Roper machine had the first known—albeit a primitive version—use of the twist grip control. The twist grip throttle control was reinvented a couple of times over the years, finally by the Indian Motorcycle Company—after this Roper design—and is still in use on today's motorcycles. Roper built more versions of his steam-powered motorcycle and in 1896, at the age of seventy-three, he showed up at a bicycle track near Harvard with a modified version of one of his designs. He was clocked at an unbelievable forty miles per hour and while slowing down the bike went into a wobble throwing Roper off the bike. Sadly, he died in this accident. Later, however, an autopsy is reported to have shown that Roper died of a heart attack and did not die from the fall.

Figure 1-3 This is a replica of what many historians consider to be the first motorcycle due to the fact that an internal combustion engine powered it. A gentleman by the name of Gottlieb Daimler created it in 1885. Copyright © 2000, 2001, 2002. Free Software Foundation, Inc. 51 Franklin Street, Fifth Floor, Boston, MA 02110-1301 USA.

Even though the Roper machine was designed years before, most historians credit Gottlieb Daimler with the invention of the motorcycle in 1885, as it was the first motorcycle in recorded history with an engine using petroleum to power its engine (Figure 1-3). Daimler designed an engine and mounted it into a wooden framed contraption in 1885.

As mentioned, it is considered by most to be the first motorcycle even though it actually had four wheels. Historians overlook the two outrigger type stabilizer wheels and consider this machine to be the grandfather to the motorcycle. Daimler's young son Paul was the first to give this machine a test ride. Daimler's machine had no pedals. Instead, the power was only supplied by the simple four-stroke engine design. Daimler later went on to build early automobiles. He left it to bicycle builders to further develop the motorcycle.

In 1892, Alex Millet invented a five-cylinder motorcycle and was the first to utilize pneumatic tires (Figure 1-4). The Millet-designed machine used a complex rotary engine built within the rear wheel. The cylinders rotated with the rear wheel while the crankshaft was actually incorporated into the rear axle.

Although short-lived due to poor design, the first motorcycle built for sale (over 200 were sold) was the Hildebrand & Wolf Mueller (Figure 1-5) in Munich in 1894. This motorcycle utilized a water-cooled twin cylinder engine that had a **direct drive system**, meaning that the wheels were directly attached to the engine, and therefore, would always be in motion if the engine were running. This made riding this motorcycle design difficult.

In 1895, the French firm of DeDion-Buton designed an engine that would allow motorcycle

Figure 1-4 In 1892, Alex Millet of France built the first motorcycle with pneumatic tires. The engine was actually a part of the rear wheel and had five cylinders. Copyright © 2000, 2001, 2002. Free Software Foundation, Inc. 51 Franklin Street, Fifth Floor, Boston, MA 02110-1301 USA.

Figure 1-5 The German company of Hildebrand & Wolf Mueller designed the first motorcycle that was produced for sale to the public. Design issues made this machine very hard to ride and therefore it had a very short life. Copyright © 2000, 2001, 2002. Free Software Foundation, Inc. 51 Franklin Street, Fifth Floor, Boston, MA 02110-1301 USA.

mass production to become a reality. The DeDion-Buton engine (Figure 1-6) design was a small high revving four-stroke single using the first battery and coil type ignition on such a small engine.

The engine was lubricated using a total loss system that dripped oil into the crankcase via a metering valve, which was then sloshed around the internals to lubricate the moving components before burning it or pumping it out onto the ground through a breather tube. While many of these engines were used, the engine design was copied by two very notable manufacturers in the United States: Indian and Harley-Davidson.

Figure 1-6 The French company of DeDion-Buton built the first internal combustion engine that was mass-produced for usage specifically in a motorcycle.

In 1900, two men, George Hendee and Carl Hedstrom, formed a partnership to manufacture a "motor-driven bicycle for the everyday use of the general public" in Springfield, Massachusetts, the Hendee Manufacturing Company. Hedstrom designed a motorcycle that debuted in 1901. A single cylinder 1.75 hp engine that was copied from the DeDion-Buton engine gave the machine the ability to travel at speeds close to twenty-five miles per hour (Figure 1-7). The real secret to this design however was its chain drive, which was superior to the belt-driven machines around at that time.

The partners picked the name "Indian" for their motorcycles; thus starting what was to become the largest production motorcycle company in the United States, building a V-Twin motorcycle with two- and three-speed gear boxes and further refined with a swing arm rear suspension. In 1914, the world's first motorcycle with electric start and a full electrical system was introduced to the industry (Figure1-8). The Hendee Special propelled Indian to be the largest motorcycle manufacturer in the world, producing over 20,000 bikes per year prior to World War I. While there have been a few attempts to revive the motorcycle brand Indian, the last true Indian motorcycle was the 1953 Indian Chief (Figure 1-9).

In 1903, 21-year-old William S. Harley and 20-year-old Arthur Davidson created for the public,

Figure 1-7 The first Indian motorcycle was produced in 1901 and was very similar in design to a bicycle. © National Museum of American History, Smithsonian Institution.

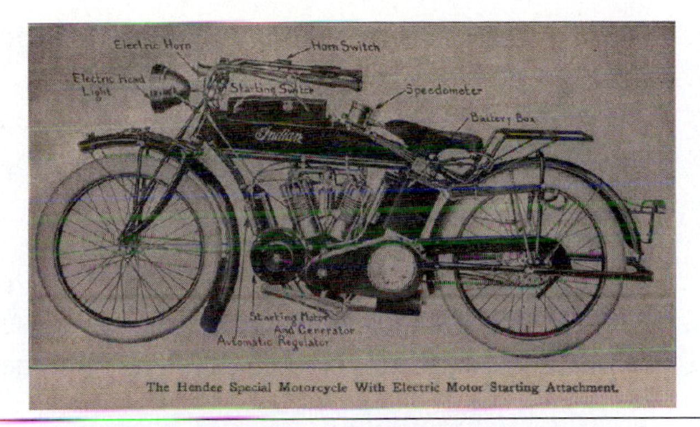

Figure 1-8 The first motorcycle that came with an electric start was an Indian Hendee Special produced in 1914.

Figure 1-9 Although many have tried to bring the name back to life over the years, the 1953 Indian Chief is arguably the last *true* Indian motorcycle.

the first production Harley-Davidson motorcycle (Figure 1-10). The factory in which they worked was a 10 x 15-foot wooden shed with the words "Harley-Davidson Motor Company" crudely scrawled on the door (Figure 1-11).

Figure 1-10 The first Harley-Davidson was produced in 1903. Courtesy of Harley-Davidson. Motor Company Archives. Copyright H-D.

Figure 1-11 An artist's rendition of the very first Harley-Davidson factory. The building was in fact an 10 x 15 foot woodshed. Courtesy of Harley-Davidson. Motor Company Archives. Copyright H-D.

Some of the earliest Harley-Davidson Motorcycles were built with racing in mind. In 1908 Walter Davidson scored a perfect 1,000 points at the Federation of American Motorcyclists (FAM) 7th Annual Endurance and Reliability event (Figure 1-12). Then, only three days later, he set the FAM economy record at over 188 miles per gallon.

The first V-Twin built by Harley-Davidson was in 1909 and had a displacement of 49.5 cubic inches and boasted seven horsepower (Figure 1-13). It was not until 1914 that the company formally entered

Figure 1-12 Family member Walter Davidson rode this machine to a perfect 1,000-point score in the Federation of American Motorcyclists Endurance and Reliability event and followed up three days later by setting an economy record of over 188 miles per gallon. Courtesy of Harley-Davidson. Motor Company Archives. Copyright H-D.

Figure 1-13 The first V-Twin built by Harley-Davidson had seven horsepower and was built in 1909. Courtesy of Harley-Davidson. Motor Company Archives. Copyright H-D.

Figure 1-14 Harley-Davidson celebrated one hundred years of manufacturing motorcycles in 2003.

into the motorcycle racing scene, but team Harley-Davidson was nicknamed the "Wrecking Crew" because of their dominance in the sport. During World War I, nearly half of all Harley-Davidsons built were used by the U.S. Military and by the end of the war it was estimated that the U.S. Army used approximately 20,000 motorcycles with the majority being Harley-Davidsons. By the year 1920, Harley-Davidson became the largest motorcycle manufacturer in the entire world with over 2,000 dealers located in sixty-seven countries. This dominance lasted until the 1950s when the British motorcycle industry came to full bloom. Today, the Harley-Davidson brand is an American icon and their products appeal to many different types of riders. The only "true" American motorcycle manufacturer still in existence from the early days, the Harley-Davidson Motor Company celebrated its centennial in 2003 (Figure 1-14).

Figure 1-15 Triumph was a strong contender in the 1950s with motorcycles like this Thunderbird.

British Motorcycles

While British motorcycle production began in the early 1900s it was the 1950s that brought motorcycles such as Triumph (Figure 1-15) and BSA (Figure 1-16) to the forefront of the U.S. motorcycle industry. Britain's motorcycle industry dominance was at its peak in 1959. Happy with their success, these companies felt somewhat invincible and failed to take note of emerging trends or replace their aging designs. Most of the

Figure 1-16 This racing version of a BSA helped the British manufacturer to create a strong consumer base with its power and looks.

engineers and company executives came from pre-war days and paid attention to only the glory days of the 1930s but unfortunately failed to look into the future.

European competition from Germany included the company BMW (Figure 1-17) who made a rapid recovery from post-war times and the Italian companies Ducati (Figure 1-18) and MV Agusta (Figure 1-19). They began to intrude on market share with their desire to build more stylish machines. The final blow to the British motorcycle industry came from the Japanese as the U.S. and European markets began to import less expensive and more reliable machines. The Japanese motorcycles showed more innovation and engineering development and the British companies were too slow to react to this competition.

Figure 1-17 German manufacturer BMW came back after World War II quickly to build very stylish (for the time) and reliable machines. Courtesy of BMW of North America.

Figure 1-18 The Italian firm of Ducati has always been regarded to have motorcycles that could be considered just as much art as functional. Machines such as this helped to capsize the British motorcycle dominance of the 1960s.

Figure 1-19 MV Agusta was another strong manufacturer that built motorcycles that the public wanted and therefore helped to reduce the British motorcycle dominance. Courtesy of Motorcycle Hall of Fame Museum Collection.

Japanese Motorcycles

The motorcycle industry throughout the world saw its biggest change in the early 1960s when a company from Japan, headed by Sochirio Honda, changed the way people looked at motorcycling. The Honda C100 Cub (Figure 1-20) was by far the most successful entry-level motorcycle and utilized the ever-popular slogan, "You meet the nicest people on a Honda" (Figure 1-21). The Cub alone

Figure 1-20 The Honda Cub was first built in the late 1950s but dominated the market in the 1960s. The 500-millionth (500,000,000) Cub was built in late 2005 making it the number one single motorcycle model ever produced by any motorcycle manufacturer in the world. Copyright by American Honda Motor Co., Inc. and reprinted with permission.

Figure 1-21 The slogan "You meet the nicest people on a Honda" helped to turn the motorcycle industry away from the "outlaw" reputation that it once was. Copyright by American Honda Motor Co., Inc. and reprinted with permission.

helped to attract annual Honda sales of over half a million dollars and more for years to come. The 500-millionth (500,000,000) Cub was built in late 2005 making it the number one single motorcycle model ever produced by any motorcycle manufacturer in the world.

In 1969, Honda introduced its CB750 (Figure 1-22), which not only boasted a four-cylinder overhead camshaft design that could reach speeds of 120 miles per hour but also included electric start, front disc brakes and a level of sophistication that was unheard of at the time. Honda also invented the All Terrain Vehicle (ATV) with its ATC90 in 1970 (Figure 1-23).

With its advanced technology and willingness to consistently strive to improve, Japan had shown to the world that it was a power to be noticed in the motorcycle industry and Japanese manufacturers Yamaha (Figure 1-24), Suzuki (Figure 1-25), and Kawasaki (Figure 1-26) soon joined Honda to dominate the motorcycle industry to this day.

Figure 1-22 The 1969 CB750 Honda brought a level of sophistication to the motorcycle industry that started the modern day motorcycle. Copyright by American Honda Motor Co., Inc. and reprinted with permission.

Figure 1-23 Honda created the ATC90 in 1970, which started the ATV revolution that has led to yearly sales well over $700,000. Copyright by American Honda Motor Co., Inc. and reprinted with permission.

Figure 1-24 This 1967 250 Yamaha Road Racer was a winner of the Daytona One Hundred Mile Lightweight class. Courtesy of Motorcycle Hall of Fame Museum Collection.

Figure 1-25 The Suzuki 250 Hustler was a big hit in the United States when it was imported in the 1960s. The Hustler had a six-speed transmission and was very fast in its day.

Figure 1-26 The Kawasaki Mach III was a 500cc triple-cylinder motorcycle that defined power when it was introduced in 1969.

TYPES OF MOTORCYCLES AND ATVS

Motorcycles are available in a wide variety of models and sizes. The model that a consumer purchases depends primarily on the intended use and personal choice of the individual rider. However, all motorcycles have basic similarities (Figure 1-27). They all contain the following common components:

- Engine
- Electrical system
- Fuel system
- Tires and wheels
- Drive train
- Handlebars with controls
- Brakes
- Seat
- Frame
- Suspension systems

These components form systems that must work together to provide a dependable and safe means of transportation. Each system performs a specific function. For example, the fuel system supplies the correct ratio of air-to-fuel mixture for proper engine operation. The ignition system provides timed electrical sparks to ignite the air-and-fuel mixture in the engine's internal combustion chamber(s). The drive train mechanism transfers the energy from the engine to the drive wheel. The brakes stop the motorcycle quickly and safely. Finally, the frame and suspension system absorb vibration and shock to give the rider a smooth, comfortable ride over varying road surfaces.

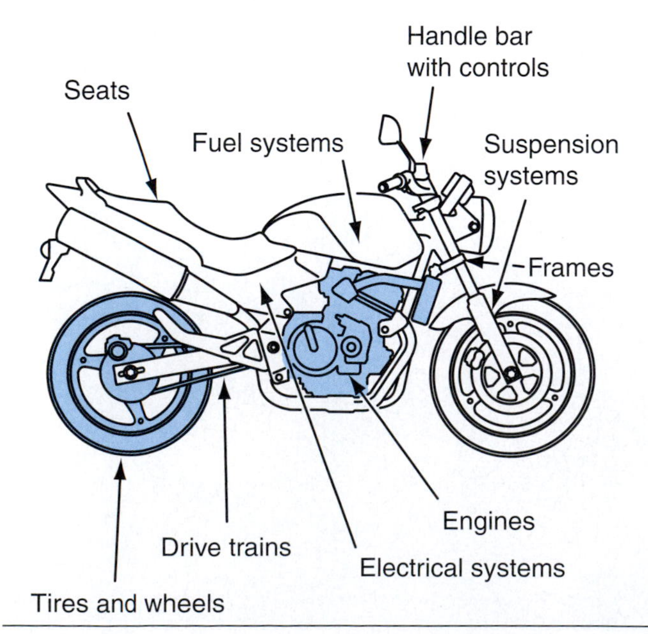

Figure 1-27 All motorcycles have basic similarities and contain common components.

Because of the numerous applications and uses, motorcycles are divided into categories by use. Each category is then subdivided by engine displacement.

Engine displacement is a size measurement given in cubic centimeters (cc). For example, a 1,000cc engine has 1,000 cubic centimeters of volume. If it is a four-cylinder engine, then each cylinder has 250cc of volume. If it's a two-cylinder engine, then each cylinder has 500cc of volume. We'll cover exactly how this measurement is calculated in another chapter.

Motorcycles are classified as:

- Street
- Dual-purpose
- Off-road

All Terrain Vehicles (better known as ATVs) are classified as:

- Three-wheelers
- Four-wheelers

Let's look at the different classifications of motorcycles and ATVs.

Street Motorcycles

Street motorcycles are available in a wide variety of designs and sizes. As the name implies, street motorcycles are designed for use on paved roadways. The **standard street motorcycle** has an upright riding position (Figure 1-28).

Japanese manufacturers initially designed today's standard street motorcycles in the late 1960s and early 1970s. They picked up the designation of **Universal Japanese Motorcycle (UJM)** after millions were built and sold. Although the current crop of standard street motorcycles is derived from the UJM version, early models of a very similar style were available from American

Figure 1-28 The standard style motorcycle has an upright riding position and is very popular. Copyright by American Honda Motor Co., Inc. and reprinted with permission.

manufacturers long before the UJM models were first imported here. These early American versions of the street motorcycles were customized by their owners into a unique class of vehicles known as custom cruisers. The standard street motorcycle is still very popular. It's available in a variety of different engine and chassis sizes to fit almost anyone's needs. The standard street motorcycle is often thought of as entry level in motorcycling circles, although many experienced riders purchase them for their daily use.

A variation of the standard street motorcycle is the **custom cruiser** (Figure 1-29). The custom cruiser design began as a set of modifications to production models of the standard street motorcycle. Custom cruisers started to evolve when riders began to customize their standard street motorcycles. It was this customizing, which began in the United States, that focused on modifying handlebar and seat design. The motorcycle manufacturers recognized the growing popularity of the customized models and started building factory versions. These manufactured custom cruisers have grown into one of the largest and most popular sales offerings of the motorcycle industry. A major difference between standard street models and custom cruisers is the riding position.

Standard street models are ridden in the upright position, and custom cruisers have a laidback riding position. Custom cruisers come in a variety of sizes and feature many different looks. **Hot-rod cruisers** are another spin-off of the standard street motorcycle (Figure 1-30), and are similar to custom cruisers. The riding position of hot-rod cruisers is slightly forward-leaning.

Figure 1-29 Custom cruisers are very popular and were designed originally by individuals based on the standard style motorcycle. Courtesy of Yamaha Motor Corporation, USA.

Figure 1-30 Hot-rod cruisers are more powerful versions of the custom cruiser and also have a more forward riding position. Courtesy of Yamaha Motor Corporation, USA.

Hot-rod cruisers are equipped with large, powerful engines. Manufacturers have redesigned the chassis and suspension systems on the hot-rod cruisers to accommodate the larger engines and the increased power common with this class of motorcycle.

From a performance standpoint, **sport motorcycles** (Figure 1-31) are by far the fastest, best-stopping, and best-handling members of the street-type motorcycles available to the public. They're available in a wide variety of sizes and power options. Many sport motorcycles are used in **road racing** (Figure 1-32) where racing is done on closed-course, street circuits.

As the name suggests, **touring motorcycles** are large, well-equipped units designed for long-distance touring travel (Figure 1-33). Many are equipped with standard and optional features such as Satellite, AM/FM and CB radios, CD players, and even cruise control. Most touring motorcycles come equipped with large fairing units (windshields and wraparound side panels) to protect riders from the elements. They're also equipped with large, weatherproof saddlebags to carry and protect luggage.

Sport-touring motorcycles (Figure 1-34) are designed for touring without some of the frills of the pure touring version. Options such as radios

Figure 1-31 If it's speed you're after then the Sport motorcycle may be just what you're looking for, as they are the fastest and best-handling, two-wheeled machines on the race track. Courtesy of Suzuki.

Figure 1-33 Touring motorcycles are chock-full of luxury items like radios, CD players, and even heated seats and hand grips. This Honda Gold Wing even comes with a GPS navigation system. Copyright by American Honda Motor Co., Inc. and reprinted with permission.

Figure 1-32 Sport motorcycles are often used in high-speed, closed-course road racing as seen here. Copyright by American Honda Motor Co., Inc. and reprinted with permission.

Figure 1-34 Sport-touring motorcycles are a combination sport bike and touring machine. They combine comfort with long-distance capabilities that allow two-up riding all day long while handling well at higher speeds when it is desired to do so. Courtesy of Yamaha Motor Corporation, USA.

and cruise control are not part of the sport-touring motorcycle design. Sport-touring motorcycles have high-performance engines, handle very well at high speeds, and provide a comfortable riding position (which is a must for long trips).

Motor scooters are another member of the street category of motorcycles. They're offered with engine displacements from 50cc (Figure 1-35) to 650cc (Figure 1-36). Because of their relative small design, they're very economical to operate and easy to park or store. The characteristics of the typical motor scooter make them excellent for daily, short-distance urban or suburban commuting. They can be equipped with saddlebags to provide limited space for carrying small parcels, books, and briefcases while some offer under-seat storage as well.

Figure 1-35 A typical small, 50cc scooter is shown here. Copyright by American Honda Motor Co., Inc. and reprinted with permission.

Figure 1-36 Scooters come in many sizes up to this 650cc model that is more than capable of riding on any highway, not to mention out performing many motorcycles as well. Courtesy of American Suzuki Motor Corp.

Dual-Purpose Motorcycles

Dual-purpose motorcycles are exactly what the name suggests (Figure 1-37). They've been designed and manufactured to perform well both on and off paved highways. Dual-purpose motorcycles share mechanical characteristics with both street motorcycles and off-road motorcycles. These motorcycles are great for riders who might desire to take the conventional route to work or school and then choose to take the road less traveled on the return trip, using trails or fire roads instead of paved highways. Dual-purpose motorcycles are equipped with more shock-absorbing suspension systems than standard street motorcycles to handle off-road conditions.

They also have special tires for improved traction when riding on the various types of road (pavement and dirt) surfaces. Dual-purpose motorcycles are becoming more popular with individuals who appreciate their versatility. Because of this growing popularity, an increasing number of specialized dual-purpose riding events, meets, and contests are held each year. These events are designed to test the versatility of the dual-purpose motorcycles and the skills and handling techniques of the riders.

Off-Road Motorcycles

As the name implies, this category of motorcycles is specifically designed for off-road use.

Figure 1-37 Dual-purpose motorcycles perform well on both paved highways and dirt roads. These machines are often ridden on trails as well. Courtesy of Kawasaki Motors Corp., U.S.A.

Off-road motorcycles (Figure 1-38) are built to handle rough terrain. The frames, suspension systems, and wheels are considerably stronger than those found on standard street motorcycles. Off-road motorcycles are often used in extreme terrains such as deserts and mountain trails. Another use for off-road motorcycles that you may be familiar with is **moto-cross** (Figure 1-39). Moto-cross is defined as a motorcycle race on a tight, closed course over terrain that includes sharp turns, hills, mud, and water.

Because of their special use, off-road motorcycles have little need for lights; consequently, the majority of models, if they have them at all, are equipped with small headlights and taillights. Certain special events require lights, but this is the exception rather than the rule. Off-road motorcycles are available in a wide variety of sizes from small to large and powerful. They're offered with both two-stroke engines and four-stroke engines. The engine sizes vary from 50cc (Figure 1-40) to over 600cc (Figure 1-41).

All-Terrain Vehicles (ATVs)

All-terrain vehicles (better known as **ATVs**) are a separate branch of the motorcycle family tree. ATVs are available in a variety of sizes and styles. They can be equipped with either two-stroke or four-stroke engines ranging from 50cc to over 700cc.

The following are two subclasses of ATVs:

- Three-wheel tricycle-style machines commonly referred to as *three-wheelers*.
- Four-wheel models commonly referred to as *four-wheelers*.

Figure 1-38 This is an example of a typical off-road motorcycle. Most off-road bikes today come with four-stroke engines. Courtesy of Yamaha Motor Corporation, USA.

Figure 1-40 This photo shows an example of a 50cc small, off-road motorcycle that has a two-stroke engine to power it. Courtesy of Yamaha Motor Corporation, USA.

Figure 1-39 A typical moto-cross motorcycle is shown here. Copyright by American Honda Motor Co., Inc. and reprinted with permission.

Figure 1-41 This is an example of a large 650cc four-stroke, off-road motorcycle. Copyright by American Honda Motor Co., Inc. and reprinted with permission.

Four-wheelers are further divided into two-wheel drive and four-wheel drive models. ATVs are versatile. They can be used for pleasure, such as off-road and backcountry touring/camping (Figure 1-42), trail riding or for racing on closed-course tracks (Figure 1-43), or for work and utility purposes (Figure 1-44). There are design differences to support the different uses. An ATV model designed for pleasure is not well suited for either racing or work. Similarly, a racing model is not designed for work.

Figure 1-42 This is an example of a recreational ATV that can be used for multiple purposes such as off-road and backcountry touring and camping. Copyright by American Honda Motor Co., Inc. and reprinted with permission.

Figure 1-43 These examples of sport and race ATVs are designed to be ridden on closed-course tracks and trails. Note the distance between the wheels and fenders for suspension clearance. Copyright by American Honda Motor Co., Inc. and reprinted with permission.

Figure 1-44 This ATV is used for utility purposes such as hunting and fishing. Note the camouflage color scheme. Courtesy of Yamaha Motor Corporation, USA.

As mentioned earlier, ATVs have been produced as both three-wheelers and four-wheelers. Although thousands are still in use, **three-wheelers** (Figure 1-45) are no longer in production. Some people feel that they have a tendency to be unstable, especially on rough and uneven terrain. For safety reasons, they've been discontinued. Owners of three-wheelers have been alerted to use caution when operating this type of ATV.

Motorcycle Industry Opportunities

The final section of this chapter introduces you to the opportunities available in the motorcycle industry. Upon completing their training, many

Figure 1-45 Three-wheeler ATVs are no longer in production by any major manufacturer but there are still thousands in use.

students obtain their first job as **entry-level motorcycle technicians** at **franchised dealerships**. A franchised dealership is authorized to sell a particular motorcycle company's products and services in a particular area. There are many positions available at a motorcycle dealership for individuals with a motorcycle repair background. Even if you're not interested in a motorcycle career at this time, this section will give you an idea of what the motorcycle industry is all about. To obtain a better understanding of the positions available at a motorcycle dealership, we will start by taking a closer look at the dealership.

Dealership Opportunities

A motorcycle dealership (Figure 1-46) is an excellent place to begin a motorcycle or ATV repair career. Often, prospective employees must be willing to start at an entry-level position and work their way up the ladder.

Most franchised and independent dealerships have the following three main departments:

- Sales department
- Parts department
- Service department

Before we discuss the service department, let's first take a look at the other two departments. It's possible that your entry job may be in the sales or parts departments. You can gain valuable experience in these departments as well. The ability to get along with people is a key requirement for working in any area of a motorcycle dealership. This is particularly true in the **sales department** (Figure 1-47) where products are displayed and sold. As a skilled salesperson, you must also be able to discuss the technical features of the different motorcycle models with customers. An education in motorcycle repair provides you with a definite advantage as a member of the sales staff. If you possess the ability to deal directly with people, the sales department is

Figure 1-46 Motorcycle dealerships come in all different sizes. This dealership sells multiple brands of bikes.

Figure 1-47 This is an example of a large motorcycle sales department.

an excellent place to learn how a motorcycle dealership operates. The sales area provides valuable exposure to business-related activities. The experience can be very beneficial, especially if you plan to run your own business someday.

The **parts department** (Figure 1-48) is also a great place to use your people skills. Parts departments sell repair parts and accessories. As a member of the parts department, you'll have constant contact with retail customers, the sales department, and the service department. You'll be dealing directly with customers, both in person and on the telephone. The parts department is more closely related to the service department than to the sales department, especially if you work as a **parts technician**. A parts technician is responsible for supplying the service department technicians with the parts that they need to complete their service and repair work.

The third department within a motorcycle dealership is the **service department** (Figure 1-49)

Figure 1-48 An example of a parts department is shown here.

Figure 1-49 Service departments come in many different sizes depending on the overall store capacity.

where motorcycles are brought in for maintenance and repairs. A small shop may have a service department that employs only one or two technicians. A medium-sized shop might employ three or four technicians plus a service manager. It's not unusual to find a significant number of employees in the service department of a large motorcycle dealership.

Larger motorcycle dealerships typically employ the following personnel in the service department:

- Lot Attendants
- Set-Up Technicians (Motorcycle Assemblers)
- Motorcycle Technicians
- Service Writers
- Service Managers

If you get a job in a motorcycle dealership service department, but not as a technician, you may be employed as a **lot attendant**. A lot attendant is usually responsible for cleaning up the display and shop areas; rearranging, cleaning, and detailing motorcycles; picking up and delivering motorcycles and supplies; and performing other related tasks. If you start out as a lot attendant, the dealership management will have a chance to evaluate your job performance before assigning you additional responsibilities.

A **set-up technician** (also known as a *motorcycle assembler*) is a step closer to becoming a motorcycle technician. The set-up technician position requires certain mechanical skills. The set-up technician uncrates and assembles all of the new motorcycles received at a dealership (Figure 1-50). The set-up activity often includes the initial service of the motorcycle (oil, gas, adjustments, and other important safety checks).

Figure 1-50 Motorcycles and ATVs (shown here) must be uncrated and assembled prior to selling. A set-up technician will normally handle this duty.

The **motorcycle technician** is frequently considered the backbone of the service department. It's not unusual to find motorcycle technicians who started in sales, in the parts department, as lot attendants, or as set-up technicians and worked their way up. As a motorcycle technician, you'll need a technical background, factory training (which the dealership can arrange for you), tools, and usually some prior mechanical experience. Some of the job assignments and responsibilities of a motorcycle technician include the following:

- Warranty service
- Preventive and scheduled maintenance
- General repair activities
- Staying current with new products, accessories, and service procedures
- Maintaining accurate repair records
- Alerting the service manager to actual or potential problems

In addition to the direct repair activity involvement of the motorcycle technician, there are other related positions available in most service departments for those who wish to try other assignments in the motorcycle service career field.

Another key employee in the service department is the **service writer** (Figure 1-51). The service writer is responsible for writing the repair orders for service work. He or she should be technically trained and should have a complete understanding of the service process. When writing a repair order, the service writer must obtain detailed failure information from the customer, verify the customer's input, and then provide the customer with an estimate of the services that might be required to correct the problem. In most dealerships, the service writer creates the repair orders, which are then distributed to the motorcycle technicians. The service writer also has a hand in job scheduling, ensuring that the repair process flows smoothly.

The **service manager** (Figure 1-51) holds the highest position in the service department. Most service managers are responsible for the following:

- Customer transactions
- Warranty claims
- Product update and information publications

Figure 1-51 The service writer and service manager work directly with customers on a daily basis.

- Technician training
- Employee hiring and dismissal
- Equipment needs
- Building maintenance
- Service policy changes
- Service files and records

Service managers usually have an extensive service background and prior management experience including:

- Technical training
- Factory service school training
- Lot attendant experience
- Set-up/assembly experience
- Motorcycle repair experience
- Customer relations skills
- Management experience

The service manager has the overall responsibility for the service department. He or she must see that everything in the service department is well organized, that all necessary parts are in stock, and that the service work is performed correctly and completed on time. The service manager must handle all customer complaints and any technical questions from both customers and technicians. The service manager needs an extensive amount of motorcycle repair experience and excellent management skills.

Finally, the top position in many motorcycle dealerships is the **general manager**. The general manager has the overall responsibility for the sales, parts, and service departments. He or she oversees the day-to-day operations of the entire business. A general manager is likely to have had experience in all of the other departments.

Other Industry Opportunities

Some individuals with motorcycle repair backgrounds (e.g., motorcycle technicians and service managers) have found challenging career opportunities as **motorcycle repair instructors** at the vocational school level. To be a motorcycle repair instructor, you must meet certain requirements. These requirements vary by locality and institution. For example, in some states, a technical instructor must be certified. To be certified in California for instance, a technician can apply for teaching credentials if he or she is qualified in one of the following ways: Seven years experience in the trade or five years experience in the trade plus two years of college (with a major in the specific trade). Because of the growing popularity of motorcycles and ATVs for sport and utility purposes, more motorcycle technical trade schools are opening every year. The demand for qualified motorcycle repair instructors is growing especially at the post-secondary school level. There are also teaching positions in most motorcycle manufacturer training schools.

Before seriously considering a career as a motorcycle repair instructor, be sure that you enjoy explaining the details of how something works, that you feel comfortable working directly with groups of people, and that you have an abundance of patience. The pay and benefits for the instructor position are usually good, but to most instructors, the most satisfying reward is watching the students develop the ability to apply their newly acquired knowledge.

If you enjoy motorcycle repair theory more than you enjoy actually repairing motorcycles, it's quite possible that you would enjoy a career as a **technical writer**. Technical writers wrote most everything that you've ever read about motorcycle repair. Technical writers in the motorcycle industry are constantly in demand, especially if they're skilled at transforming technical ideas and concepts into everyday language.

Most technical writers have the following:

- Technical training
- Higher education (college)
- Writing experience

Technical illustrators, who work closely with technical writers, create most motorcycle photographs and illustrations contained in service manuals, sales brochures, and other printed matter. Although technical illustrators create most of the illustrations, in certain cases the technical writers themselves create the illustrations. At the very least, technical writers should be able to define the illustrations or photographs needed to support the text that they've developed, and to verify that the completed illustrations support the text. There's usually a close working relationship between technical writers and the technical illustrators to develop the finished printed material. Most technical illustrators usually have the following:

- Technical training
- Some writing experience
- Photographic experience
- Technical illustration and layout experience

These are just a couple of the many possible opportunities in the area of motorcycle training, distribution, and manufacturing. Companies in the motorcycle distribution and manufacturing industry offer numerous other career opportunities as well. Although the complete list is long and varied, here is a list of these positions with a brief explanation of what the position entails:

Technical advisors assist dealer technicians with problems over the phone.

District service managers work with a group of dealers with technical-related issues within a predetermined area.

District sales managers work very much like district service managers but instead handle sales-related items.

District parts managers assist dealers with parts-related problems.

Service technical training instructors facilitate training to technicians in the same manner as motorcycle repair instructors previously mentioned.

Customer service representatives assist customers over the phone with a variety of wishes and concerns.

Quality control specialists work closely with the manufacturing engineers to ensure that the products are of the highest quality possible.

Research and development engineers help in the actual creation of the products.

Race team support technicians travel extensively and work on the factory race machines.

Advertising and marketing specialists help in selling the products to the greatest number of customers as possible.

Most motorcycle manufacturing company employees enjoy competitive salaries and generous company benefits. Before seriously being considered for a career with a motorcycle manufacturer, you will need most of the following:

- Related mechanical experience
- Employment at a dealership
- Technical training
- Factory service school training in motorcycle repair
- Higher education (college or vocational school)

We've explored several career opportunity options for someone who has the necessary skills and training in motorcycle repair. What about the person who wants to be self-employed? Are the days of the independent service technician over? Not at all! With adequate financial backing, a person with the proper skills and background could start any type of related business, including a full motorcycle dealership, a parts and accessories store, or a major service and repair business. It's possible for a trained motorcycle or ATV repair technician to start a small repair and service business with little available capital. Many of today's thriving repair and service businesses started out in the back of a garage. If self-employment is your goal, you might start out by using that spare space in your garage!

These are just some of the possibilities that await you in the exciting and challenging field of motorcycle and ATV repair. As you've discovered in this section, a wide range of career opportunities are available to qualified individuals.

Summary

- The motorcycle industry is growing at a tremendous rate and the need for trained technicians is growing rapidly.
- The history of the motorcycle dates back to the mid 1800s.
- The first two major motorcycle manufacturers in the United States were Indian and Harley-Davidson.
- British motorcycle sales were at their highest in the late 1950s.
- Japanese motorcycles came to the forefront of the industry in the 1960s.
- There are four basic motorcycle classifications: street, dual purpose, off-road and ATV.
- All motorcycles have basic similarities.
- There are numerous opportunities within the motorcycle industry, not only at the dealership level, but also at the manufacturer and distributor levels as well.

Chapter 1 Review Questions

1. Name the five positions mentioned in this chapter available in a motorcycle service department.
2. Name the three main departments in a motorcycle dealership.
3. Three things that a service writer must do when working with a customers are:
4. The three basic classifications of motorcycles are:
5. Although thousands are still being used today, what type of ATV is no longer in production by any major manufacturer?
6. Which motorcycle is often used for road racing?
   a. Hot-rod cruiser
   b. Standard
   c. Sport
   d. Sport touring
7. "UJM" stands for:
8. The first motorcycle in recorded history using an internal combustion engine was built by:
   a. Alex Millet
   b. Gottlieb Daimler
   c. William Davidson
   d. Sylvester Roper
9. The Hendee Manufacturing Company changed its name to ________ after designing its first motorcycle in 1901.
10. The first Harley Davidson V-Twin was built in:
11. In what year did Harley Davidson celebrate its 100-year anniversary?
12. In 1914, the world's first motorcycle with electric start and a full electrical system was introduced to the industry. What brand of motorcycle was this?
13. Which motorcycle usually has little to no lighting capabilities?
14. Which motorcycle is often equipped with a radio and even cruise control?
    a. Cruiser
    b. Sport touring
    c. Touring
    d. Standard
15. Which motorcycle industry professional works with a group of dealers with technical-related issues within a predetermined area?
    a. District Sales Managers
    b. District Service Managers
    c. Techncial Writers
    d. Quality Control Specialists

CHAPTER

2 Safety First

Learning Objectives

When students have completed the study of this chapter and its laboratory activities they should be able to:

- Understand the importance of safety and accident prevention in a motorcycle shop environment
- Explain the basic principles of personal safety
- Explain the procedures and precautions for safety when using tools and equipment
- Explain what should be done to maintain a safe working area in a service shop environment
- Describe the purpose of the laws concerning hazardous wastes and materials including right-to-know laws
- Describe your rights as an employee and/or student to have a safe place to work

Key Terms

Alternating Current (AC)

Carbon dioxide (CO_2)

Carbon monoxide (CO)

Class A fires

Class B fires

Class C fires

Class D fires

Conductors

Contact dermatitis

Direct Current (DC)

Earplugs

Eczema

Fire extinguishers

Fire triangle

Goggles

Halon

Hazard Communication Standard

Headset

Material handling

Material Safety Data Sheets (MSDS)

Metatarsal guards

Motorcycle Safety Foundation (MSF)

National Electrical Code (NEC)

National Fire Protection Association (NFPA)

National Safety Council (NSC)

Occupational Safety and Health Administration (OSHA)

PASS

Personal Protective Equipment (PPE)

Power tools

Respirators

Safety glasses

INTRODUCTION

Most people are concerned with safety in and around their homes. They strive to protect their families from accidents and injuries. But accidents in the workplace are often much more severe than home accidents, because workplaces contain many more potential hazards than the average home. Working on motorcycles can be fun and rewarding. But, if the proper precautions are not followed, it can be dangerous as well.

THE SAFETY ATTITUDE

Safety is more than just the absence of accidents. *Safety is an attitude* that helps you prevent injuries to yourself and others. Safe working practices should be a way of life. They should be as instinctive as putting on your seat belt or looking both ways before you cross the street. Safety is not a matter of good luck or bad luck. It's a predetermined set of mental exercises, including the following:

- Planning to work safely
- Recognizing potential safety hazards and eliminating hazards
- Following proper safety procedures at all times, particularly in your workplace

You should be aware that the **Occupational Safety and Health Administration (OSHA)** is the federal agency that publishes safety standards for business and industry. The OSHA regulations affect every business that has employees and sells its products or services. OSHA requires every employer to provide employees with safe workplaces that are free from recognized hazards. For reasons of brevity concerning the legal aspects of OSHA, it is the federal government's law enforcer for industrial-safety matters. Employers are motivated to adopt and use safe working procedures through OSHA's strict enforcement of the regulations. Safety violators receive harsh penalties and fines. You can find a complete list of OSHA's proven safety methods, practices, and regulations in one convenient resource called the Code of Federal Regulations. Even if you're a one-person operation, you should understand and follow OSHA's safety guidelines for your own protection.

In a motorcycle service department, the safety matters of primary concern are the following:

- Fire safety
- Chemical safety
- Basic electrical safety
- Ventilation of exhaust gases
- Safe operation of engines and equipment
- Good housekeeping practices
- Safe handling of heavy objects and materials
- Safe use of stands and lifts
- Proper use of personal protective equipment
- Safe riding practices

Now that we've provided you with a quick list of the safety topics and areas that OSHA's regulations cover, let's look at these important safety items one at a time.

FIRE SAFETY

A major safety consideration in the motorcycle repair business is fire prevention. Many fires occur in private garages every year and a significant number of these are started by the mishandling of gasoline, such as storing gasoline in unapproved containers or failing to clean up gasoline spills. Gasoline is the fuel for all current modern motorcycle engines. Because gasoline is one of the most flammable liquids, fire is a serious threat in any motorcycle service area.

Gasoline is not the only flammable liquid used in the service department: oils, lubricants, paints, cleaning solvents, and other chemicals can also create a fire hazard when improperly handled. Despite the fire risk, a service department can be run safely. By following basic safety practices, the danger of fire can be greatly reduced, if not eliminated entirely.

The **National Fire Protection Association (NFPA)** is the largest and most influential national group dedicated to fire prevention and protection. Its mission is to safeguard people, property, and the environment from fires. The NFPA also publishes the **National Electrical Code (NEC)**. The NEC is the national standard for all residential and industrial electrical installations in the United States and Canada. When you start planning a fire safety program for your business, check with the

NFPA. They can provide useful hints and detailed support information.

THE FIRE TRIANGLE

There are three conditions that must be present for a fire to start. These conditions are grouped together to form the **fire triangle**. The three components of the fire triangle are as follows:

- Fuel (such as wood or gasoline)
- Oxygen
- An ignition source (such as a spark)

After a fire starts, the supply of fuel and oxygen must stay at certain levels to sustain the fire. To extinguish a fire, you must remove at least one of these two legs of the fire triangle. You can put out a fire by removing the fuel source or removing the oxygen.

When analyzing fire prevention, you must always be aware of the ignition sources that could start a fire in your work area. When we consider ignition sources, most of us think of open flames, sparks, stoves, and matches. However, there are several other dangerous, but less obvious, ignition sources.

For example, a common but often overlooked source of ignition is the engine exhaust. A motorcycle's exhaust system becomes hot during operation. This heat remains in the exhaust system for a period of time after the vehicle's engine has been turned off. Therefore, if a vehicle's engine is still warm when you begin to make repairs, you must take extra precautions to prevent fires.

Another highly possible source of ignition is cigarette smoking. Smoking-related ignitions are a leading cause of fires. Sparks from lit cigarettes, heat from discarded cigarette butts, and the open flames of lighters and matches can all start fires in flammable and combustible materials. Therefore, smoking should be strictly controlled in a motorcycle service department. Smoking and nonsmoking areas should be posted with distinct, easily recognizable symbols. Smoking areas should be equipped with adequate receptacles to provide for the safe disposal of smoking materials. Smoking is prohibited in many service departments, and smokers must go to a designated outside smoking area.

Spontaneous combustion is another potential source of ignition that you should recognize. In spontaneous combustion fires, the heat for ignition is created by a chemical reaction in combustible materials. One common type of spontaneous combustion occurs when oil- or solvent-soaked rags or papers are discarded in a trash can. The decomposition of the oil or solvent often produces enough heat to ignite the rags or papers. To prevent spontaneous combustion, all oil-or solvent-contaminated rags and papers should be discarded only in designated, fireproof metal safety receptacles. Routine trash material shouldn't be discarded in these special receptacles.

THE FOUR FIRE CLASSES

Let us take a closer look at the different types of fires. The NFPA classifies fires in four categories, or classes: Classes A, B, C, and D. (Figure 2-1) Each of these four fire classes is defined by, and associated with, a different type of fuel source.

Class A fires involve the burning of wood, paper, cardboard, fabrics, and other similar fibrous materials. These materials ignite easily, burn rapidly, and produce large quantities of heat during burning. Some examples of Class A combustible materials that are commonly found in workplaces include the following:

- Paper business forms
- Company files or records
- Cleaning and polishing cloths
- Work aprons
- Dust covers
- Work area partitions

Class A fires can be extinguished with water, CO_2 (carbon dioxide), or dry chemical agents. These agents extinguish the fire by quickly cooling the burning material and lowering the temperature in the combustion zone. The symbol used to identify Class A extinguishing equipment is the letter "A" inside a green triangle.

Class B fires involve flammable liquids, gases, and other chemicals. Because many flammable and combustible liquids and solvents are used in a motorcycle service department, special care should be given to their handling, use, and storage. Some common flammable liquids include gasoline, cleaning solvents, oils, greases, turpentine, oil-

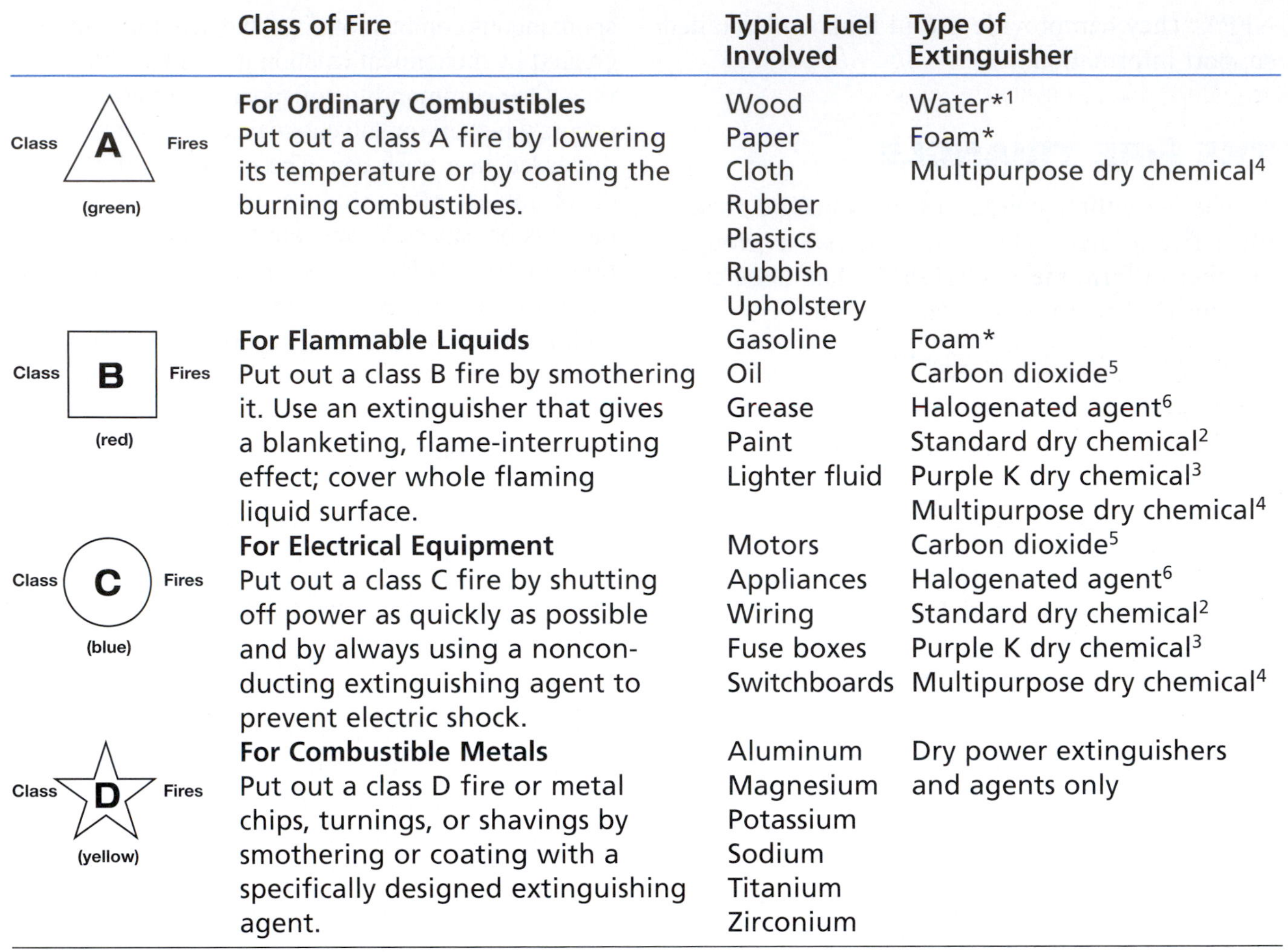

	Class of Fire	Typical Fuel Involved	Type of Extinguisher
Class A Fires (green)	**For Ordinary Combustibles** Put out a class A fire by lowering its temperature or by coating the burning combustibles.	Wood Paper Cloth Rubber Plastics Rubbish Upholstery	Water*[1] Foam* Multipurpose dry chemical[4]
Class B Fires (red)	**For Flammable Liquids** Put out a class B fire by smothering it. Use an extinguisher that gives a blanketing, flame-interrupting effect; cover whole flaming liquid surface.	Gasoline Oil Grease Paint Lighter fluid	Foam* Carbon dioxide[5] Halogenated agent[6] Standard dry chemical[2] Purple K dry chemical[3] Multipurpose dry chemical[4]
Class C Fires (blue)	**For Electrical Equipment** Put out a class C fire by shutting off power as quickly as possible and by always using a nonconducting extinguishing agent to prevent electric shock.	Motors Appliances Wiring Fuse boxes Switchboards	Carbon dioxide[5] Halogenated agent[6] Standard dry chemical[2] Purple K dry chemical[3] Multipurpose dry chemical[4]
Class D Fires (yellow)	**For Combustible Metals** Put out a class D fire or metal chips, turnings, or shavings by smothering or coating with a specifically designed extinguishing agent.	Aluminum Magnesium Potassium Sodium Titanium Zirconium	Dry power extinguishers and agents only

Figure 2-1 The symbols in this table are placed on fire extinguishers to indicate the types of fires that they are designed to be used on.

based paints, and lacquers. Common flammable gases include natural gas, propane, and acetylene.

Fires involving flammable liquids produce tremendous quantities of heat. Water is an ineffective extinguishing agent for a Class B fire. The heat from a burning flammable liquid will boil the water that's applied to the fire, turning the water into steam before it can do much good. Most importantly, almost all flammable liquids are lighter than water. The liquids float on top of the water and continue burning. This is a very dangerous situation that can cause a flammable liquid fire to spread very rapidly. The best way to extinguish a Class B fire is to smother it, removing its source of oxygen. Foams, dry chemicals, and CO_2 are the best extinguishing agents to use on a Class B fire. The symbol used to identify Class B extinguishing equipment is the letter "B" inside a red square. If you routinely keep gasoline (even in small amounts) in your workplace, you should have at least one Class B fire extinguisher in the area. You can also smother a small Class B fire with a blanket or noncombustible container. Use this method only if you can do so without risking personal injury. You should always remember that flammable liquid fires have a tendency to flare up very rapidly.

Class C fires involve live electrical equipment such as electrical boxes, panels, circuits, appliances, power tools, machine wiring, junction boxes, wall switches, and wall outlets. Some form of short circuit or overloaded circuit usually causes electrical fires. Examples of these causes include the following:

- Loose contacts or terminals
- Frayed wire insulation
- Improper installations
- Defective equipment
- Overloaded circuits

Electrical system overloads and short circuits can produce arcs, sparks, and heat. This type of electrical problem can ignite nearby combustible materials. For example, wire insulation, plastic components, and wall insulation or paneling. Water is a good conductor of electricity, and if it's applied to an electrical fire, the person holding the extinguisher could be severely shocked or electrocuted. **Carbon dioxide (CO_2)** is the most widely used extinguishing agent because it's nonconductive, it penetrates around electric equipment well, it's effective, and it leaves no residue that would have to be cleaned up afterward. Dry chemicals produce a residue that can damage electric equipment.

Halon is another extinguishing agent that's effective on all classes of fire, especially Class C. Halon is stored as a liquid under high pressure, and is released on a fire as an oxygen-depleting (smothering) gas. Although Halon is very effective, it's not readily available. Halon is a fluorocarbon compound that's classified as an ozone-depleting substance. The usage of Halon is restricted by law for environmental reasons. The symbol used to identify Class C extinguishing equipment is the letter "C" inside a blue circle.

Class D fires involve combustible metals such as magnesium, titanium, zirconium, sodium, lithium, and potassium. Flakes and fine particles of these metals can be ignited at relatively low temperatures. Metal particles are often produced by cutting or grinding operations. If cutting or grinding is done in the typical motorcycle repair shop, it's usually confined to a designated area that's uncluttered and well ventilated. The larger exposure to Class D fires is found in a *back-of-the-garage* type of operation, where space is limited and conditions might favor the start of this type of fire. You should be aware of Class D fires, and how to react to them.

Dry powder compounds and *dry chemical extinguishers* are the two primary methods to extinguish Class D fires. Dry powder compounds are completely different than dry chemical extinguishers. Dry powder compounds are usually scooped directly onto a fire. Dry chemical extinguishers apply the dry chemical charge under pressure. The symbol used to identify Class D extinguishing equipment is the letter "D" inside a yellow star. The most important reason for introducing you to the four classes of fires is to inform you of what to do and what not to do in a fire emergency. Your reaction to a fire could mean the difference between a minor incident and a major loss of property, with possible injury or death. Knowledge of the fire classes is also important when you're assessing your work area for fire hazards. Basically, most fires are preventable. It is important to remember that fire prevention isn't just a slogan. Awareness, common sense, and good work habits go a long way toward preventing fires.

Based on the nature of your work environment, the two types of fires most likely to occur in a motorcycle service department are Class A and Class B fires. But don't be negligent about the possibility of a Class C or Class D fire occurring also. Know what to do for all fire classes.

USING A FIRE EXTINGUISHER

Fire extinguishers (Figure 2-2) must be properly used to be effective on a fire. You should become familiar with the various extinguishers installed at your facility before a fire starts. This familiarization step is important for the following reasons:

- To operate a fire extinguisher safely and efficiently, you should know how to use it. You'll lose valuable fire-fighting time if you have to stop and read instructions. Be prepared!
- You could injure yourself or others by using an extinguisher improperly.
- An average fire extinguisher discharges all of its contents in only 12 to 60 seconds. You need to make the best use of all of the extinguisher contents.

Figure 2-2 Examples of different types and sizes of fire extinguishers are shown here.

To be effective, portable fire extinguishers must be readily available in a fire emergency. Extinguishers must be installed close to all potential fire hazards (Figure 2-3). The extinguishers must contain the proper type of extinguishing agent for those hazards and they must be large enough to protect the designated area. The fire hazards existing in a shop must be identified and evaluated, to verify that the proper numbers and types of fire extinguishers are installed at the correct locations.

Take the following steps before you attempt to extinguish any fire:

1. Evaluate the size of the fire. A fire in its beginning stages is called an *incipient fire*. A fire is classified as incipient (start-up) if it covers an area no larger than 2–4 feet square, has flames less than 2 feet in height, and produces low levels of smoke. Fire extinguishers can be effective for extinguishing or suppressing this size of fire. But it's not safe to use a fire extinguisher after a fire passes beyond the incipient stage. The length of time that a fire remains in the incipient stage is usually quite brief. If the fire goes beyond the start-up stage, the only course of action is to evacuate the building or facility and call the fire department.
2. Locate the exits and the escape routes you'll need in an emergency evacuation. To prevent yourself from becoming trapped in a serious situation, keep the locations of the exits in mind as you fight the fire.
3. Determine which way the flames are moving and approach the fire from the opposite direction. The flaming side of the fire radiates too much heat, and the fire could overtake you before you have a chance to escape. By attacking the fire from the opposite side, you'll be safer and you'll be able to get closer to the combustion zone of the fire.

When you've taken these preliminary steps, you're ready to use an extinguisher. To operate this type of extinguisher, perform the following:

- Remove the extinguisher from the wall.
- Grasp the handle of the extinguisher and pull out the safety pin.
- Free the hose and aim the nozzle at the fire.

Figure 2-3 Be sure to know the locations and types of fire extinguishers available in the shop.

- Squeeze the handle.
- Move the nozzle in a sweeping motion to distribute the extinguishing agent.

Here's a little hint. You can use the acronym **PASS** to help you remember how to operate a fire extinguisher. The letters in PASS stand for **Pull** (the safety pin), **Aim**, **Squeeze**, and **Sweep**.

Always direct the stream of extinguishing agent at the base of the flames. This is the fire's combustion zone. Cooling this area will extinguish the fire more quickly. Sweep the stream from side to side and work your way around the fire until it's completely extinguished. You should remember, never turn your back on a fire until you're absolutely sure that it's completely extinguished. Heat remaining inside of partially burned materials can re-ignite the fire when you're not looking. This could trap you. Portable fire extinguishers should be given a complete maintenance check annually. Maintenance may include recharging or pressure testing the extinguisher. Most fire extinguishers require pressure testing every five years. You should also inspect all fire extinguishers at least once monthly and answer the following:

- Is the extinguisher in its designated place close to possible fire hazards?
- Is the extinguisher clearly visible?
- Is access to the extinguisher free from all obstacles and obstructions?
- Is the extinguisher fully charged?

It's also a good idea (in some localities it's the law) to install battery operated smoke detectors and carbon-monoxide detectors in all buildings. Remember, smoke detectors require some maintenance. Check the detectors and their batteries routinely to ensure that they're in proper working order. Replace batteries yearly, and replace all detectors that show any signs of malfunction or incorrect operation.

The following are some important fire evacuation procedures that you should commit to memory. Never take fire safety lightly. It's recommended that you practice these procedures in regularly scheduled exercises or fire drills.

- Get out fast. Believe the alarm when you hear it. Don't waste time trying to verify that there's a fire or trying to gather things before you leave.
- Stay low to the floor to avoid smoke and toxic gases. The clearest air is found near the floor. So crawl if necessary! Cover your mouth and nose with a damp cloth, if available, to help you breathe.
- Don't open a closed door without feeling the door's surface first. If you open a door with flames on the other side, the fire could back-draft and severely burn you. Use the back of your hand (don't burn your palms), when you test for heat. If the door feels warm on your side, the temperature is probably far above the safety level on the other side. Don't open the door; use an alternate escape route!
- Never enter a burning building. This could be a fatal mistake! Professional firefighters are equipped with special protective equipment and breathing devices that allow them to enter burning buildings. They're also trained in search and rescue techniques. Leave these tasks to the professionals.
- If your clothing catches fire, don't panic! Stop, drop to the ground, and roll around to smother the flames. If a coworker's clothing catches fire, quickly wrap the person in a blanket or rug to smother the flames. Assist the person with the *stop, drop, and roll* maneuver if a rug or blanket isn't handy.

HAZARDOUS CHEMICALS

Question: Do you work with or near any hazardous chemicals? Before you answer too quickly, continue reading!

Many people are unaware that some everyday materials and products in their work areas are indeed hazardous chemicals. Materials such as paints, oils, lubricants, cleaners, degreasers, solvents, and gasoline are all potentially hazardous chemicals. Labels on chemical containers normally list the names of the chemicals and any hazards associated with them. The labels also highlight health hazards that the chemicals present, including eye and skin contact irritation characteristics, breathing and inhalation dangers, and accidental swallowing or ingestion warnings. Many labels also include first aid procedures to administer for these hazards, the name and address of the chemical manufacturer or importer, and specific ingredients of the hazardous chemical compounds. Some chemicals can cause a variety of serious health problems (or even death) if you're exposed to them.

For example, some chemicals can produce gases and vapors that are poisonous when inhaled. Other chemicals are highly flammable or explosive when exposed to sparks or open flames. And still other chemicals can cause temporary or permanent blindness if they get splashed in the eyes. Everyone should take the potential hazards of chemicals seriously. The intent here isn't to scare you. It's to alert you to possible dangers so that you'll give chemicals the respect they deserve.

Most chemicals in the workplace are not life-threatening but can cause minor injuries or illnesses. If you work with chemicals in your service department, always follow the safety precautions on the chemical's label and protect yourself with gloves and safety goggles. A common problem that chemicals can cause is a skin inflammation called **contact dermatitis** or **eczema**. Contact dermatitis usually starts as redness in the area of chemical contact or exposure and may progress into blistering, scaling, and cracking of the skin surface. People who regularly expose their hands to materials such as detergents, cleaners, degreasers, oil, and gasoline are susceptible to dermatitis. Although contact dermatitis isn't a life-threatening condition, it's very uncomfortable and could lead to more serious health complications such as infections. Contact dermatitis is much easier to prevent than it is to cure, so it's important to wear industrial rubber gloves whenever you're using chemicals that could come in contact with your skin. Exposing the skin to stronger chemicals can cause more serious injuries. For instance, exposure to acids and strong bases can cause immediate, very painful burns to the skin.

The batteries used in all types of motorcycles and ATVs contain strong and dangerous acids. The sulfuric acid found in batteries is a particularly dangerous acid you should be aware of. It can eat through clothing or burn your skin. And more importantly, it can easily cause blindness if splashed in the eyes. Be aware that storage batteries give off dangerously explosive hydrogen gas when they're being charged. Always use extreme caution when handling, storing, replacing, charging, or adding distilled water to storage batteries. Follow these safety guidelines when handling batteries:

- Keep batteries upright to prevent the acid from spilling or leaking.
- Always wear gloves, an apron, and safety goggles when handling batteries to protect your skin and eyes.
- Charge batteries in a well-ventilated location to prevent the buildup of hydrogen gas.

RIGHT-TO-KNOW LAWS

An important part of a safe work environment is knowledge of potential hazards. Right-to-know laws protect every employee in a motorcycle dealership. These laws were placed in effect when OSHA's **Hazard Communication Standard** was published back in 1983. This was originally intended for chemical companies and manufacturers that require employees to handle potentially hazardous materials in the workplace. Since then, most states have enacted their own right-to-know laws and these laws now apply to all companies that use or sell potentially hazardous chemicals or materials. The general intent of right-to-know laws is to ensure that employers provide a safe working place for their employees when hazardous materials are involved. Specifically, there are three areas of employer responsibility.

First and foremost, all employees must be trained about their rights under the legislation, the nature of the hazardous chemicals around them, and the contents of the labels on the chemicals. All of the information on each chemical must be posted in **Material Safety Data Sheets (MSDS)** and must be easily accessible to every employee (Figure 2-4). The manufacturer of the chemical must give these sheets to its customers if they are requested to do so (Figure 2-5). The sheets detail the chemical composition and precautionary information for all products that can present a health or safety risk.

ELECTRICAL SAFETY

We tend take electricity for granted so we often lose sight of its potential to cause serious injury or death. The main hazards of electricity are electrical shocks and burns. Electric shocks and burns usually result from the following:

- Faulty power tools or equipment
- A disorderly and untidy work environment
- Human error (the misuse of an electrical device)

Even if you don't work with electrical equipment on a routine basis, you should be aware of how shocks can occur. Generally, you must be in contact or in proximity with a conductor or a conductive surface. **Conductors** are normally metallic materials that readily pass or conduct electricity. If you're in proximity with a conductor and you

Figure 2-4 Right-to-know laws were created to protect employees' rights and give them access to all information regarding potentially hazardous materials in the workplace.

```
HEXANE
============================================================
MSDS Safety Information
============================================================
Ingredients
============================================================
Name: HEXANE (N_HEXANE)
% Wt: >97
OSHA PEL: 500 PPM
ACGIH TLV: 50 PPM
EPA Rpt Qty: 1 LB
DOT Rpt Qty: 1 LB
============================================================
Health Hazards Data
============================================================
LD50 LC50 Mixture: LD50:(ORAL,RAT) 28.7 KG/MG
Route Of Entry Inds _ Inhalation: YES
Skin: YES
Ingestion: YES
Carcinogenicity Inds _ NTP: NO
IARC: NO
OSHA: NO
Effects of Exposure: ACUTE:INHALATION AND INGESTION ARE HARMFUL AND MAY BE FATAL.
INHALATION AND INGESTION MAY CAUSE HEADACHE, NAUSEA, VOMITING, DIZZINESS, IRRITATION
OF RESPIRATORY TRACT, GASTROINTESTINAL IRRITATION AND UNCONSCIOUSNESS. CONTACT
W/SKIN AND EYES  MAY CAUSE IRRITATION. PROLONGED SKIN MAY RESULT IN DERMATITIS (EFTS
OF OVEREXP)
Signs And Symptions Of Overexposure: HLTH HAZ:CHRONIC:MAY INCLUDE CENTRAL
NERVOUS SYSTEM DEPRESSION.
Medical Cond Aggravated By Exposure: NONE IDENTIFIED.
First Aid: CALL A PHYSICIAN. INGEST:DO NOT INDUCE VOMITING. INHAL:REMOVE TO FRESH AIR. IF
NOT BREATHING, GIVE ARTIFICIAL RESPIRATION. IF BREATHING IS DIFFICULT, GIVE OXYGEN.
EYES:IMMED FLUSH W/PLENTY OF WATER FOR AT LEAST 15 MINS. SKIN:IMMED FLUSH W/PLENTY
OF WATER FOR AT LEAST 15 MINS WHILE REMOVING CONTAMD CLTHG & SHOES. WASH CLOTHING
BEFORE REUSE.
============================================================
Handling and Disposal
============================================================
Spill Release Procedures: WEAR NIOSH/MSHA SCBA & FULL PROT CLTHG. SHUT OFF
IGNIT SOURCES:NO FLAMES, SMKNG/FLAMES IN AREA. STOP LEAK IF YOU CAN DO SO W/OUT
HARM. USE WATER SPRAY TO REDUCE VAPS. TAKE UP W/SAND OR OTHER NON-COMBUST MATL &
PLACE INTO CNTNR FOR LATER (SU PDAT)
Neutralizing Agent: NONE SPECIFIED BY MANUFACTURER.
Waste Disposal Methods: DISPOSE IN ACCORDANCE WITH ALL APPLICABLE FEDERAL, STATE AND
LOCAL ENVIRONMENTAL REGULATIONS. EPA HAZARDOUS WASTE NUMBER:D001 (IGNITABLE
WASTE).
Handling And Storage Precautions: BOND AND GROUND CONTAINERS WHEN TRANSFERRING LIQUID.
KEEP CONTAINER TIGHTLY CLOSED.
Other Precautions: USE GENERAL OR LOCAL EXHAUST VENTILATION TO MEET
TLV REQUIREMENTS. STORAGE COLOR CODE RED (FLAMMABLE).
============================================================
Fire and Explosion Hazard Information
============================================================
Flash Point Method: CC
Flash Point Text:   9F,_23C
Lower Limits: 1.2%
Upper Limits: 77.7%
Extinguishing Media: USE ALCOHOL FOAM, DRY CHEMICAL OR CARBON DIOXIDE. (WATER MAY BE
INEFFECTIVE).
Fire Fighting Procedures: USE NIOSH/MSHA APPROVED SCBA & FULL PROTECTIVE
  EQUIPMENT (FP N).
Unusual Fire/Explosion Hazard: VAP MAY FORM ALONG SURFS TO DIST IGNIT SOURCES & FLASH
BACK. CONT W/STRONG OXIDIZERS MAY CAUSE FIRE. TOX GASES PRDCED MAY INCL:CARBON
MONOXIDE, CARBON DIOXIDE.
============================================================
```

Figure 2-5 Material Safety Data Sheets are an important part of working in a motorcycle dealership and should be readily available. They contain important information to let you know the dangers—if any—of working with chemicals.

contact a source of electricity, the electricity can pass through your body to the conductive material and then to the ground. In such a situation, your body acts like a switch, closing the electrical circuit. The sensation that you feel is an electric shock. The shock may range from mild to severe depending on the circumstances and the source of the electricity. If you're lucky, you may experience only a brief unpleasant sensation. If you're unlucky, you could be seriously injured, or worse!

There are two types of electric-power sources: **Alternating Current (AC)** and **Direct Current (DC)**. AC current flows in alternating cycles and is used to run lighting circuits and appliances in most households and workplaces. DC current is continuous and steady. DC is found in batteries and battery-powered electric systems such as motorcycle electrical systems.

Each type of electrical source has its own set of hazard characteristics. The AC found in the typical repair facility can be very hazardous if the path of the shock is from one hand to the other. This puts your heart in line with the alternating current flow, and can cause fibrillation (erratic, nonrhythmic heartbeat).

An AC shock causes the muscles to contract and relax with the alternating current cycles. This allows you to withdraw the body contact point (normally a hand) from the source of the shock when the muscles relax.

The DC found in the typical motorcycle shop can present a more serious hazard than AC because a DC shock causes the muscles to contract and freeze. There is no alternating cycles with DC; consequently, the muscles don't relax like they will with an AC shock. This makes it very difficult to let go of the shock contact point, extending the time that you're subjected to the danger of the shock. You can obtain severe burns, or worse, from a DC shock.

Always keep your work area clean and orderly to prevent electrical accidents. Keep floors clean and dry, because a wet or damp floor will make anyone standing on it more conductive to electricity. All portable electrical equipment (machines that have plugs and cords) should be inspected before each use to ensure that the equipment, cord, and plug are in good condition. If you feel the slightest tingle or shock while using electrical equipment, stop using it immediately! That slight tingle is an indication that there's a fault in the equipment's electrical wiring or ground circuit.

Avoid wearing conductive metals such as rings, watches, and chains when working with electricity. These metal objects significantly increase your risk of being shocked. You can increase your resistance to electric shock by wearing rubber gloves, standing on an insulating rubber mat, and using tools with insulated handles.

EXHAUST GAS SAFETY

When an engine is running, it creates exhaust gases that are hazardous if inhaled. The most dangerous of these gases is **carbon monoxide (CO)**. Carbon monoxide is a by-product of burning hydrocarbon fuels. It's often present in garages and around heating equipment. It's colorless, odorless, and tasteless, so you can't detect when it's present. When inhaled, carbon monoxide passes into the bloodstream and prevents red blood cells from carrying oxygen. As a result, the body suffocates from the lack of oxygen to the brain. Even small amounts of carbon monoxide can make you very ill. Adequate ventilation is necessary to prevent the buildup of dangerous carbon monoxide fumes. Also, carbon monoxide detectors should be installed in your work area. These devices are similar in appearance to smoke detectors and can be purchased at most hardware or discount stores. These detectors sound an alarm when a predetermined level of carbon monoxide is sensed in the air. They don't detect smoke, and can't be substituted for smoke detectors. Each type of device has its own function. Always follow these precautions when operating an engine:

- Never operate an engine in an enclosed area. Make sure that your workshop has proper ventilation. Use exhaust pipe extensions to direct the exhaust gases to the outside. OSHA can provide detailed information about ventilation safety requirements for buildings and work areas.
- When you're operating an engine (even if your shop is well ventilated), avoid breathing the fumes.

- Never operate an engine too close to a residential building. Exhaust gases could seep in and jeopardize the well-being of those inside.

SAFE OPERATION OF EQUIPMENT

All motorcycles and ATVs have moving parts that can be potentially hazardous. The careless operation of these vehicles when they're in for service can cause serious injuries and damage your workshop and your tools. To avoid accidents, follow these safety guidelines:

- Read the manufacturer's instruction manual carefully before operating any unfamiliar equipment.
- Never start a vehicle unless the transmission has been shifted into neutral.
- Turn off the ignition system before you start working to prevent the engine from starting accidentally while you're working on it.
- Keep your hands, fingers, and sleeves clear of all hazardous moving parts.
- Keep visitors and customers (especially children) away from all risk areas and post appropriate signs to warn customers of hazards.
- Remember that the exhaust system gets very hot during operation. Keep your hands, feet, and loose clothing away from the exhaust components whenever the exhaust system is hot.

GOOD HOUSEKEEPING PRACTICES

Now that we've discussed some of the hazards that may be present in a motorcycle service department, let's look at some of the things that can be done to prevent accidents. One of the most important considerations of any accident prevention and safety program is good housekeeping. Good housekeeping is more than just cleaning up. Housekeeping is a reflection on your organization and your work habits. A neat, well-organized work area provides the environment for good inventory control of parts and materials. A tidy work environment provides the basis for proper waste disposal. Your neatness can prevent parts, tools, work records, and other important items from being lost in the clutter and thrown out with the trash. Finally, a neat work area is less likely to contribute to accidents. The following are some specific rules related to good housekeeping:

- Keep your workbenches and your work areas clean and organized at all times. You should clean your work area at least daily, preferably after each job is completed. If a job spans several days, clean up at the end of each day. Don't allow combustible debris such as paper, cardboard, string, or rags to accumulate on or under the bench. If you must use combustible materials or flammable liquids, use only the quantity needed to complete the task and immediately return the remainder to its proper storage area. Clean up any spilled materials from floors and benches immediately. Sweep the floors every day to eliminate the buildup of dirt, dust, and other litter.
- Store flammable liquids such as gasoline and solvents in approved safety cabinets and in cool, dry areas away from ignition sources (Figure 2-6). Avoid storing flammable liquids or other chemicals in direct sunlight, heat, or humidity. Ensure that the storage area is well ventilated to prevent the buildup of potentially explosive fumes. Check all storage areas frequently for rusted containers, corroded caps or lids, and leaking containers.
- Label all flammable liquids and other hazardous materials properly. If a substance is transferred from its original container, the second container must also be properly labeled. Use portable safety cans for transporting small quantities of flammable liquids (Figure 2-7). These cans are fire-resistant and have self-closing lids. Never leave cans of flammable liquids lying around when they're not in use.
- Small amounts of flammable liquids sometimes leak or spill from machines and equipment. Place drip pans under leaking engines and vehicles to prevent the floor from becoming slippery. Drip pans should be noncombustible and should be large enough to contain any anticipated spill. If a vehicle has a persistent leak, place an absorbent, noncombustible material in the pan to soak up the liquid. Empty the drip pans regularly and dispose of the oil-soaked compounds. Oil-soaked materials are

hazardous waste that must be disposed of properly. Don't include hazardous waste with normal trash.

Figure 2-6 Always store flammable liquids in a cool, dry area away from ignition sources.

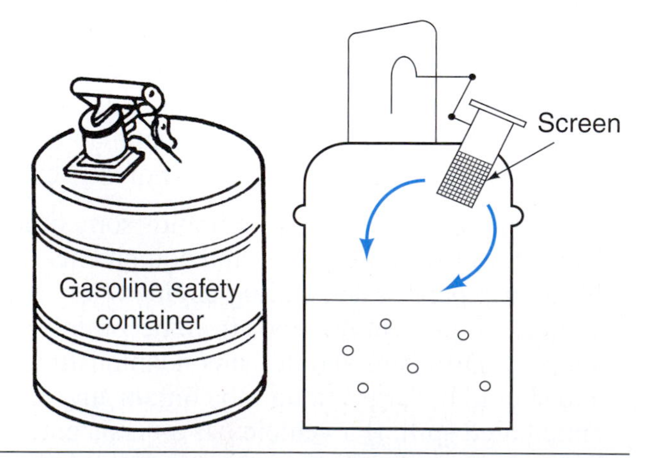

Figure 2-7 Flammable liquids should be stored in safety-approved containers. This gasoline container keeps the fumes from escaping while in storage.

- Proper garbage disposal (Figure 2-8) is another important concern for motorcycle service departments. Discard dry combustibles on a regular basis and never allow them to accumulate. Place combustibles in metal containers with lids. Lids help to contain and snuff out any fire that might start inside of the containers. When you empty the smaller containers, store the accumulated waste in a large metal refuse bin with a lid. These refuse bins should be located in a remote area away from heat sources.
- It's a good housekeeping practice to separate clean combustible wastes from dirty combustible wastes. Examples of "dirty" combustibles include papers, rags, and work clothes that are soaked with oil, grease, or solvents. These contaminated materials are more flammable than clean materials. Place all contaminated materials in separate metal containers with tight-fitting lids. Do you remember our earlier discussion in the fire safety section about spontaneous combustion? This is a good example of the use of a special discard container for disposing of contaminated material. Have oily rags and work clothes laundered by a professional industrial cleaning service. It's important to note that used liquids contaminated with dirt, grease, oils, solvents, or degreasers are classified as hazardous wastes and must be disposed of accordingly. Never empty such liquids into a sink or dump them on the ground. Federal, state, and local laws regulate the handling and disposal of hazardous wastes. Always follow authorized procedures when disposing of waste liquids.

HANDLING HEAVY OBJECTS AND MATERIALS

Material handling (moving materials from one place to another) is a concern for all occupations because this task has serious hazards associated with it. Every workplace requires some form of material handling. In a motorcycle service department, you may be required to remove or lift complete engine assemblies, packages of supplies, or pieces of equipment. Poor material-handling techniques and practices can lead to a variety of injuries including back injuries, twisted or sprained muscles and joints, hand injuries, and foot injuries. Improper material-handling procedures can also result in damaged

Figure 2-8 Place oil rags and other combustible waste in an approved container.

equipment, tools, and facilities. Because they happen so frequently, back injuries are the most costly of all injuries in terms of medical costs and lost work time. Back injuries are often the result of poor material handling and improper lifting. Most back injuries occur when workers don't know (or ignore) the proper lifting techniques (Figure 2.9). Back injuries can also result from preexisting back problems that are worsened by lifting. Some workers know how to lift heavy items correctly but ignore proper techniques to get the job done faster. To prevent injuries, always use the following lifting techniques:

- Be sure that the weight of the load isn't beyond your capacity to lift. Usually, loads of more than fifty pounds require the assistance of a second person.
- Check that the path of travel from pick-up to drop-off is clear of obstacles.
- Get a good grip on the item to be lifted. If needed, wear gloves to improve your grip.
- Stand close to the load you're going to lift.
- Bend from the knees (squat) when lifting and setting down a load. Bending from the waist places more strain on the lower back.
- Lift with a smooth, controlled motion.
- Don't twist from the waist to place the load after lifting it. Instead, turn your entire body to set a load in place.
- Use caution when placing a load above chest height or below knee height. You put more strain on the lower back in those positions.

The following are some other suggestions to prevent back injury when lifting materials:

- Use hoists, hand trucks, carts, or dollies to lift or move heavy items. These lifting devices free you from heavy lifting and protect you from injury.
- Wear a back support belt to protect the back muscles.
- Always get help to move loads that are heavy or awkward in size or shape.
- Stretch your back and arm muscles before lifting. Stretching warms up the muscles and helps prevent muscle strains, pulls, and tears.
- Keep your back and stomach muscles in good shape. A lack of good muscle tone could contribute to a severe lifting-related injury.

USING PERSONAL PROTECTIVE EQUIPMENT (PPE)

To protect yourself from injuries in the workplace, use **Personal Protective Equipment (PPE)** when appropriate. PPE includes items such as dust masks, safety glasses, gloves, and special footwear (Figure 2-10). Remember that any task can be hazardous, even if the equipment is operated properly and all safety procedures are followed. Always wear personal protective equipment wherever the potential for injury exists. The type of PPE you need varies depending on the tasks you perform.

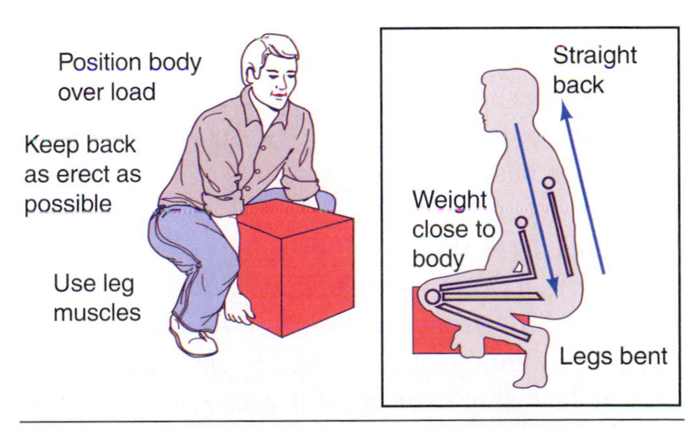

Figure 2-9 Always lift heavy loads using your leg muscles rather than your back muscles and get close to the object to make lifting easier.

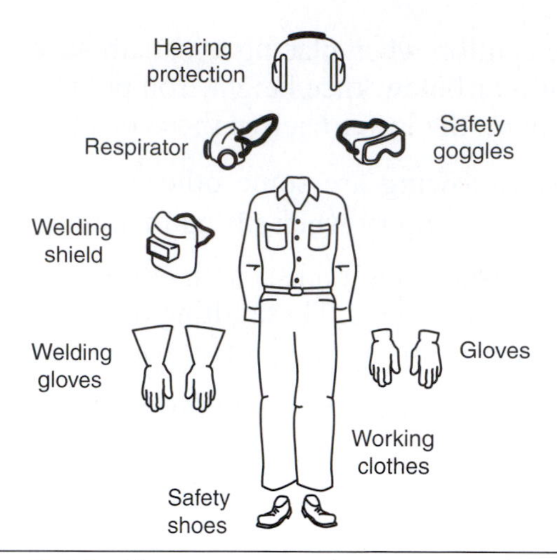

Figure 2-10 Always wear the appropriate personal protective equipment when working in a shop environment.

PROTECTING YOUR EYES AND FACE

Protective **safety glasses** and **goggles** (Figure 2-11) are available in a wide variety of styles to meet specific needs. Safety glasses with side shields provide more protection from impact and flying particles. Most safety glasses and goggles may be worn alone or over a worker's own prescription eyeglasses.

Splash goggles protect the eyes from dust, particles, and chemicals. They may contain ventilation holes to provide air circulation. Welding glasses have tinted or darkened lenses to protect the eyes from the bright flashes of welding arcs. A face shield is a cap-like device that holds a clear plastic shield over the face. The face shield protects the entire face from chemical splashes and flying particles.

Many people wear contact lenses as a replacement for glasses. Contact lens wearers must determine when it's appropriate for them to wear their contacts based on their working environment. Wearing contact lenses isn't recommended if the workplace has significant amounts of flying dirt or dust particles, or if chemical fumes are present. Remember that the only function of contact lenses is to correct your vision. They don't provide any eye protection from dust, impact, or splashes. You must still wear eye protection devices such as goggles or face shields over your eyes whenever your activity warrants such protection.

Sometimes even with the best safety practices people somehow get foreign objects in their eyes. Many shops have eye-wash stations or safety showers (Figure 2-12) that should be used whenever you or someone else has been sprayed or splashed with a chemical such as battery acid, fuel, or cleaning solvent. Have someone contact a doctor and get immediate medical attention under these conditions as well.

PROTECTING YOUR LUNGS

Respiratory-protection devices can prevent you from inhaling harmful dusts, gases, or vapors. Any

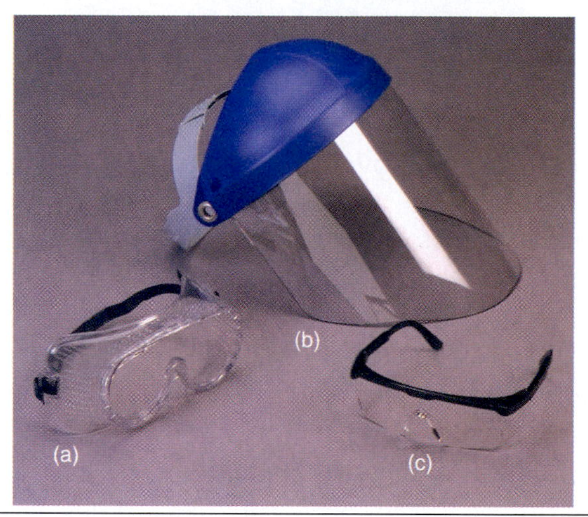

Figure 2-11 These eye protectors all have specific uses: (a) splash goggles, (b) face shield, and (c) safety glasses.

Figure 2-12 Eye-wash stations are available in properly equipped motorcycle service shops. If you need to use an eye-wash station you should follow up with a visit to a doctor to ensure no permanent damage to the eyes has occurred.

employee with an exposure to chemical fumes, dust, or any other irritants in the air should wear the appropriate respiratory protection. A typical dust mask is a small, fabric-like filter with straps that slip over the face to cover the nose and mouth. Dust masks are designed to shield the mouth and nose from dust particles. They don't filter out vapors, fumes, or gases.

Respirators (Figure 2-13) are more substantial devices than masks. Firefighters use a form of respirator device when they're called upon to enter a burning building. Respirators are made of heavy plastic, metal, and safety glass. The firefighter's version is nonflammable and insulates the user from the high temperatures of a fire. All respirators have their own oxygen supply. Because the person using the respirator doesn't breathe any of the smoke, fumes, vapors, or toxic gasses that might be present in the air, the respirator provides the best respiratory protection available.

PROTECTING YOUR HEARING

Question: How can you tell that you're in a high-noise area without using a sound-level meter? If another worker is standing three feet away and you can't have a conversation unless you shout, the work area is too noisy. Hearing protection should always be worn in areas with a high noise level. If you work eight hours a day in a high-noise environment without wearing hearing protection, you'll most likely experience a hearing loss over a period of years. If the noise level is extreme, you may suffer a hearing loss more quickly. You should always wear **earplugs** or a **headset** (Figure 2-14) in noisy areas or when you're using noisy tools.

Remember, there's no cure for noise-induced hearing loss. The prevention of excessive noise exposure is the only way to avoid hearing damage. Some earplugs are disposable (Figure 2-15), meant to be used once and thrown away. Others are intended to be cleaned and used repeatedly. A professional hearing specialist should individually fit preformed or molded plugs.

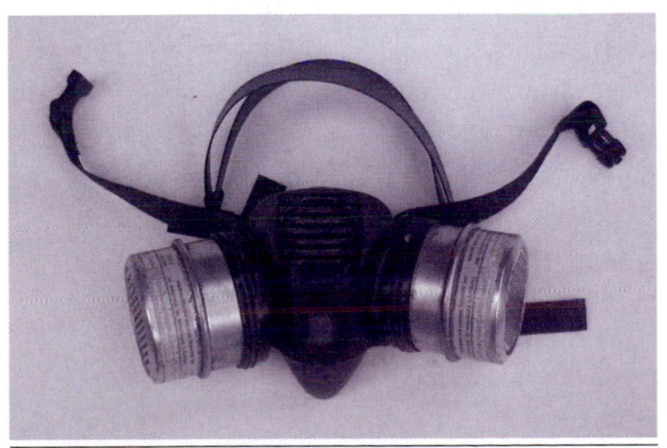

Figure 2-13 Whenever there is a danger of breathing harmful fumes, always use a high-quality respiratory breathing protector.

Figure 2-14 Protect yourself from permanent hearing damage from excessive noise by using high-quality headsets when you are working in high noise areas.

Figure 2-15 A typical pair of disposable earplugs with a small storage container is shown here. In a motorcycle shop environment these may be all that is needed to help protect your hearing in most cases.

PROPER DRESS ATTIRE

When working in a shop atmosphere, remove bracelets, necklaces, watches, and other jewelry. They can be caught in drive systems or possibly cause electrical shock if placed across an electrical circuit. To maintain a professional appearance, many shops have uniforms for their employees to wear when at work. These uniforms will include long pants and button-up shirts. Never wear shorts while working in a service shop as there will be no protection to your legs from hot surfaces such as an exhaust system.

PROTECTING YOUR FEET AND LEGS

Foot injuries are another exposure associated with material handling. These injuries usually occur when heavy materials or tools are dropped. To prevent foot injuries, it's a good idea to wear steel-toed safety shoes or boots. Various types of steel-toed safety shoes are available to provide different levels of protection. You can also attach **metatarsal guards** (special covers that go over the instep of boots) to your shoes to protect your feet. Open-toed shoes or sneakers are not considered appropriate footwear.

PROTECTING YOUR HANDS AND ARMS

Gloves, gauntlets, and sleeves protect the arms and hands from chemical splashes, heat, cuts, and tool-related injuries. In addition to standard leather and heavy-cotton construction, work gloves are also made from a variety of plastics and rubbers for use when working with solvents (Figure 2-16). Special gloves may also be designed to resist tears, cuts, and punctures. Gloves come in a variety of lengths to cover the hand, wrist, elbow, or entire arm, depending on the requirements of the job.

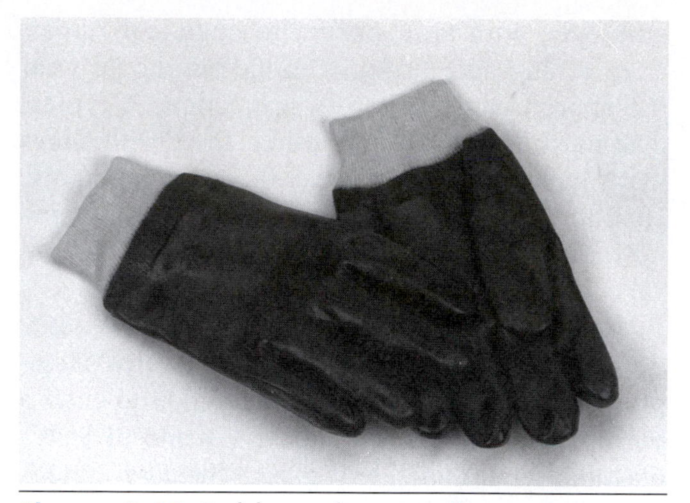

Figure 2-16 Rubber gloves will protect your hands when working with cleaning solvents.

USING TOOLS SAFELY

According to the **National Safety Council (NSC)**, more than 500,000 disabling, work-related finger and hand injuries occurred in a recent one-year period. The careless use of simple hand tools such as screwdrivers, wrenches, and hammers was the cause of many of these injuries. The most common hand and finger injuries are impact injuries (bruises, sprains, and broken bones), cuts, and puncture wounds caused by the improper use of hand tools. Improper use means using the wrong tool for the job, holding or using the tool incorrectly, or using a damaged tool.

Using Hand Tools Safely

Motorcycle technicians use dozens of different hand tools daily. By giving your tools proper care, you'll extend their useful life, and also lessen the possibility of accidents and injuries. To keep your tools in top working condition and to prevent injuries to yourself and your fellow workers, observe the following safety precautions:

- Always use the right tool for a job. Don't try to substitute one tool for another.
- Inspect hand tools often for defects. If you find a defective tool, repair it or replace it.
- When using a tool, comply with the manufacturer's instructions. Follow the instructions for the tool's use and maintenance.
- Never toss a tool to someone. Hand tools should be passed from one person to another by hand.
- Keep all tools clean. Protect the tools from corrosion. Wipe them clean when you're finished using them. Lubricate all tools with

moving parts to prevent wear and binding. Store tools in a dry and secure location.

- Keep the cutting edges on tools sharp. Sharp tools perform better and they save time.

Using Power Tools Safely

Power tools also have certain additional hazards associated with them. Common power tools that are used by a motorcycle technician include drills, power impact wrenches, and grinders. Electric power tools have three properties that contribute to their potential danger: electrical charge, high-speed movement, and momentum. The most important area of concern when using power tools is the electric charge. To avoid an electric shock when using an electric power tool, you must isolate and insulate yourself from the electric current by observing the following guidelines:

- Make sure that all electric power tools are properly grounded.
- Inspect electric power tools often for defective wiring. Visually inspect all power cords, plugs, and receptacles. Have qualified electricians replace all defective cords and plugs.
- Always unplug tools before replacing bits, blades, or grinder wheels.
- Never operate an electric power tool in a wet or damp area.
- Wear gloves to protect yourself from shocks. If you ever feel a shock or tingle when using a power tool, stop using it immediately.
- Use only extension cords that are rated to carry the current required by the power tool. An undersized extension cord can cause damage to the tool and can be a fire hazard if it overheats because of electrical overloads.

High-Speed Movement

High-speed movement is another area of concern when operating electric and air power tools. Avoid contact with any rotating tool parts because they could grab your hands, hair, or clothing. Keep all safety guards in their proper positions when operating power tools such as grinders and drills. When drilling metal, remember that friction from the high speed produces sharp, hot shavings that could cut or burn you. Also, note that if a drill becomes jammed in a piece of material, the momentum of its moving parts may cause it to spin out of control. To avoid being injured by a tool's momentum, remember the following guidelines:

- Hold all tools firmly. Pay attention to any sounds that may indicate that a tool is about to jam.
- Use only sharp cutting bits. A dull bit will frequently jam.
- Clamp or block all work pieces tightly to a firm work surface. Don't use your hands to hold the materials in position.
- Wear the appropriate personal protective equipment when operating power tools.
- Check tools for potential mechanical failure. For example, check for broken drill bits. In addition, check for faulty triggers and control switches that could cause unexpected start-ups and stops. Make sure the tools have all their guards in place.

SAFE RIDING PRACTICES

As a contributor to safe vehicle operating conditions, it's the motorcycle service technician's responsibility to verify that the set-up of each motorcycle and ATV is correct.

Each motorcycle manufacturer promotes riding safety by delivering a high-quality product. Quality ensures that the motorcycle or ATV has been designed with the rider's safety in mind. The motorcycle technician should always set up and service all vehicles with the same safety focus that the manufacturers used initially. Don't let a vehicle go out of the shop that you wouldn't be confident to ride yourself.

One of the most important elements of riding safety is the awareness and practice of safe riding habits. Riding safety also requires the proper riding apparel, and a properly maintained and serviced motorcycle or ATV.

The clothing that you wear while riding a motorcycle or ATV should provide visibility and protection. Leather jackets provide the best protection for your upper body. Denim provides some protection for your legs, but leather provides the best protection. Leather chaps worn over denim provide excellent protection for your legs. You should also wear sturdy footwear and gloves. A helmet and proper eye protection are the most important elements of riding apparel. Your helmet should fit securely and you should fasten the chinstrap snugly. No matter how experienced you are as a rider, an accident could happen at any time—while you're riding for pleasure, or while you're test-riding a motorcycle or ATV in the parking lot. Proper riding apparel doesn't guarantee that you'll be accident-free, but it will decrease the chances of serious or fatal injuries.

A good riding attitude is based on your understanding that a motorcycle or ATV is more vulnerable on the road or on the trail than a car or four-wheel-drive truck. Because motorcycles and ATVs are low-visibility vehicles and weigh less than nearly any other vehicle, the motorcyclist should be prepared to yield in all situations. Most car and truck operators have no real appreciation of how vulnerable motorcycles can be. They don't realize that stopping distances are different, or that motorcycles are less stable on gravel surfaces. Their ignorance can get the motorcycle rider in serious trouble. Always remember that fact, and ride accordingly. Always keep the odds on your side through proper vehicle maintenance and safe riding habits. You can contact the **Motorcycle Safety Foundation (MSF)** at 800-446-9227 or view their site online at *www.msf-usa.org*. Also, any motorcycle dealer will have further information on learning how to ride safely.

Chapter Summary

- It is very important to understand the importance of safety and accident prevention in a motorcycle shop environment.
- You should know the basic principles of personal safety.
- There are procedures and precautions for safety when using tools and equipment.
- You must maintain a safe working area in a service shop environment.
- There are laws concerning hazardous wastes and materials including right-to-know laws.
- You have rights as an employee and/or student to have a safe place to work.

Chapter 2 Review Questions

1. The three elements of the fire triangle are:
2. True or False? A Class B fire involves live electrical equipment.
3. Name the federal agency that publishes and enforces safety standards for business and industry.
4. What are some of the key safety areas of primary concern in a motorcycle service department?
5. What type of fire is created by a chemical reaction with combustible materials?
6. Safety is an ________ that helps you prevent injuries to yourself and others.
7. What are two types of electric power sources?
8. The letters PASS are used as a fire safety acronym. What do they stand for?
9. To prevent foot injuries, it's a good idea to wear what type of shoes?
10. When an engine is running, it creates exhaust gases that are hazardous if inhaled. The most dangerous of these gases is ________.

CHAPTER

3 Tools

Learning Objectives

When students have completed the study of this chapter and its laboratory activities they should be able to:

- Identify common hand, power, and special tools
- Know how to select the correct tool for a repair
- Understand advantages and disadvantages of various types of tools
- Identify and select the right measuring tool for different jobs
- Understand the importance of having an up-to-date service library

Key Terms

Adjustable wrenches (also called crescent wrenches)
Air ratchet
Allen wrench
Angled or remote screwdrivers
Aviation snips
Ball peen hammer
Basic hand tools
Beam-type torque wrench
Bench grinder
Bench vise
Box-end wrenches
Breaker bar
C-clamps
Center punches
Chisels
Clamps
Click-type torque wrench
Combination and rib-joint pliers
Combination wrenches
Compression gauge
Cutting pliers
Deep sockets
Dial indicator
Drill
Drill bit
Drill press
Engine Control Module (ECM)
Exhaust gas analyzer
Feeler gauges
Files
Flare-nut or line wrench
Flat chisel
Flat (or slot) tips
Flex handles
Fuel injected (FI)
Hacksaws
Hammers
Hex wrench
Hose clamp pliers
Impact screwdrivers
Locking or vise grip pliers
Mallet or soft-faced hammer
Metric
Micrometers
Multi-meter
Needle nose pliers
Offset screwdrivers
On Board Diagnostics (OBD-II)
Open-end wrenches
Phillips screwdriver tips
Pin and straight shank punches

Plastic-faced hammer
Pliers
Power impact wrenches
Pozidrive tips
Precision measuring tools
Pullers
Punches
Rechargeable cordless drills
Reed and Prince tips
Retaining-ring pliers
Rubber hammer
SAE (Society of Automotive Engineers)
Screwdriver
Screw extractor
Service manuals
Sliding t-handles
Snips or shears
Socket wrenches
Specification manuals
Speed handles
Standard wrench
Starting or aligning punches
Stubby screwdrivers
Taps and dies
Test light
Timing light
Toolbox
Torque
Torque wrenches
Torx tips
Torx wrench
Vernier caliper
Vises
Wire stripper/crimper pliers
Wrenches

INTRODUCTION

Motorcycle and ATV technicians use virtually hundreds of different tools to perform a wide variety of repair activities. In addition to the standard hand and power tools that you're already familiar with, there are specialized and precision repair tools that will probably be new to you. The assortment of tools that you'll use depends primarily on the types of vehicles you'll encounter, and the specific systems that you'll be responsible for. As an example, if you were to specialize in electrical systems, you'll need different tools than someone who specializes in internal engine repairs.

Skilled professionals, no matter what their trade or field, know how to use tools correctly and safely. Knowing exactly which tools to use for each task is essential for completing quality repair jobs quickly, safely, and efficiently. When you gain experience in the motorcycle and ATV repair field, you'll acquire the skills to do jobs faster and more efficiently. A large part of this skills development will focus on your ability to use the tools of the trade correctly.

Motorcycle and ATV repair tools can be divided into the following groups:

- Basic hand tools
- Power tools
- Special tools

It would be virtually impossible for us to discuss every type of motorcycle and ATV repair tool in this chapter. For this reason, we'll limit our discussion to the tools that you'll use most often. You've probably used many of the standard hand tools that we'll cover. But you may be totally unfamiliar with some of the specialized tools and measuring/testing instruments that we're going to introduce you to. We suggest that you take your time as you read about the various tools and instruments and familiarize yourself with their use.

BASIC HAND TOOLS

Basic hand tools are the common tools that are found in just about every workshop toolbox. Some of these basic hand tools include screwdrivers, hammers, pliers, wrenches, and socket sets. Due to the frequency of the use of these tools, most motorcycle repair technicians own a complete set of hand tools similar to the example shown in (Figure 3-1). Most of the basic hand tools are undoubtedly familiar to you.

You've probably used them, and may even own many of them. To be sure that you understand the proper use of these tools, we'll take a brief look at each of them. Let's begin with the most commonly used of all hand tools, the wrench. The largest portion of your current tool collection probably consists of different types of wrenches.

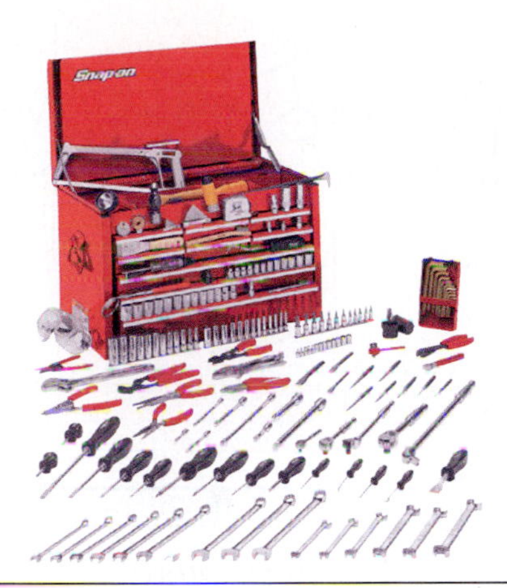

Figure 3-1 This illustration shows a typical set of hand tools that a motorcycle technician would own. This assortment of tools could be used for many different applications and is not limited to motorcycle repair.

Wrenches

Wrenches are used to tighten or loosen nut-and-bolt-type fasteners. As you probably know, wrenches come in a variety of sizes from very small to very large. The size of a wrench is determined by the width of the opening at the end of the wrench (Figure 3-2).

Metric and **SAE (Society of Automotive Engineers)** are the general classifications of the wrenches found in a motorcycle technician's toolbox. Metric wrenches are measured in millimeters, such as 10 mm, 11 mm, and 12 mm. SAE wrenches are measured in fractions of an inch, such as 1/2 inch, 9/16 inch, and 5/8 inch. As a rule of thumb, American-made motorcycles require SAE tools, and foreign-made (primarily Japanese) motorcycles require metric tools. More information on different measuring systems will be discussed in a later chapter of this textbook.

Wrenches are forged from strong, tempered steel. Each wrench size is designed to fit one particular-sized fastener. A common mistake when using wrenches is to use the wrong size (where the wrench "almost" fits). Using the wrong size wrench can in many cases damage the fastener. It could also damage the wrench. In addition to varying sizes, wrenches are also available in different styles (Figure 3-3 and Figure 3-4). The five most common styles are the following:

- Open-end wrench
- Box-end wrench
- Combination wrench
- Adjustable wrench
- Socket wrench

Open-End Wrenches

Open-end wrenches have U-shaped openings at both ends, and are designed so the length of each wrench is proportional to the size of its opening. The larger the opening, the longer the wrench's

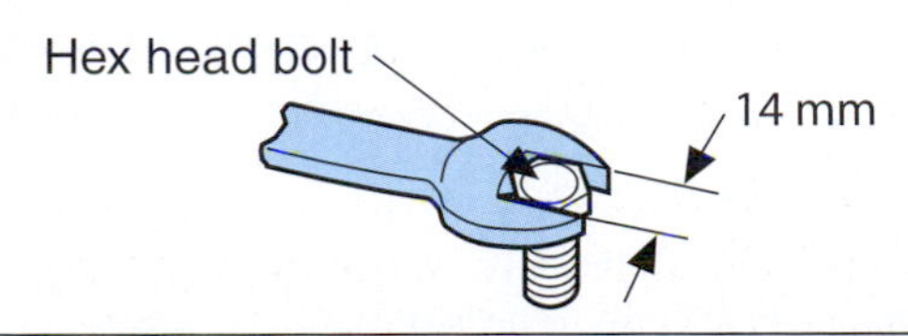

Figure 3-2 The size of a wrench is determined by the width of the opening at the end of the wrench.

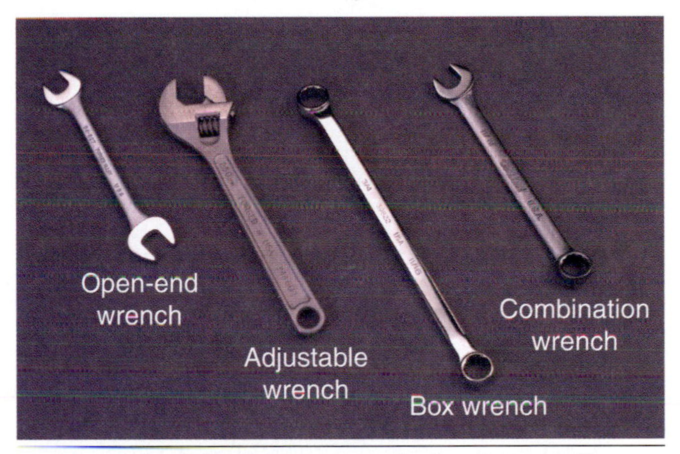

Figure 3-3 This illustration shows four different styles of wrenches.

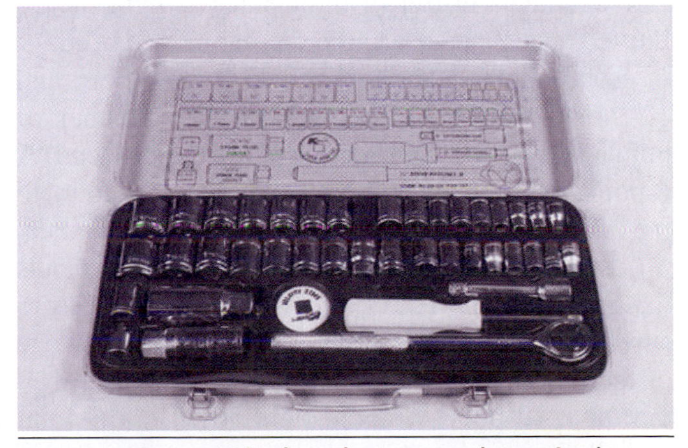

Figure 3-4 A typical socket wrench set is shown here.

handle; therefore more rotational force can be applied. Rotational force is referred to as **torque**. **Caution:** Because of the open-end wrench's design, you should never use an extension on a wrench handle to increase the torque. Using any type of handle extension could break the wrench and possibly cause an injury. There are a number of different types of open-end wrenches. Two common examples include the following:

- The **standard wrench** (Figure 3-5) has a 15-degree head angle with a different size opening on each end.
- The **flare-nut** or **line wrench** (Figure 3-6) is used for fuel and oil line fittings or brake lines. The head contacts 270 degrees of the nut for a secure grip while allowing for access around fuel, oil, and brake lines.

When using an open-end wrench, always place it squarely on the nut or bolt and pull toward you. Using this position reduces the chance of injury. Open-end wrenches have a tendency to slip when high torque is applied. If you must use a pushing motion, push with the palm of your hand. Don't grip the wrench with your fingers. You've probably heard the term "knuckle buster." Guess where that phrase came from? And always keep your wrenches clean to help prevent slipping.

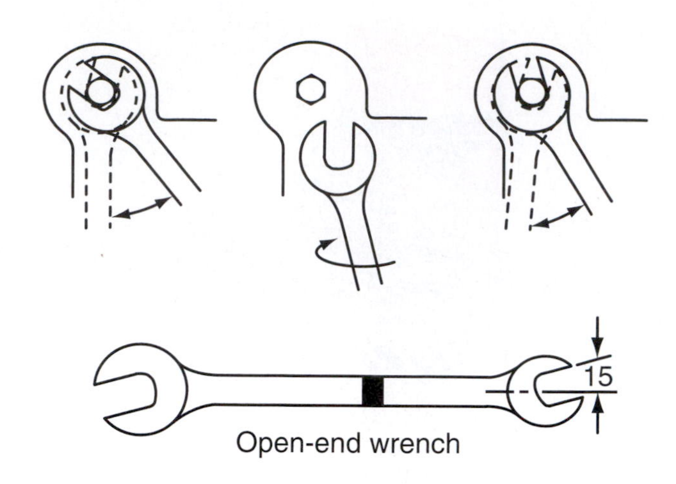

Figure 3-5 The standard wrench has a 15-degree head angle with a different size opening on each end.

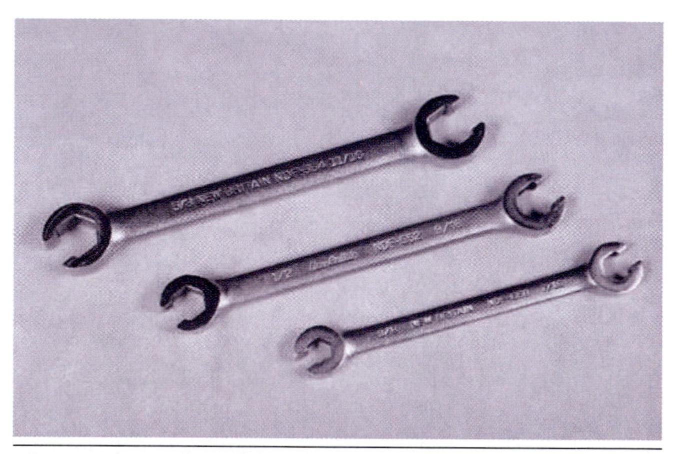

Figure 3-6 The flare-nut or line wrench is used for fuel, oil, and brake line fittings. The flare-nut wrench head contacts 270 degrees of the nut for a secure grip.

Box-End Wrenches

Box-end wrenches (Figure 3-7) should be used to loosen very tight bolts or nuts. The end of a box-end wrench encircles a nut or bolt head, providing more contact surface than an open-end wrench. Box-end wrenches have thin heads. This makes box-end wrenches useful in tight places where there's limited access space around the nut or bolt head.

Box-end wrenches are available with 6-point and 12-point openings (Figure 3-8). The number of points refers to the inside shape of the box end. A 6-point box-end wrench provides more support to the head of a bolt than the 12-point box-end wrench.

You should use a 6-point wrench on bolts and nuts that are very tight. The 6-point wrench is less likely to slip than a 12-point wrench. The main disadvantage of the 6-point wrench is that its head is thicker than that of a 12-point wrench. Furthermore, a 6-point wrench can be placed on a bolt or nut in only six different positions.

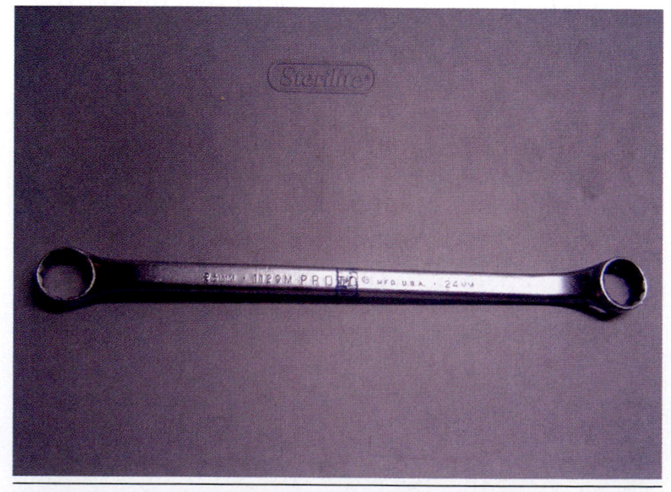

Figure 3-7 A typical box-end wrench is shown here.

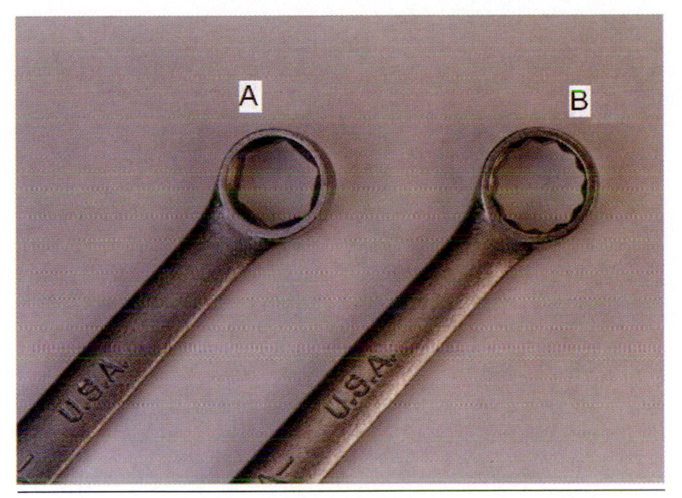

Figure 3-8 Box-end wrenches are available with (A) 6-point and (B) 12-point openings

Since its head is thinner and can be placed on the bolt or nut in twelve different positions, the 12-point style wrench is the better choice to use in tight places. You frequently can't get enough wrench travel in tight places to advance a 6-point wrench to the next turning position. With a 12-point wrench, the travel required to turn to the wrench to the next position is only half that required for the 6-point wrench. Remember, a 6-point box-end wrench is stronger and grips the fastener more securely, and a 12-point usually works better in tight spaces.

Combination Wrenches

Combination wrenches (Figure 3-9) combine the open-end wrench and the box-end wrench into one tool. In most cases, both ends of the combination wrench are the same size. This allows you to use the box-end to loosen bolts or nuts that are tight, and the open-end to quickly remove them when they're loose.

Adjustable Wrenches

Adjustable wrenches (also called **crescent wrenches**) have movable jaws that allow you to adjust the opening to fit almost any size nut or bolt (Figure 3.10). But adjustable wrenches don't grip the fastener as tightly as other types of wrenches.

Adjustable wrenches have a tendency to slip and round off the corners of nuts and bolt heads. Because of this limitation, adjustable wrenches should only be used to remove bolts or nuts that are already loosened. Adjustable wrenches can be used

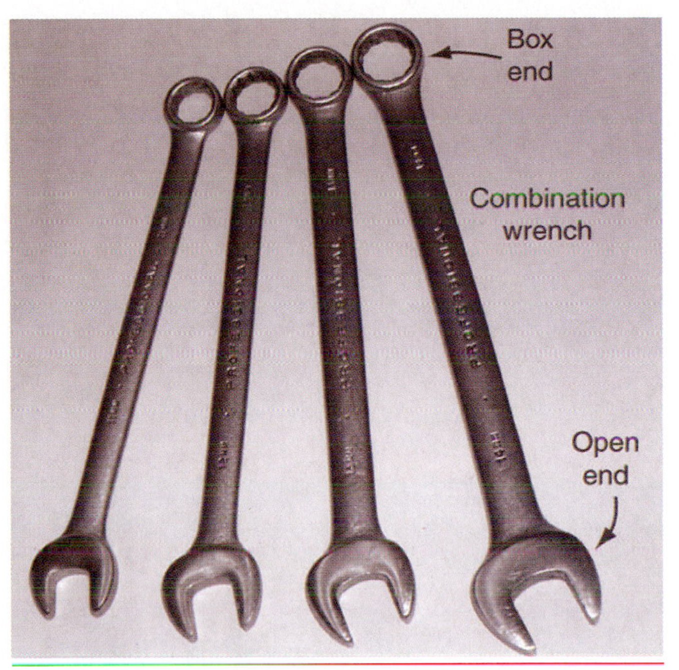

Figure 3-9 Combination wrenches are among the most popular wrenches found in a motorcycle technician's toolbox.

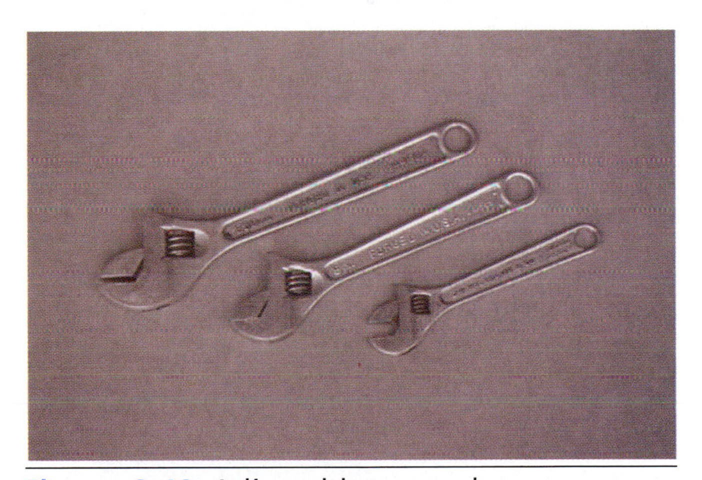

Figure 3-10 Adjustable wrenches are commonly called crescent wrenches and have movable jaws that allow you to adjust the opening to fit almost any size nut or bolt.

in a pinch, when the correct size of wrench isn't available. Remember that adjustable wrenches can slip! When using them, always pull toward you to save your knuckles, using the method shown in Figure 3-11.

Socket Wrenches

Socket wrenches are among the more frequently used tools. Socket wrenches are usually purchased in sets (Figure 3-12). The individual sockets contained in a set are in graduated sizes.

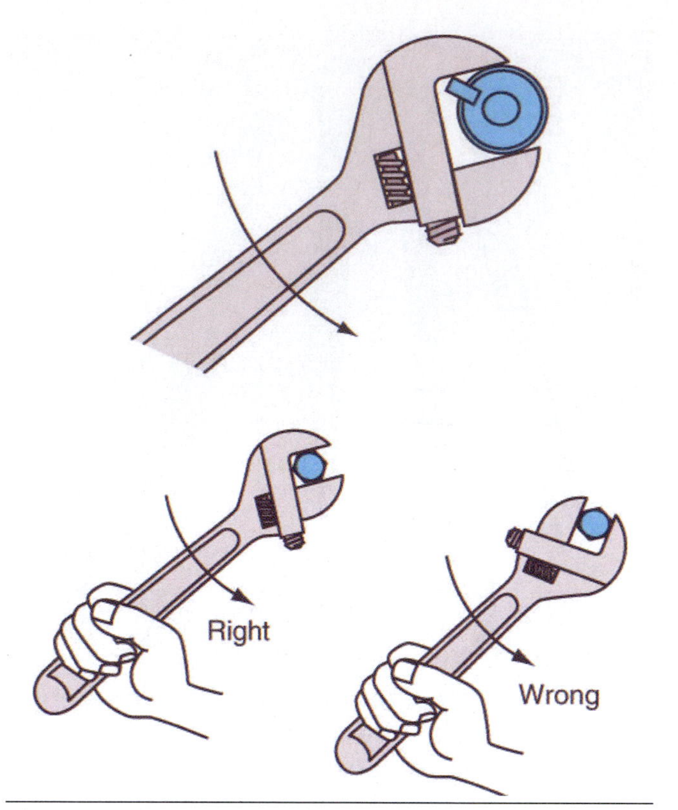

Figure 3-11 Proper adjustable wrench usage is shown here.

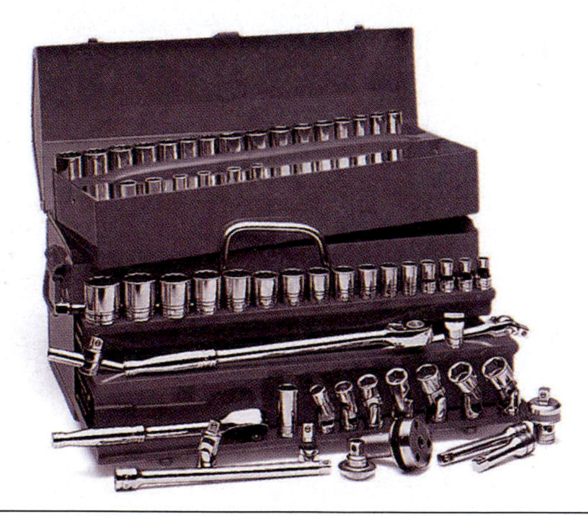

Figure 3-12 A wide range of different types of sockets is shown in this kit.

The size of the socket refers to the size of the bolt head or nut that the socket fits. Socket sizes range from small to very large, to match the largest available bolt and nut sizes.

There are two basic depths of sockets: standard and deep (Figure 3-13). **Deep sockets** allow access to recessed fasteners or nuts that are threaded onto studs.

Sockets are available in 6- and 12-point configurations just like box-end and combination wrenches (Figure 3-14). The 6-point is stronger. The 12-point is good for tight areas but is more likely to round off the nut or bolt. Some sockets have fluted corners designed to place stress on the sides of the fastener. This design tolerates higher torque with less chance of rounding off the nuts or bolt heads.

Sockets come in various degrees of hardness. The strongest are made for use with impact drivers and are usually black in color. Hand-type sockets are usually chrome-plated and are not meant to be used with air or electric impact drivers (Figure 3-15).

Sockets can be installed onto different types of handles (Figure 3-16). The drive lug on the socket handle fits into the drive hole in the socket. Sockets come with different drive hole sizes: typically 1/4 inch, 3/8 inch, and 1/2 inch. The most common handle is the reversible ratchet handle.

Figure 3-13 Examples of a standard and deep-well socket are shown here.

Figure 3-14 6- and 12-point sockets are shown here. Whenever possible it is recommended that a 6-point socket be used as it is stronger.

Figure 3-15 You can easily tell the difference between hand-type and impact-type sockets. The impact socket is black in color and the hand socket will be chrome-plated

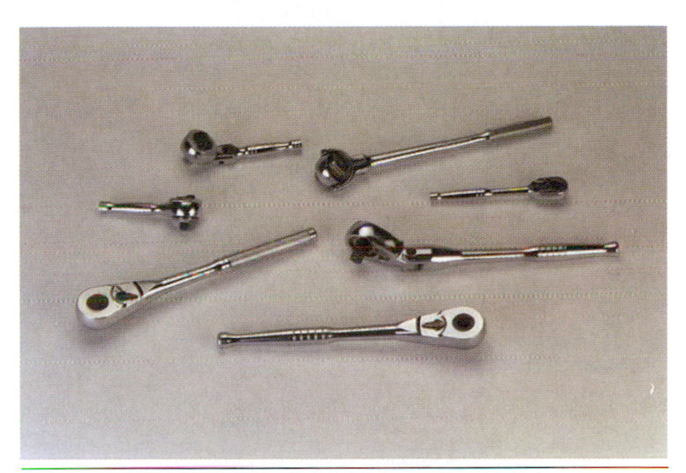

Figure 3-16 Different styles and sizes of ratchet handles for socket wrenches are shown here.

A lever on the handle allows you to change direction to either tighten or loosen fasteners (Figure 3-17). Reversible ratchet handles are great for removing and installing fasteners quickly, but they can be damaged if too much torque is placed on them.

If you're using a socket to loosen an exceptionally tight fastener, use a **breaker bar**, not a ratchet handle, to turn the socket. A breaker bar is a nonratchet solid handle with a socket drive fixed on its end. Because a breaker bar has no ratchet mechanism, it can withstand more torque than a ratchet handle.

Other popular handles include **speed handles**, **sliding t-handles**, and **flex handles** (Figure 3-18). These handles are often used to turn sockets

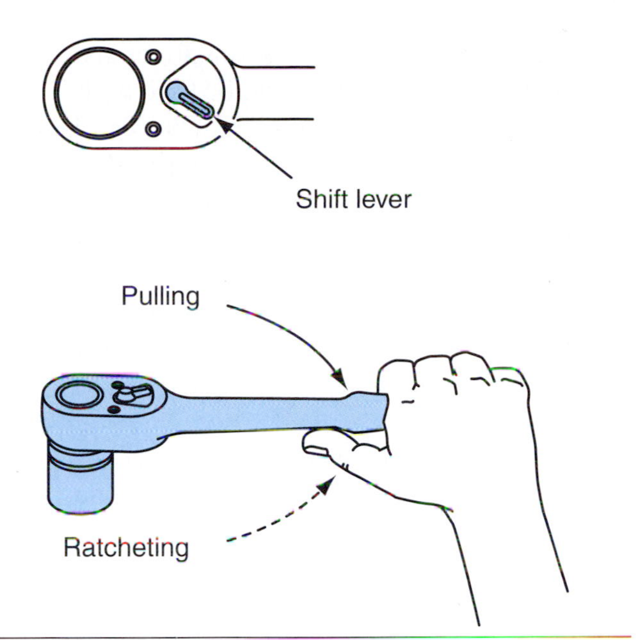

Figure 3-17 This illustration shows a reversible ratchet handle. Note that it has a shift lever that is used to change the direction of the ratchet action.

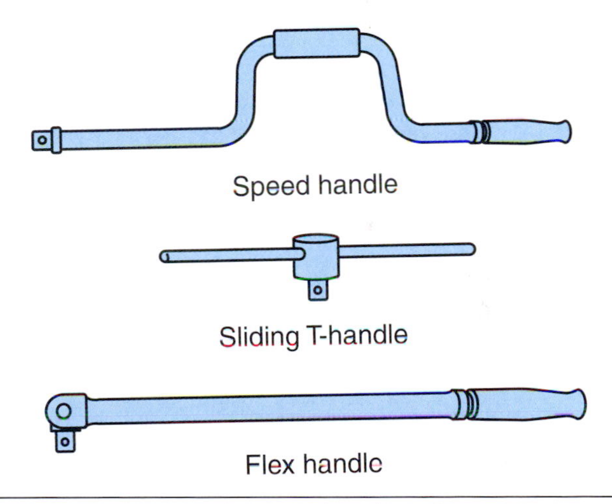

Figure 3-18 Popular types of socket drivers are shown here.

in tight spaces. Most socket sets also contain a variety of extension bars and adapters. These accessories allow the sockets to be used in many different situations.

Allen and Torx Wrenches

The **Allen wrench** or **hex wrench** is a short, six-sided rod that is used to tighten screws and bolts that contain similar six-sided (hex) indentations (Figure 3-19). A typical Allen wrench has a right-angle bend near one end. The bend forms a convenient handle. Certain special Allen wrenches are equipped with t-handles or screwdriver-style handles, and others are designed to be used with socket wrench drive handles. Allen wrenches can be purchased individually, but they're normally sold in sets that contain all of the commonly used sizes.

The **Torx wrench** is similar to the Allen wrench, except the end of the Torx wrench is star-shaped (Figure 3-19). The Torx wrench has the ability to handle more torque without slipping or stripping—Torx wrenches and bolts are used where higher fastening strengths are required. Torx bolts are ideally suited for use with impact tools. Torx wrenches are also available with screwdriver-type handles, or are attachable to socket wrench drive handles.

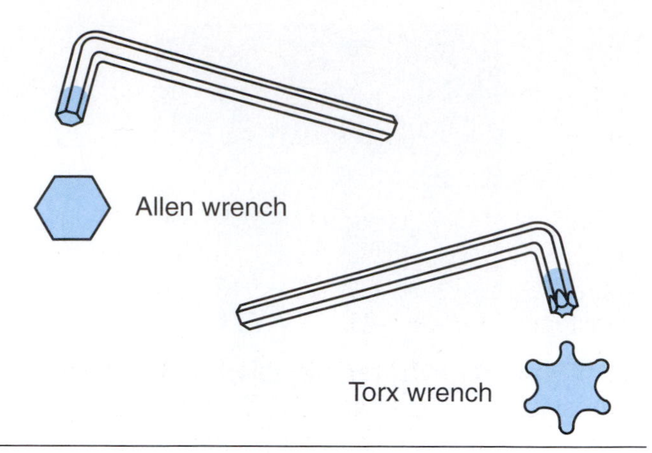

Figure 3-19 This illustration shows Allen and Torx wrenches.

Screwdrivers

Even though just about everyone is familiar with the standard **screwdriver** (Figure 3-20), let's do a quick review. A standard screwdriver has the following parts:

- A *handle,* which is usually made from plastic or wood.
- A *blade shank*, which can be round, square, or hex. If it's square or hex, you can use a wrench on the screwdriver to increase torque.
- Some screwdrivers have a *bolster* to allow you to use a wrench.
- A *tip* or *blade,* which you insert into the screwhead slot that in the most common cases is a *straight tip* or *Phillips* head.

The screwdriver is one of the most abused hand tools. Have you ever tried the following with a screwdriver:

- Used a screwdriver as a pry bar or chisel?
- Used a screwdriver handle as a hammer?
- Hammered directly on a screwdriver handle?
- Used the wrong size screwdriver?
- Used the wrong type of screwdriver?

See what we mean? There are so many opportunities to misuse screwdrivers. Almost every one of us has done one or more of these things at one time or another. If used improperly, the screwdriver can cause damage. It could even cause serious injuries to you.

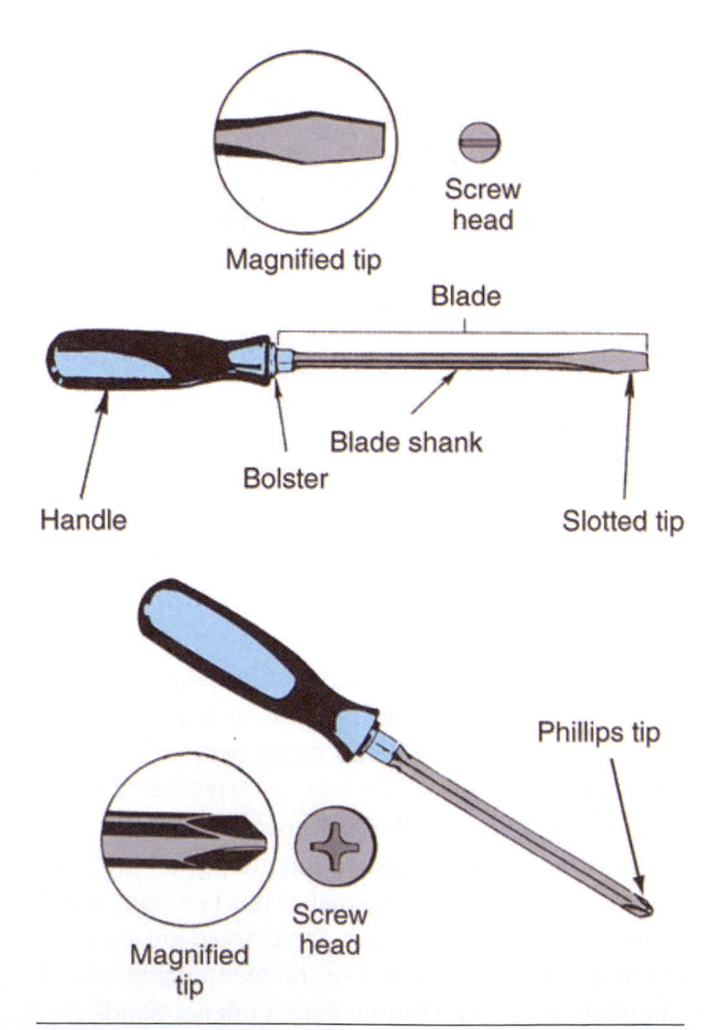

Figure 3-20 The basic parts of a screwdriver are shown here.

Types of Screwdrivers

Screwdrivers are available in a variety of shaft lengths and tips. The following are a list of the most common:

- **Flat (or slot) tips** range in size from very small (1/6-inch wide) to large (3/4 inch or wider). Be sure to always choose the correct size for the fastener (Figure 3-21).
- **Phillips screwdriver tips** (Figure 3-22) have a crossed point but a somewhat blunt end. These tips have good holding power and are less likely to slip than slotted tips. The common sizes of Phillips screw-head slots used in motorcycle work are #1, #2, and #3.
- **Reed** and **Prince tips** are similar to Phillips but have a sharper tip. Reed and Prince tips are not interchangeable with Phillips-tipped screwdrivers.
- **Pozidrive tips** are similar to the Phillips and Reed and Prince tips.
- **Torx tips** are used when higher fastening strengths are required, because they can handle more turning force without slipping.

Special Purpose Screwdrivers

Screwdrivers are also available in special shapes to provide access in restricted areas. Several of the special-shaped screwdrivers are described here:

- **Offset screwdrivers** are special purpose tools with an angled tip to allow access where space is limited. The offset screwdriver comes with either slotted or Phillips tips and is available in different sizes.
- **Stubby screwdrivers** have a short shaft and a short, fat handle. Like the offset style, the "stubby" is used in cases where there's limited access space. The stubby screwdriver is available with either slotted or Phillips tips.
- **Angled** or **remote screwdrivers** have a hollow tube with a handle on one end and a bit on the other end. They're used for jobs where you can't get hand access. Some are curved or angled. Deluxe versions may have a 1/4-inch square drive end to accommodate interchangeable tips.
- **Impact screwdrivers** are used very often on motorcycles and ATVs to remove and install fasteners where hand or wrench torque is insufficient (Figure 3-23). The drive mechanism of the impact screwdriver converts the impact force of a hammer blow to rotational torque. This torque is transferred to the screwdriver tip.

The end of the impact screwdriver is designed to accept a variety of tips. When using the impact screwdriver, always remember the following:

- Use the correct size bit
- Use a bit that was designed for impact use
- Use eye protection

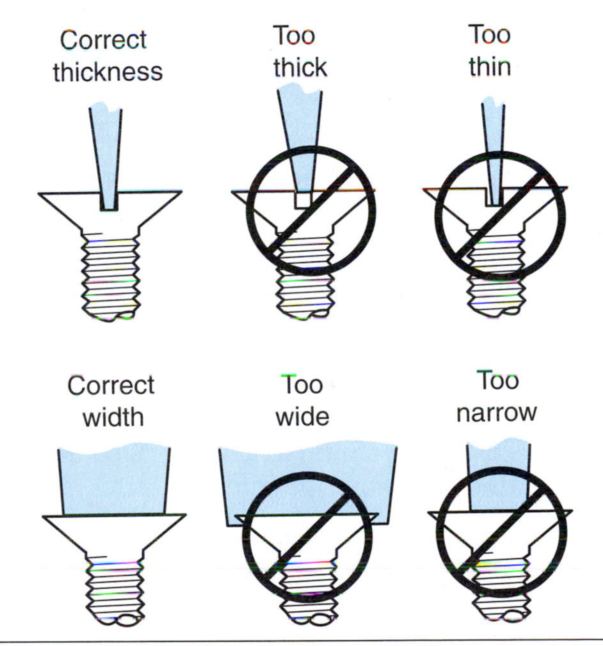

Figure 3-21 The blade of a screwdriver should fit the screw slot snugly to prevent slipping or slot damage.

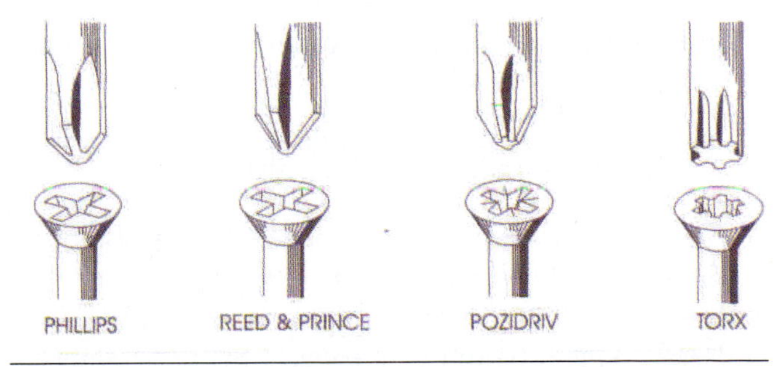

Figure 3-22 These screw head shapes are used in many applications on motorcycles and ATVs.

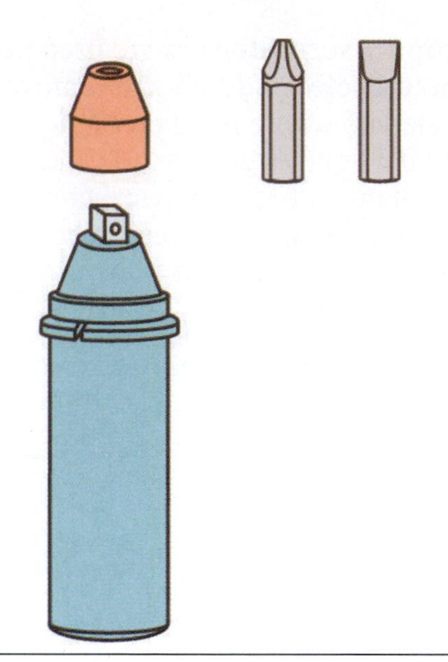

Figure 3-23 A typical impact screwdriver is shown here. Note the removable cap that would allow a socket wrench to fit on the end of the tool.

To use the impact screwdriver, install the tip onto the fastener and twist the tool in the desired direction. Hold the tool firmly and keep the tip squarely on the fastener and then strike the tool sharply with a hammer (Figure 3-24).

Observe the following rules when you're using any screwdriver:

- Always clean the slot(s) in a screw head before attempting to remove the screw.
- Always hold a screwdriver so the shaft is at a 90 degree angle to the screw slot.
- Make sure that the screwdriver blade fits a screw slot snugly to prevent slipping and possible damage to the screw and/or screwdriver.
- Never use a screwdriver to cut or remove metal, to punch holes, to pry, or for any other unintended purpose. This could damage the tool and could cause an injury.
- Never hammer on the handle of the screwdriver (except for impact screwdrivers).
- Never use a screwdriver to work on an object that you're holding in your hand. The screwdriver could slip and cause a painful injury. Use a bench vise to support the object that you're working on.

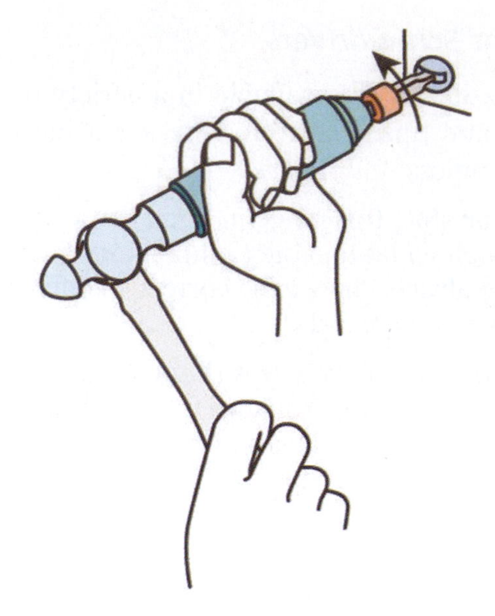

Figure 3-24 When using an impact screwdriver, hold the tool firmly and keep the tip squarely on the fastener and then strike the tool sharply with a hammer.

Pliers

Pliers are another commonly used hand tool. There are a variety of different types of pliers (Figure 3-25). They vary in size and shape, depending on their intended function. Certain pliers are designed for holding or gripping, others are designed for shaping, and others are used for cutting.

Types of Pliers

Combination and **rib-joint** (also known as *slip joint*) **pliers** are generally used to hold parts when you work on them, or to twist and bend materials. Rib-joint pliers have a slip joint that lets you open the jaws wide to grip large-diameter items. The jaws have gripping teeth. The outside end of the jaw set is for grasping flat objects, the middle is for grasping curved objects.

Locking or **vise grip pliers** are functionally similar to combination rib-joint pliers. Locking pliers are used to get a firm holding grip on items. Locking pliers can be locked in place to hold parts tightly while keeping both of your hands free. They can function as a small, portable vise. For example, you can use locking pliers to hold two metal parts in position while you install screws, washers, or bolts.

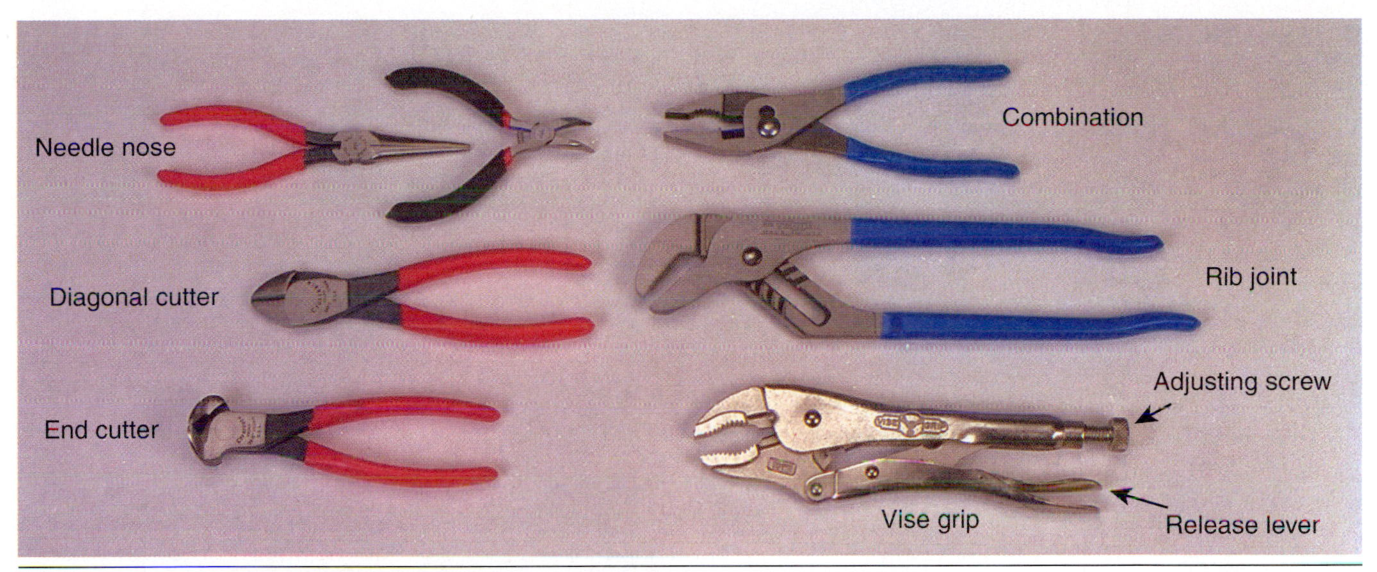

Figure 3-25 A wide variety of pliers are available. These are just a few that would typically be found in a motorcycle technician's tool collection.

Needle nose pliers are useful for gripping or twisting small parts and for reaching parts in limited access spaces. The jaws of needle nose pliers are smaller and thinner than those of long nose pliers. Needle nose pliers allow you to easily move and position parts that are too small to be handled properly with your fingers. We point out here that you can easily damage these pliers by using them for heavy work. The tips of the jaws will break if you apply too much pressure on them. Needle nose pliers are often found with bent jaws to help get to tight spaces.

Cutting pliers come in different shapes and sizes with the most popular being diagonal and the end-cutter designs.

Retaining-ring pliers (Figure 3-26) are used to spread or compress retaining rings when these rings are being removed or installed. There are two main types of retaining-ring pliers: those used for internal retaining rings, and those used for external retaining rings. These pliers are normally either long nose or angle nose to allow for varying access angles. Some models of retaining-ring pliers come with replacement tips. External retaining rings fit into grooves machined in the outer surfaces of shafts. Internal snap rings are used to retain components on the inside of hollow shafts.

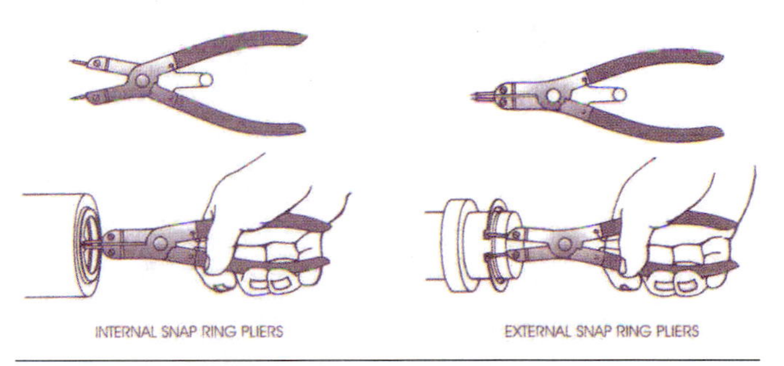

Figure 3-26 This illustration shows both internal and external retaining-ring pliers in use as well as you would see them standing alone.

Special Purpose Pliers

The following types of pliers are also often used when working on motorcycles and ATVs.

Hose clamp pliers have grooves cut into the jaws to provide positive gripping when removing or installing wire hose clamps.

Wire stripper/crimper pliers (Figure 3-27), commonly referred to as *electrical pliers*, are used to remove insulation from wire and install crimp-type electrical connectors.

Snips or **shears** are used for cutting sheet metal and metal gasket material. They come in several styles and sizes to match the material being cut.

Aviation snips are used for cutting sheet metal and come in left, right, and straight cutting styles.

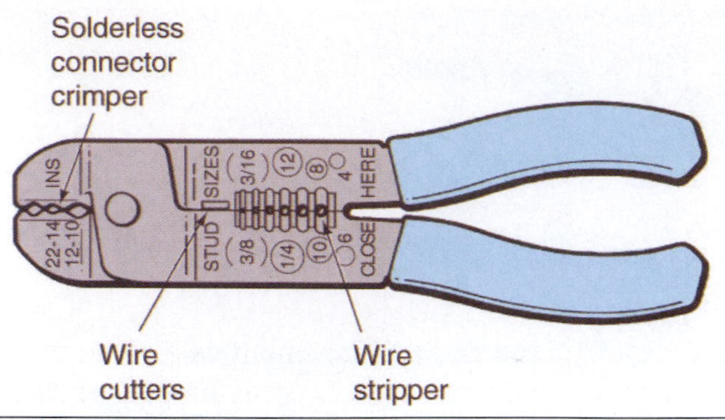

Figure 3-27 Wire stripper/crimper pliers are commonly referred to as electrical pliers and are used to remove insulation from wire and install electrical connectors.

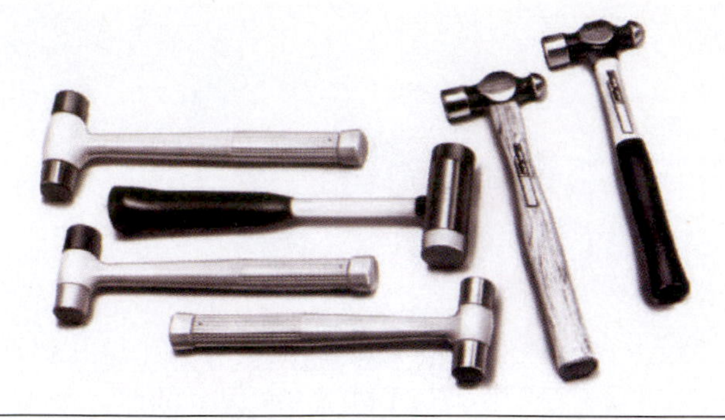

Figure 3-28 Various types of hammers are shown here.

Figure 3-29 The ball peen hammer and rubber mallet are among the most commonly found in a motorcycle technician's toolbox.

Hammers

Motorcycle technicians use a wide variety of **hammers** in their daily activities that range in size from 2 ounces to 48 ounces (Figure 3-28). Ball peen and mallet (or soft-faced) hammers (Figure 3-29) are two of the most common types.

The head of a **ball peen hammer** has two opposing striking surfaces: a flat-faced surface and a rounded surface. The flat-faced surface is used for regular hammering. The ball end is used primarily for shaping cold metal.

You should use a **mallet** or **soft-faced hammer** if a ball peen hammer might damage the part that you're working on. Several types of soft-faced hammers are commonly used in a motorcycle repair shop.

- **Rubber hammers** are used to help seat tire beads and to do sheet metal work. If you're concerned about damaging a part, use a plastic-faced hammer.
- **Plastic-faced hammers** are available with replaceable heads of varying densities (hardness). Almost all *dead blow hammers* are made from high-impact plastic. The plastic shell is filled with lead shot that helps direct the force from the hammer blow and prevents the hammer from bouncing on impact. They're frequently used where a heavy, nondamaging blow is needed, such as installing a bearing.

The following are a few things to remember when using a hammer:

- Make sure that the head is securely attached to the handle.
- Wooden hammer handles should be replaced if damaged. Be sure that the handle fits the hammerhead securely. The handle requires the correct-size wedge to properly retain the head.
- You can't repair fiberglass or steel hammer handles. Replace the hammer if the head becomes loose or damaged, or if the handle is cracked.
- Grip the hammer close to the end of the handle to provide better control of the tool and strike a stronger blow.
- Always strike the hammer face parallel with the object being hit.
- Always wear safety goggles to protect your eyes when using a hammer.

Punches and Chisels

You'll use an assortment of **punches** and **chisels** when working on motorcycles and ATVs (Figure 3-30). Punches are used for aligning and driving items while chisels are used for cutting.

Punches

Center punches have tempered ends with sharp points. They're used for punching indentation marks in metal. These marks are used as reference points for measuring or as starting points for drilling.

Starting or **aligning punches** have tapered shafts with flat tips. They're used to align holes or to start pins moving.

Pin and **straight shank punches** have no taper. They have a flat tip and are used to drive out pins. If you use the correct-size drift punch, you won't enlarge the hole and you won't damage the end of the pin.

The following are a few things to remember when using a punch:

- Hold the punch with a firm, but not overly tight grip.
- Hold the pointed end of the punch in place while striking the other end with a ball peen hammer.
- Strike the end of the punch squarely.
- Always wear approved eye protection when using punches.
- The ends of punches wear or get misshapen with use. When the end surface enlarges beyond its normal size or mushrooms, you should grind the bit end back to its original shape.

Chisels

Chisels (Figure 3-31) are another of those often misused and abused tools! They're not meant for opening paint cans, tightening or loosening screws, or prying things apart. Unfortunately, more than one chisel met an early end trying to perform one of these unintended functions! Chisels are meant to be used for cutting, shearing, and chipping.

The **flat chisel** is the most common type of chisel used in a motorcycle and ATV repair shop. The chisel has a bevel on both sides of the cutting edge. Like the punches that we just covered, the striking head on the chisel is softer than the cutting edge.

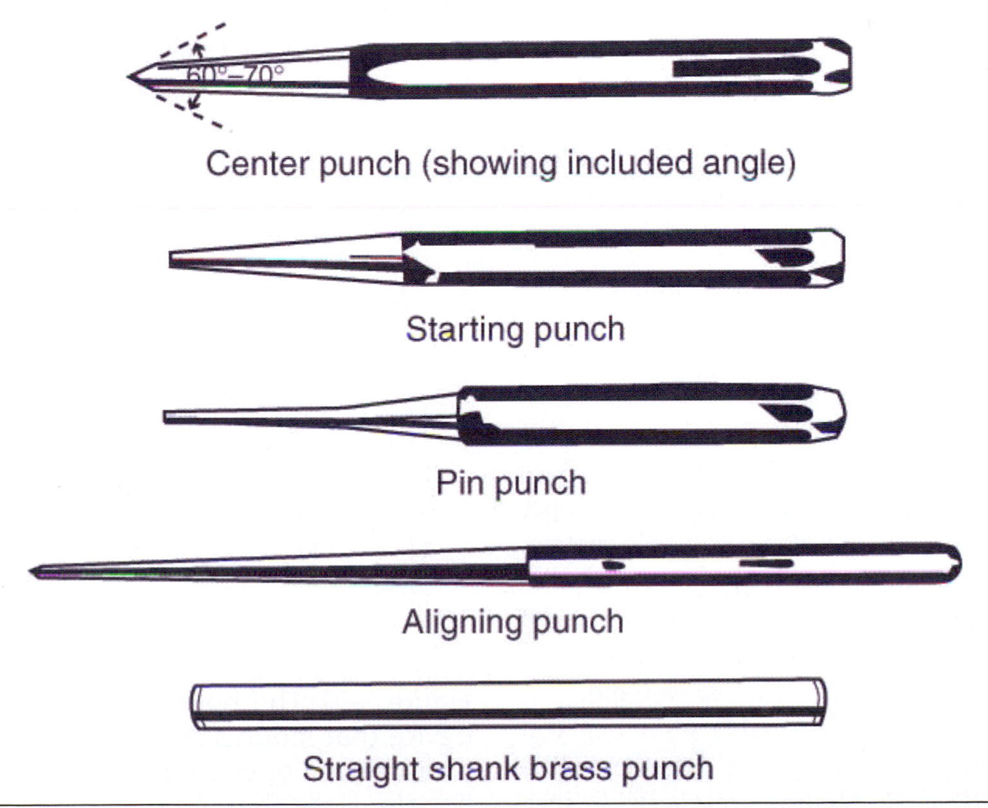

Figure 3-30 This illustration shows examples of punches used on motorcycles and ATVs.

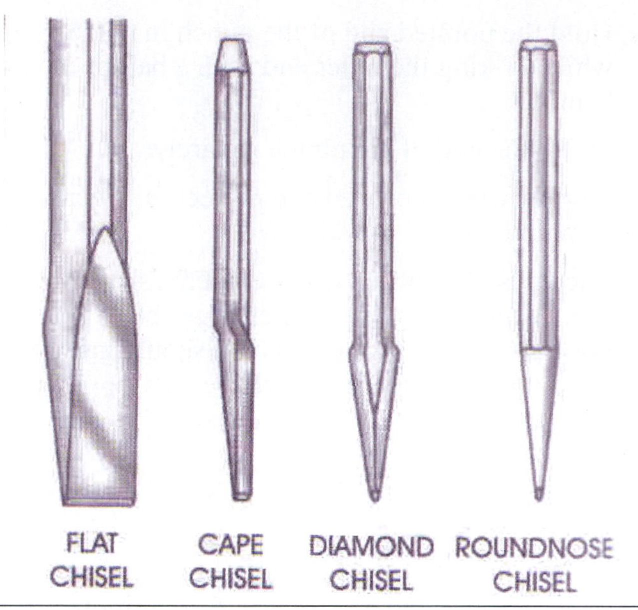

Figure 3-31 This illustration shows several different types of chisels.

The following are a few things to remember when using a chisel:

- The chisel head diameter should be approximately one half the diameter of the hammerhead that you use for striking the chisel.
- Hold the chisel the same way you would hold a punch. Don't hold it too tightly, but use a firm grip.
- Chisels wear with use. You should grind the head when it shows signs of mushrooming.
- Always keep the cutting edges of chisels sharp. Grind the cutting edges to the original contours and angles.
- Always wear eye protection when using chisels.

Clamps and Vises

Clamps and **vises** are used to hold work pieces securely. This frees up both of your hands so that you're better able to handle tools. Clamps and vises are often referred to as an extra pair of hands. Proper and timely use of a vise or clamp can help prevent injuries and eliminate damage to expensive parts and components. If you're ever in doubt as to whether you should use a vise, use it! It only takes one slip or accident to make you wish you did.

C-clamps are basically portable vises that you can use to hold pieces of material together while you work on them. C-clamps are available in a variety of sizes to accommodate a wide range of applications. Most motorcycle repair shops have a collection of C-clamps, with multiples of each size.

The **bench vise** (Figure 3-32) is a useful holding device that clamps onto a workbench or table edge. The jaws of a vise are opened and closed by turning a handle. The jaws are usually covered with soft metal. This covering protects the piece being held in the vise from scratches and dents. There are also jaw protectors made of soft materials to better protect items placed in a vise.

Cutting Tools

Various types of cutting tools will be found in a motorcycle service shop. The most common that you will find are covered in this section.

Hacksaws

Hacksaws are used to cut metal stock that's too heavy to be cut with snips or cutting pliers (Figure 3-33). Hacksaws are available with adjustable frames to accommodate different-length blades. A wing nut on the blade retention bracket is used to tighten the blade to the correct tension. An improperly tensioned blade doesn't cut well, and dulls quickly. Hacksaw blades are available in a variety of lengths, hardnesses, and tooth patterns for cutting different materials. Generally speaking, harder and thicker materials require coarser, harder hacksaw blades.

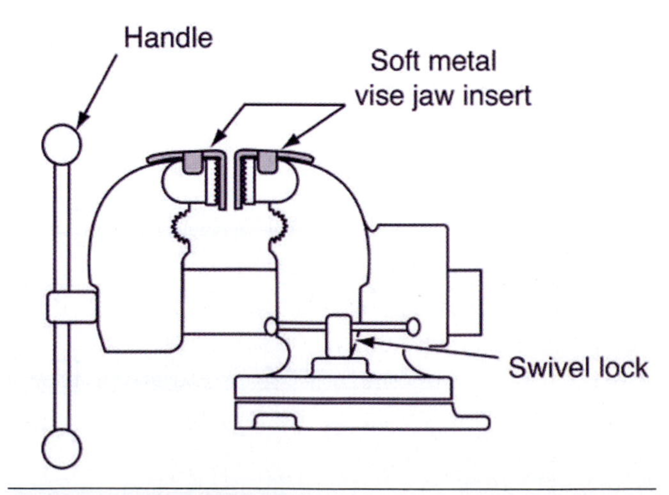

Figure 3-32 The bench vise is a useful holding device that mounts onto a workbench. The jaws of a vise are opened and closed by turning the handle. Note that the jaws are covered with a soft metal insert to protect the items being held.

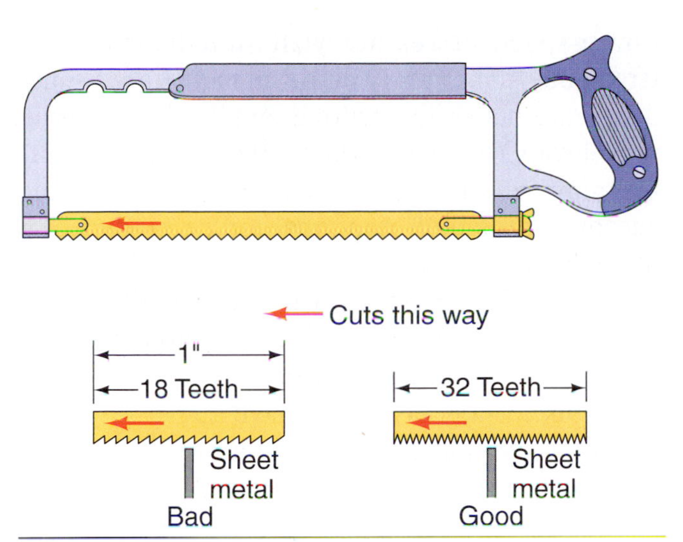

Figure 3-33 A standard hacksaw is shown here with different types of blades.

When using a hacksaw, it's important to apply pressure only on the forward stroke. Use little to no pressure on the backstroke.

Files

Files are cutting tools used to smooth surfaces and edges. Files are classified by their cross-sectional shapes and by the types of teeth (Figure 3-34). The main parts of the file are the face, edge, and tang.

Taps and Dies

Taps and **dies** (Figure 3-35) are used to cut threads in metal stock. A *tap* cuts "female" threads (equivalent to the threads in a nut). A *die* cuts "male" threads (like the threads on a screw or bolt). There are different types of taps for various uses. The tap is turned using a special wrench called a *tap wrench*. Dies are turned using a *die wrench*.

Here's a quick overview for using taps and dies. To use a tap, insert the proper-size tap into a predrilled hole in the metal stock, and screw in the tap using a back-and-forth motion until it stops. To use the die, clamp the bolt or screw stock into a vise, and screw down the die using a back-and-forth motion to cut the threads (Figure 3-36). Note that the actual use of these tools is complex, precision work that requires skill and experience. We've simplified the description so that you can distinguish between using a tap and using a die.

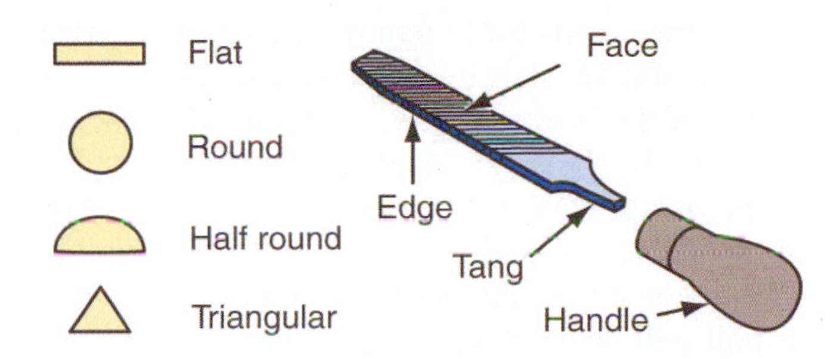

Figure 3-34 The parts of a file are illustrated here along with different types of shapes that you would see used by a motorcycle technician.

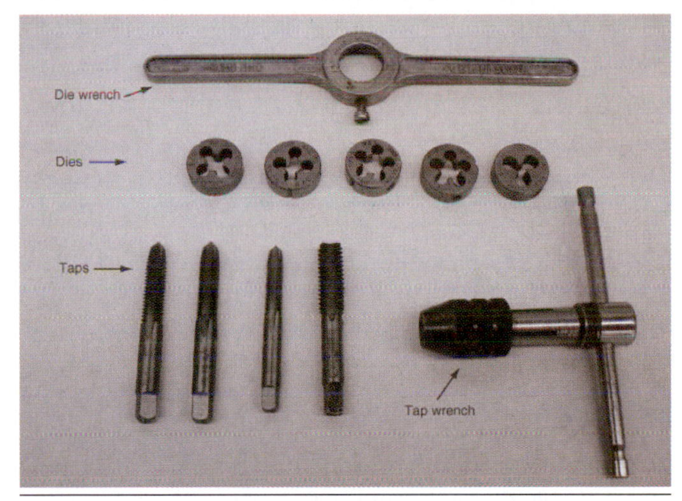

Figure 3-35 Taps and dies are used to cut threads.

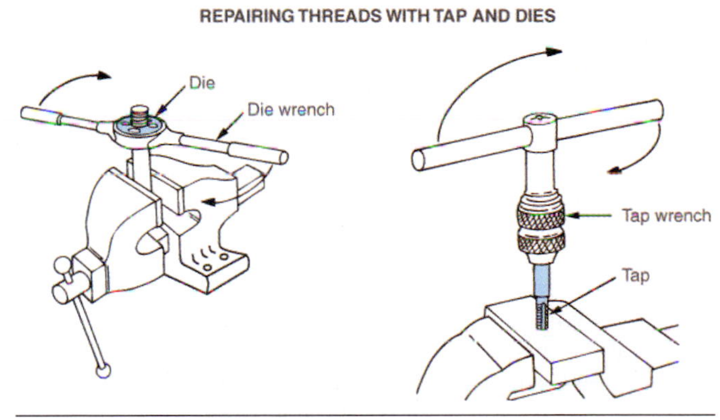

Figure 3-36 This illustration shows the use of a tap and die to repair threads in different types of fasteners.

Screw Extractors

The **screw extractor** (Figure 3-37) falls under the cutting area and is a valuable tool that's similar in operation to the tap and die. If a bolt or screw head has been accidentally sheared off, you can use a screw extractor to remove the screw. To use a screw extractor, start by drilling a small hole in the center of the broken screw. Then thread the screw extractor into the hole and twist it in a counterclockwise direction. The screw extractor self-taps into the broken screw. Continue turning in the counterclockwise direction and the broken screw or bolt will be extracted.

POWER TOOLS

Power tools are essential to the motorcycle and ATV technician. The proper use of power tools will help to make you more efficient at your job. You must exercise extreme care when using power tools. Not only can they cause severe injuries (which we'll cover soon in the section on safety), they can also break fasteners, warp covers, strip threads, and permanently damage components. The power-operated hand tools that we'll cover are either driven by air pressure or powered by electricity.

Drills

One of the more common power tools is the **drill**. A power drill, sometimes referred to as a drill motor, is a handheld device that's used to drive a drill bit.

The **drill bit** is the tool that bores through the material. The end of the bit that attaches to the power drill is the *shank*, and the power drill's socket that holds the drill bit is the *chuck*.

Power drills come in a variety of speeds. Speed is measured in rpm's *(revolutions per minute)*. You should consider rpm values before selecting a drill. Lower-speed drives are well suited for drilling large holes in certain types of metal. Higher-speed drives are better for drilling smaller holes. Drills with lower rpm values are often equipped with gripping handles. These handles are necessary because of the high rotational torque produced by these tools.

Some drills are equipped with variable-speed motors that are controlled by movable triggers. These variable-speed drills are very handy for day-to-day use. Many drills also have a reverse feature that allows you to back a drill bit out of the material you're drilling.

Power drills come in two common shop sizes, 1/2 and 3/8 inch (Figure 3-38). These measurements refer to the maximum diameter of the drill bit shank that the chuck will accept. The larger 1/2-inch power drill is used for special low-speed, high-torque situations. The 3/8-inch drill is more commonly used for general purpose applications in the shop. The 3/4-inch drills are not as common in motorcycle and ATV repair shops but are also available for drilling larger holes.

Drill bits (Figure 3-39) are manufactured from hardened steel rod stock. Drill bits are available in a variety of shapes and lengths. They're frequently sold in graduated-size sets, but each size is available separately.

Drill bits are available in both SAE and metric sizes. A well-equipped repair shop has at least one complete set of each. The drill bits used in a handheld power drill are the same as those used in a

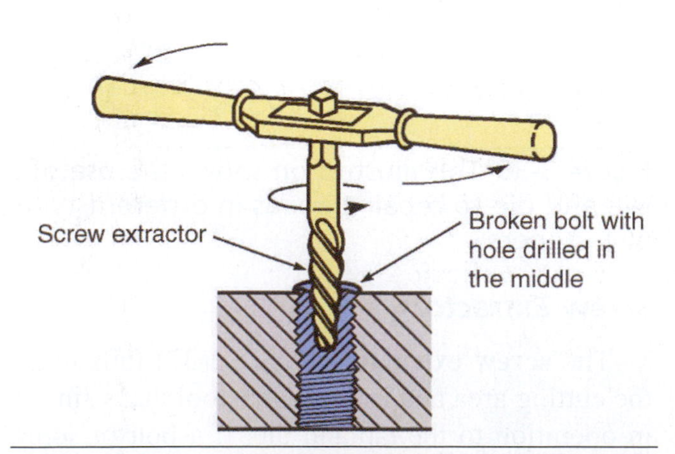

Figure 3-37 A screw extractor is a very handy tool to assist in the removal of broken fasteners.

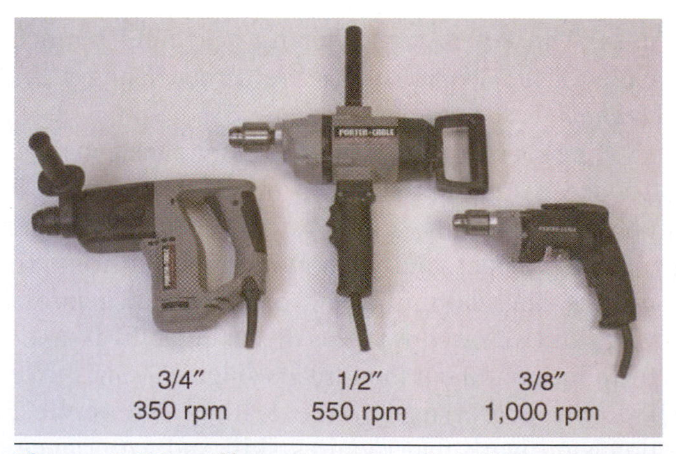

Figure 3-38 Different types of drill motors are shown here. The most common drill found in a motorcycle technician's tool collection is the 3/8-inch drill.

shop-style drill press. You must consider the type and thickness of the material you'll be drilling, and the type and size of power drill you'll be using when making a drill bit selection. Not all bits are made for cutting metal. Some are made for wood, plastics, and concrete.

Here's a safety tip when drilling metal. When the bit starts to break through the other side of the metal, the bit may catch and twist the power drill. That's why you should be sure that the piece you're drilling is properly secured and you have a firm grip on the power drill. This will help avoid damaging the part or the drill, and will reduce the risk of injury.

Rechargeable cordless drills are very popular (Figure 3-40). These drills are quite powerful and they allow you to work in difficult places without being restrained by an electrical power cord. This is particularly handy if you ever have to work in a remote location where normal power isn't available. Rechargeable batteries are the power source for cordless drills. A fully charged battery may provide several hours of drill usage. However, they must be recharged periodically; consequently, you may want to keep spare batteries handy. That way, when you're busy, the only time you'll lose is the time it takes to swap batteries. Be sure to read the instruction manual for the proper use and care of these batteries. Some types should be used until they're completely discharged, and other types may be recharged at any time during use. If you recharge the first type before it's completely discharged, the battery will be compromised and its useful life will be shortened.

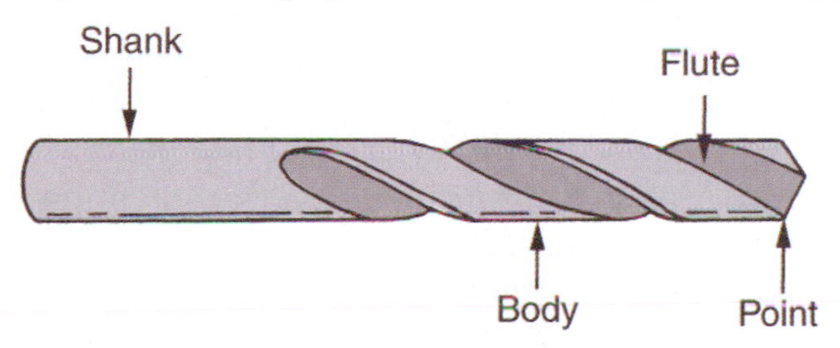

Figure 3-39 The parts of a typical drill bit are shown here.

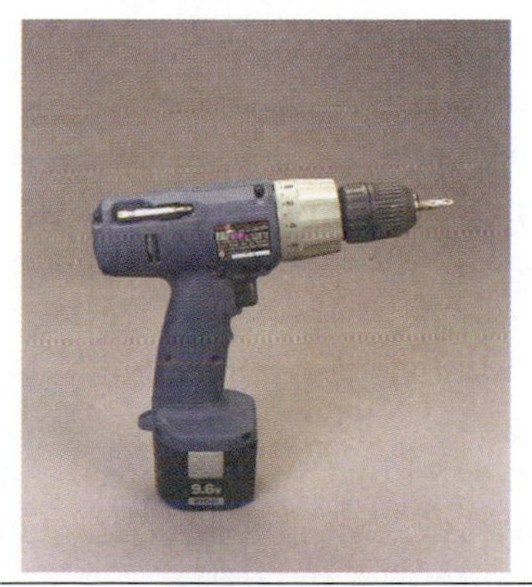

Figure 3-40 Self-powered drill motors are very popular tools and can also be used as a powered screwdriver as seen here.

Always be cautious when using power drills. Be sure to follow these guidelines:

- If you're using corded drills, use only grounded power equipment.
- Determine if there are any obstructions or wiring in the path of the bit before drilling.
- Select the correct drill-and-bit combination to perform the drilling task effectively and safely.
- Don't apply too much pressure on the drill bit.
- Use bit lubricants where required.
- Ensure that the bit is tightly clamped in the chuck.
- Never leave the chuck key in the chuck.
- Use sharp bits. A dull bit could bind, causing the drill to grab. An unexpected twist of the drill could injure your wrist.
- To drill a large hole, start by drilling a smaller hole and gradually increase the hole size by selecting increasingly larger bits.

DRILL PRESS

Another power tool you may need from time to time is the **drill press**. The drill press is a large, floor-standing device that can drill precisely located and angled holes. Because the drill press holds the drill bit in an exact position, it can cut at precision angles and depths. Drill presses use many of the same drill bits that are used with the handheld power drill.

BENCH GRINDER

The **bench grinder** is another useful tool in the motorcycle and ATV service department. The typical bench grinder (Figure 3-41) has two rotating wheels, one with an abrasive grinding surface and the other with a buffing wire surface. The wheels can be used to sharpen tools, buff or polish metal, and remove rust from parts.

It cannot be overstated that caution should be used every time you use a bench grinder! Always use eye protection. Always keep a firm grip on the piece you're grinding or buffing. Finally, make sure that all guards, guides, and shields are in place before you operate a bench grinder.

Air Tools

The **air ratchet** is similar to the standard hand ratchet wrench but is larger than the hand ratchet (Figure 3-42). The size of the air ratchet drive restricts the places where it can be used directly. With a selected set of shaft extensions and universal joint couplers, the air ratchet can be used for most applications. There are 1/4-inch and 3/8-inch drive lug models. The air ratchet turns at a much higher speed than a handheld ratchet and lets you remove nuts and bolts more quickly.

Power impact wrenches are driven by compressed air. They come in either 3/8-inch or 1/2-inch sizes (Figure 3-43). Power impact wrenches are very useful for component disassembly but aren't recommended for reassembly. It's very easy to apply too much torque using a power impact wrench. The excess torque could easily shear the fastener and/or damage the part or assembly that you're working on.

Figure 3-41 A typical bench grinder has a grinding wheel as well as a wire brush attachment.

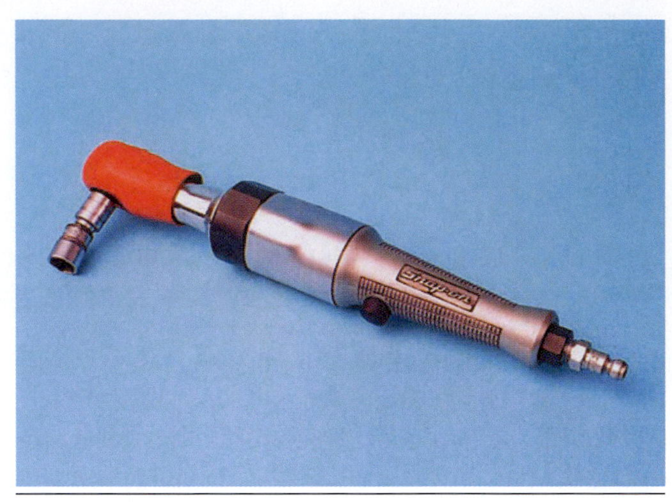

Figure 3-42 An air ratchet can aid in the speed of removal of fasteners, but should not be used to tighten fasteners.

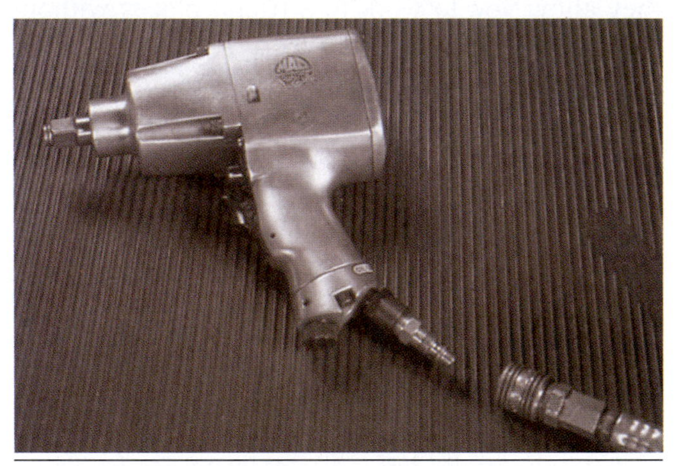

Figure 3-43 A typical air-powered impact wrench is seen here. These types of tools should only be used to remove items and not used to install as they are so powerful that they could break or over-torque the fastener in question.

As previously mentioned, special heavy-duty impact sockets are designed for use with air or impact wrenches. If you recall, they're the ones that normally have a black finish. Here's a handy hint: Never start a bolt or nut with a powered impact wrench. The fasteners can cross thread, causing damage to the fastener or component. Remember to always use approved eye protection when using any power tools. Additionally, it is important to note that when using compressed-air power tools, never exceed the recommended air pressure rating of the tools.

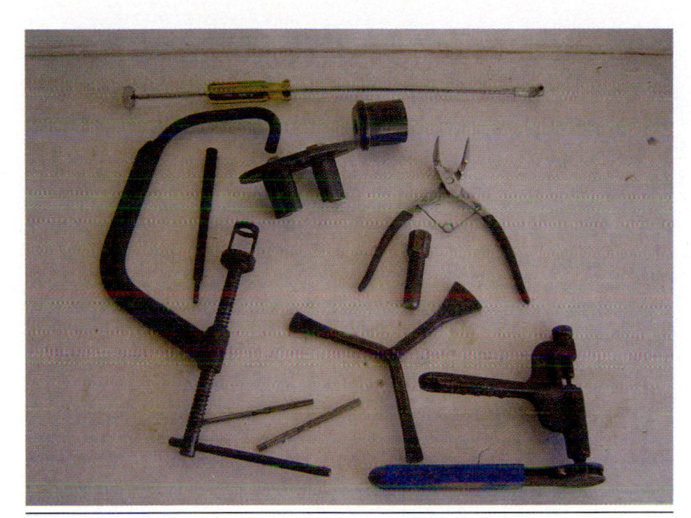

Figure 3-44 An assortment of special tools used on motorcycles is shown here.

SPECIAL TOOLS

The basic hand tools that we described are often used to repair motorcycles and ATVs. But many repair activities involve procedures that can't be performed with standard tools alone. You'll need various special tools, especially when rebuilding engines (Figure 3-44). Special tools are also used to work on brake and suspension systems. As you learn more about motorcycle and ATV repair, you'll discover that a special tool is made for almost every purpose.

You should be aware that some special tools are designed to be used on only one make or model of vehicle, while other special tools can be used on a variety of vehicles. In most cases, these special tools are owned by the service department and controlled by the service manager, but many technicians opt to purchase all of the special tools that they utilize to ensure that they know that they have the right tool at hand whenever they are in need of it. We'll review the names and uses of some special motorcycle and ATV repair tools now. We don't expect you to understand how to use these tools yet. We just want you to become somewhat familiar with them. In later chapters, we'll cover the use of specialized tools in detail. Now, let's take a look at some of the more common special tools.

Pullers

Pullers are used to safely remove gears, flywheels, and various components from shafts. A puller can easily separate machined parts that are tightly pressed together without damaging the parts. There are many different types of pullers used on motorcycles and ATVs (Figure 3-45). For safety reasons and to protect precision parts, you should use pullers whenever possible to separate parts.

Precision Measuring Tools

Precision measuring tools are used to make exact measurements of various parts or distances between parts. Measuring tools are used primarily to check for wear of components and specific part-to-part clearances when reconditioning or rebuilding an engine. The measuring instruments used during motorcycle repair can range from standard rulers to precision instruments. They're designed to measure thickness, clearances, pressure, and fastener tightness. Let's take a quick look at some of the common measuring tools.

Vernier Caliper

The **Vernier caliper** is one of the most commonly used precision measuring tools in motorcycle repair work (Figure 3-46). They can be used to measure inside, outside, or depth measurements, which makes them very versatile. Vernier calipers can be quite accurate.

To operate Vernier calipers, slide the jaws of the caliper around the part (Figure 3-47). The caliper indicates the size of the object on its display scales. You can use the jaws of most sliding calipers to measure both outside and inside dimensions.

Individual styles of Vernier calipers display their measurements differently.

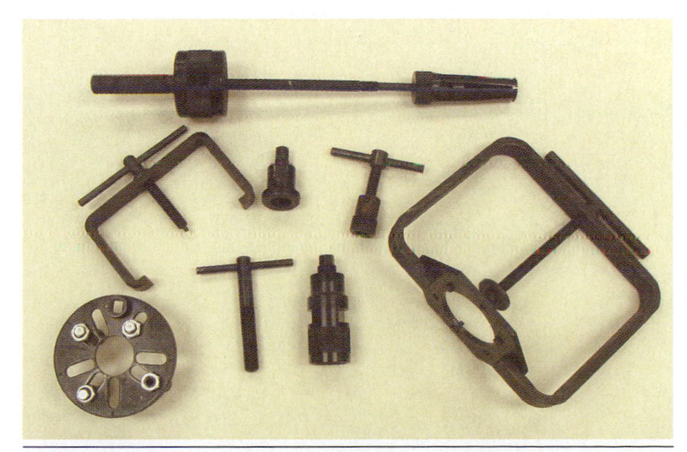

Figure 3-45 Different types of motorcycle- and ATV-related pullers are shown here.

Some of the more common styles include the printed beam and Vernier scale, the dial gauge, and the digital display. Our caliper example has a printed beam and Vernier scale. You must read the numbers on the scale and perform some calculations to obtain the actual measurement.

In contrast, digital calipers (Figure 3-48) are considered by many to be much easier to use. The caliper displays the true measurement directly on the digital display screen and most will show in SAE or Metric measuring systems.

Micrometer

As mentioned, Vernier calipers can be quite accurate, but they're less accurate than **micrometers**. In those cases when exact precision is needed, micrometers should be used. The micrometer is used to measure the size of a part or component. The basic components of a micrometer are shown in Figure 3-49.

Micrometers come in a variety of sizes and styles (Figure 3-50). Some micrometers are designed to measure the outer dimensions of an object, while others are made to measure inner dimensions, such as the inside diameter of a cylinder. Another type of micrometer measures depth. Using a micrometer is a common procedure while working on motorcycles and ATVs and therefore an understanding of their use is important. Many micrometers use an electronic readout, which allows quick and easy-to-read measurements, while many micrometers can be read using a Vernier scale on the tool itself. Just like

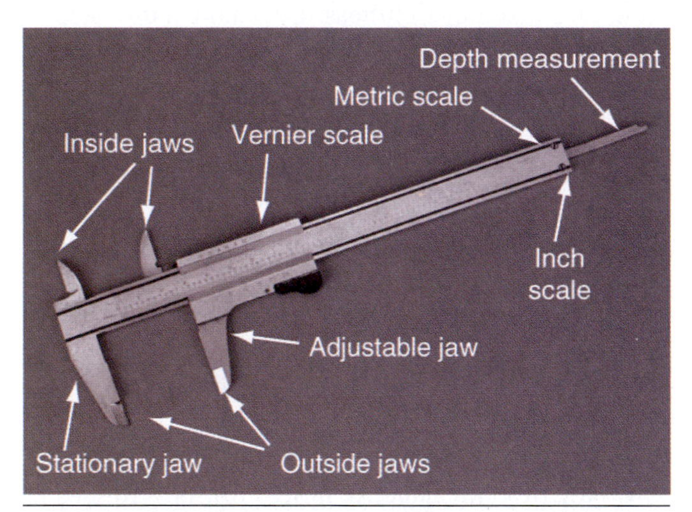

Figure 3-46 The Vernier caliper is a versatile tool that can measure inside and outside diameters as well as depth. The different parts of a Vernier caliper are shown here.

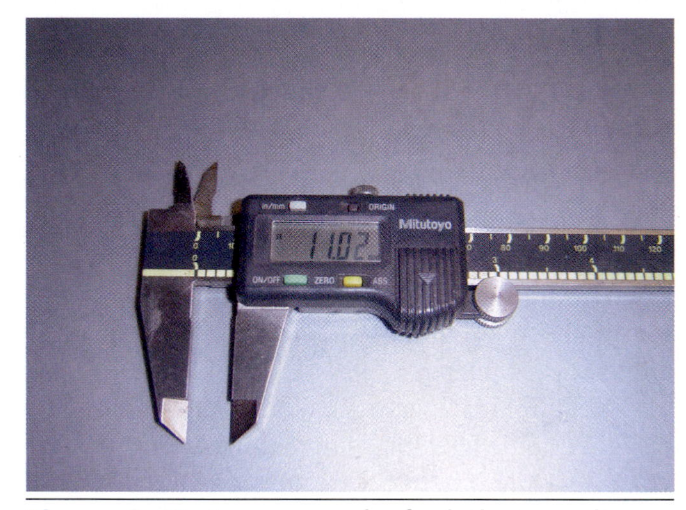

Figure 3-48 Some people feel that reading a digital caliper is easer than the standard version. As you can see here, that point is hard to argue!

Figure 3-47 A Vernier caliper in use.

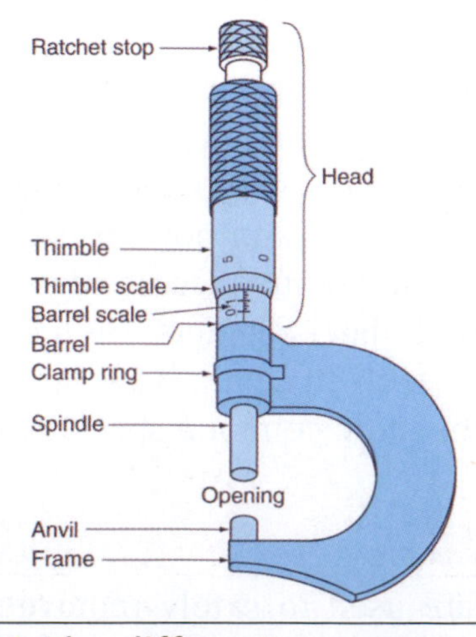

Figure 3-49 The different components of a typical micrometer are shown here.

Vernier calipers, micrometers can be purchased with digital displays, which are easier to read and can also be easily converted from SAE to Metric denominations (Figure 3-50A).

To use a micrometer, place the object between the *anvil* and the *spindle*, and then turn the *thimble* until the anvil and spindle contact the object with a light resistance. You can read the measurement on the thimble and sleeve of the gauge (Figure 3-51).

Micrometers are often used during the engine-rebuilding process. During a rebuild, you'll completely disassemble an engine and inspect and measure all of the engine parts to determine if they're worn or damaged. Since the micrometer can measure thickness so precisely, it can easily detect small changes in part sizes that indicate wear. Using a micrometer is a typical procedure performed during engine reconditioning.

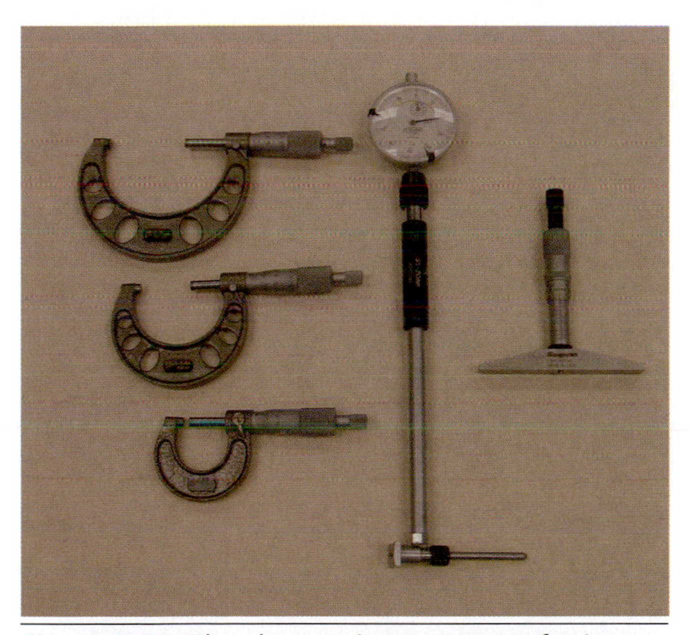

Figure 3-50 The three primary types of micrometers used on motorcycles are shown here.

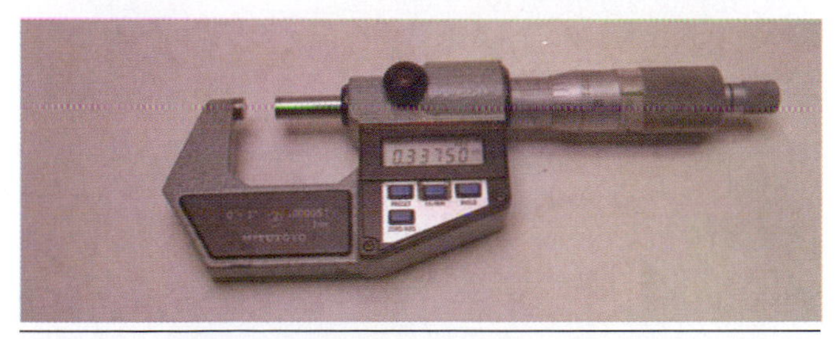

Figure 3-50A Just like the Vernier caliper, micrometers also come in versions that read digitally.

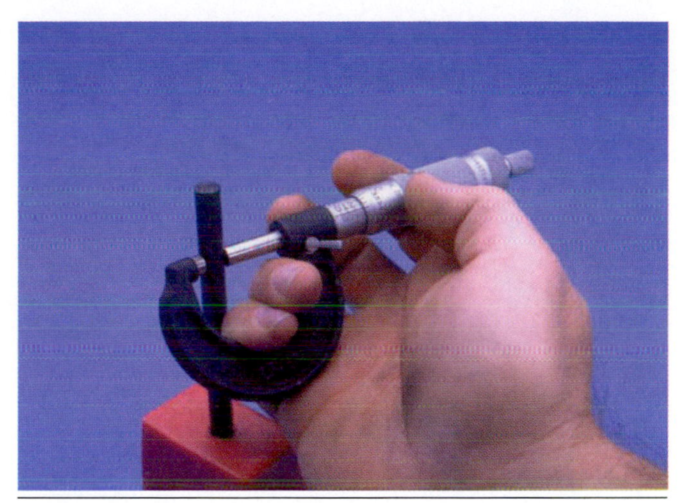

Figure 3-51 An outside micrometer is shown being used here.

Dial Indicator

So far with precision measuring tools, we've looked at tools that measure the sizes of parts. However, there are times when you'll need to measure the distance that a part moves, such as the in-and-out movement (travel) of a shaft. The most common way to measure this type of movement is with a tool called a **dial indicator**. A dial indicator is simply a dial gauge with a plunger that sticks out from one side (Figure 3-52) and can be positioned in any one of numerous positions. Dial indicators are held in place by a magnetic base or by the use of a flexible shaft that can be locked in place once the desired location has been determined.

To use the dial indicator, attach it to a solid object (i.e., a magnetic base) next to the item you're going to measure (Figure 3-53). Position the dial indicator so that the plunger contacts the object to be measured. Move the object back and forth. The dial indicates the distance that the plunger travels. Dial indicators are often used during engine and transmission rebuilding.

Torque Wrench

Accurate tightening of fasteners is a key consideration in the proper operation of engines and other motorcycle components. In shop repair and service manuals, manufacturers specify the exact amount of torque that should be used to tighten fasteners, such as head bolts. Because it's almost impossible to accurately tighten a bolt by hand to a specified torque, special tools called **torque wrenches** are used.

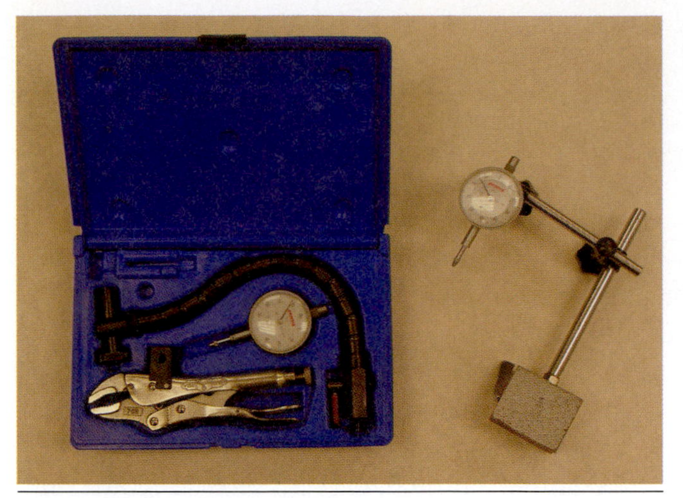

Figure 3-52 Two different dial indicators are shown here. The indicator on the right uses a magnetic base to hold it in place while the one on the left uses a flexible shaft that locks in place.

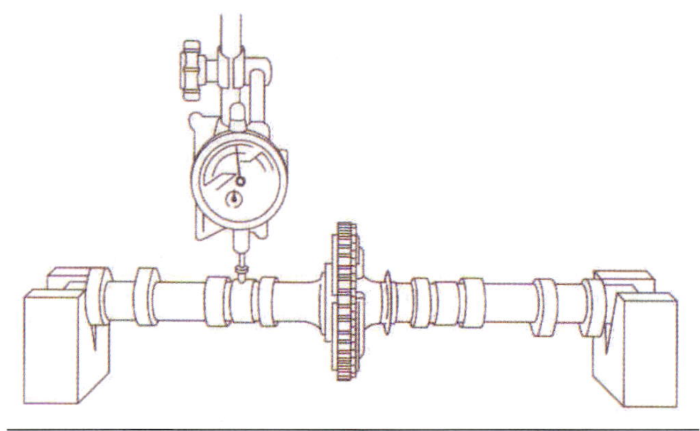

Figure 3-53 A dial indicator in use is illustrated here measuring a camshaft from a motorcycle.

A torque wrench allows you to apply the exact amount of tightening force (torque) to a fastener. Frequently, the torque wrench is a modified socket drive handle that has a torque-measuring device built into it. This allows you to use standard interchangeable sockets when you're setting the torque on fasteners.

A torque wrench contains a measuring dial or scale. As you tighten a nut or bolt with the wrench, the dial or scale indicates how much torque you're applying to the fastener. These scales are usually calibrated in SAE foot–pounds (ft–lb), inch–pounds (in–lb), or Metric Newton-meters (N/m). There are two common types of torque wrenches used in the motorcycle service shop (Figure 3-54).

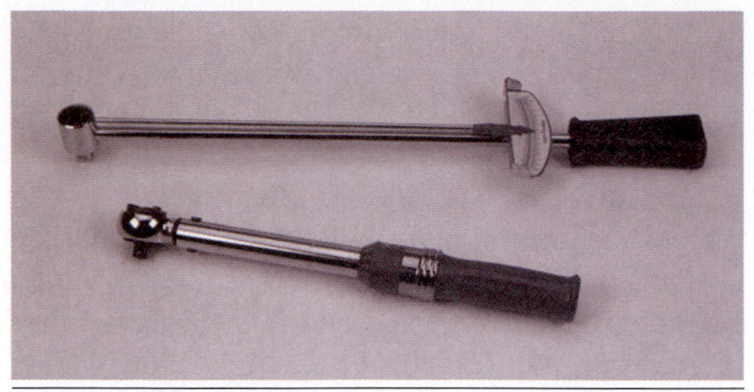

Figure 3-54 The two most popular types of torque wrenches found in a motorcycle technician's tool collection are the beam and click types as seen here.

- The **beam-type torque wrench** contains a metal pointer rod. As you tighten a bolt, the rod points to the measured torque value.
- The **click-type torque wrench** is somewhat easier to use than the beam type. With a click-type torque wrench, you preset the desired amount of torque on a calibrated dial before tightening the bolt. When you reach the preset torque value, the wrench clicks and no more torque is needed to be applied to the fastener.

Feeler Gauges

Feeler gauges are another type of precision measuring tool. Feeler gauges are typically used to measure very small spaces between two parts, such as spark plug gaps. These small spaces are often called *clearances*. A typical feeler gauge is actually a set of gauges, made up of a large selection of metal blades that can spread open like a fan (Figure 3-55). The blades vary in thickness to

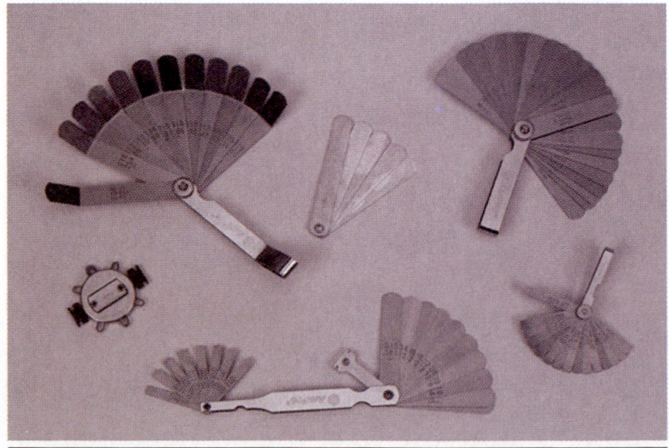

Figure 3-55 A wide array of feeler gauges is shown here.

provide a complete range of very precise measurements. Each blade is marked with its thickness, in some cases marked in both metric and SAE sizes.

To measure the distance between two parts, insert one blade at a time between the parts until you find the blade that's an exact fit. The marked size of that blade indicates the measured clearance between the two parts. In some cases, a combination of two or more blades is used to precisely measure the clearance. In those cases, you must add the total of the blade thickness to determine the clearance.

Test Instruments

Troubleshooting and repairing motorcycles requires a wide variety of special test instruments. These test instruments are designed to test the condition of various systems. For example, there are several different instruments that you can use to test an engine's ignition and electrical systems. Other testing devices are used to measure compression and pressure. Let's take a brief look at some of the more common test instruments.

Multi-meter

You can use several different special instruments to test or measure electrical circuits within a motorcycle's electrical system. The most common electrical testing instrument is the **multi-meter** (also called a *volt-ohmmeter* or VOM). This one instrument can measure voltage, current, and resistance. The multi-meter is a box-like device that has two flexible wire test leads connected to it (Figure 3-56). The ends of the wire leads hold probes that are used to contact the electrical circuitry.

The probes can be either needle-point or alligator-clip design. Most are a combination to allow you to either probe, or connect the test leads, depending on what component is being tested. The probes are touched to different areas of an electrical circuit to make electrical measurements. The multi-meter has a display face to provide the circuit measurement information to the user. The display can be either an analog type that uses a moving needle (Figure 3-57) or a digital display type. Both types of multi-meters are popular in the motorcycle technician's tool collection.

The circuit information displayed by the multimeter helps you determine where problems such as broken wires, faulty connections, or defective components may exist in an electrical system. Note that this is a very brief and very basic description of the multi-meter's operation. The actual operation of a multi-meter is much more complex. You must always observe electrical-safety precautions when using the multi-meter. You could damage the multi-meter and the electrical circuit if you use it improperly. More importantly, you could receive a severe electrical shock. We'll discuss in detail where to use a multi-meter in a later chapter.

Fuel Injection Diagnostic Testers

Quite some time ago, all motorcycles came with carburetors to mix air and fuel to make an engine run. Ignition points made a spark occur at a precise time, which allowed the fuel and air mixture to burn properly to make the engine run smoothly. Back

Figure 3-56 A typical digital multi-meter is shown here.

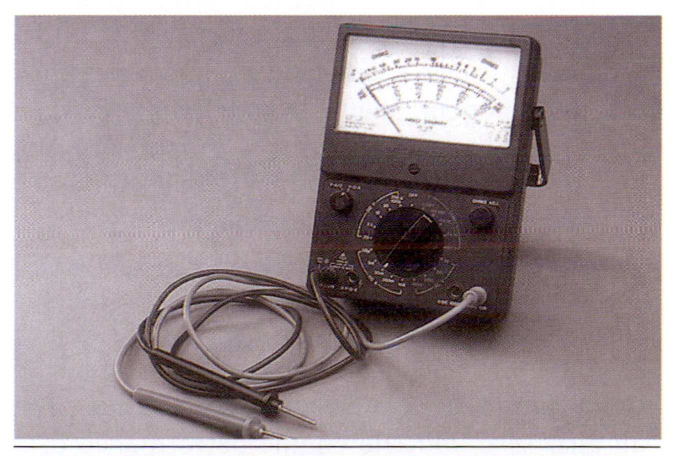

Figure 3-57 A typical analog multi-meter is shown here.

then, a technician diagnosed engine problems by listening, feeling, and smelling. Then came electronic engine controls that regulated fuel mixtures and ignition timing with great precision. Plus, with these controls came the ability to effectively change how the engine ran and adapted to changing temperatures and elevations. As time progressed, it was apparent that you could not fix problems when they occurred just by listening, feeling, and smelling these new systems. So, engine control systems had limited self-diagnostic abilities, and would flash lights in code to reveal internal problems—but more was needed.

In the early 1970s, the first on-board diagnostic systems were integrated into American cars to help keep emissions within limits. In 1996, **On Board Diagnostics (OBD-II)** became a standard for measuring automobile engine operating parameters, while most manufacturers had additional diagnostics for measuring and tracking other vehicle information. It takes a computer to diagnose a computer, and OBD-II automotive diagnostic systems are available from hundreds to thousands of dollars, depending on capabilities for automobiles. This is the technology that has trickled down to motorcycles today.

Today, more and more motorcycles and even ATVs are being designed with **fuel injected (FI)** systems that utilize computers and electronic sensors. These systems require special tools (Figure 3-58) to assist the technician in diagnosing problems and can even help in determining the overall health of the FI system. These are highly specialized tools and most require special training to understand and decipher the information that they are transmitting.

Diagnostic tools cannot actually pinpoint a faulty component; they can only tell you when parameters that the **Engine Control Module (ECM)** reads do not fall within normal values. It's up to you as the technician to find out why.

Beyond just reading trouble codes, most FI diagnostic tools can read the actual values of the sensors and signals sent from the ECM, like how long the fuel injectors are open. This data can help you to zero in on a problem. On some testers, you can also save and file engine performance data to create permanent performance records of motorcycles using these special tools. You can also easily email performance data to manufacturer help lines for help analyzing unusual problems.

Timing Light

Although not used on many motorcycles and ATVs today, you can use a special device called a **timing light** (Figure 3-59) to determine if the timing of the spark impulse is correct. The timing light is usually connected between the spark plug and the ignition system.

To use this type of tool, you connect it between the spark plug and the plug wire, start and run the engine, and watch the timing light flash against a rotating timing mark. Each time the spark plug fires, the timing light will produce a flash of light (strobe) that freezes the mark. The illuminated timing mark is compared with a fixed reference mark to verify correct timing. If the timing is wrong, the two marks

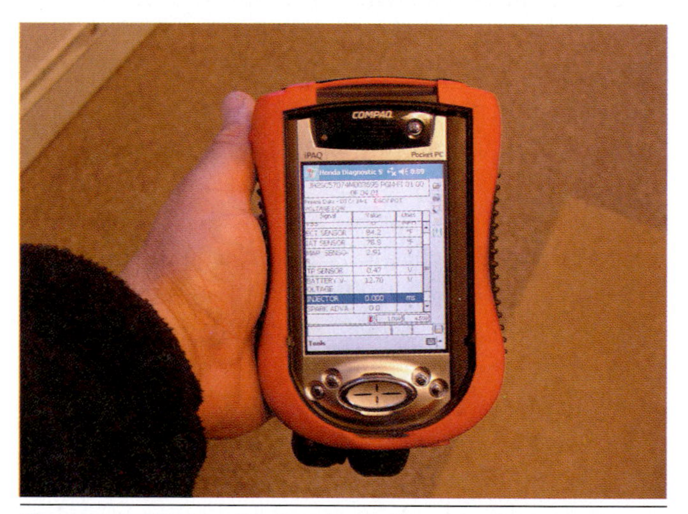

Figure 3-58 Fuel injection diagnostic testers are becoming more popular with the advent of more and more fuel injected motorcycles and ATVs.

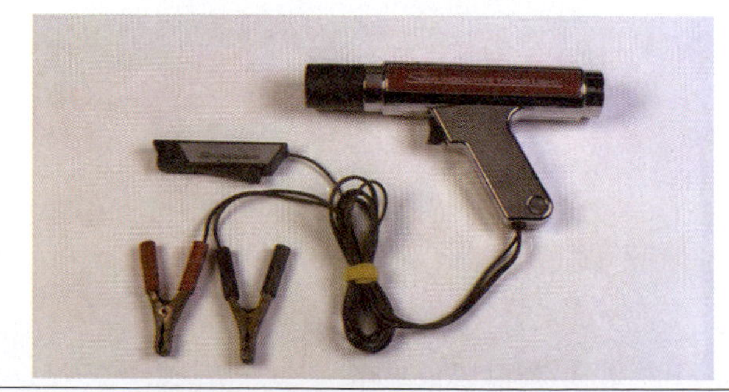

Figure 3-59 Although not used (or needed) as often today as they once were, timing lights are still a very popular tool that is used to determine if the engine's spark timing is correct.

(fixed and rotating) will not line up. You then adjust the timing until the marks come into alignment. Most of today's motorcycles and ATVs do not have adjustable timing; therefore, if the timing is not correct, there is a problem somewhere in the system that will require a component to be replaced.

Test Lights

A **test light** (Figure 3-60) is a tool that is used to determine if electrical power is available to a particular circuit and is often used in diagnosing failed lighting systems. Test lights come in different styles: one is self-powered, which verifies that a circuit is complete and the other is non-self-powered, which verifies that power is being supplied to the circuit. Both types have various uses in the service shop environment.

Compression Gauge

The **compression gauge** is used to measure pressure in an engine cylinder (Figure 3-61). A mixture of air and fuel is compressed inside of the engine's cylinder before the mixture is ignited. The higher the cylinder compression, the better the fuel mixture will burn. As engine components wear, the compression in the cylinder will decrease. By measuring the cylinder's compression pressure, you can determine if the engine parts are worn sufficiently to require an engine rebuild.

To use the compression tester, unscrew the spark plug from the cylinder head and install the compression tester gauge adapter into the cylinder head in place of the spark plug (Figure 3-62). When you turn the engine over a number of times, the amount of pressure that's developed in the cylinder is displayed on the gauge.

Exhaust Gas Analyzer

An **exhaust gas analyzer** is used to measure the levels of certain emission gases in an engine's exhaust. Most modern exhaust analyzers are designed to measure the levels of hydrocarbons and carbon monoxide in the exhaust (Figure 3-63). The word *hydrocarbons* refer to raw, unburned fuel in the exhaust. *Carbon monoxide* is a toxic, odorless gas that's produced when fuel doesn't burn completely in an engine. To analyze an engine's exhaust, insert the exhaust analyzer probe into the tailpipe to sample the exhaust gas. The analyzer will produce readings that indicate the levels of various gases in the exhaust.

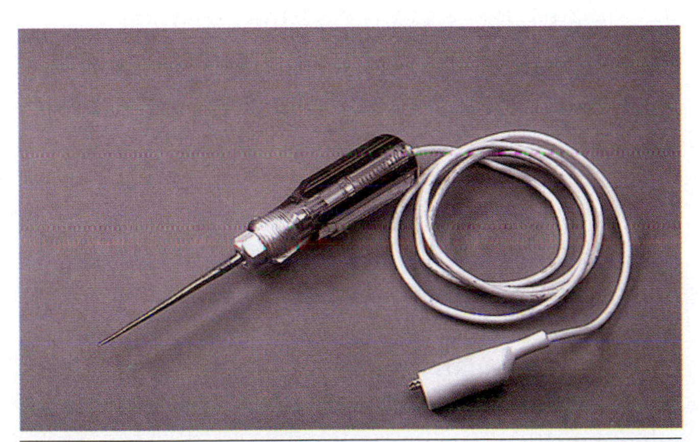

Figure 3-60 A test light is a very handy tool when diagnosing a problem in an electrical system.

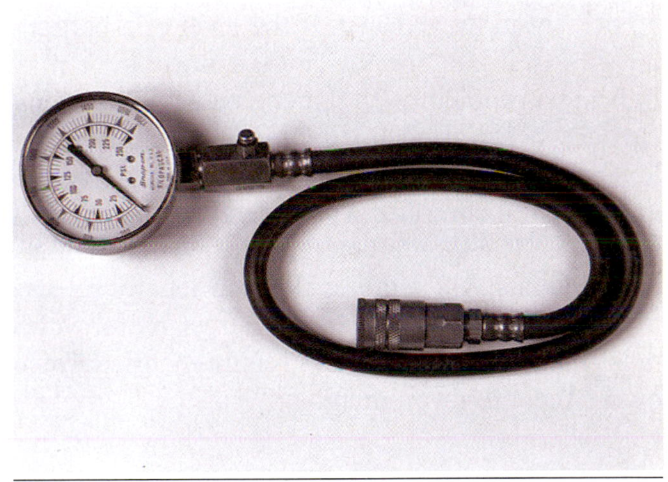

Figure 3-61 Compression testers measure the amount of pressure inside an engine and are used often on all types of motorcycles and ATVs.

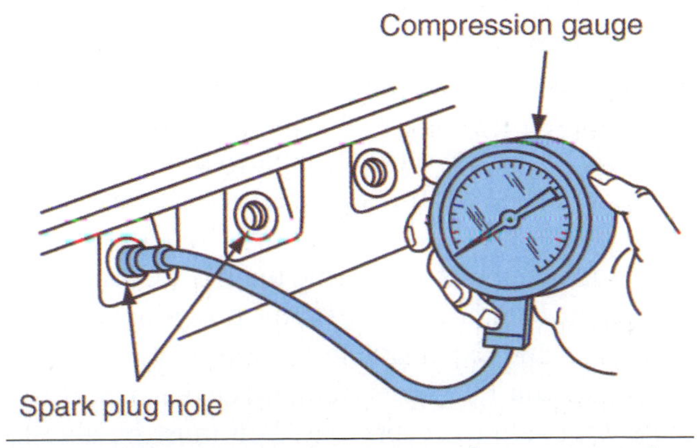

Figure 3-62 Proper placement of a compression tester is shown here.

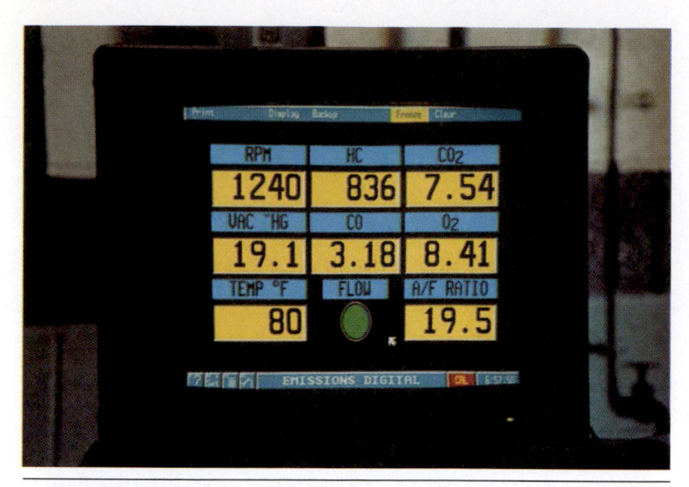

Figure 3-63 An exhaust gas analyzer is used to determine the amount of emissions being produced from an engine.

You can compare these readings with an analyzer diagnostic chart to make an analysis of the engine's condition. If the engine is working properly, certain percentages of each of the gases will be present in the exhaust. If the level of a particular gas is higher or lower than normal, it may indicate a problem with the engine or fuel system. Besides being used for diagnosing engine problems, a vehicle's exhaust may also be analyzed to ensure that it complies with federal and state emission laws. These emission laws define the amounts of pollutants that motorcycles can legally release into the environment. All new street motorcycles must meet specific exhaust requirements defined by the federal government.

PURCHASING TOOLS

Very few motorcycle technicians own every type of available repair or diagnostic tool. This is true because most technicians don't perform every type of repair. Just what tools you'll need depends on your repair specialty area and the availability of equipment at the service department where you work. In most motorcycle service departments, technicians are required to buy their own basic hand tools and possibly some diagnostic tools. But the service department usually provides specialized tools and expensive test equipment.

When you buy tools, remember that you'll use these tools almost every day. You must be able to depend on them. For this reason, the tools that you buy should be of high quality to provide you with many years of service. It's a very good idea to buy professional quality, brand name tools. These tools are usually of the highest quality and are often backed with lifetime warranties. This means that the tools should provide quality service for your entire working lifetime. If a tool under warranty is damaged or fails in normal use, you can return it to the manufacturer for a replacement. You can purchase most tools individually or in sets. If you intend to enter the motorcycle repair field, you may want to consider purchasing a complete starter set rather than dealing with individual tools. In most cases, purchasing tools in sets is less expensive than buying them individually. But if you already own a reasonable assortment of quality tools, you may need to purchase only a few additional items to meet your tool needs.

As you gain experience and start performing more repairs, you can add more tools to your starter set. Remember that it's better to avoid buying too many tools until you're reasonably established in your field. Before purchasing an assortment of tools that you may never use, stop and analyze your situation. When you determine the types of repairs that you'll be doing routinely, you can purchase extra tools based on your specific needs. This planning will help you avoid spending a lot of money on tools that you may never need in the future. Tools aren't cheap, but they're a sound investment in your future. Invest wisely!

STORING TOOLS

Proper tool maintenance and storage should be an important part of your daily routine. Professional technicians always take proper care of their equipment. Because you'll probably own a large collection of tools one day, it's very important to establish good work habits and housekeeping practices early in your career. One of the most important first steps is to organize your tools so that you can find them easily. You don't want to spend a lot of time looking for a tool each time you need it. And equally important, tools are an expensive investment. You should keep them locked up when you're not using them. The best way to store and organize your tools is to keep them locked in a

Figure 3-64 A typical portable toolbox is shown here.

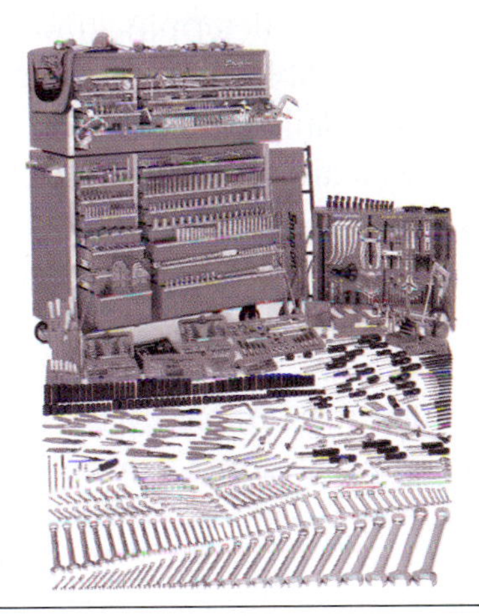

Figure 3-65 Many long-term motorcycle technicians go all out and purchase tool boxes such as this one.

sturdy, professional-quality **toolbox**. Toolboxes come in a variety of sizes and price ranges. They can be small and portable (Figure 3-64), or large, floor-standing units with wheels (Figure 3-65). When choosing a toolbox for storage, always consider your future needs and plan ahead! As a professional technician, you'll continue to add tools to your collection as the need arises. So, when you choose a toolbox or cabinet, select one that will allow room for future expansion.

SERVICE INFORMATION LIBRARY

Whenever you're repairing or rebuilding a motorcycle or ATV, the correct service reference material is a must. Because each make and model is different, you'll frequently need to look up manufacturer's information and specifications about the particular machine you're working on. An up-to-date service information library (Figure 3-66) is as essential to your work as the proper tools. Franchised motorcycle dealerships are required to keep a variety of reference books and **service manuals** on hand (Figure 3-67). Service manuals can be printed copy or on a CD that is used in a computer. These collections of service documentation grow every time new models and features are introduced to the market. Service manuals contain a great deal of helpful information, such as engine identification information, engine reconditioning specifications, and recommended repair procedures.

Most manufacturers also send out **specification manuals** known as "spec" manuals. Our usage of the term "specification" focuses on certain precision measurements made on parts of the vehicle, particularly in the engine. The exact measurements are determined by the manufacturer and are listed in the vehicle's service manual or spec manual. When you're doing a major rebuild of a motorcycle engine, many of the engine specifications must be checked with precision measuring instruments that we mentioned earlier in this chapter. These measurements are compared with the specifications listed in the service manual or spec manual. Any deviations from the listed acceptable tolerance limits indicate a problem that should be corrected as part of the repair process. An engine must conform exactly to its specifications to operate properly. Some common engine specifications include the following:

- Spark plug gap, which is the width of the gap between the spark plug's electrodes
- Cylinder-bore—the inside diameter of the engine's cylinder
- Torque specifications—tightness of fasteners, usually measured in foot–pounds

Some technicians attempt to make repairs to a vehicle without using service manuals, but this is not a good practice for several reasons. First, even the most experienced technicians can't remember every specification for every make and model vehicle. If you were to work on only one motorcycle

Figure 3-66 A service library will contain service manuals and specification manuals. Some manufacturers even put their manuals on CD with interactive sections within them to aid the technician in making repairs.

Figure 3-67 A properly stocked franchised motorcycle dealer service library will contain at least five model years of service manuals.

make and model, you could probably handle most repairs without the use of outside references. But in almost all motorcycle service departments, technicians work on a large variety of models and often from many different manufacturers. Secondly, manufacturers make changes and improvements to their vehicles from year to year. Each year, more new vehicles are introduced and more features are added to existing models. An up-to-date service library will help you stay current with the latest changes to the vehicles you service. And finally, service manuals are often necessary in those cases where you weren't the person who disassembled a particular component. You may be picking up the handiwork of someone else. With a service manual at hand, you can quickly determine how the component should be assembled and how it should be adjusted once it's installed.

Service manuals are useful tools, but they can't tell you everything. Most manuals are written with the assumption that you already know the basics of motorcycle repair, such as how engines operate, how the various motorcycle systems operate, how to disassemble and reassemble engines, and how to use the proper tools and measuring instruments. Most service manuals concentrate on specifications and repair procedure sequences. Therefore, service manuals can never take the place of good training. Service manuals are simply additional tools to help you make repairs correctly and efficiently.

Summary

- You should be able to easily identify common hand, power, and special tools.
- It is very important to know how to select the correct tool for a particular type of repair.
- Using the correct types of tools when doing a job to ensure that you are being the most efficient you can be will help you to be the most effective at your job as a technician.
- There are various different special tools needed to properly disassemble and reassemble motorcycles and ATVs.
- There are many different types of precision measuring tools and each has a specific purpose.
- A proper service information library is among the most important tools that you can use.

Chapter 3 Review Questions

1. A ________ is a test instrument that's commonly used to measure volts, ohms, and other electrical measurements.
2. Motorcycle tools can be classified into these three categories:
3. A ________ is used to tighten a nut or bolt with an exact required amount of pressure.
4. One of the most commonly used precision measuring tools in motorcycle and ATV repair can measure inside and outside diameters as well as depth. This tool is called a:
5. The ________ measuring system uses millimeters.
6. A short, six-sided rod used to tighten screws with similar indentations is the:
7. Which type of wrench would be a better choice in tight places?
   a. 6-point box wrench
   b. 12-point box wrench
8. *True* or *False*. An air ratchet is suitable for tightening fasteners.
9. Franchised motorcycle dealerships have service libraries that include what two types of manuals?
10. What type of wrench combines the open-end wrench and the box-end wrench into one tool?

CHAPTER

Measuring Systems, Fasteners, and Thread Repair

Learning Objectives

When students have completed the study of this chapter and its laboratory activities they should be able to:

- Identify the basic measuring systems used on motorcycles and ATVs
- Identify common fasteners used on motorcycles and ATV's
- Describe the four most important bolt dimensions
- List the three basic types of threads used on fasteners
- Determine and record bolt grade/tensile strength
- Determine the appropriate tightening torque for a threaded fastener
- Describe common ways to remove damaged fasteners
- Describe how to clean and repair damaged threads

Key Terms

Acorn nut or cap nut
American National Standards Institute (ANSI)
Axial tension
Beam-type torque wrench
Bolt
Bolt diameter
Bolt head markings
Bolt length
Castle-headed nut
Clamping force
Click-type torque wrench
Combination bolt (CT)
Cone-type lock washer
Conventional system
Dial-type torque wrench
DR-type bolts
Elastic range
Flat washer
Head size
Heli-Coil®
International Standards Organization (ISO)
International System of Units (SI)
Metric system
Nut
Plastic range
Preload
Self-locking nut
Shear
Society of Automotive Engineers (SAE)
Split-ring-type lock washer
Stake-type lock nut
Stretched bolt
Stud
Tensile strength
Thread pitch
Tongued lock plate washer
Torsion
Torx bolts
UBS bolts
Well nuts
Yield point

INTRODUCTION

Motorcycles and All Terrain Vehicles (ATVs) use two systems of weights and measures side by side. Some motorcycles and ATVs made in the United States utilize the U.S. or conventional system of weights and measurement. Virtually all motorcycles and ATVs built around the rest of the world use the metric system, which is considered by many to be an easier system to use.

All motorcycles and ATVs have hundreds of parts held together with threaded fasteners. Unlike permanent fastening methods like welding, riveting, or gluing, threaded fasteners are essential as a nonpermanent connection that can be disassembled whenever necessary.

A competent motorcycle technician must know the different measurement systems as well as about fasteners, including how to properly install and tighten the various fasteners to correct manufacturer specifications and how to repair and remove fasteners when they break. A technician must be able to use a torque wrench to complete most tightening procedures to insure a quality reassembly. The technician must also use factory "torque" values and the recommended tightening methods and sequence. An improperly tightened fastener can fail by loosening (backing out) or breaking to cause a dangerous condition for the operator of the motorcycle or ATV. This chapter will review measurement systems, fastener identification and classification, removing broken and seized fasteners, thread repair procedures, and fastener reinstallation guidelines.

MEASUREMENT SYSTEMS

For motorcycle systems to operate properly, the parts must fit together securely. Fasteners, parts, and the tools needed to work on them are made to specific sizes, or measurements.

Motorcycle manufacturers use the two most common weights and measurement systems—the conventional system and the metric system.

The Conventional System

The measurement system found often on products made in the United States is the **conventional system**, which is also known as the standard and United States Customary (USC) system. Measurements used in this system require knowing different combinations of numbers. For example: one foot contains 12 inches. Each inch can be divided into equal units as in halves (1/2"), quarters (1/4"), and eighths (1/8") of an inch. Rulers and other measuring instruments divide each inch into units ranging from 16, 32, 64, or 128 equal parts.

Sizes must be known for fasteners, tools, and parts. Sizes are stated in inches or parts of inches. The parts of inches are expressed either as fractions or as decimal numbers. For example, a fastener diameter of one-half inch can be written as: 1/2" or as a decimal, 0.5". A 1/8" fastener diameter would be 0.125" as a decimal number.

The Metric System

The **metric system** is also called the **International System of Units (SI)**, because it is used in virtually all countries around the world. The metric system is easier to use than the conventional system because it uses a simple decimal system to determine different base units of measurement. There is no need to memorize that 12 inches make a foot, 3 feet make a yard, and so forth. Motorcycles and ATVs generally use the following metric size increments:

- A millimeter equals one-thousandths (1/1,000) of a meter.
- A centimeter equals one hundredth (1/100) of a meter.
- A meter is the base point measurement of the metric system.

It is easy to remember that each metric unit of measurement is ten times the size of the previous unit. For example, 10 millimeters (10 mm) equal 1 centimeter (1 cm). A meter (39.37 inches) equals 1,000 mm or 100 cm. Decimal numbers are always used in the metric system. Units smaller than 1 meter are listed as 0.01 for centimeter and 0.001 for millimeter. Each smaller metric unit of measurement is 1/10 (0.10) the size of the previous unit.

FASTENERS

Fasteners are devices that hold the parts of a motorcycle or ATV together. Hundreds of fasteners are used in today's modern vehicles. Figure 4-1 illustrates some of the most common types.

The Nuts and Bolts of Bolts and Nuts

A **bolt** is a metal rod with external threads on one end and a head on the other. When a bolt is threaded into a part other than a nut, it can also be called a cap screw. A **nut** has internal threads and usually is made with a six-sided outer shape. Figure 4-2 shows examples of bolts and nuts. When a nut is threaded onto a bolt, a powerful clamping force can be produced.

In motorcycle technology, many bolts and nuts are named after the parts they hold. For instance, the bolts holding the cylinder head on the block are called cylinder head bolts. The bolts on an engine connecting rod are called connecting rod bolts.

When more turning force, or torque is applied, more force is exerted on the nut. The force creates a tension in the bolt, which clamps the mating part together. **Preload** is the technical term for the tension caused by tightening the fastener that holds the parts together. Producing sufficient preload force is the key to strong and reliable bolted joints that will not loosen or break under load. Figure 4-3 illustrates the forces that act on a bolted component. We will discuss these forces later in this chapter.

Figure 4-1 Many of the common types of fasteners used on today's motorcycles are shown here.

Fastener Anatomy

Bolts and nuts come in various sizes, grades (or strengths), and thread types. A good technician must be familiar with the differences. The four most critical bolt dimensions (Figure 4-4) are:

- **Bolt diameter**—the measurement of the outside diameter of the bolt threads.
- **Head size**—the distance across the flats or outer sides of the bolt head. This is the same as the wrench that would be used to install or remove the bolt.
- **Bolt length**—measured from the bottom of the bolt head to the threaded end of the bolt.
- **Thread pitch**—sometimes referred to as thread coarseness. Thread pitch for metric fasteners is the distance from the top of one thread to the

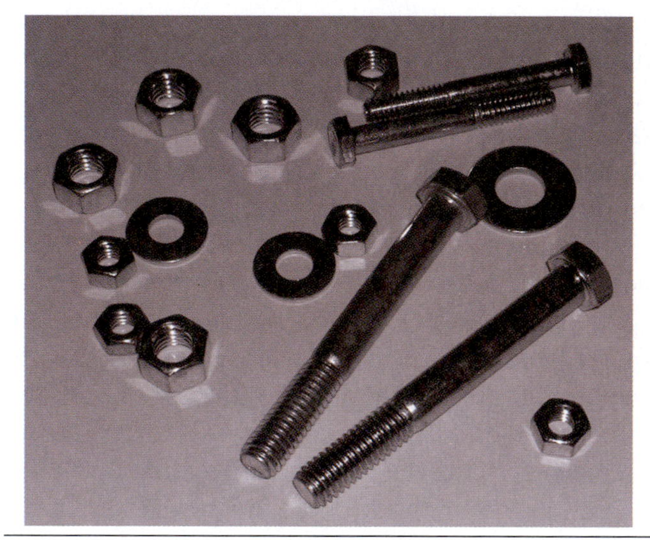

Figure 4-2 Examples of nuts and bolts are shown here.

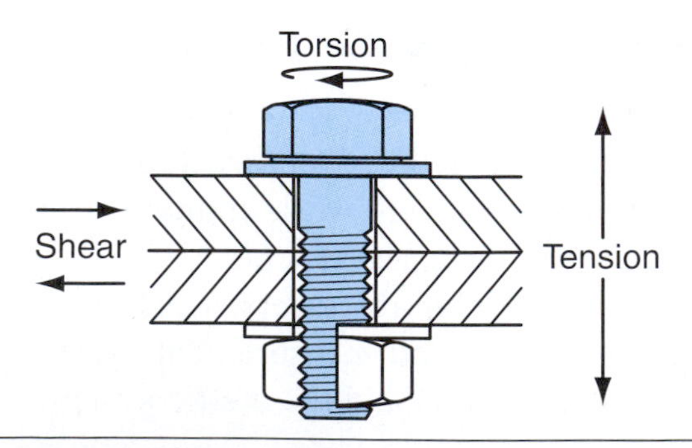

Figure 4-3 Preload is the technical term for the tension caused by a bolt that has been tightened. Copyright by American Honda Motor Co., Inc. and reprinted with permission.

top of the next (Figure 4-5). The International Standards Organization (ISO) sets these specifications. This is also the distance the bolt moves in one complete revolution. Standard or U.S. Customary (USC) fastener thread pitch is determined by the number of Threads Per Inch (TPI). Refer to Figure 4-6 for an example of thread pitch measurement using the conventional system.

Thread Types

In respect to thread standards, there is a metric thread (SI), a parallel thread for piping (PF), a taper thread for piping (PT), and a unified thread (UNC, UNF).

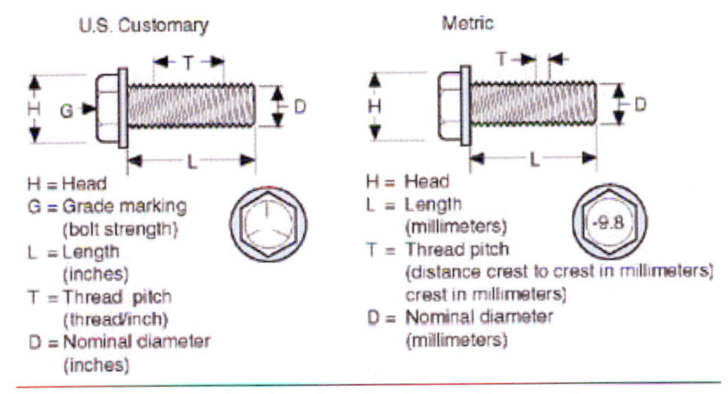

Figure 4-4 Critical bolt dimensions are shown here.

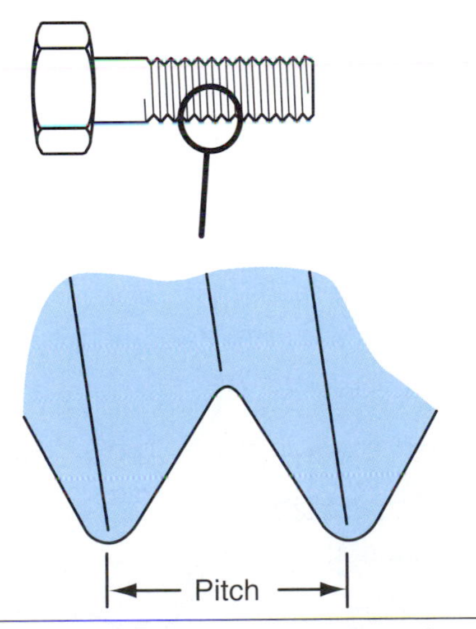

Figure 4-5 Thread pitch for metric fasteners is the distance from the top of one thread to another. Copyright by American Honda Motor Co., Inc. and reprinted with permission.

The three most common types of threads used on fasteners are:

- Coarse threads (UNC–Unified National Coarse)
- Fine threads (UNF–Unified National Fine)
- Metric threads (SI–System International)

An additional thread is used on motorcycles, but in limited applications. Threads on oil pressure switches, coolant temperature sensors, etc., may utilize the following:

- Tapered pipe thread (PT)
- Parallel thread (PF)

Take note that PT and PF threads are *not* compatible with each other or any other threads previously mentioned.

Never substitute thread types or thread damage will result. To prevent damage to fasteners, always thread the bolt or nut by hand (or fingers) for the first three to five complete turns. If the fastener only threads a turn or two then starts to bind, there may be a mismatch. Do not continue to tighten using a big wrench, air tool, or impact wrench, as damage is liable to occur.

Bolts and nuts come in right- and left-hand threads. With right-hand threads, the fastener must be turned clockwise to tighten (this rhyme may help you to remember: *righty tighty, lefty loosy*). This is the most common style of thread.

A left-hand thread requires turning the fastener in a counterclockwise direction to tighten.

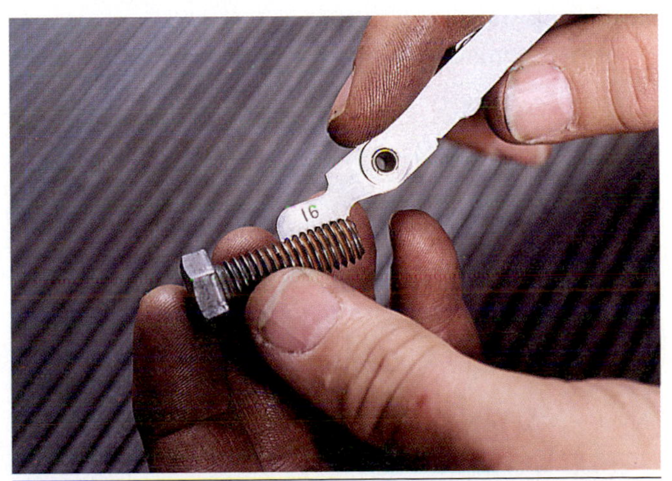

Figure 4-6 A thread pitch gauge that measures thread per inch is shown here.

Left-hand threads are not common. The letter "L" may be stamped on fasteners with left-hand threads.

Bolt Grades

Tensile strength, or grade, refers to the amount of pull a fastener can withstand before breaking. Bolts are made of different metals. Some are better than others.

Tensile strength of bolts can vary. **Bolt head markings**, also called grade markings, specify the tensile strength of the bolt. Standard or U.S. Customary (USC) bolts are identified with *lines* or *slash marks* (Figure 4-7). Count the lines and add two to determine the strength of the bolt. These are known as **Society of Automotive Engineers (SAE)** type and are evaluated by the **American National Standards Institute (ANSI)**, which is used to measure a fastener's tensile strength. SAE is the Technical Standards Board that issues and recommends industry standards. ANSI is an organization in the United States that sets technical standards.

Metric bolts are *numbered*. The higher the bolt number, the greater the strength of the bolt (Figure 4-7). The **International Standards Organization (ISO)** defines fastener quality in terms of tensile strength and yield strength. This standard is used for metric fasteners and may someday eventually replace all other grading standards. Metric fasteners classified as 8.8 or higher are required to have the markings. The ISO is a worldwide federation of national standards bodies from each of 140 countries.

When replacing any bolt, *always* replace with the same grade markings. A weaker bolt may easily snap, causing part failure and a dangerous condition. Replacing a bolt with a stronger one is not always safe. Harder bolts tend to be more brittle and may fail in specific applications.

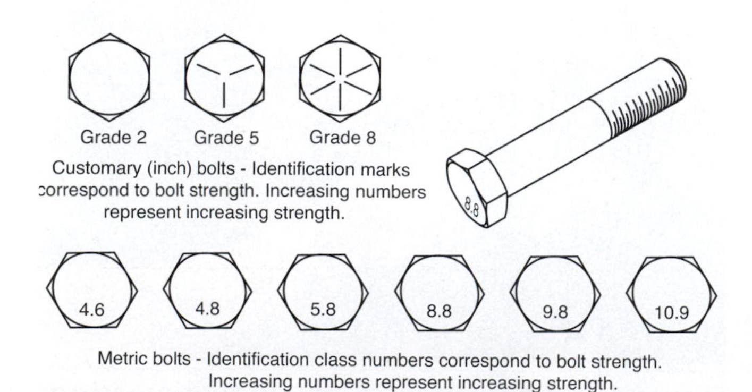

Figure 4-7 Typical bolt grade markings.

Bolt Types Used on Motorcycles and ATVs

There are various types of bolts found on motorcycles and ATVs. This section will cover a few of the most commonly found.

Deep Recess (DR) Type Bolts

DR-type (deep-recess) **bolts** without strength markings (flange bolts with hex heads and weight reduction recesses in them) are classified by outer flange diameters (Figure 4-8). Their outer flange diameters are larger than standard bolts with a flanged head. Be careful to install these bolts in the correct location and apply the correct torque.

Uniform Bearing Stress (UBS) Bolts

UBS bolts (Figure 4-9) are designed to resist loosening. A small angle of 5–60 degrees is rolled or forged on the underside of the bolt head. This causes the flange to flex as the bolt is tightened and provides additional friction to hold the bolt tight. They can be identified by an undercut radius under the bolt head. UBS bolts may or may not be marked with strength marks.

Torx Bolts

Torx bolts (Figure 4-10) are characterized by a 6-point star-shaped pattern that is patented. This design will carry a greater torque from the socket to the bolt. These bolts come in both external and

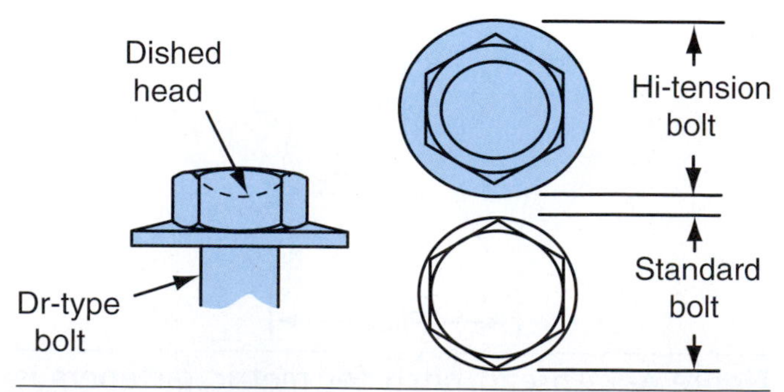

Figure 4-8 A DR-type bolt is shown here. Copyright by American Honda Motor Co., Inc. and reprinted with permission.

internal types. The outside torx is classified as an E type and the internal as a T type. These are used on many ATV differentials.

Combination (CT) Bolt

The **combination bolt (CT)** is a type of self-tapping bolt (Figure 4-11). It forms the female threads when it is screwed into the unthreaded pilot hole by deforming the walls of the hole. The lower half of the CT bolt features the combination of the standard threads and the low threads. Few chips and shavings are produced from the CT bolt.

Note: When the CT bolt is reused, tighten the bolt with care to prevent damage to the existing threads. When replacing a CT bolt (Figure 4-12), use a new CT bolt or a standard bolt of shorter length (L sb). Do not use a standard bolt of length (L ct).

Studs

A **stud** (Figure 4-13) is a fastener with external threads on each end. To hold parts together, a stud first is tightened into a threaded hole in a part (such as an engine case). A second part (a cylinder head) fits over the exposed end of the stud. Finally, a nut is fitted to the exposed end of the stud and is tightened. Many studs use two different thread sizes so be sure to know which side of the stud should be inserted into the threaded hole.

Tension Bolts

Certain areas of the motorcycle or ATV are subjected to repeated and severe external forces such as vibration or expansion from heat. Special bolts (Figure 4-14) with greater elasticity are used in these areas. These fasteners can stretch further than common bolts without permanently stretching. They are used on cylinder heads, connecting rods, and crankcases.

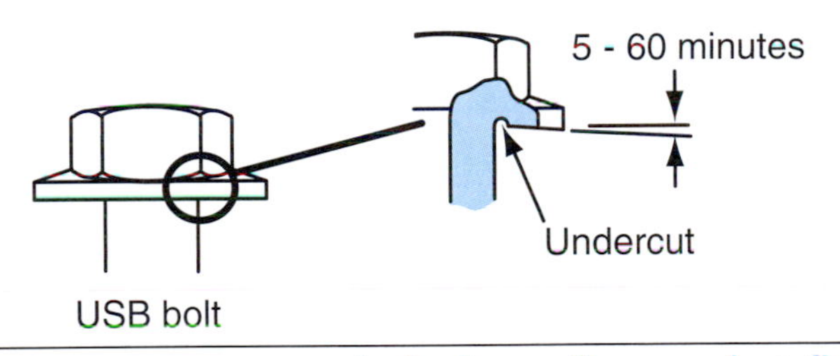

Figure 4-9 UBS-type bolts have flanges that flex to resist loosening. Copyright by American Honda Motor Co., Inc. and reprinted with permission.

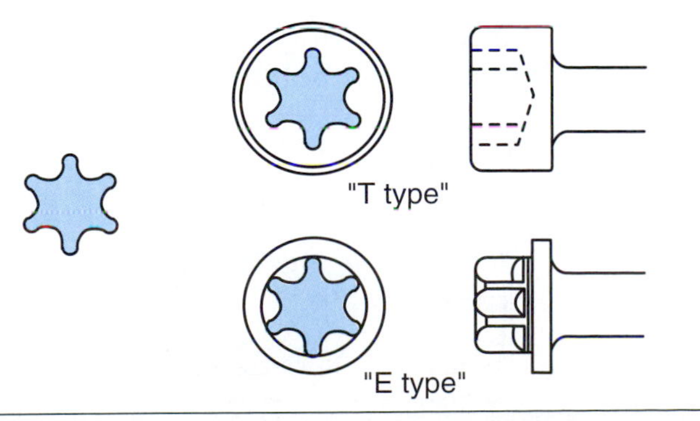

Figure 4-10 The two types of TORX bolts are shown here. Copyright by American Honda Motor Co., Inc. and reprinted with permission.

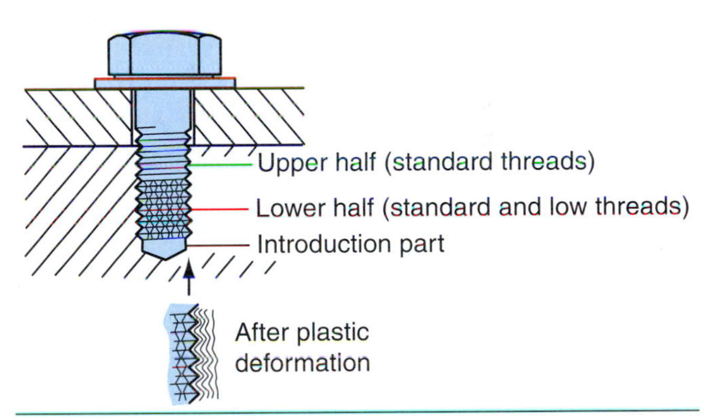

Figure 4-11 CT-type bolts are self-tapping. Copyright by American Honda Motor Co., Inc. and reprinted with permission.

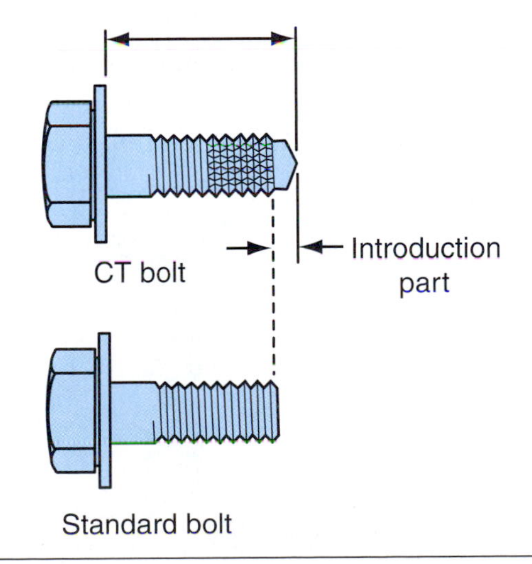

Figure 4-12 If replacing a CT bolt with a standard type, be sure to use a shorter bolt or else you could damage the component. Copyright by American Honda Motor Co., Inc. and reprinted with permission.

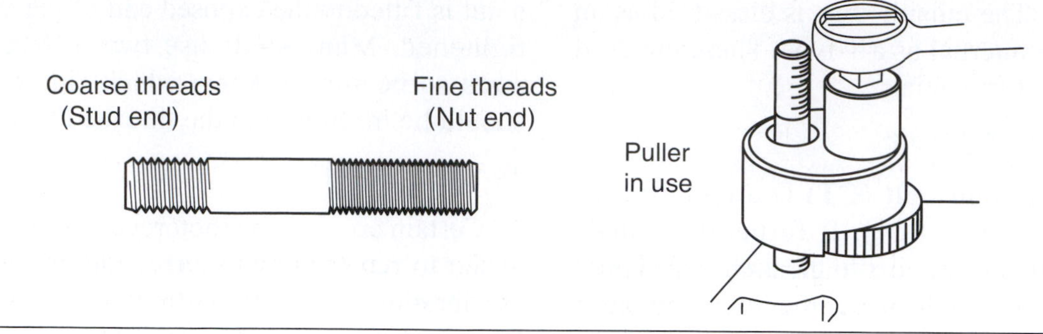

Figure 4-13 Studs have external threads on both ends.

Nuts

A **nut** has internal threads and usually a six-sided outer shape. The grade of nut used must match the grade of bolt that it is used with. Manufacturers use several different nut markings to denote the grade identification (Figure 4-15). Some marks are on the top and others are marked on the sides. Several types of nuts are commonly used on motorcycle and ATV applications.

Self-Locking Nuts

A **self-locking nut** (Figure 4-16) has a spring plate on the top. This spring plate presses against the thread, making it difficult for the nut to loosen. After removal, this type of nut can be used again. This type of nut can be used on the frame, in suspension pivot bolt/nut applications, and as axle nuts. The following are two basic points to consider when working with self-locking nuts:

- The bolt head must be held during nut installation and removal due to the resistance of the nut spring plate against the bolt.

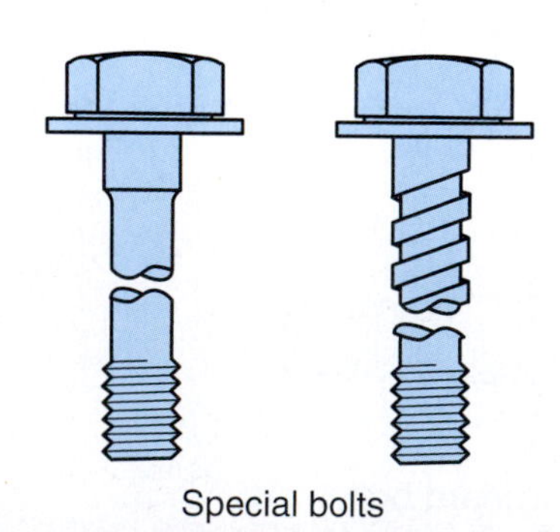

Figure 4-14 Special bolts are used in various places on a motorcycle. Be sure to use the correct bolt when reassembling. Copyright by American Honda Motor Co., Inc. and reprinted with permission.

- If the bolt length is too short, the spring plate portion of the lock nut will not engage with the thread fully and therefore the nut will not lock.

Castle-Headed Nut

The **castle-headed nut** (Figure 4-17) allows a cotter pin to be installed through a nut and bolt to prevent loosening. Applications include important safety points on the frame, axle nut, and brake torque rod. The following are a few points to consider when working with castle nuts:

- Always use new cotter pins during assembly.

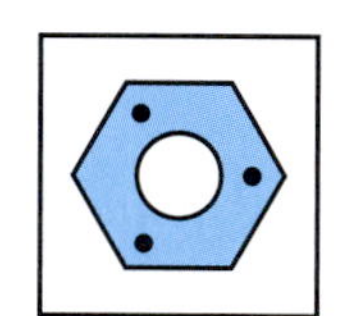

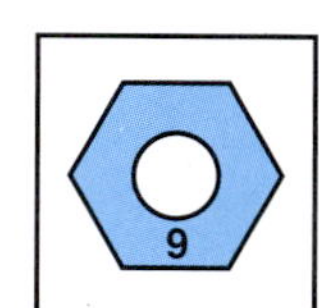

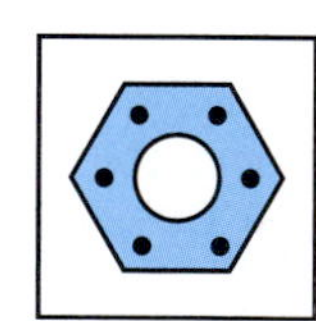

Figure 4-15 Grade markings on nuts are used in the same fashion as on bolts.

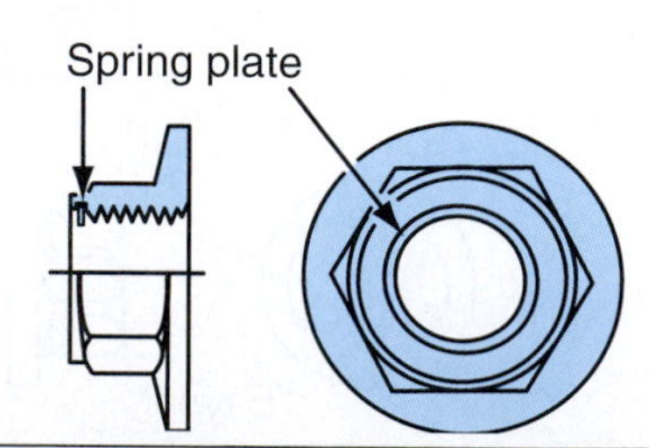

Figure 4-16 A self-locking nut is shown here. Copyright by American Honda Motor Co., Inc. and reprinted with permission.

- Tighten the nut to the specified torque. Align the next possible pin hole while tightening the nut just beyond the specified torque.
- Do not align the holes in a position where the nut torque is less than the specified torque.

Stake Nut

The **stake-type lock nut** (Figure 4-18) incorporates a metal collar at the top of the nut. A punch is used to stake (bend or indent) the collar of the nut to match a groove in the shaft that it is being used on. Applications include the clutch center lock nut, a shift drum stopper plate, or wheel bearing retainers. The following are a few points to consider when using stake nuts:

- During disassembly, eliminate the staking point before the nut is loosened.
- Replace the nut if the old staked area of the nut aligns with the groove of the shaft after tightening the nut to the specified torque.
- After tightening the nut to the specified torque, stake the nut collar using a drift punch. Ensure that the staking point has entered into the grove at least two-thirds of the groove depth.

Well Nut

Well Nuts are rubber fasteners with a brass threaded sleeve installed in the center (Figure 4-19). A screw normally threads into the brass insert. When tightened, the insert causes the rubber to expand and hold it in place. Well nuts are used on body panels, fairing parts, and windshields/windscreens. Do not use excessive force to tighten.

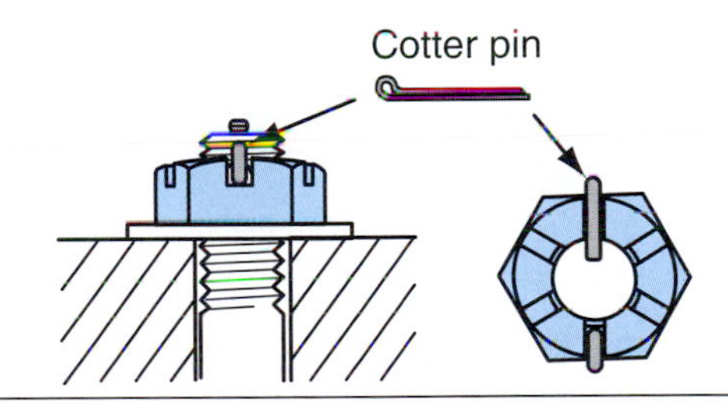

Figure 4-17 Castle-headed nuts are used with cotter pins to prevent loosening. Copyright by American Honda Motor Co., Inc. and reprinted with permission.

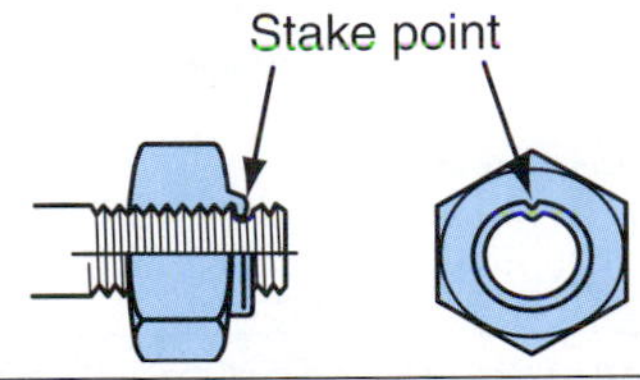

Figure 4-18 Stake-type lock nuts are usually used inside of engines. Copyright by American Honda Motor Co., Inc. and reprinted with permission.

Acorn (Cap) Nuts

An **acorn nut** or **cap nut** (Figure 4-20) is a decorative nut with a finished or plated surface and is often used to cover the threaded end of a bolt or stud. Many are made of stainless steel to prevent corrosion or chipping. Bolts (or studs) must be of the proper length or the nut may bottom out on the stud before the clamping action can occur. Applications include head-nuts and engine covers.

Washers

There are several types of washers common to the motorcycle industry.

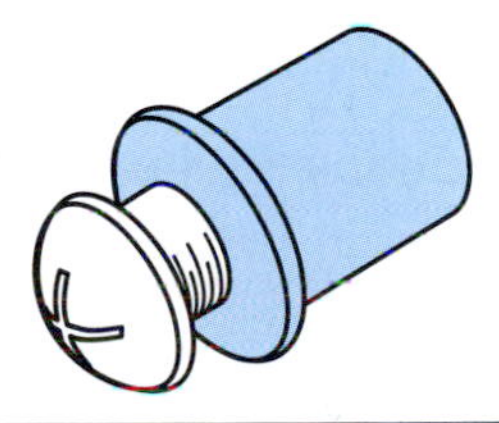

Figure 4-19 Well nuts are used mainly on plastic parts like windshields. They expand when tightened. Copyright by American Honda Motor Co., Inc. and reprinted with permission.

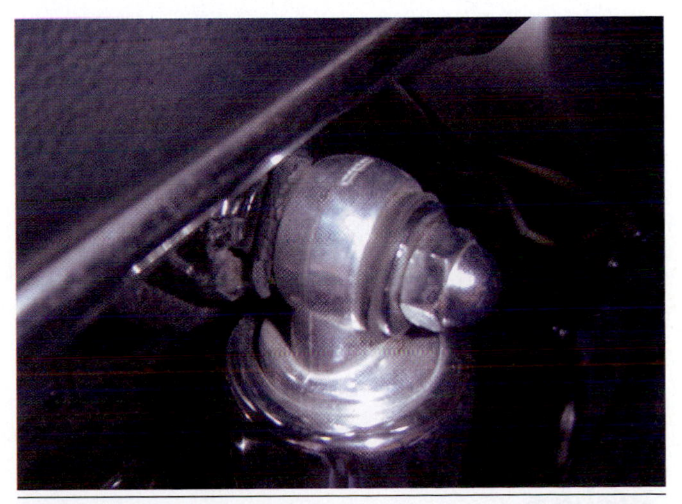

Figure 4-20 Acorn or cap nuts are used to cover the threaded end of a bolt or stud. Be careful to install them correctly or the nut will bottom out and not tighten properly.

Flat Washer

A plain **flat washer** is used under bolt heads and nuts. Using the proper washer is also necessary to achieve the correct load on a bolt. A washer will also increase the clamping surface under the fastener and prevent the bolt or nut from digging into the part. This is important when the material to be clamped is soft, such as aluminum, magnesium, or other metal softer than steel.

Split-Ring-Type Lock Washer

A **split-ring-type lock washer** (Figure 4-21A) is compressed under the bearing surface (of the nut) pressure and the elasticity of the spring. The edges of the ring ends prevent loosening. Application includes various points on the frame and bolts—incorporating washers are also used. When using with a plain washer, always put the lock washer between the nut and plain washer (Figure 21B).

Cone-Type Lock Washer

The **cone-type lock washer** (Figure 4-22) is a dished washer made from spring steel. When installed, the center of the washer sets up from the surface. The bearing surface (of the nut) presses on the cone-type washer and the spring reaction presses against the nut to prevent it from loosening. Applications include the clutch lock nut and primary gear lock nut as well as the drive sprocket center bolt.

Tongued Lock Plate Washer

The **tongued lock plate washer** (Figure 4-23) serves the purpose of washer and locking device. Simply bend the tongue (claw) to the flat face of the nut or into the groove of the nut to lock the nut or bolt head. Applications include the clutch locking nuts as well as important safety points on the frame including steering head bearing top nut and driven sprocket nuts. Replace the lock plate with a new one whenever the lock plate is removed.

Thread-Locking Agents and Sealers

Under some conditions, special chemical compounds called thread-locking agents may be needed to help threaded fasteners do their job. Thread-locking agents (Figure 4-24) can be used where vibration would cause the fastener to loosen. These compounds are anaerobic adhesives that set up in the absence of air. Thread-locking agents come in several grades depending on the desired strength, with the most popular being high-strength Red and medium-strength Blue. Applications include frame components, fork socket bolts, brake disc bolts, and fasteners inside the engine such as the stator coil bolts, bearing retainer bolts, and shift drum stopper plate bolt.

Take the following into consideration when using thread-locking agents:

- Application of a locking agent increases loosening torque. Take care not to damage the bolt during removal. However, sometimes heat (using a heat gun) is needed to soften the

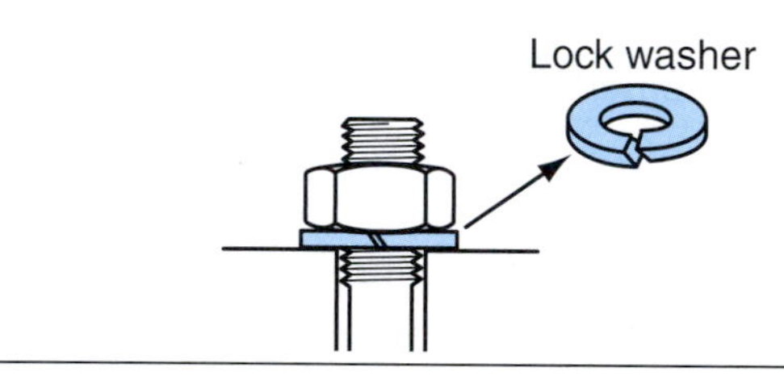

Figure 4-21A A typical split-ring lock washer is shown here. Copyright by American Honda Motor Co., Inc. and reprinted with permission.

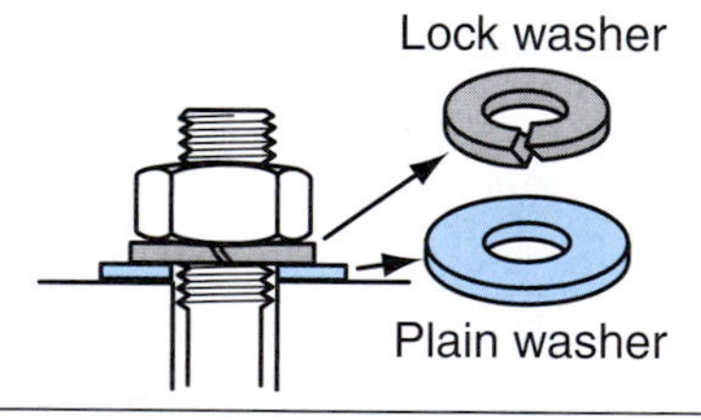

Figure 4-21B The correct installation of a flat and lock washer is shown here. Copyright by American Honda Motor Co., Inc. and reprinted with permission.

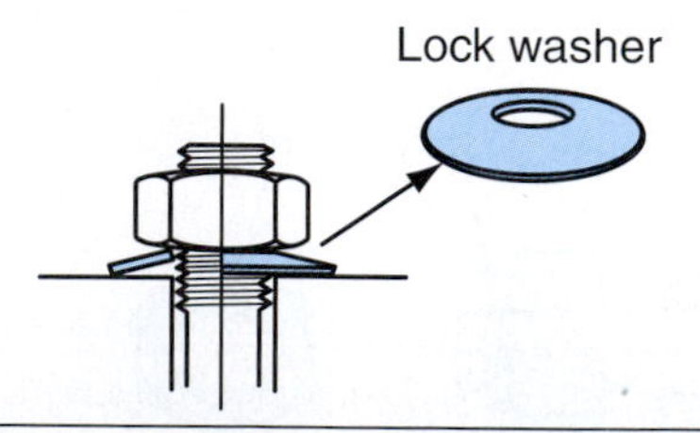

Figure 4-22 Cone-type washers must be installed correctly to function properly. Copyright by American Honda Motor Co., Inc. and reprinted with permission.

material when higher strength compounds are used.

- Before applying locking agent clean off all oil and/or residual adhesive remaining on the threads and dry them completely.
- Applying a small amount of adhesive to the end of the bolt threads distributes the adhesive as it is threaded in.
- Excessive adhesive may, during loosening, damage the thread or cause the bolt to be broken.
- Locking agents may cause plastic parts to crack. Do not let the locking agent touch plastic parts.

Along with thread-locking agents thread sealants are used in the same manner in locations where the threads of a bolt or other fastener protrudes into an area where liquids such as a coolant or oil could cause corrosion or leak past the threaded area.

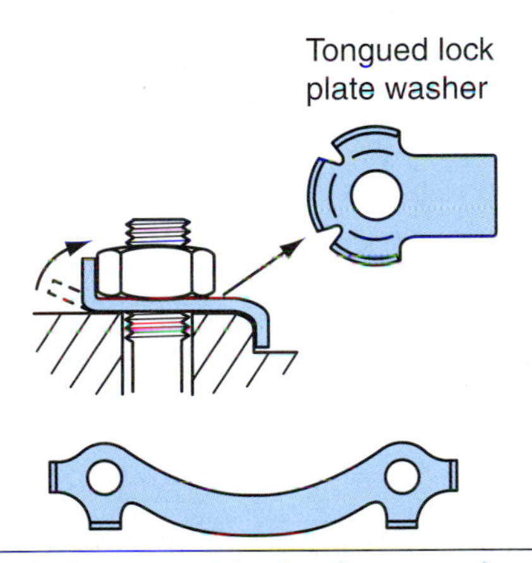

Figure 4-23 Tongued lock plate washers can be seen often on the rear sprocket of a motorcycle. Copyright by American Honda Motor Co., Inc. and reprinted with permission.

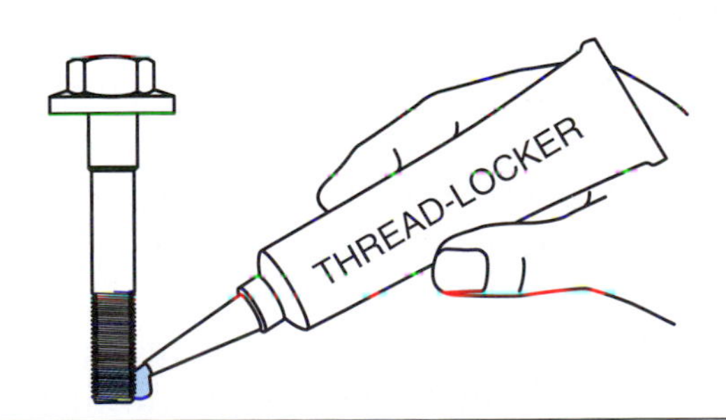

Figure 4-24 Thread-locking agents are often used on motorcycles and ATVs to help prevent loosening of the fastener. Copyright by American Honda Motor Co., Inc. and reprinted with permission.

INSPECTION, CLEANING, AND REPAIR OF THREADED FASTENERS

Before any reassembly process, threaded fasteners must be inspected, cleaned, and sometimes repaired to be made ready to function normally. It is critical to review the motorcycle or ATV service manual to determine if specific bolts, nuts, or washers can be reused, or must be replaced. Failure to replace specific fasteners such as crankshaft connecting rod bolts/nuts or cylinder head bolts or nuts as recommended by the manufacturer could cause future fastener failure that would be expensive as other damage would certainly occur to the related components.

Fastener Inspection and Cleaning

Always clean fasteners thoroughly. Installing fasteners with dirt or other foreign matter on the threads or on the bolt- or nut-bearing surfaces will result in improper tension even if the proper torque value was applied (Figure 4-25). As the dirt of foreign matter breaks down due to vibration and the attached parts working against each other, the fastener will soon work its way loose.

Solvent cleaning can remove surface grime such as oil and grease. It is sometimes necessary to use a power wire wheel to remove rust and other materials from the head and threads of the fastener. Wear gloves and eye protection when using a power wire wheel. Using a thread die to chase (clean) the threads (Figure 4-26) will insure that the thread is both clean and free of nicks that could cause excessive friction when installed.

If a bolt has been over-tightened, it will stretch (Figure 4-27). A **stretched bolt** can be identified

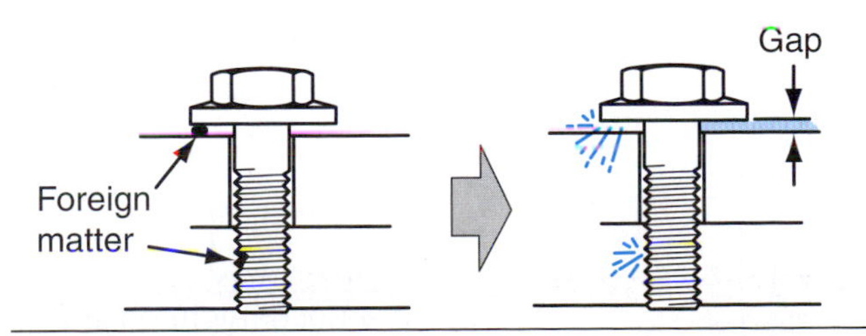

Figure 4-25 If debris is in or around a fastener, it will not have the ability to fully function. Copyright by American Honda Motor Co., Inc. and reprinted with permission.

Figure 4-26 Cleaning a bolt using a wire wheel is shown here. Be sure to use proper safety apparel when using a wire wheel. Can you note what safety apparel is missing in this picture?

when a nut threads down the bolt easily then binds when it reaches the stretched area. At that point, it will become hard to turn. A stretched bolt must be discarded and replaced.

Threaded holes in various components must be chased (cleaned) using a thread tap (Figure 4-28) to insure that the internal threads are clean and free of contamination. If damaged threads are found, it is possible to make repairs by replacing the threads. Thread replacement will be covered in detail later in this chapter.

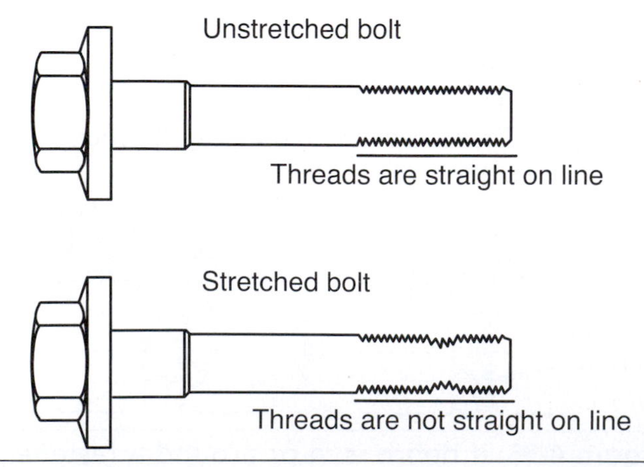

Figure 4-27 An example of a stretched bolt (bottom) is shown compared to an un-stretched bolt (top).

Figure 4-28 A technician is "chasing" threads here with a tap.

STRESSES ON THREADED FASTENERS

Earlier in this chapter, we mentioned that preload is the technical term for the tension caused by tightening the fastener that holds the parts together. The forces acting on a threaded fastener include axial tension, torsion, and shear (Figure 4-29).

- **Axial tension** is the stretching force applied to a bolt when it is tightened into a case or a nut is tightened onto it. Axial tension is the most important point in the tightening of a fastener.
- **Torsion** is a twisting force applied to the head of a bolt when it is tightened.
- **Shear** is the force exerted at 90 degrees to the center line of a bolt.
- **Clamping force** is the force applied by a bolt holding two parts together.

When a fastener is tightened, an axial tension is applied to it. This stress stretches the fastener and reduces its diameter slightly. As we tighten the fastener more, we reach the **yield point** (Figure 4-30), which is the maximum tensile stress that may be impressed upon a material without straining it beyond the elastic limit. Continuing to tighten the fastener moves it into the **plastic range** in which we are permanently stretching the fastener. If we continue to tighten the fastener, we will reach the ultimate tensile strength of the fastener and shortly thereafter, the bolt will break.

Preload

For any fastener to work properly, it must be stretched sufficiently to produce a static preload (clamping force) that is greater than the expected external loads. This is what the torque applied to the fastener does. Bolts and screws stretch like a spring and then return to their original length when tightened in the **elastic range**. It is critical that the fastener be tightened the right amount. Too little and they will loosen, too much and they may break or damage threads. Fasteners tightened into the **plastic range** will be permanently stretched and should be replaced. Fortunately, the engineers have taken care of these design characteristics. All we have to do is prepare the fasteners correctly and tighten them to the service manual specified torque. Preparing the fasteners includes cleaning, inspecting, and blowing them dry as discussed earlier.

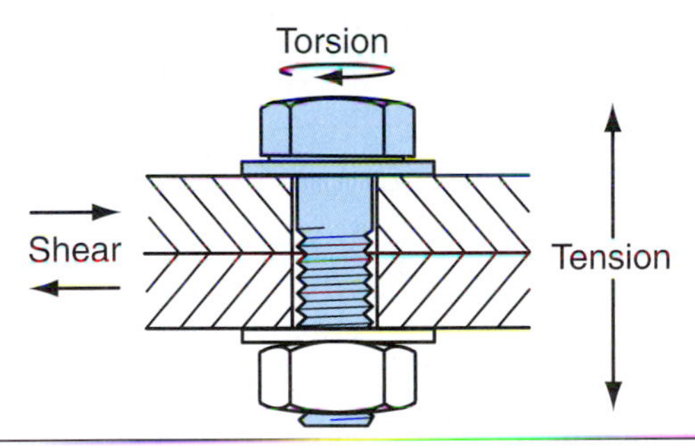

Figure 4-29 The forces acting on a threaded fastener include axial tension, torsion, and shear as shown. Copyright by American Honda Motor Co., Inc. and reprinted with permission.

There are some variables that can affect tightening torque. Tightening torque can decrease over time, from external forces or vibration.

Proper tightening forces are specified according to fastener strength, strength of fastened parts, and intensity of external forces. Tightening must be carried out to the service manual specification, especially at critical fasteners. As an example, tightening a connecting rod-bearing cap with a higher torque than specified will reduce the oil clearance for the bearing to less than specified, which may lead to premature bearing seizure. A low torque, on the other hand, may allow the nuts or bearing caps to loosen and fall off during engine operation, leading to serious engine damage.

TIPS FOR WORKING WITH THREADED FASTENERS

- Always clean bolts, screws, and threads in engine cases and blow them dry. Dirt or foreign matter under the head or on the threads will prevent a proper torque from being applied and risk loosening.

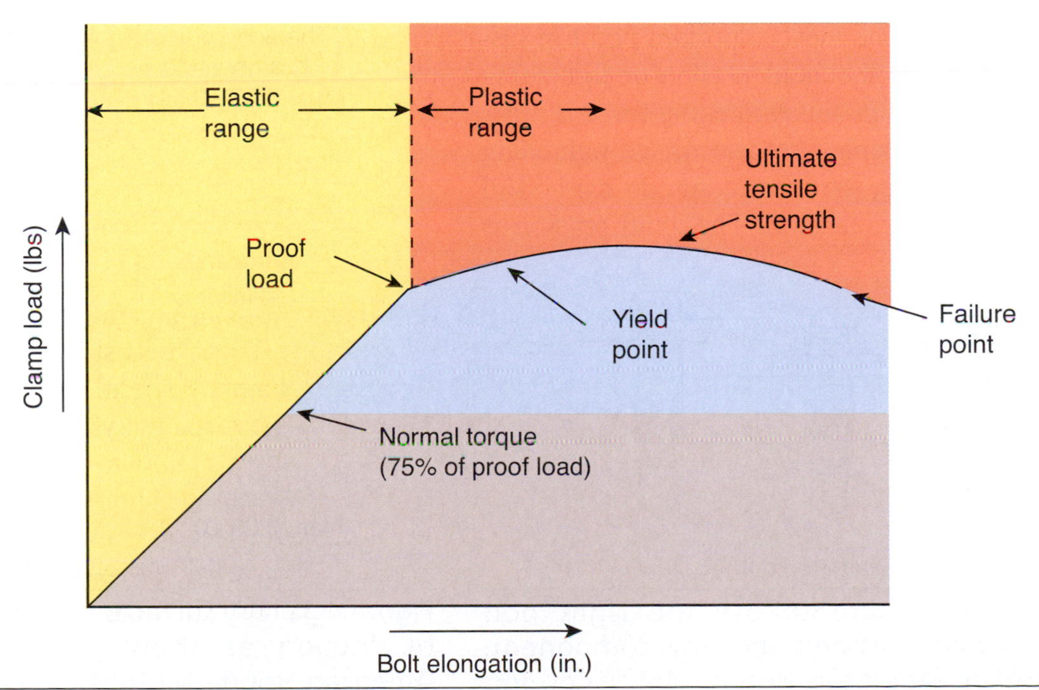

Figure 4-30 This graph shows the different levels of tension that can be given to a fastener.

- Do not lubricate fasteners unless specified in the service manual.
- Bolt or screw lengths can vary for engine covers and cases (Figure 4-31). These different lengths must be installed in the correct locations. If you are not sure, place the bolts in their holes and compare the exposed lengths. Each should be exposed the same amount. A bolt that is too long will bottom out and break or strip threads. A bolt that is too short will pull the threads out of the case. The exposed length of bolts should be at least 1.5 times their diameter.

To prevent warping important components and ensure proper gasket sealing, multiple-sized fasteners should be torqued as follows:

- During disassembly, always loosen the small fasteners first.
- Tighten all fasteners to hand-tight.
- Torque largest fasteners before smaller fasteners.
- Torque the bolts in sequence to half the specified torque and then repeat the sequence to the full-specified torque.
- If no sequence is given, torque in a crisscross pattern from inner to outer.

TIGHTENING AND TORQUE

As mentioned, the most important point in fastener tightening is the axial tension or tightening force. The problem is, this tightening force is difficult to measure. Using a predetermined tightening torque is, therefore, the most common method of controlling fastener tension. The axial tension is proportional to the torque applied in certain conditions; accordingly, the most common condition being clean, dry threads.

Friction at the threads uses up 40 percent of the applied torque. The friction between the bolt head and mating surface uses 50 percent of the torque. That leaves 10 percent to tighten the bolt. Dry surfaces have the highest friction (Figure 4-32). When a lubricant is applied to fasteners, the friction is decreased. The μ symbol indicates the coefficient of friction. The lower this number, the less friction. Dry threads have the high est friction; oiled threads have the lowest.

If the threads are lubricated, more of the torque is applied to the axial tension. This means the parts are held together with a *greater* force and the fastener is stressed more. The graph shown in Figure 4-33 gives some examples of how much the friction is reduced when kerosene or oil are applied to the threads. With the same tightening torque the axial tension increases greatly. This graph can be compared to the images in Figure 4-32.

Some manufacturers specify that oil be applied to the threads and to the underside of the head on certain fasteners. It is important that these parts be oiled before tightening. If they are assembled and tightened dry they will not apply the correct pre-

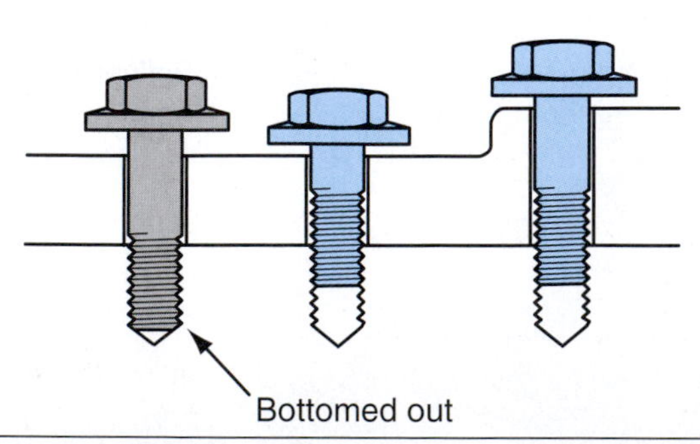

Figure 4-31 Be sure to use the right bolt lengths when reassembling any component. Copyright by American Honda Motor Co., Inc. and reprinted with permission.

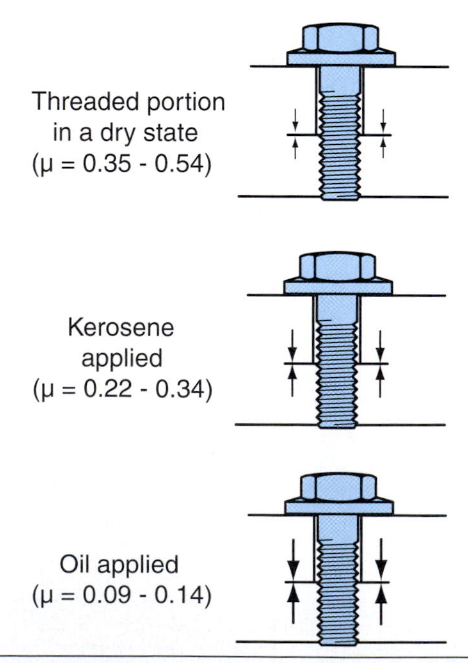

Figure 4-32 Dry surfaces have the highest levels of friction as shown here. Copyright by American Honda Motor Co., Inc. and reprinted with permission.

load (clamping force), which could cause the bolt to loosen or a joint to leak. Also of great importance: Do not over oil; excessive oil could cause a hydraulic lock and damage a part.

All other threaded fasteners must be assembled and tightened dry. Lubrication of these bolts may cause them to break.

Torque

Torque values are determined according to fastener size and strength, and to the strength of the parts that are fastened together. In earlier service manuals torque values are specified within a certain range. In current service manuals, a single value is given for the torque. This is equivalent to the middle of the range in earlier manuals. Service manuals specify torque in Newton meters, kilogram–meters, and foot–pounds.

As seen in Figure 4-34, torque is simply a force applied to a lever of a specific length. One kilogram attached to the end of a 1-meter arm gives 1 kilogram-meter (kg-m) of torque. A weight of 5 kg applied to an arm 1/5 of a meter gives the same torque.

Torque = Force + Length

Torque Wrenches

As mentioned in Chapter 3, there are two common types of torque wrenches used by technicians to tighten fasteners to specifications. There is an additional type of torque wrench that is not found as often but we will cover it as well in this section.

Beam-Type Torque Wrench

The **beam-type torque wrench** (Figure 4-35) is the least expensive and works by the beam bending in response to the torque applied. It is simple, reliable, and accurate. When tightening a bolt, apply force in the center of the handle. This allows the beam to bend in the manner designed to indicate the correct torque.

- Do not over-torque the wrench or the beam may bend permanently.
- Rough handling can bend the pointer arm. If bent, it can be bent back to the center without loosing accuracy.
- If the beam is bent it cannot be bent back.

Click-Type Torque Wrench

The **click-type torque wrench** (Figure 4-35) is popular with technicians due to ease of use and easy-to-store profile. A click-type torque wrench works by preloading a "snap" mechanism with a spring to release at the specified torque. When the mechanism releases the ratchet head makes a "click" noise. The torque is set by rotating the micrometer-style handle to the appropriate torque setting. To use, pull on the handle until you feel (or hear) the click.

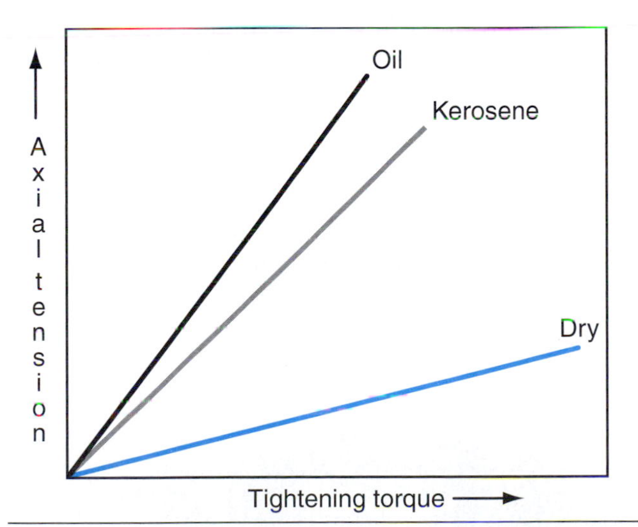

Figure 4-33 This graph shows different levels of friction when assembled dry and when using kerosene or oil. Copyright by American Honda Motor Co., Inc. and reprinted with permission.

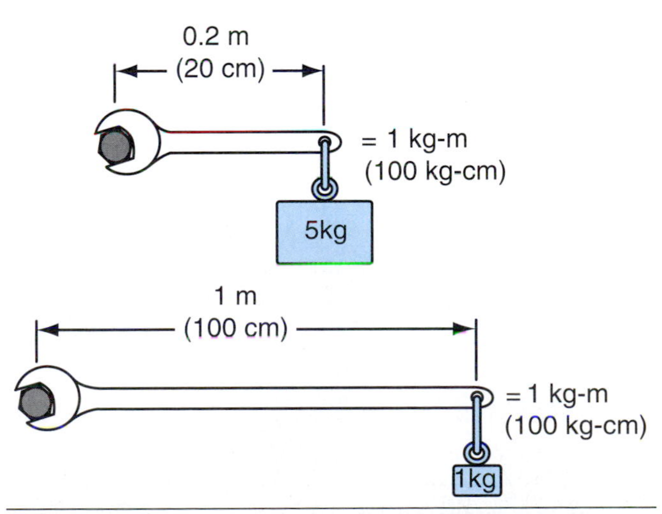

Figure 4-34 A graphic example of torque is shown here. Copyright by American Honda Motor Co., Inc. and reprinted with permission.

- A ratchet head makes it easy to use in a confined space.
- Do not use this torque wrench to loosen tight fasteners as it may damage the calibration.
- When finished, always return the wrench to its lowest setting before putting it away. This will prevent the internal spring from taking a set.
- The actual click from a torque wrench is considered a "cycle" of that wrench. Most manufacturers recommend basic calibration after 8,000 to 12,000 cycles.

Dial-Type Torque Wrench

The **dial-type torque wrench** (Figure 4-36) is not as popular as the previous two types of torque wrenches mentioned but it provides easy operation and the ability to rotate the bezel face to "zero" to calibrate the wrench before use. The pointer will move up the scale as the fastener is tightened. Some technicians will turn the bezel face to the torque setting, then pull the wrench back to zero to obtain the proper torque.

- Be sure to "zero" the tool before use.
- The glass/plastic bezel cover can break or scratch; use care when storing the tool.
- Calibration should be checked every twelve to eighteen months.

Hints for Using Torque Wrenches

The following guidelines will ensure precise accuracy and reliability when using a torque wrench:

- Even the most expensive torque wrenches lose accuracy. It is a good idea to have your torque wrench calibrated periodically to maintain accuracy.
- Always work on CLEAN threads.
- For an accurate torque wrench reading, the final turn of the nut must be tightened with the torque wrench.
- Use your torque wrench for **TIGHTENING ONLY**!
- Always loosen or remove bolts, nuts, or studs with a standard wrench (not with the torque wrench).
- If using a dial-type torque wrench be **SURE** the dial and pointer are set correctly before applying torque.
- Wear **PROPER SAFETY GEAR** when using a torque wrench.

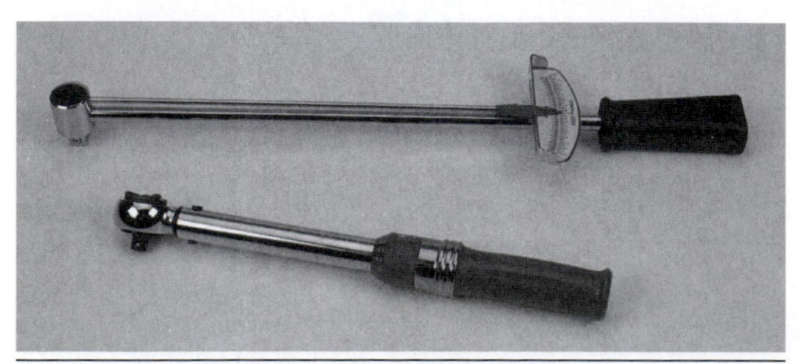

Figure 4-35 A beam-type torque wrench is shown on top and a click-type torque wrench is shown on bottom.

REPAIRING AND REPLACING BROKEN FASTENERS

Sooner or later every technician has to remove a broken bolt or repair a damaged thread. It may take just a few minutes to repair or it may take hours. It may cost a few cents for a new bolt or screw, or hundreds of dollars to repair or replace the component made unusable by a repair job gone bad.

A bolt that has broken off above the surface (Figure 4-37A) may be removable with vise grip pliers (Figure 4-37B). The following are steps to assist in removing a broken fastener using this method:

- Tap the top of the broken fastener with a ball peen hammer a few times to loosen or break the grip between the threads.
- It may also be helpful to apply a penetrating oil to loosen any rust or corrosion.
- Another technician trick is to heat the surface (with a heat gun or propane torch) to expand the metal and loosen the grip on the broken bolt.

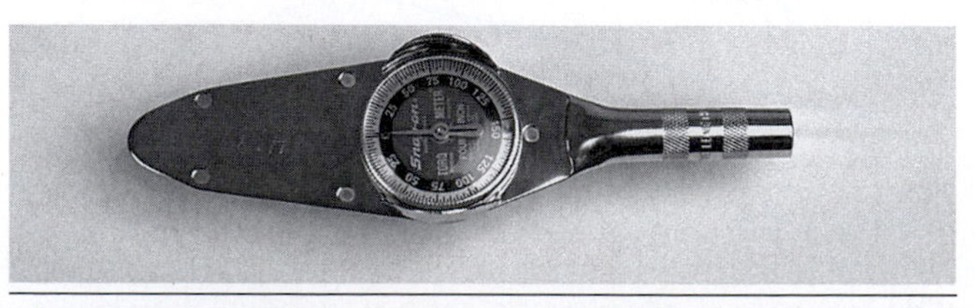

Figure 4-36 Dial-type torque wrenches are used on motorcycles as well as on ATVs.

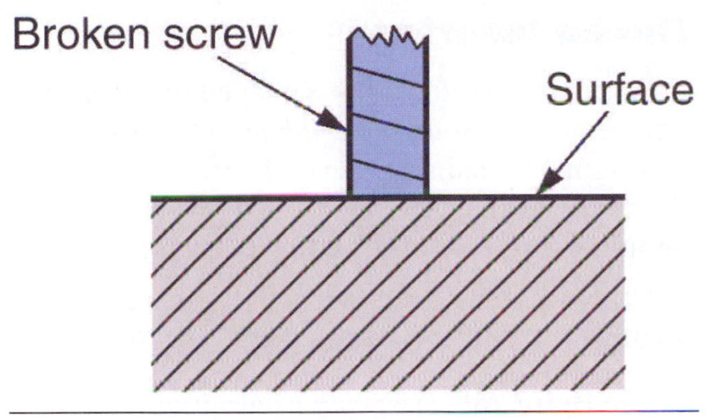

Figure 4-37A In most cases, a bolt that has broken off above the surface can be removed easier than one that has been broken off flush.

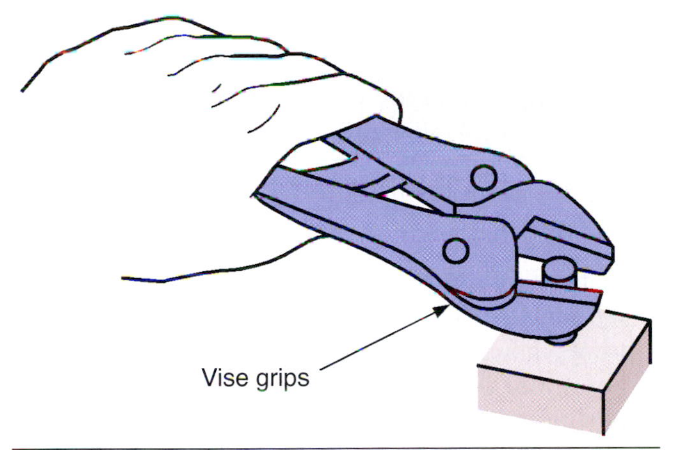

Figure 4-37B Using a vise grip wrench can be used to remove fasteners that have broken off above the surface.

Screw Extractors

A fastener broken flush with the surface may require the use of screw extractors. Some technicians may try using a sharp punch or chisel to tap the broken fastener counterclockwise to turn it out of the threaded material (Figure 4-38). This sometimes works but at times other means are necessary. Also, using "left-hand drills," to drill into the broken fastener may actually remove the bolt as you drill the larger hole. The drill motor must turn counterclockwise to work properly when using left-handed drill bits, therefore a reversible drill will be needed in these situations.

There are two common types of screw extractors (Figure 4-39). The easy out has large, widely spaced, sharp edges and tapers to a rounded tip. The fluted extractor has multiple sharp-edged groves along its length and does not have any taper.

Both types of extractors are usually packaged as kits to be used on several sizes of broken bolts, screws, or studs. To remove a broken fastener using an easy out, perform the following steps:

- Use a center punch to mark the center of the fastener.
- Drill a pilot hole all the way through the bottom of the fastener using the smallest drill bit. Then, drill the hole out using the recommended size drill bit for the tool being used.

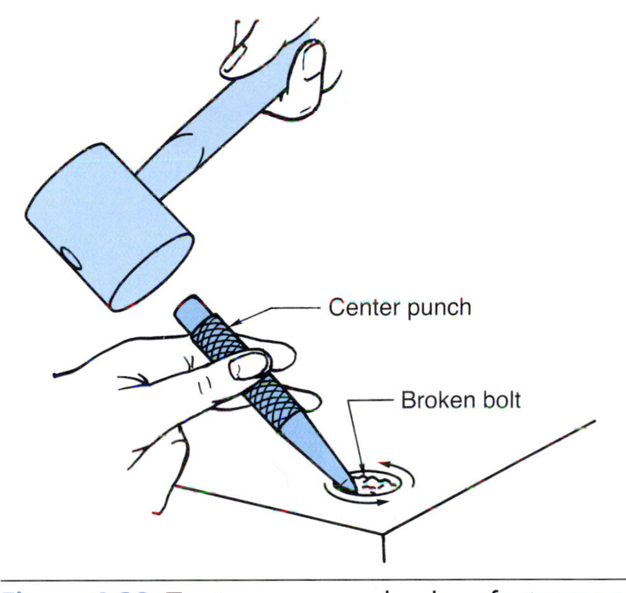

Figure 4-38 Try to remove a broken fastener off flush first by using a sharp punch or chisel and a hammer.

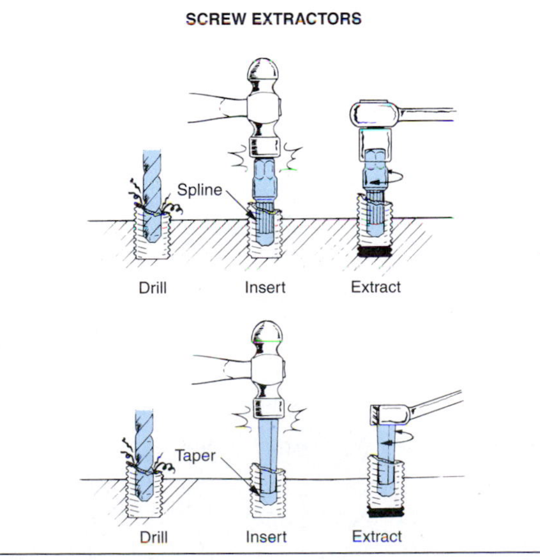

Figure 4-39 The two most common screw extractors are shown here.

- Insert the easy out, attach the handle, and turn counterclockwise to remove the damaged fastener (Figure 4-40).

The fluted extractor set includes several sizes of fluted extractors, the correct size drill bits, splined hex nuts, and drill guides. To remove a broken fastener using a fluted extractor, the same methods as mentioned for the easy out should be applied.

Extractors are hardened, which makes them difficult to remove if they break. A broken extractor cannot be drilled with a common drill. Caution should be used not to apply too much force when removing the damaged fastener.

If a drill, tap, or easy out does break off (somewhere it's not supposed to be), there are other methods of removal. A process known as Electro Discharge Machining (EDM) can remove the broken piece(s). The process is quite expensive and requires the complete disassembly of the item to be repaired. The process is like welding but in reverse. The results are very good but finding a shop in your area that can use this system of fastener removal may be a little tricky.

Drilling and Retapping

Occasionally a broken fastener can be repairedby drilling it slightly smaller than the original diameter. The hole is then carefully tapped to the original size. This procedure sometimes works and may be well worth the effort.

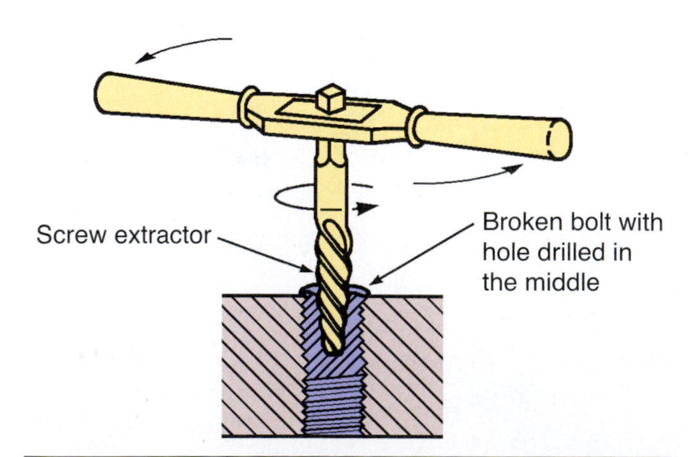

Figure 4-40 An example of using an easy out is shown here.

Thread Inserts

When a screw thread is damaged or stripped, it may be necessary to use a thread insert to restore the original thread size. Several varieties of thread inserts are available. They all require drilling and tapping a slightly larger hole. The insert is then threaded and locked into place.

Heli-Coil®

The **Heli-Coil®** is an oversized spring coil made of stainless steel (Figure 4-41). Heli-Coil® kits are available for Standard and Metric thread sizes. When installed, the inside thread of the Heli-Coil® restores the original thread size and pitch, and provides a repair that is just as strong and sometimes even stronger than the original thread.

To install a Heli-Coil®:

- Use the kit drill to drill an oversized hole.
- Tap the hole using the tap included with the kit.
- Install the insert using the special tool provided with the kit. The insert is installed slightly lower than the surface. A drive tang at the bottom of the Heli-Coil® locks to the tool during installation. The tang is notched for easy removal once the insert is installed.
- Break the driving tang off the bottom of the Heli-Coil® with a hammer and punch. Remove the tang before installing the bolt.

Solid-Threaded Inserts

In applications that require high strength, such as repairing stripped head bolt/stud holes, solid-threaded inserts are used. The insert shown in

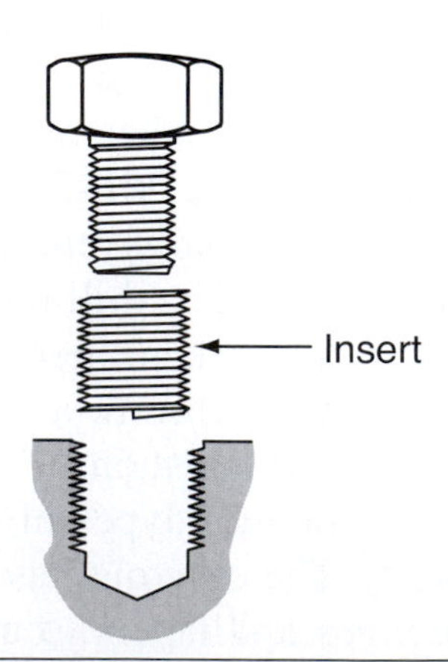

Figure 4-41 Heli-Coil® are used often when thread repair is required.

Figure 4-42 has bottom external threads that are cold rolled. The installation tool locks the insert into place by forcing the bottom threads to expand the mating external threads into the threads cut into mating material (engine, transmission, etc.).

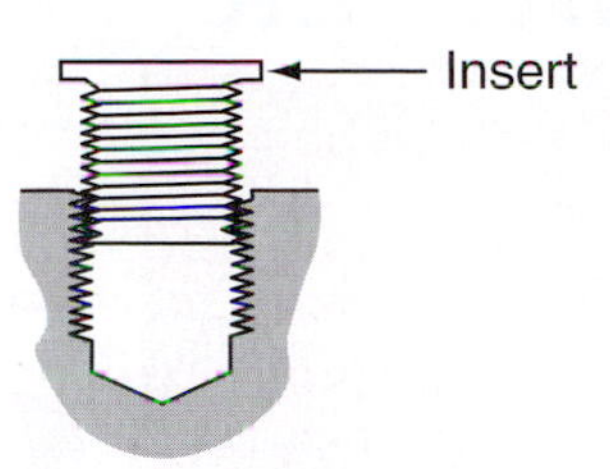

Figure 4-42 Solid-threaded inserts are also widely used when repairing threads in a component.

Summary

- Motorcycles and All Terrain Vehicles (ATVs) are held together with hundreds of threaded and nonthreaded fasteners.
- There are two common systems used to classify the measurement systems used on motorcycles: the conventional system and the metric system.
- Bolts and nuts come in various sizes, grades (strength), and thread types. Bolt dimensions provide bolt size, bolt head size, bolt length, and thread pitch.
- Quality standards for fasteners are established by the American National Standards Institute (ANSI) and by the International Standards Organization (ISO).
- A variety of washers serves to achieve the correct load on a bolt and to prevent loosening of threaded fasteners.
- The proper tightening of fasteners is a critical skill. Using a torque wrench correctly is a critical skill needed to correctly reassemble components on a motorcycle.
- Torque wrenches used to tighten threaded fasteners will vary in accuracy and for best results should be calibrated regularly.
- Broken bolts or studs can be difficult to remove or replace. Proper methods must be used to remove broken fasteners.
- Thread inserts are used to restore threads in engine or transmission cases, and even to replace spark plug threads.

Chapter 4 Review Questions

1. Motorcycles and ATVs use two common measurement systems, the ________ and the ________.
2. ________ is the technical term for the tension caused by tightening the fastener that holds the parts together.
3. The four most critical bolt dimensions are: bolt size, bolt head size, bolt length, and ________.
4. The three most common types of threads used on fasteners are ________, ________ and ________.
5. The thread pitch of a bolt is determined by the number of threads per inch when using the ________ of measurement.
6. A ________ has internal threads and usually a six-sided outer shape.
7. The two common tools used to torque fasteners are the ________ and ________ torque wrenches.
8. During disassembly, always loosen the ________ first.
9. When a screw thread is damaged or stripped, it may be necessary to use a ________ to restore the original thread size.
10. Don't lubricate fasteners unless specified to do so in the ________.

CHAPTER

5 Basic Engine Operation and Configurations

Learning Objectives

When a students have completed the study of this chapter and its laboratory activities they should be able to:

- Learn about reciprocating engines
- Understand the basic differences between a two-stroke engine and a four-stroke engine
- Understand how engines are rated
- Distinguish the primary components found in a two-stroke and four-stroke engine
- Learn about the different cooling systems used on motorcycles and ATVs
- Understand the different engine configurations found on motorcycles and ATVs

Key Terms

Bottom-dead center (BDC)
Bore
Brake horsepower (bhp)
Clutch
Combustion chamber
Combustion chamber volume (CCV)
Compression
Compression ratio
Constant
Counter-balancer
Crankcases
Cylinder
Dynamometer (dyno)
Electric start
Engine displacement
Exhaust
Forced draft
Gaskets
Horsepower (hp)
Intake
Kick starter
Mechanical work
Open draft
Overhead camshaft
Piston
Power
Power
Ratio
Reciprocating engines
Recoil start
Revolutions per minute (rpm)
Stroke
Top-dead center (TDC)
Torque
Transmissions
Valves
Water jacket

INTRODUCTION

BASIC ENGINE OPERATION

In this chapter, you will look at the different motorcycle and ATV engine designs and configurations. We will start by giving you a very simple overview of engines in general and then give you a basic overview of the different types of engines found on motorcycles and ATVs. Virtually all motorcycles and ATVs have either a four-stroke engine or a two-stroke engine. Each type of engine has advantages and disadvantages. We will cover these in detail in future chapters.

For now, just keep in mind the primary difference between these two engine designs.

- The four-stroke engine has a power stroke for every two turns (720 degrees rotation) of the crankshaft, which is every **four piston strokes**.
- The two-stroke engine has a power stroke every full turn (360 degrees rotation) of the crankshaft, which is every **two piston strokes**.

Reciprocating Engines

Motorcycles and ATVs use **reciprocating engines**. A reciprocating engine has a piston that moves alternately *up* and *down* inside a cylinder. All reciprocating engines—from tiny model airplane engines to large truck engines—have a number of common components.

The first illustration shows a cylinder with a piston positioned inside it (Figure 5-1). A **cylinder** is a circular tube that's closed at one end. The **piston** is a circular plug that moves up and down inside the cylinder. The closed end of the cylinder is sealed by a cylinder head. When a piston is at its lowest position in the cylinder, it's said to be at **bottom-dead center (BDC)**. When the piston is at its highest position in the cylinder, it's said to be at **top-dead center (TDC)**.

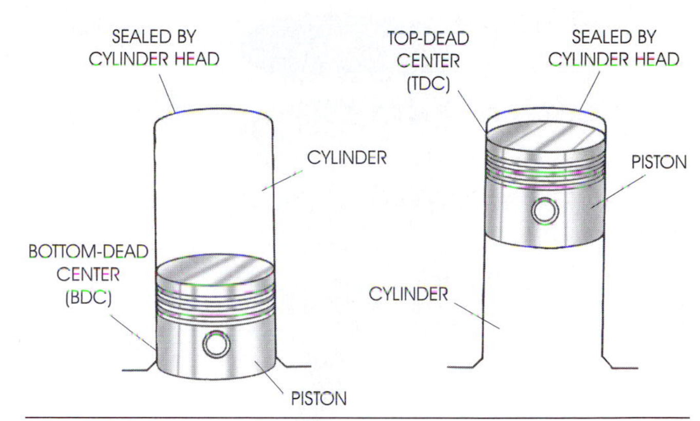

Figure 5-1 A simplified drawing of a cylinder and piston. Note the position of the cylinder at top-dead center (TDC) and bottom-dead center (BDC).

When the piston is at TDC, a small amount of space remains in the cylinder head. This small space above the top of the piston and below the cylinder head is called the **combustion chamber** (Figure 5-2). In the combustion chamber, a mixture of fuel and air is burned to produce power. When the air-and-fuel mixture burns in the combustion chamber, it produces a release of energy. This energy release is strong enough to force the piston downward into the cylinder. When the piston moves up or down, the distance it travels is the piston **stroke**.

The bottom of the piston is attached to a connecting rod and crankshaft (Figure 5-3). When the piston is forced downward in the cylinder, the piston's motion is transferred to the connecting rod and crankshaft. The connecting rod and crankshaft convert the up-and-down (reciprocating) motion

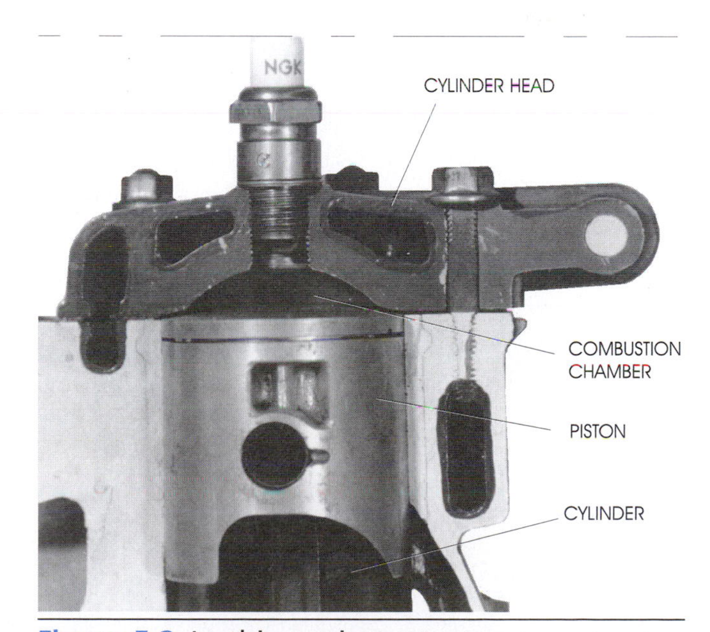

Figure 5-2 In this engine cutaway, you can see the piston, cylinder, and cylinder head. The small space between the top of the piston and cylinder head is called the *combustion chamber*.

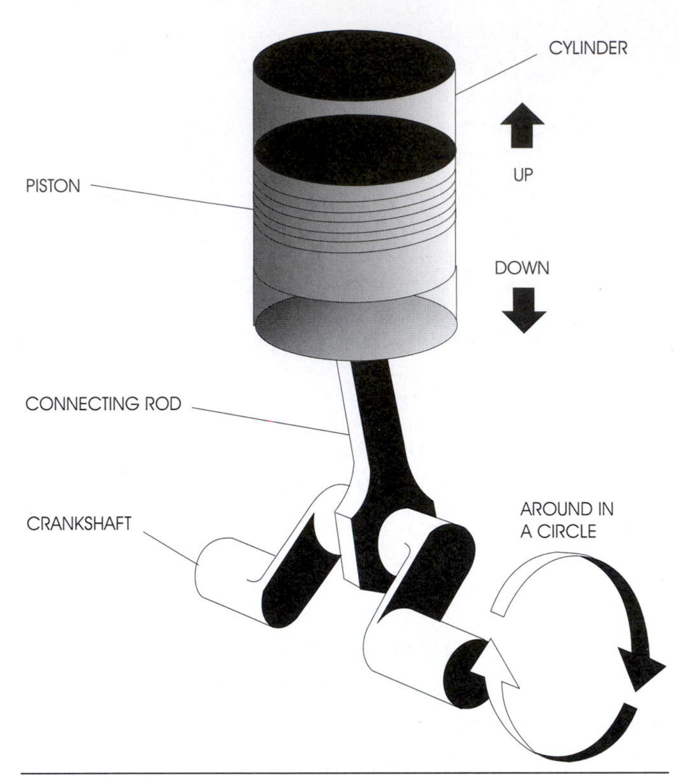

Figure 5-3 The connecting rod connects the piston to the crankshaft. The reciprocating (up-and-down) motion of the piston is changed to rotary (circular) motion at the crankshaft.

of the piston into a circular (rotary) motion. You can compare this conversion of reciprocating motion into rotary motion. This motion is similar to pedaling a bicycle. When you pedal a bike, the up-and-down motion of your legs is converted into circular motion that drives the rear wheel.

Four-Stroke Basic Engine Operation

Today, four-stroke engines are found on most motorcycles and ATVs in the United States. The primary reason for this is that the four-stroke engine design is more environmentally friendly than any alternative engine designs. This will be discussed in more detail in later chapters.

The operation of a typical four-stroke engine is divided into four stages: intake, compression, power, and exhaust (Figure 5-4). Here is a brief description of each stage.

Stage 1: **Intake**. The piston moves downward and draws an air-and-fuel mixture into the cylinder.

Stage 2: **Compression**. The piston rises and compresses the air-and-fuel mixture into the combustion chamber.

Stage 3: **Power**. The air-and-fuel mixture is ignited. The release of energy from the ignited air-and-fuel mixture pushes the piston back down in the cylinder. The downward motion of the piston is transferred to the rod and crankshaft.

Stage 4: **Exhaust**. The piston rises and pushes the exhaust gases out of the cylinder.

When a four-stroke engine is operating correctly, it continually runs through these four stages.

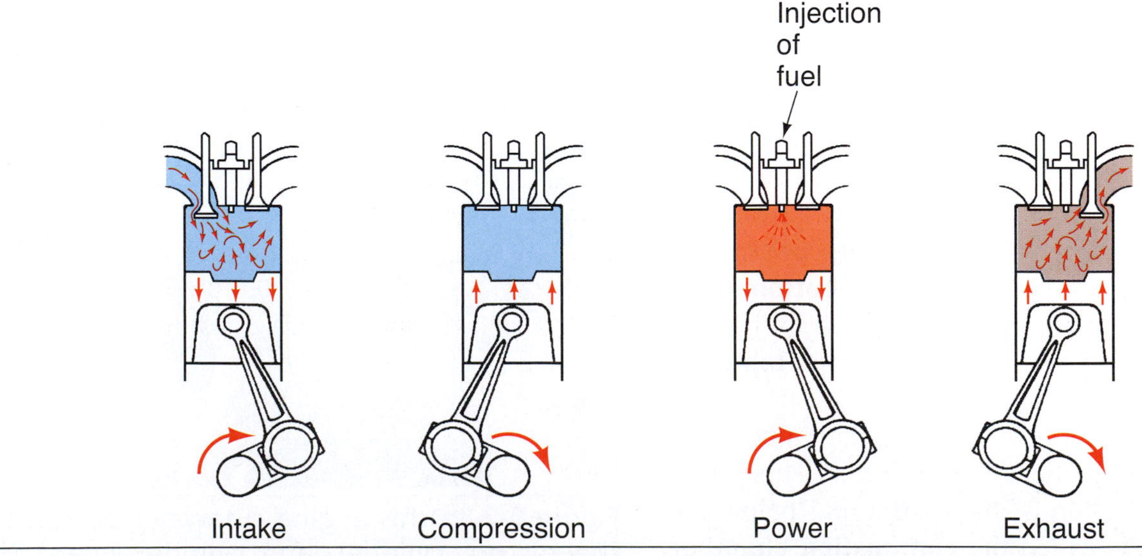

Figure 5-4 This illustration shows the four stages of the four-stroke engine in their proper order: intake, compression, power, and exhaust.

Two-Stroke Basic Engine Operation

The two-stroke engine operates somewhat differently from the four-stroke engine. In a four-stroke engine, the four stages (intake, compression, power, and exhaust) require four piston strokes. In a two-stroke engine, these stages are accomplished in only two piston strokes.

The two-stroke engine has inlet and outlet holes called ports located at different heights in the sides of the cylinder (Figure 5-5). As the piston moves up and down in the cylinder, the ports are covered (closed) or uncovered (opened) at different times by the piston.

The air-and-fuel mixture in the two-stroke engine enters below the piston into the area around the crankshaft. The air-and-fuel mixture typically contains oil to lubricate the crankshaft and related components of a two-stroke engine. The air-and-fuel mixture is delivered from the crankshaft area to the combustion chamber through a port above the piston. As the air-and-fuel mixture enters the combustion chamber, exhaust gases are forced out an exhaust port. As the piston rises, all ports are eventually sealed off. The air-and-fuel mixture is compressed and ignited, producing a power stroke that forces the piston downward and transfers motion to the connecting rod and crankshaft.

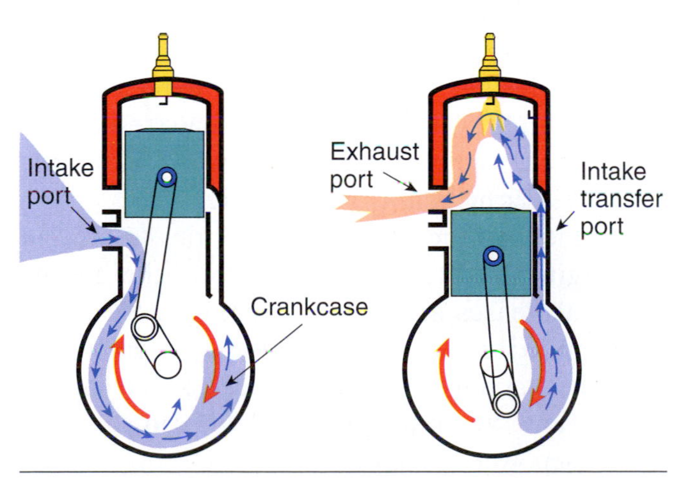

Figure 5-5 The two-stroke engine uses ports to allow the intake and exhaust gasses to flow through it. The piston opens and closes these ports as it goes by them.

ENGINE RATINGS

Now that we know the stages of basic engine operation, let's take a look at how motorcycle engine manufacturers rate and classify their engines. Motorcycle and ATV engines are normally classified in one of the following ways:

- By the size of the engine
- By the amount of power it produces

Before we begin to discuss how engine power is measured, let us define a few basic terms that we'll be using in this discussion.

Work

We're all familiar with the term work. People work in some way or another every day of their lives. However, when we refer to mechanical work, we can actually measure the amount of work that's done. By definition, **mechanical work** is a force that's applied over a specific distance. We can calculate the amount of work that's being performed by a device (or a person) by multiplying the amount of force being applied by the distance over which it's applied. Therefore, the formula for work is: Work = Distance x Force (W=D x F). Using this formula, if the amount of force applied is measured in pounds and the distance is measured in feet, the amount of work performed is measured in units called *foot–pounds* (ft-lb).

Let's look at a simple example. Suppose you want to move a box from the floor to a shelf. The box weighs 10 pounds (lbs), and the shelf is located 5 feet (ft) from the floor. If you lifted the box and placed it on the shelf, you performed a certain amount of work. You can calculate the amount of work by using the formula mentioned. The box weighs 10 lbs, so the amount of force you applied was 10 lbs. You lifted the box 5 ft off the floor, so the distance is 5 ft. Substitute these values into the formula and solve.

- Work = Distance x Force
- Work = 5 ft x 10 lb
- Work = 50 ft-lb

Thus, the amount of work you performed in this example was 50 ft-lb.

Note that this work formula does not mention time. The same amount of work is performed whether you took ten seconds or ten minutes to move the box from the floor to the shelf. The time required to perform the task is directly related to the strength of the person doing the job. If we do not consider the amount of time it took to perform a task, we are unable to determine the strength of the person who did the work.

The same idea applies to motorcycle engines. We can use the formula to calculate how much work an engine can perform, but without a time factor we can't determine the true strength of the engine. In order to calculate the engine's strength, we have to figure in the time the engine takes to complete a job.

The rate at which work is accomplished is called **power**. In other words, power is work per unit of time. The following formula is used to calculate power:

- Power = Work / Time

In this formula, we'll divide the amount of work (ft-lb) by the amount of time in seconds (s). The amount of power in this case will be measured in units called foot–pounds per second (ft-lb/s).

Let's return to our earlier example of the box and the shelf. We calculated that 50 ft-lb of work was required to move the 10-lb box from the floor onto the 5-ft-high shelf. Suppose that you completed this job in 2 seconds. We can use the power formula to calculate the power involved in doing the job.

- Power = Work / Time
- Power = 50 ft-lb / 2 s
- Power = 25 ft-lb/s (foot–pounds per second)

Therefore, from our calculations you can see that you required 25 ft-lb/s of power to complete your task.

Now that we have looked at the basic concepts of work and power, let us see how we apply this information specifically to motorcycle and ATV engines.

Horsepower

Many years ago, when internal combustion engines were invented, no one knew how to express the amount of work they could do. At that time, horses provided most of the transportation and power. As a result, a gentleman by the name of James Watt made some observations and concluded that the average horse of the time could lift a 550-pound weight one foot in one second, thereby performing work at the rate of 550 foot–pounds per second, or 33,000 foot–pounds per minute (m). Mr. Watt then published those observations and stated that 33,000 foot–pounds per minute of work was equivalent to the power of one horse, or, one **horsepower (hp)**. Nobody argued his calculation—hence the term horsepower was created.

- 1 hp = 550 ft-lb/s
- 1 hp = 33,000 ft-lbs/m

Even though horses are seldom used to perform work nowadays, we still use the standard unit of horsepower to describe the power output of motorcycle engines. So, the next time you're looking at a motorcycle in a showroom and the salesperson tells you it has 90 hp, you can just imagine that motorcycle being pulled by 90 horses. That's pretty impressive!

Today, we rate almost all engines by their horsepower output. Stronger engines produce more horsepower. Now that you understand how to calculate horsepower, let's look at how we measure it.

There are several different ways to calculate horsepower. The most common is to measure the horsepower output of an engine as it runs on a device called a **dynamometer (dyno)**. A dynamometer is a measuring instrument designed to measure power.

During a typical dyno test (Figure 5-6), the technician places the motorcycle on the dyno and runs it at full throttle. The dyno places a load (resistance) on either the output shaft or rear wheel. Usually, the load is either hydraulic or electronic.

As the load increases, its force tries to prevent the engine from turning. Therefore, the engine speed decreases as the load increases. Since the load applied is a known value, the dyno can determine the amount of torque produced by the engine. If we know the torque produced, we can calculate the horsepower of the engine.

Because this type of test involves the slowing or braking of the engine, the type of horsepower measured this way is commonly referred to as the **brake horsepower (bhp)**. The brake horsepower

Figure 5-6 A dynamometer (also known as a dyno) is used often in the motorcycle industry to measure horsepower and torque.

rating is the maximum power output of the engine. You will usually see the specifications for an engine given in bhp.

In practical use, a motorcycle or ATV engine is normally operated at a level well below its maximum power output. If the engine was always run at maximum horsepower, it would have a very short life span.

You can compare an engine that's running at its maximum rated power to a person running at top speed. That person would not be able to keep up the pace for long, and neither would an engine running at its maximum horsepower.

Torque

Torque is a measurement of twisting or rotational force. Remember that an engine's output is in the form of rotational motion. The power output from the crankshaft is used to turn the rear wheel of the motorcycle. You can compare the torque produced by an engine to the twisting force a person exerts when opening a jar lid. Engine torque is usually measured in foot–pounds (ft-lbs) and can be measured on a dynamometer.

As you've probably figured out by now, the ideal engine would have high horsepower and lots of torque. Unfortunately, we don't see this combination too often in real life. In a typical motorcycle or ATV engine, the horsepower and torque that are developed will vary with the speed of the engine (Figure 5-7). This speed is measured in **revolutions per minute (rpm)**. RPM is a measurement of how many complete turns (360 degrees) the crankshaft makes in one minute.

In a typical engine, horsepower generally increases as the rpm increases. Remember that power is related to the rate (speed) that work can

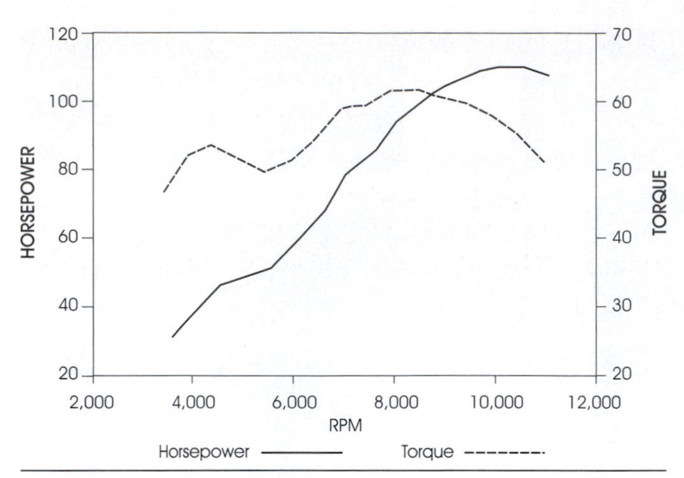

Figure 5-7 A typical dynamometer chart shows horsepower and torque over a wide range of engine revolutions (rpms).

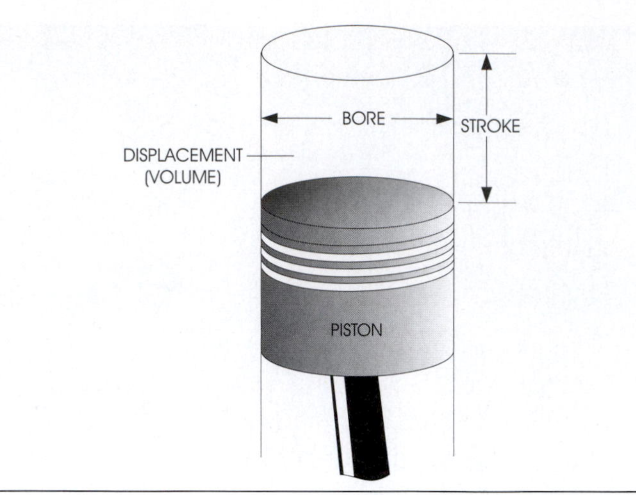

Figure 5-8 In this illustration, you can see the bore and stroke of an engine. This information is used to determine the engine displacement.

be done. Therefore, the maximum horsepower develops near the maximum rpm limit of the engine. Torque, in contrast, is produced somewhat differently. The maximum torque is normally produced at a lower rpm range and then declines as the rpm increases. This means that the maximum torque and the maximum rpm don't usually occur at the same time. So, when manufacturers design motorcycles and ATV engines, they must compromise. The design usually depends on the particular application.

More torque usually means better acceleration and more low-end power. Higher horsepower usually means higher top speed capabilities. The amount of horsepower and torque an engine develops depends on many design factors. The displacement, compression ratio, fuel mixture, engine design, ignition timing, and valve timing (on four-stroke engines) all affect horsepower and torque.

Engine Displacement

When you hear people refer to the size of an engine, they don't mean the overall size of the engine but rather the size of the area inside the engine where the air-and-fuel mixture is burned. The size of this area is known as the engine displacement.

By definition, **engine displacement** is the volume of space that the piston moves as it moves from BDC to TDC. The distance that the piston travels up or down in a cylinder is called the **stroke** of the engine (Figure 5-8). The diameter of the cylinder is called the **bore**. Displacement on motorcycle engines is usually measured in *cubic centimeters (cc)*; but it may also be measured in *cubic inches (ci)*. The displacement of an engine is usually stated in the service manual and stamped on the engine itself.

You can calculate the displacement of an engine if you know the diameter of the cylinder and the length of the stroke of the engine. The displacement of an engine can be calculated by using the following formula:

- Displacement = B x B x 0.7854 x S x N

In the formula, the letter B stands for the diameter (bore) of the cylinder.

The number 0.7854 is a constant. A **constant** is a number used in a formula that never changes. The letter S stands for the length of the stroke of the engine. The letter N stands for the number of cylinders in the engine.

To see how this formula works, let's look at an example. Suppose an engine has one cylinder with a diameter of 54.0 millimeters (mm). The stroke length of the engine is also 54.0mm. To calculate the displacement, we must first convert millimeters into centimeters (cm). Most motorcycles are rated in cubic centimeters. Simply moving the decimal point one space to the left does this: 54.0mm equals 5.4 cm (remember that one centimeter equals 10 millimeters). Substitute these values into the formula to determine the size of the engine.

- Displacement = B x B x 0.7854 x S x N
- Displacement = 5.4 cm (bore) x 5.4 cm (bore) x 0.7854 5.4 cm x (stroke) x 1 (number of cylinders)
- Displacement = 123.672cc

This particular example is a very common bore and stroke for a 125cc off-road motorcycle engine. Note that the motorcycle manufacturer will round off the displacement number to describe the size of the engine.

We can use this same formula to determine the displacement of an engine in cubic inches. For instance, if we have a four-cylinder engine with a bore of 2.76 inch and a stroke of 1.91 inch, the displacement equation would look like this.

- Displacement = 2.76 in. (bore) x 2.76 in. (bore) x 1.91 in. (stroke) x 0.7851 x 4 (number of cylinders)
- Displacement = 45.709ci (cubic inches)

This is a common-sized engine used in today's four-cylinder motorcycles.

Displacement is the most common way of describing a motorcycle or ATV engine. Motorcycle engine displacements range from 50cc to over 1,800cc, and ATVs have a general displacement range of 50cc to 800cc.

An engine's displacement has an effect on the power that the engine develops. In most cases, the larger the displacement, the more power the engine will develop. However, this doesn't mean that a smaller engine can never develop more horsepower than a larger one. Many factors besides displacement affect an engine's power. In general, however, an engine with a larger displacement will develop more horsepower.

Compression Ratio

You have learned that the displacement of an engine is the volume of space that a piston moves as it travels up and down in a cylinder. When a piston is at BDC, the cylinder volume (CV) is at its largest. When the piston is at TDC, the cylinder volume is at its smallest. The volume of space used when the cylinder is at its smallest volume position is also known as **combustion chamber volume (CCV)**. The ratio of the largest cylinder volume to the smallest cylinder volume is called the **compression ratio** (Figure 5-9). A **ratio** is simply a comparison between two values. The compression ratio can be calculated by using the following formula:

- Compression Ratio = CV / CCV

The volumes at BDC and TDC can be determined by using a combination of mathematical calculations and special test instruments. The typical technician will never be required to measure these volumes, so we won't get into the details of how the volumes are determined. However, you should be aware that the compression ratio of an engine affects the amount of power that the engine develops.

Let's examine the compression ratio in a typical motorcycle engine. For example, the volume of a cylinder (CV) at BDC is 100cc, and the volume of a combustion chamber at TDC (CCV) is 10cc. The compression ratio of this engine is therefore 10 to 1. This ratio may be written as 10 to 1 or abbreviated as 10:1.

The compression ratio is important in a motorcycle engine because it determines how effectively fuel is burned in the cylinder. As you learned earlier, fuel burns inside the cylinder to produce power.

An engine's compression ratio determines how much the fuel mixture will be compressed when the

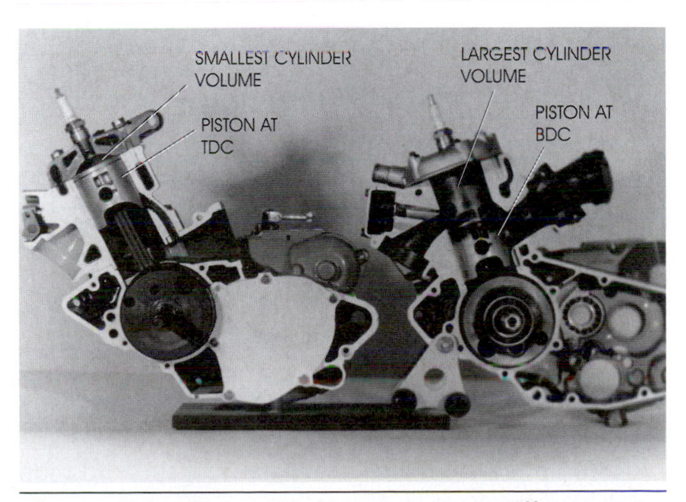

Figure 5-9 These engine cutaways allow us to see an engine at top-dead center (TDC) as well as bottom-dead center (BDC). Using these values allows us to determine compression ratios.

piston rises. The higher the compression ratio, the more the mixture is compressed. Our example engine had a compression ratio of 10:1. This means that when the air-and-fuel mixture first enters the cylinder, the mixture has a potential volume of 100cc.

When the piston is at TDC, the 100cc of mixture is compressed into a 10-cc space. When this compression occurs, the pressure of the mixture increases dramatically. This large increase in pressure helps the mixture burn more completely and produce more power when it's ignited. The compression ratio is important, but every engine has its limitations.

If an engine's compression ratio is too high, the excessive pressure can damage the engine. If the compression ratio is too low, the engine may not develop much power. Different engines have their own specified compression ratios. Most modern motorcycle engines have a compression ratio in the range between 8:1 and 12:1.

BASIC FOUR-STROKE ENGINE DESIGN

Now that we have established the required components of a simple internal combustion engine, let us discuss ways of making it work for us as a suitable motorcycle power source.

To best explain the basic design of a typical four-stroke engine used in a motorcycle or ATV we will cover the components found on a single-cyclinder four-stroke engine. Most motorcycles found with single-cylinder engines use smaller displacement (under 500cc) engines. But there are some single-cylinder engines rated at 650cc displacement or larger. However, most ATVs will be found with single-cylinder four-stroke engines. The single-cylinder four-stroke engine is the least complex of all four-stroke engine configurations (Figure 5-10).

Figure 5-10 A single-cylinder four-stroke engine is shown here.

Four-Stroke Engine Bottom End

The bottom end of a four-stroke engine contains the heart of the engine—the crankshaft. We will begin from the bottom of the engine and then work toward the top.

Crankshaft

The four-stroke crankshaft may be made up of two flywheel halves (Figure 5-11), or it may be a one-piece design (Figure 5-12). It will have a connecting rod, rod (crank) pin (if it's a multi-piece design), and a connecting rod bearing. At least one bearing on each end supports the crankshaft and allows it to rotate freely. The crankshaft is located in the engine crankcases.

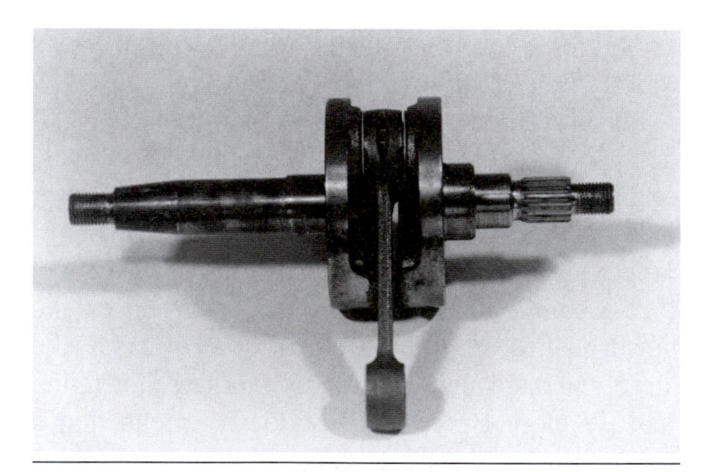

Figure 5-11 A multi-piece single-cylinder four-stroke crankshaft is shown here.

Figure 5-12 A one-piece crankshaft is shown here without the connecting rods.

Crankcases

The engine **crankcases** are used to hold all of the engine components together and provide the main engine mounting points (Figure 5-13). There are two crankcases: center and side. The center crankcases hold the major components. The side crankcases enable you to gain access to various parts of the engine without having to fully disassemble it. The side crankcases are sometimes also known as side covers.

Seals

Seals are used to protect rotating shaft bearings. The seals are typically located at the ends of the rotating shafts. These seals prevent gases and oil from escaping from the crankcases and also prevent outside substances from getting into the crankcases.

Bearings

Bearings are found primarily in the crankcases of the four-stroke engine. Bearings are designed to reduce friction and to allow shafts to rotate freely under various engine loads. You will also find a four-stroke single-cylinder engine's transmission in the crankcases.

Transmissions

The transmission found on a four-stroke engine (Figure 5-14) typically consists of gears, shafts, and shifting mechanisms. These components work together to transmit power from the crankshaft to the rear wheel and to help keep the engine running in the desired rpm range. The clutch in the four-stroke engine (Figure 5-15) is used to engage and disengage the transmission and the rear wheel from the crankshaft power output.

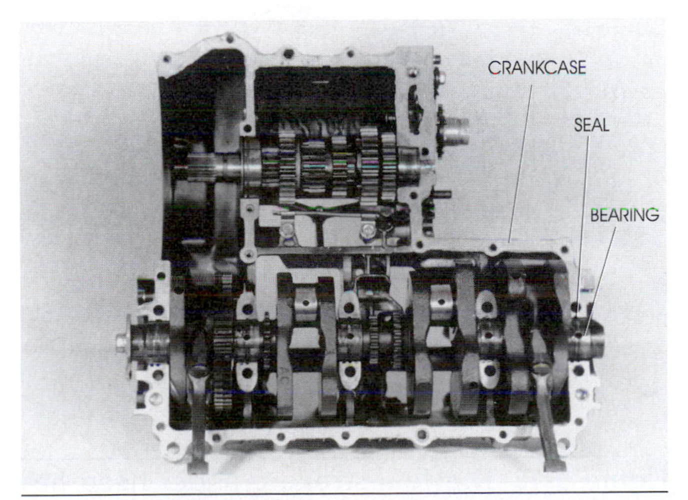

Figure 5-13 In this picture, you can see a typical crankcase and the location of a seal and bearing.

Figure 5-14 A typical motorcycle transmission is shown here and consists of gears, shafts, and shifting mechanisms.

Gaskets

Gaskets are used to seal the mating surfaces of various parts of the engine. The two surfaces are usually both metal. Gasket material may be rubber, cork, or metal.

Starting Devices

The last part of the bottom end is the *starting mechanism*. The four-stroke engine uses one of three different starting mechanisms:

- A **kick starter** has an external lever on the side crankcase. The rider pushes the lever down with his or her foot to turn the crankshaft. Turning the crankshaft starts the engine.
- The **electric start** mechanism is similar to an automobile's starter mechanism. It uses an electric motor with a reduction gear to turn the motorcycle engine's crankshaft. The rider energizes the starter motor with a push-button or key-lock mechanism.
- The **recoil start** mechanism uses a pull cord that's attached to a pulley on the crankshaft. The rider pulls the cord to turn the crankshaft, similar to starting a gasoline-powered lawnmower. The cord has a return spring to rewind it into the housing.

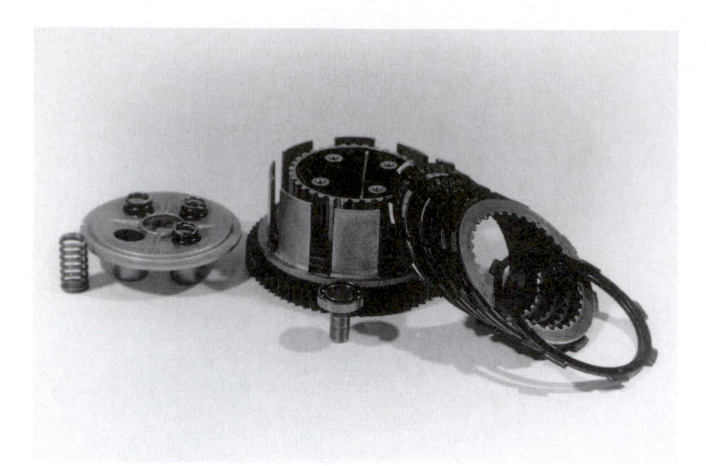

Figure 5-15 A typical motorcycle clutch is shown here.

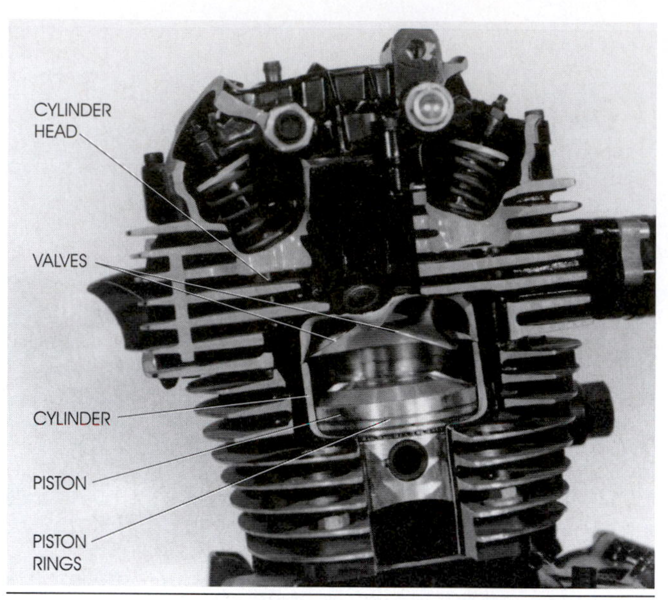

Figure 5-17 This picture shows a four-stroke engine cylinder and related components.

All of the items mentioned up to now are found in the bottom end of a single-cylinder four-stroke engine. Now let us look at the top end.

Four-Stroke Engine Top End

Piston

The piston is attached to the crankshaft connecting rod. The piston is held in place by a wrist pin. Clips prevent the pin from moving. The wrist pin usually has a bushing between it and the connecting rod. Piston rings are located in the slots on the outer diameter of the piston to make a seal between the piston and cylinder wall (Figure 5-16). The piston travels up and down in the cylinder, which is usually positioned vertically (Figure 5-17).

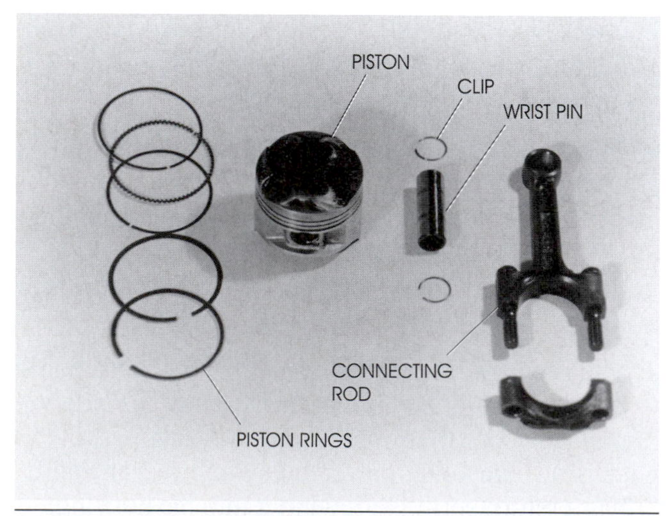

Figure 5-16 This picture shows a four-stroke piston and its related parts.

Cylinder Head

The cylinder head is attached to the cylinder to seal the top of the cylinder from the outside of the engine. The cylinder head of the four-stroke motorcycle and ATV contains holes that are called ports (Figure 5-18). These ports are opened and closed with *valves*.

The **valves** in a four-stroke engine control the air-and-fuel mixture that's drawn into the cylinder and the exhaust gases that are expelled. These valves are actuated by one (Figure 5-19), or two camshafts (Figure 5-20). Camshafts are normally

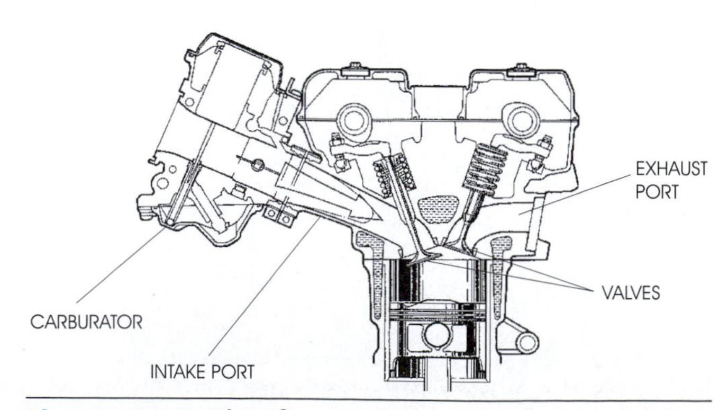

Figure 5-18 The four-stroke cylinder head has ports and valves to allow the flow of intake and exhaust gases.

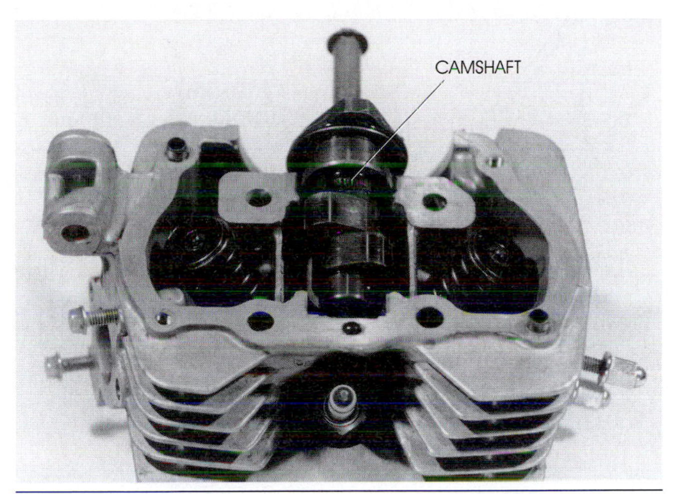

Figure 5-19 This picture shows a single overhead camshaft cylinder head.

Figure 5-20 This picture shows a double overhead camshaft cylinder head.

located on top of the cylinder head. This engine style is known as an **overhead camshaft**. The camshaft is connected to the crankshaft via a chain, belt, or set of gears.

Counter-Balancer

Many single-cylinder engines use a counter-balancer to help keep the engine running smoothly. The **counter-balancer** is a device that balances the power pulses created by the power strokes (Figure 5-21). Many manufacturers of single-cylinder engines use a gear or chain-driven counter-balancer to offset the uneven forces that

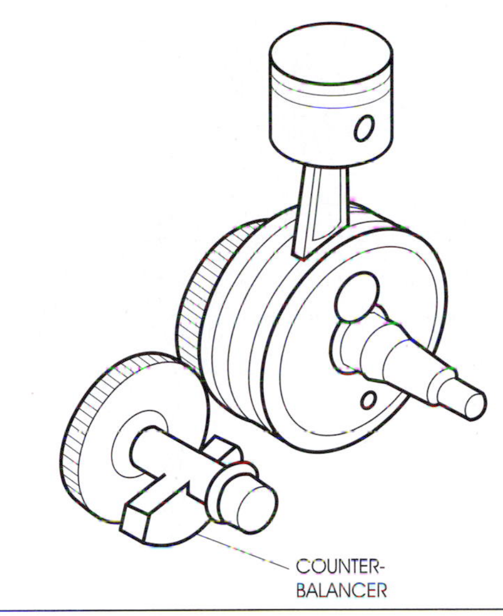

Figure 5-21 Counter-balancers are used to help some four-stroke singles and twins run smoothly.

create vibration. This system requires additional parts but very little maintenance.

BASIC TWO-STROKE ENGINE DESIGN

Now we'll cover the two-stroke engine's major components and their layout. Of all the engine designs used on motorcycles and ATVs, the two-stroke engine is the least complex. In later chapters, we'll give you a more detailed explanation of how these components work. For now, however, let's concentrate on the basics.

The single-cylinder two-stroke engine is the least complex of all two-stroke engines (Figure 5-22). The cylinder may be positioned at any angle but most modern engines position the piston upright or vertical to the motorcycle frame. The major components are located in the top and bottom ends of the engine. Let us start with the bottom end.

Two-Stroke Engine Bottom End

Just as with the four-stroke engine, the bottom end of a two-stroke engine contains the heart of the engine—the crankshaft as well as crankcases, seals, bearings, transmission, clutch, gaskets, and a starting mechanism with only minor differences to the actual components.

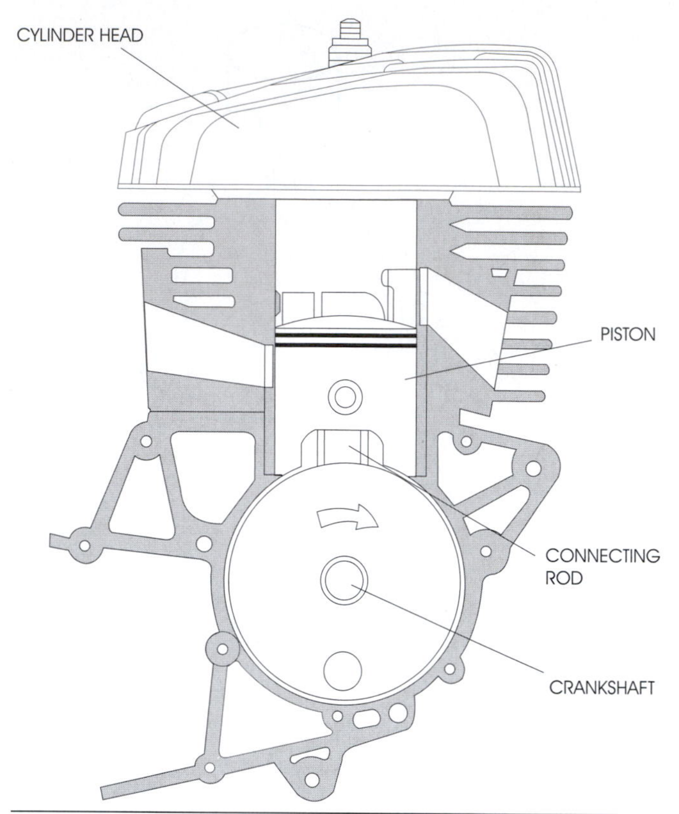

Figure 5-22 The single-cylinder two-stroke engine is the least complex of all two-stroke designs.

Crankshaft

The two-stroke crankshaft is made up of two flywheel halves, a connecting rod, a connecting rod pin, and a connecting rod bearing (Figure 5-23). A bearing on each end supports the crankshaft and allows it to rotate.

Crankcases

The crankshaft is located inside the engine crankcases. The engine crankcases are used to hold

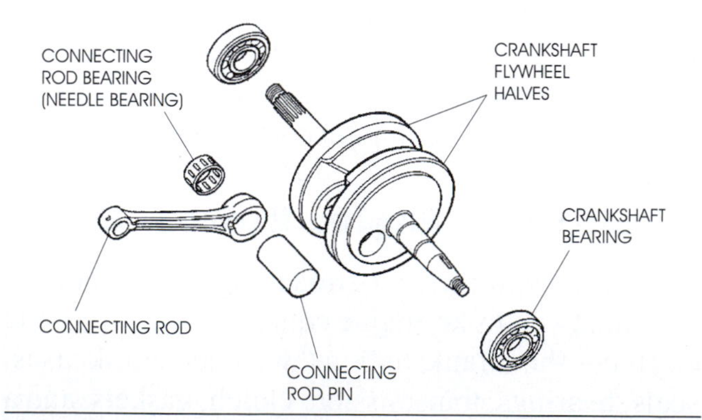

Figure 5-23 The parts of a two-stroke single-cylinder crankshaft are illustrated here.

Figure 5-24 The engine crankcases are used to hold all of the various components together.

all of the engine components together and supply the main engine mounting points (Figure 5-24). Just as with the four-stroke engine, there are two crankcases—center and side. The center crankcase on a two-stroke engine is also pressurized around the crankshaft as the fuel-and-air mixture is on both the top and bottom of the piston. The side crankcases enable you to gain access to the various parts of the engine without having to fully disassemble it. The side crankcases are also known as side covers.

Seals and Bearings

Seals are used to protect rotating shaft bearings (Figure 5-25). The seals are typically located at the ends of the rotating shafts. These seals prevent the loss of gases and oil from the crankcases. Their sealing action also prevents outside substances from getting into the crankcases.

Bearings are found primarily in the crankcases of the two-stroke engine. Bearings are designed to reduce friction and to allow shafts to rotate freely under various engine loads.

Transmission and Clutch

You'll also find the two-stroke single-cylinder engine's transmission in the crankcases. Transmissions consist of gears, shafts, and shifting mechanisms. These components work together to transmit power from the crankshaft to the rear wheel and to help keep the engine running in the desired rpm range.

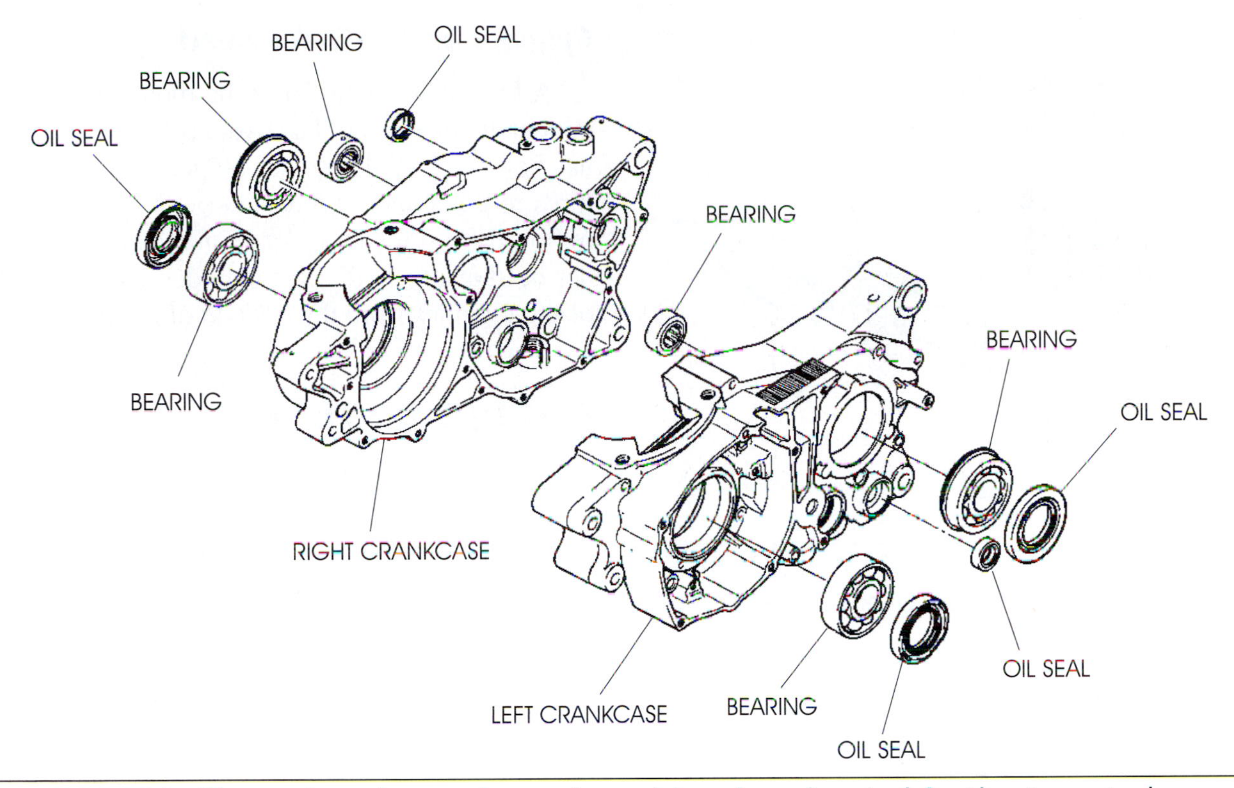

Figure 5-25 This illustration shows the seals and bearings located in the two-stroke engine crankcases.

The clutch in the two-stroke engine is used to engage and disengage the transmission and the rear wheel from the crankshaft power output just as found on the four-stroke engine.

Gaskets

Gaskets are used to seal the mating surfaces of various parts of the engine (Figure 5-26). These surfaces are usually both metal. Gasket material may be rubber, cork, or metal.

Starting Devices

Just as with the four-stroke engine, the two-stroke engine uses one or more of the following methods for starting—kick start, recoil start, or electric start. Now, let us look at the top end of the two-stroke engine.

Two-Stroke Engine Top End

Piston

The piston is attached to the crankshaft connecting rod (Figure 5-27). The piston is held in place by a piston pin (also known as a wrist pin). Clips prevent the wrist pin from moving laterally.

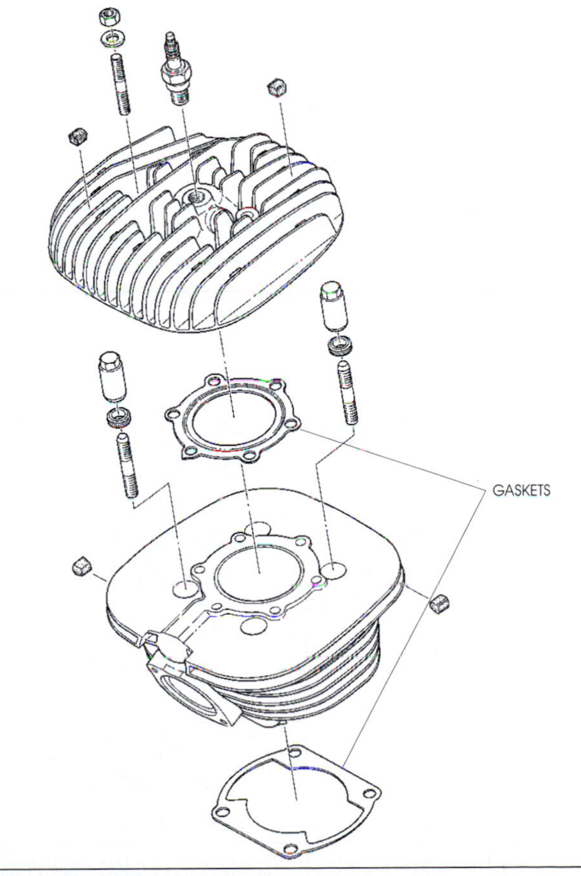

Figure 5-26 This illustration shows typical gaskets found on engine mating surfaces.

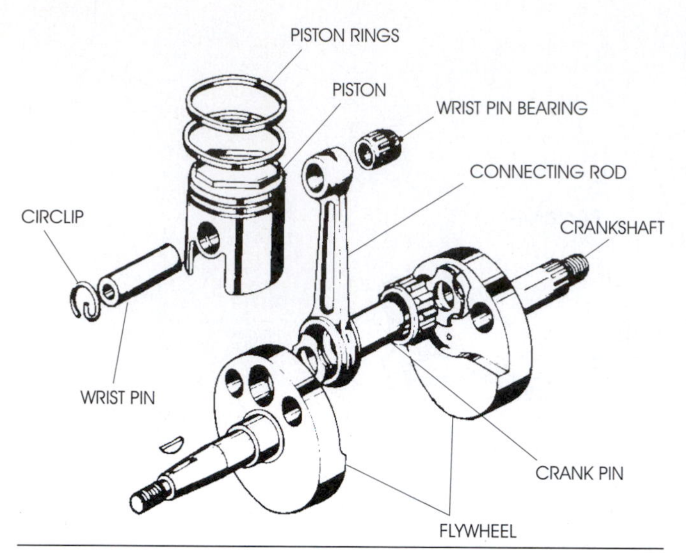

Figure 5-27 The piston and its related parts are attached to the crankshaft connecting rod.

There's usually a bearing where the piston pin attaches to the connecting rod. Each piston has one or more rings around its outside surface. These rings are called piston rings. The rings form a seal between the piston and the cylinder surfaces to improve the compression and exhaust functions.

Cylinder and Cylinder Head

A two-stroke cylinder is the most complex component on the engine. The two-stroke engine cylinder has holes in its side ports. The ports allow fuel mixtures to enter the cylinder and exhaust gases to be removed from the cylinder (Figure 5-28). The cylinder head is attached to the cylinder and seals the top of the cylinder from the outside of the engine.

ENGINE COOLING

A significant amount of heat is generated in any internal combustion engine during the combustion stage of the engine's operation. All engines must have a way to dissipate this heat. Excessive heat will damage the components. Engines on motorcycles and ATVs use one of two ways of maintaining ideal operating temperature—*air-cooled* and *liquid-cooled* temperature control.

Air-Cooled Engines

Air-cooled engines use cooling fins on the cylinder block and the cylinder head to remove any excess

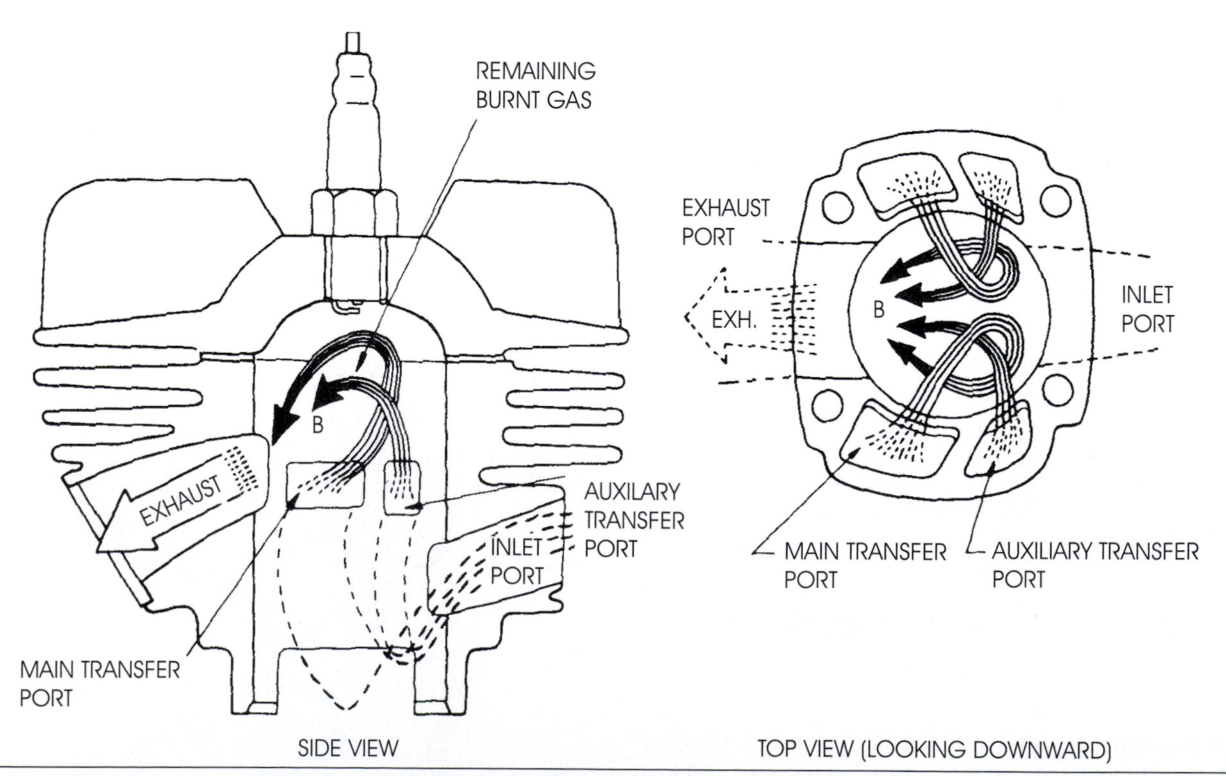

Figure 5-28 The ports in a two-stroke cylinder allow the flow of gases in and out of the engine using the piston to open and close them.

heat from the engine. This is done using either the open draft design or the forced draft design.

Open Draft Cooling Systems

Most air-cooled motorcycles and ATVs use the **open draft** design. This engine cooling system uses the movement of the open air over the fins while the motorcycle is moving (Figure 5-29). The heat is drawn away from the engine through the fins to keep the engine temperature at a safe level.

Forced Draft Cooling Systems

The **forced draft** design uses air from an engine-driven fan to move cool air through ducts. These ducts, called *shrouds*, surround the engine and keep it cool by forcing the air in toward the cylinder and head fins (Figure 5-30). Forced draft cooling systems are mostly found on scooters where the engine is covered from the passing outside air flow, therefore needing an alternative means to cool the engine.

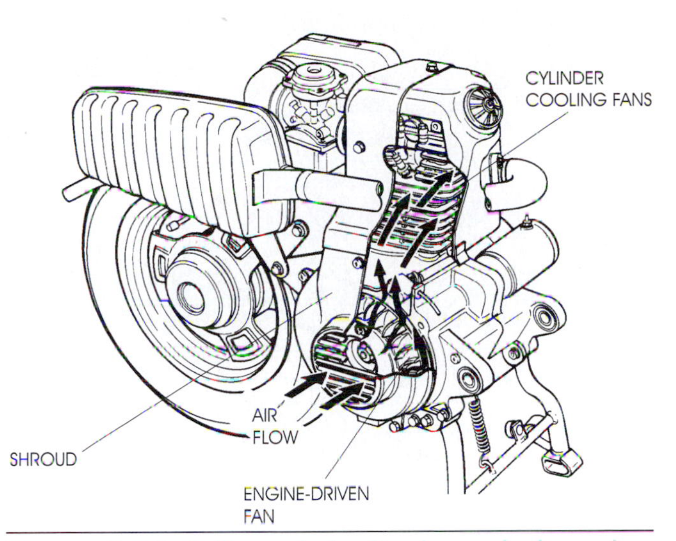

Figure 5-30 A forced draft air-cooled engine uses an engine-driven fan to move air.

Liquid-Cooled Engines

The main difference between an air-cooled engine and a liquid-cooled engine (Figure 5-31) is the use of a liquid instead of air to maintain proper engine operating temperature.

This liquid is usually made up of a 50/50 mixture of distilled water and anti-freeze (ethylene glycol). The cylinder and cylinder head have water jackets. A **water jacket** (Figure 5-32) is a series of passageways surrounding the cylinder and combustion chamber. As the liquid circulates through these passageways, the heat is transferred from the metal to the liquid, which helps to control the internal heat of the engine.

Other components such as the radiator, water pump, thermostat, hoses, and a reservoir tank assist with circulation and cooling of the liquid coolant (Figure 5-33).

A prime advantage of liquid-cooling is the ability to keep the engine at a constant temperature. Another advantage with liquid-cooled engines is that the engines run quieter because the

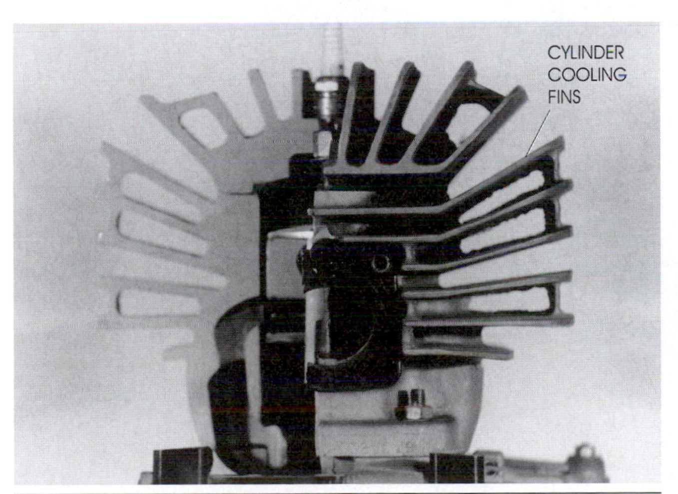

Figure 5-29 An open draft air-cooled cylinder and head are pictured here. Note the cylinder fins that are used to dissipate heat away from the engine.

Figure 5-31 Side by side, the differences are easily seen between a liquid-cooled engine on the left and an air-cooled engine on the right.

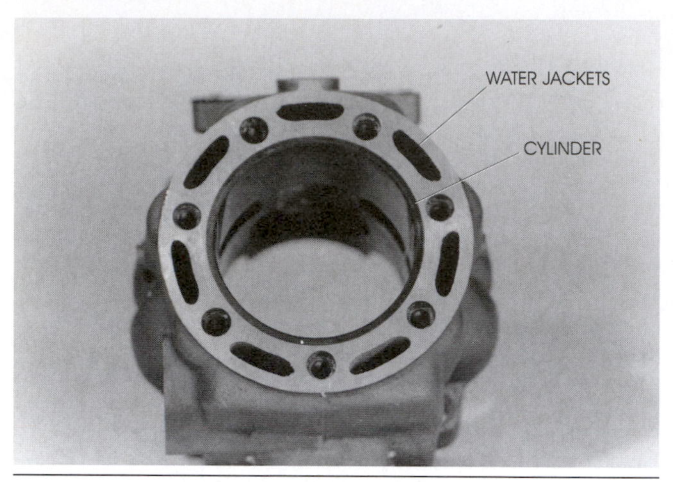

Figure 5-32 The liquid-cooled engine has water jackets surrounding the cylinder.

coolant provides sound dampening to the internal engine noises.

ENGINE CONFIGURATIONS

Now that we have given you a basic understanding of basic engine design, let's take a look at the different engine configurations that you might see in a motorcycle or ATV. The obvious difference between a single-cylinder engine and a multi-cylinder engine is the number of cylinders and pistons. Multi-cylinder engines run more smoothly because there are more power strokes per 360 degrees of crankshaft rotation. In most cases more cylinders also means more displacement

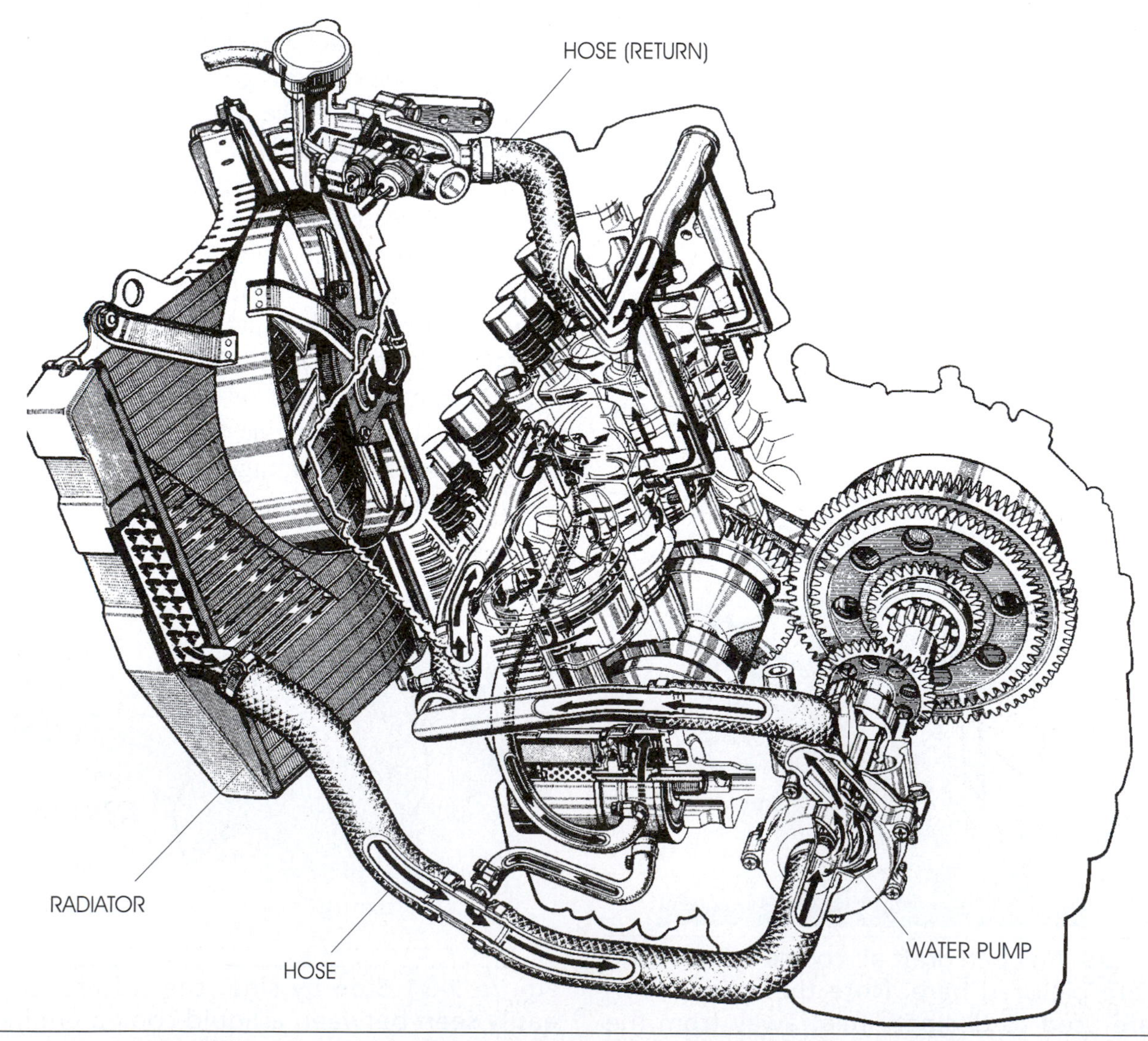

Figure 5-33 A typical liquid-cooled engine and its components are shown here.

and more power. Multi-cylinder engines may be either air-cooled or liquid-cooled. Here are the main types of engine configurations that can be found on motorcycles.

Parallel Twin-Cylinder Engines

Parallel twin-cylinder motorcycle engines use either a 180-degree or 360-degree crankshaft design (Figure 5-34). If one piston is up while the other is down, the engine is considered to have a 180-degree crankshaft. If both pistons rise and fall together, the engine is considered to have a 360-degree crankshaft.

Horizontally Opposed Twin-Cylinder Engines

On horizontally opposed twin-cylinder motorcycle engines, the crankshaft is in line with the motorcycle frame (Figure 5-35). The crankshaft in this engine configuration will always be a 180-degree design. Because the cylinders oppose each other, the pistons move in and out at the same time and keep the engine in balance. This design allows the engine to be mounted lower in the frame, creating a lower center of gravity and improving weight distribution.

V-Twin Engines

The V-twin engine design (Figure 5-36) allows for the greatest amount of engine displacement in the smallest overall area. This type of engine design is most often found on cruiser-type motorcycles, although it has also been adapted for use on sport-type motorcycles.

In-Line Multi-Cylinder Engines

In-line multi-cylinder engines are very popular and can be found in three-, four-, and even six-cylinder designs. The cylinders may be vertical, horizontal, or positioned at any angle in between. The cylinders are normally transversely positioned from left to right as you sit on the motorcycle but can also be found longitudinally positioned parallel with the frame. An in-line multi-cylinder engine

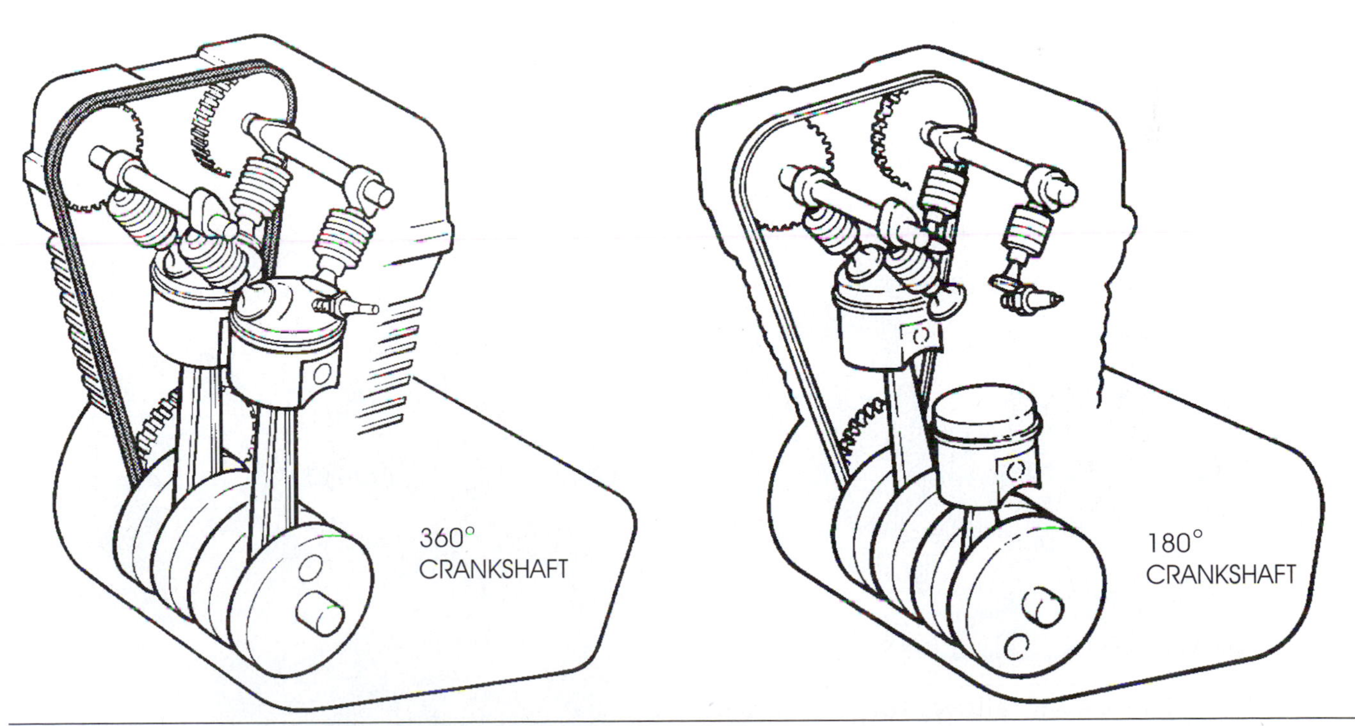

Figure 5-34 A parallel twin-cylinder engine configuration is illustrated here. A 180-degree crankshaft design has one piston at TDC while the other piston is at BDC as can be seen on the right as compared to a 360-degree crankshaft design on the left.

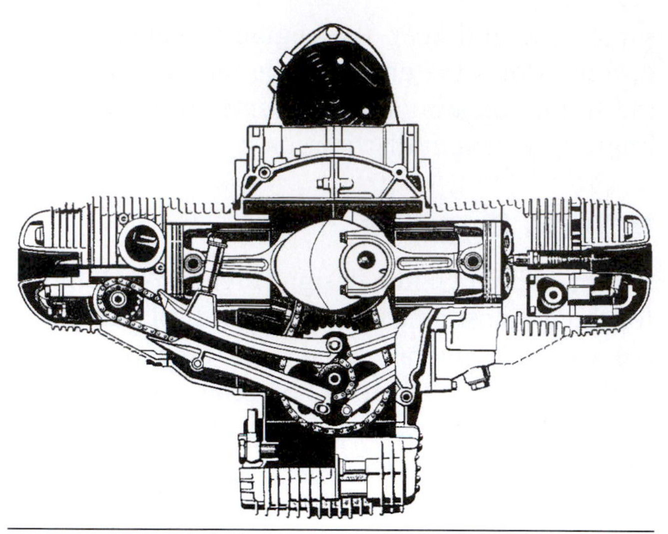

Figure 5-35 A horizontally opposed twin is illustrated here.

is usually designed so that the power strokes are spaced to occur at an equal number of degrees apart. In a four-cylinder engine, four power strokes occur in two full turns of the crankshaft (720 degrees of rotation). The crankshaft and camshaft (on four-stroke engines) are designed so that there's 180 degrees between each power stroke. This engine design creates a very smooth-running engine.

V-Four Engines

The V-four engine design is more compact than the in-line four-cylinder engine and produces minimal vibration (Figure 5-37). It may have either a 180-degree or 360-degree crankshaft design (Figure 5-38). A negative with V-four designs is the cost involved to manufacture the engine. It is more complex and costly because it uses essentially twice as many cam drive components as a comparable in-line four-cylinder engine.

Horizontally Opposed Multi-Cylinder Engines

Horizontally opposed multi-cylinder engines (Figure 5-39) are similar to the flat twin design with additional cylinders on each side. They come in both four- and six-cylinder configurations. Because the cylinders lie side by side, most engines using this design are liquid-cooled to ensure adequate rear-cylinder cooling. The opposed multi-cylinder engine is mounted lower in the frame than an in-line engine to lower the center of gravity and improve weight distribution. The Honda Goldwing is a popular model of this design.

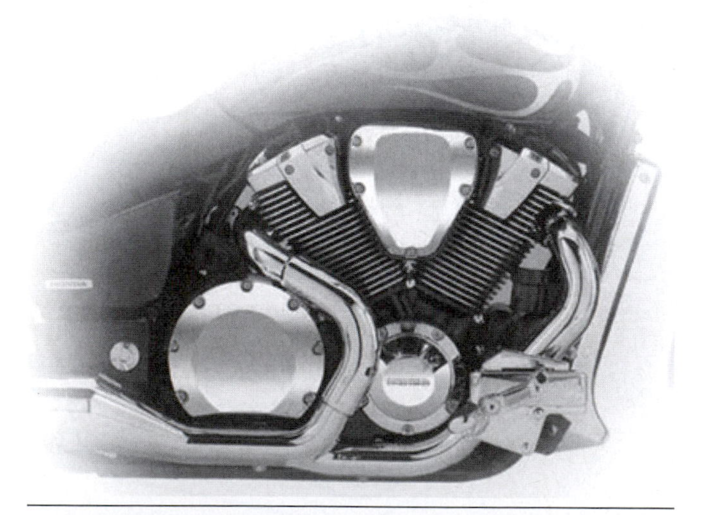

Figure 5-36 The V-twin allows for the largest displacement in the smallest overall space. It is found most often on custom cruiser motorcycles.

Figure 5-37 The V-four motorcycle engine is very complex in nature.

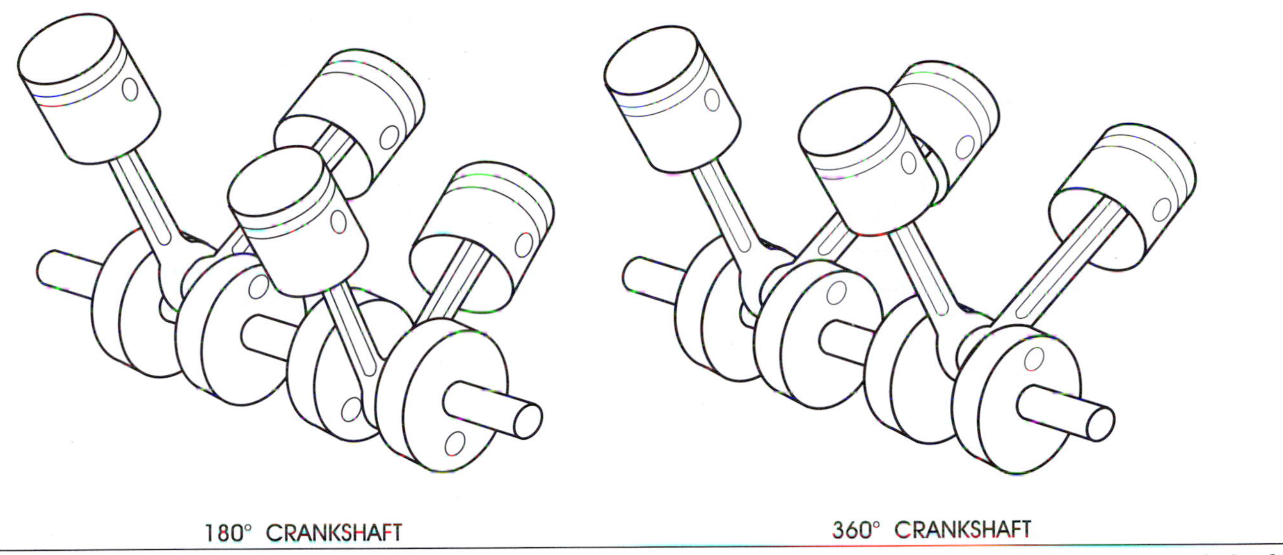

Figure 5-38 This illustration shows the V-four, four-stroke, 180-degree and 360-degree crankshaft designs.

Figure 5-39 The horizontally opposed multi-cylinder engine is very wide and is usually liquid-cooled to prevent the rear cylinders from overheating.

Summary

- Virtually all motorcycles and ATVs use reciprocating engines as a source of power.
- There are two different types of engines used on motorcycles and ATVs. Two-stroke and four-stroke. While these engines share many of the same design features there are basic differences found between them as well.
- Engines are rated by the amount of horsepower and torque that they produce.
- There are distinguishable components found in a two-stroke and four-stroke engine.
- There are three basic types of engine cooling systems used on motorcycles and ATVs.
- There are many different engine configurations found on motorcycles and ATVs.

Chapter 5 Review Questions

1. The single-cylinder four-stroke engine has a power stroke every ________ degrees of crankshaft revolution.
2. During which stage of engine operation does the burning mixture of air and fuel force the piston downward in a four-stroke engine?
   a. compression
   b. power
   c. intake
   d. exhaust
3. The four basic four-stroke engine operation stages, in their proper order, are
   a. compression, intake, power, exhaust
   b. intake, compression, power, exhaust
   c. power, compression, intake, exhaust
   d. intake, power, compression, exhaust
4. By definition, work is a force that is applied over a specific ________.
5. Work / Time = ________.
6. What would the compression ratio be if a cylinder has a volume of 250cc and a combustion chamber volume of 20cc?
   a. 0.08:1
   b. 8:1
   c. 10:1
   d. 12.5:1
7. When the piston is at its highest point, it's said to be at
   a. TDC
   b. BDC
   c. ABC
   d. BBC
8. What is the name of the device used to measure engine torque and horsepower?
   a. manometer
   b. compressor
   c. dynamometer
   d. Vernier caliper
9. A measurement of the twisting or rotational force that an engine can produce is called:
   a. rotary distance
   b. horsepower
   c. torque
   d. compression ratio
10. The connecting rod and crankshaft convert up-and-down (reciprocating) motion into
    a. a small, contained explosion
    b. rotary motion
    c. a release of exhaust gases
    d. movement in the cylinder head

CHAPTER

6 Internal-Combustion Engines

Learning Objectives

When students have completed the study of this chapter and its laboratory activities they should be able to:

- Explain the physical laws associated with motorcycle and ATV engines
- Describe the operation of a basic internal-combustion engine
- Explain how fuel and air are used to make an engine operate
- Identify the component parts used in a four-stroke engine
- Describe the theory of operation for a four-stroke engine
- Visually identify the component parts used in a two-stroke engine including the piston, crankshaft, cylinder head, and cylinder
- Explain the theory behind the operation of the two-stroke engine
- Understand the different induction systems used on the two-stroke engine
- Describe how a two-stroke engine physically differs from a four-stroke engine
- Understand both the advantages and disadvantages of both the two- and four-stroke engines used in the modern motorcycle and ATV

Key Terms

Active combustion
Active energy
Atomized liquid
Base circle
Boost port
Bottom-dead center (BDC)
Boyle's law
Cam ground
Camshaft
Camshaft drive tensioner
Camshaft lift
Carbon dioxide (CO_2)
Carbon monoxide (CO)
Catalyst
Chamfered
Clearance ramps
Combustion
Combustion chamber
Combustion lag
Compression event
Compression ring
Compression stage
Compression stroke
Crankcase reed valve
Crown
Cylinder
Cylinder reed valve
Duration
Dykes piston ring
Energy

Environmental Protection Agency (EPA)
Exhaust event
Exhaust port
Exhaust port
Exhaust stage
Exhaust stroke
Exhaust valve
Hydrocarbons (HC)
Keystone piston ring
Ignition
Induction
Intake event
Intake port
Intake stage
Intake stroke
Intake valve
Internal-combustion engine
Labyrinth seal
Law of action and reaction
Law of inertia
Momentum
Multi-piece crankshaft
One-piece crankshaft
Oil control ring
Oxides of nitrogen (NO_x)
Piston
Piston port/crankcase reed
Piston ring
Piston ring lands
Piston skirt
Poppet valves
Ports
Post combustion
Potential energy
Power band
Power event
Power stage
Power stroke
Primary area
Ring grooves
Rocker arm
Scraper ring
Secondary area
Shim and bucket
Squish area
Squish band
Standard piston ring
Stellite
Stroke
Top-dead center (TDC)
Transfer event
Transfer port
Transfer ports
Valve closing devices
Valve guides
Valve overlap
Valve seats
Vaporized liquid
Viscosity
Water (H_2O)
Wrist pin boss

INTRODUCTION

In Chapter 5, we learned about the various motorcycle and ATV engine configurations. Now we'll focus on how motorcycle and ATV engines operate. We'll begin by discussing certain physical laws that pertain to engines. Next we'll describe the theory of operation for a basic internal-combustion engine.

After you understand basic engine operation, we will focus on the four-stroke engine. We'll discuss the basic components used in a four-stroke engine and then take an in-depth look at how the four-stroke engine operates. We will then look at the two-stroke engine. We will identify the components found in the two-stroke engine and then learn how the two-stroke engine operates. We will then describe the different types of induction systems. After completing our look at the two-stroke engine, we'll discuss the differences between two-stroke engines and four-stroke engines. At the conclusion of this chapter, we will look at the advantages and disadvantages of each of these engine designs.

GENERAL AND SCIENTIFIC TERMS

Before we go into detail on how engines operate, it is important that we understand certain terms and principles related to gasoline engines and the combustion process.

Matter

Matter can be described as any substance that occupies space and has weight—that is, a substance of which physical objects are composed. Matter can't be created or destroyed but can be changed from one form to another by a chemical or physical process. An example of matter changing from one form to another is a block of ice. The ice turns to water if not kept at a freezing temperature. Furthermore, if enough heat is applied, the water can be changed to steam.

Matter can be in the form of a solid, liquid, or gas. The block of ice in the previous example is considered to be solid matter because it has three dimensions (length, width, and depth) that can be measured.

When the ice melts, it changes from a solid form into a liquid form (water). A liquid has no definite shape and conforms to the shape of the container holding it. A liquid has the ability to transmit pressure but can't be compressed. Another interesting fact about a liquid is that it won't burn! The following are two terms describing liquids that relate to engine operation:

- An **atomized liquid** consists of liquid drops suspended in air. An example of an atomized liquid is an early morning fog. Because an atomized liquid is still a liquid, it will not burn.
- A **vaporized liquid** is a liquid that's converted to a gaseous state through a heating process. A vaporized liquid has the ability to burn. Vaporized liquids are used to make an engine run.

Steam is a gas or gaseous matter. Keep in mind that we aren't talking about gasoline when we talk about gas. Gasoline is a liquid. A gas has no definite shape and, like a liquid, conforms to the shape of its container. A gas can transmit pressure but is lighter than a liquid when compared in equal volumes. Unlike a liquid, a gas is highly compressible.

An excellent example of a gas is the air we breathe. Air is made up of approximately 78 percent nitrogen, 21 percent oxygen, and 1 percent inert or inactive gases. The oxygen in the air keeps us alive and helps an engine run at its best.

Air density can be described as the amount of oxygen per given volume of space—or in other words, the thickness of air. The air all around us is actually compressed. At sea level the air pressure is 14.7 pounds per square inch (psi). Air density decreases as we increase altitude or when the temperature rises. When the air density decreases there are fewer oxygen molecules in the air. It's more difficult for you to work at the same level of intensity at 10,000 feet above sea level than at 1,000 feet above sea level. It's also more difficult to work at the same level of intensity on a very hot and humid day than on a cool and dry day. The same changes affect how an engine runs! As air density decreases, there are fewer oxygen molecules in the air for you and your engine to breathe.

Viscosity

A liquid will flow through a path such as a water hose. Its path of flow affects how fast a liquid flows. For example, a liquid won't flow uphill without some sort of pressure behind it. The temperature of a liquid also affects its ability to flow. As the temperature of a liquid increases, the liquid has a tendency to get thinner. This change is known as viscosity. **Viscosity** is the measure of a liquid's resistance to flow. You'll normally see the word viscosity when referring to the oil used in engines. A viscosity with a high number has a greater resistance to flow as compared to a low viscosity number.

Boyle's Law

As we stated earlier, a gas, similar to air, can be compressed. There's a physical law known as **Boyle's law**, which states, "the product of the pressure and the volume of a given mass of gas is constant if the temperature isn't changed." Boyle's law tells us that when a gas is compressed, its temperature and pressure increase. The more a gas is compressed, the greater its temperature and pressure. Each time you decrease the volume of a gas by one half, you double the pressure.

Pressure Differences

Pressure differences result in movement of a gas from a high-pressure area to a low-pressure area, moving any matter that may be in the way along with it. High pressure always seeks low pressure. A common example of pressure differences can be seen every day in the weather around us. In engines, carburetion and the intake or induction phase take advantage of pressure differences for operation.

Momentum

Velocity is the speed of an object. Mass is the weight of any form of matter. **Momentum** is the driving force that's the result of motion or movement. Momentum is determined by multiplying mass (weight) times velocity. (Momentum = Mass × Velocity)

Laws of Motion

Two laws of motion concerned with understanding engine operation are the law of inertia and the law of action and reaction. The **law of inertia** states that anything at rest or in motion tends to remain at rest or in motion until acted upon by an outside force. The **law of action and reaction** states that for every action there's an equal and opposite reaction.

Energy

Energy is the ability to do work. If we have lots of energy, we can do lots of work. Energy itself can't be seen; however, the results of energy can be seen. An example is lifting a box and setting it on a table. Only the physical movement of the box can be seen.

Energy exists in many forms and can be changed from one form to another. For example, a battery changes chemical energy to electrical energy. No conversion of energy is 100 percent efficient. An example is your home heating system. As your heating system burns fuel, most of the heat is used to warm your home; however, some of the heat is lost up the chimney. The same is true in a motorcycle engine. Burning fuel in the engine provides the energy to move the motorcycle, but some energy is lost in friction and heat produced by the engine.

There are two types of energy:

- **Potential energy** is stored energy, such as in a charged battery or a can of gasoline.
- **Active energy** (or *kinetic energy*) is energy in use or in motion, such as when a battery is used to light a lamp or when gasoline is used to run an engine.

BASIC INTERNAL-COMBUSTION ENGINE OPERATION

In Chapter 5 we discussed the basic components and operation of an engine. Now we will discuss these topics again but in detail.

Motorcycle and ATV engines use a principle called combustion to operate. **Combustion** is the rapid combining of oxygen molecules with other elements. Chemical changes can result in the release of heat when elements are combined with oxygen. Heat usually speeds up any chemical changes and can act as a catalyst. A **catalyst** speeds up the chemical reaction of something without undergoing any change itself. Cold usually slows down most chemical changes. This is why engines tend to run better when warmed up than when they are first started.

Types of Combustion Engines

The engines found in all modern motorcycles and ATVs are internal-combustion engines. In an **internal-combustion engine**, compressed fuel and air is burned inside the engine to produce power. The internal-combustion engine produces mechanical energy by burning fuel. In a motorcycle engine, fuel is sent to the engine through an induction system, where it's burned inside to produce the power that's used to help make the engine run. In contrast, an external-combustion engine burns fuel outside of the engine. A steam engine with a boiler is an example of an external-combustion engine. We will not discuss the external-combustion engine design here because it is not relevant to motorcycle or ATV use.

Construction of the Internal-Combustion Engine

Let's begin by reviewing some of the common parts in an engine (Figure 6-1). The **cylinder** is a hollow metal tube. The top end of the cylinder is sealed by the cylinder head. The cylinder head is bolted onto the top of the cylinder. The **piston** is a can-shaped metal component that can move up and down inside the cylinder. The piston is the main moving part in an engine.

The area above the piston and below the cylinder head is called the **combustion chamber**. In this chamber, a mixture of air and gasoline is compressed and burned to produce power. A spark plug is screwed into a threaded hole in the cylinder head. The end of the spark plug protrudes through the cylinder head and into the combustion chamber. The spark plug is used to ignite the compressed air-and-fuel mixture in the cylinder and cause it to burn. The sparking action of the spark plug is controlled by the engine's ignition system, which we will discuss in detail in a later chapter.

When the air-and-fuel mixture burns in the combustion chamber, it releases energy. The expansion of the gases due to this energy release is enough to force the piston downward into the cylinder. The bottom end of the piston is attached to a connecting rod and crankshaft assembly. When the piston is forced downward in the cylinder, the piston's downward motion is transferred to the rod and crankshaft. The rod and crankshaft then convert the up-and-down (or reciprocating) motion of the piston into circular or rotary motion.

Now, let's take a few minutes to review a few terms that were presented earlier. When a piston is at its highest position in the cylinder, it's said to be at **top-dead center (TDC)** (Figure 6-2). When the piston is at its lowest position in the cylinder, it's said to be at **bottom-dead center (BDC)**. The total distance that the piston moves from the top of the cylinder to the bottom of the cylinder is called the **stroke**.

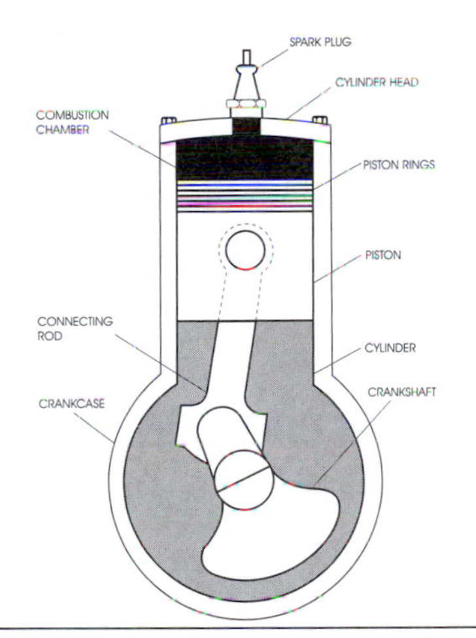

Figure 6-1 This drawing shows a simplified cutaway view of an engine. During operation, the piston moves up and down inside the cylinder. This movement is transferred to the crankshaft through the connecting rod.

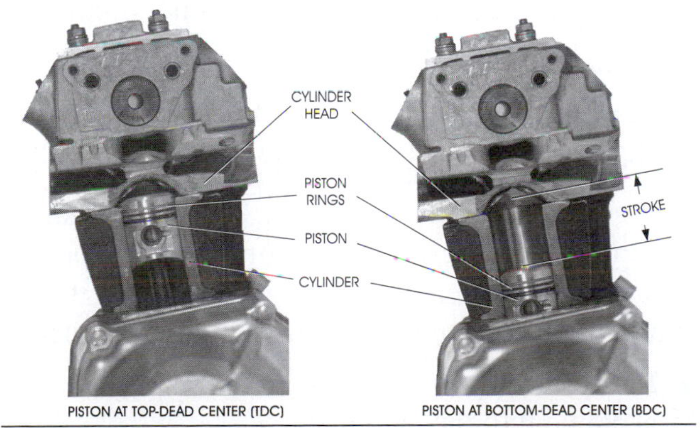

Figure 6-2 This illustration shows a cutaway view of an actual engine with the piston located at top-dead center (left) and at bottom-dead center (right).

The outside surface of a piston has several horizontal grooves cut into it. Each groove holds a metal ring called a piston ring. A **piston ring** is a metal ring that's split at one point and is designed to apply pressure against a cylinder to separate one side of the piston from another. The piston rings slip over the outside of the piston and fit into the piston ring grooves. Once they're in place, the rings stick out like ridges on the surface of the piston. When a piston is inside a cylinder, the piston rings press outward against the walls of the cylinder. This helps form a tight seal between the piston and the cylinder, which is necessary for proper engine operation.

Methods of Internal Combustion

There are two methods of initiating normal combustion in an engine. The first method is **ignition**, which is the contact of a fuel with a spark. The second method is by reducing the space combining oxygen and a combustible material, which produces heat. Motorcycle and ATV engines use a combination of these two methods. The air-and-fuel mixture is compressed into a very small space. An ignition spark begins the combustion process. As the air-and-fuel mixture burns, the hot expanding gases push the piston down. The combustion process changes potential energy in the form of the air-and-gas mixture (chemical energy) to active (kinetic) energy in the form of heat. The following are the three phases of combustion that occur during the power stroke of the internal-combustion engine: combustion lag, active combustion, and post combustion.

Combustion Lag

The first phase of the internal-combustion engine process starts after the piston compresses the air-and-fuel mixture. The spark plug ignites a small portion of this mixture. A ball of fire spreads outward and begins to consume the remaining compressed air-and-fuel mixture. However, this ball of fire that initiates combustion doesn't immediately spread outward. Before the chain reaction spreads to the outside area of the combustion chamber, a short period of relatively slow burning time takes place. The time interval between the time that the spark occurs and the energy release of the air-and-fuel mixture is called **combustion lag**.

Active Combustion

The second phase of the internal-combustion engine process begins when the initial combustion lag is overcome and the chain reaction begins to spread quickly outward. A rapid temperature and pressure buildup occurs as the charge is consumed. The chain reaction of burning molecules accelerates and the chemical conversion causes heat to be released very quickly. This increase in temperature causes the pressure in the cylinder to increase. This phase is known as **active combustion**.

Post Combustion

As the piston moves down and the volume inside the cylinder increases, the pressure drops and the power is then absorbed by the piston. The cylinder now eliminates spent gases to prepare for the next cycle of fresh air-and-fuel mixture. All engines begin to release exhaust gases out of the cylinder well before the piston reaches bottom dead center. This is known as **post combustion**.

Results of Combustion

The heat and power generated within the combustion chamber produce work, which is realized through the crankshaft and eventually through the drive system of the motorcycle or ATV. Although most four-stroke engines run at lower temperatures, cylinder head temperatures can be as high as 300–375 degrees Fahrenheit (F). Combustion chamber gas temperatures within the engine are known to be as high as 4,000 degrees F. This relates to cylinder pressures reaching 800–1,000 pounds per square inch (psi). The heat produced expands the gases in the combustion chamber and pushes the piston towards BDC.

Chemical changes occur during combustion, which convert the fuel-and-air mixture into the following chemicals:

- **Carbon monoxide (CO)** results from partially burned fuel or fuel that's not completely burned during the combustion process. As mentioned in a previous study unit, remember that carbon monoxide is a colorless, odorless, poisonous, and deadly gas.
- **Hydrocarbons (HC)** result from unburned or raw fuel.
- **Carbon dioxide (CO_2)** is the result of complete combustion.
- **Oxides of nitrogen (NOx)** are forms of oxidized nitrogen resulting from extremely high combustion temperatures.
- **Water (H_2O)** also results from complete combustion. Unbelievably, for every gallon of fuel burned, approximately one gallon of water is produced in a vaporized form.

In the United States, the **Environmental Protection Agency (EPA)** has developed emission standards for street-legal motorcycles. Since 1978, motorcycles designed for street use must comply with EPA emission standards. The two key emissions produced by motorcycles and monitored by the EPA are HC and CO. The EPA also monitors noise emissions.

INTERNAL-COMBUSTION ENGINE OPERATION

So far, we've looked at the basic internal-combustion process of a typical engine. Now, let's take a look at how combustion is used to allow the engine to operate. In order to work, all internal-combustion engines must do four basic things. The four basic things an engine must do are the following:

- Take in air and fuel
- Squeeze or compress the air-and-fuel mixture
- Ignite and burn the mixture
- Get rid of the burned gases

The engine actions we've just described are the four stages of engine operation. The proper names for these stages are intake, compression, power, and exhaust. When an engine is operating, it continually runs through these four stages, repeatedly.

- **Intake stage** Air has been mixed with fuel and is drawn into the cylinder.
- **Compression stage** The piston rises and compresses the air-and-fuel mixture trapped inside the combustion chamber.

- **Power stage** The air-and-fuel mixture is ignited and energy is released. The release of energy of the ignited fuel pushes the piston back down the cylinder. The downward motion of the piston is transferred through the connecting rod to the crankshaft.
- **Exhaust stage** The exhaust gases are released from the cylinder.

The four stages then begin all over again.

One **engine cycle** is a complete run through all four stages of operation: intake, compression, power, and exhaust. Keep in mind that the four stages of operation we've described occur very quickly, and they repeat continually for as long as the engine is running. All motorcycle and ATV engines operate in these same four basic stages, and all the stages must occur in order for the engine to run properly. To understand how an engine works, one of the most important things you can do is memorize the four stages of engine operation. Once you understand these four stages, everything else we discuss about engine operation will fall into place.

BASIC FOUR-STROKE ENGINE COMPONENTS

Now that we've reviewed the basics of engine operation, let's look at the four-stroke motorcycle and ATV engine components in more detail. We'll cover all the basic parts of a four-stroke engine and explain their purpose. Many of the components are found in both four-stroke engines and two-stroke engines. Refer to the illustrations for reference as we discuss each component. Be aware that not all engines look exactly alike. However, the illustrations provided are typical of many motorcycle and ATV four-stroke engines you will see.

Four-Stroke Cylinder Heads

Cylinder heads are constructed of aluminum alloy or cast iron (Figure 6-3). The four-stroke cylinder head holds the intake and exhaust valve train components and the spark plug. The cylinder head also seals the top end of the cylinder for compression of the air-and-fuel mixture under the spark plug to increase combustion efficiency. Holes in the cylinder head called ports provide for the air-and-fuel mixture to come into the combustion chamber and also remove the spent gases after the combustion process. Cylinder heads also aid in the transfer of heat from the engine by the use of fins on air-cooled engines or by using water jackets on liquid-cooled engines.

Many modern four-stroke motorcycles and ATVs use multi-valve cylinder heads. Cylinder heads may have anywhere from two valves to eight valves per cylinder. The intake valve area is usually larger than the exhaust valve area on the four-stroke engine.

Poppet Valves

Four-stroke engines use mechanical valves called **poppet valves** to control the gases coming into and going out of the engine. Poppet valves, which are tulip-shaped, open and close every other crankshaft revolution. The poppet valve may be made from stainless steel, carbon steel, or titanium. The various parts of the valve are shown in Figure 6-4.

- The valve tip is the part of the valve that rides against the valve-opening device. Most valve tips are stellite-plated for wear. **Stellite** is an extremely hard metal alloy that resists wear and won't soften at high temperatures.
- The keeper groove is where keepers lock the valve-and-spring retainer in place.

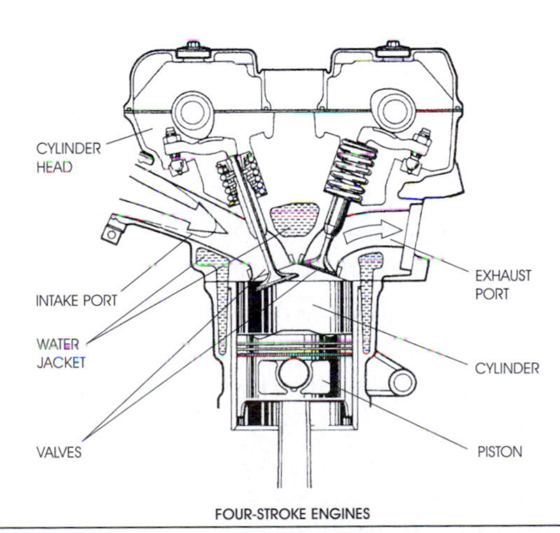

Figure 6-3 This illustration shows a typical liquid-cooled engine cylinder head attached to the cylinder.

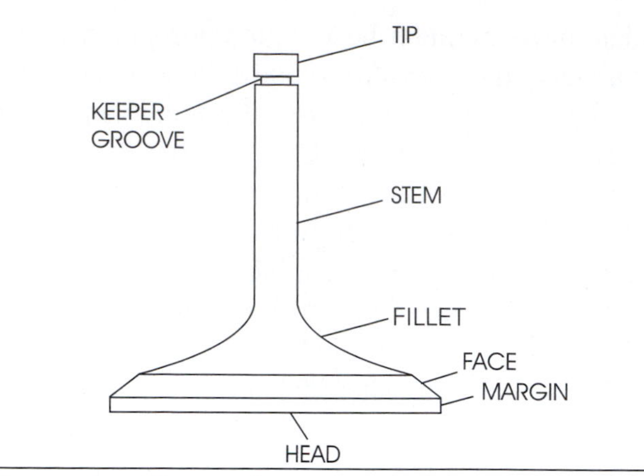

Figure 6-4 The various parts of a typical poppet valve are shown here.

- The valve stem is the thrust surface for the valve guide and is considered to be a major wear area. If the stem is worn, excessive amounts of oil can pass between the stem and guide into the combustion chamber. If oil leaks into the combustion chamber, smoke appears in the exhaust.
- The valve fillet is the sloped area of the valve that connects the valve stem to the valve head.
- The valve face mates with the cylinder head valve seat to seal gases in the combustion chamber and aid in heat transfer. The valve face is often coated with stellite to reduce wear and prolong the life of the valve.
- The margin supports the valve face and shields the face from high combustion temperatures.
- The valve head is the bottom portion of the valve and forms a part of the combustion chamber.

Common wear areas of the valve are the tip, face, stem, and keeper groove.

Valve Seats

Cylinder head **valve seats** (Figure 6-5) are stationary in the cylinder head and are the sealing surface for the valve face. There are normally at least three angles cut into the valve seat to allow for better air-and-fuel flow into the cylinder through the valve opening.

Valve Guides

Valve guides (Figure 6-5) are installed in the four-stroke cylinder head. Valve guides provide a bushing surface for the valve stem.

Valve Stem Seals

Valve stem seals (Figure 6-6) are installed on the valve guides and are used to prevent excessive oil from entering between the inside of the valve guide and valve stem. This is particularly important on the exhaust side, as excessive oil passing by the valve stem seal will cause smoking out of the exhaust.

Valve-Closing Devices

Valve-closing devices keep the valve closed when required. The most common method to close a valve is with the use of coil springs (Figure 6-6) attached between the valve and the cylinder head. The springs are held in place with valve spring retainers and valve keepers that fit into the valve keeper grooves. While many modern four-stroke engines use only one valve spring, there are usually two coil springs per valve to reduce the chance of valve float. Valve float is the point at which the valve doesn't stay in constant contact with the valve train. Valve float can occur when the valve springs are weak or during excessively high engine speed. Other devices have been used in the past, such as torsion bars and hairpin springs, but are not seen commonly in today's modern motorcycle or ATV engines.

Valve-Opening Devices

The main valve-opening device is the cam follower. The cam follower contacts the valve tip and is used to transfer motion from the camshaft to the valve. There are two common types of valve-opening devices used on the four-stroke motorcycle and ATV—rocker arms and shim and bucket.

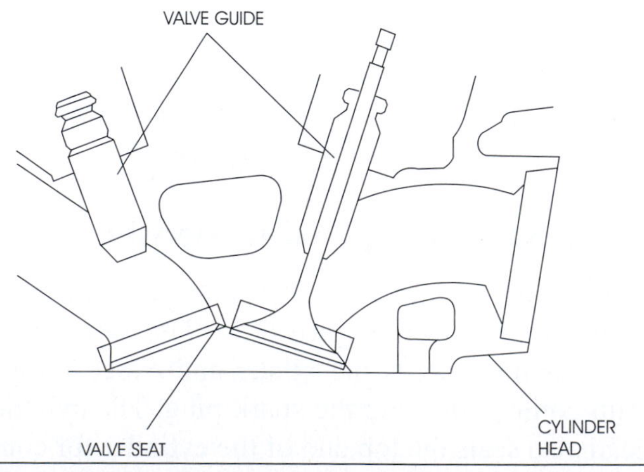

Figure 6-5 Valve seats and guides are located in the cylinder head.

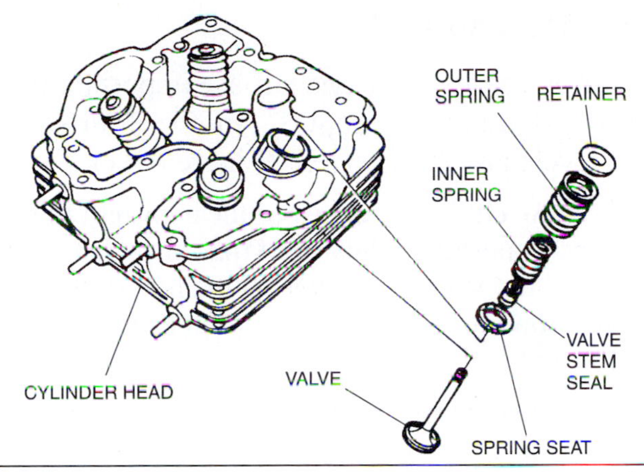

Figure 6-6 This illustration shows the valve stem seal and common components used to close the valve.

Rocker arm The rocker arm is a lever that can gain a mechanical advantage and change the direction of force applied to it (Figure 6-7). The rocker arm can also open more than one valve. There are various rocker arm designs used on the four-stroke engine but they all perform the same function. A disadvantage of rocker arms is that they create side loads on valve stems and guides, which can cause excessive wear.

Shim and bucket The second common type of valve-opening device is the **shim and bucket** (Figure 6-8). The bucket is located above the valve in the cylinder head. The shim may be located either above or below the bucket. This type of valve-opening device doesn't create side loads and allows for a lighter mass in the valve train because it has fewer moving parts. This type of valve-opening device is most popular in higher revving motorcycles such as sport bikes.

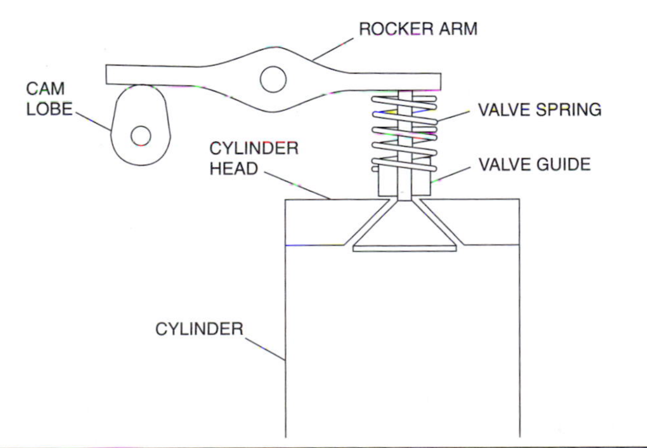

Figure 6-7 This illustration shows a simple diagram of a rocker arm valve-opening device.

Valve Clearance

Valve clearance or lash (Figure 6-9) is necessary to allow for heat expansion, oil clearance, and for proper sealing of the valve. Too little clearance causes improper sealing of the valve and excessive heat. Too much clearance causes excessive wear and noise.

There are different types of valve lash adjusters used to adjust valves on motorcycles and ATVs. Three of the most popular types are as follows:

- The screw and lock nut (Figure 6-9) uses a screw that can be turned in or out to change the clearance. After the adjustment has been made, a lock nut holds the screw in place. The screw and lock nut may be located on the rocker arm, on a push rod, or on a valve lifter.
- The shim and bucket (Figure 6-8) is used for both a valve-opening device and an adjustment device. The shims are used to adjust the valves for proper clearances. Changing the size of the shim changes clearances. The two popular types of shim-and-bucket adjusters are shim-over-bucket, where the shim rests on top of the bucket and shim-under-bucket, where the shim rests under the bucket. You must remove the camshaft to replace a shim with this design.
- Hydraulic valve lash adjusters (Figure 6-10) automatically adjust for the proper clearance by using oil pressure to maintain zero lash at all engine temperatures and speeds.

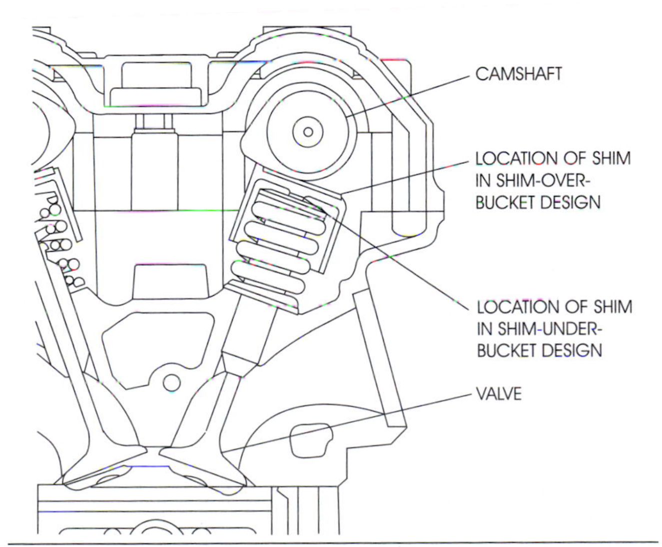

Figure 6-8 This illustration shows the location of the shims in the shim-over-bucket design and the shim-under-bucket design.

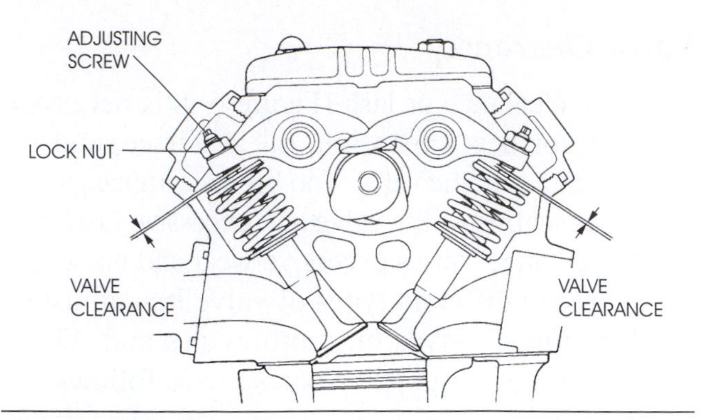

Figure 6-9 This illustration shows valve clearance with a screw and lock nut for adjustment. Valve clearance is important to allow for heat expansion, oil clearance, and proper sealing of the valve.

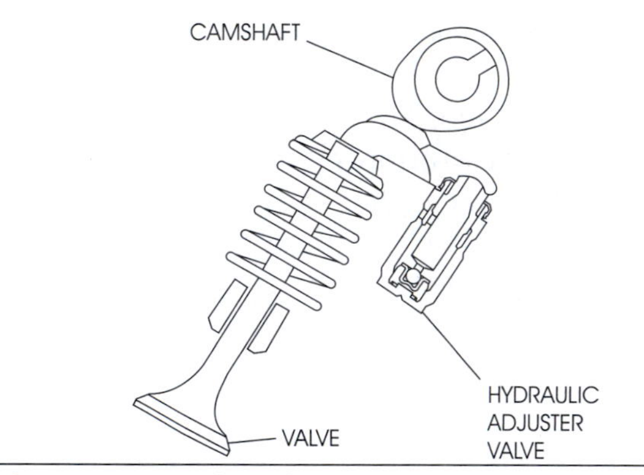

Figure 6-10 This illustration shows a hydraulic adjuster valve that automatically adjusts for proper valve clearance.

Four-Stroke Camshafts

The purpose of the **camshaft** (also known as the "cam") is to change rotary motion to reciprocating motion. The camshaft is also a mechanical valve timer that controls the following:

- When to open
- How fast to open
- How far to open
- How long to stay open
- When to close
- How fast to close
- How long to stay closed

Parts of a Camshaft

Various parts of a camshaft are important to its ability to function properly in the four-stroke motorcycle and ATV engine (Figure 6-11).

- The **base circle** is the area of the camshaft that forms a constant radius from the centerline of the journal to the heel. The *heel* is the part of the cam that allows the valve to seat onto the cylinder head and seal off the combustion chamber.
- **Clearance ramps** take up the valve clearance and open and close the valve. These ramps act similarly to a shock absorber and are used to gently (relatively speaking) open and close the valve.
- The flanks of the cam determine and control the acceleration of the opening and closing of the valve.
- The nose is the area of the camshaft where the valve is opened the greatest distance from the cylinder head area; it controls lift dwell. *Lift dwell* is the amount of time in crankshaft degrees that the valve stays open at maximum lift.
- **Camshaft lift** is a measure of the difference between the base circle and the nose. Depending on the type of engine, this may or may not translate into the actual valve lift, which is the distance that the valve actually moves away from the cylinder head.

The **duration** of a camshaft is a measure of how long the valve is held open. Duration is measured in crankshaft degrees.

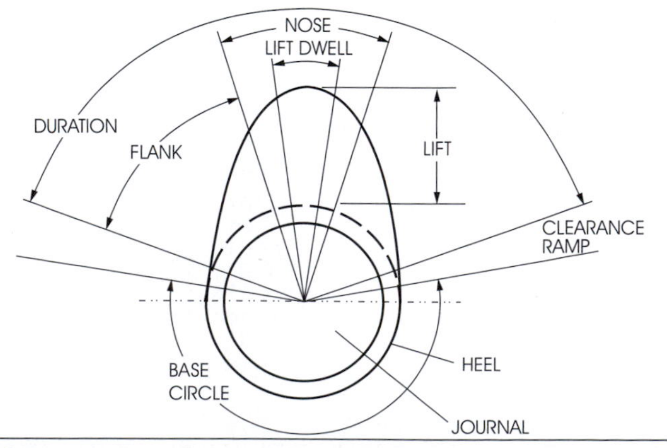

Figure 6-11 The various parts of a camshaft are shown here.

Most camshafts have what's known as valve overlap built into them. Valve overlap occurs between the exhaust and intake strokes. **Valve overlap** is the time that both valves are open simultaneously and is measured in crankshaft degrees. High-performance engines generally have more valve overlap to allow for more air-and-fuel mixture to be packed into the cylinder combustion chamber. While this allows for higher peak power, an engine with a camshaft exceeding 30 degrees of valve overlap will lack efficiency in the low- and mid-range power areas of the engine.

Camshaft Drives

The camshaft rotates at one-half the speed of the crankshaft to properly time the intake and exhaust valves with the piston as it moves up and down the cylinder. There are three methods used to drive a camshaft on a four-stroke motorcycle or ATV engine:

- The chain type of camshaft drive normally operates in an oil bath to maintain lubrication (Figure 6-12).
- The gear type of camshaft drive also operates in an oil bath for lubrication and to reduce noise (Figure 6-13).
- The toothed-belt type of camshaft drive runs on pulleys with teeth. This system is very quiet and requires no lubrication; however, proper alignment and tension are critical (Figure 6-14).

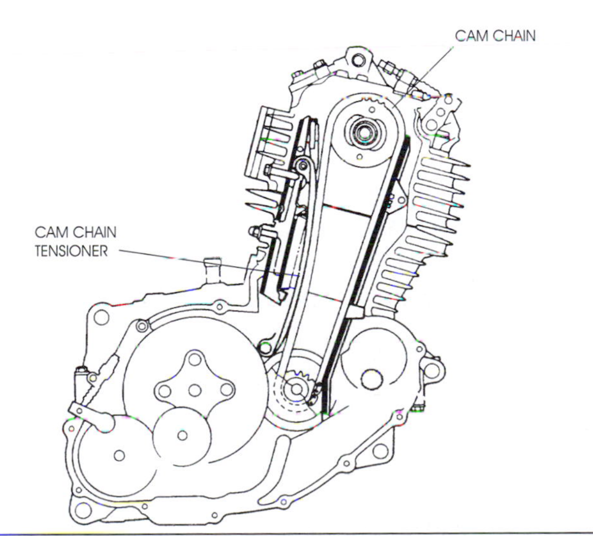

Figure 6-12 Example of a chain-driven camshaft and tensioner.

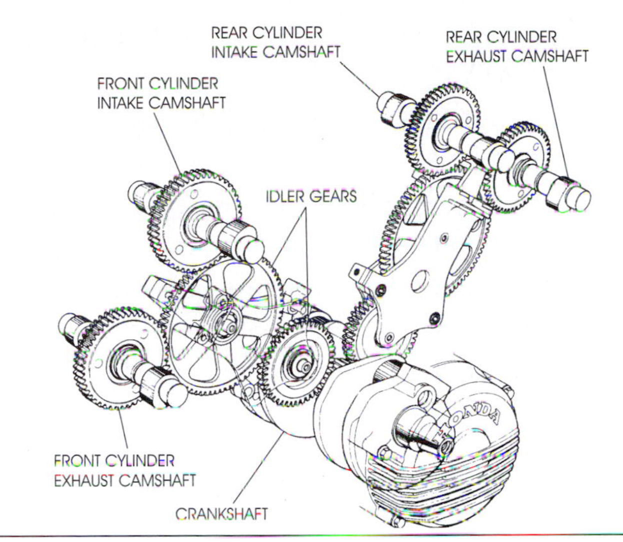

Figure 6-13 Example of a gear-driven camshaft.

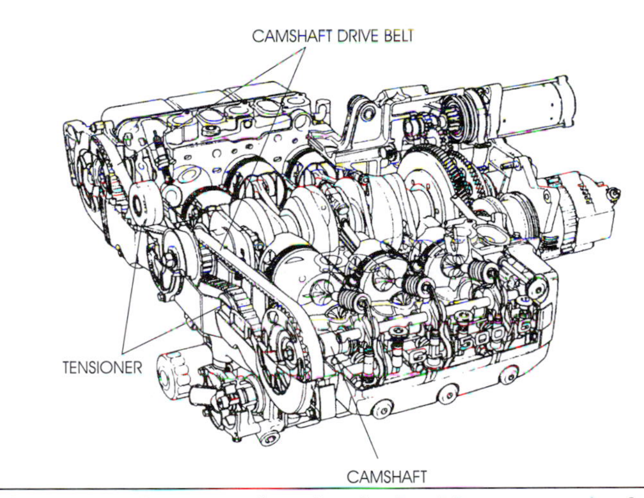

Figure 6-14 Example of a belt-driven camshaft and tensioner.

Camshaft Drive Tensioners

The purpose of a **camshaft drive tensioner** is to keep the proper tension on the cam chain or cam belt (Figure 6-12 and Figure 6-14). There are manual tensioners that require adjustment and automatic tensioners, which automatically adjust the tension while the engine is running.

Valve Train

The valve train consists of the valve components in the cylinder head and the valve drive components previously discussed. There are four common types of valve trains found on motorcycles and ATVs:

- A pushrod with rocker arm is shown in Figure 6-15A.
- A single overhead cam with rocker arms is shown in Figure 6-15B.
- A dual overhead cam with rocker arms is shown in Figure 6-15C.
- A dual overhead cam with shim and buckets is shown in Figure 6-15D.

Four-Stroke Cylinders

The purpose of the four-stroke engine cylinder is to guide the piston as it travels up and down. The cylinder helps to transfer engine heat and may be either air cooled or liquid cooled (Figure 6-16).

There are different types of materials used in the construction of a cylinder. Each material has its advantages and disadvantages.

- Cast-iron cylinders have a one-piece design and can be fit with oversize pistons by boring to a larger size. When a cylinder is bored, material is removed from the cylinder to enlarge the hole. A larger "oversized" piston is then used in place of the previous piston. The cast-iron cylinder is inexpensive to manufacture but has poor heat transfer characteristics when compared to other materials used to construct cylinders. Cast-iron cylinders are also very heavy.

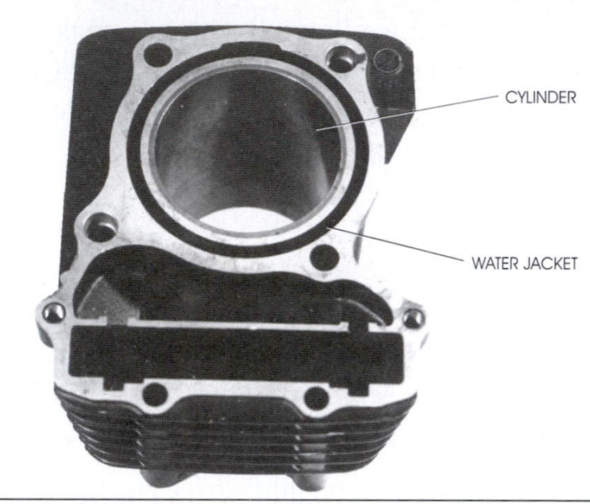

Figure 6-16 This is a typical liquid-cooled cylinder. A liquid-cooled cylinder uses water jackets as opposed to the cylinder fins used on an air-cooled cylinder.

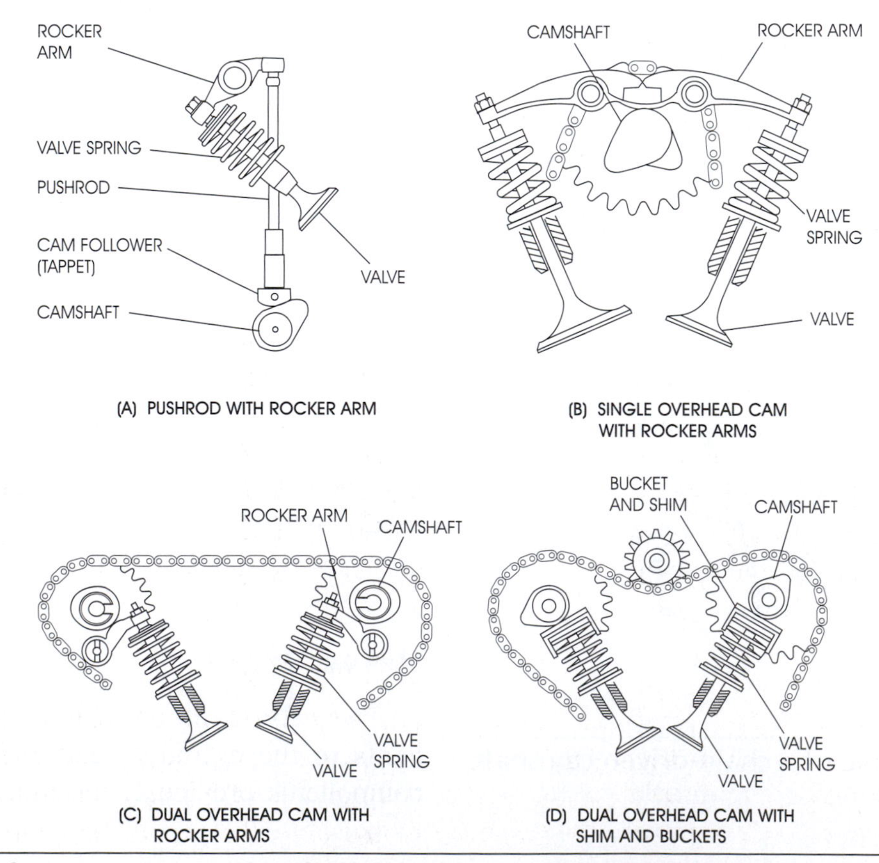

Figure 6-15 The four common types of valve trains are illustrated here.

- Aluminum cylinders with cast-iron or steel sleeves have much better heat transfer abilities than cast-iron cylinders and are much lighter in weight. These cylinders can also be bored to a larger diameter. In most cases, the sleeve can be replaced if needed.
- Plated-aluminum cylinders, which are also called Nikasil or composite cylinders, have the best heat transfer characteristics of any cylinder produced today. They're the lightest-weight cylinders available and when properly maintained are the longest-lasting cylinders. The disadvantage of plated-aluminum cylinders is that they can't be bored to a larger diameter and therefore must be replaced when damaged. These cylinders are expensive to replace when compared to the other types of cylinders.

Cylinders have tiny scratches purposely cut into them called *crosshatching*. Crosshatching is accomplished by honing the cylinder wall with a tool called a cylinder hone. The purpose of honing is to help seat the piston rings and retain a very thin layer of oil on the cylinder walls to keep them properly lubricated.

Cylinders must be round from top to bottom to work properly. They shouldn't have any taper or out-of-roundness. We'll discuss how to measure cylinders later.

Pistons

The purpose of the piston is to transfer power produced in the combustion chamber to the connecting rod. The piston is manufactured in a way that makes it directional. This makes it necessary to install the piston in the specific manner indicated in the service manual.

Pistons are tapered from top to bottom. The top of the piston is smaller than the bottom to allow for different heat expansion rates of the piston. To allow for further heat expansion, pistons are **cam ground** so they're oval in shape when cold. When the piston reaches operating temperature, it becomes round to match the cylinder.

There are two common piston-manufacturing methods: cast aluminum and forged. Cast-aluminum pistons are the more common of the two.

Forged pistons use aluminum alloy forced into a die under extreme pressures. This manufacturing method produces a stronger piston, but makes the piston more expensive.

A piston has several parts (Figure 6-17).

The **crown** is the top of the piston and acts as the bottom of the combustion chamber. The crown is the hottest part of the piston—due to combustion chamber temperatures. The crown area expands more than the rest of the piston because it's hotter and has more mass. The piston crown may have a positive dome, flattop, or negative dome. There also may be notches in the crown to allow for valve relief.

The **ring grooves** allow for installation of piston rings. The bottom ring groove of the four-stroke piston has holes or slots for oil return that help remove oil from the cylinder wall. This also helps to lubricate the wrist pin. The **piston ring lands** support the piston rings.

The **wrist pin boss** is where the piston attaches to the small end of the connecting rod. A hardened tool-steel *wrist pin* attaches the piston to the rod. The wrist pin is normally held in place with retaining clips to prevent it from contacting the cylinder wall.

The **piston skirt** is the load-bearing surface of the piston. The piston skirt contacts the cylinder wall and is the primary wiping surface for the cylinder wall. The largest diameter of the piston is usually at or close to the bottom of the skirt and 90 degrees from the wrist pin. This is where the piston is normally measured.

Piston Rings

The purpose of **piston rings** is to aid in heat transfer from the piston to the cylinder wall, seal in

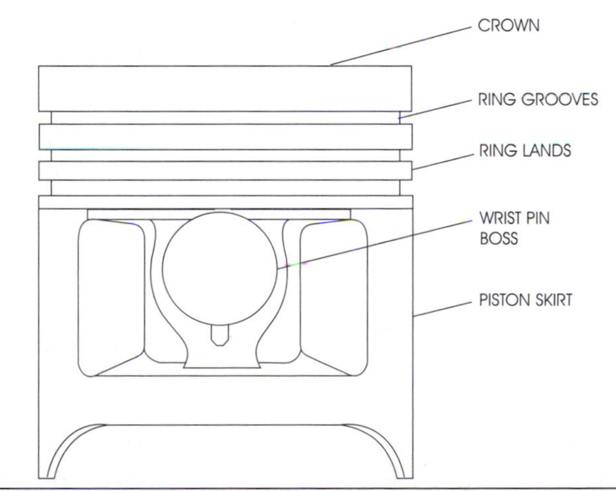

Figure 6-17 The parts of a typical four-stroke piston are shown here.

the combustion gases, and prevent excessive oil consumption. Generally, there are three types of piston rings used on the four-stroke piston (Figure 6-18).

The **compression ring** is closest to the piston crown and is used to seal most of the combustion chamber gases. The compression ring is usually made of cast iron and may be chrome-plated, Teflon, or moly-coated.

The **scraper ring** is the middle ring and aids in sealing the combustion chamber gases. The scraper ring scrapes excessive oil from the cylinder wall. Like the compression ring, the scraper ring is made of cast iron, but in most cases has no coating. Many modern engines no longer use the scraper ring to reduce friction, therefore the engine's horsepower is increased.

The **oil control ring** is the ring closest to the piston skirt. The oil control ring removes the oil from the cylinder walls left behind by the piston skirt.

Piston rings have an end gap to allow for heat expansion. The end gap is measured by using a feeler gauge after fitting the ring squarely in the cylinder (Figure 6-19).

Crankshafts

As we discussed earlier in Chapter 5, the purpose of a crankshaft is to change the reciprocating motion from the piston into a rotary motion. The main parts of a crankshaft include journals and counterweights (Figure 6-20).

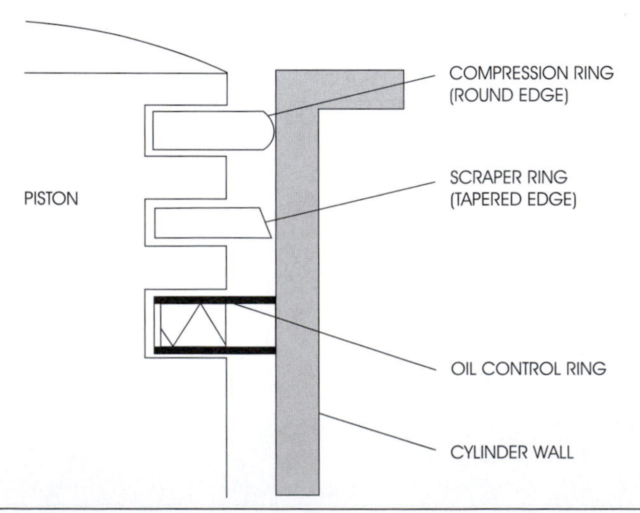

Figure 6-18 Three common types of piston rings found on four-stroke engine pistons.

Main journals support the mass of the crankshaft and are located at the center of the rotating axis.

Connecting-rod journals support the connecting rods and are offset from the main journals.

Counterweights add momentum to the crankshaft. The counterweights assist in keeping the crankshaft rotating and the engine running smoothly by counterbalancing the reciprocating masses from the piston and connecting rod. Some engines use remote counterweights, which are located on a separate shaft and are chain or gear driven. Remote counterweights must be timed properly with the crankshaft.

There are two different types of crankshafts used in four-stroke engines.

The **one-piece crankshaft** is cast or forged as one part (Figure 6-20). The one-piece crankshaft is the stronger in design but must be used with a

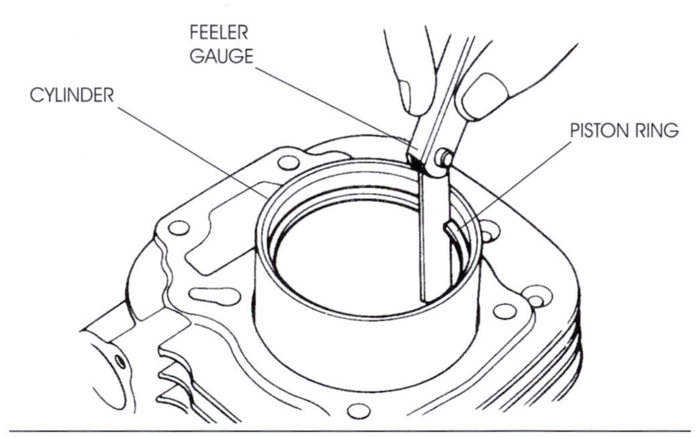

Figure 6-19 The piston ring end gap is measured using a feeler gauge. The ring end gap is required to allow for heat expansion.

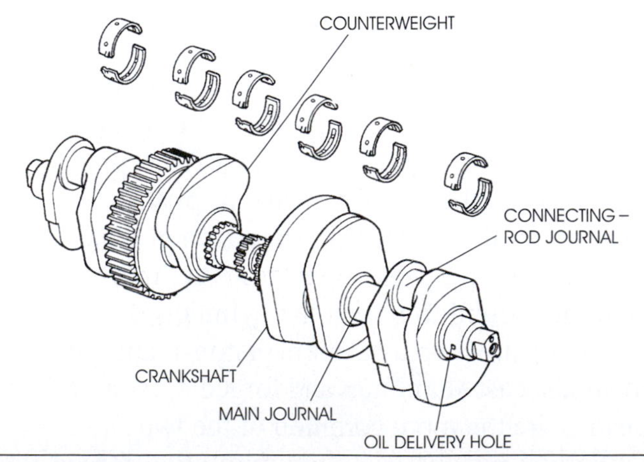

Figure 6-20 A one-piece crankshaft used on a four-stroke engine is illustrated here.

multi-piece connecting rod. Most one-piece crankshafts are cross-drilled for oil delivery to the connecting-rod journals and use plain bearings at the main and connecting-rod journals. Most one-piece crankshafts require high oil pressure to the bearings. Generally, you cannot rebuild one-piece crankshafts.

The **multi-piece crankshaft** uses crankshaft halves that are cast or forged (Figure 6-21). The connecting-rod journal is a pin (crankpin) that's press-fit into the halves. A one-piece connecting rod uses a roller bearing at the connecting-rod journal. The multi-piece crankshaft generally uses ball bearings on the main journals. Most multi-piece crankshafts can be rebuilt. Figure 6-22 shows an assembled multi-piece four-stroke engine crankshaft with connecting rod.

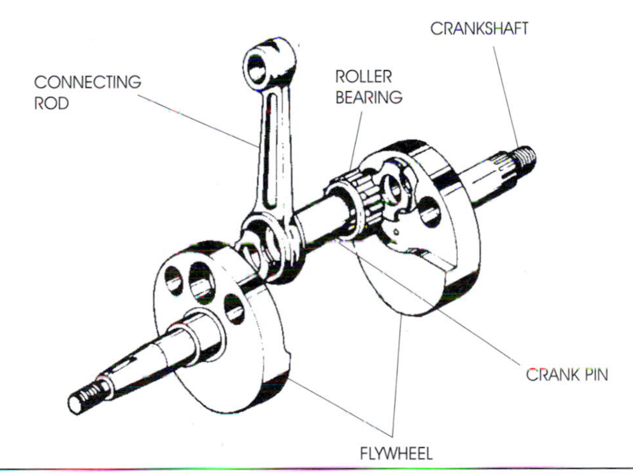

Figure 6-21 A multi-piece crankshaft used on a four-stroke engine is illustrated here.

Figure 6-22 An assembled multi-piece four-stroke engine crankshaft is pictured here.

Multi-Cylinder Crankshafts

Multi-cylinder crankshafts use different offset positions for each cylinder, depending on the design (Figure 6-23).

- In the 360-degree design, both pistons move up and down together. This design requires more counterweight and tends to vibrate at higher engine speeds.
- In the 180-degree design, the pistons move in opposite directions. This design requires less counterweight and has less vibration at higher engine speeds. When this design is used on an in-line four-cylinder engine, a pair of 180-degree crankshafts is used.
- In the 120-degree design, the pistons move 120 degrees apart from each other. This design is used on three- and six-cylinder engines. Six-cylinder engines use a pair of 120-degree crankshafts.

Connecting Rods

The connecting rod is a lever that transfers power from the piston to the crankshaft. Connecting rods are usually made of forged steel, titanium, or aluminum and use an I-beam construction. There are two types of connecting rods used on four-stroke motorcycle and ATV engines.

The one-piece connecting rod (Figure 6-24A) is the stronger in design. It uses a roller bearing at the larger end and must be used with a multi-piece crankshaft. The one-piece connecting rod normally has holes or slots on both ends for added lubrication.

The multi-piece connecting rod (Figure 6-24B) is somewhat weaker in design when compared to the one-piece rod. The parts of the multi-piece connecting rod consist of the connecting rod, connecting-rod end cap, and connecting-rod bolts and nuts. The multi-piece connecting rod uses a two-piece plain bearing at the larger end, which normally requires high oil pressure for proper lubrication. The multi-piece connecting rod is normally used with a one-piece crankshaft.

Crankcases

The purpose of the crankcase is to contain and support the major engine components. These components can include the crankshaft, oil pump, cylinder,

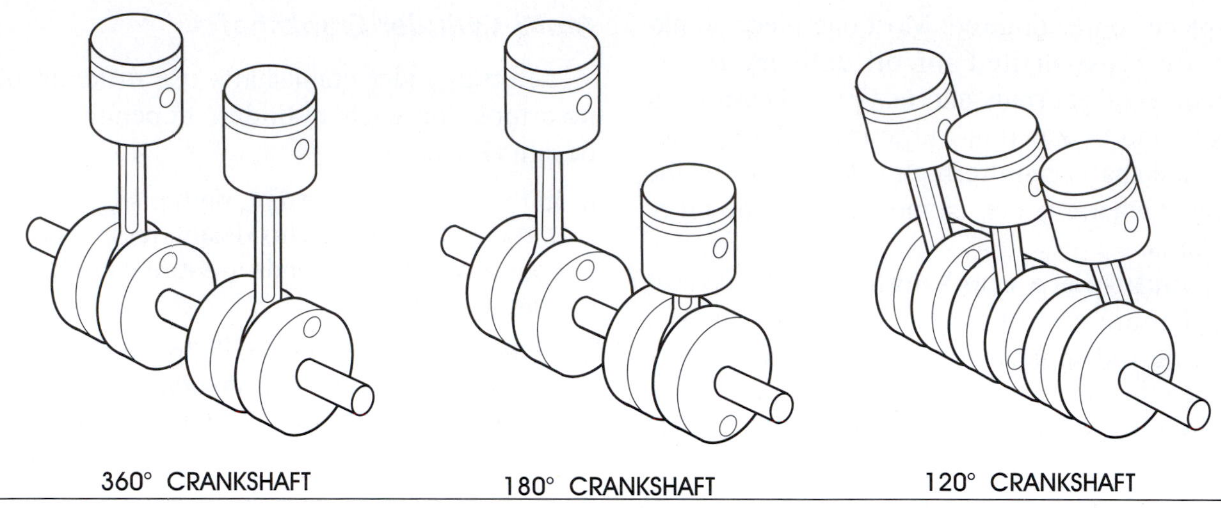

Figure 6-23 Three different multi-cylinder crankshaft designs are shown here.

Figure 6-24 This illustration shows the two types of connecting rods. Figure A shows a one-piece connecting rod with the piston attached. Figure B shows a multi-piece connecting rod with the piston detached.

primary shaft, primary drive, transmission, and camshaft. The four-stroke crankcase must be vented to the atmosphere to prevent excessive pressure from building up inside the engine. Today, government regulations require that crankcase ventilation re-circulate back through the combustion chamber.

There are three types of crankcases on four-stroke engines.

- A one-piece crankcase consists of a case that's a single-piece construction with a separate access cover to remove parts. This type of crankcase is rarely found on an engine in a modern motorcycle or ATV.
- Vertically split crankcases consist of two case halves that separate vertically (Figure 6-25). This design requires the removal of the cylinders before the case halves can be split.
- Horizontally split crankcases consist of two case halves that separate horizontally. This design allows the bottom halves to be removed with the cylinder(s) still attached to the top half.

FOUR-STROKE ENGINE THEORY OF OPERATION

Although the four-stroke engine is somewhat complex in design due to the parts necessary for it to function, it's relatively simple in terms of operation. The engine runs by repeatedly completing a cycle of operation. Each cycle of operation consists of two crankshaft revolutions in which four piston strokes occur. Each of the four piston strokes performs a distinct operation. The four operations (or stages) that are required for the engine to produce power are intake, compression, power, and exhaust. These operations must occur in the proper order for the engine to run correctly.

Valve Operation

As we mentioned earlier, the four-stroke engine uses mechanical valves: the intake valve and the exhaust valve (Figure 6-26). These valves move up and down to open and close during engine operation. The **intake valve** opens to allow the air-and-fuel mixture to flow into the combustion chamber. The **exhaust valve** opens to allow exhaust gases to flow out of the combustion chamber after the air-and-fuel mixture is burned.

The intake and exhaust valves are mechanically lifted to make them open and close. A valve-lifting device that rests on the lobes of the camshaft opens the valves. As the camshaft turns, the lobes open the valves in a timed sequence to match up properly with the up-and-down motion of the piston. The valve springs hold the valves closed when they are not being forced open by the camshaft and lifting device.

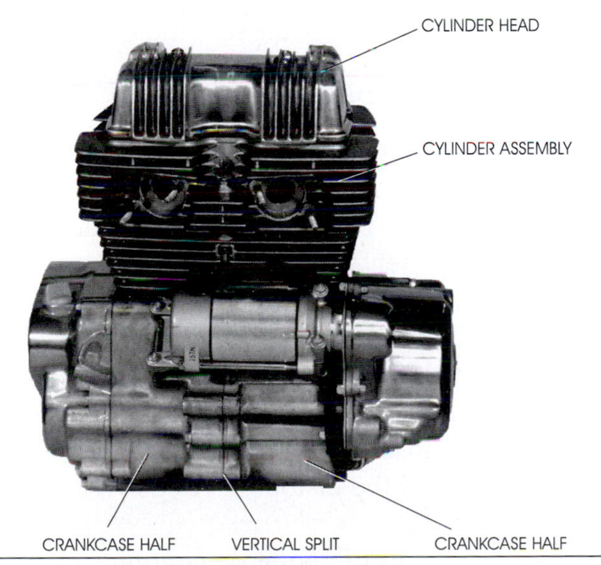

Figure 6-25 This engine has a vertically split crankcase design. The cylinder and cylinder head assemblies must be removed to split the crankcase halves.

Fuel Induction

In order to burn properly in an engine, fuel must be mixed with air. This is done via a carburetor or a fuel injection system generally known as a fuel induction system. Fuel moves from the fuel tank into the induction system, where it's atomized and mixed with air. The air-and-fuel mixture is then transferred out of the induction system and into the cylinder through the intake valve, where it's vaporized. We'll discuss the specific functions and details of various induction systems in a later chapter.

The Strokes of a Four-Stroke Engine

Now, let us take a closer look at the individual operations, known as strokes, that occur in the four-stroke engine (Figure 6-27).

The Intake Stroke

During the **intake stroke**, the air-and-fuel mixture enters the cylinder. The intake sequence starts when the intake valve begins to open. As the piston moves downward in the cylinder away from the cylinder head, the volume of the cylinder above the piston expands. This increase in volume creates a low-pressure area, which develops less-than-atmospheric pressure inside the cylinder. With the

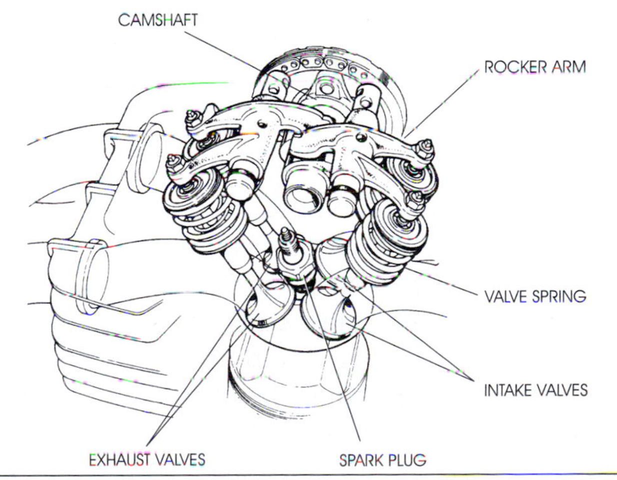

Figure 6-26 This illustration shows the intake and exhaust valves. These mechanical valves open and close to allow the air-and-fuel mixture into the engine, and the exhaust gases to leave the engine. The camshaft rotates with the crankshaft to push the valves open and allow them to close at the proper times.

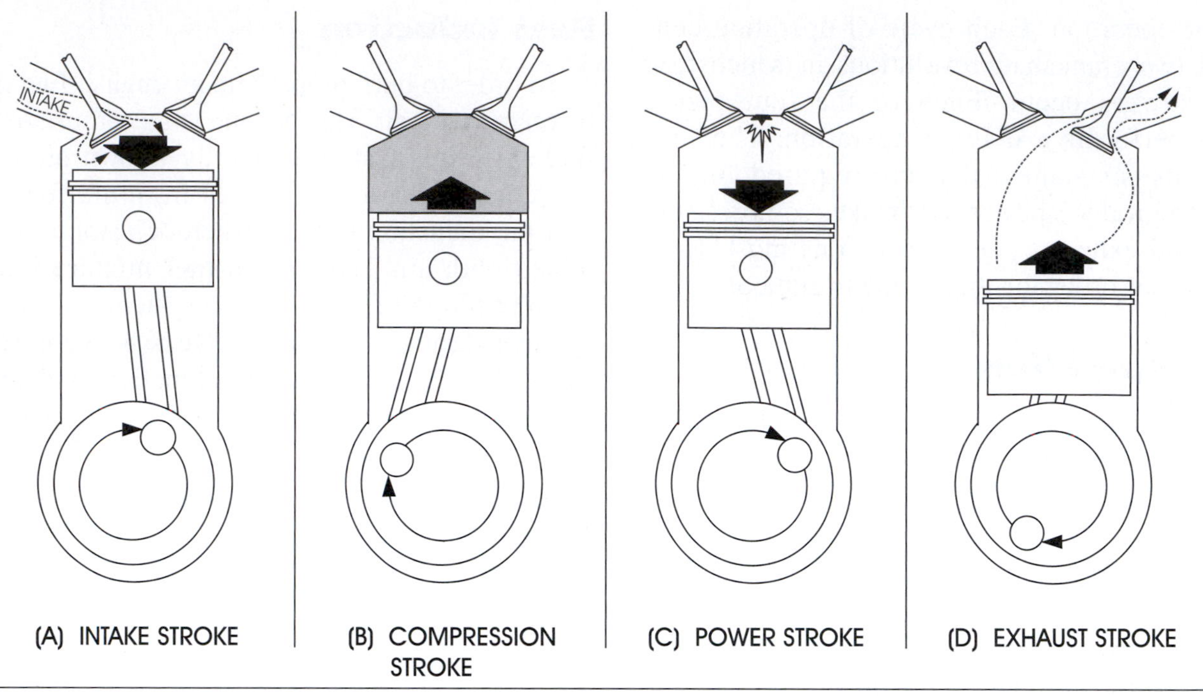

Figure 6-27 This illustration shows the four piston strokes in the sequence needed to complete one cycle of operation. An engine runs by repeatedly completing this cycle.

intake valve open, a path is completed through the intake manifold and carburetor. In an effort to balance the pressure difference between the atmospheric pressure of the outside air and the less-than-atmospheric pressure inside the cylinder, the outside air moves through the carburetor toward the cylinder. (Remember, a high-pressure area will always seek a low-pressure area.) The intake valve closes and seals the combustion chamber when the piston is near the bottom of its stroke near the crankshaft.

The Compression Stroke

When the piston approaches bottom-dead center, both valves are closed. The air-and-fuel mixture is now trapped inside the sealed combustion chamber. At this point the piston begins to rise, which compresses the air-and-fuel mixture very tightly in the combustion chamber. This is known as the **compression stroke**.

The Power Stroke

Just before the piston reaches TDC during the compression stroke, the engine's ignition system fires the spark plug. That is, the ignition system produces an electric current that causes a spark to jump across the two electrodes of the spark plug. When the spark is applied to the compressed air-and-fuel mixture, a contained explosion occurs and the compressed air-and-fuel mixture is burned.

When gases explode, they expand rapidly. The force of this contained explosion pushes the piston down in the cylinder. The connecting rod, which is connected between the piston and crankshaft, causes the downward motion of the piston to force the crankshaft to rotate. This is known as the **power stroke**.

The Exhaust Stroke

As the piston moves downward during the power stroke, the exhaust valve opens. By the time the piston reaches bottom-dead center, the exhaust valve is completely open. As the piston moves up again, it pushes the burned gases out the exhaust valve. This is known as the **exhaust stroke**.

Once the exhaust stage is completed, the four stages of operation begin again. The movement of the camshaft closes the exhaust valve and opens the intake valve, and the piston moves down to begin a new intake stage.

The four stages of operation continue as long as the engine is operating. Keep in mind that these cycles are repeated at a very high rate of speed. An average motorcycle engine crankshaft rotates anywhere from 1,000 to 13,000 revolutions every minute. This means that these engine cycles are repeated thousands of times.

TWO-STROKE ENGINES

We will now describe the components found in the two-stroke engine and explain how the two-stroke engine operates as well as learn about the different types of induction systems (the way that the air-and-fuel mixture passes through the engine) found on a two-stroke engine.

You've already learned that an engine is classified according to the number of strokes its piston takes to complete one full engine cycle. You've also learned that, in order for any engine to operate, it must run through four stages of operation: intake, compression, power, and exhaust. The four-stroke engine accomplishes these four stages in four piston strokes—one stroke for each stage. Now we will learn how the two-stroke engine operates.

As you're aware, in a two-stroke engine the piston takes only two strokes to complete one full operational cycle. When the piston in a two-stroke engine moves in an upward direction, it completes the intake and compression stages. When the piston moves downward, it completes the power and the exhaust stages.

Two-stroke engines are much simpler in design than four-stroke engines. The basic two-stroke engine has only three moving parts: the piston, the connecting rod, and the crankshaft. Note that there's no camshaft used to operate valves for the flow of the air-and-fuel mixture or exhaust gases on the two-stroke engine.

TWO-STROKE ENGINE COMPONENTS

Before you learn how two-stroke engines operate, we'll discuss the component parts used on two-stroke engines. You will notice that many of the parts used on the two-stroke engine are similar, if not identical, to those used on the four-stroke engine.

As we've mentioned before, not all engines look exactly the same. The engines illustrated here are typical of many of the two-stroke engines you will see.

Two-Stroke Engine Cylinder Heads

The primary difference between the two-stroke engine cylinder head and the four-stroke engine cylinder head is that there are no ports (and therefore no valves) in the two-stroke cylinder head. This makes the two-stroke cylinder head much simpler to produce.

One main purpose of the two-stroke cylinder head is to create a combustion chamber by sealing the area between the cylinder and the cylinder head (Figure 6-28). A second purpose is to hold the spark plug. The **squish area** of the combustion chamber forces the air-and-fuel mixture into a tight pocket under the spark plug to increase the combustion efficiency. This squish area, or as it's also known, **squish band**, is more critical in the two-stroke engine as compared to the four-stroke engine.

The modern two-stroke cylinder head is constructed of aluminum alloy. Like four-stroke cylinder heads, two-stroke cylinder heads also aid in the transfer of heat from the engine by the use of fins on air-cooled engines or water jackets on liquid-cooled engines.

Two-Stroke Cylinders

The main difference between the two-stroke cylinder and the four-stroke cylinder is that the two-stroke cylinder has **ports** located in the cylinder wall (Figure 6-29). These ports serve the same

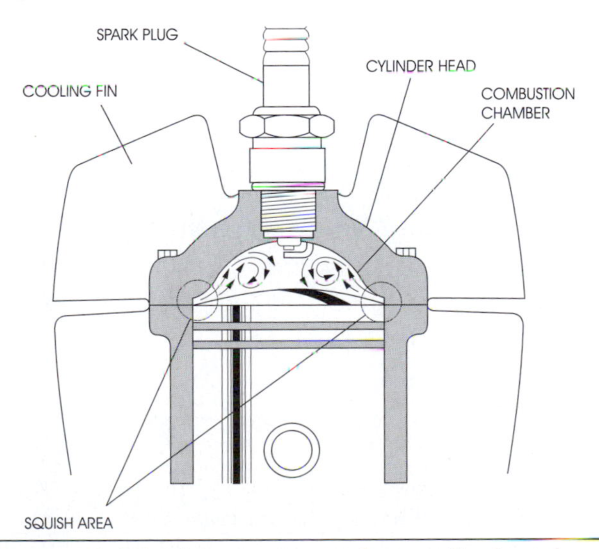

Figure 6-28 The two-stroke cylinder head attaches to the cylinder.

purpose as the ports in the cylinder head of the four-stroke engine.

Ports allow the air-and-fuel mixture to enter the cylinder and exhaust gases to leave the cylinder.

The ports in a two-stroke cylinder may be bridged. Bridged ports are used on very wide ports to prevent the piston ring from catching on the edge of the port and utilize a piece of metal between the ports that acts like a bridge. Both the upper and lower edges of the ports are chamfered. When ports are **chamfered**, the sharp edge of the port is removed to help keep the piston ring from catching as it moves up and down in the cylinder. The ports that may be found in the two-stroke cylinder are:

- The **exhaust port**, which is used to allow the exhaust gases to escape.
- **Transfer ports**, which are used to transfer the intake gases from the bottom of the cylinder to the combustion chamber through the two-stroke cylinder.
- The **intake port**, which is used to allow the gases to enter the engine.

Like the four-stroke cylinder, the two-stroke engine cylinder guides the piston as it travels up and down. The cylinder also aids in transferring engine heat and may be either air cooled or liquid cooled. Also, just as with the four-stroke engine cylinder, there are different types of materials used in the construction of a two-stroke cylinder. Each material has its own advantages and disadvantages. You may wish to review the four-stroke cylinder section of this chapter for a refresher on the different materials and their advantages and disadvantages.

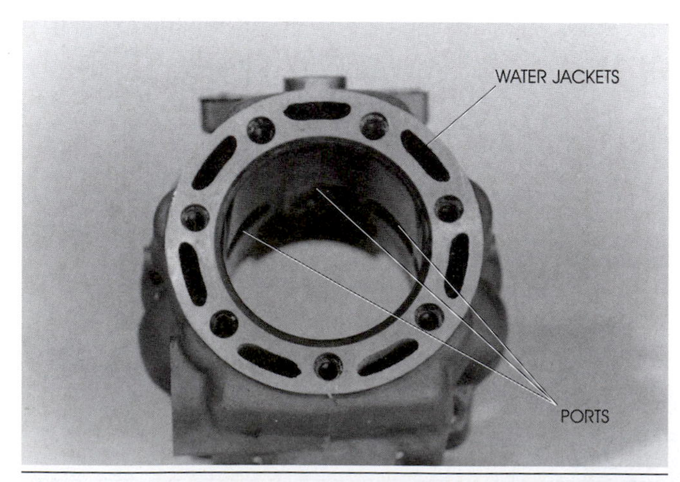

Figure 6-29 Ports are located in the cylinder wall of the two-stroke engine. Water jackets help cool the cylinder.

Two-Stroke Pistons

Just as with the four-stroke piston, the purpose of the two-stroke piston is to transfer the power produced in the combustion chamber to the connecting rod. The two-stroke piston has pins to prevent the piston rings from rotating around the piston (Figure 6-30).

The two-stroke piston is somewhat different in design from the four-stroke piston. Figure 6-31 illustrates the parts of the two-stroke piston.

The crown, which acts as the bottom of the combustion chamber, is the top of the piston. The crown is the hottest part of the piston, due to combustion chamber temperatures. The crown area expands more than the rest of the piston because it's hotter and has more mass. The piston crown on the two-stroke engine normally has a positive dome, but may have a flat top or even a negative dome or dish. The piston crown on a two-stroke engine also controls the duration of the exhaust and transfer ports. **Duration** is the time that the ports are open and is measured in crankshaft degrees.

The ring grooves have pins installed in them to prevent the piston rings from rotating. The two-stroke piston normally has no more than two piston ring grooves. Piston ring lands support the piston rings.

The wrist pin boss is where the piston attaches to the small end of the connecting rod. A hardened tool-steel wrist pin attaches the piston to the rod. The wrist pin is normally held in place with retaining clips to prevent the wrist pin from contacting the cylinder wall.

The piston skirt is the load-bearing surface of the piston. The piston skirt contacts the cylinder wall and is the primary wiping surface for the cylinder wall. The largest diameter of the piston is usually at or close to the bottom of the skirt, 90 degrees from the wrist pin. This is where the piston is normally measured. Two-stroke engines that use a reed valve induction system have a cutaway, or holes, machined on the intake side of the piston skirt. Two-stroke induction systems will be discussed later in this chapter.

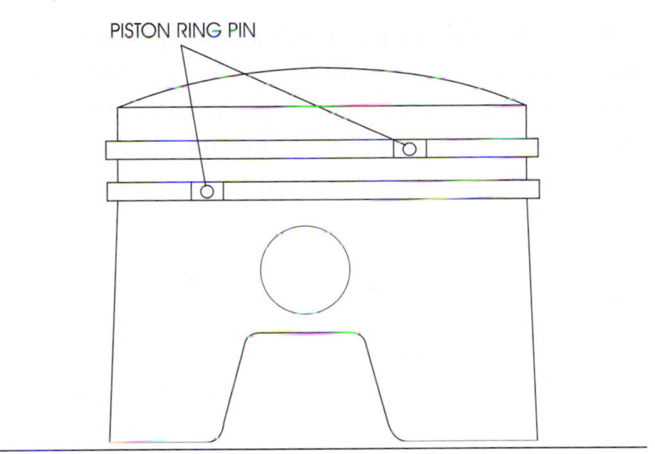

Figure 6-30 Pins installed in the two-stroke piston prevent the piston rings from rotating.

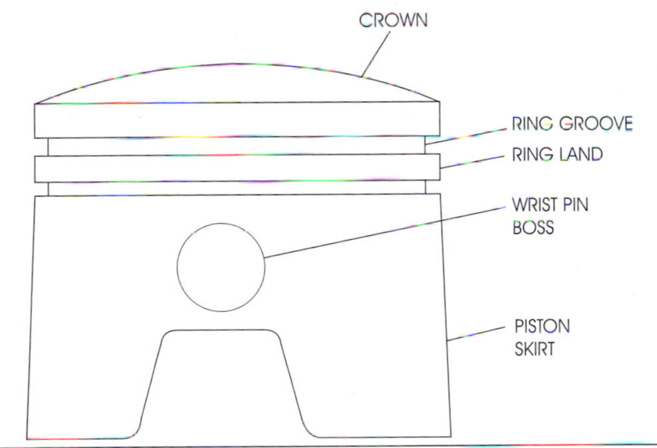

Figure 6-31 The parts of a typical two-stroke piston are shown here.

Two-Stroke Piston Rings

The purpose of the two-stroke piston ring is to aid in heat transfer from the piston to the cylinder wall and to seal in the combustion gases. There are three different types of piston rings used on the two-stroke piston (Figure 6-32).

The **standard piston ring** is rectangular in shape and is the most popular ring found in the two-stroke engine. Standard rings are usually made of cast iron and are chrome-plated.

The **keystone piston ring** is a wedged-shaped ring that seals better than the standard piston ring. However, the keystone ring is more expensive to manufacture and requires a special wedge-shaped piston groove.

The **Dykes piston ring** is an L-shaped ring that's only used as a top ring on the piston. This type of piston ring expands outward when the combustion gases force the piston downward. Although

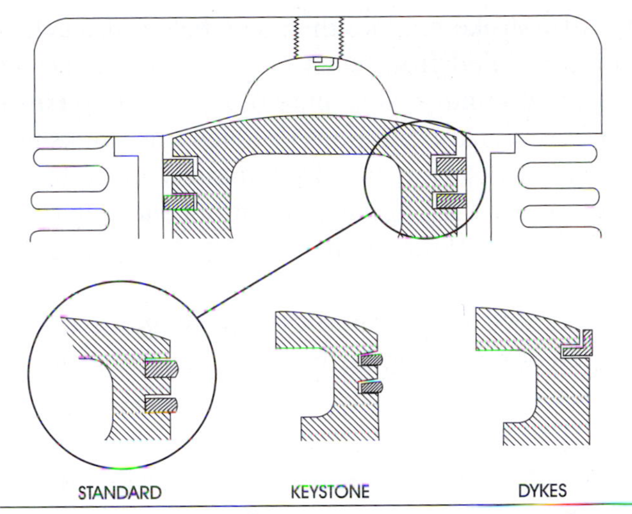

Figure 6-32 The three common types of two-stroke piston rings are illustrated here.

the Dykes ring is the most expensive piston ring to produce, it's also the best-sealing ring found on a two-stroke piston.

All piston rings must have an end gap to allow for heat expansion. As we discussed earlier with the four-stroke piston ring, the piston ring end gap is measured using a feeler gauge (blade) after fitting the ring squarely in the cylinder. In the two-stroke engine, the ring end gap fits around the piston pin that prevents the ring from rotating around the piston.

Two-Stroke Crankshafts

The two-stroke engine normally uses a multi-piece crankshaft similar to that used on the four-stroke engine. The crankshaft halves are cast or forged. The connecting-rod journal is a pin (crankpin) that's press-fit into the crankshaft halves. A one-piece connecting rod uses a roller bearing at the connecting-rod journal. The multi-piece crankshaft generally uses ball bearings on the main journals. Most multi-piece crankshafts found on a two-stroke engine can be rebuilt.

Two-Stroke Multi-Cylinder Crankshafts

As with four-stroke engine designs, two-stroke multi-cylinder crankshafts use different offset positions for each cylinder, depending on the design. However, a major difference between the four-stroke multi-cylinder crankshaft design and the two-stroke multi-cylinder crankshaft design is that in a two-stroke engine, the areas below the cylinders are sealed from each other. This is done by using a labyrinth seal (Figure 6-33). The **labyrinth seal** is stationary and doesn't touch the crankshaft journal, although it has an extremely close tolerance. When the engine is running, the labyrinth area fills with fluid that forms a seal to separate the cylinders.

Most two-stroke multi-cylinder engines utilize the 180- or 120-degree design depending on the number of cylinders.

Two-Stroke Engine Connecting Rods

The connecting rod is a lever that transfers power from the piston to the crankshaft. Connecting rods are usually made of forged steel or aluminum and use an I-beam construction. The connecting rods found on two-stroke engines are a one-piece design. This design uses a roller bearing at the big end and a needle bearing at the small end of the rod. The one-piece connecting rod normally has holes or slots on both the small and large ends for added lubrication.

Two-Stroke Engine Crankcases

The purpose of the two-stroke crankcase is the same as that of the four-stroke crankcase—to contain and support the major engine components. These components can include the crankshaft, cylinder, primary drive, and transmission. Unlike the four-stroke engine, which requires ventilation, the two-stroke crankcase must be sealed from the atmosphere and from the transmission area to allow pressure to build up inside the engine. The two-stroke engine crankcase is assembled as a vertically split or horizontally split crankcase.

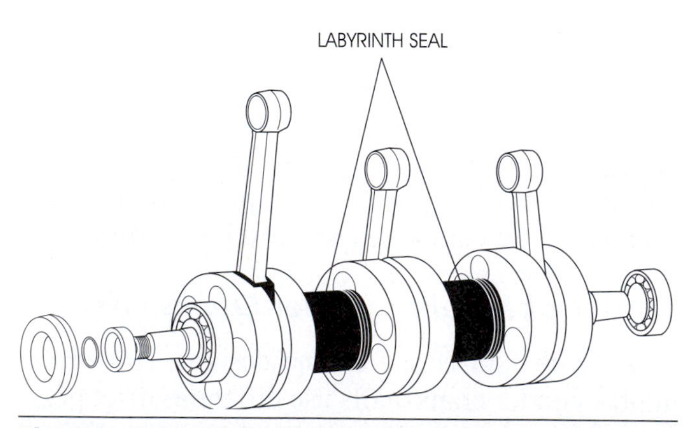

Figure 6-33 This illustration shows a three-cylinder two-stroke crankshaft with two labyrinth seals.

Exhaust Port Power Valves

The basic two-stroke engine has a limited **power band**. This means that the engine makes usable power within only a very small rpm range, especially when compared to the four-stroke engine. Motorcycle and ATV manufacturers of two-stroke engines have developed variations of components known as exhaust power valves to help increase the power band and improve engine performance. These two-stroke exhaust valve designs help to use the engine's speed to change the port timing and therefore change the power characteristics of the engine as it's running. There are many different designs of power ports used on motorcycles and ATVs. We will briefly discuss some of the most popular systems.

One design uses a cylindrical valve that's incorporated into the exhaust port. This valve matches the shape of the port and rotates to increase or reduce the exhaust port height, which changes the exhaust port timing. By changing the exhaust port timing, we can change where the engine makes the most power.

Another design uses two sliding guillotine valves activated by rocker arms. The rocker arms are controlled by centrifugal force, which allows the valves to open and close at a predetermined engine speed.

Still another power port system changes the actual volume of the exhaust system using a sub-chamber. By changing the volume of the exhaust system's expansion chamber, we can also change where the engine makes the most power and therefore change the effective power band.

Finally, there's a system that uses both a valve to alter the exhaust port height and a device to change the exhaust chamber volume. This system obtains optimum performance at all engine speeds and creates a very wide power band.

Two-Stroke Exhaust System Expansion Chambers

The two-stroke exhaust system is a tuned chamber that operates from sonic (sound) waves created by the engine. The shape of the chamber has a direct effect on the performance of the two-stroke engine's operational characteristics. The parts of the expansion chamber that create the shape and allow for differences in power delivery are shown in Figure 6-34. The shape of the expansion chamber also aids in the scavenging of residual exhaust gases and allows for an adjustment of the power band characteristics of the engine.

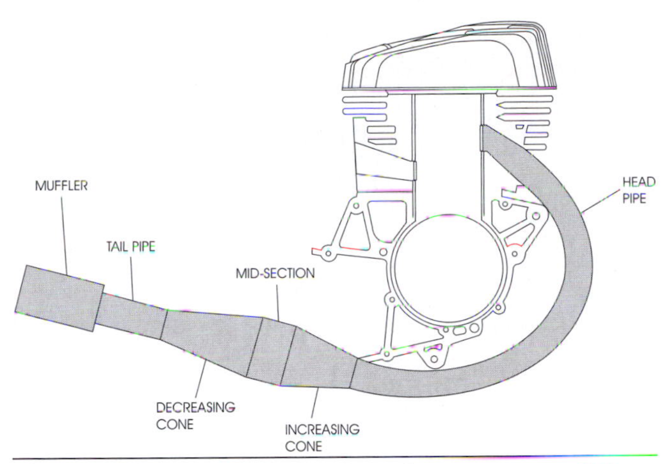

Figure 6-34 Expansion chambers are used in a two-stroke engine. The various parts are used to tune the performance of the engine design.

TWO-STROKE ENGINE THEORY OF OPERATION

Although a two-stroke engine has many of the same components as a four-stroke engine, its method of operation is very different. You'll remember that in a four-stroke engine, one power stroke occurs every two revolutions of the crankshaft. In a two-stroke engine, one power stroke occurs for each crankshaft revolution. Two-stroke engines are much simpler in design than the four-stroke engine. The basic two-stroke engine has only three moving parts: the piston, the connecting rod, and the crankshaft. However, the two-stroke engine is much more complex in its operation.

The moving parts in all engines must be lubricated with oil to prevent wear. Although we will discuss the lubrication systems used on two-stroke engines at a later time, it should be noted that the two-stroke engine mixes the oil used for lubrication of the engine components with the fuel supply.

The Four Stages of Engine Operation

Remember, the two-stroke engine must go through the same four stages of engine operation as any internal-combustion engine—intake, compression, power, and exhaust. However, where the four-stroke engine uses one piston stroke to accomplish each stage, the two-stroke engine accomplishes the four stages in just two piston strokes (Figure 6-35). Each time the piston moves upward, it completes the intake and compression stages. Each time the piston moves downward, it completes the power and exhaust stages. Because two stages of engine operation occur for each piston stroke, the operation of the two-stroke engine is more complex when compared to the four-stroke engine.

Two-Stroke Engine Areas

The two-stroke engine is split into two different areas (Figure 6-36):

- The **primary area** is the area below the piston crown including the crankcase. The crankcase in a two-stroke engine must be sealed to allow for the compression of the intake gases while they are in the primary area.
- The **secondary area** is the area above the piston crown including the combustion chamber where the air-and-fuel mixture is compressed to prepare for ignition.

Two-Stroke Engine Ports

Two-stroke engines don't use the same mechanical valves in the combustion chamber as the four-stroke engine. Instead, the two-stroke engine has holes in the cylinder walls that are called *ports* (Figure 6-37). These ports control the flow of the air-and-fuel mixture and exhaust gases. As the piston moves up and down in the cylinder, it covers and uncovers these ports, allowing the air-and-fuel

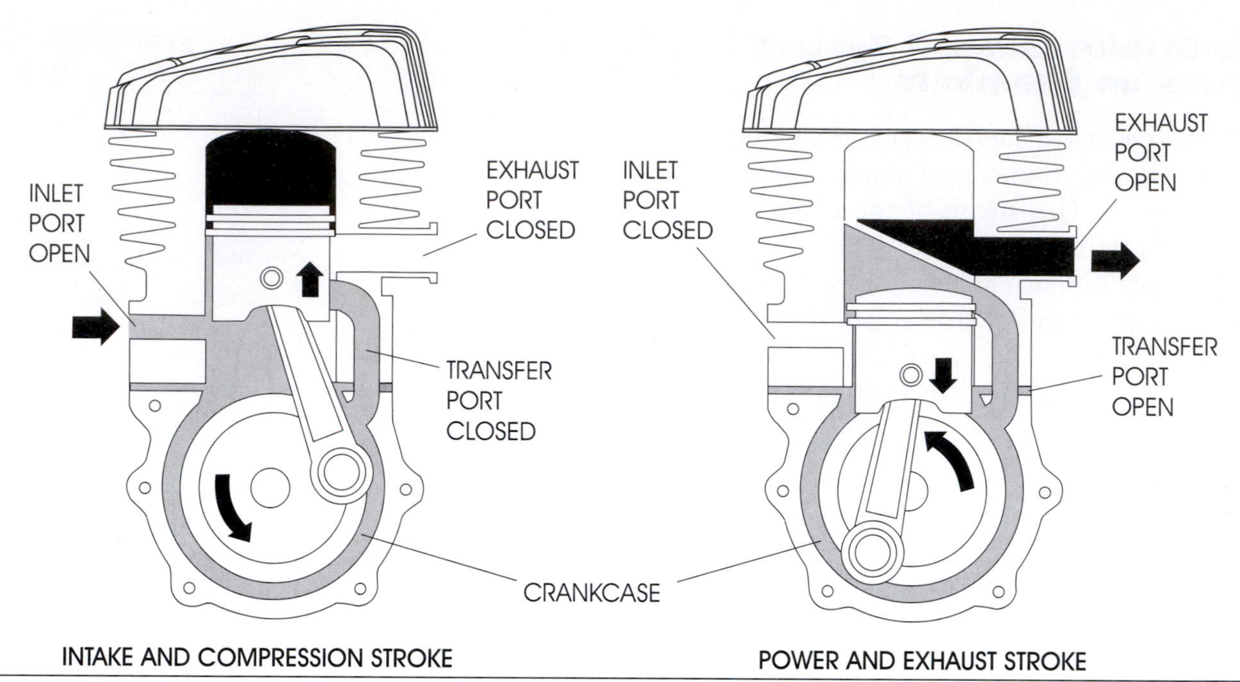

Figure 6-35 In a two-stroke engine, as the piston moves upward, it completes the intake and compression stages of operation. As the piston moves downward, it completes the power and exhaust stages of operation.

mixture to enter while also allowing the removal of the exhaust gases.

Two-stroke engines have different types of ports to allow for the flow of intake and exhaust gases.

The intake port is used to control the flow of fresh air and fuel into the primary area (crankcase area). Depending on the induction system used, the intake port either is the lowest port in the cylinder or is in the crankcase.

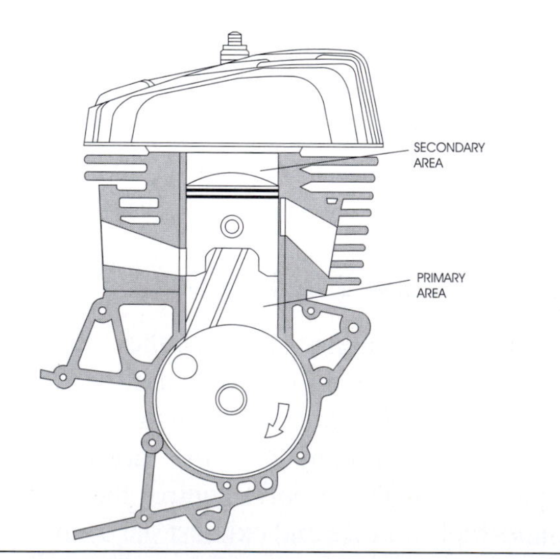

Figure 6-36 The area below the piston is called the primary area. The area above the piston is called the secondary area.

The **boost port** isn't used on all two-stroke cylinders. It's found primarily on two-stroke engines using reed valves. When boost ports are used, there may be one or more, which are located at the rear of the cylinder, opposite the exhaust port. The purpose of a boost port is to allow an extra amount of the air-and-fuel mixture to flow into the combustion chamber directly from the intake port area. This directly bypasses the crankcase and transfer ports to help fill the secondary area with additional fresh air and fuel to produce more power.

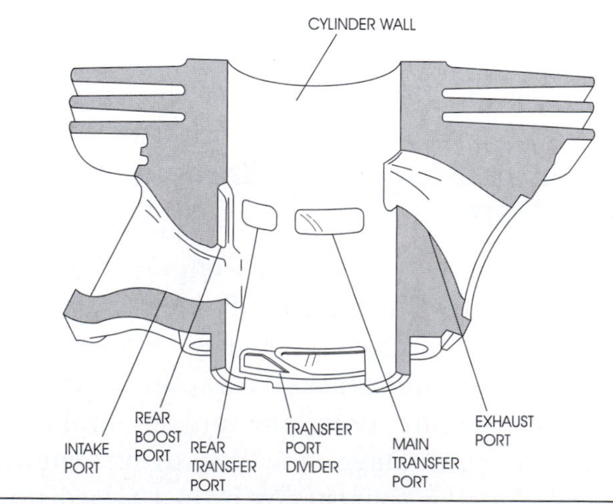

Figure 6-37 This cutaway view of a two-stroke cylinder shows the ports.

The **transfer ports** are used to control the transfer of the air-and-fuel mixture from the primary area to the secondary area. The transfer inlet is located at the bottom of the cylinder where it meets the crankcase assembly. The transfer outlet is located in the middle of the cylinder, attached to the crankcase through a transfer tube, which is cast into the cylinder. The transfer ports are controlled by the position of the piston crown. The number of transfer ports varies from engine to engine. When more than two transfer ports are used, the extra ports are called auxiliary ports.

The **exhaust port** controls the flow of the exhaust gases from the cylinder. The exhaust port is the highest port in the cylinder. The opening and closing of the exhaust port is controlled by the position of the piston crown.

Two-Stroke Engine Events

There are actual events that occur in each engine cycle of a two-stroke engine (Figure 6-38). To complete all five events, it takes only two strokes of the piston, which is one revolution of the crankshaft.

The **intake event** begins when the piston moves toward top-dead center (TDC). The primary area located below the piston increases in size, which causes the pressure to decrease. Because of the pressure difference, fresh air and fuel are pushed into the primary area through the intake port. The intake event has the longest port duration of all of the two-stroke events.

The **compression event** (also known as secondary compression) occurs as the secondary area decreases above the piston. The air-and-fuel mixture that was previously brought into the cylinder is compressed while the piston is still moving toward TDC. At a precise time, the ignition fires and creates a spark at the spark plug.

The **power event** begins after TDC when the expanding combustion gases caused by the ignition force the piston downward. The power event ends when the exhaust port is uncovered by the piston (opens).

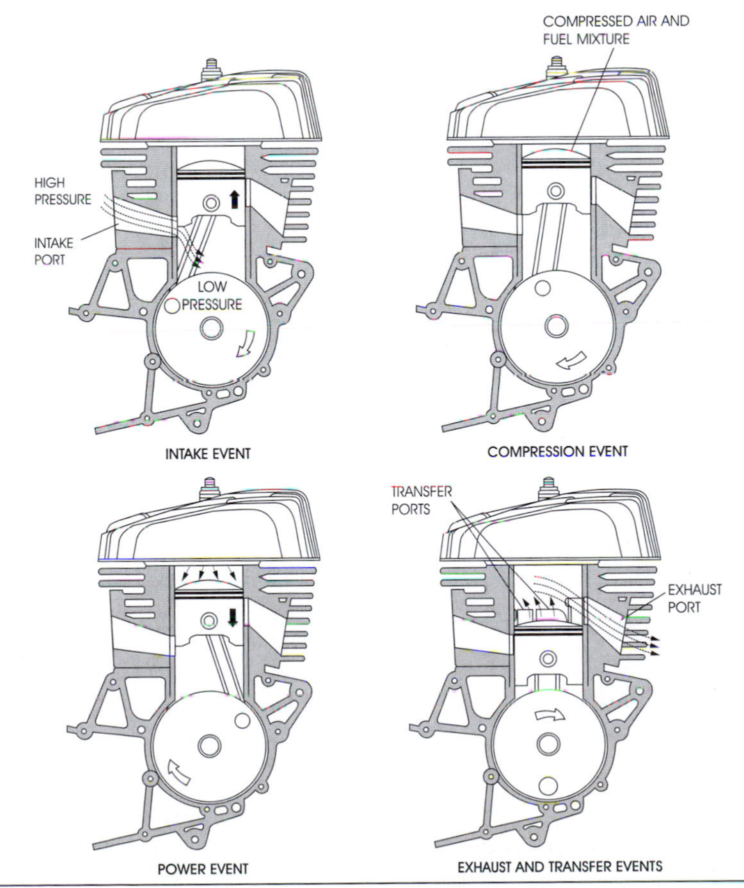

Figure 6-38 This illustration shows the five events that occur in one crankshaft revolution on a two-stroke engine.

The **exhaust event** begins when the piston crown uncovers (opens) the exhaust port while moving down toward bottom-dead center (BDC). Because of the high pressure in the cylinder, the exhaust gases are pushed into the exhaust system.

The **transfer event** also occurs, as the piston is moving toward BDC. While the piston is traveling downward, the primary area is decreasing, which increases the primary-area pressure. This is known as primary compression on a two-stroke engine. While this is occurring in the primary area, the secondary-area pressure is decreasing. Because of the pressure difference between the primary and secondary area, the fresh air-and-gas mixture located in the primary area is pushed through the transfer ports into the secondary area. The transfer event occurs during the exhaust event, which helps scavenge (similar to four-stroke valve overlap) residual exhaust gases by pushing the remaining exhaust gases out the exhaust port. The transfer event has the shortest port duration time. The transfer event uses what's known as loop scavenging, in which the transfer ports are angled away from the exhaust ports. The angle of the transfer ports directs the fresh air-and-fuel mixture up and away from the exhaust port to prevent the mixture from directly flowing out of the port.

Two-Stroke Expansion Chamber Theory

The two-stroke engine exhaust systems used on motorcycles and ATVs provide much more than just a simple escape path for exhaust gases. As the piston moves downward on the power stroke, the exhaust port opens (Figure 6-39A). A high-pressure wave of exhaust escapes from the combustion chamber and travels down the head pipe to the expansion chamber. The wave expands and slows as it enters the large portion of the expansion chamber. As the wave expands, a vacuum is created behind it, helping to draw out any remaining exhaust gases from the combustion chamber.

The wave flows through the center section of the expansion chamber, then compresses and accelerates as it reaches the tapered exhaust end. A portion of the high-pressure wave is reflected back toward the exhaust port as it contacts the tapered end of the expansion chamber (Figure 6-39B). The reflected pressure travels back up the head pipe and helps keep the fresh air-and-fuel mixture from escaping out the exhaust port (Figure 6-39C).

The remainder of the high-pressure wave travels out the tailpipe and is exhausted to the atmosphere. As the wave exits the tailpipe, it creates a low-pressure wave that moves back up the tailpipe behind the reflected high-pressure wave. The low-pressure wave is timed to arrive at the exhaust port to draw or scavenge the exhaust gases after the next power stroke. The expansion chamber must be properly tuned to permit the pressure waves to cause the desired effect.

Exhaust pressure waves have a direct relationship to engine speed; therefore, exhaust-pulse scavenging isn't always effective. The scavenge effect works only within a narrow range (or power band) of engine speed. This means that different exhaust-system designs are required for different motorcycle and ATV applications.

Be aware that a damaged exhaust system affects exhaust scavenging and results in a loss of power. Likewise, if the exhaust system is allowed to get excessively dirty, performance is also affected.

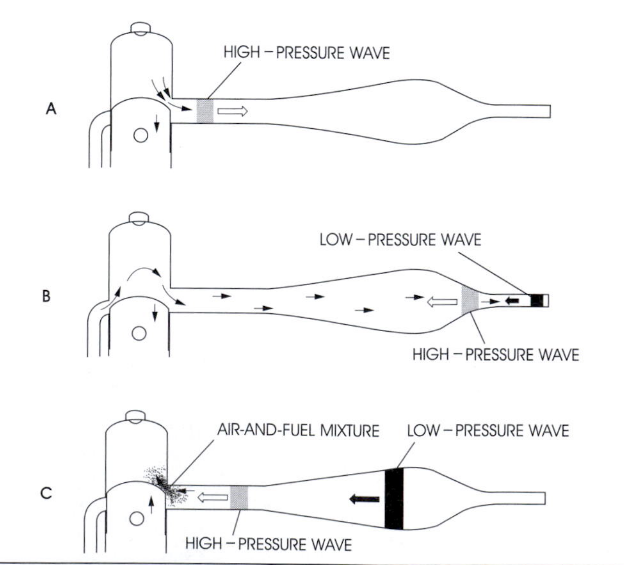

Figure 6-39 This illustration shows how a two-stroke engine expansion chamber functions.

TWO-STROKE ENGINE INDUCTION SYSTEMS

As we've discussed, the intake air-and-fuel mixture flows through ports inside the engine. Two-stroke engines use different methods of controlling the intake flow by what's known as induction. **Induction** is the method used to pass the air-and-fuel mixture through the intake port of the engine. Two-stroke motorcycle and ATV engines use three different types of induction systems.

Piston Port Induction

The piston port engine is the oldest and simplest type of two-stroke engine. This engine design contains all three-engine ports (intake, transfer, and exhaust) in the cylinder walls. As the piston moves up and down, it covers or uncovers the ports.

The piston skirt opens and closes the intake port. As with all two-stroke engines, the piston port engine has a sealed crankcase. As the piston moves upward, low pressure is created in the crankcase. The intake port is uncovered and the air-and-fuel mixture is drawn into the crankcase. As the piston continues to move up the cylinder, the exhaust and transfer ports are covered and the air-and-fuel mixture that's already in the combustion chamber is compressed.

When the piston approaches TDC, the spark plug fires and ignites the air-and-fuel mixture in the combustion chamber. The piston is forced downward by the expanding gases and the exhaust gases flow out the exhaust port. As the piston is moving downward, the air-and-fuel mixture in the crankcase is compressed. The transfer port is still closed at this time. As the piston continues downward, the piston uncovers the transfer port. When the transfer port is uncovered, the air-and-fuel mixture moves from the crankcase area, through the transfer port, and into the combustion chamber side of the piston. As the air-and-fuel mixture enters the combustion chamber, it helps to remove the remaining exhaust gases. When the piston starts to rise, the intake and compression stage begins again.

The piston port engine has the narrowest power band of all two-stroke engines. This engine design may be tuned to run at low speed or high speed, but not both. If the piston port engine is tuned to run at high speed but is operated at low speed, the carburetor tends to allow fuel to come back out the carburetor. This occurrence is called spit back.

Reed Valve Induction

To aid in keeping the crankcase sealed and prevent a loss of pressure as the piston moves downward, most modern two-stroke motorcycle and ATV engines use small one-way valves called *reed valves* (Figure 6-40). A reed valve opens during the intake-and-compression stage and then closes tightly during the power-and-exhaust stage to seal the crankcase area and prevent any of the fuel mixture from escaping back into the carburetor. This prevents spit back.

A reed valve is generally placed between the carburetor and the intake port of the engine. As the piston moves upward during the intake-and-compression stage of operation, the air-and-fuel mixture is pulled through the reed valve into the crankcase. When the piston reaches TDC, the reed valve closes to prevent the air-and-fuel mixture from flowing back through the carburetor. In this way, the air-and-fuel mixture is compressed more completely in the crankcase, which allows it to be pushed more forcefully into the combustion chamber as the

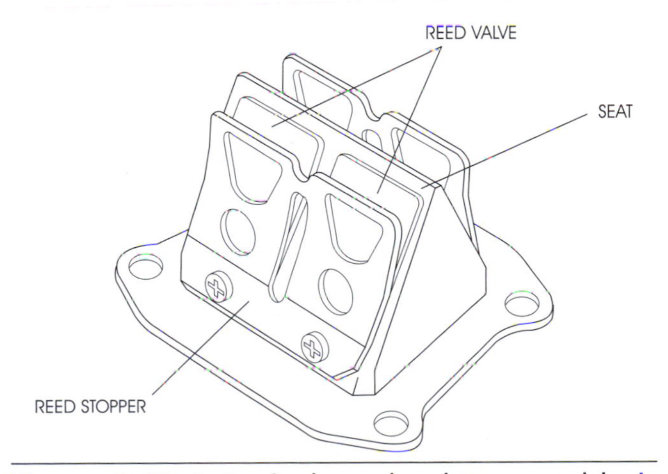

Figure 6-40 A typical reed valve assembly is shown here.

piston reaches BDC. Reed valves are made from one of two materials: stainless steel or fiber resin material (fiberglass or carbon fiber).

The piston of a reed valve engine has either a cutaway or holes on the intake side of the piston skirt to allow the flow of the intake mixture at all possible times (Figure 6-41).

Common Reed Valve Induction Systems

There are three different types of reed valve induction systems commonly used on motorcycle and ATV two-stroke engines.

The **cylinder reed valve** induction design (Figure 6-42) has the intake port in the same location as the piston port engine (located on the cylinder). With a cylinder reed valve engine, the intake port never closes. The purpose of the cylinder reed valve engine is to broaden the standard piston port engine power band. By using reed valves on a piston port engine, we can tune the engine to run at high speed while the valve prevents spit back through the carburetor at lower speeds.

The **crankcase reed valve** induction system (Figure 6-43) has the intake port located directly on the crankcase. This design can develop a very wide power band because it allows a shorter and more direct path to the crankcase and permits approximately one-third more transfer-port area. This lets more air and fuel enter the combustion chamber.

The **piston port/crankcase reed** induction system (Figure 6-44) takes the benefits of both the piston port induction system to control the lower-speed range of the engine and the crankcase reed induction system for high-speed operation.

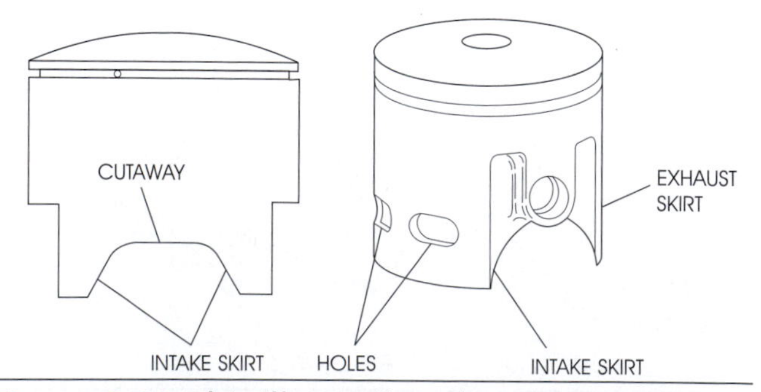

Figure 6-41 This illustration shows two types of reed valve pistons. The piston on the left uses a cutaway on the intake side of the piston skirt. The piston on the right has holes on the intake side of the piston skirt.

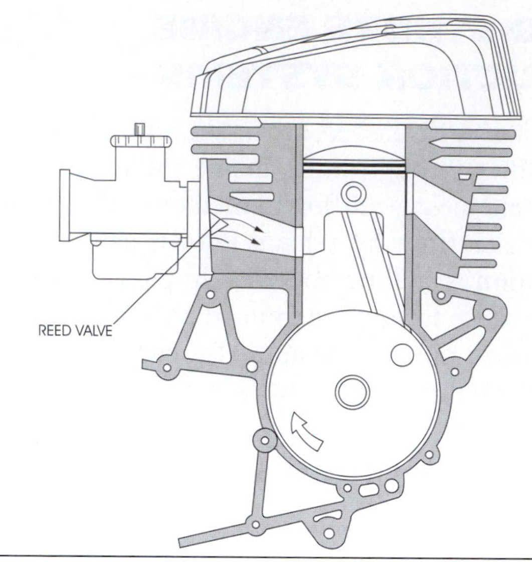

Figure 6-42 This illustration shows the location of the reed valve in a cylinder reed valve induction system on an air-cooled two-stroke engine.

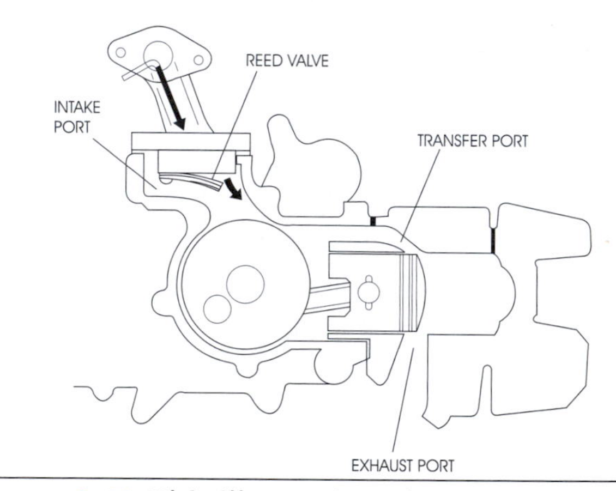

Figure 6-43 This illustration shows the location of the reed valve in a crankcase reed valve induction system.

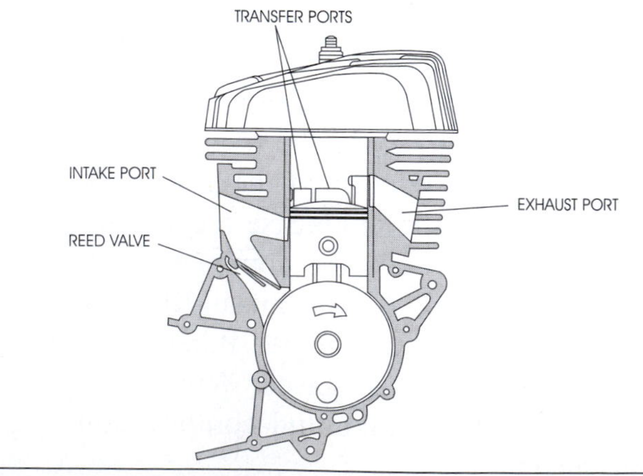

Figure 6-44 This illustration shows the location of the reed valve in a piston port/crankcase reed induction system.

Rotary Valve Induction

The rotary valve induction system has the intake port located on the crankcase of the engine (Figure 6-45). A rotary disk that covers and uncovers the intake port controls the opening and closing of the port. The disk is attached directly to the crankshaft. When the cutaway opening on the disk aligns with the intake port, fuel flows into the crankcase. In a rotary valve assembly, the rotary plate rotates between two fixed plates, or between a fixed plate and the crankcase. The two fixed plates, or fixed plate and crankcase, also contain openings. Fuel enters the crankcase only when all three holes line up during the rotation of the rotary plate. At all other times, the rotary valve blocks the passage from the carburetor to the crankcase.

The rotary valve engine design is seldom used on today's motorcycles due to its size. Rotary valve engines have the carburetors attached to the crankcase near the crankshaft and each cylinder has its own disk, which requires that the engine be wider than other designs. This engine design generally has the widest power band because it has the shortest and most direct intake path into the primary area of the crankcase.

COMPARING TWO-STROKE TO FOUR-STROKE ENGINES

There are many advantages and disadvantages when comparing two-stroke and four-stroke engines. Because each engine is specifically designed to produce good power over a broad range of engine speeds, you would think that manufacturers would tend to make only one type of engine. Motorcycle and ATV manufacturers are constantly working on building better, longer-lasting, and more powerful engines. Much progress has been made in a very short period of time. For example, in the early 1980s it was almost unheard of to see a stock motorcycle of any size (including the large 1,000-cc engines) with 100-horsepower engines. Today, many 600 cc engines develop over 100 horsepower.

So, which engine is better—the two-stroke engine or the four-stroke engine? Let us look at the two-stroke engine and compare it to the four-stroke engine design.

Two-Stroke Engine Advantages

The most noticeable advantage of the two-stroke engine over the four-stroke engine is that the two-stroke engine has fewer internal moving

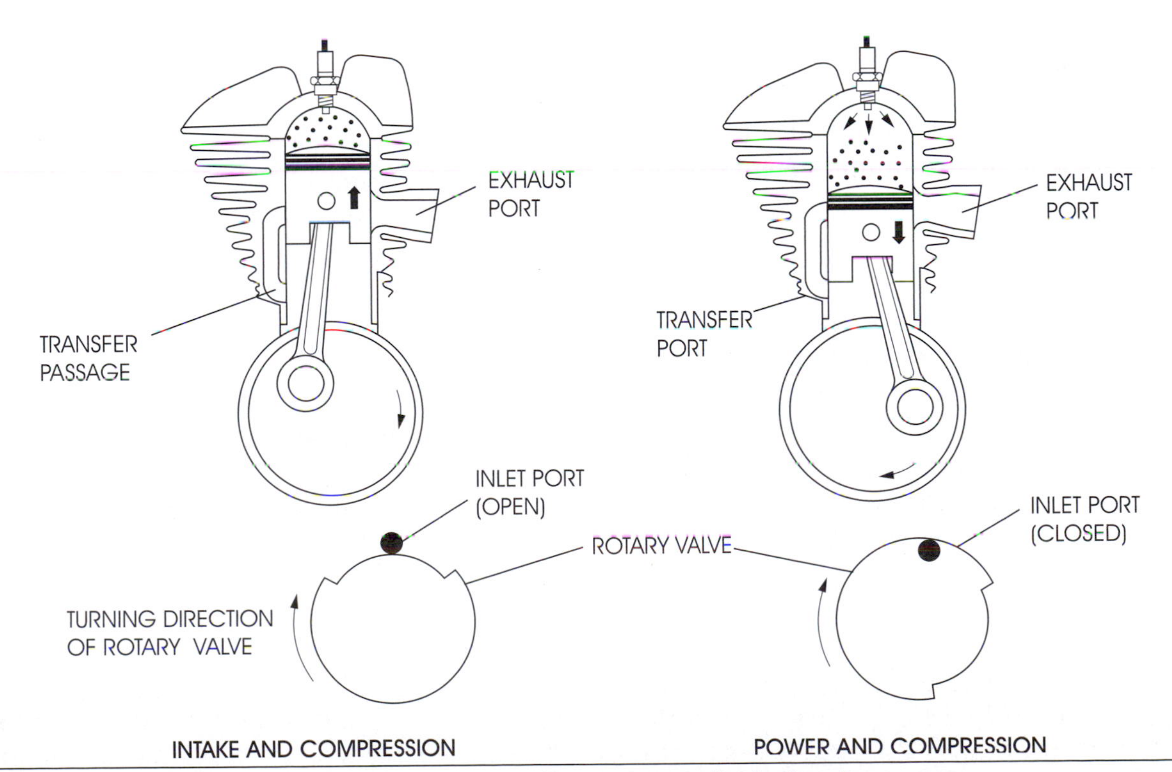

Figure 6-45 This illustration shows a rotary valve two-stroke induction system.

parts. This allows the two-stroke engine to be smaller in physical size and made lighter in weight than the four-stroke engine, as shown in Figure 6-46. The engine on the left is a liquid-cooled two-stroke engine and the engine on the right is a liquid-cooled four-stroke engine. Both engines have the same engine displacement (250cc). Note the physical size difference between the two. The four-stroke engine is taller than the two-stroke because of the valve train needed to open and close the ports in the cylinder head. In almost all cases, the two-stroke engine will be lighter in weight than the four-stroke engine when comparing equal displacement engines.

Another advantage of the two-stroke engine is that it normally produces more horsepower compared to a four-stroke engine of equal size. This is because the two-stroke engine has twice as many power strokes in the same given period of time as the four-stroke engine, resulting in better mechanical efficiency.

With these advantages, why aren't all motorcycle and ATV engines two-strokes? To answer that question, let us look at the disadvantages of the two-stroke engine.

Two-Stroke Engine Disadvantages

The primary disadvantage of the two-stroke engine is that it emits a high amount of hydrocarbon emissions (unburned fuel) from the exhaust, making it a very high air-polluting engine design. This occurs during the transfer and exhaust events of operation. While the intake gases are being transferred from the primary area to the secondary area of the engine, some of the raw, unburned fuel mixture escapes directly out the exhaust system and into the atmosphere. Even with a properly tuned expansion chamber, some raw fuel escapes from the exhaust port at certain engine speeds.

Other disadvantages of the two-stroke engine are directly related to one of its primary advantages.

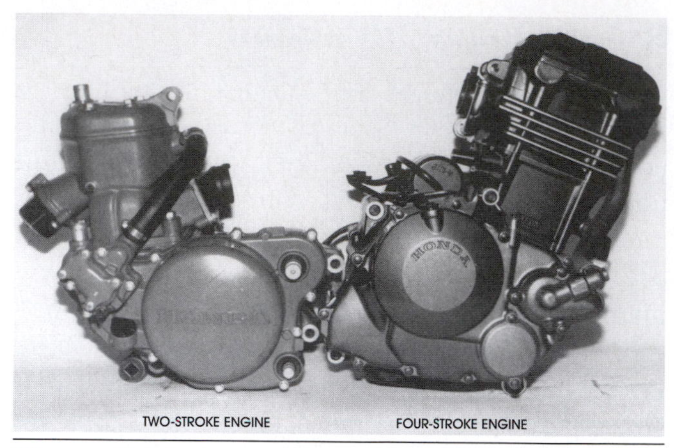

Figure 6-46 A comparison of like-sized two-stroke and four-stroke engines is shown here.

Because the two-stroke engine creates a power stroke every time the crankshaft makes one revolution, it burns more fuel than the four-stroke engine in the same time period. This generally results in poorer fuel economy. The two-stroke engine also runs at hotter temperatures compared to four-stroke engines for this same reason. Because the two-stroke engine runs hotter, it's prone to wear out internal parts sooner than the four-stroke engine. This tends to make the two-stroke engine less reliable and require more frequent service than the four-stroke engines used in today's motorcycles and ATVs.

Finally, as we mentioned earlier, the basic two-stroke engine has a narrow power band when compared to the four-stroke engine's generally wide power delivery over a range of speeds. For these reasons, most motorcycle and ATV manufacturers primarily build four-stroke engines.

Comparison Tables

Table 6-1 summarizes our discussion of the advantages and disadvantages of the two-stroke engine design.

Table 6-2 summarizes the advantages and disadvantages of the four-stroke engine design.

Table 6-1: TWO-STROKE ENGINE ADVANTAGES AND DISADVANTAGES

Two-Stroke Advantages	Two-Stroke Disadvantages
Fewer internal parts	High HC emission (air pollution)
Lighter weight	Poor fuel economy
More horsepower	Hotter engine temperatures
Better mechanical efficiency	Generally less reliable
	Requires frequent servicing
	Narrow power band

Table 6-2: FOUR-STROKE ENGINE ADVANTAGES AND DISADVANTAGES

Four-Stroke Advantages	Four-Stroke Disadvantages
Lower HC emissions (less air pollution)	More moving internal parts
Better fuel economy	More weight
Cooler engine temperatures	Less horsepower
Generally more reliable	Lower mechanical efficiency
Less vibration (smoother running)	
Wider power band	

Summary

- There are certain physical laws associated with motorcycle and ATV engines.
- All internal-combustion engines run using the same basic operational methods.
- A mixture of fuel and air is used to make an engine operate.
- There are many different components used in a four-stroke engine to allow it to operate properly.
- There are four piston strokes that must be in sequence to complete one full cycle in a four-stroke engine.
- There are many different components used in a two-stroke engine to allow it to operate properly. Many of them are very similar to the components found on four-stroke engines.
- While simple in concept, the theory behind the operation of the two-stroke engine is somewhat complex.
- There are different types of induction systems used on the two-stroke engine.
- A two-stroke engine has various physical differences when compared to a four-stroke engine.
- There are advantages and disadvantages with both a two-stroke and four-stroke engine used in the modern motorcycle and ATV.

Chapter 6 Review Questions

1. If you travel from sea level to 10,000 feet above sea level, what happens to the density of the air?
2. If a gas is compressed, what happens to its temperature?
3. What type of liquid has the ability to burn?
4. What are the stages of engine operation?
5. What type of combustion engine is found on all modern day motorcycles and ATVs?
6. The camshaft in a four-stroke engine rotates at ________ the speed of the crankshaft.
7. The one-piece connecting rod is used on multi-piece crankshafts. True /False
8. On a four-stroke engine, if the piston is rising toward TDC and the valves are closed, the engine is on the ________ stroke.
9. Hot gasses are released from the engine during the ________ stroke.
10. The cylinder used on a two-stroke engine contains ________ to allow for the flow of gasses through the engine.
11. The three basic moving parts in a two-stroke engine are the ________, ________, and ________.
12. The ________ opens and closes the ports in a two-stroke engine.
13. The ________ induction system is the oldest and simplest two-stroke induction design.
14. Two-stroke engines generally generate more hydrocarbon emissions than four-stroke engines. True / False
15. Two-stroke engines are heavier than four-stroke engines. True / False

Chapter

7 Lubrication and Cooling Systems

Learning Objectives

When students have completed the study of this chapter and its laboratory activities they should be able to:

- Define the four key purposes of lubrication
- Describe the types of oil and how oil is classified
- Explain why bearings, bushings, and seals are needed in an engine
- Identify the different types of bearings used in motorcycles and ATVs
- State the purpose of both two- and four-stroke engine lubrication systems
- Identify the different types of lubrication systems used in both two- and four-stroke motorcycle engines
- Describe how cooling systems work and why they are used
- Identify the components of motorcycle cooling systems
- Identify the various specialty lubricants used in lubrication system maintenance

Key Terms

Additives
Air-cooled
American Petroleum Institute (API)
Axial
Ball bearings
Bearings
Bleeding
Bushing
Centrifugal oil filter
Coolant
Dry lubricants
Dry-sump
Engine seizure
Fiber oil filter
Forced-draft cooling
Fraction distillation
Friction
Gear-type oil pump
Grease
Hydrometer
Hydronamic lubrication
Internal-oil cooling
Liquid cooling
Multi-viscosity oil
Needle bearings
Open-draft cooling
Oil filter bypass valve
Oil injection system
Oil pressure relief valve
Paper oil filter
Petroleum-based oil
Plain bearings
Plunger-oil type pump
Premixed
Press fit
Pressure testing
Push fit
Radial
Radiator
Radiator cap
Roller bearings
Rotor-type pump
Screen or wire-mesh oil filter
Seals

Society of Automotive Engineers (SAE)
Sump
Synthetic-based oil
Tapered roller bearings
Telltale hole
Thermostat
Twin-sump
Viscosity
Viscosity index
Water pump
Wet-sump

INTRODUCTION

Both two-stroke and four-stroke engines used in motorcycles have many moving internal parts that are machined to extremely close tolerances. To the naked eye, these parts have a smooth, fine finish to optimize wear inside the engine. However, if you were to look at these same parts under a microscope, you would see that these seemingly smooth parts are actually quite rough. To reduce the friction that occurs if two or more of these surfaces make contact, it's necessary to maintain a thin layer of lubrication between them (Figure 7-1).

A thin layer of lubricant between all internal engine parts effectively separates the parts from each other and provides a slight cushion for them to rest against. A lack of lubrication between these parts causes an immediate buildup of excessive heat, and in extreme cases causes the parts to melt together. When two or more parts inside an engine are hot enough to melt together in this manner, it's known as an **engine seizure**.

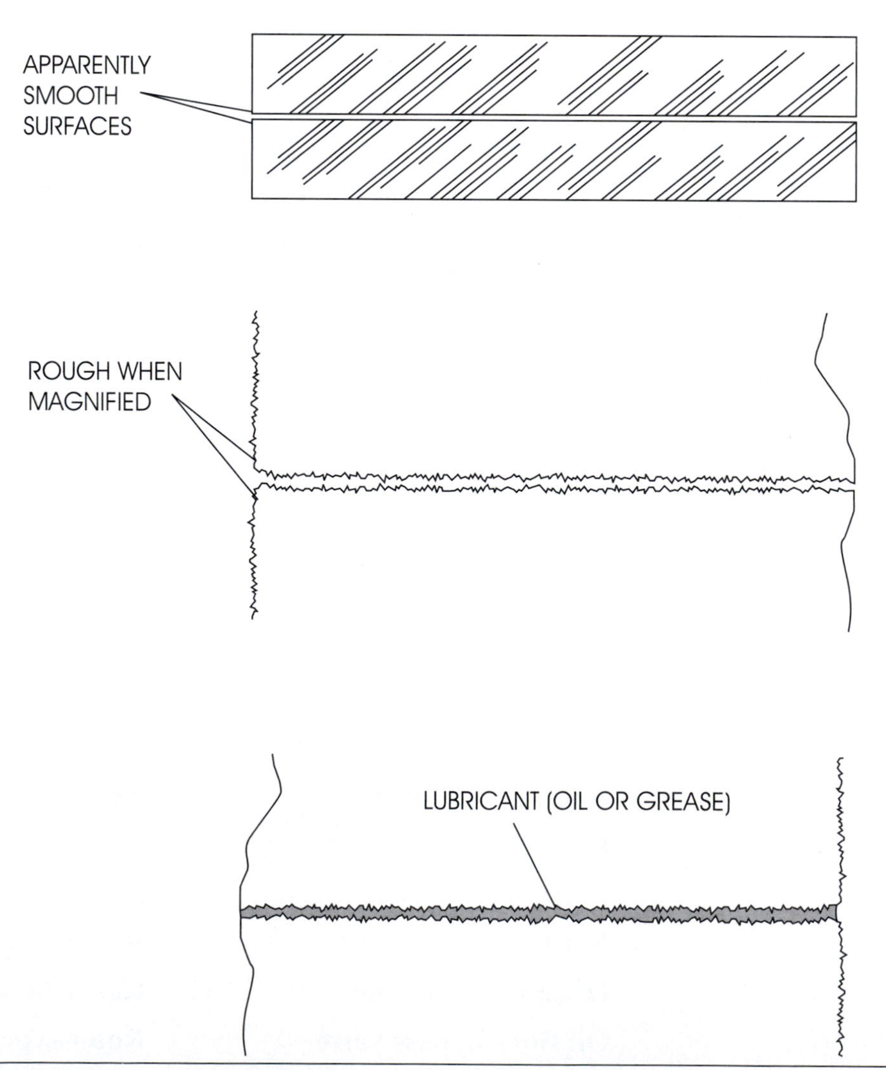

Figure 7-1 Engine parts that seem smooth may actually be quite rough when viewed under magnification. Lubricants provide a cushion to prevent the rough edges from making contact with each other.

In addition to its role as a lubricant, oil inside an engine performs many other duties. The oil film that coats each of the internal parts keeps air and moisture from the parts and prevents the buildup of corrosion. Also, when used in the four-stroke engine, oil is constantly recirculated, carrying away contamination and trapping it in an oil filter. It's important to note that oil filters must be replaced or cleaned on a regular basis. Oil is also used to aid in the creation of seals of both two- and four-stroke engine parts that require additional sealing. Finally, oil helps to disperse the heat generated in high-temperature areas such as the piston, cylinder, transmission, and combustion chamber.

Although motorcycle engines built today are very efficient, they still waste a considerable amount of energy. Ideally, a 100-percent efficient engine would convert all of the heat energy it produces into mechanical energy. Unfortunately, there's a considerable amount of heat energy produced in engines that's unable to be converted, thus creating unwanted engine heat. It's the job of the lubricants and the engine lubrication and cooling systems to remove the unwanted heat, thus preventing engine damage.

Now, let us learn in detail about the lubricants, lubrication systems, and cooling systems used on the engines found on motorcycles and ATVs.

The importance of an engine's lubrication system and the lubricants used in motorcycle and ATV engines cannot be overemphasized. If the proper lubricants are missing or the engine's lubrication system is operating improperly, the moving parts inside the engine will get hot enough to actually melt together. These internal engine components can score or even lock together in a matter of minutes. The buildup of heat, caused by friction, is among one an engine's worst enemies.

This chapter will give you an understanding of the types of lubricants and lubrication systems used in motorcycles and ATVs. It covers both the two-stroke and four-stroke engine designs. You'll learn how bearings, bushings, and seals help control and reduce friction. In addition, we'll discuss lubrication requirements for components of motorcycles and ATVs. You'll also learn about the specific cooling systems used on motorcycles and ATVs. Everything you will learn about motorcycle engine lubrication systems applies to ATVs as well.

From now on, we'll refer only to motorcycles in this chapter. Unless stated otherwise, you can assume that the information applies to both motorcycles and ATVs.

LUBRICANTS AND LUBRICATION

An engine has two main enemies: friction and heat. The main purpose of lubrication is to reduce friction. What is friction? **Friction** is the resistance to motion created when two surfaces move against each other, or when a moving surface moves against a stationary one. Friction can occur in many places in a motorcycle engine. Some of these places are on the following:

- Cylinder wall where the piston and piston rings rub wrist pin and mating surfaces of the piston and connecting rod
- Crankshaft pin, bearing, and connecting rod
- Teeth of any gear inside the engine
- Camshaft lobes, tappets, valve guides, and valve stems (on four-stroke engines)

If these surfaces are not protected by some form of lubrication, they will quickly heat up and wear.

The great amount of heat created inside the engine by friction places a strain on vital engine components. To prevent these parts from overheating, modern motorcycle engines require high-quality lubricants. Lubricants and engine lubrication systems used on modern motorcycles have been greatly improved over the years due to the research done by motorcycle and lubricant manufacturers. The useful life of an engine without proper lubrication might be measured in minutes; the service life of a properly lubricated engine can be hundreds of thousands of miles!

Lubrication serves four key purposes. When used properly, a lubricant will do the following:

- Cool
- Clean
- Seal
- Lubricate to reduce friction

Cooling

When oil in the engine flows through oil passageways, it helps cool the internal engine components by absorbing heat away from the metal parts. The circulating hot oil is then returned to the engine's crankcase where it is cooled.

Cleaning

While the oil is moving around and through the internal engine components, it's also cleaning the engine's parts. Combustion and normal wear and tear of engine parts produce tiny metallic particles and other contaminants. Today's motorcycle engine oils contain special additives that help hold these contaminants in suspension until they can be removed. The lubrication system removes contaminants as it passes oil through the engine's oil-filtering system. The oil filter system can't remove all the contaminants in the oil, so those remaining are removed by draining the engine oil. This is why oil turns darker in color after many hours of use.

Sealing

Another function of the engine oil is to help the piston rings seal in engine compression and combustion pressures. A thin oil film between the piston rings and cylinder wall is essential for the rings to seal properly. Oil between the piston ring groove and piston rings also aids in preventing combustion pressure leakage. Oil provides yet another sealing function for the engine seals located on sliding or rotating shafts.

Lubricating

It should now be clear that motorcycle engines consist of many internal components and parts that contact each other as they move. As a result of this contact, a certain amount of friction and heat is always present. The main purpose of oil in a lubrication system is to help reduce friction by keeping a thin layer of oil between all of the engine parts. This thin film of oil helps to prevent excessive metal-to-metal contact and unwanted friction.

Keep in mind that friction is the resistance to movement between two surfaces. A lubricant helps to reduce friction and the heat that friction produces. Lubricants thus reduce engine component wear.

Types of Engine Oil

In today's high-revving motorcycle engines, selecting suitable lubricating oil is very important. Motorcycle and oil manufacturers have worked hard to develop oils that meet and even exceed the high demands of motorcycle engines. There are three basic types of oils used in the modern motorcycle.

Petroleum-Based Oils

The first type of oil is a standard mineral-based oil, more widely known as **petroleum-based oil**. Petroleum-based oil starts out as crude oil, located in large underground pools all over the world. After this oil is removed from the ground, it's heated in a process known as **fractional distillation**. This process separates the needed lubricating oil from other elements within the crude oil. The oil is then blended with other additives to create the desired oil viscosity. **Viscosity** refers to the measure of a fluid's resistance to flow. It's determined by the rate of oil flow under certain controlled conditions. High-viscosity oil is thicker at room temperature than low-viscosity oil.

Standard petroleum-based oils perform poorly without additives. **Additives** are selected and used in the manufacturing of oils to improve the oil's operating qualities. Additives don't change the basic characteristics of oil, they just add new properties to it. Several additives are used in oil today; each is selected for a specific purpose. Table 7-1 lists the types of commonly used additives and the characteristics they impart to oil. One example of an engine oil additive is sulfur, which is used to improve the oil's extreme-pressure properties. Another example is zinc, which is added to oil to increase its shear strength. Some desirable characteristics that result from the proper use of oil additives are a higher film strength, resistance to foaming of the oil, resistance to oxidation of the oil, and the ability to keep oil contaminants in suspension. One important and interesting fact about standard mineral-based oils is that the base oils themselves don't wear out, but the special additives do!

Table 7-1: IMPROVING AN OIL'S OPERATING QUALITIES

Additive	Characteristics Imparted to Oil
Oxidation inhibitor	Increased life, less sludge
Corrosion inhibitor	Protection against chemical attack
Viscosity index improver	Improved viscosity–temperature characteristics
Pour-point depressant	Low-temperature fluidity
Oiliness agent	Increased load-carrying ability
Extreme-pressure additive	Lubrication under extreme pressures
Anti-foam agent	Resistance to foam
Detergent/dispersant	Ability to suspend contamination

Synthetic-Based Oils

Another type of oil is **synthetic-based oil**. Synthetic oils were developed during World War II, when petroleum products were not widely available. Today's synthetic oils operate more efficiently and over a larger range of temperatures than standard mineral-based oils. When synthetic oils are manufactured, a variety of different types of synthetic additives are added to help increase the oil's effectiveness. Advantages to using synthetic-based oils include the ability to handle higher engine temperatures before breaking down, and less viscosity change with change in temperature. The main disadvantage to using synthetic-based oils is that it may not be compatible with some petroleum-based oils and it can cause clutch slippage in many instances as it is too slippery. Synthetic-based oils are also more expensive than petroleum-based oils.

Petroleum/Synthetic Blended Oils

Blended oils are very popular today. They combine a petroleum base stock with synthetic additives, instead of petroleum additives. This combination greatly increases the quality of the oil. For this reason, blended oils are widely used in the engines of today's motorcycles.

Oil Classification

The internal-combustion engine has become more sophisticated and technologically advanced over the years, thus increasing the operating requirements of engine oil. As a result, two general automotive agencies were created to test, standardize, and classify lubricating oils (Figure 7-2). The first agency is the **American Petroleum Institute (API)**. The second agency is the **Society of Automotive Engineers (SAE)**. These agencies classify oil based on the following criteria:

- Manufacturer requirements regarding additives
- Intended use of the oil
- Viscosity ratings

Letter Classification Codes

Oil classifications use a double-letter code to indicate their intended use and manufacturer requirements. Oils used in motorcycle internal-combustion engines have a code that begins with

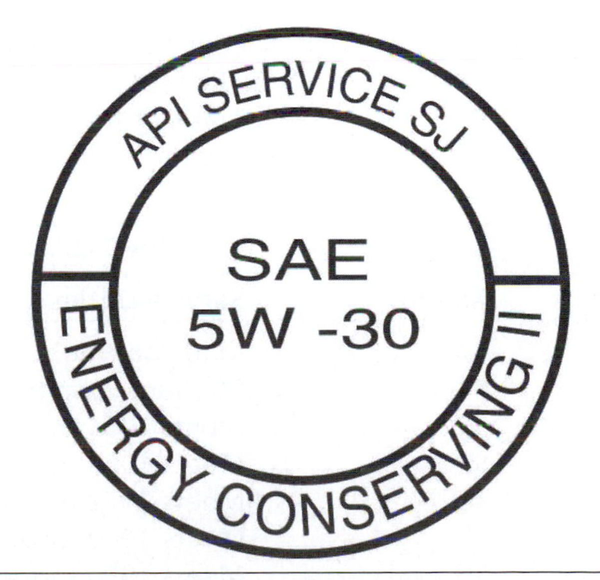

Figure 7-2 All oils have labels showing a classification determined by the American Petroleum Institute (API) and viscosity ratings by the Society of Automotive Engineers (SAE).

the letter S. The lowest-grade oil for use in the gasoline engine has a classification of SA. Rarely used today, this type of oil is a straight petroleum-based oil containing no additives. This oil classification isn't recommended for use in any motorcycle. Today's highest-classification of oil is rated SL. Oils rated SE, SF, SG, SH, or SJ are also recommended for use in motorcycle engines. Most manufacturers today recommend oil rated at "SG or higher" in their engines.

Table 7-2 shows the various classifications used by the API. The API also classifies diesel engine oil. The letter code for diesel oil is C. Although you wouldn't use oil designed for a diesel engine in a motorcycle, there are times when the same oil can be used for either engine. When this is the case, it's indicated on the API label. It's important to understand which oil to use in a particular motorcycle engine. The motorcycle service manual and owner's manual contain the manufacturer's recommendation for oil.

Engine Oil Weight Classifications

Let's now discuss oil weight and how the Society of Automotive Engineers measures it. The weight of oil is a reference to its thickness, from extremely thin oils of 5-weight, up to 90- or 140-weight oils. It's important to understand that a lower-number weight, such as 10-weight oil, is thinner than a higher-numbered oil, such as 40-weight oil. The oil weight-numbering system tells us that, all conditions being equal, 10-weight oil flows at a faster rate than 40-weight oil when poured through holes of the same size at the same temperature.

To determine the viscosity of oil, it is poured through an orifice at a predetermined temperature. Two temperatures are used to test oil viscosity. For winter usage, oils are tested at 0 degrees Fahrenheit and show the letter W after the number to indicate "winter." All other oils are tested at a temperature of 210 degrees Fahrenheit or higher. Keep in mind that the weight numbers indicate the oil's viscosity rating only. A higher-numbered oil, such as 50-weight oil, doesn't indicate a better lubricant than say, 10-weight oil. The weight of an oil has nothing to do with its quality; it's a designation used for comparison purposes only.

So, what does this really mean? Different temperatures have a direct effect on oil viscosity. Oil tends to become thicker at lower temperatures, therefore, a lower-weight number or a thinner oil should be used in cold weather conditions. Extremely hot temperatures require a higher weight number, or heavier, thicker oil.

Multi-Viscosity Oil

The majority of engine oils used in modern motorcycles have designation numbers such as 10W40 or 20W50. This type of oil is called **multi-viscosity oil**. These designations indicate oil viscosity that's suitable for use under many different climatic and driving conditions. For example, a 10W40-rated oil gives proper lubrication in both cold and warm temperatures. The 10-weight rating indicates the oil will flow when the temperature is cold (recall that the W stands for winter test). Protection is also provided as the temperature increases, as indicated by the 40-weight rating.

Table 7-2: API OIL CLASSIFICATIONS

Rating	Service Duties
SA	Mild conditions, no additives (base oil only)
SB	Medium conditions, uses anti-foaming and detergent additives
SC	Meets 1964 through 1967 automotive manufacturer requirements
SD	Meets 1968 through 1971 automotive manufacturer requirements
SE	Meets 1972 through 1979 automotive manufacturer requirements
SF	Meets 1980 through 1989 automotive manufacturer requirements
SG	Meets 1990 through 1993 automotive manufacturer requirements
SH	Meets 1994 through 1996 automotive manufacturer requirements
SJ	Meets 1997 through 2000 automotive manufacturer requirements
SL	Meets 2001 through current automotive manufacturer requirements

Multi-viscosity oils contain additives that allow the oil to thicken at higher temperatures to improve the viscosity index. The **viscosity index** is the number used to indicate the consistency of the oil with changes of temperature. An oil labeled 10W30 is a 10-weight oil at 0 degrees Fahrenheit, but has the viscosity of a 30-weight oil at 210 degrees Fahrenheit. It is important to note that oils such as these are specially designed; therefore, combining a straight 10-weight oil and 40-weight oil does not have the same effect as the factory-prepared 10W40 multi-viscosity oil.

The type or grade of lubricant best suited for proper maintenance of motorcycle engine components depends on many factors. Considerations include the type of motorcycle riding conditions; on- or off-road riding; hard-and-fast riding versus moderate-to-easy riding; and dry, wet, or extremely hot weather conditions. Motorcycle manufacturers recommend a certain type or grade oil or lubricant for each specific component need. It's best to follow the manufacturer's recommendations. If these recommendations aren't readily available, check with the manufacturers themselves or ask your local motorcycle dealership. They can usually give advice that will help determine product suitability for the job at hand.

Specialty Lubricants

Now that you have a basic understanding of engine oils, let's briefly discuss other types of lubricants used on motorcycles. The type of specialty lubricant used depends on the component to be lubricated.

Grease

Grease is a lubricant that's suspended in gel. It's often used on non-engine-related components such as wheel bearings, swing arm bearings, and for the lubrication of steering-head bearings. Grease is designed for long-term lubrication.

Dry Lubricants

Dry lubricants are used to lubricate without attracting contaminants. These lubricants use an evaporating solvent as a carrier. Dry lubricants are often used on cables and areas that are in the open atmosphere.

Drive Chain Lubricants

There are many different brands of drive chain lubricants. Just remember that there are two basic types of chain lubricants. Regular chain lubricants are used for standard drive chains; specialty chain lubricants are available specifically for O-ring drive chains.

Other Lubricants

You'll use many other types of specialty lubricants, such as silicone spray, penetrating oils, and multipurpose lubricants, on various parts of motorcycles. These products are widely available and are used for everything from helping to loosen rusted nuts and bolts to lubricating squeaky parts.

FRICTION-REDUCING DEVICES

The purpose of friction-reducing devices in an engine or anywhere on a motorcycle is, as the name indicates, to reduce friction. These devices, called **bearings**, are also used to reduce free-play between engine shafts, allow for proper spacing, and support different types of loads. Bearings can use either a rolling motion or a sliding motion to reduce friction. Examples of bearings that use a rolling motion are the following:

- Ball bearings
- Roller bearings
- Tapered roller bearings
- Needle bearings

Examples of bearings that use a sliding motion are the following:

- Plain bearings
- Bushings

Keep in mind that the purpose of any bearing is to help reduce the buildup of friction between moving parts that are carrying a load.

Ball Bearings

Ball bearings are the most popular bearing used in motorcycle engines because they provide the greatest amount of friction reduction and have the ability to handle both **axial** (side-to-side) and **radial** (rotating) loads.

Ball bearings consist of spherical balls contained in a cage and held in place by inner and outer races (Figure 7-3). The cage ensures the balls don't touch one another. Ball bearings require very little lubrication. They are used to support transmission shafts in both the two- and four-stroke engine cases, allowing the shafts to rotate.

Roller Bearings

Roller bearings are similar in design to ball bearings, except they use cylindrical-shaped rollers instead of spherical balls (Figure 7-3). The roller bearing is capable of withstanding higher radial loads than the ball bearing due to its greater available surface contact area.

Tapered Roller Bearings

Figure 7-3 also shows a variation of the roller bearing, known as a **tapered roller bearing**. The diameter of each roller of a tapered roller bearing is larger at one end than at the other. Tapered roller bearings are normally used in pairs with opposing angles, such as in steering mechanisms.

Needle Bearings

Needle bearings are yet another variation of the roller bearing (Figure 7-3). The needles or rollers of a needle bearing are usually several times longer than their diameters. Needle bearings normally have an attached cage, which keep the needles from making contact with one another. Needle bearings often have only an outer race. In this case, the needles

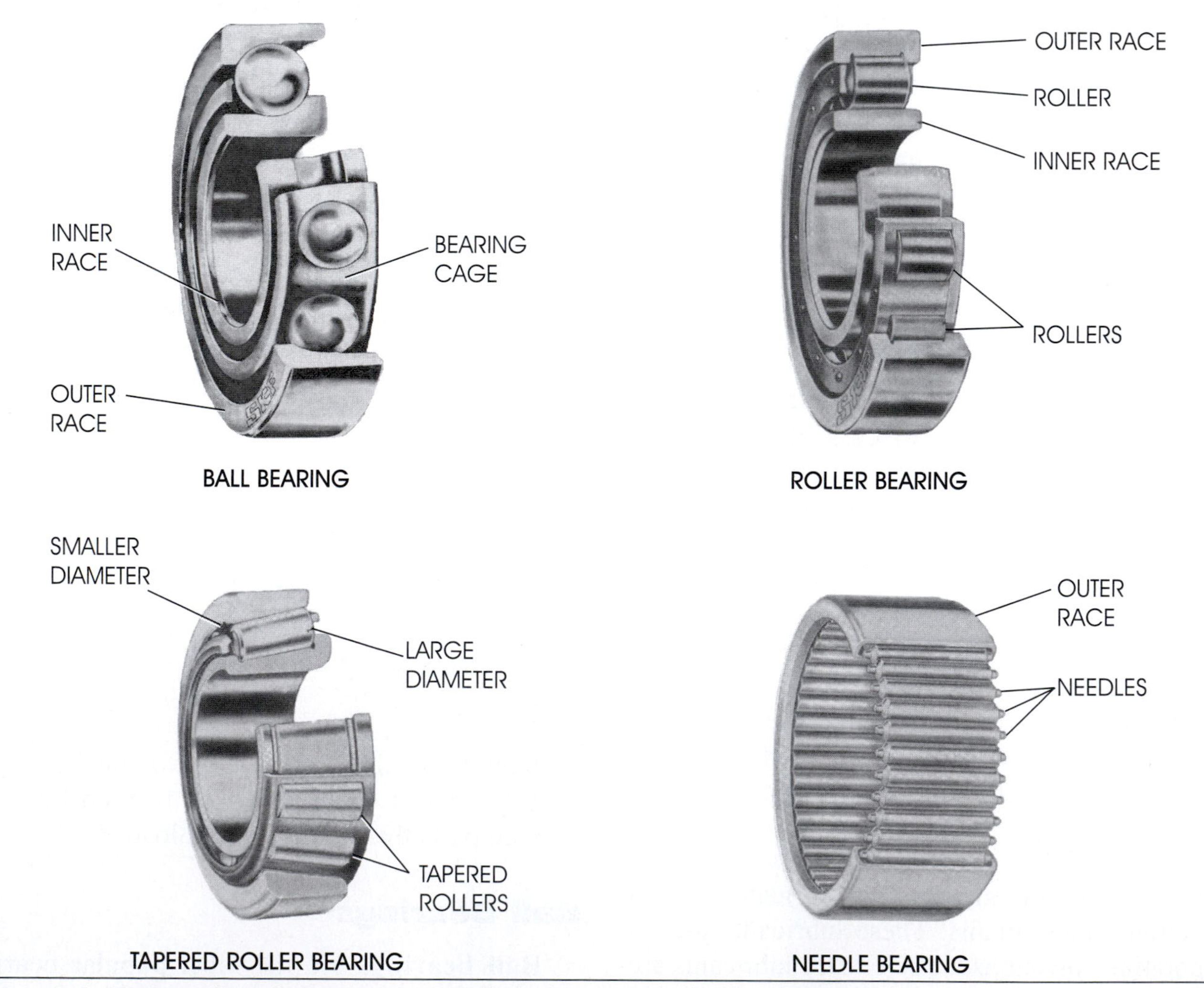

Figure 7-3 Various bearings commonly found in motorcycle engines.

make direct contact with the shaft surface they're supporting. Needle bearings can be found on some transmission shafts, swing arms, camshafts, and in two-stroke engine connecting rod small ends.

A key advantage of both the ball bearing and the roller bearing is that they can survive with minimal lubrication. However, if there is a total lack of lubrication, either bearing will be destroyed due to the extreme heat that occurs with friction.

Plain Bearings

Precision insert bearings, more widely known as **plain bearings**, are typically made in the shape of a cylindrical sleeve and are designed to withstand extremely heavy loads (Figure 7-4). They're used almost exclusively in the four-stroke engine. These bearings normally come in two separate pieces, but may also be of a one-piece design.

Plain bearings have a large surface area, which provides the ability to handle high radial loads. However, they don't reduce friction as efficiently as a ball bearing. Plain bearings can be found in two-piece connecting rod big ends, or they may be found as two-piece bearings supporting a crankshaft in the engine's crankcases.

Plain bearings require constant high oil pressure to produce a lubricating film known as **hydrodynamic lubrication**. This oil film, located between the rotating shaft and the bearing, is required at all times during engine operation to prevent unwanted metal-to-metal contact. The plain bearing receives the majority of its wear during the start phase of engine operation, due to the lack of lubrication at the time of engine start-up. Hydrodynamic lubrication can be further defined as a system of lubrication in which the shape and relative motion of the bearing surfaces causes the formation of a fluid film having sufficient pressure to separate the surfaces.

Bushings

Like plain bearings, the purpose of a **bushing** is to support large radial loads, and occasionally, axial loads. Bushings are cylindrical in design with a lining made of a soft alloy such as brass, aluminum, plastic, or silicone bronze (Figure 7-4). Most bushings are press-fit into place and are normally replaceable. A **press fit** is a force fit that's accomplished using a press. In contrast, a **push fit** is a force fit that is accomplished manually.

Seals

Although not used to reduce friction, **seals** are devices that are designed to prevent leakage used on transmission shafts and other rotating shafts

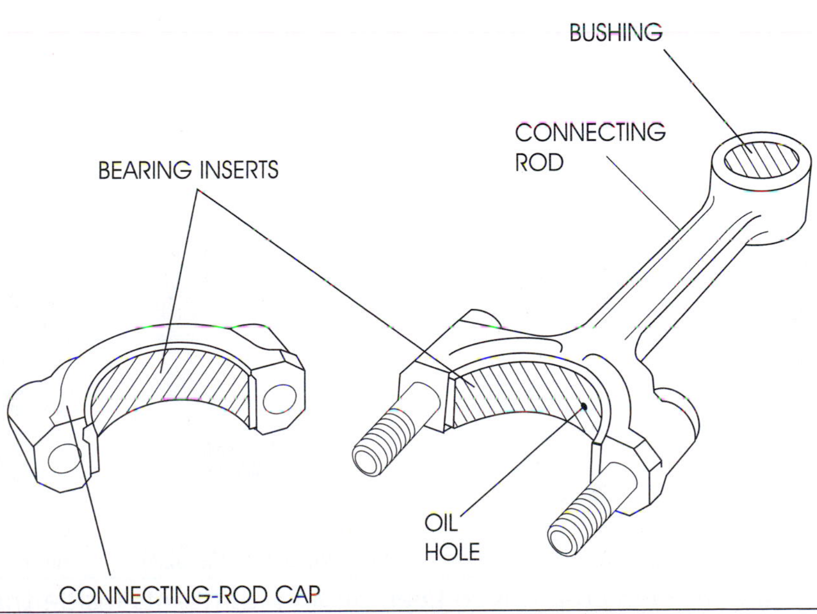

Figure 7-4 Plain bearings and bushings can support large radial loads.

within a motorcycle engine to perform the following functions:

- Prevent oil loss from the engine and bearings
- Keep contaminants from entering the engine and bearings
- Seal out atmospheric air when necessary

There are many types of seals, as shown in Figure 7-5. Seals are usually held in place by press or push fit. In most cases, the seal lip applies tension against the shaft by the use of a spring, which creates a predetermined amount of sealing pressure. The engine oil keeps the seal lubricated for durability.

TWO-STROKE ENGINE LUBRICATION

As mentioned in Chapter 6, the lubrication of a two-stroke engine is different from that of a four-stroke engine. In the two-stroke engine, lubrication is accomplished by mixing fuel with a recommended two-stroke oil, then introducing the mixture to the internal moving parts of the engine. When the oil-and-fuel mixture enters the engine, the oil lubricates the piston and other moving parts. The mixture also enters the combustion chamber, where it's ignited by a timed ignition spark. Because the oil does not burn as well as the fuel, much of it exits out of the exhaust system.

Another distinguishing fact about two-stroke lubrication systems is that all two-stroke motorcycle engines separate the transmission from the crankcase where the crankshaft, piston, and ring(s) are located.

The oil used to lubricate the transmission is different from the oil used to lubricate the piston, piston rings, cylinder, and crankshaft. Although some four-stroke engines do this as well it is not as common as on the two-stroke. A standard multi-grade lubricant or a special blended transmission gear oil is used to lubricate the transmission in a process known as an oil bath splash. As the gears rotate, they pick up the lubricant and splash the oil onto other moving components in the transmission.

Oils used to lubricate two-stroke engines are specially prepared and recommended by the manufacturer. They help reduce piston-to-cylinder wall scuffing and reduce excessive carbon buildup in the cylinder combustion chamber, exhaust ports, and exhaust systems. It's essential to combine the special two-stroke oil and fuel before the mixture enters the engine, to ensure that all of the internal engine components receive the correct amount of lubrication. The oil uses the fuel as a carrier to get into the engine, then separates itself from the fuel. Although it does eventually burn, the oil is not designed to burn with the fuel in the combustion chamber, but to lubricate the moving parts of the engine.

There are two methods used to provide two-stroke engine lubrication.

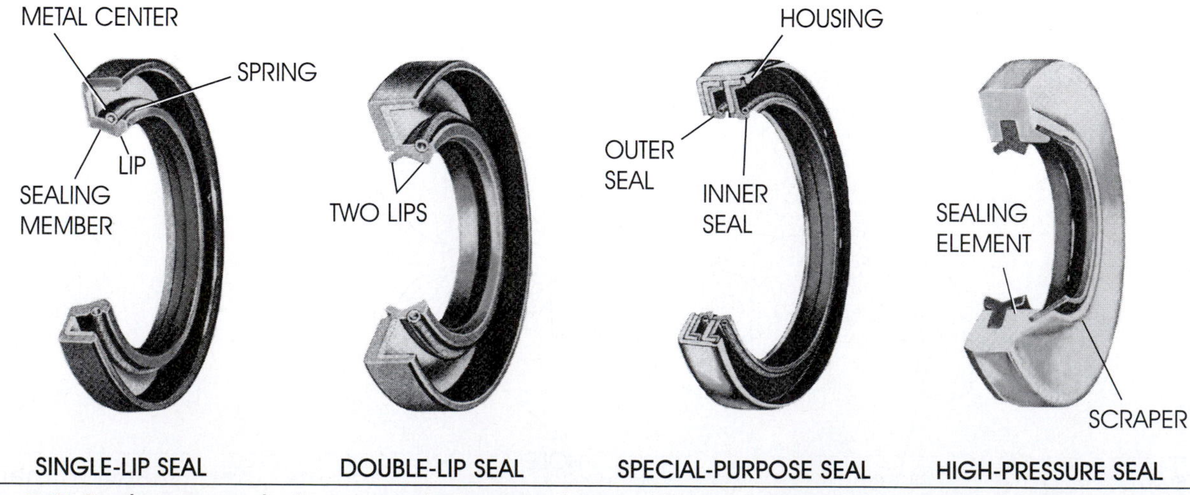

Figure 7-5 Seals are made in many designs to keep oil and lubricants inside the component.

Premixed Fuel and Oil

The **premixed** method of lubrication in a two-stroke engine requires the use of a specified ratio of gasoline and oil mixed together in a recommended container. The container can be the fuel tank of the motorcycle, but it's usually recommended that this procedure be completed in a separate fuel container to ensure proper mixing. Figure 7-6 shows that the fuel and oil have already been combined prior to entering the engine. It is important always to shake the mixture well before using it, to ensure that the oil and gas are completely mixed.

Use the following steps to determine how much oil to mix with the fuel to obtain the proper fuel-to-oil ratio:

1. Divide the number 128 (the number of ounces in a gallon of gas) by the manufacturer's recommended fuel-to-oil ratio.
2. Multiply the result of Step 1 by the number of gallons of gasoline to be used. The result is the number of *ounces* of oil you need to add to the gas.

As an example, let us determine how much oil must be added to 5 gallons of fuel when the manufacturer recommends a 40:1 fuel-to-oil ratio.

Step 1: Divide 128 by 40. The result is 3.2, which is the number of ounces of oil that must be added to each gallon of fuel.

Step 2: Multiply 3.2 by 5. The result is 16. Therefore, 16 ounces of oil must be added to 5 gallons of fuel.

The recommended ratio of fuel to oil may vary from 16:1 (16 parts of fuel to every one part of oil) to 50:1 (50 parts of fuel to one part of oil). Variations are based upon the manufacturer's recommendations and the brand of oil used. Many different brands of oil are available, each with different lubrication qualities. Be sure to investigate each product carefully and choose one based on the manufacturer's recommendations. More importantly, choose one that best protects the two-stroke engine's moving parts.

One disadvantage of premixing the fuel and oil is that there's no way to adjust the amount of oil entering the engine with the fuel as the engine

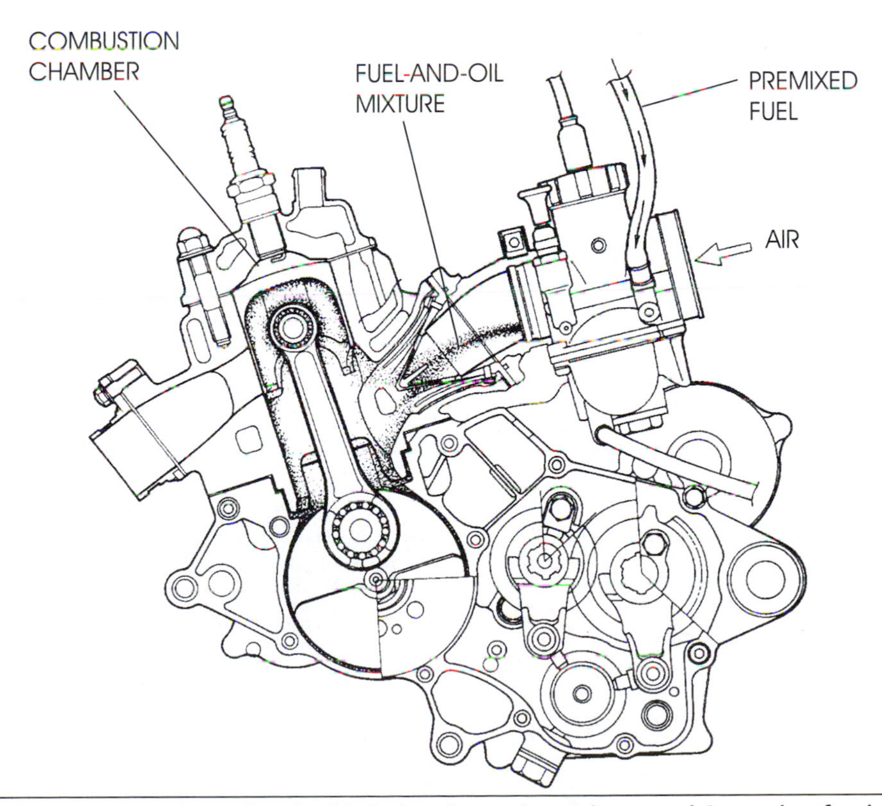

Figure 7-6 With the premixed method of lubrication, the rider combines the fuel and oil manually, prior to the mixture entering the engine.

operates. At slow engine speeds, the engine doesn't work as hard; thus the proportion of oil may be greater than required to lubricate the engine components. The result may be excessive oil in the engine. As the engine speed is increased, excess oil exits through the combustion chamber, causing exhaust smoke as the mixture burns. Excessive oil can also cause the spark plug to foul, or misfire. At higher engine speeds, on the other hand, the proportion of oil may not be adequate enough to supply sufficient lubrication needed to reduce friction. These are reasons why it is so important to use the manufacturer's recommended oils and oil ratios.

Oil Injection

The method of injecting oil into the engine instead of premixing it with the fuel, requires the use of a pump. An oil pump measures and feeds the oil from a separate storage tank to all of the two-stroke engine's components that require lubrication. With an **oil injection system** (Figure 7-7), the mixing of oil and gas occurs automatically. Increasing or decreasing the speed of the engine regulates the amount of oil pumped to the engine's components.

Oil injection systems offer several advantages. Not only does the motorcycle rider not have to mix the oil with the gas, but perhaps more importantly, oil is supplied to the engine's internal components in the correct amount required to provide the best protection at different engine speeds. Most two-stroke engines using oil injection have internal oil passages to help feed more oil to those internal engine components that require extra lubrication. Examples of these areas are the bearing of the connecting rod big end and the crankshaft main bearings. A smaller quantity of oil is fed to those engine components requiring less oil. Examples include the piston wrist pin and rings.

Oil injection systems can reduce the total oil consumption of the engine because the oil is used only as needed, based on the engine's operating speed. As the engine speed increases, more oil is injected into the engine; at slower speeds, less oil is injected. This helps reduce oil consumption, spark plug fouling, and excessive smoke, and can help to increase the engine's life. More importantly, oil injection pump systems supply the proper amount of oil to the moving parts, even when the carburetor slide is suddenly closed down and the engine speed remains high. This prevents

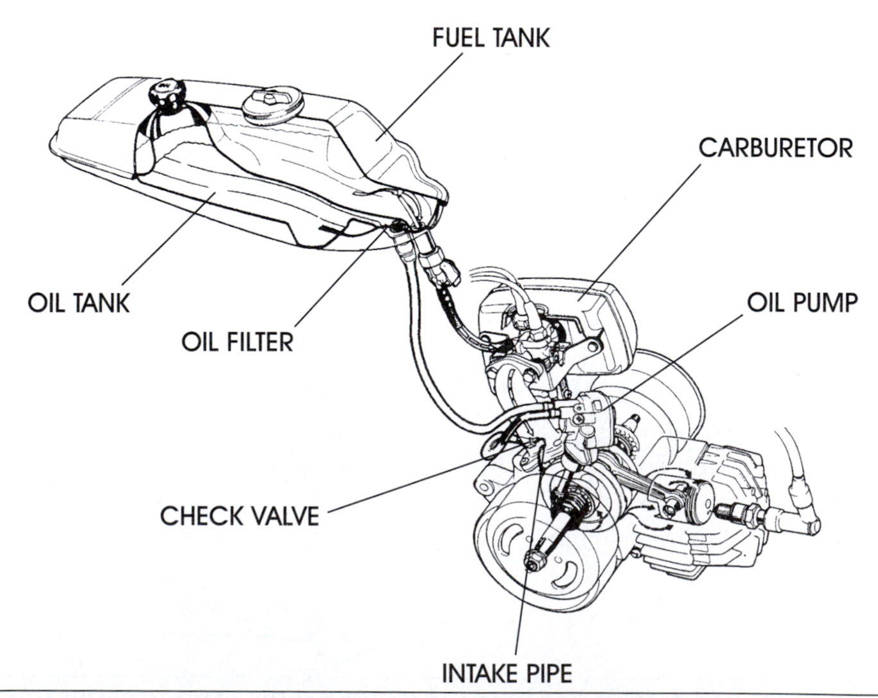

Figure 7-7 With the oil injection method of lubrication, the mixing of oil and gas occurs automatically, assisted by an oil pump that feeds the oil from a storage tank.

engine seizure that would otherwise be caused by lack of lubrication. Remember, with the premixed method, oil must enter the two-stroke engine with the fuel. When the carburetor slide is closed on deceleration, the oil supply to the engine's internal components is drastically reduced. An oil injection system, on the other hand, does not have this effect.

A cable connecting the throttle housing to the oil pump controls oil pump output (Figure 7-8). Turning the throttle twist grip causes the carburetor throttle valve and oil pump to open. This allows the oil pump to automatically increase its oil output in proportion to the air/fuel supply from the carburetor. Most two-stroke engines synchronize the oil pump to the carburetor throttle valve. That is, the pump is set to supply a quantity of oil to the engine in direct relation to the quantity of air/fuel mixture supplied to the engine. As a rule, the pump lever to which the cable is connected is adjusted to move to a predetermined point when the throttle is wide open and to return to its original position when the throttle is closed.

Different methods are used to indicate the oil pump lever position in relation to the carburetor throttle valve opening. Usually a punch mark on the throttle valve is aligned with a mark on the carburetor body when the pump lever is in the proper position. This is a common procedure, but you will need to check the motorcycle or ATV service manual for further instructions on the type of pump design and recommended adjustment.

In oil injection systems, the oil storage tank must not be allowed to run dry. If this happens, or if the oil pump is removed, air enters the oil lines. Air in the system won't allow the oil to flow properly to the components in need of lubrication. The air must be removed to ensure that proper lubrication takes place. The method used to remove air bubbles from oil lines is called **bleeding**. A common way to bleed air from the system is to allow gravity to do the job. This is done by removing the oil line located at the oil pump. After this line is removed, the oil flows out the hose, bringing the air bubbles along with it. You should allow the oil to drip into an oil pan that you've placed under the pump. How does this really work? The oil storage tank is located physically higher than the oil pump. When the oil line is removed from the pump, gravity lets the oil flow freely from the oil storage tank to the oil pump. Manufacturers suggest various methods for bleeding a system. You should follow their instructions very carefully.

Two-Stroke Transmission Lubrication

As discussed in Chapter 6, two-stroke crankcases must be sealed. Therefore, another means of lubrication must be provided for the transmission and

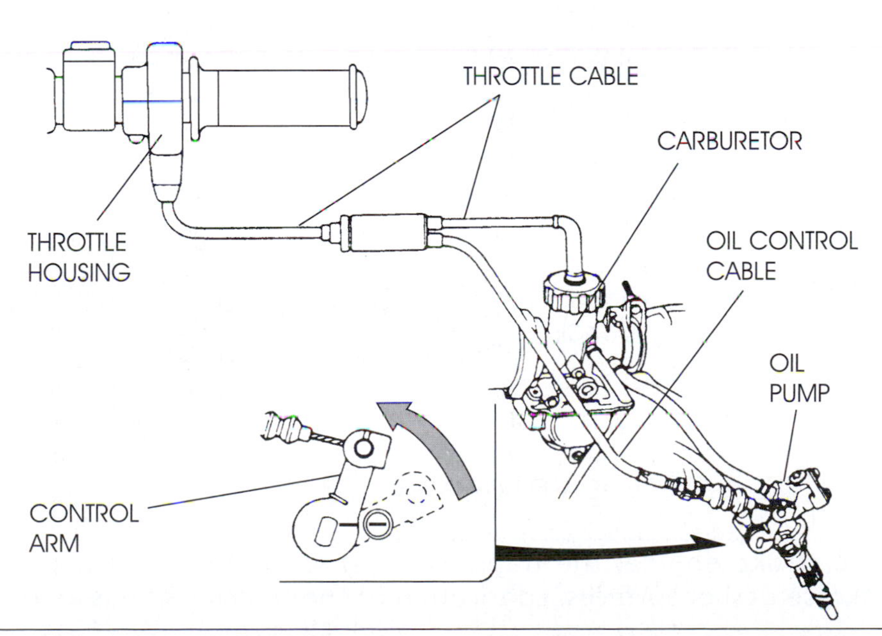

Figure 7-8 The throttle cable controls the oil-injected two-stroke engine oil pump.

other primary components, such as the clutch and starter. An oil bath splash method is used to provide lubrication for the separately sealed transmission and components (Figure 7-9). As the gears rotate, they pick up the lubricant and splash the oil onto other moving components in the transmission.

Special motorcycle oils including multi-grade oils and special gear oils are recommended for transmission lubrication purposes. You should use the oils recommended by the model-specific manufacturer. The amount of oil to be used is specified in the model-specific service manual.

FOUR-STROKE ENGINE LUBRICATION

Unlike the two-stroke engine, the lubrication system for four-stroke motorcycle and ATV engines requires the engine oil and gas to be kept separate from each other. Therefore, the four-stroke engine doesn't mix the oil with the gas, nor does the oil intentionally enter the combustion chamber. Consequently, oil is not burned but is recirculated throughout the engine.

The earliest four-stroke engines for motorcycles used what was known as a total-loss lubrication system. A total-loss lubrication system worked by using an oil tank full of oil and oil lines that allowed the oil to drip from the lines onto the bearings and then be splashed onto the piston and cylinder walls by the rotating movement of the bearings. The oil was then allowed to leak out of the crankcase. In some engine designs, the oil leaking out of the crankcase was routed so that it dripped onto the chain that powered the rear wheel.

Modern four-stroke engine lubrication systems are much more sophisticated than the total-loss systems of old. They consist of the following major components:

- Engine sump (The word **sump** refers to the lowest portion of the crankcase cavity.)
- External oil storage tank, used by some four-stroke engines

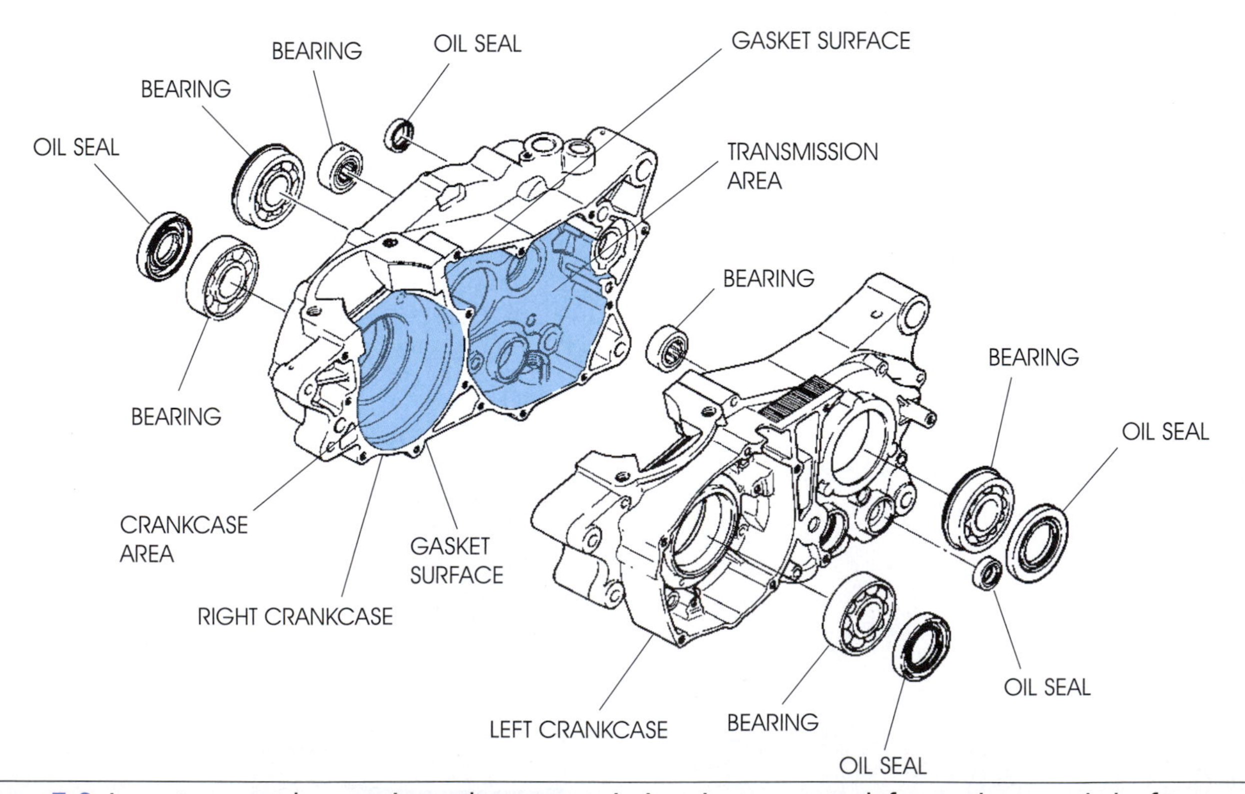

Figure 7-9 In a two-stroke engine, the transmission is separated from the crankshaft area by means of the crankcase gasket surfaces. Lubrication of the transmission is accomplished using an oil bath splash.

- Oil pump, used to pressurize or force the oil through oil lines and passageways to the engine components that require lubrication
- Pressure relief valve, used to control excessive oil pressure

We will discuss these components in more detail later.

The two types of four-stroke engine lubrication systems used on motorcycles are the wet-sump and the dry-sump. A **dry-sump** engine stores its oil supply in a separate oil storage tank, while the **wet-sump** system stores its oil in the bottom of the engine's crankcase.

Dry-Sump Lubrication

The components of a typical dry-sump lubrication system are the oil storage tank, oil strainer, oil feed line, oil pumps, engine oil passageways and, an oil return line (Figure 7-10). In this system, two oil pumps are typically used. One pump acts as an oil pressure feed; the other is an oil return pump. Oil in the oil storage tank is gravity-fed to the pressure-feed side of the oil pump. The oil strainer keeps relatively large pieces of dirt and debris from entering the feed line. The pump then forces oil, under pressure, through oil passages in the engine. This lubricates the moving internal engine components, which would otherwise be damaged by the heat created by friction. Oil that's thrown off the pressure-fed parts lubricates other internal engine components using the splash lubrication method. Finally, the excess oil collects in the sump and is returned to the oil storage tank by the oil return pump.

One advantage of the dry-sump lubrication system in four-stroke engines is that the oil has a better place to cool by being stored in a separate storage tank away from the hot engine. A one-way check valve prevents the oil in the storage tank from leaking back into the sump. What would happen if the anti-leak valve were to malfunction? Engine crankcases

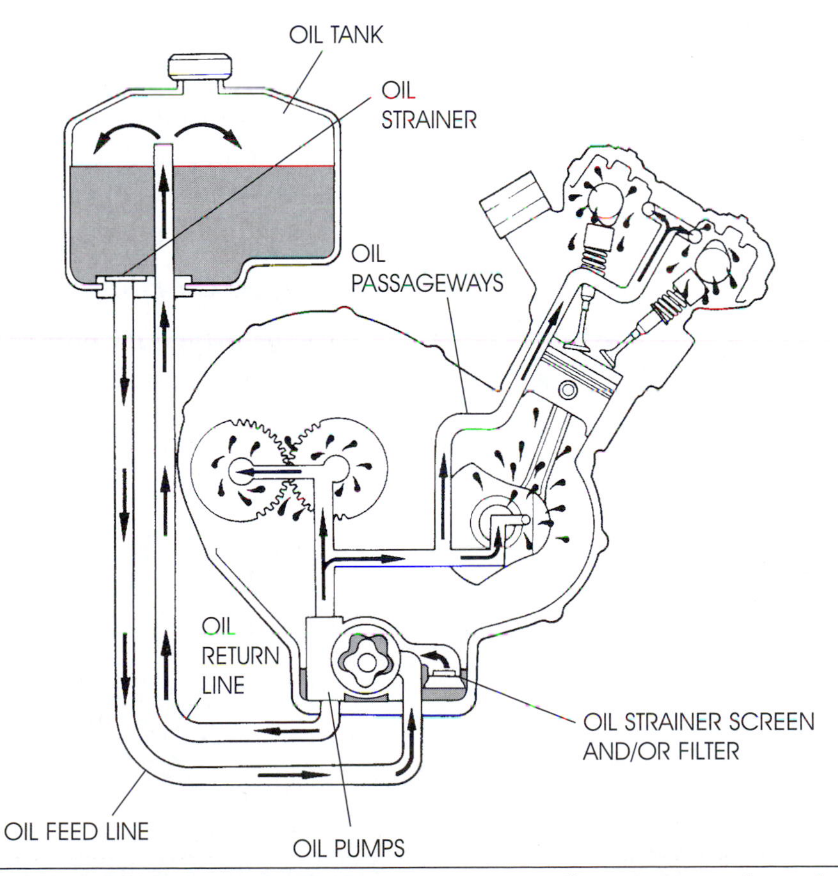

Figure 7-10 A dry-sump lubrication system uses a separate oil storage tank and typically uses two oil pumps.

aren't designed to contain large amounts of oil. The oil in the storage tank would fill the engine sump, causing the engine to smoke excessively. Such smoking could possibly foul the spark plugs.

Wet-Sump Lubrication

Wet-sump lubrication systems differ from dry-sump lubrication systems in two major ways. First, in wet-sump lubrication systems, oil is stored in the engine crankcase, not in a separate oil tank (Figure 7-11). Second, only one oil pump is used in wet-sump four-stroke engines. Similarly to dry-sump lubrication systems, oil is pressure-fed to all areas in need of lubrication. Examples of areas needing lubrication are cam bearing areas, pistons and cylinders, and main crankshaft bearing areas. The oil that's pressure-fed to these areas is thrown off the rotating components and drained back to the sump, where the oil pump picks it up once again and recirculates it to those high-friction areas that need lubrication.

One advantage of a four-stroke wet-sump lubrication system is that it has fewer components and no external oil lines carrying the major supply of engine oil. This results in the wet-sump system being less prone to oil leaks. Another advantage of wet-sump systems is a lower center of gravity, because the engine oil is contained in the bottom of the engine. This helps to improve the handling of the motorcycle.

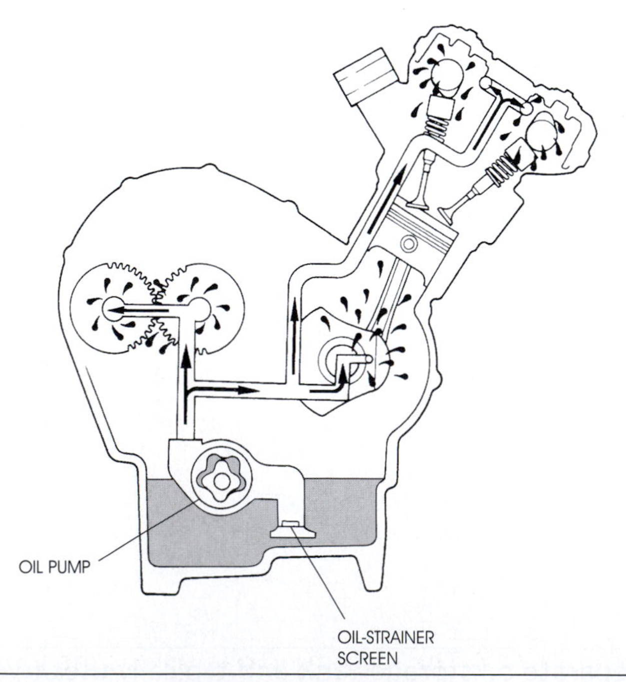

Figure 7-11 A wet-sump lubrication system stores oil in the engine crankcase.

Twin-Sump Lubrication Systems

Some four-stroke engines are now equipped with **twin-sump** lubrication systems (Figure 7-12), which are designed to separate the crankshaft piston and valve train from the clutch and transmission. This design is very simlar to the two-stroke lubrication system and is utilized in some high performance single-cylinder four-stroke off-road motorcycles and ensures a cool supply of oil to the clutch for longevity and also prevents clutch and transmission contaminants from entering the engine oil areas. This system also reduces the amount of circulating oil and required size of the oil pump.

Because there are separate oil supplies with the twin-sump design, the crankcase oil level and transmission oil level are checked independently.

Oil Pumps

Four-stroke motorcycle engines use one of three basic types of oil pumps. They are the gear type, plunger type, and the most commonly used type—the trochoid or rotor type. It is important to understand the operation of each type.

Gear-Type Oil Pumps

The **gear-type oil pump** consists of a housing and two spur gears: a drive gear attached to the oil

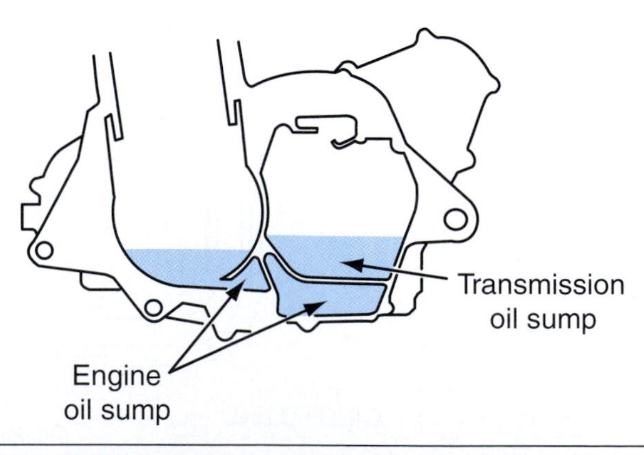

Figure 7-12 A twin-sump lubrication system stores oil in the engine crankcase and the transmission much like the two-stroke engine.

pump shaft and a driven gear (Figure 7-13). The teeth of the gears are meshed together and move oil as they rotate. As the oil is picked up at one side of the oil pump, it's forced to the other side by the gear teeth. This oil pump design produces only moderate oil volume and pressure.

Plunger-Type Oil Pumps

The **plunger-type oil pump** consists of a set of check valves, a piston, and a cylinder (Figure 7-14). When the piston moves up in the oil pump's cylinder, oil is drawn in past the inlet check ball. As the piston moves back down the cylinder, the inlet check ball closes and oil is forced past the outlet check ball. This pump design is capable of high pressure, but produces only low-volume oil flow.

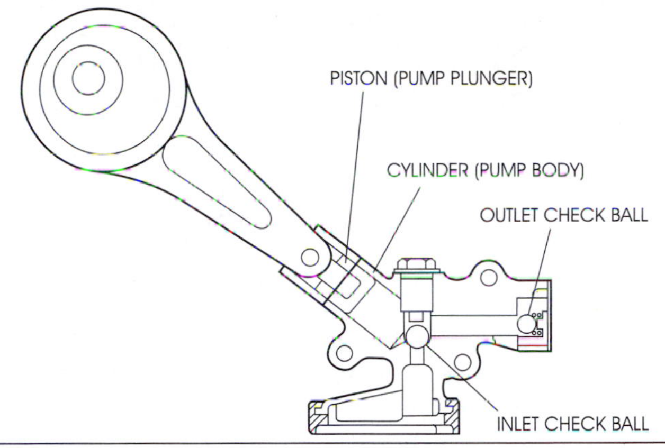

Figure 7-14 A plunger-type oil pump is shown here.

Rotor-Type Oil Pumps

The most commonly used oil pump in four-stroke motorcycle engines is the trochoid or **rotor-type pump** (Figure 7-15). The rotor pump consists of a pair of rotors: an inner rotor and an outer rotor. The inner rotor is shaft driven, while the outer rotor is moved by the inner rotor and is free to turn in the housing. The lobes on the rotors squeeze oil through passages in the pump body. As the inner rotor rotates, oil is constantly picked up from the inlet side, transferred, and pumped through the outlet side. Oil pressure is created when the oil is squeezed between the inner and outer rotors. The rotor-type oil pump design is capable of creating both high volume and high pressure. Most of these pumps have a spring-loaded valve used to bleed off excessive oil after the oil pressure reaches a predetermined level.

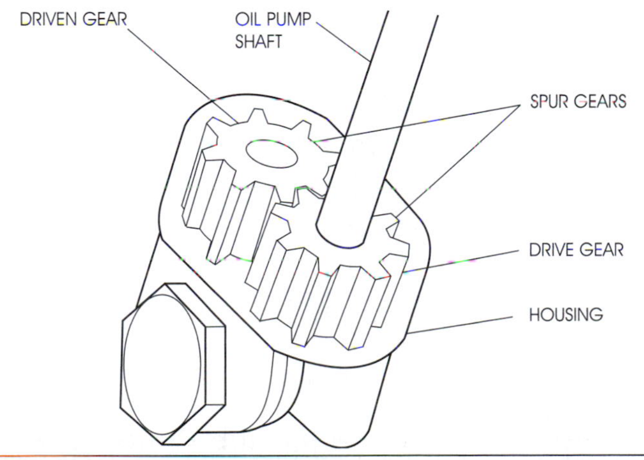

Figure 7-13 A gear-type oil pump is illustrated here.

Oil Pressure Relief Valve

The **oil pressure relief valve** is usually located near the oil pump (Figure 7-16). Its purpose is to prevent excessive oil pressure from building up by bleeding excessive oil back into the crankcase. The oil pressure relief valve operates during cold starts when the oil is thick or when the engine is run at excessive rpm's.

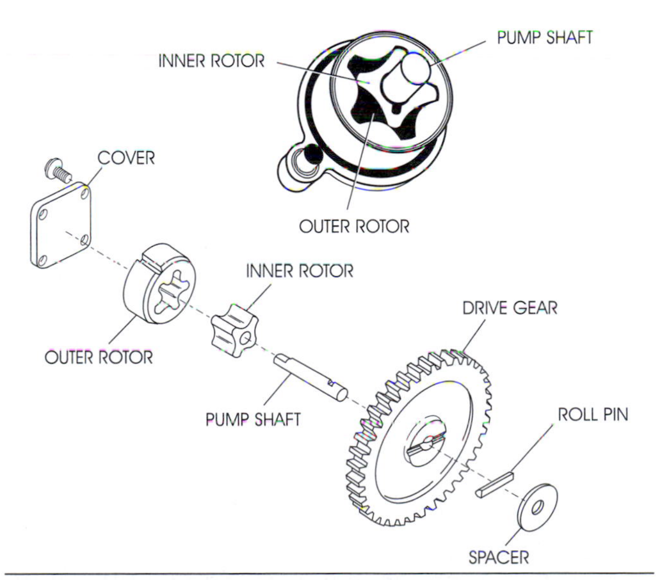

Figure 7-15 The rotor-type oil pump is the most widely used pump in the modern four-stroke motorcycle engine.

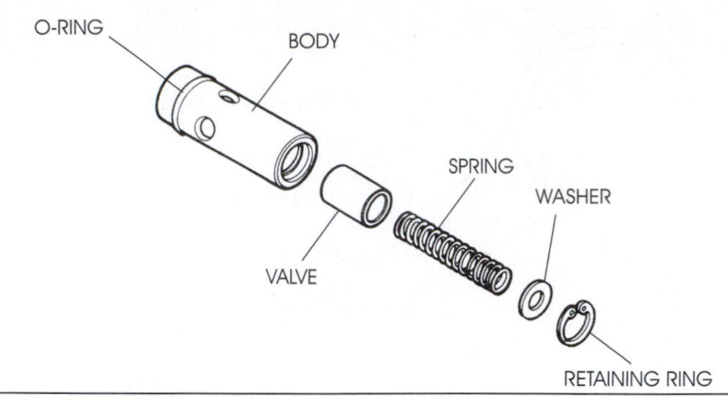

Figure 7-16 The components of the oil pressure relief valve are shown here.

Oil Filters

The four types of oil filters used on four-stroke engines are as follows:

- The **paper oil filter** uses treated pleated paper; it may be a spin-on type or a cartridge type (Figure 7-17).
- The **fiber oil filter** traps contaminates throughout the filter, not just on the surface area. This type of filter was used for many years in dry-sump lubrication systems, but is no longer popular.
- The **screen** or **wire-mesh oil filter** traps large contaminants.
- The **centrifugal oil filter** spins the oil at crankshaft speed. Because dirty oil is heavier, it is trapped within a canister that surrounds the filter assembly.

Both the screen and centrifugal types of oil filters are found primarily on smaller four-stroke motorcycle engines.

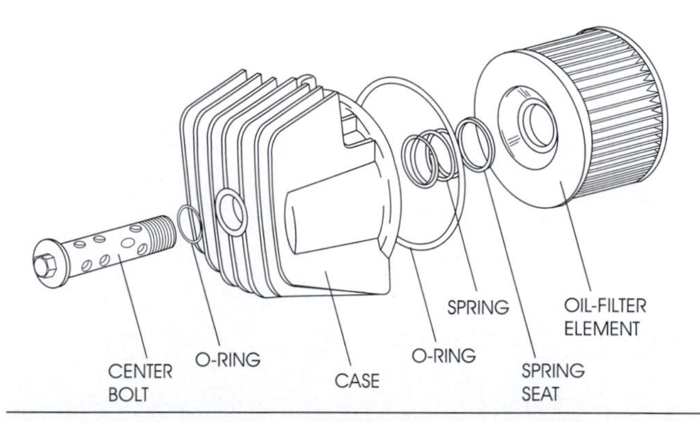

Figure 7-17 A paper-style cartridge oil filter, with components that complete the oil filter assembly is shown here.

Oil Filter Bypass Valve

Most spin-on oil filters and cartridge-type oil filter bolts include an **oil filter bypass valve**. When the oil flow through the filter is restricted from an excessively dirty filter or extremely cold running conditions, the oil filter bypass valve allows the oil to bypass the filter, thus providing essential lubrication to critical engine components. Notice that the oil filter bypass valve illustrated in Figure 7-18 has a bolt, which contains a spring and a ball. When pressure is excessive in the oil filter housing, the ball pushes the spring, allowing the oil to bypass the filter. Unfiltered oil is better than no oil at all.

Oil Passages

The flow of oil for a typical four-stroke motorcycle is shown in Figure 7-19. The illustration provides evidence of the many different duties performed by the oil used for lubrication in this engine. Oil passages, or pipes, deliver the oil from the pump to the crankcase, camshaft, valves, and all other internal parts in need of lubrication. Most crankcases have holes machined into them to supply oil to the crankshaft and transmission shafts. Some models use an external oil line to deliver oil to the top end, while some four-stroke engines simply pump oil up one or more cylinder studs. These studs are sealed with O-rings, which must be replaced any time the engine is disassembled.

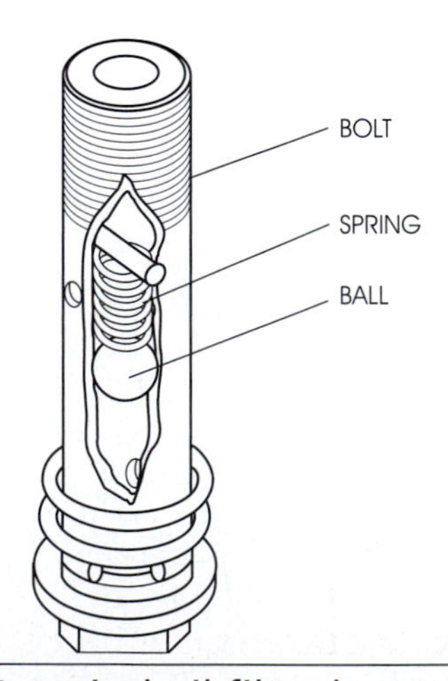

Figure 7-18 A typical oil filter bypass valve and its components are shown here.

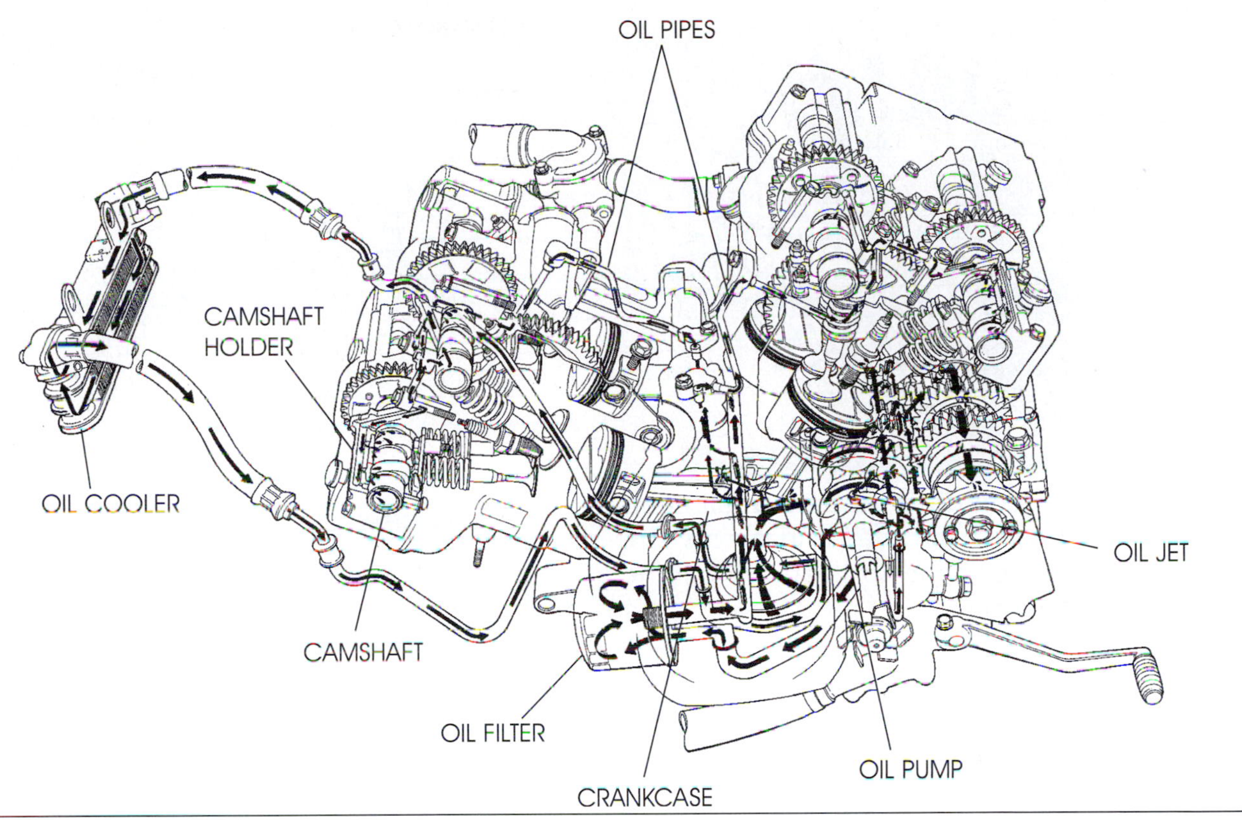

Figure 7-19 Four-stroke engines use many paths for the oil to get to the numerous lubrication points.

COOLING SYSTEMS

Motorcycle engine cooling systems assist in the removal of excess heat produced by the engine. They're designed to allow the engine to operate at a temperature predetermined by the manufacturer. There are three types of cooling systems found on motorcycle engines: internal-oil cooling, air cooling, and liquid (coolant) cooling.

Internal-Oil Cooling

All motorcycle engines use **internal-oil cooling** in their engine designs. The components of internal cooling are the oil, oil coolers (used on some motorcycles), oil pumps, and oil filters, along with the oil passages and oil lines. As the oil is circulated throughout the engine, heat is transferred to the oil from the engine components with which the oil has come in contact.

Air Cooling

Air-cooled engine designs use fins on the cylinder head and cylinder to dissipate heat to the surrounding air. The following are two air-cooling methods used on motorcycles:

- **Open-draft cooling** uses the movement of the motorcycle to force air over the fins, removing the excess heat from the engine (Figure 7-20).
- **Forced-draft cooling** uses an engine-driven fan, which draws air through ductwork called *shrouds* (Figure 7-21). Shrouds surround the cylinder and the cylinder head.

Liquid Cooling

Although creating a slightly heavier motorcycle because of the extra needed components, **liquid cooling** is extremely popular with motorcycles being built today. A liquid-cooled engine gives the

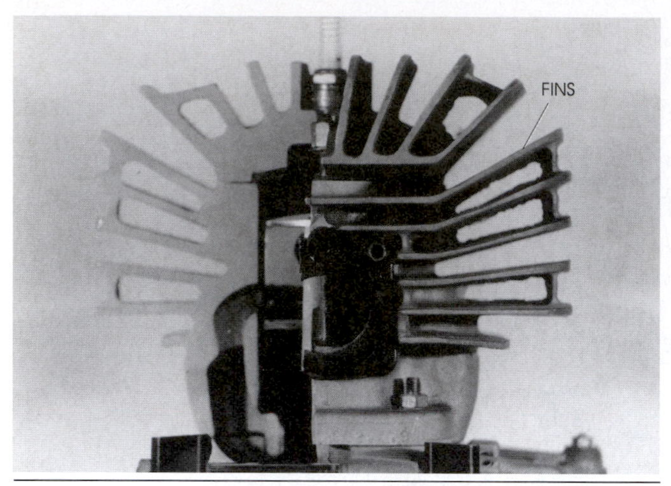

Figure 7-20 The open-draft method of air cooling uses the flow of air over and across the cylinder and cylinder head to keep the engine cool when in motion.

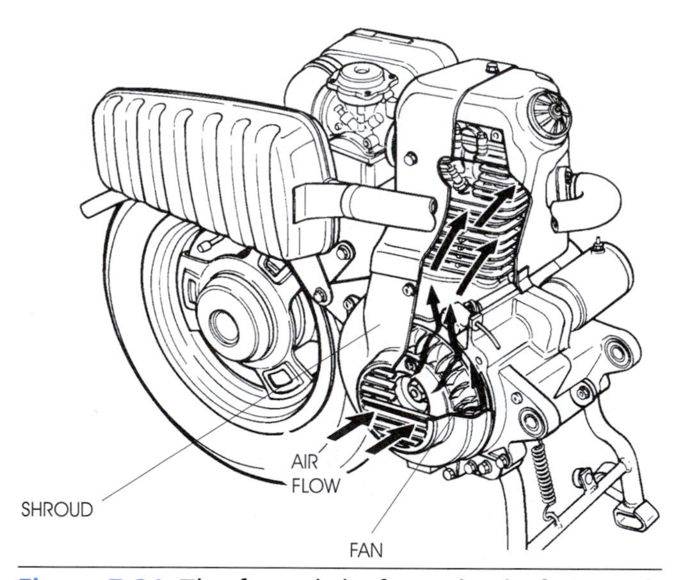

Figure 7-21 The forced-draft method of air cooling uses an engine-driven fan to draw air through the engine shrouds.

motorcycle manufacturer the ability to better control the engine temperature. Liquid-cooled motorcycle engines contain the following various components:

- Water pump
- Radiator
- Thermostat
- Radiator cap
- Radiator fan

A typical liquid-cooling system flow path is illustrated in Figure 7-22.

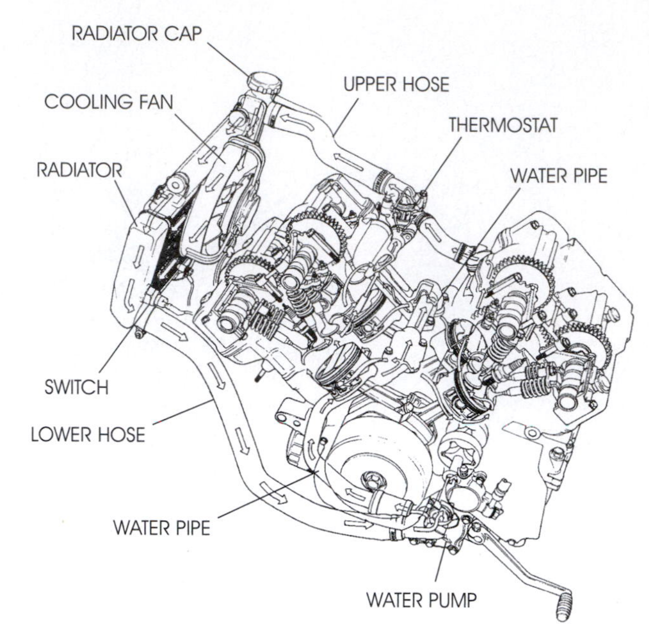

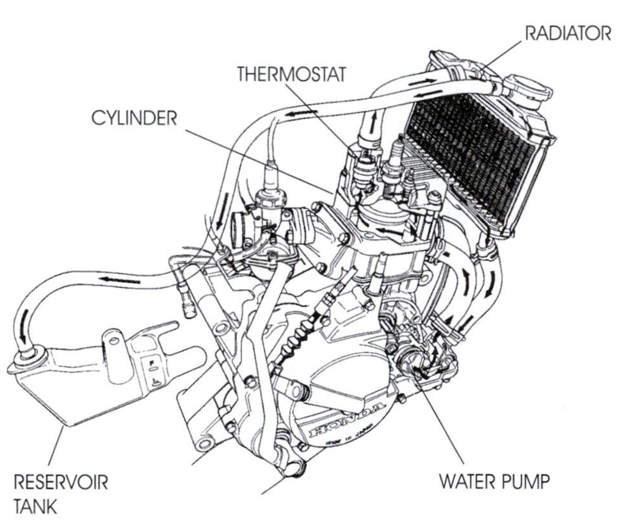

Figure 7-22 Liquid-cooled engines are a popular choice for modern motorcycle engine designs. Two examples are shown here using both a two-stroke and four-stroke engines.

Water Pump

The liquid-cooled engine's **water pump** purpose is to circulate the coolant. The water pump is driven by the engine (Figure 7-23). It draws coolant through the inlet pipe and discharges it into the engine's water jackets. The water pump ensures that the coolant is sent to all needed areas in a uniform manner. The pump consists of a pump shaft, impeller, bearings, mechanical seal, oil seal, and housing.

The water pump housing includes a drain hole, known as a **telltale hole**. Coolant leaking out the

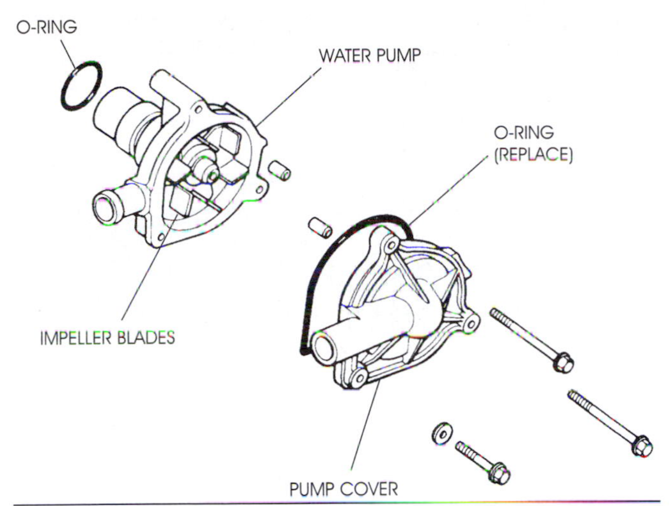

Figure 7-23 A typical water pump assembly is shown here.

telltale hole is an indication that the mechanical seal is leaking (Figure 7-24). This is the most common problem found in a motorcycle's liquid-cooling system. Some water pumps may be rebuilt, but in most cases, the pump is replaced if the mechanical seal fails. If engine oil appears to be leaking out the telltale hole, the oil seal is at fault. The oil seal is normally replaceable and does not require replacement of the water pump.

Radiator

The **radiator** is a cooling device that allows for rapid heat removal. The radiator cools the moving liquid inside the cooling system as it's pumped through the engine. A radiator is also known as a heat exchanger. The radiator consists of small tubes or passages surrounded by very thin cooling fins (Figure 7-25). On most motorcycles, the radiator is constructed of aluminum alloy. As the coolant temperature rises, the coolant expands. Most radiators also include a reservoir, or reserve tank, which allows expansion and contraction of the coolant. If a radiator is damaged, it normally needs to be replaced. Although it's possible to repair radiators on motorcycles, the cost to repair them may be higher than the cost to replace them.

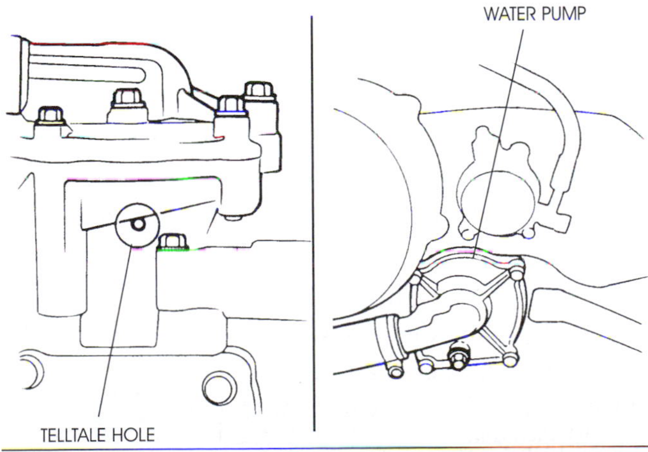

Figure 7-24 The telltale hole in a water pump gives warning when the pump needs repair or replacement.

Thermostat

The **thermostat** is a temperature-sensitive flow valve. Its purpose is to provide a quicker engine warm-up time. It also ensures that the engine operates at a predetermined temperature. When the engine is cold, the thermostat is in the closed position. This allows the coolant to flow through the engine only and not into the radiator. When the engine reaches its predetermined operating temperature, the thermostat opens, permitting the coolant to flow through the radiator (Figure 7-26).

The thermostat may be tested by suspending it in heated water and checking the temperature with

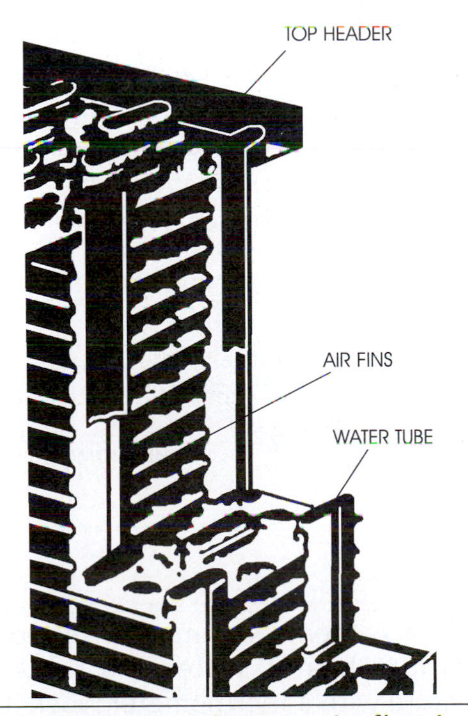

Figure 7-25 In a radiator, air flowing past the fins cool the coolant as it flows through the water tubes.

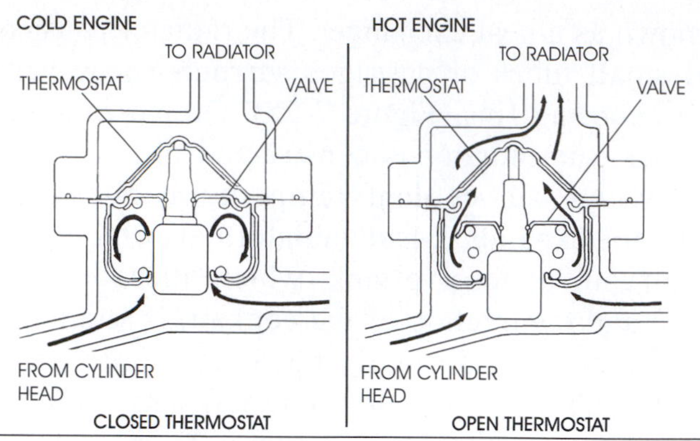

Figure 7-26 The thermostat remains closed when the engine is cold, keeping the flow of coolant from reaching the radiator. This provides for quicker engine warm-ups. When the engine reaches a predetermined operating temperature, the thermostat opens, allowing coolant to pass through the radiator.

a thermometer when it opens (Figure 7-27). If the thermostat doesn't open at the correct temperature given by the manufacturer, doesn't open at all, or doesn't close, it must be replaced.

Radiator Cap

The **radiator cap** seals the cooling system from the outside atmosphere. It's also used to limit the cooling system's operating pressure. The boiling point of a liquid is increased by 3 degrees Fahrenheit for each 1 psi of pressure. Motorcycle radiator caps are usually designed to hold 12–17 psi of pressure. They may be tested using a pressure tester (Figure 7-28). If the cap fails the test, it must be replaced.

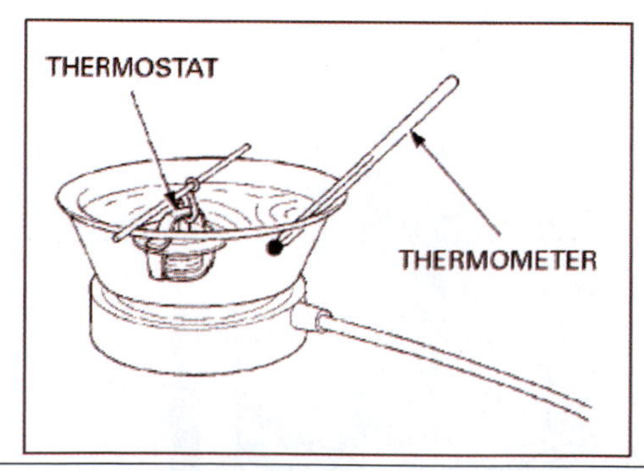

Figure 7-27 A thermostat may be tested by suspending it in heated water and checking the temperature with a thermometer when it opens.

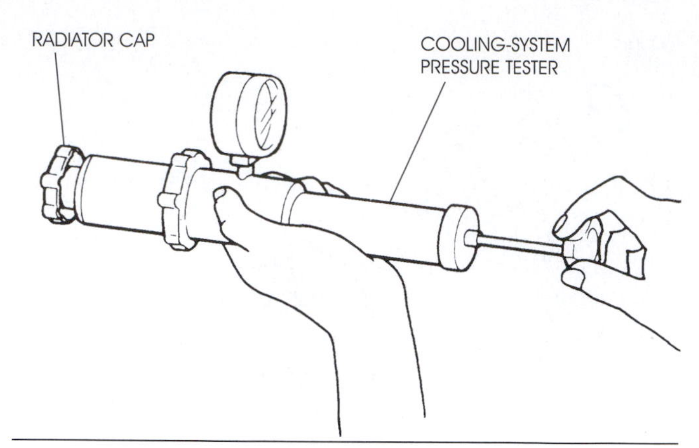

Figure 7-28 A cooling system tester can be used to perform a radiator cap pressure test.

Coolant

Coolant for a liquid-cooled motorcycle usually consists of a 50/50 mixture of distilled water and antifreeze (ethylene or propylene glycol). One part water is used for each part of antifreeze because water has much better heat transfer capabilities than pure antifreeze. For the purpose of engine protection, distilled water is better than plain tap water because it doesn't contain mineral deposits that can cause corrosion. You must use the type of coolant recommended by the manufacturer to ensure that the engine is protected from damage due to corrosion. You should never use 100 percent ethylene glycol in any cooling system. Ethylene glycol is a very poor coolant when used by itself. The purpose of using antifreeze in a liquid-cooled engine is to lower the freezing point and raise the boiling point of the liquid (water). By mixing antifreeze with water in the proper proportion, antifreeze lowers the freezing point of water to less than 30 degrees Fahrenheit. (Water normally freezes at 32 degrees F.) At the same time, antifreeze raises the boiling point of water to more than 225 degrees Fahrenheit. (Water normally boils at 212 degrees F.)

Antifreeze also contains lubricants, anti-foaming additives, and corrosion inhibitors that help protect the engine. The antifreeze-and-water coolant is pumped along water jackets through the cylinder head and cylinder. The heated coolant then flows through the radiator, where heat is dissipated to the surrounding air.

Radiator Fan

The fan used with the liquid-cooled engine system helps to move air through the radiator when the motorcycle isn't in motion or when it's moving too slowly to get the correct amount of air through the radiator. The fan usually uses DC voltage to operate and is controlled by a temperature-sensitive sensor. It is designed to operate only when the engine reaches a predetermined temperature. The fan may be designed to continue running even if the motorcycle is shut off.

Liquid-Cooling System Testing

Liquid-cooled systems can be pressure-tested to check for leaks (Figure 7-29). **Pressure testing** verifies that the system can hold a specified pressure for a required period of time. Specifications for this test are available in the manufacturers service manual for the specific model. If the system fails the pressure test, check the hoses, pipe connections, the water pump installation, and seals.

Another common test for liquid-cooled engines is a *specific-gravity test*. This test uses a **hydrometer**, which measures the weight of liquid as compared to the weight of water (Figure 7-30). When antifreeze is added to water, the hydrometer measures the weight change.

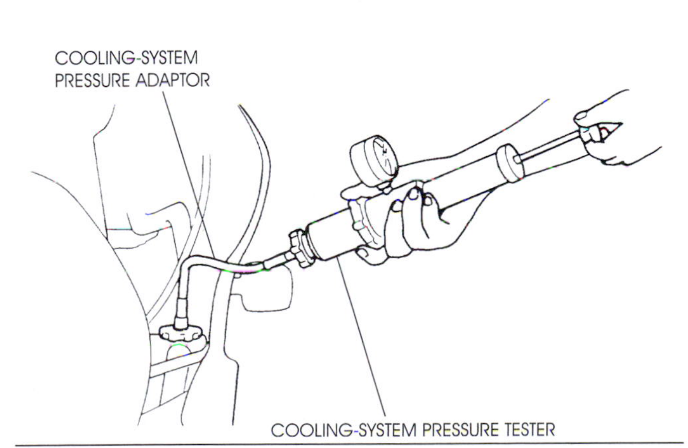

Figure 7-29 Cooling systems must be pressure tested to verify that the system is operating correctly.

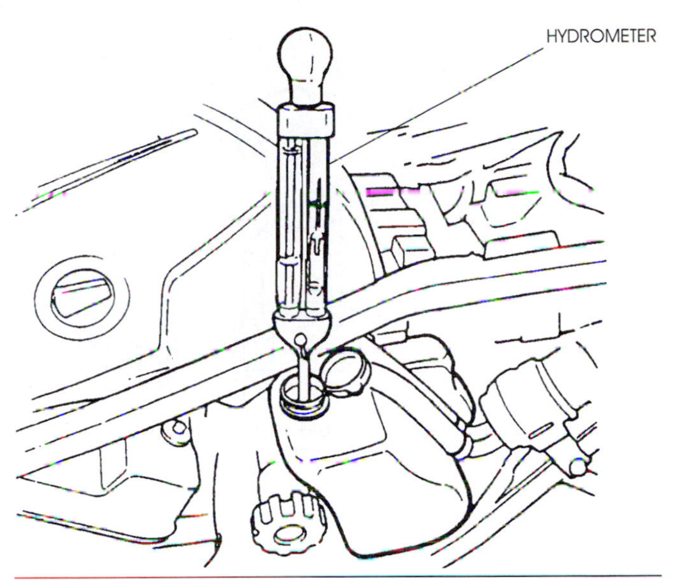

Figure 7-30 The hydrometer measures the specific gravity of the coolant to verify that it's correctly mixed.

LUBRICATION SYSTEM MAINTENANCE

Although Chapter 19 covers maintenance in detail, we will briefly cover lubrication system maintenance now to help you become familiar with it.

Two-Stroke Engine

Only minor maintenance is required with the two-stroke engine lubrication system. If you're using the premixed method of lubrication, you must ensure a correct ratio of oil to fuel. If you have an oil injected lubrication system, you must verify that there is an adequate supply of oil in the oil tank.

Two-Stroke Transmission and Clutch

The two-stroke transmission and clutch have a separate oil drain plug, normally found on the bottom of the engine crankcase (Figure 7-31). It is necessary to drain and replace the oil on a regular basis as described in the appropriate owner's or service manual.

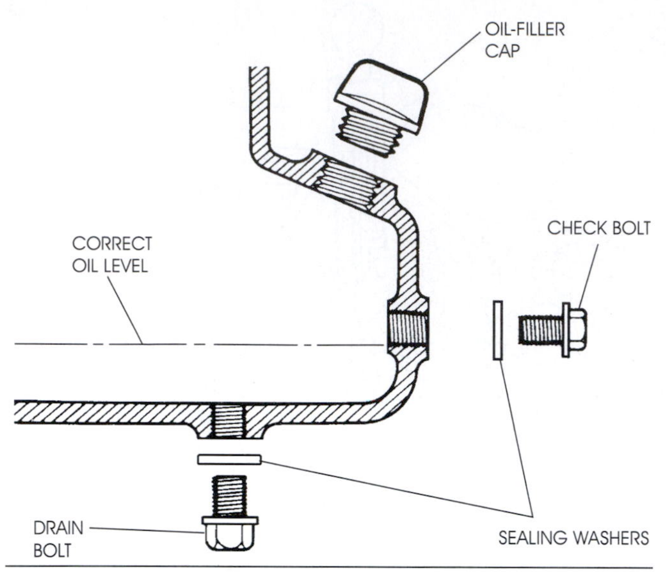

Figure 7-31 Draining oil from a two-stroke engine crankcase.

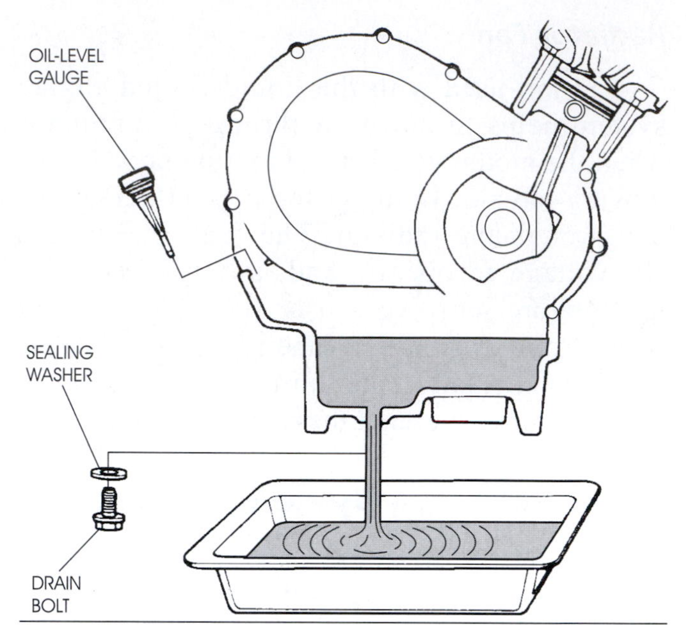

Figure 7-32 Draining oil from a four-stroke engine crankcase.

Four-Stroke Engine, Transmission, and Clutch

Most four-stroke motorcycle engines use the same oil to lubricate the engine as they use for the transmission and clutch. These engines usually have just one drain plug for removing oil from the engine crankcase (Figure 7-32). In most cases, there's also an oil filter that needs to be replaced at the same time the engine oil is changed. Changing the oil filter is as important as changing the engine oil because the filter contains most of the dirt and contaminants from the engine.

Other Components

While the engine must have a source of oil to operate properly, lubrication of other parts of the motorcycle is equally important. This is usually accomplished by applying grease or oil at regular time intervals to points of wear. Note that the grease used for wheel bearing lubrication is different from grease used on other chassis components. Use caution to not over-lubricate components, as this may cause dirt to collect. The component outside the engine that requires the most attention (as far as lubrication is concerned) is the chain. This is especially true for off-road motorcycles and ATVs. Dirty conditions cause considerable wear to the chain and sprockets. Figure 7-33 points out lubrication points on a typical ATV. Lubrication points are similar on motorcycles.

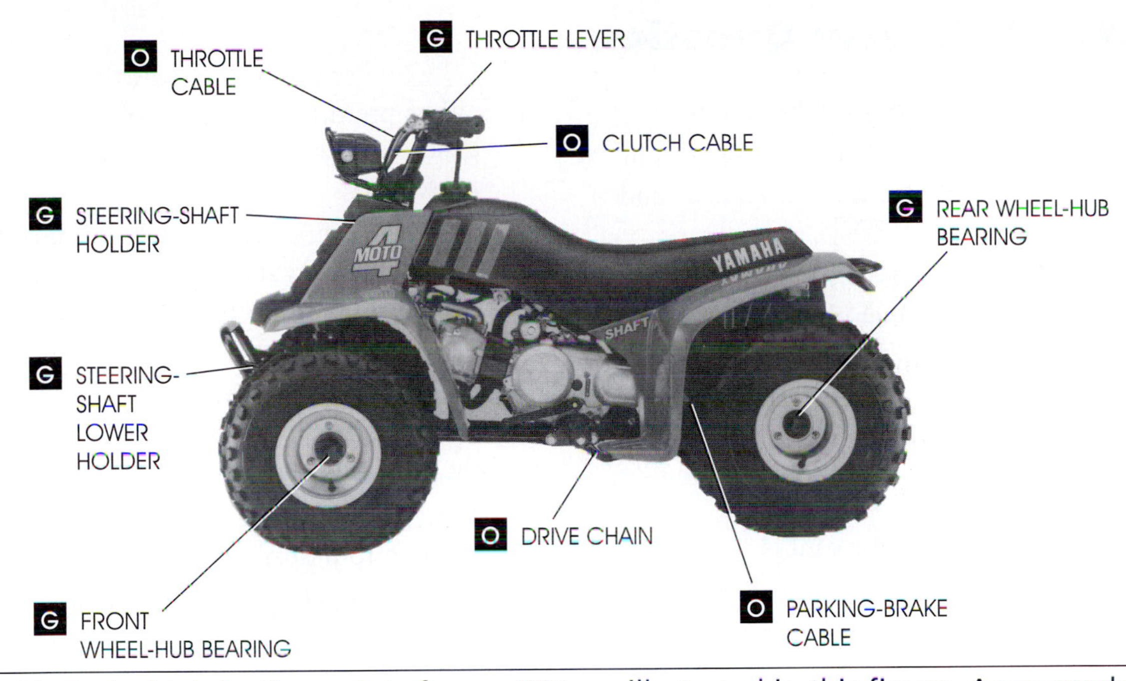

Figure 7-33 Typical lubrication points for an ATV are illustrated in this figure. Areas marked with O require oil lubrication; areas marked with G require grease lubrication.

Summary

- There are four key purposes of lubrication: cooling, cleaning, sealing, and lubricating to reduce friction.
- There are various types of oil, which are classified by letter codes and viscosity ratings.
- Bearings, bushings, and seals are needed in an engine to reduce friction.
- There are six different types of bearings used in motorcycles.
- Both two- and four-stroke engine lubrication systems have specific purposes.
- There are two different types of lubrication systems used in two-stroke engines and three types of lubrication systems found on four-stroke engines.
- All motorcycles include internal cooling, and can also be air-cooled or liquid-cooled to prevent overheating.

Chapter 7 Review Questions

1. Name four purposes of oil lubrication.
2. The three basic types of engine oils used in motorcycles are ________, ________, and ________.
3. ________ is the process that separates lubricating oil from other elements in crude oil.
4. ________ is the term used to refer to the thickness of oil.
5. The abbreviation API stands for ________.
6. The abbreviation SAE stands for ________.
7. What special type of lubricant is manufactured in gel form?
8. An oil that has the letter W after its number rating indicates that it was tested at what temperature?
9. The ________ bearing is similar in design to the ball bearing, but is capable of withstanding higher radial loads because it has more surface area.
10. The ________ bearing can come in either one piece or two pieces.
11. Two methods used to provide proper lubrication of a two-stroke engine are ________ and ________.
12. When premixing a two-stroke engine's gasoline and oil, what does the ratio 50:1 represent?
13. Using the steps described in this chapter, what's the correct amount of oil to premix with 5 gallons of gasoline using a 20:1 fuel-to-oil ratio?
14. The oil used in a two-stroke motorcycle transmission is the same oil used to lubricate the piston crankshaft and cylinder. True/False
15. Which type of oil pump is most commonly used on four-stroke motorcycle engines?
16. What are the two types of paper oil filters found in four-stroke lubrication systems?
17. Which type of oil filter spins the oil at crankshaft speed, trapping contamination in a canister?
18. Of the three types of cooling systems, the ________ cooling system is found in all engines.
19. The type of air-cooled engine that uses an engine-driven fan and shrouds is called a ________ system.
20. The component used as a heat exchanger on a liquid-cooled engine is called the ________.

CHAPTER 8 Fuel Systems

Learning Objectives

When students have completed the study of this chapter and its laboratory activities they should be able to:

- Define fuel octane ratings and state the factors that affect these ratings
- Explain the primary principles of carburetor operation
- Identify various fuel delivery systems used on motorcycles
- Identify the components of each type of carburetor
- Describe the operation of the circuits in each type of carburetor
- Explain the concept of carburetor synchronization
- Understand the purpose of fuel injection
- Identify the components of an electronic fuel-injection system

Key Terms

Accelerator pump
Air cutoff valve
Air filters
Air mixture screw
Atomization
Base carburetor
Bleed-type needle jet
Butterfly throttle valve
Bypass ports
Carburetor
Carburetor synchronization
Choke plate cold start system
Circuits
Cold start systems
Computerized Fuel Injection (CFI)
Constant velocity
Detonation
Electric fuel pump
Electronic Fuel Injection (EFI)
Enrichment cold start system
Ethanol alcohol
Flammable
Float chamber
Foam air filters
Fractional distillation
Fuel filters
Fuel injection
Fuel injector
Fuel lines
Fuel mixture screw
Fuel pumps
Fuel tank
Fuel valves
Gasoline
Gauze air filters
Hydrocarbon
Idle air bleed passage
Idle circuit
Idle fuel jet
Idle outlet port
Injector discharge duration
Jet needle
Main jet
Main jet circuit
Malfunction Indicator Light
Manual fuel valves
Mechanical slide

MTBE
Needle jet
Octane rating
Oxygen
Oxygenated fuels
Paper air filters
Pilot air screw
Pilot jet air passageway
Primary-type needle jet
Programmed Fuel Injection (PGM-FI)
Sequential fuel injection
Slide cutaway
Tickler system
Vacuum fuel pump
Vacuum-operated fuel valves
Vacuum valve with electrical assist
Vent hoses
Venturi principle
Volatile

INTRODUCTION

All motorcycle and ATV engines require a fuel system and a carburetion system to operate. For motorcycle and ATV repair, it's important to have a good understanding of both of these systems. In this chapter, you'll first learn about fuels used for motorcycle and ATV engines. You'll then learn about the principles of carburetion. We'll discuss the types of fuel delivery systems used to get the fuel from the fuel tank into the engine. We'll also describe the different types of carburetors found on motorcycles and ATVs. Finally, we will discuss the use of multiple carburetors and fuel-injection systems.

All of what you'll learn about motorcycle fuel and carburetion systems applies to ATVs as well. From now on, we'll refer only to motorcycles in this chapter. Unless stated otherwise, you can assume that the information applies to both motorcycles and ATVs.

FUEL

When we speak of fuel in relationship to the internal-combustion engine, we're referring to gasoline. **Gasoline** (or a gasoline-and-oil mixture in the two-stroke engine) is the fuel used in most standard motorcycles. Gasoline is a **volatile** (evaporates easily), **flammable** (burns easily), **hydrocarbon** (chemical compound of carbon and hydrogen), liquid mixture used as a fuel. Oxygen must be present for gasoline to combust (burn). You may be aware that some racing machines use highly specialized fuels. These fuels won't be a part of this discussion of the basic principles of standard carburetion.

Gasoline is removed from crude oil by a process called fractional distillation, which is the process used to separate a mixture of several liquids, based on their different boiling points. The **fractional distillation** process is based on the fact that each hydrocarbon boils or vaporizes within a certain temperature range. Thus, crude oil is heated in stages until all the various hydrocarbon classes have been individually vaporized and collected. Additives are then blended with the gasoline to give it distinct properties.

The purpose of fuel is to give satisfactory engine performance over a wide range of conditions. Fuel is rated by a method known as fuel octane rating or knock rating. **Octane rating** is the measure of a fuel's ability to resist detonation. The higher the octane rating, the higher the fuel's resistance to detonation. **Detonation**, also known as engine knocking, is the explosion of highly compressed air and the un-burnt fuel mixture in the cylinder. Excessive detonation can cause catastrophic damage to the inside of an engine by breaking a piston or cracking a cylinder head. Today, the chemicals isooctane and heptane are the main additives used in gasoline to resist detonation.

There's no advantage to using gasoline of a higher rating than what the engine needs to operate detonation free. There are factors that can influence the octane rating needs of a motorcycle engine:

- Higher engine and air temperature encourages detonation.
- Higher altitudes discourage detonation.
- An air-and-fuel mixture that is too lean encourages detonation.

The method of riding the motorcycle also affects detonation. The heavier the load the rider applies to the engine, such as riding up a steep hill with the machine overloaded with excess weight, allows a greater chance that detonation will occur.

As you can see, different factors can influence the octane rating needs of a motorcycle engine.

The most important thing to remember is to use a gasoline with an octane rating that meets the motorcycle manufacturer's minimum requirements. This information is provided by every motorcycle manufacturer and can be found in the machine owner's manual.

Oxygenated fuels have an oxygen-based component such as alcohol or ether that contains more oxygen than normal. Adding oxygen to fuel helps the fuel reduce harmful engine carbon monoxide emissions. The two most popular oxygenated additives and the maximum amounts in which they can safely be used in gasoline are:

- **MTBE** (methyl tertiary butyl ether)—up to 15 percent
- **Ethanol alcohol** (also known as gasohol)—up to 10 percent oxygen

OXYGEN

Oxygen is a tasteless, odorless, colorless gas that makes up about 20 percent of the air we breathe. Oxygen in the air that's drawn into the engine has the ability to combine with gasoline to form a combustible vapor. Pure oxygen has the ability to explode if submitted to extreme compression. Ignited oxygen produces a very high temperature and a great amount of energy. However, because engines don't receive pure oxygen, and the compression ratio used in an internal-combustion engine is too low to cause the amount of oxygen that's present to ignite on its own, a fuel mixture is combined with the intake air. The air-and-fuel mixture permits combustion to take place at a compression ratio lower than that required for pure oxygen to burn. Therefore, a combination of air (oxygen) and fuel (gasoline) is necessary to obtain the explosive characteristics required to operate an internal-combustion engine.

THE CARBURETOR

The amount of power produced by an engine is directly related to the heat energy put forth by the air-and-fuel mixture. The more combustible the mixture becomes, the greater the amount of heat that's generated. Motorcycle engines have the ability to transform heat energy into usable power. The greater the amounts of productive heat produced by combustion, the more power you can expect from the engine.

The **carburetor** is a device used to mix the proper amounts of air and fuel together in such a way that the greatest amount of heat energy is obtained when the mixture is compressed and ignited in the combustion chamber of the engine.

The function of the carburetor is to mix the correct amount of fuel with sufficient air so the fuel atomizes (breaks up), allowing it to become a highly combustible vapor. When this vapor enters the combustion chamber of the engine and is compressed by the action of the piston, a spark ignites it, creating the power to operate the engine. To obtain the maximum amount of power from the fuel supply, the exact proportions of air and gas must be mixed and must reach the combustion chamber of the engine in a vapor of precisely the right consistency.

The proper amounts of air and fuel, as they pertain to different engine running conditions, are shown in Table 8-1. Keep in mind that the ratios are the weight of the air and fuel entering the engine, not the volume.

Table 8-1: The air-to-fuel ratios as compared to various engine conditions are shown here.

AIR-AND-FUEL MIXTURES	
ENGINE CONDITION	**A/F**
Starting, cold engine	10 to 1
Acceleration	9 to 1
Idling (no load on the engine)	11 to 1
Partly open throttle	15 to 1
Full load, wide-open-throttle	13 to 1

You learned earlier that liquids do not burn. Gasoline is a liquid. Oxygen, on the other hand, is a gas and has the ability to burn. The most efficient combustion of gasoline and oxygen occurs only when they're combined and turned into a vapor from the heat produced by the engine. This is a delicately balanced mixing process accomplished by the carburetor. Two primary principles are involved in the carburetion operation:

- The principle of atomization
- The Venturi principle

Let us look at each of these principles in detail.

Principles of Atomization

Atomization is the process of combining air and fuel to create a mixture of liquid droplets suspended in air.

As the piston begins the intake stroke, the air pressure in the cylinder is reduced. The pressure difference causes the higher-pressure outside air to flow through the air filter and carburetor, and into the engine (Figure 8-1). Atomization takes place when the carburetor meters gasoline into the fast-moving air passing through it. A primary function of the carburetor is to atomize the fuel to create an air-and-fuel mixture.

The Venturi Principle

Carburetor design is based on the Venturi principle. The **Venturi principle** simply states that a gas or liquid that's flowing through a narrowed-down section (venturi) of a passage will increase in speed and decrease in pressure compared to the speed and pressure in wider sections of the passageway. The Venturi principle is shown in Figure 8-2.

A venturi has a particular shape—a modified hourglass figure, you might say—so that air passing through the carburetor on its way to the combustion chamber passes through the venturi. The hourglass shape of the venturi causes the stream of air to increase in speed and decrease in pressure, creating a pressure difference in the venturi. This pressure difference is important, as it allows fuel to be drawn into the air stream and atomized.

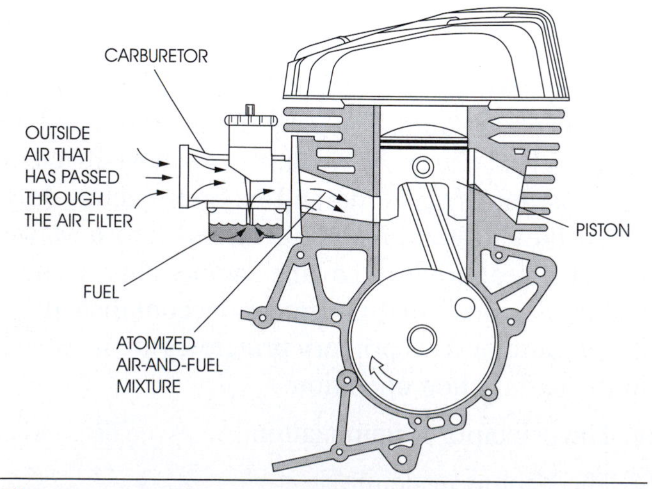

Figure 8-1 The principle of atomization is shown here.

The major air passage in the carburetor body is called the carburetor bore. The air entering the carburetor bore is controlled by its speed and by the size of the venturi. A typical main carburetor bore may have a diameter of 41mm. compared to a venturi diameter of 26mm. When air rushes to fill the cylinder, the speed of the air is faster if it must pass through a small opening than if it must pass through a large opening. As mentioned earlier, as air speed increases, air pressure decreases. The speed of air as it passes through the carburetor is an important factor in the breaking up (or atomization) of the fuel, as well as controlling the amount of fuel that's delivered into the venturi. You can see, in Figure 8-3, that air is drawn into the carburetor through the venturi, where it gains considerable speed. This increase in air speed is directly related to a fall in air pressure in the venturi, which then draws fuel from an outlet. The fuel is atomized under the influence of atmospheric pressure as it is mixed with the incoming air.

Venturi size and shape are of considerable importance. If a venturi is too large, the flow of air is slow and won't atomize sufficient fuel to make a balanced mixture. On the other hand, if the venturi is too small, not enough air passes through to fill the vacuum inside the cylinder created by the engine. A

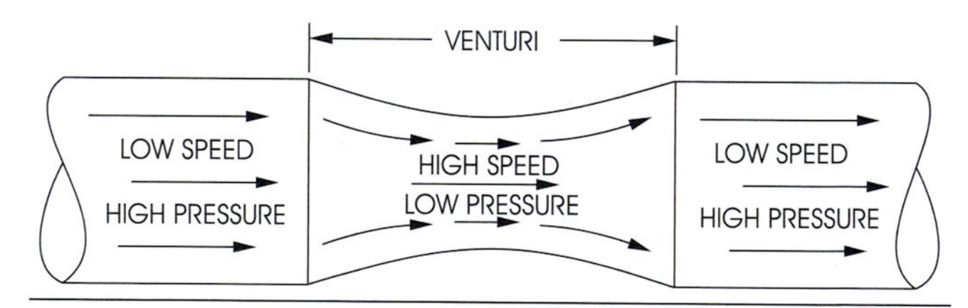

Figure 8-2 This image demonstrates the Venturi principle.

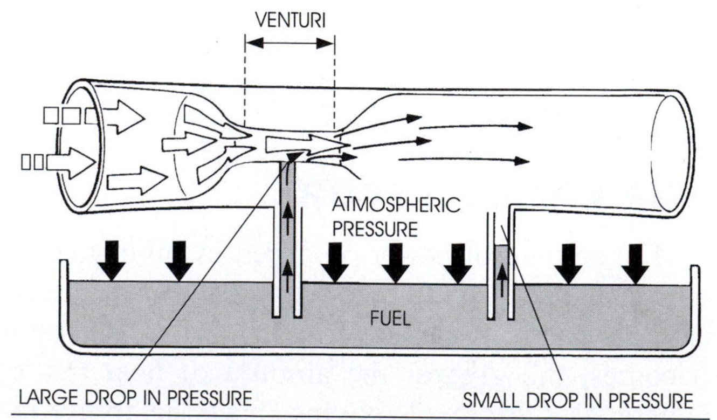

Figure 8-3 Effect of low pressure in a venturi is demonstrated here.

large engine that creates a great vacuum uses a carburetor with a large venturi. A small engine requires a smaller venturi to be most effective.

Carburetors are equipped with mechanisms for regulation of the air and fuel volumes that are allowed to pass through the venturi. All carburetors have a venturi that operates on the same basic principle. Variations are in size, method of attachment, or in the system used to open and close the venturi. However, the principle of operation is the same for all carburetors.

The slide-type carburetor is the most popular type of carburetor used on motorcycle engines and has a venturi whose size is adjusted by a throttle slide (Figure 8-4). The throttle slide, or throttle valve as it's also called, is simply a piston that is raised and lowered in a cylinder. The method used to raise and lower each type of slide is fully explained later in this chapter. By its change in position, the throttle slide controls the venturi opening size. When the throttle slide is raised, the size of the venturi is enlarged and the amount of air allowed to enter the engine is increased. This causes the engine speed to increase. When the throttle slide is lowered, the venturi size is reduced. That is, the air passage through the venturi shrinks and engine speed is decreased. The process of atomization and the Venturi principle are why carburetors work.

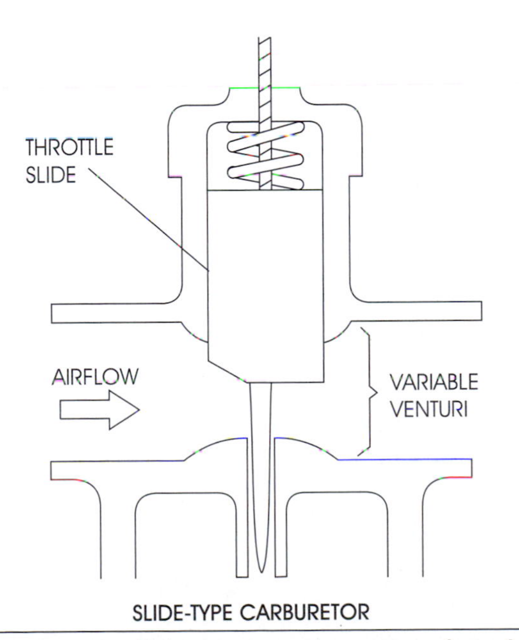

Figure 8-4 A slide-type carburetor showing a variable venturi. Note that the slide can move up and down to change the size of the venturi.

FUEL DELIVERY SYSTEMS

The fuel delivery system of most motorcycles consists of many components, each of which is discussed in this section. Servicing fuel delivery systems is very important and involves inspecting and cleaning or replacing many of these components.

Fuel Tank

The **fuel tank** is designed to store fuel (gasoline). Fuel tanks can be made of steel, aluminum, plastic, or even fiberglass. Most modern street motorcycle fuel tanks are made of a light, thin steel, while many off-road motorcycle fuel tanks are made of plastic. The important thing to remember is that the fuel tank is a reservoir that safely stores a supply of fuel for the carburetion system (Figure 8-5). In many cases, the fuel tank uses a gravity feed system to allow fuel to flow into the carburetor. The fuel tank will always be higher than the carburetor when using the gravity feed system.

Typically, the fuel tank is vented to the atmosphere, but some states (California, for example) require fuel tanks to be vented into a charcoal canister (Figure 8-6). This canister retains the hydrocarbon vapors, keeping them from entering the air we breathe.

Figure 8-5 The fuel tank is used to store gasoline, and is mounted above the engine and carburetor in this photograph.

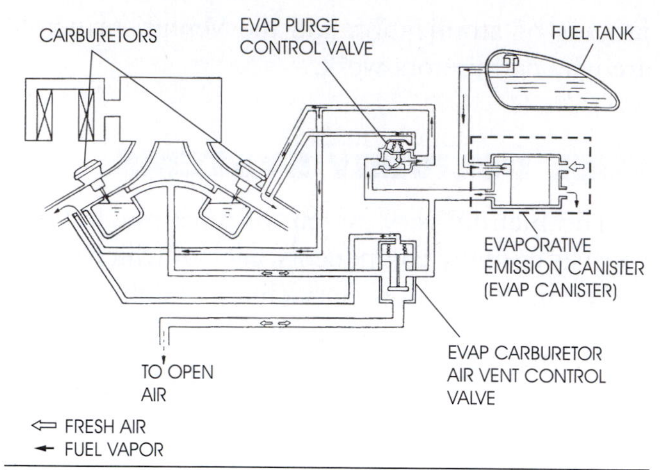

Figure 8-6 A charcoal canister (called the Evaporative Emissions Canister or EVAP canister in this illustration) is used in many motorcycles to help reduce hydrocarbon emissions.

Fuel Valves

Fuel valves, also known as fuel petcocks, are on/off valves that control the flow of gasoline from the fuel tank to the carburetion system. There are different types of fuel valves used on motorcycles: manual fuel valves, vacuum-operated fuel valves, and vacuum fuel valves with electric assist.

Manual Fuel Valves

Manual fuel valves allow the rider to control the fuel flow by turning the valve to one of three positions: on, off, or reserve (Figure 8-7). When turned to the on position, fuel flows to the carburetor from the main fuel supply. When turned to the off position, the flow of fuel stops. The reserve position serves as a reminder to the rider that the fuel tank needs to be filled. This position is an important warning device because, unlike an automobile, most motorcycles don't have a fuel gauge. When in the reserve position, fuel is drawn from a reserve section of the fuel tank. At this point, there is usually less than one gallon of fuel left, so it would be wise to seek a gas station soon!

Vacuum-Operated Fuel Valves

Vacuum-operated fuel valves also have levers with three positions: on, reserve, and prime. The on and reserve positions allow the fuel to flow only when the engine is running and engine vacuum is present. The engine vacuum pulls on a diaphragm inside the fuel valve, allowing fuel to flow freely to the carburetor. When the lever is in the prime position, fuel flows at all times. This position doesn't require engine vacuum to allow fuel to flow. The prime position is usually used only when the carburetor has been drained of all fuel, after long storage, or following disassembly. As you think about this, the prime position is very helpful. If the carburetor has no fuel, the engine won't start, so there's no vacuum available for the fuel valve diaphragm to operate, thus no fuel flows to the carburetor. The prime position overrides the vacuum diaphragm, allowing fuel to flow without the engine running. The vacuum fuel valve normally has two hoses, as shown in Figure 8-8, one hose for fuel delivery, and another smaller hose for engine vacuum.

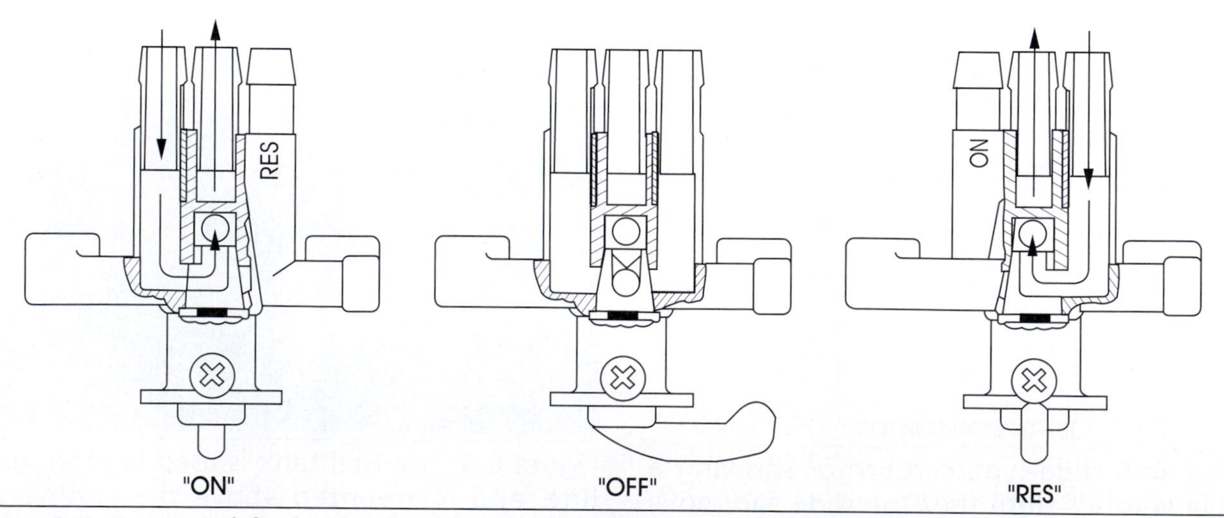

Figure 8-7 A manual fuel valve in the on, off, and reserve positions is shown here.

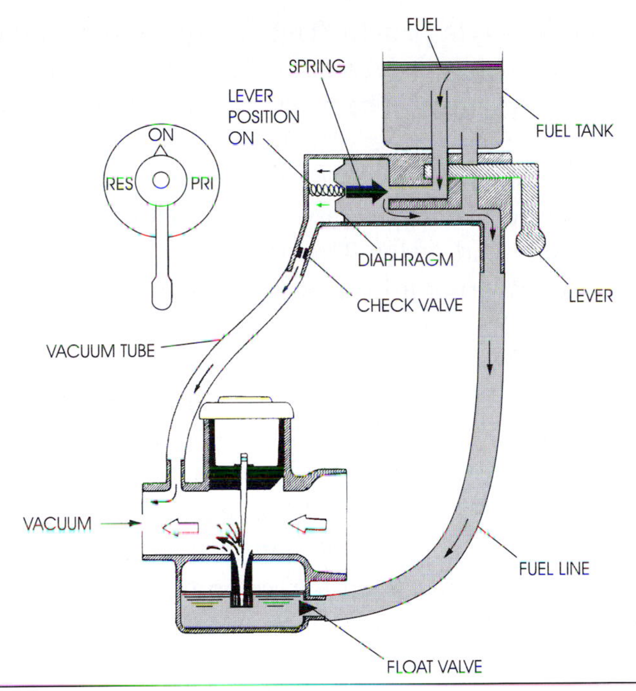

Figure 8-8 A vacuum-operated fuel valve uses engine vacuum to allow fuel to flow by use of a diaphragm as shown here.

Vacuum Valve with Electric Assist

The third fuel valve system is a **vacuum valve with electric assist**. This fuel valve is the same as the vacuum-operated fuel valve, except it has a float gauge inside the fuel tank. When the fuel level reaches a predetermined level, the float gauge signals an electrical switch that automatically switches the fuel valve from the on position to the reserve position.

Fuel Lines

Fuel lines are used to flow gasoline from the fuel valve to the carburetion system and are usually made of neoprene. It's important to use the manufacturer recommended fuel line, because some hoses can be affected or damaged by today's gasoline. It is always important to route the fuel line away from hot engine parts and carburetor linkages.

Fuel Pumps

Many motorcycles use a fuel pump. The purpose of a fuel pump is to deliver fuel from the fuel tank to the carburetion system (Figure 8-9). Fuel pumps are always used on fuel-injection systems. Fuel injection is a type of carburetion discussed later in this chapter. A fuel pump is also required when the motorcycle's fuel tank is lower than the carburetor. The fuel pump supplies fuel under pressure to keep the carburetor filled with fuel.

There are two types of fuel pumps used on motorcycles: vacuum and electric.

Vacuum Fuel Pumps

The **vacuum fuel pump** uses a diaphragm that's moved by the pressure differences of engine vacuum and atmospheric pressure. It works in the same manner as the vacuum fuel valve explained earlier in this section. A vacuum fuel pump system is illustrated in Figure 8-10.

Electric Fuel Pump

The **electric fuel pump** is operated electronically by the use of an electric solenoid, or relay, that pumps the fuel from the fuel tank to the carburetor (Figure 8-9). An electric fuel pump operates only when the motorcycle is running, unless it is bypassed.

Vent Hoses

Vent hoses are used on fuel tanks and carburetors to permit atmospheric air pressure to enter into certain important areas within the fuel system. If

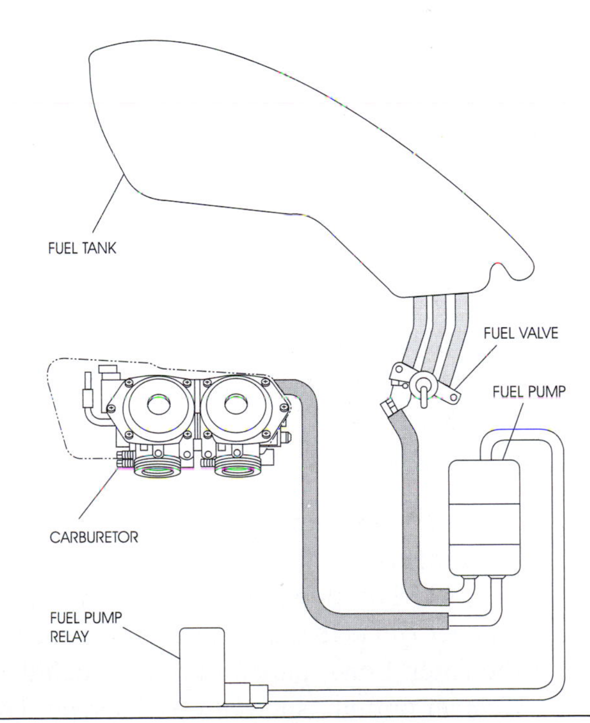

Figure 8-9 A typical electric fuel pump system.

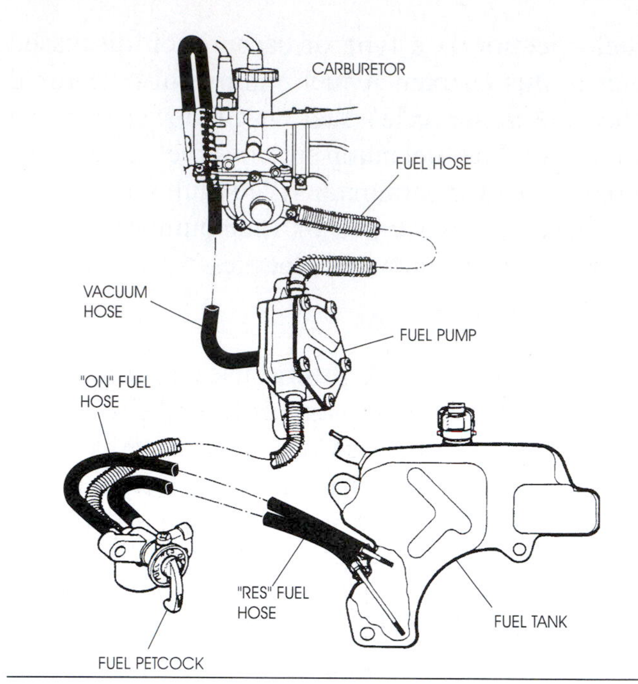

Figure 8-10 A typical vacuum fuel pump system is illustrated here.

these hoses become plugged, twisted, or curled, the fuel will not flow correctly.

Fuel Filters

Fuel filters help remove contaminants from the fuel before they reach the carburetor. Common locations are on the top of the fuel valve or petcock in the fuel tank, in the fuel valve, in line with the fuel hose, or on top of the seat in the carburetor.

Air Filters

Air filters are designed to filter the incoming air to the carburetor. Air filters are very important to the life of an engine. If dirt or other contaminants are allowed to go through the carburetor with the air-and-fuel mixture, they will damage the engine quickly.

Paper Air Filters

The first and most commonly used air filter in street motorcycles is the **paper air filter** (Figure 8-11A). The paper air filter consists of laminated paper fibers that are sealed at the ends or sides of the filter. Some paper air filters include a supportive inner or outer shell of metal screen. The paper used in these air filters is molded in an accordion-style pattern. This design increases the surface area and decreases the restriction of air passing through it. The paper air filter design must be kept dry and free of oil. If it becomes excessively dirty or has oil in it, the paper air filter must be replaced. Don't try to clean a paper air filter with soap and water. This will damage the paper fibers, rendering it incapable of doing its job.

Foam Air Filters

Foam air filters (Figure 8-11B), which are used on most off-road motorcycles, uses a special foam coated in oil to aid in trapping dirt and other contaminants. The foam filter usually fits over a metal apparatus to help hold its shape. These filters work by slowing down the incoming air and collecting particles of dirt as the air passes through the filtering material. The dirt sticks to the filter and remains there until the filter is serviced. When the filter becomes dirty, it can be cleaned in a warm, soapy water solution, then rinsed and dried. After drying, it must be oiled using special foam air filter oil (Figure 8-12).

Gauze Air Filters

The **gauze air filter** is very similar to the paper air filter. Surgical gauze is used to trap the dirt as the air passes through it. The main difference compared to the paper filter is that when dirty, you can clean this filter in warm, soapy water, then rinse and dry it. You must also use special gauze filter oil when servicing this type of air filter.

Figure 8-11A The paper air filter is molded in an accordion-style pattern to provide greater surface area.

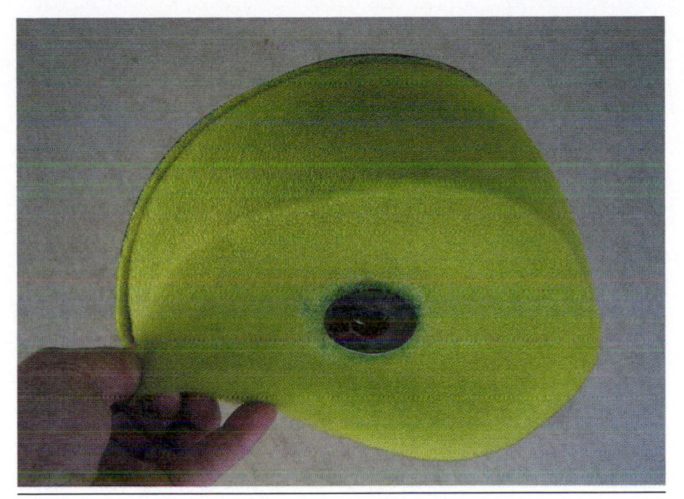

Figure 8-11B The foam filter uses a special foam coated in oil to trap dirt from getting into the engine.

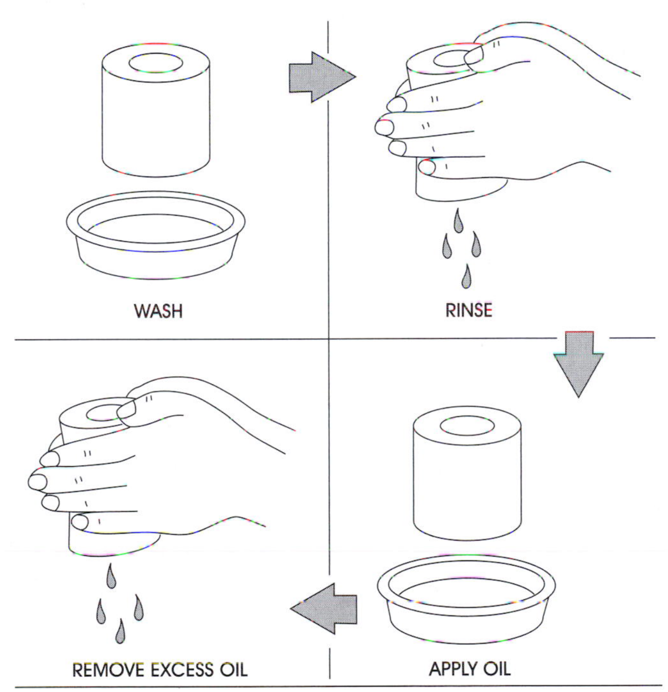

Figure 8-12 When cleaning a foam air filter, wash it with soap and water, rinse well and dry, and apply the proper oil as stated by the filter manufacturer and finish by removing the excess oil.

Intake Manifolds

The purpose of the intake manifold is to help deliver the air-and-fuel mixture to the engine. After the mixture has passed through the carburetor, the intake manifold delivers it to the engine's cylinders, allowing the air and fuel to continue mixing during the delivery. The intake manifold also secures the carburetor to the engine. The intake manifold can be either clamped or bolted to the cylinder or cylinder head. The size and form of the intake manifold varies depending on the particular motorcycle. Intake manifolds can be made of neoprene or aluminum. Once again, this depends on the particular engine design. There are three types of intake manifold mounting methods: spigot, flange, and clamp.

Spigot

The spigot-type mount allows the carburetor body to fit inside a rubber-like intake manifold, while a clamp is used to hold it in place (Figure 8-13).

Flange

Through the flange-type mount, the carburetor body has mounting points cast into it (Figure 8-14). These mounting points bolt to the intake manifold.

Figure 8-13 A spigot-type intake manifold mount is shown with the carburetor attached and removed.

Clamp

With the clamp-type mount, the carburetor body has a clamp cast into it. The carburetor body fits over the intake manifold as shown in Figure 8-15. This intake manifold style is the least likely to be used because of the high cost of the machining process required to manufacture it.

CARBURETOR SYSTEMS AND PHASES OF OPERATION

Now we'll learn what happens inside a carburetor during its various phases of operation. But first, recall that all carburetors work using the same basic principle. The carburetor has the task of combining the air and fuel into a mixture that produces power for the engine. First, the engine draws in air. The pressure difference between the outside atmosphere (higher pressure) and the inside of the cylinder (lower pressure) forces the air to pass through the carburetor. The air mixes with a predetermined amount of fuel that's also moved by pressure differences into the air stream of the carburetor venturi. Carburetors use several different fuel metering systems, which supply fuel for the air-and-fuel mixture in regulated amounts. These metering systems are called **circuits**, and their operating ranges overlap. We'll discuss these circuits as well as the operation of the various carburetor systems used on all carburetors. We will begin the discussion at the start of the fuel process—that is, the fuel tank.

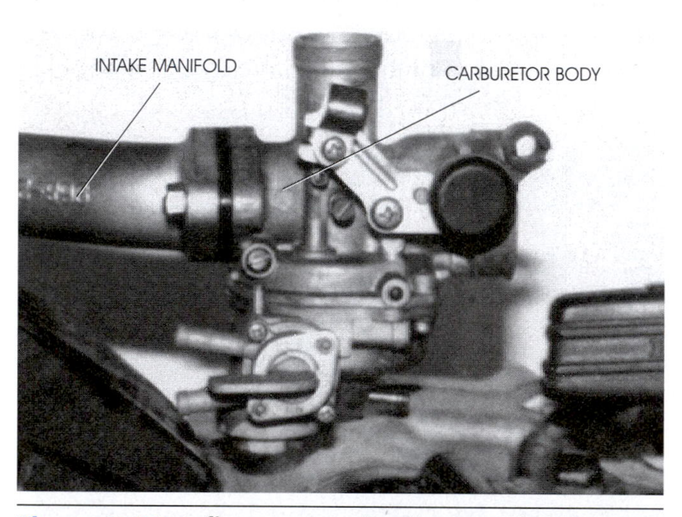

Figure 8-14 A flange-type intake manifold mount has the carburetor bolted solid to the manifold and is mainly seen on older motorcycles.

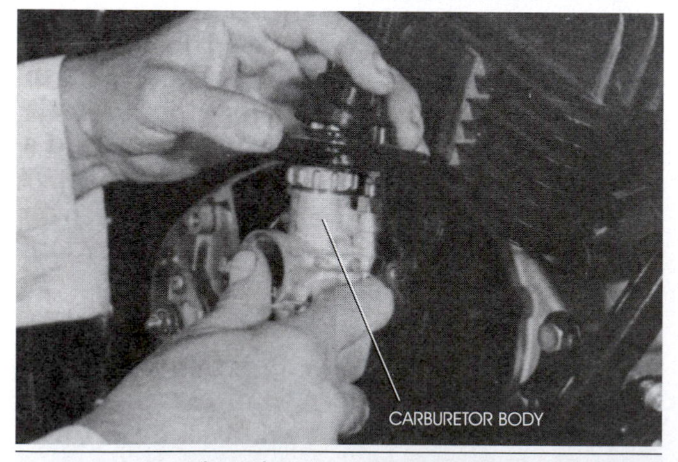

Figure 8-15 The clamp-type intake manifold is typically used when the carburetor is mounted in an engine case and is least used due to the high manufacturing cost.

Fuel Feed System

As mentioned earlier, the fuel tank is used to store the fuel for the engine. Fuel is delivered from the fuel tank to the carburetor using either a gravity feed method or a fuel pump.

Float System

The **float chamber** is located in the carburetor body in most cases (Figure 8-16). It's designed to hold a constant level of fuel for the engine. As the fuel is used in the engine, the fuel level becomes low and allows a float valve to open and fill the float bowl to a specified level. This causes the float, which is attached to the float valve, to rise. When the specified level is reached, the float valve closes and stops the fuel from entering. This operation repeats continuously as the engine is running. To keep the float bowl at atmospheric pressure, an air vent passage connects the float bowl to the outside air of the carburetor. Also, an overflow tube is provided to drain any excess fuel to the outside of the carburetor.

The fuel level can be modified on many carburetors by adjusting a small tang that rests on the float valve. Other floats aren't adjustable. In this case, the float and float valve must be replaced if out of specification.

The float valve contains a small spring and valve pin that is used to depress the valve so it

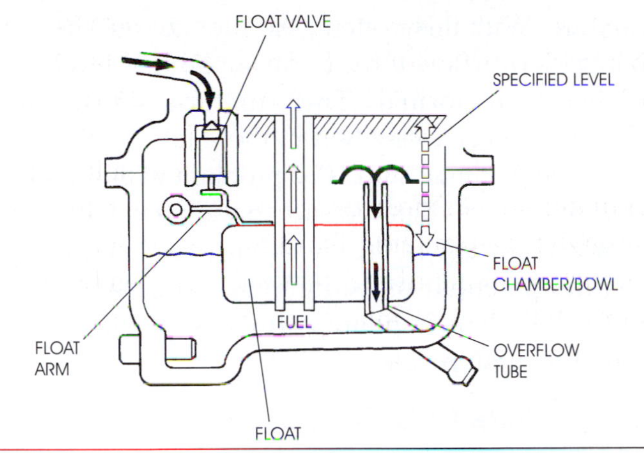

Figure 8-16 A typical float chamber is illustrated here.

doesn't become dislodged from the seat when closed (Figure 8-17). Vibration of the running engine could otherwise cause the float valve to become dislodged.

Accelerator Pump System

When the throttle slide is suddenly opened, the mixture of air and fuel being drawn into the engine is lean. This is due to the sudden increase of airflow into the cylinder. When this occurs, the fuel can't be drawn into the venturi quickly enough to keep the engine running properly. Therefore, the engine is in need of assistance in the transition from one carburetor phase to the next. To avoid this condition, some carburetors use an **accelerator pump** to temporarily enrich the mixture. As the throttle slide opens, a diaphragm located in the pump is depressed by a rod (Figure 8-18). Fuel is supplied to the main bore of the carburetor via the accelerator nozzle. As the throttle valve closes, spring action returns the accelerator's diaphragm to its original position.

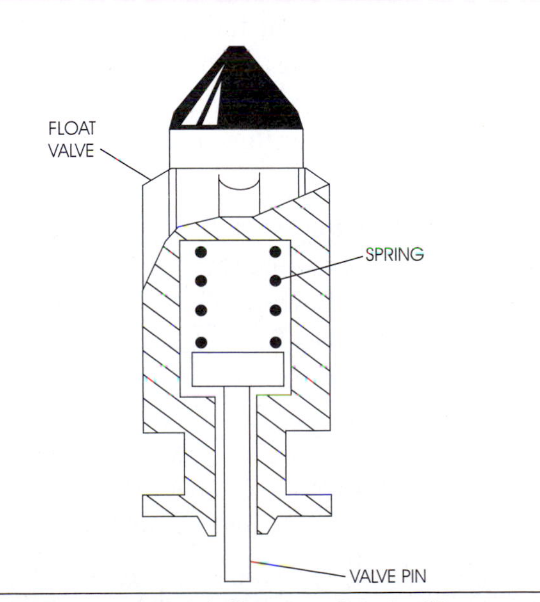

Figure 8-17 A typical float valve is illustrated here.

Air Cutoff Valve System

When the throttle slide is suddenly returned to the closed position after the engine has been run at high rpm, engine braking is applied. When this occurs, the fuel mixture becomes lean and the result can be an engine that backfires or pops on deceleration. Under these conditions, the **air cutoff valve** shuts off a portion of the air to the idle

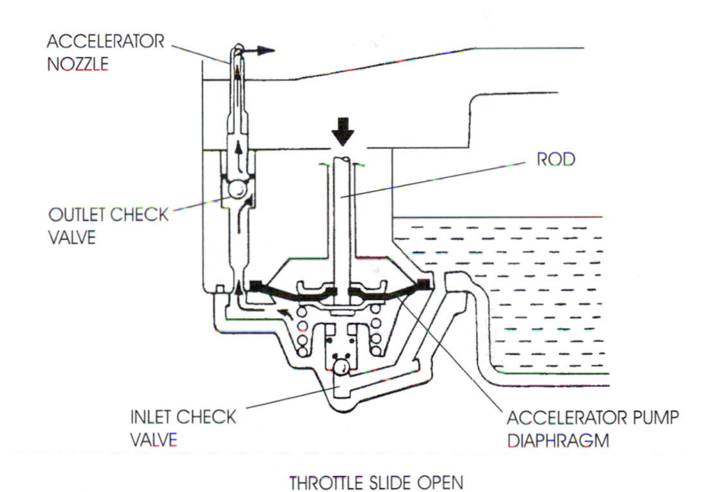

Figure 8-18 When using an accelerator pump, fuel is allowed to spray into an engine to temporarily enrich the air/fuel mixture that prevents a lean mixture during sudden throttle openings.

circuit of the carburetor (Figure 8-19). The reduction of air to this circuit allows for a temporarily rich mixture. The air cutoff valve is controlled by engine vacuum.

Cold Start Phase of Operation

For the cold start phase of engine operation, a rich fuel mixture is needed because the engine metal is cold. When the engine is cold, the air-and-fuel mixture is also cold and won't vaporize or combust readily. To compensate for this reluctance to burn, the amount of fuel in proportion to the amount of air must be increased. This is accomplished by the use of a cold start system. **Cold start systems** are designed to provide and control a richer than normal air-and-fuel mixture necessary to quickly start a cold motorcycle engine. Most carburetor cold start mixtures are designed to operate at a ratio of approximately 10:1, that is, 10 parts of air to 1 part of fuel. Carburetors manufactured today usually include one of three types of cold start devices.

Tickler Cold Start System

In the early days of motorcycling, a **tickler system** was used on many European carburetors to aid in the quick start of a cold engine. This system used a pin and spring-loaded rod that, when pushed down, caused the carburetor's float needle to allow an excessive amount of fuel to enter the float bowl. The result was a richer air-and-fuel mixture. With this system, the fuel ran out the carburetor's overflow tube, because the fuel level was higher than normal. This, in turn, caused the engine to receive raw fuel in the intake port for an easier cold engine start. The problem with this cold start design was that the engine was easily flooded. More important is the fact that overflowing fuel has the potential of being very dangerous. With today's higher standards, this system is rarely found on motorcycles.

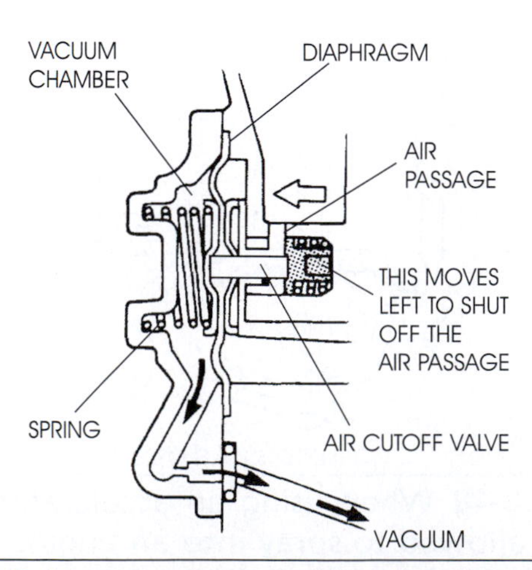

Figure 8-19 An air cutoff valve is used in many carburetors to prevent backfiring under deceleration.

Choke Plate Cold Start System

The **choke plate cold start system** is an air restriction system that controls the amount of air available during a cold engine start. This system uses a rider-controlled plate, called a choke valve, to block air to the carburetor venturi at all throttle openings (Figure 8-20). This plate has either a small hole cut into it or a spring-loaded *relief valve*, to allow some air into the carburetor venturi. This gives the engine enough air to run, but creates a very rich mixture in comparison to the mixture created if the plate were in the open position. The choke valve is located on the air filter side of the carburetor.

Enrichment Cold Start System

The **enrichment cold start system** is the most commonly used system on today's motorcycles with carburetors. With this system, an enrichment device feeds additional fuel into the carburetor via the starting enrichment fuel jet (Figure 8-21). The incoming air combines with the extra-rich fuel and moves quickly toward the engine's intake tract. The enriched mixture ignites readily and the engine

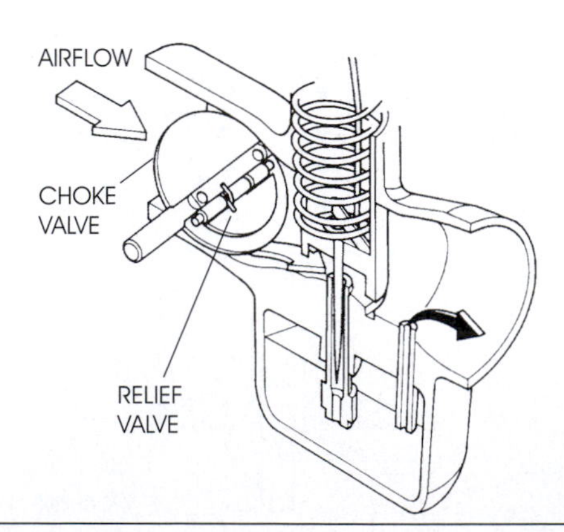

Figure 8-20 The choke valve cold start system closes off air to the engine.

starts. This system is somewhat like having a carburetor within a carburetor as the enrichment system acts under the same basic principles of a standard carburetor. When using this cold start system, it's important to remember to keep the carburetor's throttle valve closed. If the throttle valve is open too far, too much air enters the carburetor venturi. This makes the cold start mixture too lean, and thus, ineffective.

Slow-Speed Phase of Operation

Now that the engine is started and is getting warm, it doesn't require as rich a mixture as in the cold start phase. To maintain an idling speed, the engine needs a continuous flow of the air-and-fuel mixture, but the quantity required is just enough to keep the engine turning over. The mixture of air and gas is moderate. When the carburetor is in this idling phase, the throttle slide is almost completely closed, permitting only a small amount of air to pass through the main venturi. The major portion of the air is inducted through the **pilot jet air passageway** (Figure 8-22). This very small air passage in the pilot jet is sometimes called the primary venturi. The purpose of the **pilot air screw** is to adjust the amount of air flowing through. This can be regulated by turning the pilot air screw in or out slightly. As this tiny, fast-moving stream of air enters through the pilot jet, it picks up fuel from the float bowl and continues on into the main venturi. A very small volume of air in the main venturi, which slips past the slightly opened throttle slide combines with the mixture coming from the primary venturi. This is the final mixture that's drawn into the engine through the intake manifold. This slow-speed phase is in use from idle until the throttle slide is approximately one-quarter open.

Mid-Range Phase of Operation

The mid-range phase is used in cruising speeds, when the throttle slide is approximately 1/4 to 3/4 open. In this phase, the throttle slide is moved upward to permit a larger quantity of air to pass through the main venturi (Figure 8-23). Raising the throttle slide raises a **jet needle**, which fits through the opening at the top of the needle jet. Notice that the jet needle is tapered. As it moves upward, it permits more gas to flow from the float bowl up through the main jet, out the needle jet, and into the main venturi. The amount of fuel that's permitted to be drawn through the needle jet is controlled by the jet needle. As the jet needle is raised upward, more fuel is allowed to pass through the needle jet than when the jet needle is down.

Many jet needles have grooves cut into them for the placement of a retaining ring (Figure 8-24). Raising the retaining ring on the needle lowers the needle in the needle jet, creating a leaner mixture. On the other hand, lowering the retaining ring on the needle raises the needle in the needle jet, thus creating a richer mixture. In this way, the amount of fuel can be metered and atomized by the flow of the air entering through the venturi.

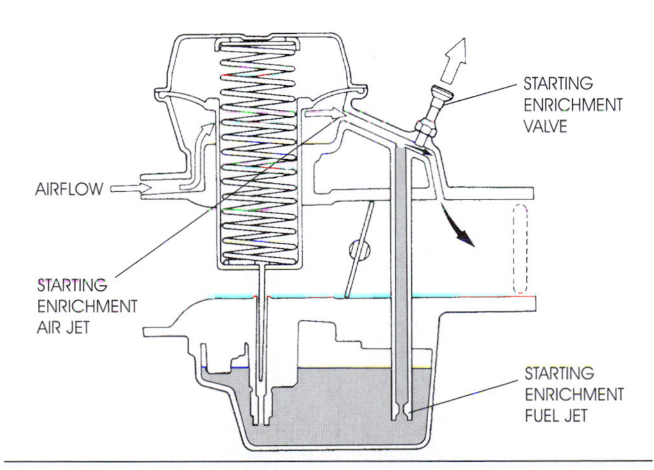

Figure 8-21 The enrichment valve permits the flow of more fuel to the engine and acts as a carburetor within a carburetor.

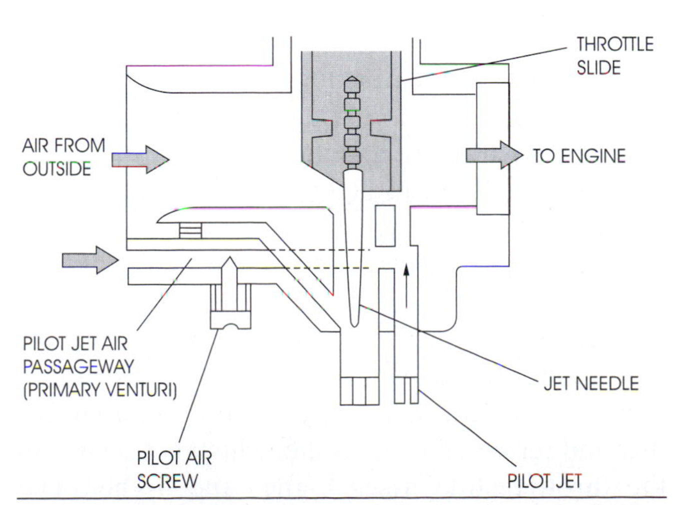

Figure 8-22 The slow-speed phase of carburetion can be seen in this illustration.

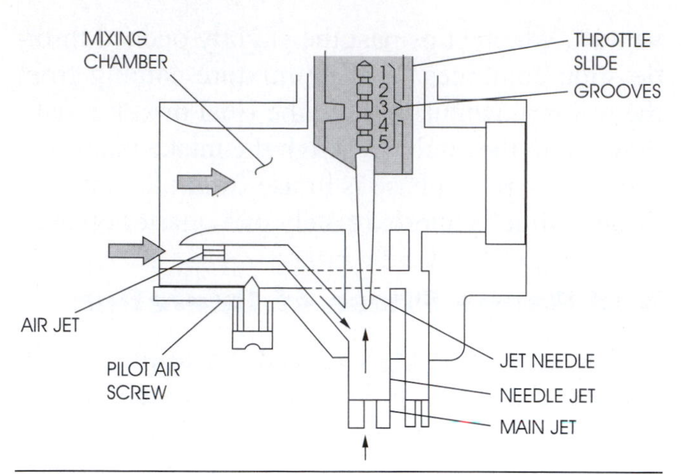

Figure 8-23 This illustration shows the mid-range phase of carburetion.

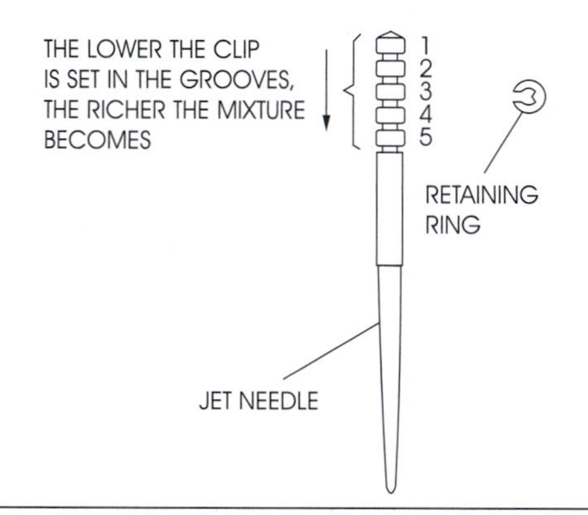

Figure 8-24 Many jet needles have grooves in them that allow for adjustment. Lowering the clip will raise the needle, which makes for a richer mixture while lowering the needle will tend to lean the mixture as fuel is blocked from flowing into the engine by the needle.

The object of the **main jet** is to control the amount of fuel allowed to pass through the needle jet. A hole located at the bottom of the main jet holder permits fuel to enter the needle jet during the mid-range phase of operation. The size of the main jet is not important at this point because fuel flow is partially restricted by the jet needle in the needle jet.

During the mid-range phase, the small quantity of fuel that enters the main venturi through the pilot jet is of minor consequence. As the speed of the engine increases, the demand for more fuel increases as well. As the throttle slide is raised further and further, the jet needle, which is fastened to the throttle slide, rises higher and higher. The tapered part of the jet needle moves up, permitting the amount of fuel entering the carburetor to be increased. Thus, in the mid-range phase of carburetor operation there's a wide range of possible positions that the needle and throttle slide can assume in order to accommodate variations in engine cruising speed.

High-Speed Phase of Operation

In the high-speed phase of operation, the throttle slide is 3/4 open to wide open. The only difference that can be noticed between the high-speed phase and the mid-range phase is that the throttle slide is closer to its absolute wide-open position (Figure 8-25). The jet needle is, therefore, in its highest position, allowing for the greatest amount of fuel flow into the engine. The jet needle barely fits through the needle jet, allowing the flow of gas through the main jet to be virtually unobstructed. The amount of fuel entering the carburetor venturi area is now totally controlled by the size of the main jet. While in the high-speed phase of operation, the engine is operating at maximum rpm.

A key point in all of these phases is that all phases of carburetor operation overlap with one another, as can be seen in Figure 8-26.

TYPES OF CARBURETORS

There are several carburetor designs, but as you've learned, the fundamental operation is the same for each design. Carburetors must atomize the fuel before the fuel reaches the engine. Proper atomization ensures that the air-and-fuel mixture is vaporized and the engine performs at its best. Carburetors

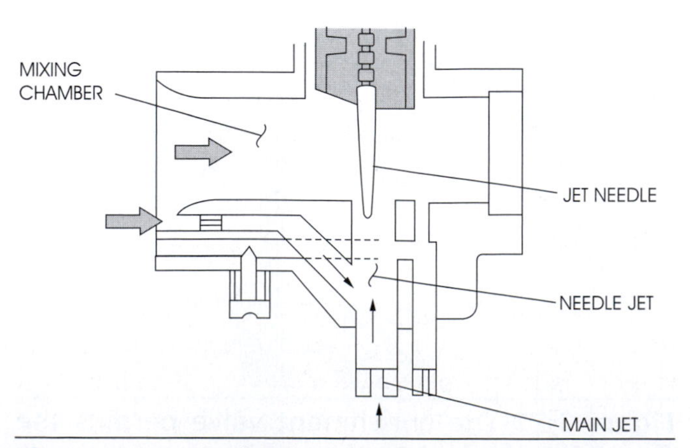

Figure 8-25 The high-speed phase of carburetion is illustrated here.

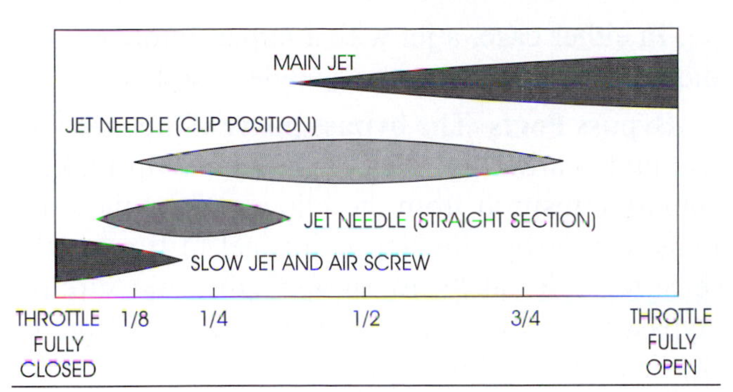

Figure 8-26 An important fact to remember is that all phases of carburetor operation overlap.

used on motorcycles can be grouped into two categories: variable venturi and fixed venturi.

Variable Venturi Carburetors

The variable venturi carburetor is, by far, the most popular carburetor used on today's motorcycles. There are two basic designs of the variable venturi carburetor. The **mechanical slide** variable venturi carburetor uses a rider-controlled slide to vary the size of the venturi. The **constant velocity** variable venturi carburetor, better known as the CV carburetor, uses engine vacuum to control the size of the venturi. Most of the internal components of these two carburetor designs are exactly the same, and serve the same function. The key difference between the mechanical slide and CV carburetors is in the way the venturi size is changed.

Fixed Venturi Carburetors

As the name implies, the fixed venturi has no way of changing the physical size of its venturi. The amount of air and fuel entering the engine is controlled by a throttle plate. This type of carburetor is not used on modern motorcycles; the fixed venturi carburetor was used on some older motorcycles, as well as on watercraft and snowmobiles.

Mechanical Slide Carburetors

As mentioned earlier, the venturi of the mechanical slide motorcycle carburetor is controlled by the rider. The rider operates the throttle, which in turn raises and lowers the throttle slide. This procedure controls the air volume, velocity, and pressure in the carburetor venturi. To provide the proper air-and-fuel mixture at all throttle openings, the mechanical slide carburetor has several fuel metering circuits, which we'll discuss in this section.

The Idle Circuit

The idle circuit may also be called a slow-speed or pilot circuit. The **idle circuit** meters the air-and-fuel mixture at engine idle and up to approximately 1/4-throttle opening. The idle circuit allows fuel to flow at all times during engine operation but has little effect after approximately 1/4 throttle. Because this circuit also has the smallest fuel jets, it's usually the first carburetor circuit to become restricted due to fuel contamination. Figure 8-27 illustrates the components of the idle circuit.

Idle Air Bleed Passage The **idle air bleed passage** is located at the air intake side of the carburetor. The purpose of the idle air bleed passage is to aid in the atomization of the idle circuit by mixing air with fuel in the circuit. Some air bleed passages may use a removable air jet. These air jets may have a number stamped on them to indicate their airflow rate. A larger number indicates a larger jet. A larger air jet allows a greater amount of airflow over a smaller air jet; thereby creating a slightly leaner carburetor condition.

Idle Outlet Port On a mechanical slide variable venturi carburetor, the **idle outlet port** is located in front of the carburetor slide close to the

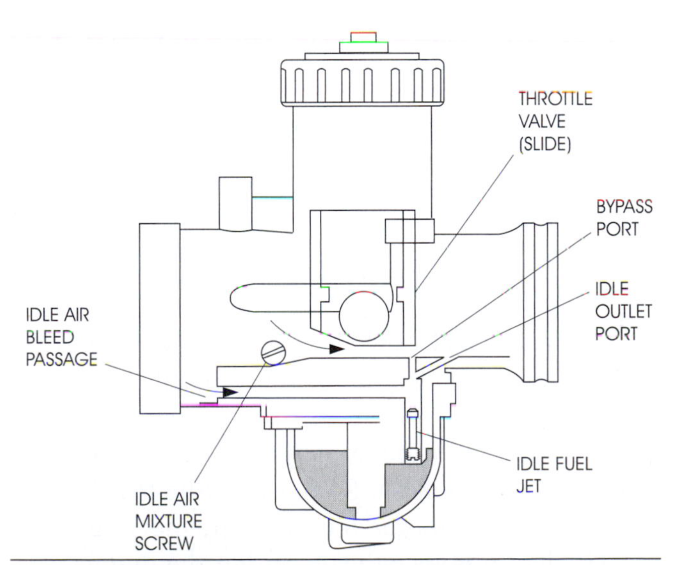

Figure 8-27 The idle circuit in a mechanical slide carburetor consists of an air bleed passage, an idle outlet port, a mixture screw, and a fuel jet.

intake port. The idle outlet port is the only means of fuel flow while the engine is idling.

Idle Mixture Screw Motorcycle carburetors have either an air mixture screw or a fuel mixture screw to help with idle circuit adjustment. Some motorcycle carburetors may have both, but this is extremely rare. Adjustments made to an air mixture screw or a fuel mixture screw require different procedures and have different effects.

Air mixture screw On a mechanical slide carburetor, the **air mixture screw** is always located on the air filter side of the carburetor slide. It varies the flow rate of the air bleed passageway. To create a richer air-and-fuel mixture, you turn the air mixture screw in (or clockwise). This reduces the amount of air allowed to enter the circuit, creating a rich mixture. To create a leaner air-and-fuel mixture, you turn the air mixture screw out (or counterclockwise). This increases the amount of air allowed into the circuit, causing a leaner mixture.

Fuel Mixture Screw The fuel mixture screw is always located on the engine side of the carburetor slide. The **fuel mixture screw** controls the amount of fuel exiting the idle outlet port. The fuel mixture screw changes the amount of fuel entering the carburetor while the engine is at idle. When this screw is turned out (or counterclockwise), it enriches the air-and-fuel mixture by increasing the amount of fuel allowed to enter the circuit. When the fuel mixture screw is turned in (or clockwise), the mixture becomes lean by reducing the amount of fuel allowed to enter the circuit.

When a fuel mixture screw is used on a street-legal motorcycle, the carburetors have factory-installed anti-tamper plugs over the screw to prevent the untrained consumer from changing the factory- and EPA-approved setting.

The Idle Fuel Jet The **idle fuel jet** is used to meter the amount of fuel flow through the entire idle circuit. This fuel jet is usually made of brass and may be removable or may be a pressed-in, non-removable component. Removable idle fuel jets are numbered. The larger the number, the larger the jet. The jet number indicates either the diameter of the jet's opening or the flow rate of fuel through the jet. In either case, a jet with a larger number flows more fuel than a jet with a smaller number.

Bypass Ports The **bypass ports** of the mechanical slide carburetor are designed to help make a smooth transition from the idle circuit to the mid-range circuit by allowing more fuel to flow as the throttle slide starts to rise off the idle circuit. Bypass ports are located directly under the front of the mechanical slide's carburetor slide.

Carburetor Slides The slide of a mechanical slide carburetor is controlled by the rider either by twisting the throttle grip or by pushing on a thumb lever. The slide raises and lowers, controlling the size of the venturi and providing more or less engine power as the rider wishes. The slide is one of three basic designs: round, flat, or radial flat (half round and half flat). The idle adjustment screw is installed in a location such that it can push the slide up or down to adjust the engine idle setting to factory specifications (Figure 8-28).

Slide Cutaway The slide cutaway is located on the air filter side of the slide bottom and is cut at an angle (Figure 8-29). The cutaway portion of the slide always faces the air filter side of the carburetor venturi. The cutaway in the slide controls the airflow rate allowed to pass through the engine, primarily between 1/8- and 1/4-throttle opening. When a larger cutaway is used, the mixture of air and fuel is leaner. The purpose of the slide cutaway is to aid in transition from the idle circuit to the mid-range carburetor circuit.

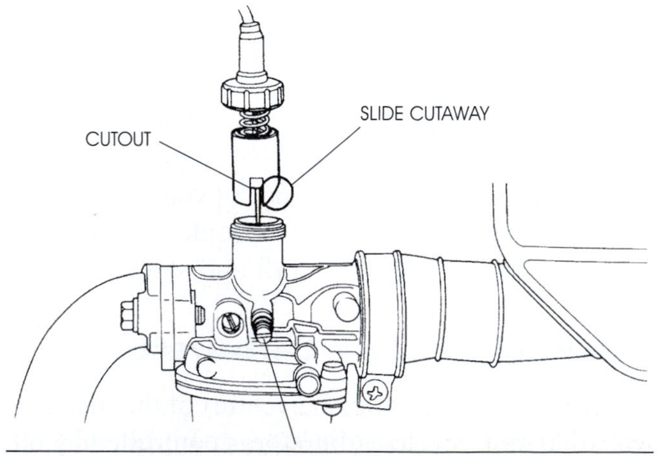

Figure 8-28 The idle adjustment screw on a typical mechanical slide carburetor.

The Mid-Range Circuit

The mid-range circuit is most effective between 1/4 to 3/4 throttle opening. The mid-range circuit contains two primary components: the needle jet and the jet needle.

Needle Jet The **needle jet** is a stationary jet that's located in the carburetor body. This jet is numbered for its flow rate. A larger number indicates a richer mixture for the entire mid-range circuit. The needle jet is located in series with the main jet (see Figure 8-23 shown earlier in this chapter). The needle jet contains the air bleed tube for both the mid-range and main jet circuits. There are three types of needle jets used on mechanical slide carburetors.

Bleed-Type Needle Jet The **bleed-type needle jet** is identified by several air bleed holes located on its sides (Figure 8-30). The top of the bleed-type needle jet sits flush with the venturi floor. The bleed-type needle jet allows for good atomization of fuel over a wide range of conditions.

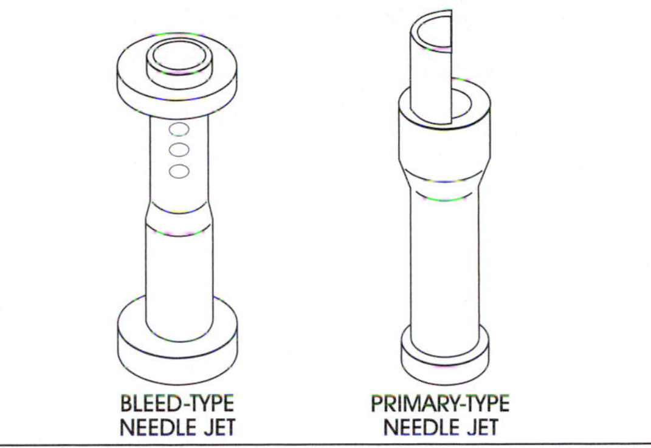

Figure 8-30 Different types of needle jets are used on mechanical slide carburetors depending on the intended use.

Primary-Type Needle Jet The **primary-type needle jet** has a hood that protrudes into the throat of the carburetor. The hood, which acts similar to an airfoil, is designed to help increase the negative pressure above it. This design has only one air bleed hole and is found primarily in two-stroke motorcycle engines.

Primary Bleed-Type Needle Jet This jet is a combination of the bleed and primary types. It's hooded and also has several air bleed holes. The primary bleed needle jet allows for good atomization along with increased throttle response over the primary-type jet. This needle jet may have an air bleed passage or an air bleed jet. Although needle jets have different designs, they all have the same responsibility of helping to atomize the fuel for both the mid-range circuit and the main circuit.

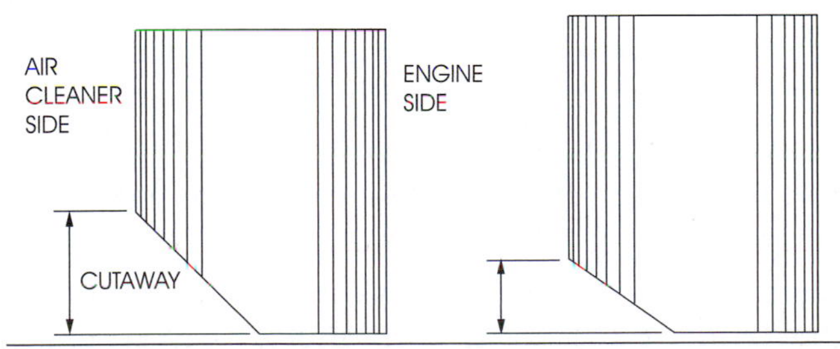

Figure 8-29 The slide cutaway on the left allows more airflow than the slide cutaway on the right. The left slide creates a leaner mixture in the carburetor than the right slide.

Jet Needle The jet needle is a long, tapered needle that moves up and down with the throttle slide as the slide opens and closes. The jet needle varies the amount of fuel flow as it flows through the needle jet. At 1/4-throttle opening, the needle restricts the flow of fuel through the needle jet more than at 3/4-throttle opening (Figure 8-31).

As mentioned earlier and shown in Figure 8-24, many jet needles have up to five grooves cut into them for a retaining ring, allowing for adjustment of their static height positions. The number one or top groove is the leanest jet needle setting, as it lowers the needle down into the needle jet. On the other hand, the number five or bottom groove is the richest setting, as it raises the needle higher in the needle jet, thus allowing greater fuel flow.

The Main Jet Circuit

The **main jet circuit** controls the range of 3/4 to wide-open throttle. As mentioned earlier, the main fuel jet is located in series with the needle jet, and works in conjunction with the needle jet and jet needle. At 3/4 throttle to full throttle, the needle jet becomes virtually unrestricted. The purpose of

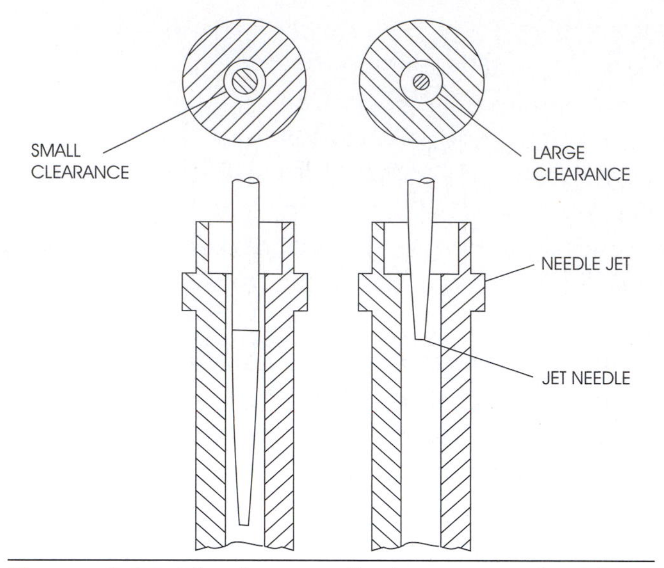

Figure 8-31 The jet needle lets more fuel into the carburetor venturi as it rises in the slide.

the main jet is to regulate the amount of fuel that flows into the needle jet at this throttle opening range. Without a main jet in place, the engine would allow far too much fuel to flow into the venturi of the carburetor. This would create an excessively rich air-and-fuel mixture and flood the engine. It is very important to remember that although the throttle opening controls the operation of all the circuits, there is an overlap from one circuit to the next to allow for a smooth-running engine over a very wide range of engine power needs.

Constant Velocity Carburetors

The constant velocity, or CV, carburetor is also a variable venturi carburetor that's widely used on motorcycles. The constant velocity carburetor is very similar to the mechanical slide carburetor. The major difference is that the venturi size isn't controlled by a throttle cable that raises and lowers the throttle slide, instead, on a CV carburetor, the venturi is opened when the throttle slide is raised by pressure differences created by the engine as it's operating. The air-and-fuel mixture is actually controlled by the needs of the engine. If the engine is running at a slow speed, it won't create enough vacuum to raise the slide and, therefore, won't draw in an excessive amount of air and fuel. In contrast, a cable-controlled mechanical slide carburetor can be opened fully by the rider regardless of engine needs, creating the potential for drawing in an excessive amount of air at low engine speeds.

Another difference between a CV carburetor and a mechanical slide carburetor is that the CV carburetor has a rider-controlled butterfly throttle valve. The butterfly throttle valve regulates the flow of air into the engine intake tract. Other than these two differences—the procedure that moves the throttle slide and the use of a butterfly throttle valve—the components and functions of the CV carburetor are nearly identical to those used on the mechanical slide carburetor. Let's learn more about the two components that are different on the CV variable venturi carburetor.

Butterfly Throttle Valve

The rider-controlled **butterfly throttle valve** is a thin, flat disc that fits in the venturi between the throttle slide and the intake manifold (Figure 8-32). The butterfly throttle valve's job is to open and close the body of the carburetor, thus increasing or decreasing the flow of air into the engine intake tract. The rider controls the action of the butterfly valve by manipulating a cable connected to the handlebar twist grip (throttle). An idle adjustment screw is installed in a location that allows it to open the butterfly throttle valve and adjust the engine idle setting to factory specifications.

CV Carburetor Slide

The CV carburetor slide looks very similar to the mechanical throttle slide. The difference is that the CV-type slide has a rubber-like (usually neoprene) diaphragm or a piston-like apparatus that separates the vacuum chamber from the atmospheric pressure area beneath the vacuum chamber (Figure 8-32). The differences in pressure control the movement of the CV carburetor.

Recall what happens when the throttle is opened on a mechanical slide carburetor. When the rider twists the throttle control on a mechanical slide carburetor for a quick increase in engine power, the effect on the carburetor is to instantly raise the throttle slide. This briefly upsets the proportion of the air-and-fuel mixture, creating a rapid increase in the airflow through the carburetor venturi. The effect is a temporary lean mixture until the carburetor system catches up and returns to a balanced mixture.

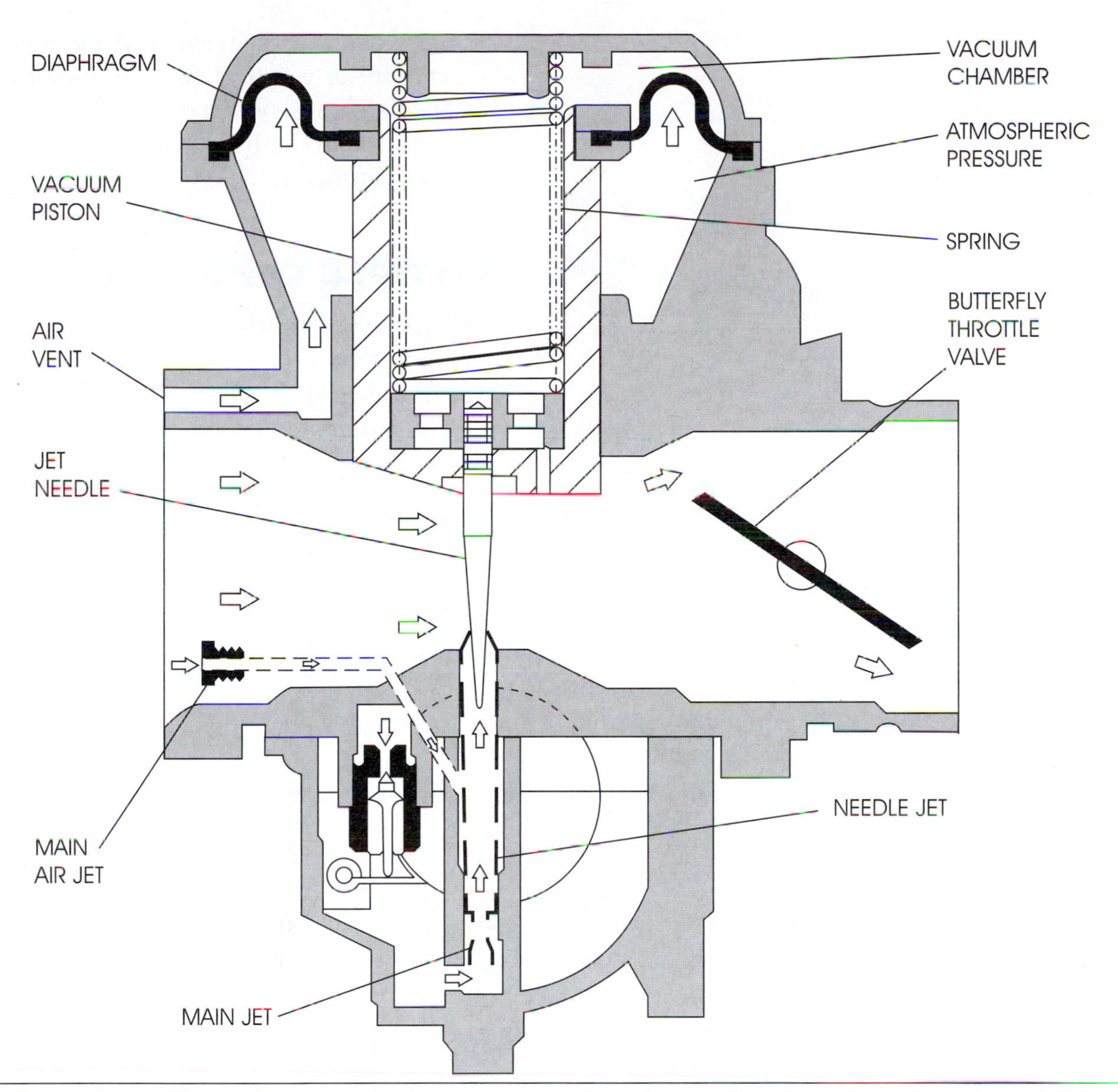

Figure 8-32 On a CV carburetor, the rider controls a butterfly throttle valve. In addition, a diaphragm separates the vacuum chamber from the atmospheric pressure area beneath the vacuum chamber. These pressure differences control the raising and lowering of the CV carburetor slide.

In contrast, with motorcycles equipped with CV carburetion, when the rider turns the throttle control for a sudden increase in speed, the butterfly throttle valve opens between the carburetor and the engine. The airflow in the main bore of the carburetor exerts a strong negative pressure on the lower section of the vacuum piston. At this point, air is drawn out the carburetor's vacuum chamber and the pressure in the chamber drops. The atmospheric pressure beneath the vacuum piston is greater than the pressure in the vacuum chamber, and this allows the slide to rise. This process can be seen in Figure 8-33. A low pressure area is created above the top part of the slide, which causes the slide to rise. As the slide rises, the tapered jet needle lifts up in the needle jet and permits an increased amount of fuel to enter and become atomized with the incoming air stream. When the throttle slide is returned to the closed position, the process is reversed. The airflow in the main bore is closed off. The pressure within the venturi and carburetor bore rises because of a decrease in the airflow, which slows down the air speed. The throttle slide is lowered by the force of a spring located within the vacuum chamber.

The CV carburetor slide can be thought of as an on-demand carburetor system, where the slide opens

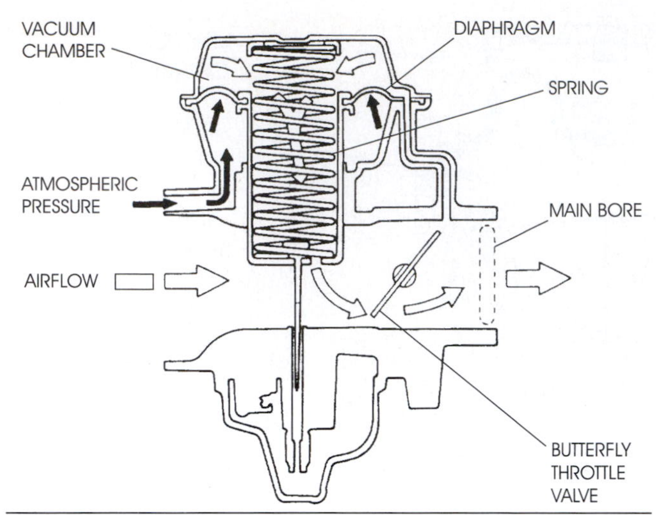

Figure 8-33 Atmospheric pressure pushes the CV carburetor slide upward as the pressure in the vacuum chamber decreases.

in keeping with the engine's demands. A decrease in speed means a decrease in demand. When the rider returns the throttle to an idle, the butterfly valve between the carburetor and the engine closes, the vacuum to the top side of the diaphragm decreases, and the throttle slide eases downward. With the decrease in demand, the quantity of the air-and-fuel mixture decreases. Changes in speed are made more smoothly when a CV carburetor is used.

As mentioned earlier, the circuits and parts used in the CV carburetor are the same as in the mechanical slide carburetor. If needed for review, see the previous discussion on mechanical carburetors in this chapter.

Comparison of Mechanical Slide and CV Carburetors

The CV carburetor offers a number of efficiency advantages over the mechanical slide carburetor. In addition, both types of variable venturi carburetors are better suited for use on motorcycles than the fixed venturi carburetor. The differences between the two variable venturi carburetors are minimal. In practice, the selection of a particular variable venturi carburetor depends upon the engine design and overall cost of manufacturing the machine. Generally, the CV carburetor costs more to manufacture than the mechanical slide carburetor. On many special purpose machines, such as moto-cross racing motorcycles, the simplicity and responsiveness of the mechanical slide carburetor is felt to be more important than improved efficiency. Therefore, both CV and mechanical slide carburetors are used in a wide range of motorcycles.

MULTIPLE CARBURETORS

Almost every multi-cylinder motorcycle engine uses one carburetor for each cylinder. In these cases, the engine is using multiple carburetors. Multiple-carburetor designs can be the mechanical slide or CV types and are identical to the carburetors we've discussed in this chapter. The only difference is that there are more of them.

Base Carburetor

Multiple carburetors are connected together as banks in such a way that one carburetor controls the opening of all the others (Figure 8-34).

The controlling carburetor is known as the **base carburetor**. The base carburetor can be easily spotted as the carburetor with the idle adjustment screw attached to it as seen in Figure 8-35. Carburetors are connected together by the use of linkages and plates to allow each carburetor to open with the base carburetor.

Multiple Carburetor Disassembly

When disassembling multiple carburetors to clean or rebuild them, they should *never* be separated from their banks unless an individual carburetor within the set needs to be replaced, which is extremely rare. Multiple carburetors may be cleaned and repaired as a set. The primary reason for not separating multiple carburetors from their banks is due to the many very small springs and linkages between them as seen in Figure 8-36.

Carburetor Synchronization

Carburetor synchronization is the process of making the internal systems within a set of two or more carburetors operate the same in terms of the amount of air-and-fuel mixture drawn through each one. Carburetor synchronization is measured by the engine's vacuum at each of the carburetor

Figure 8-34 A typical bank of four carburetors.

intake manifolds. The operating temperature, smoothness, and response of a multi-cylinder motorcycle engine depend greatly upon proper synchronization. This is especially vital to performance on a multi-cylinder engine having one carburetor per cylinder. Although the physical linkages and synchronization methods vary from model to model, the basic principles of carburetor synchronization are the same for all multiple-carburetor engines. These principles are covered thoroughly in Chapter 19.

Remember, no matter how complex carburetors may seem due to the number of carburetors used, the basic fundamentals of carburetion always remain the same. A carburetor is simply an atomizer that supplies an internal-combustion engine with a combination of vaporized fuel, mixed with air in amounts that will burn most efficiently.

FUEL INJECTION

Fuel injection is a modern means to carburetion on the majority of today's motorcycles. The purpose of fuel injection is to allow an extremely precise metering of air-and-fuel mixture ratios at any given engine condition. This results in each cylinder getting only the amount of fuel it needs at all times instead of a preset amount being delivered at all times as with traditional carburetors. Other than the method of getting fuel into the engine, the basic components of this system aren't much different from a standard carburetor engine. In today's motorcycle engines, fuel injection is very popular especially on street-going models. Fuel injection makes it much easier to adhere to the strict guidelines of the United States' Environment Protection Agency (EPA). This helps to ensure that the air we breathe remains as clean as possible. How? By making our modern motorcycles are

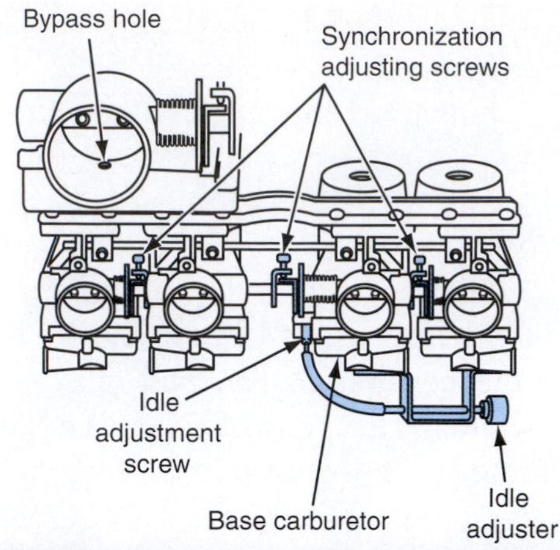

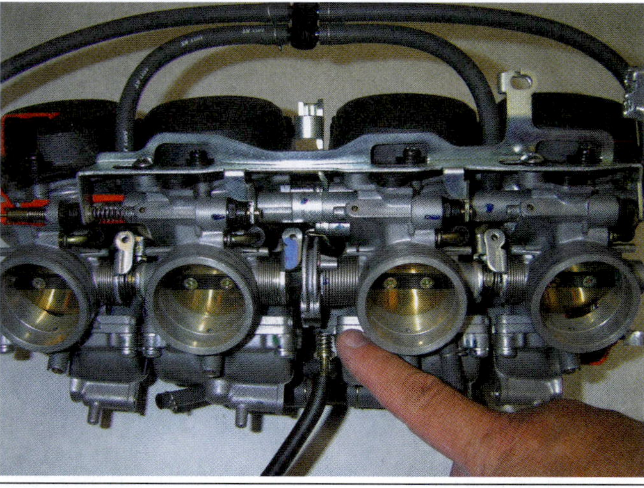

Figure 8-35 The base carburetor is the one with the idle adjustment screw attached to it. The synchronization screws are used to adjust each individual carburetor to operate equally.

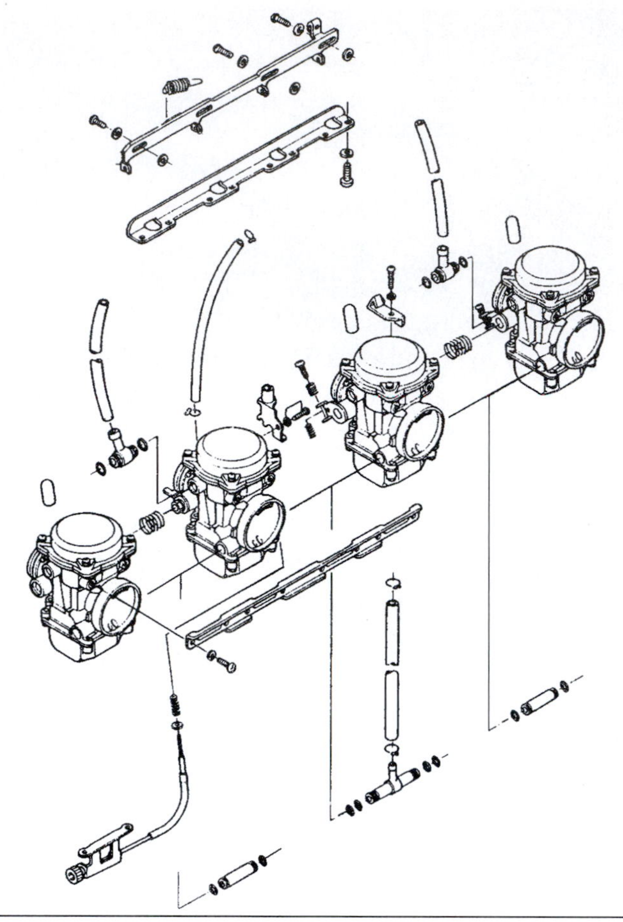

Figure 8-36 There are many small parts to a bank of carburetors. A bank of carburetors should be kept intact. Most repairs such as cleaning can be done without separating the carburetors.

as fuel efficient - combustion-wise - as possible while remaining at a performance level unlike anything we have seen in the past. We are also seeing more and more application of fuel injection on ATVs and off-road motorcycles.

The primary advantage of fuel injection over traditional carburetion is the ability for a fuel-injected engine to automatically adjust to the constantly changing atmospheric conditions that it runs in. Conditions such as temperature, humidity, and altitude all have affects on traditional carburetion that would cause a carbureted motorcycle to run differently unless one was to make physical adjustments to the carburetor settings. With a fuel-injected machine, all of these conditions are compensated for by the use of sensors found within the fuel-injection system.

There primary type of fuel injection found on today's motorcycles is called indirect fuel injection. There is also another type of system known as direct fuel injection.

Direct Fuel Injection

With the direct fuel-injection system, fuel is injected directly into the combustion chamber. This type of fuel injection is found primarily on diesel engines, and not generally found on motorcycle engines. The direct system injects an extremely fine mist of fuel into the combustion chamber just prior to top-dead center (TDC) of the engine's compression stroke.

Indirect Fuel Injection

Indirect fuel-injection systems are found on virtually all motorcycle engines that utilize fuel injection. When an indirect fuel-injection system is used, fuel is injected into the intake tract before the intake valve. All modern fuel-injected motorcycle engines use **Electronic Fuel Injection (EFI)**, although the name that some manufacturers use may be different such as **Computerized Fuel Injection (CFI)** or **Programmed Fuel Injection (PGM-FI)**. All of these systems use an Electronic Control Unit (ECU) to control the amount of fuel being delivered to the engine.

Indirect electronic fuel-injection systems give motorcycles the ability to meet EPA standards—standards that are getting tougher to comply with each year. At the same time, modern EFI systems provide excellent performance.

Electronic Fuel-Injection System Components

Figure 8-37 shows an overview of the systems and components found in a typical EFI system on a modern motorcycle. We will discuss EFI-related system components beginning with the area of fuel delivery.

Fuel Pump

Fuel pumps used with electronic fuel-injected motorcycles have three primary requirements. They must be electric powered; they must have the ability to handle a high volume of fuel; and they must have the ability to supply high pressure to the injectors. Numerous modern motorcycle EFI fuel pumps are located inside the fuel tank of the motorcycle to save space (Figure 8-38) as well as preventing vapor lock, a condition that is caused when gasoline overheats and begins to actually boil within the fuel pump. The ECM controls the operation of the fuel pump. The fuel pump will normally operate for a couple of seconds after the key is first turned on to pressurize the fuel injectors.

The fuel pump consists of an electric armature that spins between two magnets and turns an impeller that draws fuel in and through the pump (Figure 8-39). A check valve is incorporated to maintain pressure at the fuel injectors to allow for quick engine starts. Fuel is sealed in this system and therefore cannot evaporate or deteriorate during long periods of storage such as during the winter months. A relief valve is also located within the fuel pump and is opened to send fuel back into the fuel tank if a fuel line were to become restricted and cause excessive pressure build up.

Fuel Filters

There are normally at least two **fuel filters** used on EFI systems. Before fuel enters the fuel pump, it

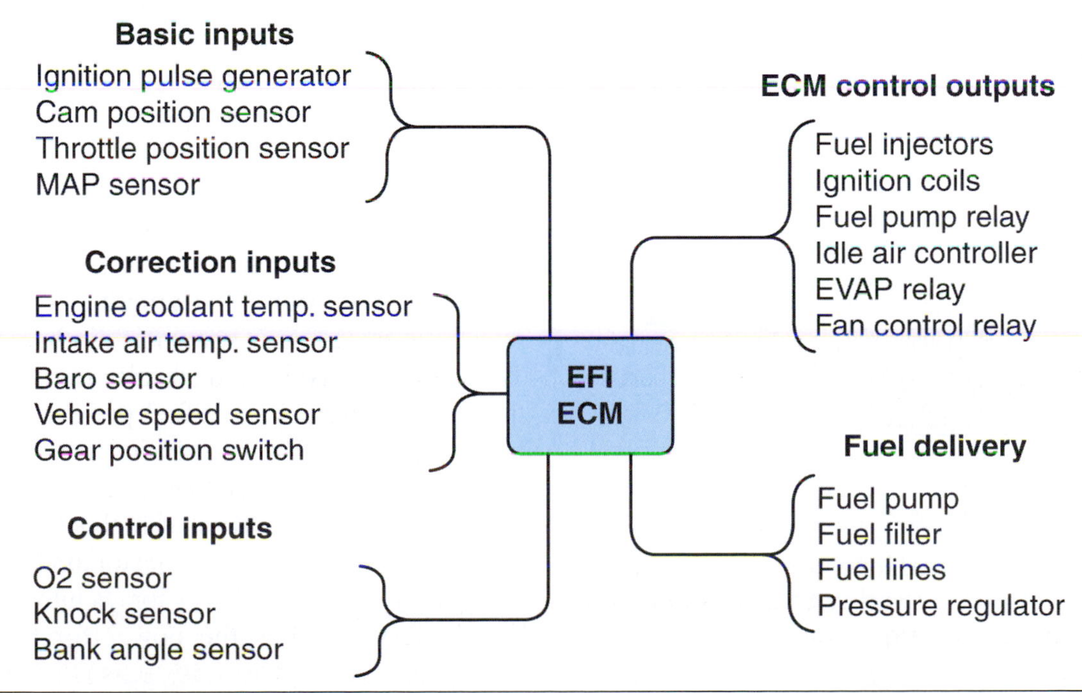

Figure 8-37 Components of an electronic fuel-injection system. Copyright by American Honda Motor Co., Inc. and reprinted with permission.

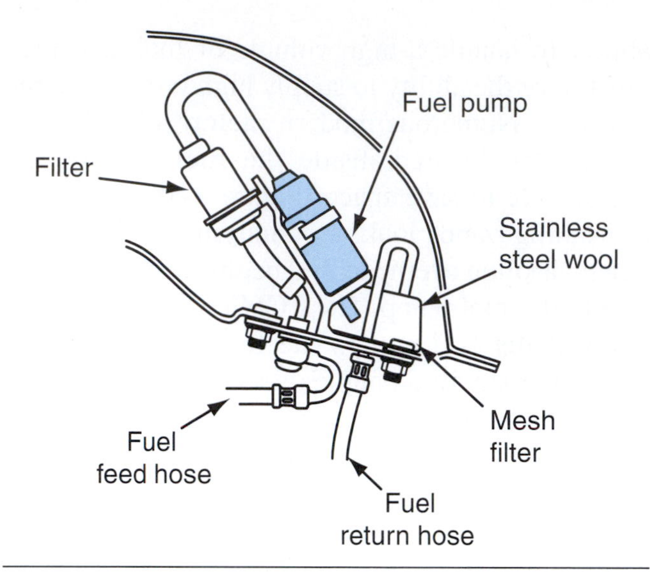

Figure 8-38 Locating the fuel pump inside the fuel tank saves space and also prevents vapor lock.

must first go through a mesh filter that prevents grit and rust from entering the pump and damaging it. Another filter used is a large inline type and can be mounted inside the fuel tank or outside of the tank (Figure 8-39 and Figure 8-40). The operation of fuel filters is critical in a fuel-injected system because clogged fuel injectors won't function properly.

Fuel Lines

EFI machines use special high-pressure fuel lines from the fuel pump to the injectors and can

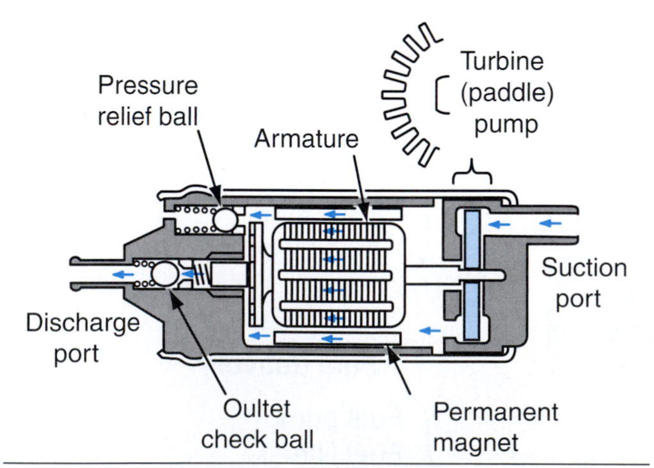

Figure 8-39 The various internal components of an electric fuel pump used with a fuel-injection system is illustrated here.

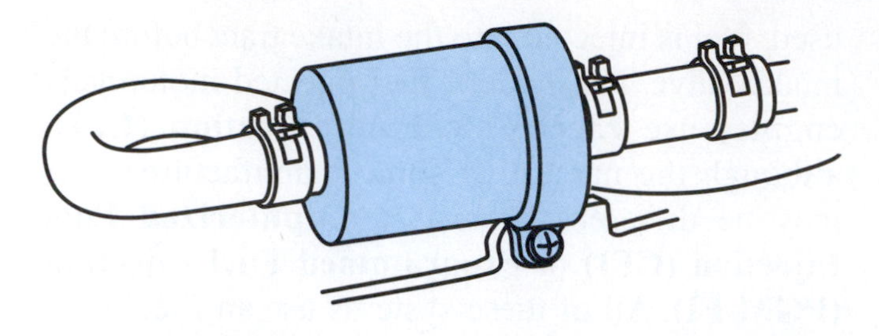

Figure 8-40 A typical externally mounted fuel filter is illustrated here.

be damaged if mishandled by excessive bending or stretching. The damage in many cases will be internal and therefore you will not see it until the line breaks under pressure. When servicing EFI motorcycles be sure to adhere to the appropriate service manual to avoid damaging the fuel lines.

Fuel Pressure Regulator

The fuel pressure regulator maintains correct fuel pressure and keeps it above the pressure of the intake manifold. Excessive pressure is returned to the fuel tank by a separate return hose. A typical fuel pressure regulator can be seen in Figure 8-41.

Fuel Injectors

The **fuel injector** is an electronically operated solenoid that turns fuel on and off (Figure 8-42). They are normally closed and are either fully closed or fully open. The electronic control module tells the fuel injector when to turn on and off. The control unit also determines how long the injector must stay on, therefore telling the injector how much fuel is injected into the engine. This is known as **injector discharge duration**. The length of time the fuel injector is turned on is known as discharge duration. Three things cause fuel to atomize in an EFI system: the shape of the injector, fuel pressure, and turbulence in the air intake tract. Inside the injector there's a spring-loaded plunger that closes against a valve seat. Once seated, the flow of fuel is blocked. When the solenoid coil within the injector assembly lifts the plunger, the pressurized fuel sprays into the cylinder. The battery supplies the power for the solenoid

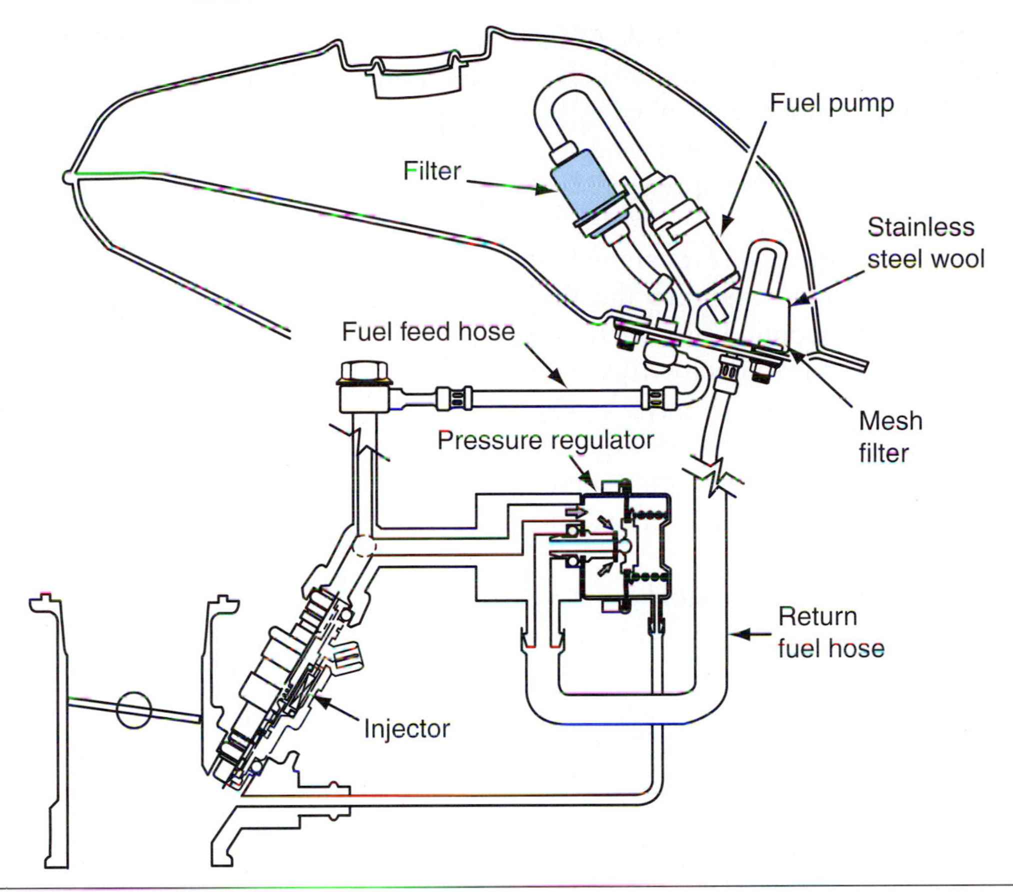

Figure 8-41 In this illustration you can see that the fuel filter is located inside the fuel tank.

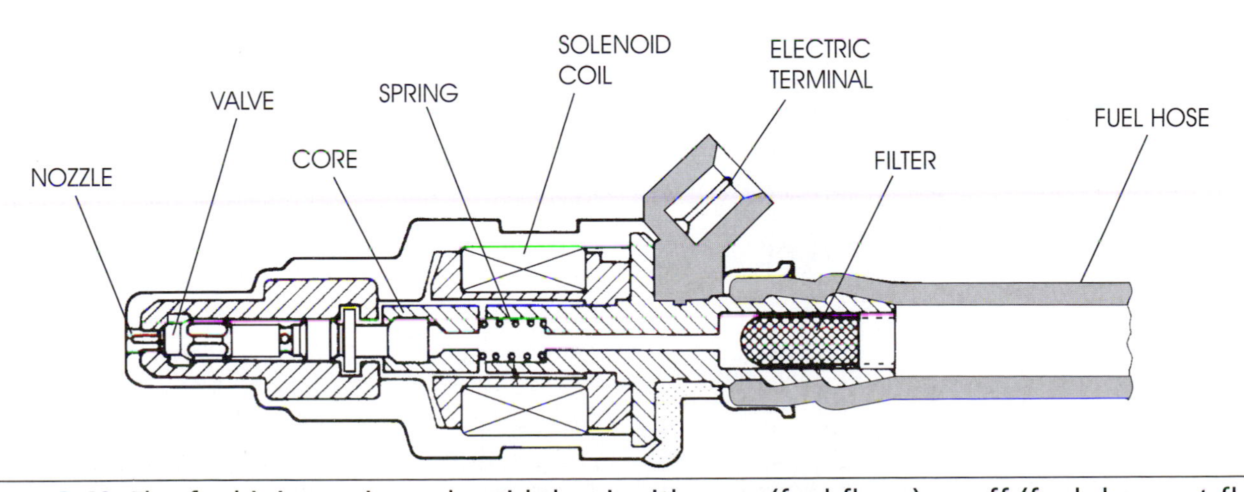

Figure 8-42 The fuel injector is a solenoid that is either on (fuel flows) or off (fuel does not flow).

coil and the ECM controls the ground side of the injector, therefore making the injectors "switch to ground circuits." Each injector is controlled by the ECM and fuel is delivered only to the cylinder as it is needed. This is known as **sequential fuel injection**.

Fuel injector tip openings are designed to provide a spray pattern that atomizes the fuel to help it mix with the incoming air. There are many different types of fuel injector tips with different outlet hole designs (Figure 8-43). Different outlet designs are used to vary the spray pattern to the manufacturers requirements for different performance levels as well as cost of manufacturing.

Fuel Injector Failures Fuel injectors can have two types of failures: electrical and mechanical.

Figure 8-43 Various types of tips can be found on a fuel injector. Decisions on the type of injector used can be based on intended use as well as cost. Copyright by American Honda Motor Co., Inc. and reprinted with permission.

The following are three possible electrical failures for an injector:

- High resistance
- An open
- A short

Some motorcycles can detect electrical failures while the machine is running. Others can only detect a failure when starting the motorcycle.

The following are two possible mechanical failures for an injector:

- Leaking fuel (partial or complete)
- Blocked fuel discharge (partial or complete)

Possible fuel leakage is indicated by the following:

- Dark spark plug color
- Fuel-fouled spark plugs

Blocked fuel discharge is indicated by the following:

- A cold exhaust pipe on that cylinder

Electronic Control Module

The Electronic Control Module (ECM) uses a microcomputer to process data and control the operation of the fuel injectors, ignition spark, and timing as well as the fuel pump. The ECM receives information from basic input sensors and determines what, when, why, and how long the various operation steps need to be controlled. Depending on the manufacturer an ECM can also be called an Electronic Control Unit (ECU).

ECM Inputs and Outputs

The engine control module has the following three types of inputs:

- Basic
- Correction
- Control

The basic inputs (Table 8-2) provide information that the ECM needs to select a particular mixture control map (most EFI systems have at least two maps). The ECM then selects the basic fuel discharge duration from the chosen map. These inputs include ignition pulse, camshaft position sensor, throttle position sensor, and the vacuum pressure in the intake manifold (MAP sensor).

The correction inputs (Table 8-3) provide the information the ECM needs to adjust the basic fuel discharge duration. Typical correction inputs would include engine temperature, intake air temperature, barometric pressure (BARO), and vehicle speed.

The control inputs (Table 8-4) provide the information the ECM needs to adjust engine operation. These inputs would be the oxygen sensor and knock sensor. A bank angle sensor is often used on motorcycles to cut off electrical power to the ECM in the case of the machine tipping over. Bank angle sensors are designed to stop the engine.

ECM outputs include the fuel injection and ignition spark as well as the operation of the fuel pump and cooling fan on liquid-cooled machines.

Sensors

Various sensors that can be found back in Figure 8-37 monitor the engine and atmospheric conditions such as throttle position, engine rpm, engine and intake air temperature, vehicle speed, Manifold Absolute Pressure (MAP, which is calculated into air density), coolant temperature, and piston position. These sensors assist in all aspects of EFI and send information to the ECM to allow the engine to run as efficiently as possible.

Throttle Body

Most models with PGM-FI have one throttle valve for each cylinder. The throttle body contains the injector as well as a butterfly valve (Figure 8-44). Motorcycles with EFI do not need to depend on the venturi effect due to the fuel injector's delivery of a precise amount of fuel at any given time, unlike a carbureted motorcycle that will receive the same amount of fuel at a given throttle opening.

EFI Self-Diagnostics

Most modern motorcycles have a self-diagnostic system incorporated to assist technicians when problems arise. Various components on EFI are monitored continuously by the self-diagnosis function. If the ECM notices a fault, a **Malfunction Indicator Light (MIL)** switches on (Figure 8-45) and depending on the severity of the fault may only give a warning to the rider. In other cases, the engine may go into a fail-safe operation mode. This allows the engine to continue to run but at a reduced performance level, or stop completely depending on the severity of the fault such as when an electrical-related problem is detected by the system's sensors. The MIL is used to detect and assist in diagnosing any EFI-related electrical failure.

Table 8-2: Typical basic inputs for an electronic fuel-injection system are shown here. Copyright by American Honda Motor Co., Inc. and reprinted with permission.

Sensor	Measures	Signal	Component	Comment
Crankshaft Position Sensor	Engine Speed	Spiked Wave (AC)		The EFI system cannot operate without this input. Measure this voltage signal with a peak voltage tester.
Camshaft Position Sensor	TDC of Cylinder No. 1	Spiked Wave (AC)		The EFI system cannot operate without this input. Identifies cylinder 1 for sequential injection during each intake stroke. Measure voltage signal with a peak voltage tester.
Throttle Position (TP) Sensor	Throttle Setting	Analog (DC) 4.5V IDLE WOT 0.5V IDLE → WOT		Mounted on the end of the throttle shaft. Sensor voltage varies with throttle opening. Idle: about 0.5 volts. Full open: about 4.5 volts. Sensor is not adjustable. Throttle body must be replaced if it fails.
Manifold Absolute Pressure (MAP) Sensor	Vacuum pressure in the intake manifold	Analog (DC) 4.5V 0.5V LOW → HIGH PRESSURE		Voltage from the MAP sensor increases with absolute pressure. With key on and engine off, voltage from the map sensor is 2.3 to 3.0 volts. Compare this output with the BARO sensor output, with key on, engine off. They should be very similar. Note: Some models also use the map sensor on startup to measure atmospheric pressure. No BARO sensor is used.

Table 8-3: Typical correction inputs for an electronic fuel-injection system are shown here. Copyright by American Honda Motor Co., Inc. and reprinted with permission.

Sensor	Measures	Signal	Component	Comment
Engine Coolant Temperature (ECT) Sensor	Engine Temperature	Analog (DC)		This sensor measures engine coolant temperature. The output voltage decreases as the temperature increases. At 70 degrees F the output voltage is 2.8 to 3.0 volts. Compare this output voltage to the IAT sensor when the engine is cold. They should be similar.
Intake Air Temperature (IAT) Sensor	Temperature in air cleaner case	Analog (DC)		This sensor measures temperature in the air cleaner case. The IAT sensor has a plastic case that allows it to respond quickly to changes in air temperature. At 70 degrees F this output voltage is 2.8 to 3.0 volts. Compare this to the ECT sensor when the engine is cold. They should be similar.
Barometric Pressure (BARO) Sensor	Atmospheric Pressure	Analog (DC)		This sensor measures atmospheric pressure. Its input allows the ECM to adjust the air/fuel mixture for changes in altitude. The output voltage with the key on and engine off is 2.3 to 3.0 volts. Compare this sensor's output with the MAP sensor under the same test conditions. They should have similar output voltages. A MAP sensor part is frequently used as the BARO sensor.
Vehicle Speed Sensor	Road Speed	Digital (DC)		The vehicle speed sensor input is used to switch the cooling fans off at 12 mph on the GL1800.
Gear Position Switch	Gear Position	N/A		Gear position is supplied to the ECM in order for it to select the correct ignition map for the gear selected. Motorcycles have separate ignition maps for neutral, 1st, 2nd, 3rd, and above.

Table 8-4: Typical control inputs for an electronic fuel-injection system are shown here. Copyright by American Honda Motor Co., Inc. and reprinted with permission.

Sensor	Measures	Signal	Component	Comment
Oxygen (O_2) Sensor	Oxygen in exhaust	Analog (DC) 1V 0V		An oxygen sensor produces its own voltage: Nearly a full volt if it senses very little oxygen, when the mixture is rich. Close to zero when the mixture is too lean and there is excessive oxygen. About a half volt when the mixture is just right.
Knock Sensor	Detonation			The knock sensor detects detonation and from this input the ECM retards the ignition timing.
Bank Angle Sensor	Vehicle Position	N/A	IGNITION SWITCH TRANSISTOR TO ENGINE STOP RELAY LATCH-UP CIRCUIT Upright	The bank angle sensor controls the operation of the engine-stop relay and the fuel-cut relay. The sensor consists of a reed switch, latch-up circuit, and an oil-dampened pendulum with two magnets.
				When the ignition switch is turned on, power flows through the latch-up circuit turning the transistor on, then the current flows through the coil of the engine-stop relay to ground.
			TO ENGINE STOP RELAY IGNITION SWITCH TRANSISTOR MAGNET LATCH-UP CIRCUIT LEAD SWITCH Tipped Over	When the motorcycle and sensor are tipped more than 50 degrees the magnet sensor pendulum closes the reed switch. This causes the latch-up circuit to turn off the transistor, opening the circuit between the engine-stop relay and ground. This stops power to the ECM.
				To reset the sensor, turn the ignition key switch off before attempting to restart the engine.

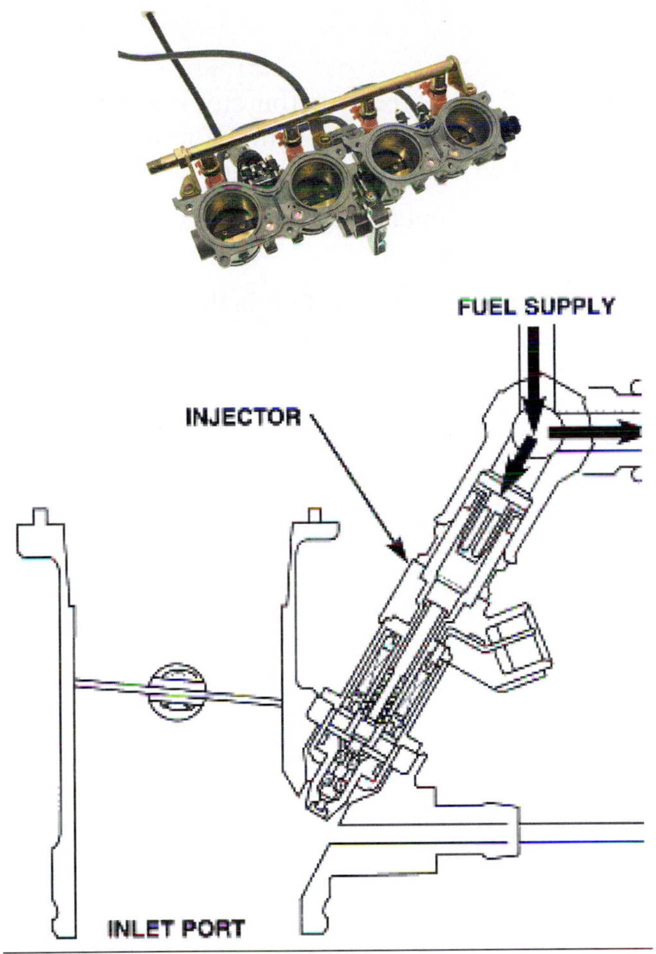

Figure 8-44 A throttle body for an electronic fuel-injection system is shown here along with an illustration of a fuel injector and the inlet port of the throttle body. Illustration Copyright by American Honda Motor Co., Inc. and reprinted with permission.

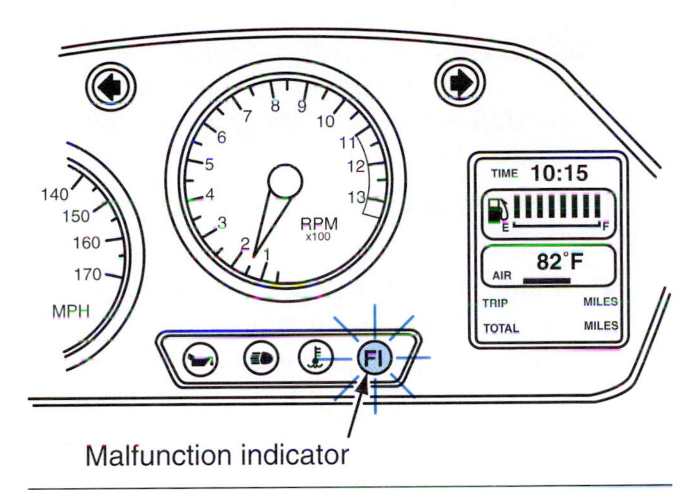

Figure 8-45 The malfunction indicator light (also known as the "MIL"), will let a rider know if a failure is detected in the EFI system. Copyright by American Honda Motor Co., Inc. and reprinted with permission.

Summary

- Fuel has different octane ratings and various factors affect these ratings
- The primary principles of carburetor operation are:
  - Atomization: The process of combining air and fuel to create a mixture of liquid droplets suspended in air.
  - The Venturi Principle: A gas or liquid that is flowing through a narrowed-down section of a passage will increase in speed and decrease in pressure compared to the speed and pressure in wider sections of the passageway.
- Fuel delivery systems consist of many separate components and servicing fuel delivery systems involves inspecting and cleaning or replacing many of these components.
- Each different type of carburetor has many different components that function in similar ways when compared to one another.
- Describe the operation of the circuits in each type of carburetor
- Carburetor synchronization is the process of making two or more carburetors operate the same in terms of the amount of air-and-fuel mixture drawn through each one.
- The purpose of fuel injection is to allow an extremely precise metering of air-and-fuel mixture ratios at any given engine and atmospheric condition.

Chapter 8 Review Questions

1. True or False: Gasoline by itself as a liquid will not burn
2. The air we breath contains ________ which is used to help ignite the fuel mixture in an engine
3. The ________ rating is defines as the fuels ability to resist detonation in an engine
4. If the needle clip is raised on a carburetor, the mixture becomes ________
5. The choke plate cold start system controls the amount of ________ entering the carburetor
6. True or False: Carburetor fuel delivery systems overlap one another.
7. The venturi on a CV type carburetor is controlled by ________
8. The ________ mixture screw is always located oni the air filter side of the carburetor slide.
9. The ________ controls the airflow from idle to wide open on the CV type carburetor
10. The mid-range fuel circuit works in conjunction with the ________ circuit.
11. The carburetor ________ circuit has the greatest effect between the range of idle and 1/4 throttle opening
12. The fuel pump on an electronic fuel injection system is ________ powered.
13. The amount of time that the fuel injector is open is known as ________.
14. Sensors on a fuel injection system monitor the throttle position, air volume and other important engine information transmitting this information to the ________
15. The ________ is most likely to have an anti-tampering plug installed to prevent adjustments.

CHAPTER

9 Drives, Clutches, and Transmissions

Learning Objectives

When students have completed the study of this chapter and its laboratory activities they should be able to:

- Identify the different gears used in transmissions
- Calculate gear and drive ratios correctly
- Identify the functions of the primary drive systems
- Identify the different components that make up the primary drive systems
- Understand and identify the different clutch types
- Identify the different clutch release mechanisms
- Identify the different types of transmissions and shifting components
- Identify and understand the different types of final drive systems

Key Terms

Backlash
Belt-driven final drive
Belt-driven primary drive
Centrifugal clutch
Chain-driven final drive
Chain-driven primary drive
Clutch
Clutch bearing
Clutch center
Clutch drag
Clutch friction discs
Clutch outer
Clutch plates
Clutch pressure plate
Clutch push rod
Clutch release mechanisms
Clutch slippage
Direct-drive gear ratio
Direct-drive transmission
Dual-range transmission
Final drive ratio
Fixed gear
Freewheeling gear
Gear-driven primary drive
Gear ratios
Improper cable adjustment
Improper release rod adjustment
Indirect-drive transmission
Lever movement
Lifter rod
Over-drive gear ratio
Overall gear ratio
Primary drive
Primary drive ratio
Shaft-driven final drive
Sliding gear
Sprag
Sprockets
Transmission gear ratio
Under-drive gear ratio
Variable-ratio clutch

INTRODUCTION

This chapter focuses on the transmissions and related components used in motorcycle engines. Here you will learn about the different types of gears that will be found in a motorcycle engine. You will also learn how the primary drive and different clutch systems function in an engine. We will then discuss how and why transmissions are used in motorcycle engines and will finish this chapter by discussing the different types of final drive systems found on motorcycles.

All motorcycles use some type of a drive system to propel the vehicle. The primary drive, clutch, transmission, and final drive systems work together to transfer the power that's produced by the engine crankshaft to the rear wheel. To use the power made at the engine's crankshaft, a motorcycle may use gears, belts, clutches, sprockets and a chain, or a combination of these items.

Before you learn about the clutches, transmissions, and final drives used in motorcycles and ATVs, you must first understand some basic information related to gears and gear ratios.

GEARS

In basic terms, a gear is one or more rotating levers. There are five primary purposes for using gears in engines:

1. To transmit power from one shaft to another
2. To change the direction of rotation
3. To increase torque, which results in a decrease in the revolutions per minute (rpm), or speed, of the gear; note that the force behind a moving gear is the torque of the drive system.
4. The increase of rpms will result in a decrease in torque
5. To time properly certain components in the engine, such as engine balancers and camshafts on four-stroke engines

There are several basic types of gears used in a motorcycle's engine. Gears are also used in many other areas of a motorcycle as well. Because you will see many different types of gears, it is important that you can identify and understand their purpose (Figure 9-1).

Spur Gear

A spur gear, or straight-cut gear, has teeth that are straight, which allows a tooth to mesh entirely with another spur gear tooth. The spur gear is the most common gear used in engines and is commonly used in the transmission of a motorcycle engine. The spur gear is the simplest gear to manufacture, is very durable under various strenuous loads, and, because of its simplicity, is the least expensive gear to manufacture. However, this type of gear also makes the most noise as it meshes with another spur gear.

Offset Spur Gear

An offset spur gear is essentially two spur gears that are attached side by side. The teeth of the offset spur gear are offset by one-half of a tooth. The gears are normally attached to one another by a rubber damper, which allows the offset spur gear to engage its mating spur gear. The offset spur gear is used to reduce the amount of free movement between the teeth of two meshed gears. This free movement is called backlash. **Backlash** can be defined as the play or loose motion in a gear due to the clearance between two opposing gears.

Helical Gear

A helical gear has teeth that are angled, which allows it to be much quieter than a spur gear. Unlike the spur gear, the teeth of two helical gears, when meshed, don't fully contact each other. The lack of full contact between the gears causes the gears to be quieter. When engaged and under load, helical gears create side loads that tend to force the gears to one side.

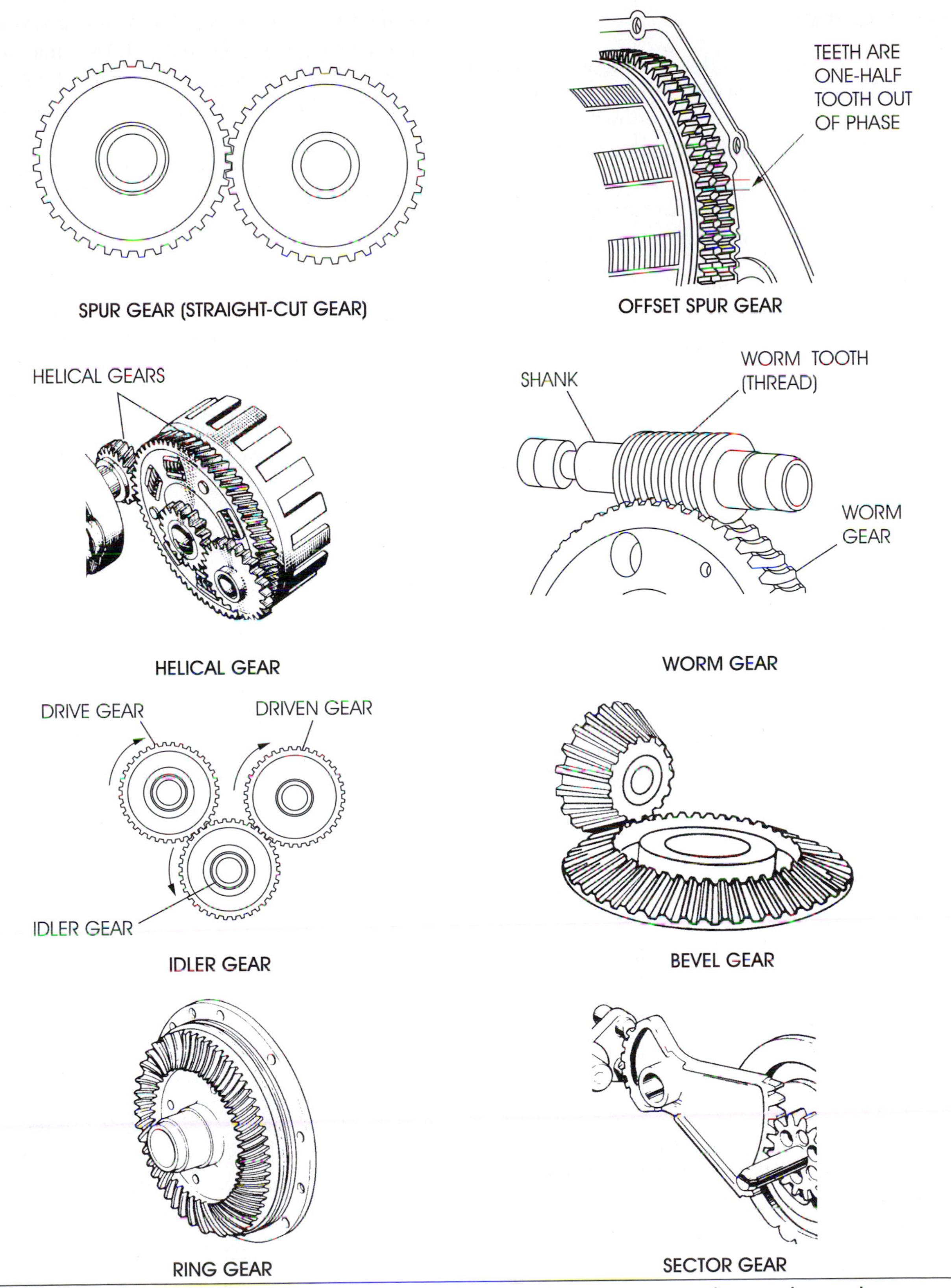

Figure 9-1 The different types of gears that will be found on a motorcycle are shown here.

Worm Gear

A worm gear has teeth that are cut at an angle to be driven by a shank that has at least one complete tooth (thread) spiraled around the surface. A worm gear is used to connect nonparallel, nonintersecting shafts. A worm gear allows for a very high gear reduction ratio. Worm gears are most commonly found in speedometer and tachometer drives.

Idler Gear

An idler gear is a gear that's situated between a drive gear and a driven gear to transfer motion without a change of direction between the drive gear and driven gear. An idler gear does not change the gear ratio of the gear set. We will discuss gear ratios later in this chapter.

Bevel Gear

A bevel gear can be a spur bevel with straight-cut teeth or a spiral bevel with curved teeth. The bevel gear is used to transmit power at 90 degree angles. The bevel gear is the most common design found on the pinion gear (drive gear) of a final shaft-drive system and also on some camshaft drives. We will discuss final drive systems later in this chapter.

Ring Gear

A ring gear is a metal wheel that has teeth around the inner edge of the gear. The ring gear is most commonly seen in the design of gears used for the final driven gear on shaft-drive final drive systems. The ring gear is normally used in conjunction with a bevel gear.

Sector Gear

A sector gear is a pie-shaped segment of a helical or spur gear. The sector gear is used in kick-start mechanisms and shift linkages. A sector gear allows partial movement of certain components.

GEAR RATIOS

Gear ratios are used to alter the speed of a rotating component to a useful rpm. A gear ratio is a numerical comparison of the number of revolutions of the drive gear as compared to one revolution of the driven gear. This is really simpler than it sounds.

An example of a gear ratio would be five to one, or, as stated in numerical terms, 5:1. This ratio states that the drive gear makes five revolutions for every one revolution of the driven gear (Figure 9-2). The formula used to determine gear ratios is as follows:

Gear ratio = (# of teeth on driven gear) divided by (/) (# of teeth on drive gear)

Let's try an example using the above formula to determine a gear ratio. If a driven gear has 40 teeth and a drive gear has 10 teeth, what's the gear ratio? You simply take the driven gear (40) and divide it by the drive gear (10) to get 4.

Driven gear (40) / Drive gear (10) = 4

Therefore, these gears have a gear ratio of four to one, or 4:1. In other words, the drive gear rotates four times for every one revolution of the driven gear. In many cases, gear ratios are rounded off to a thousandth, so 4:1 could read 4.000:1.

Gear ratios can be categorized into one of three groups (Figure 9-3).

1. An **under-drive gear ratio** is a gear ratio that's greater than 1:1. In this situation, the drive gear always has fewer teeth than the driven gear. An under-drive gear ratio increases the torque output of the driven gear but decreases the rpm of the driven gear.

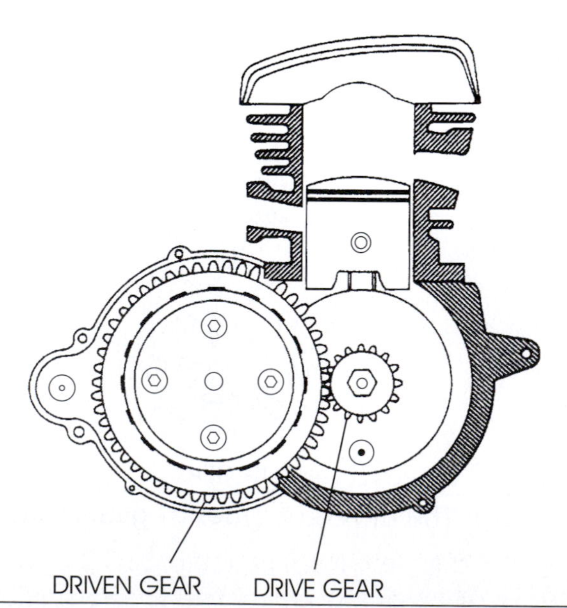

Figure 9-2 A drive and a driven gear from a motorcycle engine are illustrated here.

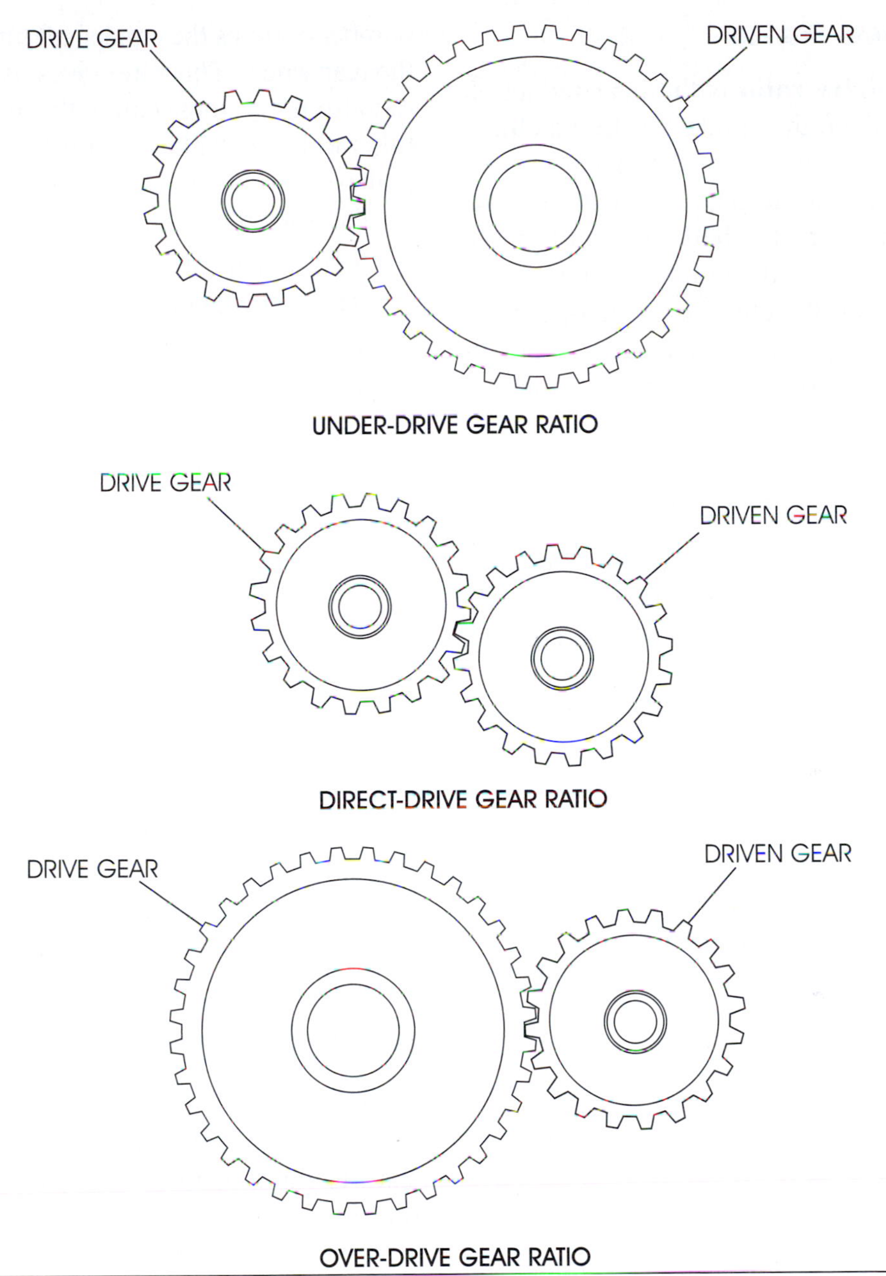

Figure 9-3 The three different types of gear ratios are shown here.

2. A **direct-drive gear ratio** is a gear ratio that's exactly 1:1. In this situation, the drive gear always has the same number of teeth as the driven gear. A direct-drive gear ratio will not change the torque or rpm of the driven gear.

3. An **over-drive gear ratio** is a gear ratio that's less than 1:1. In this situation, the drive gear always has more teeth than the driven gear. An over-drive gear ratio increases the rpm but also decreases the torque of the driven gear.

There are many different gear reductions used inside a motorcycle engine. The proper gear reduction must be used to allow an engine to operate at its full potential as designed by the engineers. You will find the following four types of ratios used in motorcycle engines:

1. Primary drive ratio
2. Transmission gear ratio
3. Final drive ratio
4. Overall ratio

Primary Drive Ratio

The **primary drive ratio** is the gear reduction that's determined from the crankshaft to the clutch of the engine. When computing the primary drive ratio, the crankshaft output gear is the drive gear. The clutch has a gear attached to it and is the driven gear. For example, if a crankshaft gear has a tooth count of 41 and the clutch-attached gear has a tooth count of 89, the primary drive ratio is 89 (driven) divided by 41 (drive). The result (rounded off to the thousandth) is 2.171:1.

Transmission Gear Ratios

The transmission transmits power from the clutch, which is attached to the mainshaft (drive side of the transmission) of the engine, to the countershaft (driven side of the transmission) of the engine. **Transmission gear ratios** represent how many turns of the mainshaft cause one revolution of the countershaft and are needed to allow the motorcycle to roll from a stop and then to increase its speed while staying within a given engine rpm range. There may be as few as two to as many as six different transmission gear ratios used on a motorcycle. For example, a main shaft gear with 22 teeth and a countershaft gear with 28 teeth have a gear ratio of 1.273:1. This is determined by dividing the driven countershaft gear teeth (28) by the drive main shaft gear teeth (22), which results in a gear ratio of 1.273:1.

Final Drive Ratio

The **final drive ratio** is a numeric comparison between the countershaft of the transmission and the rear wheel. In the case of a chain final drive the transmission's countershaft has a sprocket attached to it, which drives the rear wheel through a roller chain. An example of a final drive ratio is an output shaft sprocket size of 16 teeth and a rear wheel sprocket size of 46 teeth. In this case, you would divide 46 by 16. The result of the ratio (rounded off to the thousandth) is 2.875:1.

Overall Gear Ratio

There's one other ratio that directly relates to the ratios that we just covered. This ratio is known as the **overall gear ratio**. This ratio compares the number of times the crankshaft turns to the turning of the rear wheel. This ratio gives us the numerical gear ratio from the crankshaft to the rear wheel. Generally, this ratio is calculated with the transmission in its highest gear, but it can be calculated for any gear.

The formula to calculate the overall ratio is:

Overall gear ratio = (primary) × (transmission) × (final drive) ratios

If we take the figures obtained in each of the examples previously mentioned, we can calculate the overall gear ratio (rounded off to the thousandth) as follows:

$$2.171 \times 1.273 \times 2.875 = 7.946:1$$

This tells us that for every one revolution of the rear wheel, the crankshaft rotates 7.946 times.

It is important to understand that gears and gear ratios have a very important use in engines. Engines with high horsepower can use a lower gear ratio. Engines with a low horsepower rating require a higher gear ratio to allow for the full use of the available horsepower. An engine with a higher overall gear ratio will not have as much top speed capability but will reach its maximum available speed quicker than an engine that has a lower overall gear ratio.

PRIMARY DRIVES

All engines require a gear reduction system that's used to transfer the power from the crankshaft to the transmission, and then from the transmission to the rear wheel. The gear reduction system used for transferring the power from the crankshaft to the clutch is called the **primary drive**. As you now understand, gear reduction is necessary to allow the engine to remain in the appropriate range of rpm while maintaining various speeds at the rear wheel. In other words, we need a gear reduction system so that the engine crankshaft can turn at one speed while the rear wheel turns at another speed. A clutch system is needed to engage and disengage the power from the crankshaft to the transmission.

Before we learn about the different types of clutches, we'll first discuss the types of primary drive systems found in today's motorcycle and ATV engines. A primary drive system transfers power from the crankshaft to the clutch by using gears, a chain, or a belt (Figure 9-4). These are the three basic methods of connecting the engine to the clutch and transmission.

Gear-Driven Primary Drive

Most motorcycles use a **gear-driven primary drive**. This drive system may utilize spur, offset spur, or helical gears to transfer power from the crankshaft to the clutch system. With a gear-driven primary drive system, the two gears turn in opposite directions because each gear is on a separate shaft. This makes the clutch turn in the opposite direction of the crankshaft. Gear-driven primary drive systems must be lubricated to prevent excessive heat caused by the friction of the gears as the engine operates.

Chain-Driven Primary Drive

The **chain-driven primary drive** uses a chain and two gears or **sprockets** to transfer power from the crankshaft to the clutch system. Sprockets are the teeth on the outside of a wheel that engage the links of a chain. With a chain-driven primary drive, both sprockets turn in the same direction and use either a roller chain or a Hy-Vo chain design. The Hy-Vo chain design is the most commonly used as it's a much stronger design and quieter than the roller chain. Just like with a gear-driven primary drive, you must keep chain-driven primary drive systems well lubricated.

Belt-Driven Primary Drive

The **belt-driven primary drive** system uses a toothed belt known as a "Gilmer-type belt" and two pulleys with teeth attached to them. Just like the chain-driven primary drive system, the belt-driven type has both pulleys turning in the same direction. This type of primary drive is very quiet because it uses a belt instead of gears or a chain. Unlike the other primary drive systems, you must keep a belt-driven primary drive dry.

CLUTCH SYSTEMS

Now that we have a gear reduction system in place for the power at the crankshaft, there must be a way to interrupt that power flow when we need to stop the motorcycle. We also need a way to permit a gradual engagement of the motorcycle when you decide to have the machine move again. This power interruption is done by the use of a **clutch**. A clutch is a mechanism for transmitting rotation, which can be engaged and disengaged.

The clutch is normally placed between the primary drive of the engine and the transmission. Clutch actuation can be divided into two different types:

- Manual clutch actuation, which is controlled by the rider
- Centrifugal clutch actuation, which is controlled by the engine

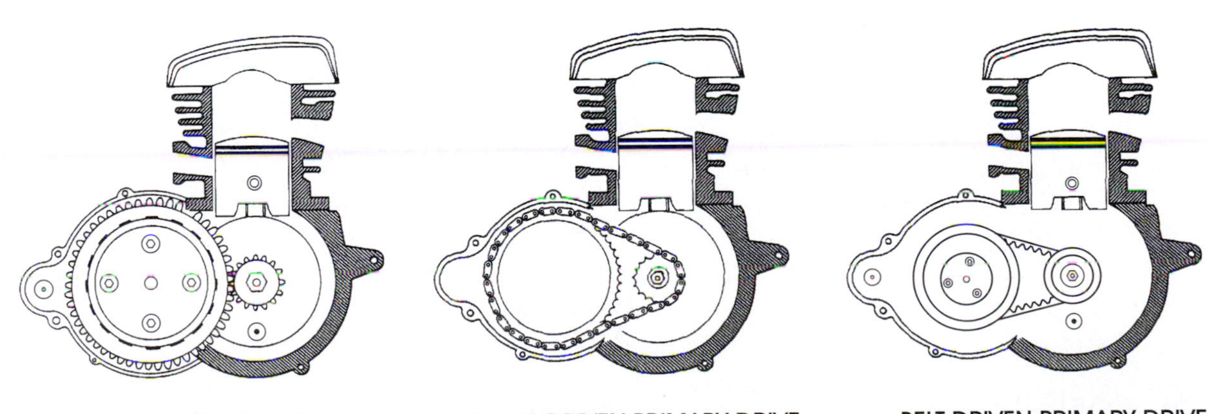

Figure 9-4 The three types of primary drives are illustrated here. Note that the gear-driven drive gears will turn in opposite directions while the chain- and belt-driven drives turn the drive and driven gears in the same direction.

The following are the three basic clutch designs used in engines:

1. Manual clutch
2. Centrifugal (also referred to as automatic) clutch
3. Variable-ratio clutch

Manual Clutch

A manual clutch allows the rider to control the engagement and disengagement of the power flow from the crankshaft to the transmission. The manual clutch is the most conventional type of clutch used on motorcycles. A manual clutch may be wet or dry meaning that a wet clutch is cooled with oil and a dry clutch is cooled with air. Both wet and dry manual clutches are virtually identical in both design and function although the wet clutch is much quieter when the engine is operating. Manual clutch actuation is normally done by cable or hydraulics but can also be done by electronics, which will be discussed later in this chapter.

As just mentioned, the main difference between wet and dry clutches is the way they're cooled. A wet clutch operates in an oil bath to keep its components cool. The oil carries the heat generated by the clutch away from the clutch to help keep it cool. A dry clutch uses airflow to keep the clutch cool. There are two styles of manual clutches used: single-plate and multi-plate. A single-plate clutch, which is very similar to the clutch system used for automobiles, is normally found in conjunction with a dry clutch. A multi-plate clutch is the most conventional style of clutch found on motorcycles. All manual clutches consist of the same basic components. We will concentrate primarily on the multi-plate manual wet clutch because dry clutches are not as popular on motorcycles but again they are both nearly identical in both design and function.

Multi-Plate Manual Clutch

A multi-plate clutch has as few as two plates and up to as many as twelve plates. Most modern motorcycles have between four and eight plates. By using more plates, the clutch can be made smaller in diameter while keeping the same amount (or more) of friction material for engagement purposes. Although terminology varies, all multi-plate clutches have the same basic components (Figure 9-5).

- The **clutch outer**, which may also be called the clutch basket, contains all of the components of the clutch and has the driven gear for the primary drive attached to it. This component freewheels on the transmission main shaft and rotates whenever the engine is running. Most clutch outers use springs or rubber dampers attached to the driven gear to absorb excess power pulses so

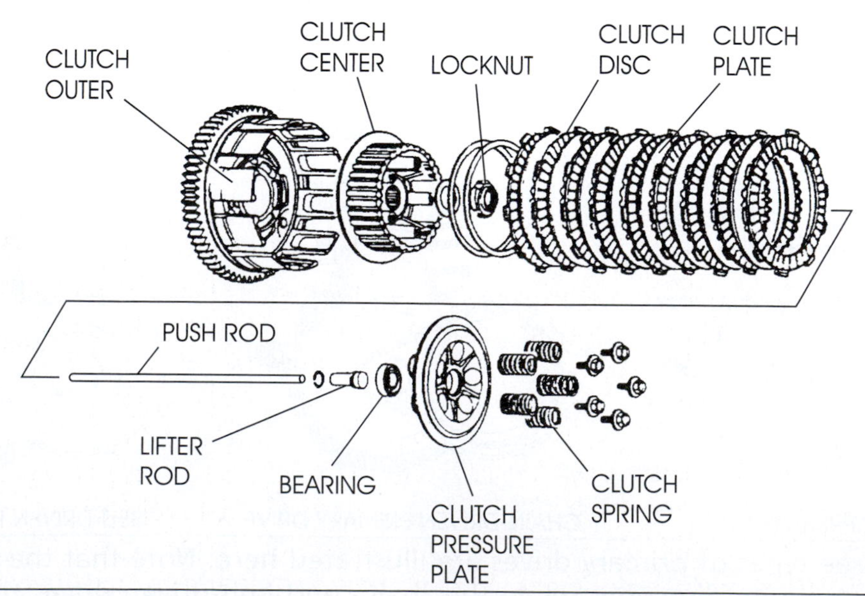

Figure 9-5 The components of a typical multi-plate manual clutch are illustrated here.

that the pulses aren't transmitted through the rest of the driveline. The clutch outer is driven by the engine and is connected to all of the clutch friction discs by the use of slots or fingers that surround the plates.

- The **clutch center**, which may also be called the inner clutch hub, is splined to the transmission main shaft and secured with a locknut or a circlip. The clutch center has splines machined into it and is driven by the clutch plates. When the clutch plates rotate, the clutch center also rotates. The clutch center is connected directly to the transmission.
- **Clutch friction discs** transmit the power of the clutch outer to the clutch center and are made of a high-friction material. Clutch friction discs are also known as friction plates. The materials used to make friction plates are well-kept secrets by the makers of different clutches, but it's known that some of the materials used to produce friction discs are cork, neoprene, and Kevlar. Friction discs are connected to and driven by the clutch outer. Clutch friction discs have tabs on their outer edge that fit into the slots, or fingers in the clutch outer. Like the clutch outer, the clutch discs rotate whenever the engine rotates. This being a wet clutch, the friction material is kept cool by the oil in the lubrication system. As the clutch friction material wears down, it begins to contaminate the oil in the transmission.
- **Clutch plates** are connected to and used to transfer the power from the clutch friction discs to the clutch center. When the clutch plates rotate, the transmission shaft also rotates as they are attached to the clutch center via inner tabs that fit into the splines of the clutch center. Clutch plates can be made out of steel or aluminum depending on the manufacturer's desired performance and longevity.
- The **clutch pressure plate** applies pressure to the clutch plates and friction discs, which prevents the clutch from slipping. When pressed together, the clutch plates and discs form one unit and allow power to flow through them. The clutch pressure plate is pushed by the clutch release mechanism, which releases the pressure applied to the plates and discs to separate them from each other and to disengage the clutch. Clutches are designed to have an inner pressure plate or an outer pressure plate (Figure 9-6). An outer pressure plate will have pressure pushing from inside the clutch to the outside of the clutch to disengage the engine from the transmission. An inner pressure plate will have pressure pushing from outside the clutch to the inside of the clutch to disengage the engine from the transmission. Both types have the same results although disassembly procedures are slightly different for each type.
- Clutch springs hold the clutch pressure plate firmly against the clutch plates and discs.
- The **lifter rod** applies pressure to the clutch pressure plate to release the clutch.
- The **clutch bearing** reduces the friction of the lifter rod as it applies pressure to the clutch pressure plate.
- The **clutch push rod** pushes the lifter rod, which in turn releases the clutch pressure plate. Push rods may be installed through the main shaft as illustrated in Figure 9-5 or they may be attached to the outside of the clutch.

Single-Plate Manual Clutch

The single-plate manual clutch uses the same basic components as the multi-plate clutch. As the name implies, the main difference between the two clutches is that the single-plate manual clutch uses only a single friction plate. This type of clutch is normally a dry-type clutch using air to keep it cool instead of oil from the lubrication system.

Manual Clutch Release Mechanisms

Clutch release mechanisms are used to disengage the power flow from an engine's clutch to the transmission. When the clutch is disengaged, the release mechanism pushes against the clutch pressure plate. This pressure separates the clutch drive plates from the driven plates. The clutch must be engaged gradually to prevent a sudden grabbing of the clutch, which would cause the motorcycle to lurch forward. There are several types of manual clutch release mechanisms and all use a cable that's attached to the handle bar to actuate them. You will also see hydraulic clutch actuation release mechanisms that will use hydraulic fluid in place of the cable.

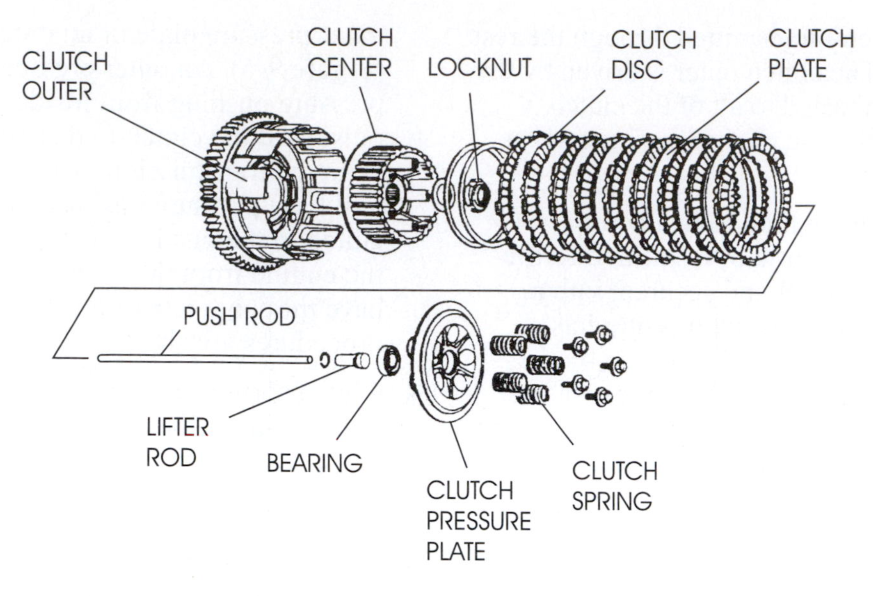

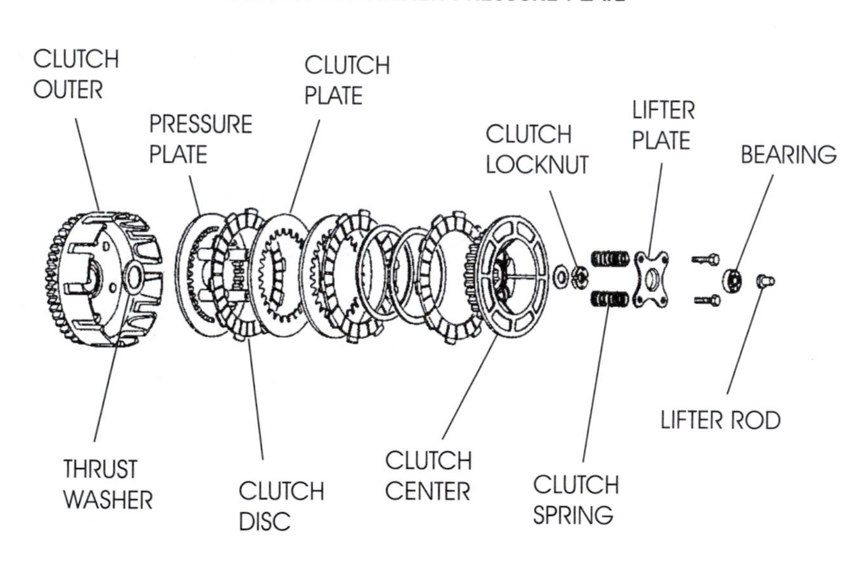

Figure 9-6 Clutches can have inner pressure plates or outer pressure plates. The result is the same for each clutch but disassembly procedures are different for each type.

Multi-Plate Manual Clutch Operation

A clutch is either engaged or disengaged (Figure 9-7). When the clutch lever is pulled, the clutch push rod pushes against the clutch lifter rod. The clutch lifter rod applies pressure to the clutch pressure plate, resulting in a gap between the clutch friction discs and clutch plates. This separates the power of the crankshaft from the rear wheel and allows the clutch to slip. This places the clutch in a disengaged position.

When the transmission is shifted into gear and the clutch lever is released, the clutch friction discs and clutch plates become caught between the pressure plate and the clutch center. This now prevents the clutch from slipping and the power of the crankshaft is transmitted to the rear wheel.

Centrifugal Clutch

The **centrifugal clutch** (Figure 9-8) is among the simplest clutch designs and uses the engine's rpm to engage and disengage the crankshaft power

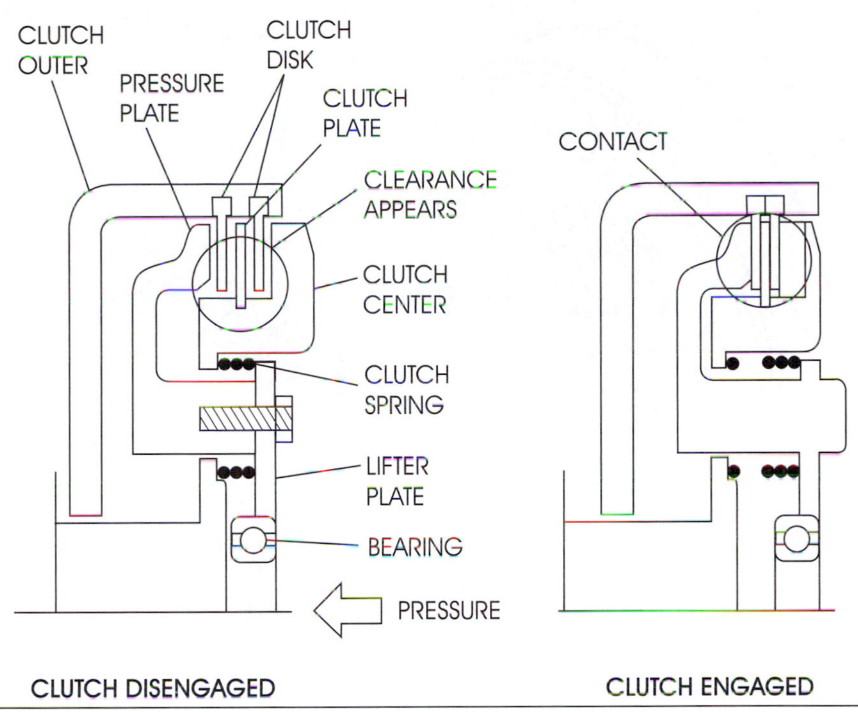

Figure 9-7 When all of the clutch plates are contacting each other, the clutch is engaged. When they are separated, the clutch is disengaged.

to and from the engine's final drive. Since this clutch system relies on the engine for engagement, there's no need for a handlebar-mounted clutch lever. The centrifugal clutch uses weights and springs to determine the rpm's necessary to disengage and engage the clutch. When the engine's rpms are high enough to overcome the tension of the springs, centrifugal force throws the weights outward against the clutch drum. This action engages the crankshaft with the clutch outer.

In many ATVs, this type of clutch is often used in conjunction with a manual multi-plate clutch, as shown in Figure 9-9. The combination of these two clutch systems allows the engine to remain in gear while idling to allow the rider to come to a complete stop without having the engine stall.

Variable-Ratio Clutch

Variable-ratio clutch systems (Figure 9-10) are often seen on motorscooters and also can be found on some ATVs. Although named a "variable-ratio clutch," this system is actually a transmission system that provides a variable drive ratio between the engine and the rear wheel. The variable-clutch system consists of a drive pulley that's attached to the engine crankshaft and a driven pulley that's attached to a shaft, which may also incorporate a centrifugal clutch. The two pulleys are connected by a drive belt. Many variable-ratio clutch systems also have a final gear reduction between the driven pulley and the rear wheel to provide an extra increase in torque when needed.

Variable-Clutch Drive Pulley Operation

The drive pulley consists of a fixed and a movable face. The movable face has the capability to slide toward the drive face. There's also a ramp plate incorporated in the drive pulley, which pushes weight rollers against the drive face as can be seen in Figure 9-10. As engine speed increases, centrifugal force pushes the weight rollers outward. This force pushes the movable face toward the fixed face, which in turn pushes the drive belt upward toward the top of the drive pulley (Figure 9-10A). This reduces the drive ratio by forcing the drive belt to ride on a pulley of larger diameter. As the engine speed decreases, the belt is pulled back into the drive pulley, which

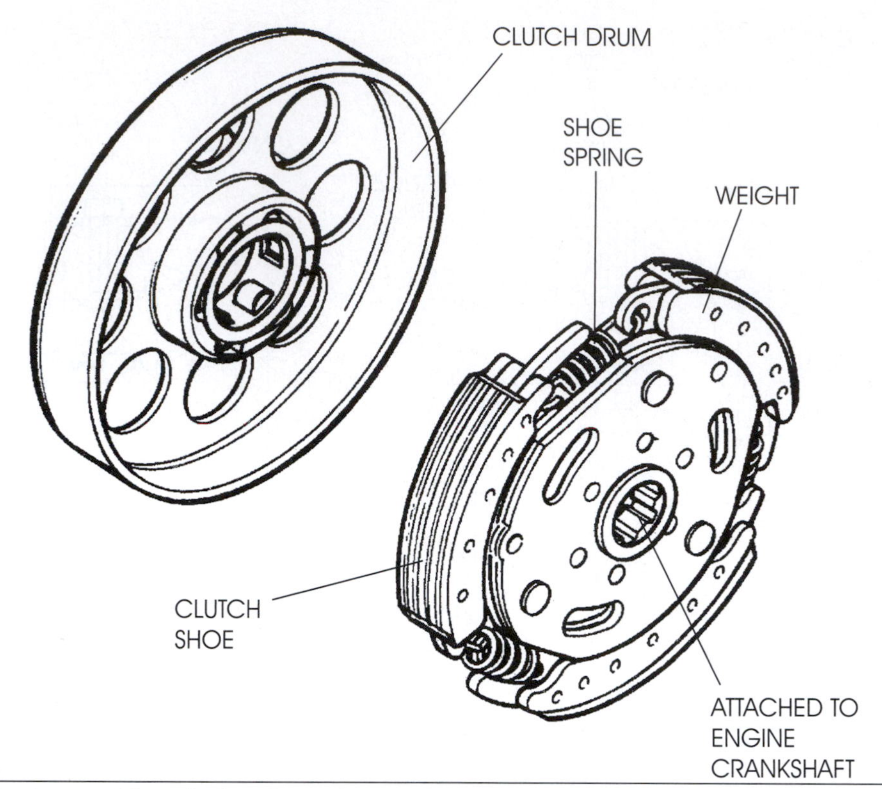

Figure 9-8 A typical centrifugal clutch is shown here.

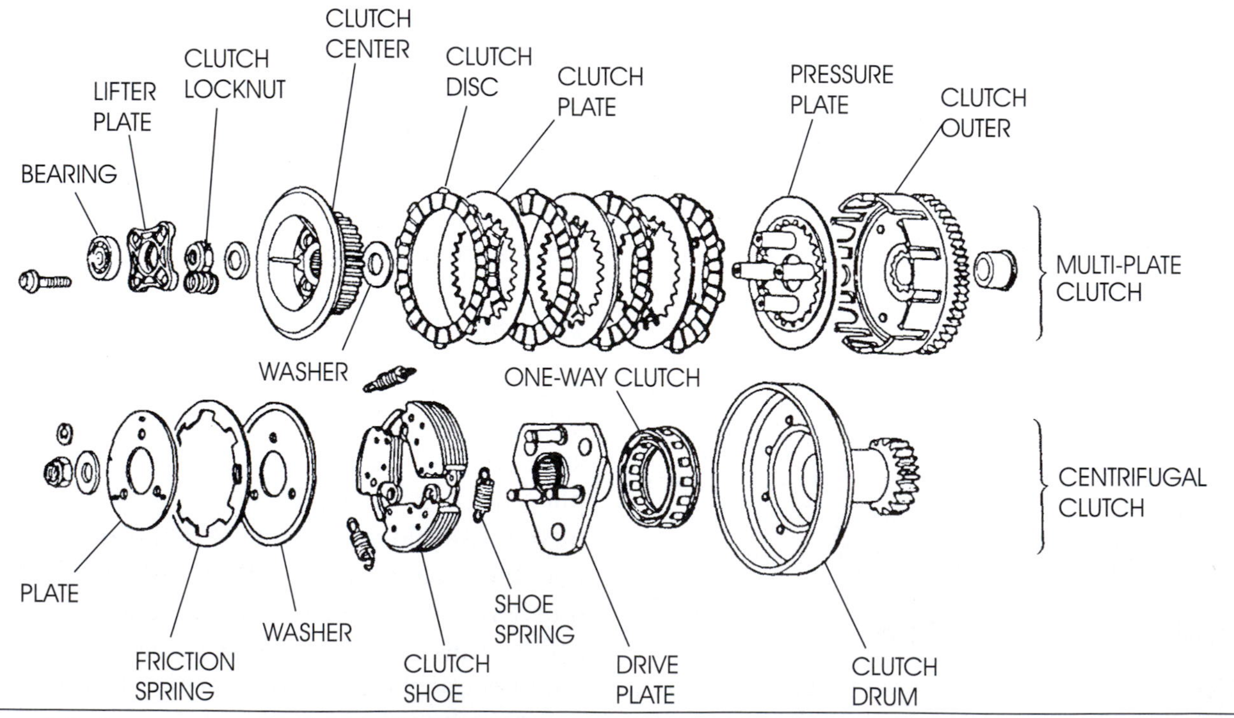

Figure 9-9 Some motorcycles and many ATVs use both a multi-plate and centrifugal clutch that allows the engine to idle in gear without moving.

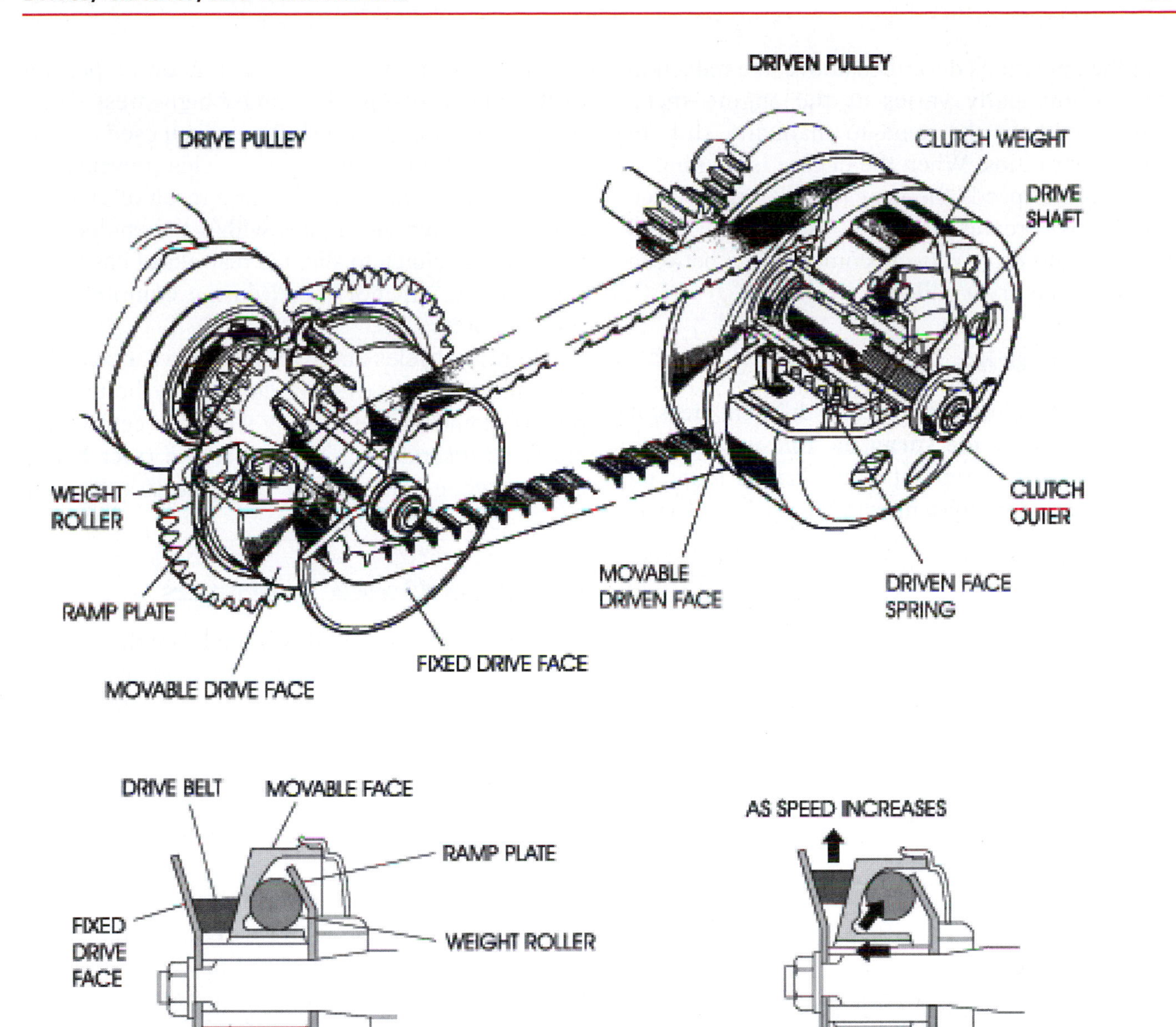

Figure 9-10 The variable-ratio clutch can be seen on many motorscooters and some ATVs as well.

increases the drive ratio by allowing the belt to ride on a pulley of smaller diameter.

Variable-Clutch Driven Pulley Operation

Because the belt remains constant in length, the driven pulley reacts to the drive pulley by allowing the drive belt to be pulled in toward the center of the driven face. By doing this, the diameter of the belt on the driven face decreases at higher engine speeds (Figure 9-10B). When the engine speed decreases and the belt is pulled back into the drive pulley, the driven face spring moves the movable driven face toward its resting and original position. This pushes the drive belt back to the circumference of the driven pulley.

In the operations described above, the reduction ratio automatically varies as the engine speed changes without the need to manually shift to change gear ratios. When the engine is running at lower engine speeds the gear ratio is high, which allows for greater torque. As engine speed increases, the drive ratio becomes lower between the driven and drive pulleys.

Sprag Clutch

Another type of clutch used in different areas of the motorcycle is the **sprag** or "one-way" clutch (Figure 9-11). This type of clutch can be found on some primary drive systems as well as many electric starting systems. The sprag clutch allows a portion of the clutch to slip when under high-stress situations such as rapid downshifting, when used in conjunction with the primary drive. This prevents the loss of traction that can occur as a result of extreme compression-braking forces within the engine. By allowing the clutch to slip, the rear wheel has better traction and will not "hop" during compression-braking forces.

On motorcycles and ATVs that have a centrifugal clutch system, a Sprag clutch may be used to provide some engine braking when the centrifugal clutch is disengaged. This gives the rider better control when bringing the machine to a stop or going down hills.

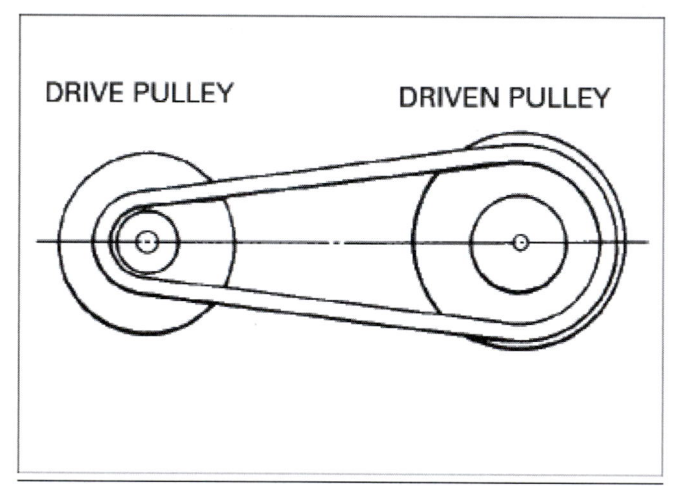

Figure 9-10A The drive belt position at a low engine speed condition.

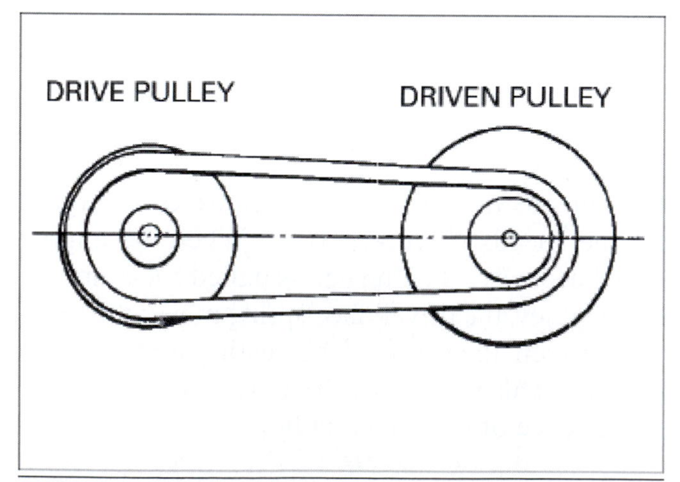

Figure 9-10B At higher engine speed, the drive belt is pushed up on the drive pulley and forced down into the driven pulley, changing the ratio of the pulleys.

Common Clutch Problems

You'll find two common clutch problems with any of the previously mentioned clutches in this chapter. The first problem is a clutch that slips. **Clutch slippage** occurs when the clutch doesn't have the ability to transfer all of the engine's power flow. Improper adjustment, weak clutch springs, or worn clutch friction discs and/or plates may cause this problem.

The second common problem you will find with a clutch is known as **clutch drag**. Clutch drag occurs when the clutch is unable to fully disengage. Clutch drag is evident when the engine power can't be disengaged from the rear wheel. An example of this condition is when you squeeze the clutch lever

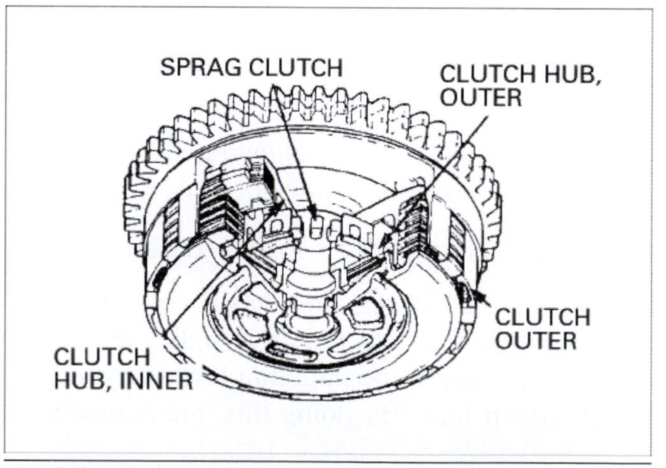

Figure 9-11 A sprag clutch is designed to allow power to flow in one direction and slip in the opposite.

and the motorcycle still tries to move forward. This is also known as "creeping." Possible causes of this condition include warped or binding clutch plates, a worn clutch outer or clutch center, improper clutch adjustment, or a worn release mechanism.

Clutch Service and Adjustment

Most clutch problems originate from an improperly adjusted clutch. Whenever there's an indication of a clutch problem, it's important to verify that the clutch is adjusted properly. In most cases, a simple adjustment can resolve either a slipping or a dragging clutch if caught early on. Most manual clutches use a clutch cable to link the handlebar lever to the clutch release mechanism, which is attached to the engine. Most motorcycles that have a clutch cable also have a clutch cable length adjuster and a clutch release adjuster—both of which must be adjusted properly to ensure a long-lasting clutch. A clutch basket can also wear and cause the plates to stick (Figure 9-12).

The cable length adjuster is used to make minor adjustments in the free movement or "play" in the cable. The manufacturer has the correct specification for the lever in the service manual and the owner's manual. You can adjust the amount of play in the cable simply by turning the adjuster in or out (Figure 9-12). Too much play can cause a clutch to drag and too little play can cause a clutch to slip.

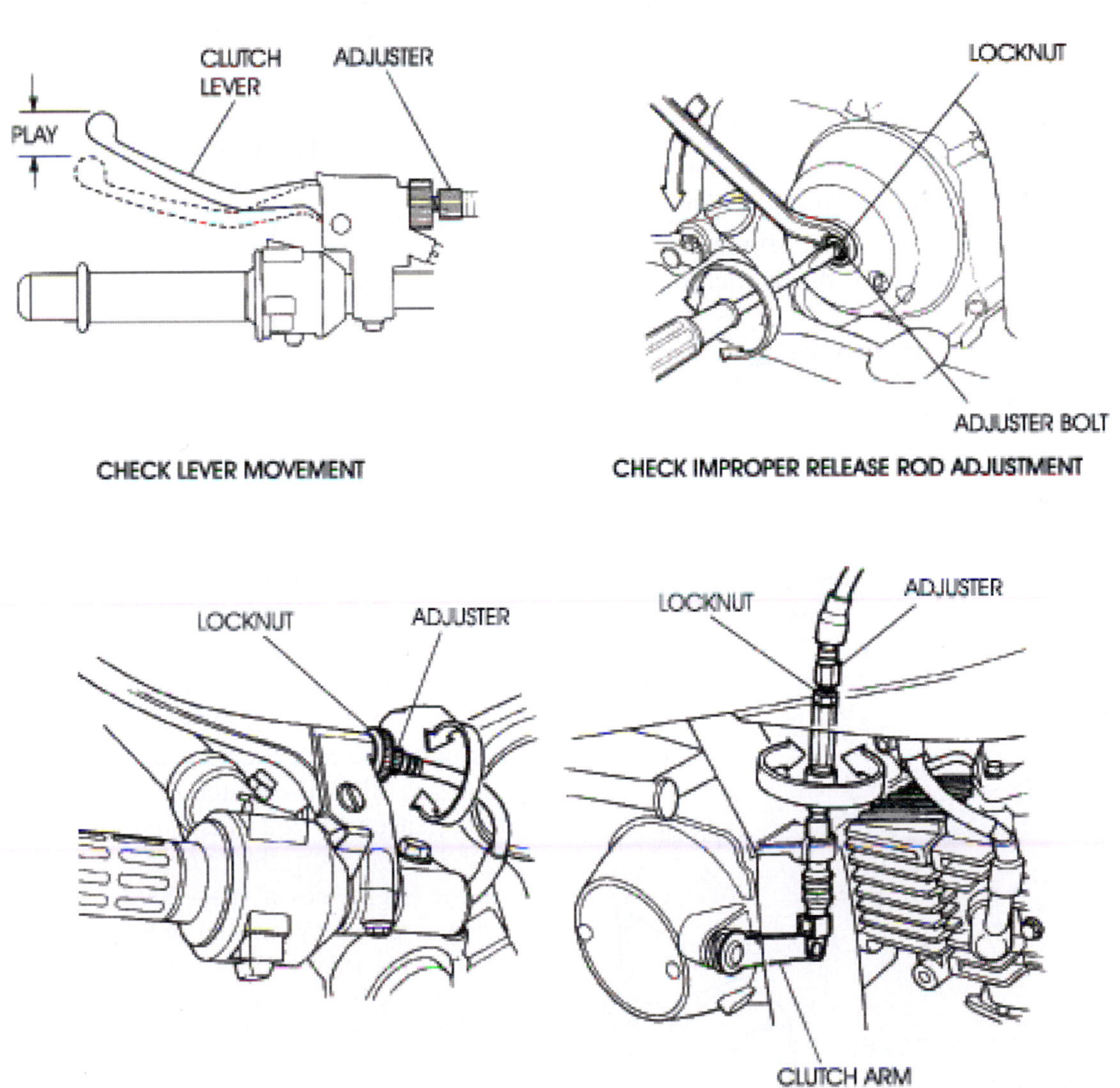

Figure 9-12 The clutch release adjuster is normally found on the engine, while the cable length adjuster can be found by the clutch lever on the handlebar.

The clutch release adjuster (Figure 9-13) is used to verify that there's a proper amount of clutch rod movement for correct disengagement. Normally, you should not need to adjust this side of the cable unless a new cable is being installed or if clutch repairs have been made. Because there are many different types of adjusters and each has a specific method of adjustment, you'll need to refer to the manufacturer's service manual for the correct procedure. For any adjustments thereafter, you will only need to remove excess slack from the clutch cable as it stretches.

Manual Clutch Adjustment Sequence

The proper sequence for clutch adjustment is as follows:

1. Adjust the release mechanism at the engine.
2. Adjust the cable play at the clutch lever.

As discussed before, clutches that drag (don't fully disengage) will prevent smooth shifting of the transmission gears; clutches that slip (don't fully engage) can not transmit all of the available engine power to the transmission. Sometimes either one of these conditions can be corrected by making simple clutch adjustments. The following may cause these problems:

- Not enough lever movement at the handlebar
- Improper cable adjustment
- Improper release rod adjustment

These problems may be interrelated. The following items must be checked to ensure correct adjustment of each part:

- **Lever Movement** You should be sure that the clutch lever on the handlebar isn't bent and has as much movement as it was designed to have. You should also be sure that the lever doesn't contact the handlebar grip before the inner clutch cable has been pulled to its maximum extension.
- **Improper Cable Adjustment** You must adjust the cable so that it has the correct (manufacturer's specified) free play. That is, so that the lever on the handlebar can be pulled a pre-described amount before the pressure of the clutch spring resisting release can be felt. Take the measurement from where the cable attaches to the handlebar lever. Adjust the cable by turning the adjuster at the handlebar. Turn the adjuster "in" for more play and "out" for less play.
- **Improper Release Rod Adjustment** Many motorcycles have a release rod that is separate from the clutch cable. The release rod adjustment must allow enough play in the release rod to allow the springs to expand, but not too much play or the rod won't move far enough to relieve the pressure from the clutch plates to allow the clutch to disengage. To adjust the release rod, you must first disconnect the clutch cable from the release rod lever located on the engine cover case. Next, you adjust the release rod screw so that pressure is applied against the springs. Then you must readjust the adjustment screw to allow for the manufacturer's recommended play of the release rod lever. You should be able to move the end of the release rod lever a certain amount from fully open before the clutch spring pressure resists the lever movement. Each manufacturer will offer the correct specification for release rod adjustment in their service manual.

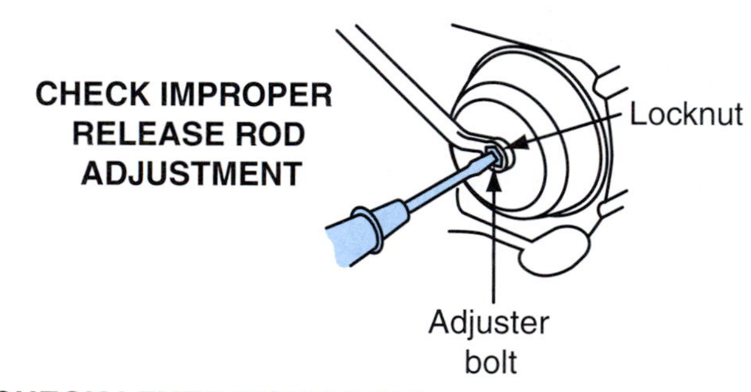

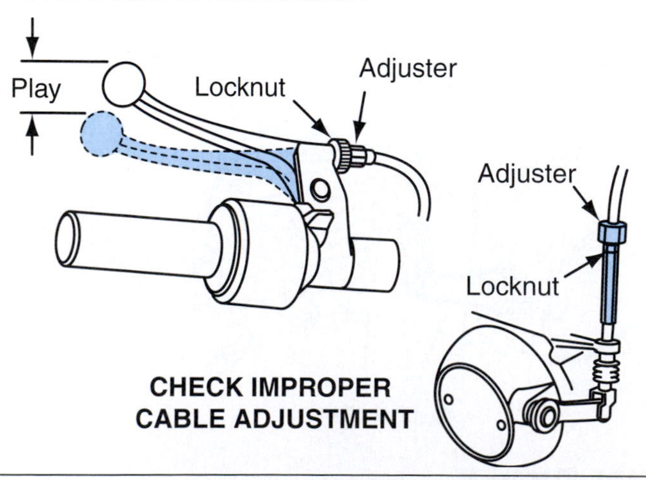

Figure 9-13 Clutch outer basket damage is caused by the friction plates wearing into the fingers of the basket, as can be seen here.

TRANSMISSIONS

Now that you understand how and why primary drive systems and clutches are used on motorcycles, we can move on to explain about engine transmissions. All motorcycles require a method of transmitting the power from the engine to the rear wheel. Transmissions are used to change the speed of the machine while keeping the engine within its usable power band. A transmission consists of shafts and gears, which are arranged to provide different gear ratios to the rear wheel. Transmissions allow the rider to increase and decrease the wheel speed while maintaining a constant engine speed.

Transmission types vary from one manufacturer to another. However, if you can understand the basic principles of how the most popular transmissions operate, you'll be on your way to becoming a well-informed motorcycle technician. In this section, we're going to help you get a thorough understanding of how transmissions work.

In its simplest form, a transmission consists of a centrifugal clutch attached to the crankshaft. A chain will be connected between the clutch and the rear wheel via a sprocket. As the engine speed increases, the clutch activates and propels the rear wheel. This type of transmission is an example of a single-speed transmission, which is far from the most efficient system available.

Transmission Gears

Most transmissions contain a combination of gears (Figure 9-14). The gears that may be found in a transmission include fixed gears, freewheeling gears, and sliding gears.

Fixed Gears

A **fixed gear** is one that doesn't move on the shaft to which it's attached. Fixed gears are attached to a shaft in one of three ways. The gear may be machined as part of the shaft, splined to the shaft, or pressed onto the shaft. A fixed gear rotates at shaft speed.

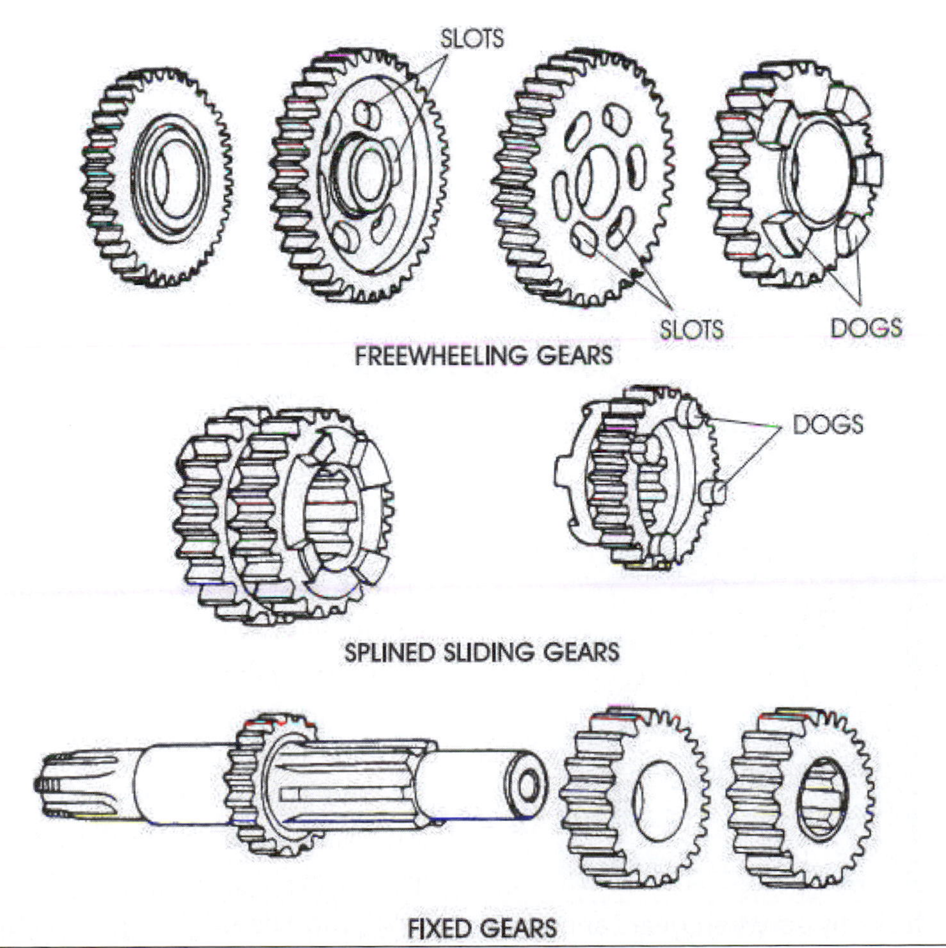

Figure 9-14 The different types of transmission gears are shown here.

Freewheeling Gears

A **freewheeling gear** moves freely on its shaft and is usually held in place by retaining clips. A freewheeling gear does not have to rotate at shaft speed and has slots or protrusions (called dogs) on its sides, which allow the gear to engage a sliding gear. Virtually all modern motorcycles have a freewheeling gear opposed to a fixed or sliding gear on the opposite shaft. This type of transmission is called a constant mesh transmission and will be discussed a little later in this chapter.

Sliding Gears

A **sliding gear** is one that can slide across the axis of the shaft. A sliding gear is splined to the shaft and rotates at shaft speed. The purpose of this type of gear is to engage and disengage transmission gears. Sliding gears are moved left or right across the axis of the shaft by a shift fork (Figure 9-15). A sliding gear has dogs on its sides that are designed to engage a freewheeling gear.

Constant-Mesh Transmissions

The most common transmission design found on motorcycles today is the constant-mesh transmission design (Figure 9-16). With a constant-mesh transmission, the teeth of all gears mesh with their mate on the opposing shaft at all times. At the beginning of this chapter you learned that there are several different types of gears found on motorcycles. The most common gears used in a typical constant-mesh transmission are spur (straight-cut) gears. There are two types of constant-mesh transmissions—indirect-drive and direct-drive.

Figure 9-15 Shift forks fit between gears and are used to slide the gears from one side to the other.

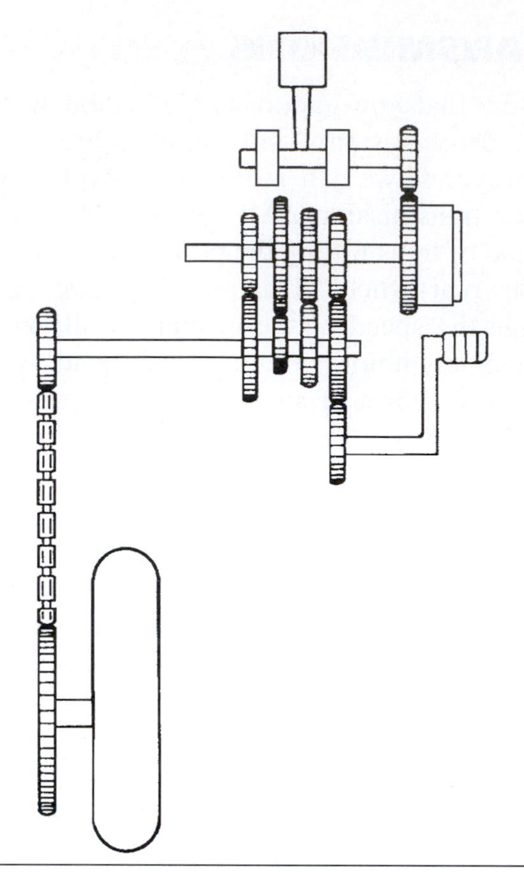

Figure 9-16 A simple constant-mesh indirect-drive transmission is illustrated here.

Indirect-Drive Transmissions

Indirect-drive transmissions have power entering on one shaft and exiting from another shaft on a different axis (Figure 9-17). The mainshaft of an indirect-drive transmission is splined to the clutch center. Power is transmitted through the clutch to the main shaft. The main shaft rotates opposite the countershaft. Power flow generally exits on the countershaft. The transmission countershaft turns the countershaft sprocket, which is connected to the rear wheel.

The following facts apply to indirect-drive transmissions:

- All gears on the main shaft are drive gears.
- The smallest gear on the main shaft is a fixed gear and is part of low gear. The largest gear on the main shaft is part of high gear.
- All gears on the countershaft are driven gears.
- The largest gear on the countershaft is part of first gear. The smallest gear on the countershaft is part of high gear.

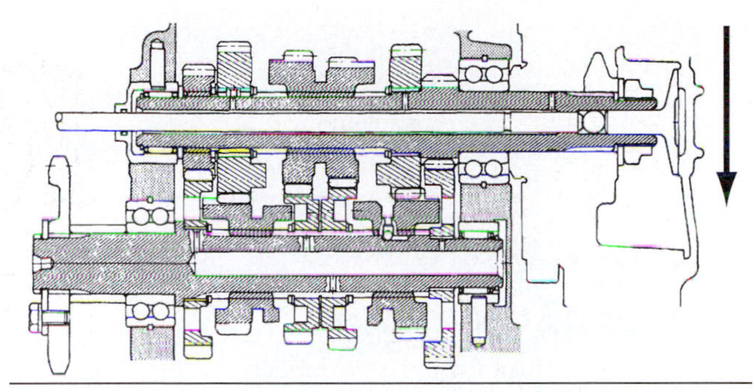

Figure 9-17 Indirect-drive transmissions have power entering on one shaft and exiting from another shaft on a different axis.

Direct-Drive Transmissions

With a **direct-drive transmission**, the power flow enters on one shaft and leaves on another shaft of the same axis, instead of a different axis as found with the indirect-drive transmission. Top or high gear with the direct-drive transmission always has a ratio of 1:1, meaning that the rear wheel is turning at the same rpm as the clutch, hence the name "direct-drive." The direct-drive transmission is often housed in a separate gearbox, which is not attached directly to the engine.

This type of transmission was widely used on older European motorcycles and is still seen on some American-made motorcycles.

Dual-Range Transmissions

Another type of transmission, the **dual-range transmission**, can also be called a sub-transmission. This type of transmission is found in some street motorcycles, dual-purpose motorcycles, and ATVs as well. A dual-range transmission is normally a two-speed (high and low range), auxiliary transmission that's placed into the power flow between the transmission output side and the final drive system. Usually, a dual-range transmission has a manual shifting arrangement. The dual range allows a motorcycle to offer two different overall gear ratios. For instance, this type of transmission allows a trail bike to have both on-road (high range) or off-road (low range) gearing, and a street motorcycle to have the option of a low range for quicker acceleration as well as a high range for a lower engine rpm at higher speeds.

Shifting Transmission Gears

When you move the gearshift lever on a constant-mesh transmission, you're also moving mechanical connections, which are linked to certain gears within the transmission. Each movement of the shift lever locks one set of gears into position, and this set of gears is then engaged. At the same time, all other gears within the transmission are disengaged. Thus, shifting gears is the way you control the position of gears in the transmission and select different gear ratios that best suit the riding conditions. There are several different components that make up the gear-shifting mechanism used on motorcycle and ATV transmissions. Later in this chapter, we will look at some of them more closely.

The shifting sequence for a transmission is as follows:

- Your foot moves the shift lever, which is attached to a shift linkage.
- The shift linkage moves a shifting mechanism.
- The shifting mechanism moves a shift fork, which in turn moves a sliding gear to engage and disengage the transmission gear.

Let's now trace the power flow through a simple four-speed indirect-drive transmission. Although the gear arrangement and number of gears vary from four to six in most motorcycles, the power flow of indirect-drive transmissions is similar to that of the following basic four-speed motorcycle transmission example.

Indirect-Drive Transmission Power Flow

Neutral

Let's begin with the gear selection where there's no power transmitted to the rear wheel—neutral. Using Figure 9-18 as a reference, let's find out why power can't be delivered when the gears are in this position:

- Gear #1 is a fixed gear and turns with the main shaft. It meshes with gear #5 on the countershaft, causing it to spin. Because gear #5 is a freewheeling gear, no power is transmitted to the countershaft.

- Gear #2 is also a freewheeling gear, so power cannot be transmitted to sliding gear #6 when the main shaft rotates.
- Gear #3 is a sliding gear and turns with the main shaft. It meshes with freewheeling gear #7, so power cannot be transmitted in this position.
- Gear #4 is a freewheeling gear and cannot transmit power to fixed gear #8 when arranged in this neutral position.

If you could see inside the transmission with the engine running and the gears in this neutral position, you would see the main shaft and gears #1 and #3 turning gears #5 and #7 with no rotation of the countershaft.

First Gear

Comparing Figure 9-18 (neutral gear) with Figure 9-19, you'll notice that gear #6, a sliding gear, has moved to the right. This gear change from neutral to first gear connects freewheeling gear #5 with sliding gear #6, which is splined to the countershaft. By following the arrow, you can trace the power flow from the mainshaft and gear #1 to gears #5 and #6, and through the countershaft, which is now rotating to transmit power in first gear or the lowest gear ratio of the transmission.

Second Gear

Shifting from first gear into second gear causes two major changes in the arrangement of the gears. When comparing Figure 9-19 to Figure 9-20 you'll notice that sliding gear #6 has moved to the left, out of engagement with gear #5. Also, sliding gear #3 has moved to the right, into engagement with freewheeling gear #2. You can trace the power flow in the figure by following the solid line arrow

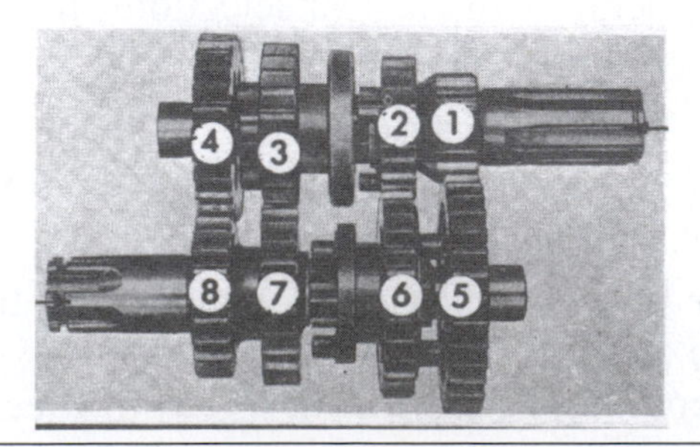

Figure 9-18 When in neutral, the engine does not turn the rear wheel.

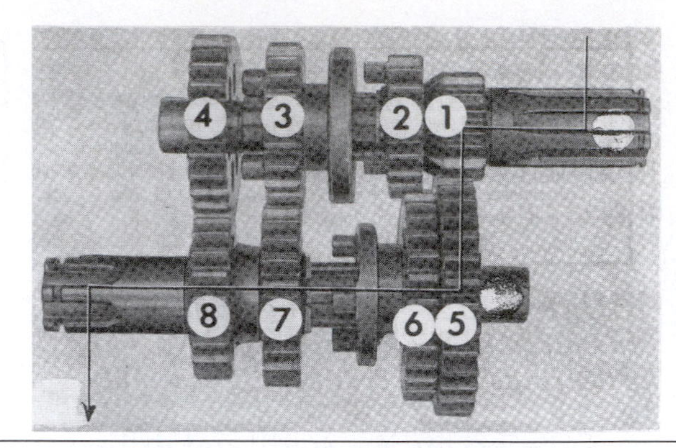

Figure 9-19 The power flow of first gear is shown here.

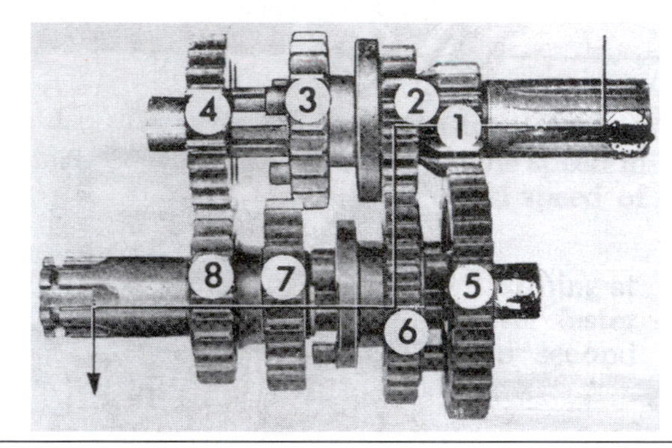

Figure 9-20 The power flow of second gear is shown here.

from the mainshaft and gear #2 to gear #6 (a sliding gear), which transmits power and rotation to the countershaft.

Third Gear

In Figure 9-21, notice that in shifting from second gear to third gear, sliding gear #3 has moved to the left, out of engagement with gear #2, and now meshes with gear #7. In addition, notice that sliding gear #6 has moved to the left to lock freewheeling gear #7 to the countershaft. As the arrow indicates, power flows through the mainshaft and gear #3 (splined) into gear #7 and through the countershaft.

Fourth Gear

In a four-speed transmission, fourth gear is also called top gear or high gear. In Figure 9-22, notice that in shifting from third gear to fourth gear, gear #6 has shifted to the right, out of engagement with gear #7. Also notice that sliding gear #3 has moved to the left to connect with

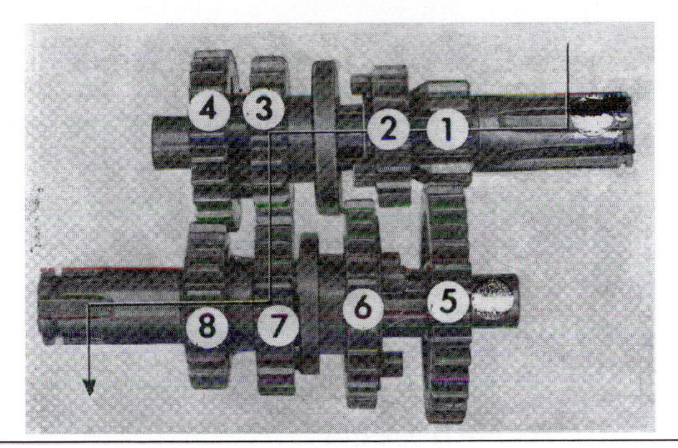

Figure 9-21 The power flow of third gear is shown here.

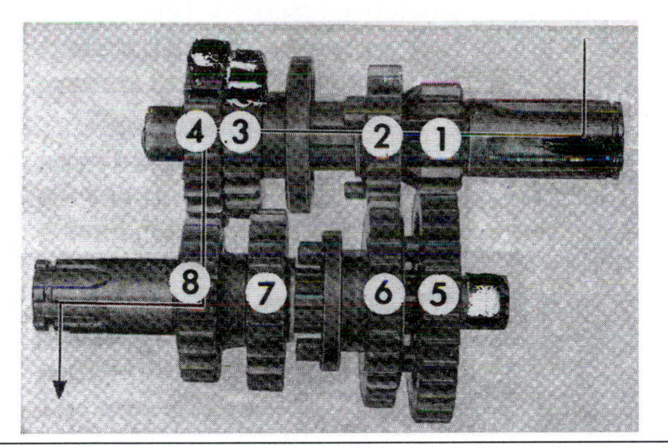

Figure 9-22 The power flow of fourth gear is shown here.

freewheeling gear #4. As the arrow indicates, power flows through the mainshaft and gears #3 and #4 and then into gear #8 (a fixed gear) and out through the countershaft to the final drive system. Final drive systems are covered later in this chapter.

Shifting Mechanisms

Manufacturers of motorcycles use many different mechanical systems to change the internal transmission gears from one gear ratio to another. Let's look at how the most common shifting mechanisms work. Once you gain an understanding of these mechanisms, learning how the variations operate will become easy.

The most common shifting mechanism in a constant-mesh transmission uses a component called a shift drum along with shift forks (Figure 9-23). You'll recall that shift forks fit into the grooves of sliding gears and allow movement

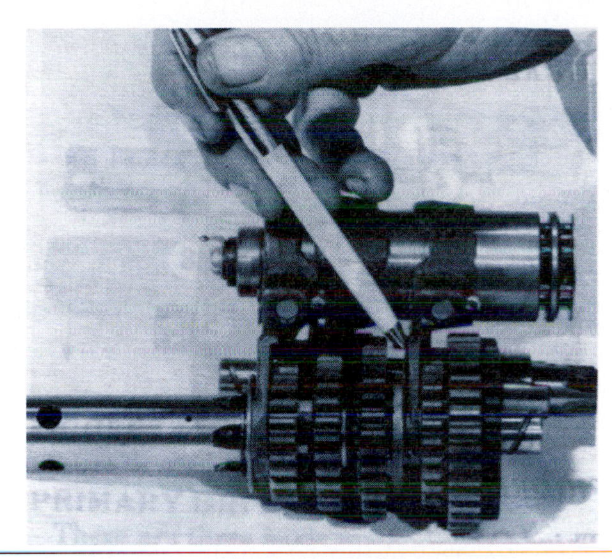

Figure 9-23 A shift drum turns and moves the shift forks back and forth in a timed manner to change from one gear to another.

to the right or left on the transmission shaft. Now let us learn what makes them move.

Figure 9-24 shows a typical entire shifting-mechanism assembly. When the rider's foot moves the change pedal (shift lever) downward, gearshift arms A and B move forward. The prongs on the end of gearshift arm B push on the gearshift drum pins, which causes the shift drum to rotate forward. Notice the machined grooves (slotted guides) in the shift drum. If you locate the shift forks, you'll notice that their cam follower pins fit into the machined grooves. When the shift drum rotates, the machined groove causes the shift forks to move in a timed sequence to the right or to the left to engage and disengage the sliding gears.

When the rider shifts to a higher gear ratio by moving the change pedal upward, the gearshift arms pull and rotate the shift drum in the opposite direction. This motion causes the shift forks to move the gears into engagement with the gear selected by the rider.

Shift detents are used to help in locating the next gear as the shift drum rotates. A component called a shift drum stopper lever applies spring-loaded pressure to the shift drum to ensure that the shift drum does not go past the intended gear and helps to lock in the desire gear. There are different variations of shift detents used within transmissions, but they all serve the same purpose.

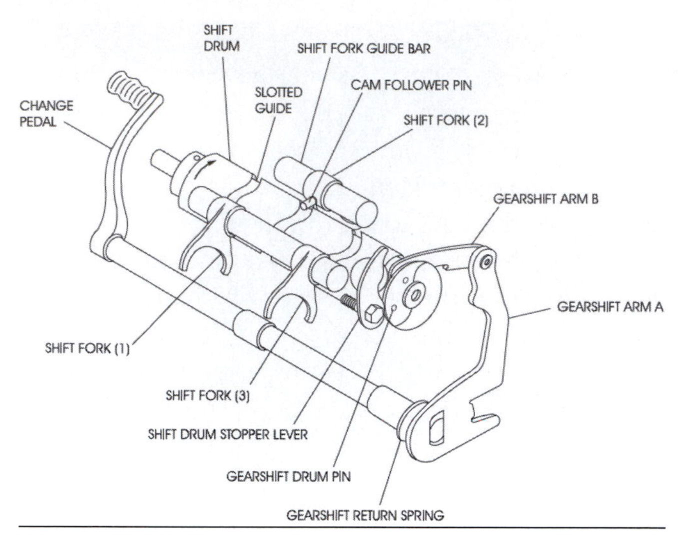

Figure 9-24 The various components of a typical shift mechanism are shown here.

Transmission Problem Symptoms

Because each part in the transmission does a certain job, when a failure occurs, you can usually tell which part is at fault by the symptoms. The following items discuss some common malfunctions of transmissions and how to recognize the malfunctions.

Difficulty Shifting

When there is a difficulty in shifting gears in a transmission, it may indicate either a clutch problem or a transmission problem. If the clutch is at fault, you'll notice symptoms such as grinding gears when you shift into first gear. As we discussed earlier, in many cases if the clutch is at fault, a simple adjustment may very well solve the problem. If the transmission is at fault, you'll notice difficult shifting between gears while the motorcycle is moving. This problem may indicate a bent shift fork or seized gear on one of the transmission shafts. When a shift fork is damaged, it will no longer fit properly in the grooves of the gear and will not fully engage into its mating gear. To fix this problem, disassembly of the engine will be required to replace the shift fork. When a gear has seized on the transmission shaft, the problem is usually caused by a lack of proper lubrication. As with the shift fork, you'll also need to disassemble the engine to repair the problem. But also take care to ensure that you have found out why there was a lubrication problem. For instance, was there a lack of oil in the transmission? Or, is there another lubrication issue such as a clogged oil jet somewhere inside the transmission? These are questions to answer before reassembly to make sure you do not have to do the repair twice.

Inability to Shift Gears

Occasionally you'll find a machine that shifts into one gear, but won't shift into the next gear. This problem is often caused by the shift return spring, which returns the shifting lever to its original position. You can see this spring in Figure 9-24 and can usually repair this type of problem by simply replacing the spring. The spring is normally located behind the clutch assembly and therefore you will need to remove the clutch, but in most cases, you will not need to disassemble the engine to repair the problem.

Strange Sounds

Occasionally you will have customers who may complain of strange sounds coming from the transmission of their motorcycle. Unusual sounds may range from a low growl to a high-pitched whine to clunking noises. Below we'll describe the two most common noise complaints, which may indicate a transmission problem. Most confirmed unusual noises that are not part of a manufacturing design that come from the transmission of a motorcycle would likely require you to disassemble and carefully inspect for worn or broken parts.

You should always try to compare a noise complaint to another machine of the same model to ensure that there is indeed a problem to begin with as it is very difficult to explain to a customer that there is no problem once you have stated that the noise was not one that should be there in the first place. Remember, you are considered to be the expert and once you let a customer know that you believe that there is indeed a problem, it will be very difficult to explain that there is not a problem after all if it is determined that the noise was not an actual problem.

Constant Growling Sound A growling sound from the transmission usually indicates a bearing failure. When a bearing fails, it may cause a transmission shaft to move slightly out of position. When this occurs, the gears don't mesh properly and can produce a low growling noise. As the transmission speed increases the low growling sound may turn into a high-pitched whine. In these cases, not only will you need to replace the bearing but after a close inspection of all of the gears, you will most likely have to replace some gears as well.

To verify that the transmission is indeed where the sound is coming from you will need to test ride the motorcycle to listen for the sound. Determine whether the sound increases in volume or pitch as you ride faster and continues at high speed even if you pull the clutch lever in.

Clunking Noises An excessive clunking sound when the engine is in a particular gear while under a load usually indicates broken teeth on one or more gears. In this case, you will need to disassemble and inspect all of the transmission components.

Jumping Out of Gear

When the gear dogs and/or slots become excessively rounded, the gears will tend to slip out of the holes when the engine rpm increases, and the engine "jumps out of gear." If this happens only occasionally, it may be that the transmission was not fully engaged when it was shifted into the gear in question, but if shifted incorrectly enough times, the gears will wear and the forks may become bent. Since the shift forks may also become damaged from the excessive pressure they encounter as the transmission jumps out of gear it will be very important to inspect them carefully along with the shift drum. The slots in the drum may become chipped and if not replaced with the rest of the worn parts, it will not shift the transmission correctly. Therefore, when a transmission is jumping out of gear, you will need to inspect closely all of the gears, shift forks, and the shift drum.

Automatic Shift Transmissions

Motorcycles may utilize different styles of automatic transmissions as well as the systems already mentioned. Most motorcycles and ATVs with a branded automatic transmission in reality are using an electrical type of shifting mechanism that shifts the engine with electronics (by pushing an up or down shift button) instead of using a footshifter mechanism. Honda uses this type of system on many of their ATVs.

In recent years, Yamaha has also introduced an automatic type of shifting mechanism for a motorcycle utilizing the same basic internal components of an indirect-drive system but utilizing electronics to disengage the clutch and shift the motorcycle into a different gear. This system features an electric-shift transmission that eliminates the clutch lever entirely. With the Yamaha system, either shifting occurs through the traditional foot-operated gearshift lever or bar-mounted paddle switches.

Yamaha calls this system "Yamaha Chip Controlled Shifting" (YCC-S). The YCC-S system uses two electronically controlled actuators—one for the clutch and one for the gear shift. The ECU at the heart of the YCC-S system is programmed to evaluate and compute data being transmitted by various sensors that constantly monitor engine rpm, mainshaft rpm, running speed, gear position, and throttle position. All of this information is used by the ECU to calculate the precise timing of every shift request in order to ensure smooth upshifting and downshifting.

Whenever the rider chooses to change gears, the system actuates the clutch push rod mechanism to disengage the clutch at the appropriate time, while in the shift system a shift actuator and shift rod engages the gears.

Some manufacturers also use "true" automatic transmissions for some of their ATVs as well as motorcycles. Two common types of this style of automatic transmission is the hydrostatic transmission and an automotive-style transmission. Since these types of transmissions are not used on many different types of motorcycles, they are certainly worth mentioning.

Hydrostatic Transmissions

The hydrostatic type of transmission can be found on not only some Honda ATVs but also on some motorcycles as well. This is a hydro-mechanical, continuously variable transmission housed within a fully sealed self-contained assembly. The engine drives a hydraulic pump that forces hydraulic fluid (in this case, engine oil) through pistons from the pump side to the other side of the pump called the motor side, which is attached to the output shaft of the machine. By the use of an angled plate called a swash plate, the transmission can obtain variable ratios by moving the plate from one angle to another allowing for a continuously variable shifting ability.

Automotive-Style Automatic Transmissions

The automotive-style automatic transmissions are found mainly on ATVs and controls gear selection by use of a small automotive-style torque converter and multiple shifting clutches that are attached to each gear set within the transmission. To shift this type of machine, control valves allow engine oil to flow in and out of the respective clutch packs to engage and disengage transmission gears through an electronic control module (ECM). Gears are changed either automatically or by the ECM using input from various sensors such as engine rpm, vehicle speed, and load resistance.

STARTING SYSTEMS

A motorcycle may utilize a kick-start lever or an electric starter to start the engine. We will discuss both of these types of starting systems, which will complete our discussion of transmissions.

Electric-Starting Systems

An electric-starting system uses an electric motor, which uses a gear or a chain to turn, in most cases, the crankshaft. Electric-starting systems work much like the starting system on a car, and are very popular on both on-road as well as off-road motorcycles. As a matter of fact, most every street motorcycle employs an electric-start system. We will discuss electric starters in detail in a later chapter.

Kick-Starting Systems

Many motorcycles have an engine kick-starting system that's actually a part of the transmission. Kick-starting systems fall into one of two categories—primary drive or direct drive.

In general, when a kick starter is used a lever is depressed, which activates a mechanism that turns a gear. This gear rotates another gear that's either attached directly to the primary drive system or to the main shaft. The difference between a direct drive and primary drive is that the primary drive system attaches directly to the primary drive of the engine and has the ability to start the motorcycle in neutral and any other gear as long as the clutch lever is pulled in. This is very handy on off-road motorcycles as it is common to try to restart the motorcycle in gear after a fall on the trail or track. The direct drive kick starter will only start in the neutral position due to the fact that the starter gears mate directly with a gear on the mainshaft and if tried to start with the engine's clutch pulled, will not engage as the clutch is not engaged to the crankshaft at that time.

FINAL DRIVE SYSTEMS

Now you should have a good understanding of how power is transferred from the crankshaft to the clutch and through the transmission. Next we will discuss the final drive system, which is a gear reduction system that takes the power from the transmission and allows it to flow to the rear wheel. There are three types of final drive systems —chain, belt, and shaft.

Chain-Driven Final Drive

The **chain-driven final drive** system (Figure 9-25) is the most common final drive system found on motorcycles. A chain-driven final drive consists of two sprockets, one attached to the countershaft of the transmission and one attached to the rear wheel. A chain is used to connect the sprockets. Different-sized chains are used depending on the power and size of the machine. By using a chain-driven final drive one can easily change gear ratios simply by replacing the existing sprockets with different-sized sprockets.

The sprockets and chain in a chain-driven final drive system wear out over time and require frequent maintenance to make them last. The drive chain also needs to be serviced more often than any component found on other final drive systems. Correct adjustment includes proper alignment of

Figure 9-25 The chain-driven final drive system is the most common final drive system found on motorcycles.

Figure 9-26A A chain that can be pulled away from its sprocket should be replaced.

Figure 9-26B A chain that is in good condition will not pull away from its sprocket.

the chain as well as proper chain free play. Also, proper lubrication of the final drive chain will help to prolong the life of the chain as well as the sprockets. There are two common ways to check a chain for wear:

- Try to lift the chain at various points around the rear sprocket. At a point midway between the top and bottom of the sprocket, try to pull the chain away from the sprocket. If you can pull the chain so that one-third of the sprocket tooth shows below the chain (Figure 9-26A), the chain should be replaced and the sprocket should also be closely inspected for wear. If the chain does not pull out of the sprocket teeth (Figure 9-26B), it does not require replacement.
- Lay the chain on a flat surface and measure the length of the chain when it's compressed (pushed together) to its shortest length and measure it. Pull the chain to stretch it out as far as possible. If the chain stretches in excess of 1/4 inch per foot, it should be replaced.

The chains used on chain final drive systems are composed of pin links and roller links (Figure 9-27). Pin links are composed of two plates and two pins. Roller links are composed of two plates, two bushings, and two rollers. The links are connected together by a master link or are considered to be an endless chain and have no master link. Many chains used on motorcycles use O-rings in between the rollers and plates to help protect the chain and to keep lubrication inside the roller.

The sprockets on chain final drive systems are flat metal plates with teeth around the outside edges. The chain fits around the sprockets with the teeth of the sprockets fitting into the open spaces between the rollers of the chain. Worn sprockets will ruin a chain. Sprocket wear is visible and the condition of a sprocket can be judged by comparing it to a new one (Figure 9-28).

Belt-Driven Final Drives

Belt-driven final drive systems are used on a few select models of motorcycles. These systems use a "Gilmer-type" belt that has teeth molded into it that mesh with a pair of toothed pulleys. The belt requires no lubrication and actually must be kept

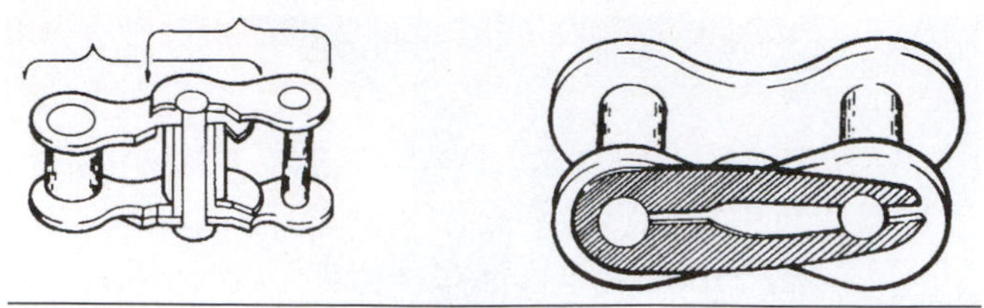

Figure 9-27 A drive chain link and master link are illustrated here.

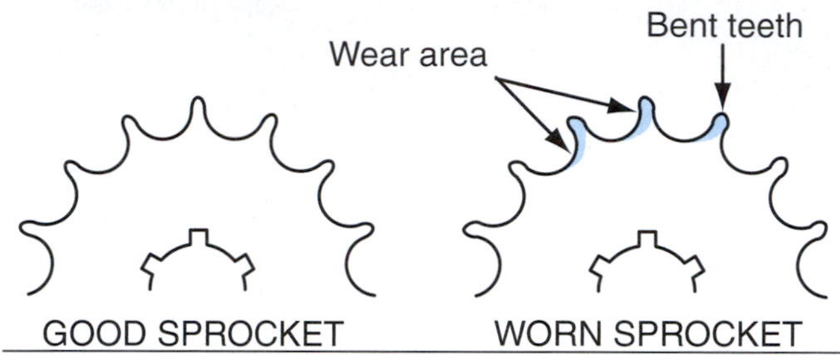

Figure 9-28 Different wear patterns of a worn sprocket as compared to a good sprocket. An incorrectly adjusted chain will make a sprocket wear out quickly.

dry. The toothed pulleys are like sprockets in many ways but are not as easy to change if you wanted to the change the final gearing due to availability. This system has certain maintenance requirements such as proper alignment of the belt and pulleys. Belt tension is extremely critical with this type of final drive system.

Shaft-Driven Final Drives

Although it has the least mechanical efficiency, a **shaft-driven final drive** system is deemed to be the best system available for a final drive as these systems are strong, clean, and require very little maintenance. Shaft drives are the least likely final drive system to have a failure, and will most likely last longer than the motorcycle on which they're used when they're properly maintained. There are many parts and gears included in a shaft-driven final drive system, which is the main reason that it is the least efficient mechanically (Figure 9-29). This system uses bevel gears to transfer power at 90° angles. Even though it is less efficient than a chain final drive system, the shaft-driven final drive systems are becoming popular on ATVs. The reason is that the complete system is self-contained and will not wear out as quickly as a chain or belt system.

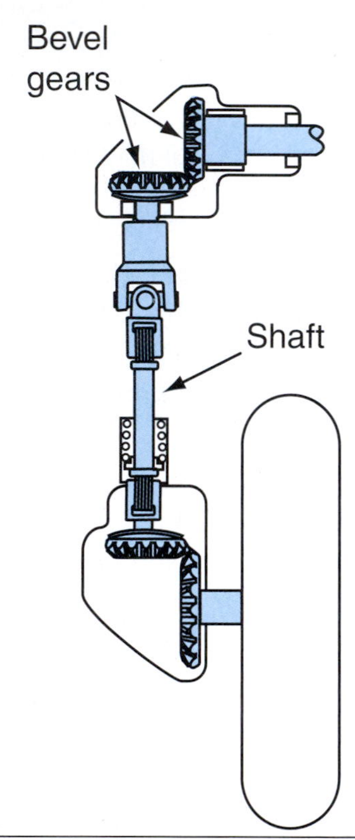

Figure 9-29 Shaft-driven final drive systems are becoming more common on motorcycles as well as ATVs as they require very little maintenance. They are, however, less mechanically efficient than a chain-and-sprocket design as there are many moving parts working in different directions to allow power from the countershaft to get to the rear wheel.

Summary

- There are several basic types of gears used in a motorcycle's engine. Gears are also used in many other areas of a motorcycle as well. Because you will see many different types of gears, it is important that you can identify and understand their purpose.
- Gear ratios are used to alter the speed of a rotating component to a useful rpm. A gear ratio is a numerical comparison of the number of revolutions of the drive gear as compared to one revolution of the driven gear.
- The gear reduction system used for transferring the power from the crankshaft to the clutch is called the primary drive system.
- There are three basic types of clutches used on motorcycles: manual, centrifugal, and variable ratio.

Chapter 9 Review Questions

1. The ________ gear is the most commonly used gear on motorcycles.
2. If a crankshaft drive gear has 24 teeth and the clutch outer-basket driven gear has 68 teeth, the ratio (rounded to the thousandth) for this set of gears is ________.
3. A gear on a main shaft has 28 teeth and the countershaft has 33 teeth; the ratio (rounded to the thousandth) for this set of gears is ________.
4. If the countershaft sprocket has 15 teeth and the rear wheel sprocket has 42 teeth, the ratio (rounded to the thousandth) for this set of gears is ________.
5. By using the ratios from questions 2, 3, and 4, the overall gear ratio for this motorcycle is ________.
6. The ________ is used to engage and disengage the power flow from the engine to the transmission.
7. When the ________ primary drive system is used, the crankshaft rotates in the opposite direction as the clutch.
8. The ________ on a manual clutch are normally made of a high-friction material.
9. A clutch that won't fully disengage can be the cause of a motorcycle or ATV that has a tendency to creep even though the clutch lever is fully depressed. True/False
10. The two most common problems found in clutches are ________ and ________.
11. The proper measuring tool to check for the thickness of the clutch friction plates is a ________.
12. The ________ transmission is the most common type of transmission found in motorcycle engines today.
13. For a constant-mesh transmission to operate correctly there must be a ________ gear opposed to a fixed or sliding gear on the opposite shaft.
14. A ________ is used to move a sliding gear.
15. When an engine is in a gear selection where there is no power being transmitted to the rear wheel, it is considered to be in ________.

Chapter

10 Two-Stroke Engine Top-End Inspection

Learning Objectives

When students have completed the study of this chapter and its laboratory activities they should be able to:

- Understand the importance of engine problem diagnosis
- Understand the necessity of proper engine component inspection
- Describe the concept of the motorcycle two-stroke engine power valve

Key Terms

Compression check
Cylinder head
Diagnosis
Honda CR125R
Piston
Piston ring end gap
Piston ring grooves
Piston ring side clearance
Piston-to-cylinder clearance
Ring lands
Symptoms
Wrist pin

INTRODUCTION

While becoming less prominent in the motorcycle industry due to the advancement of the four-stroke design and more stringent emission control regulations, there are many two-stroke motorcycles being used today. Two-strokes on today's motorcycles are usually off-road motorcycles and almost all are of the single-cylinder liquid-cooled variety.

Top-end engine disassembly is a process in which all the parts of an engine above the engine crankcases are removed. An engine may be disassembled to make needed repairs, or the disassembly may be the first step in a complete rebuild. During an engine rebuild, an engine's components are replaced to a "like new" condition.

DIAGNOSTICS

If the motorcycle has an engine-related problem, before any components are disassembled the technician must diagnose the condition. **Diagnosis** is the process of determining what's wrong when something isn't working properly by checking the symptoms. **Symptoms** are the outward, or visible, signs of a malfunction. For example, a knock or a slipping clutch is a symptom. The actual cause might be a broken, worn, or malfunctioning part. Of course, this is only one example and to assist with diagnosing a problem you may need to use other testing equipment to help find the problem.

Often, complete diagnostics cannot be confirmed until the machine is actually disassembled. For instance , if a two-stroke engine develops a rattle in the top end, an experienced technician may recognize the sound and tentatively conclude that the piston is worn out. That would be the diagnosis. The technician would not be able to confirm the diagnosis until disassembling the engine and actually seeing that the piston and cylinder are worn or damaged.

Correctly diagnosing problems is the most difficult and important part of a technician's job. Diagnosis is difficult because the technician often can't see the faulty part before disassembly. Correct diagnosis is important because the technician must not waste time disassembling and inspecting parts that have not failed.

To diagnose problems, experienced technicians mentally divide a motorcycle into sections. For example, suppose the rear wheel won't turn. The technician first inspects the wheel and tire to be sure that they're free to rotate. Then, the technician inspects the brake assembly to ensure that the brake isn't locked up. Finally, the technician checks that the final drive system is in proper order. Thus, the technician inspects each part that is connected to the problem following a logical order of possible malfunctions.

As this chapter covers the top-end of the two-stroke engine, we're going to present some engine problems and give you the possible diagnoses. In diagnosing, you should have a mental picture of the parts connected with the problem. Table 10-1 lists some common problems you might encounter.

Troubleshooting will be discussed later in the chapter, but as you will see, you can handle quite a few common engine repairs by disassembling the top end of the two-stroke engine.

GENERAL TIPS BEFORE BEGINNING ENGINE REPAIRS

- Be sure the motorcycle is clean. Dirt or other foreign particles cause damage to internal working mechanisms.
- Use the correct tools. Use a six-point socket whenever possible. The second choice is a box-end wrench. Use extreme care when using open-end wrenches as they may spread at the jaws, which can round the heads of the fasteners.

REPAIR PROCEDURES

The procedures in this chapter are general in nature and their purpose is to familiarize you with the types of activities you'll encounter when working on a two-stroke top end. Always refer to the appropriate motorcycle service manual for disassembly information. The service manual contains all the information to do the job correctly, including detailed instructions for the specific model of motorcycle, special tools, and service tips. Above all, the service manual contains the appropriate safety information.

Table 10-1: DIAGNOSING ENGINE PROBLEMS

Symptom	Check	Notes
Engine won't start	Fuel flow	Loosen the carburetor drain screw. Fuel should flow as you loosen the screw. If fuel doesn't flow, verify that there's fuel flowing from the fuel tank to the carburetor.
	Ignition	Be sure the spark plug is firing at the correct time. Ignition systems are covered in Chapter 16.
	Compression	Hold a compression gauge in the spark plug hole while the engine is rotating at cranking speed. Engine compression should be within manufacturer's specification. You'll find the actual specification in the appropriate service manual. The most common causes of low compression are excessive piston wear or excessive piston ring wear, which will be covered later in this chapter.
Engine rotates and may run but has a loud, heavy knock	Broken piston skirt	If the piston-to-cylinder clearance is too large, the piston rocking on the bore may break the skirt.
	Big-end connecting rod bearing failure	See Chapter 11 on lower-end assembly for connecting rod problems.
Engine runs but has a light, rapid tapping noise	Piston ring wear	Excessive piston ring-to-cylinder clearance likely means that dirt or other debris has been getting into the engine through the air filter side of the engine.
	Piston pin wear	Unless a lack of lubrication is found, excessive wear on the piston pin is unusual because the pin is made from high-quality steel.
Engine won't rotate	Piston seizure	Piston seizure means that the piston is stuck in the cylinder. Possible causes of piston seizure are: • Improper fuel/oil mixture • Improper air/fuel mixture • Incorrect spark plug or ignition timing • Insufficient piston-to-cylinder clearance
	Failure of the lower-end assembly	The engine may not rotate due to the failure of the lower-end assembly, including clutch or transmission problems. These items are covered in Chapter 11, which relates to the lower end of the motorcycle two-stroke engine.

TWO-STROKE TOP-END DISASSEMBLY AND INSPECTION

The components of a typical two-stroke engine top-end assembly are illustrated in Figure 10-1. This exploded view of the top-end assembly of a liquid-cooled two-stroke engine is what you would see at the beginning of most manufacturer service manuals. The top-end assembly consists of the following parts:

- Cylinder and head fasteners
- Cylinder head
- Cylinder head dowel pins
- Cylinder head gasket
- Power valve assembly and covers
- Cylinder and cylinder head studs
- Cylinder
- Cylinder base gasket
- Cylinder dowel pins
- Piston and piston rings
- Piston wrist pin
- Piston wrist pin bearing
- Piston wrist pin retaining clips

As was discussed in an earlier chapter, the focus was on the differences between an air-cooled two-stroke engine and a liquid-cooled two-stroke engine. Before moving forward, you may wish to review this information before working on a two-stroke engine.

In this chapter, you'll see images of a liquid-cooled, single-cylinder two-stroke engine as it is the most common design found in today's motorcycles. Both air- and liquid-cooled two-stroke engine designs use similar approaches to disassembly and assembly procedures.

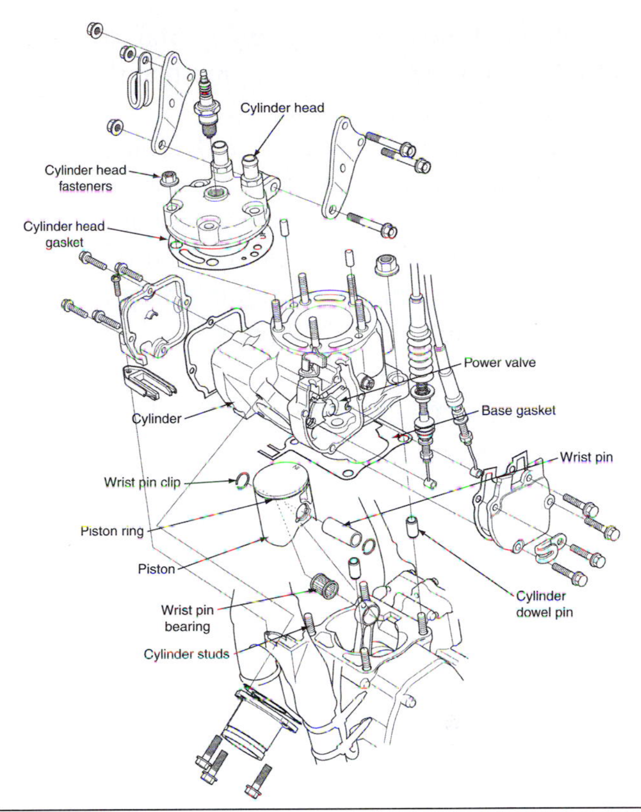

Figure 10-1 The components found in a typical two-stroke liquid-cooled single-cylinder motorcycle engine are shown in this exploded view. Copyright by American Honda Motor Co., Inc. and reprinted with permission.

Disassembling the Top End of the Engine

As there are many different motorcycle manufacturers building two-stroke engines, we will not show you how to disassemble the top end of a two-stroke engine. We strongly recommend that you utilize the appropriate manufacturer's service manual and follow their procedures when disassembling an engine. Incidentally, for this chapter, we're using the engine from a liquid-cooled **Honda CR125R** motorcycle for illustrative purposes (Figure 10-1). The procedures for the disassembly of a two-stroke ATV are the same as for a motorcycle.

Figure 10-1 This chapter will use images of the Honda CR125R for illustrative purposes. The basic design elements of this motorcycle are used in most other manufacturer's two-stroke engines.

Repairs of the top-end assembly of a two-stroke motorcycle normally do not require that the engine be removed from the chassis.

As you disassemble the top end, look closely at all of the parts and record your observations. Note any possible sources of damage to any of the parts. Note any marks on the piston and the rings. Your notes will be valuable as you complete your inspection on the individual parts.

TWO-STROKE ENGINE TOP-END INSPECTION

Once you have the two-stroke engine's top end apart, it is important to clean and inspect each component before reassembling. This inspection process is as follows:

- Inspecting the cylinder head
- Determining if the piston is reusable
- Fitting new piston rings to the cylinder
- Checking cylinder wear to see if the bore is within manufacturer specifications
- Cleaning and inspecting the power valves
- Checking the reed valves

Part of this inspection is measuring the top-end components. The purpose of these measurements is to determine when a part is excessively worn. You should always measure the following two-stroke motorcycle top-end components:

- Cylinder head
- Cylinder
- Piston
- Piston rings

Knowledge of certain measuring tools is essential to completing some of these inspections. We explained these tools in Chapter 3 of this textbook.

The special measuring tools you will use to complete the top inspection are the following:

- Dial bore gauge
- Outside micrometer
- Inside micrometer
- Feeler gauge
- Compression gauge

Cleaning Top-End Components

Cleaning and visual inspection of all parts and top-end components is very important. By carefully cleaning all the components included in the two-stroke top end, you'll find potential problems before they occur. You'll need degreasing solvent to remove carbon deposits and oil deposits on the piston. Brush gently and use special care not to damage any parts as many may be fragile. You may need a wire brush and a scraper to remove residual gasket materials from gasket surfaces and excess carbon from the cylinder exhaust port and the cylinder head. Wash all parts in approved solvent and dry with regulated, compressed air pressure and a lint-free shop rag. Remember that most cylinder heads and cylinders are made of aluminum, which is a relatively soft metal, so be careful you don't dig into them with a scraper or brush when cleaning. Remove any remaining old gasket material from the cylinder and crankcase joint-mating surfaces. That is where the cylinder joins the crankcase.

Never use gasoline or other highly flammable liquids to wash parts. Uncontrolled fires are easy to start, but hard to stop.

Cylinder Head Inspection

The **cylinder head** is the component that seals the upper end of the cylinder. It's normally attached to the top of the engine's cylinder by several fasteners. A gasket between the head and the cylinder helps create an airtight seal. Since the cylinder head must seal off the cylinder, the head must be in good condition and free of cracks and distortion. Before inspection of the head, be sure that it is clean.

Once the cylinder head is thoroughly cleaned, you can check it for any visible signs of damage. Check for small cracks or other damage in the combustion chamber area (Figure 10-2). Also, if any cooling fins are broken or water jackets are damaged, the head will have to be replaced. In most cases, cylinder heads are very reliable, and you won't find any damage on them. The most common problem you'll see is damage to the threads in the spark plug's hole from incorrectly installing the spark plug.

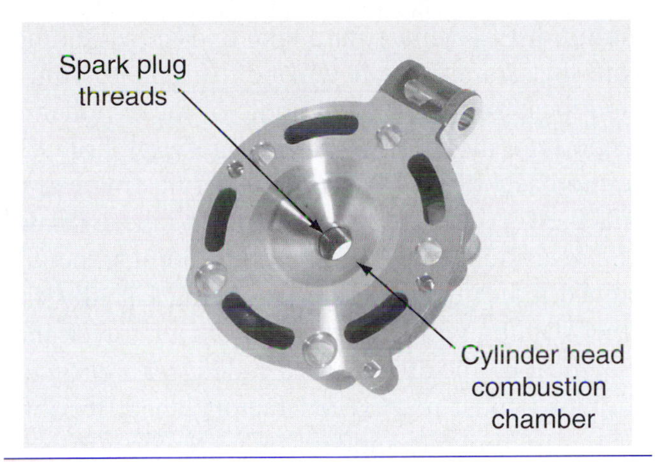

Figure 10-2 Check the cylinder head combustion chamber for cracks and look closely for damaged spark plug threads. Copyright by American Honda Motor Co., Inc. and reprinted with permission.

Cylinder Head Distortion Inspection

After the cylinder head has been cleaned and if it appears to be free from damage, you can move on to check for distortion of the surface where the new head gasket will be installed. Remember that the cylinder head must seal tightly to the top of the cylinder. The gasket between the head and the cylinder can compensate for some variation in flatness, but the surface of the head must still be quite flat or the seal will fail. The manufacturer's service manual will tell you the maximum amount the surface of a usable cylinder head can be distorted or warped.

Check for cylinder head surface distortion using a straightedge. Place the cylinder head so that the surface on which the gasket is installed is facing up. To check for distortion, place the straightedge across the surface of the head. Try to insert the blades of a feeler gauge to measure the distortion at that point (Figure 10-3). The thickness of the blade that fits the clearance is the amount of cylinder head distortion at that particular location on the head.

Because the straightedge is very narrow, move it about the surface of the head and measure for distortion in several locations. In most cases, placing the straightedge diagonally across the head's surface gives the best indication of the head's flatness. Generally, you compare the dimension you measure with the manufacturer's specification to find the maximum amount of distortion at any point on the surface.

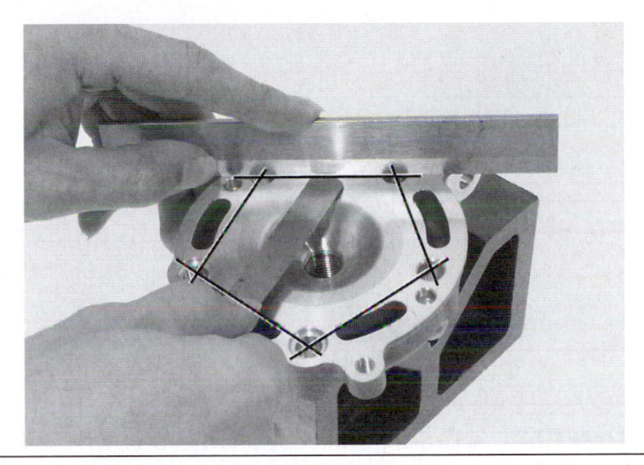

Figure 10-3 A technician is using a straightedge and a feeler gauge to check for cylinder head distortion. Copyright by American Honda Motor Co., Inc. and reprinted with permission.

If the cylinder head's distortion is within specifications, the head-mating surface area is acceptable, and the head is ready to be reinstalled and used. However, if you find that the cylinder head is distorted (warped) more than is allowed, the head will have to be replaced. In some cases, you can have the head surface machined but by doing so, the compression ratio will become higher as the distance between the cylinder head combustion chamber and the piston will be reduced. If too much surface of the cylinder head is removed, you may cause further damage to the engine. Therefore, it is recommended that you replace the cylinder head instead of attempting to repair a warped head surface.

Cylinder Inspection

Check the cylinder-to-cylinder head-mating surface for distortion the same way you checked the cylinder head (Figure 10-4).

Cylinder Wall Inspection and Measurement

After checking for distortion, visually inspect for obvious cracks, scratches, and scoring marks on the cylinder. Once it has been verified that there is no visual damage or wear, measure the bore of the cylinder with a dial bore gauge to check for excessive wear (Figure 10-5). Note that due to the large ports used in the modern two-stroke engine it is sometimes difficult to measure a point in the middle of the cylinder. Refer to the manufacturer's service manual to be assured that

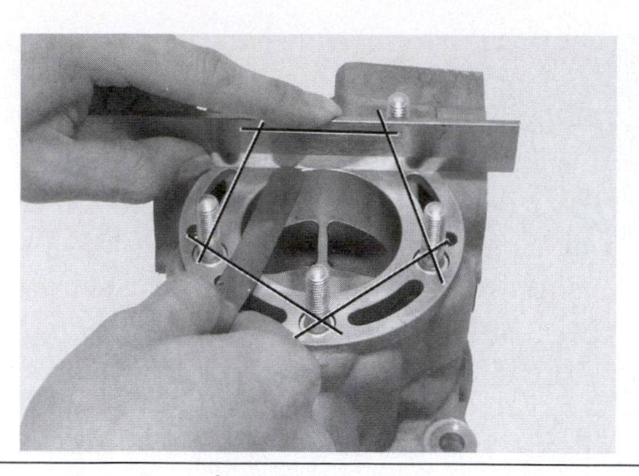

Figure 10-4 A technician is using a straightedge and a feeler gauge to check for cylinder distortion here just as done with the cylinder head. Copyright by American Honda Motor Co., Inc. and reprinted with permission.

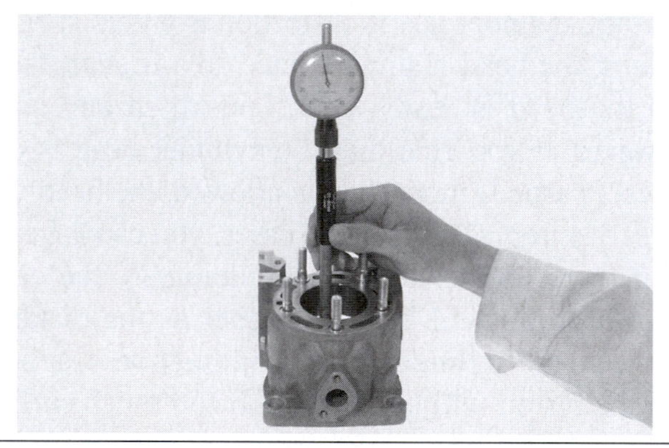

Figure 10-5 A technician is shown here measuring a cylinder with a dial bore gauge. Copyright by American Honda Motor Co., Inc. and reprinted with permission.

you measure all points of the cylinder as specified by the manufacturer.

This is to determine the amount of wear within the cylinder wall. Movement of the piston and rings within the cylinder causes cylinder wear. The area of greatest wear is the area in which the rings travel. This is because the rings must press tightly against the cylinder wall in order to seal the compression into the cylinder. In the area of ring travel, more heat is generated on the front and backside of the cylinder wall. By front and back, we mean at a 90-degree angle to the wrist pin.

Insert the dial bore gauge into the cylinder at a point near the top of the ring travel. Take a gauge reading. Move the dial bore gauge to a point near the top of the exhaust port and take another reading, then move to the bottom of the exhaust port. Finally, move the dial bore gauge to a point near the bottom of the cylinder and take one more reading. Note that sometimes it will be difficult to find a good location toward the middle of the cylinder as the ports in modern two-strokes are very large. Be sure to check the manufacturer's service manual to find the recommended locations to take cylinder bore measurements. Compare the readings; the difference indicates the amount of wear. Usually, the larger reading will be in the area of ring travel: that is, at the top. The difference is called *cylinder taper*. The taper must not exceed factory specifications for the motorcycle model you're repairing. Each model will have different specifications. If the "taper" exceeds allowable limits, the cylinder must be replaced or, if allowable, bored to a new size and fitted with a new, oversized piston and rings. Take the same readings at 90 degrees for the first three readings to determine if the cylinder is out of round (Figure 10-6).

Boring Cylinders

Boring a cylinder requires special machine tools and is therefore performed by a specialist. Cylinder reboring can be done at some motorcycle dealerships, but it's normally done at a machine shop that's set up to handle such a job.

If you must re-bore a cylinder, first obtain a correct oversized piston. Then, bore the cylinder to fit the piston. Many two-stroke cylinders use a plated bore (Figure 10-7) fused to the cylinder

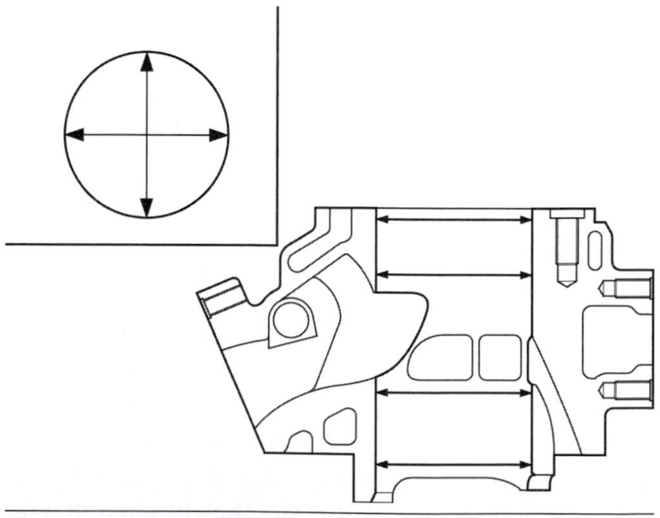

Figure 10-6 This image points out typical points to measure a cylinder for wear. Copyright by American Honda Motor Co., Inc. and reprinted with permission.

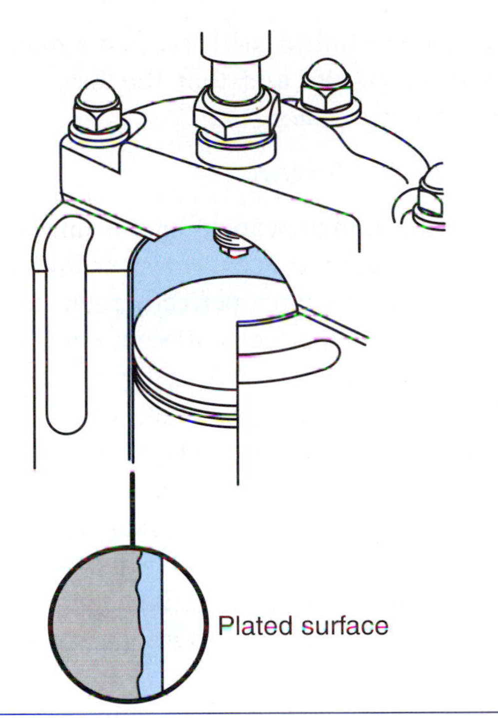

Figure 10-7 This image shows a plated cylinder. Plated cylinders last a long time if properly maintained, but most may not be bored oversize.

wall. While this design greatly reduces weight and friction and improves heat transfer, in most cases the cylinder cannot be bored and must be replaced if its measurements are outside the manufacturer's specifications.

Inspecting the Piston Assembly

As you'll recall, the **piston** is the cylinder-shaped component that moves up and down in the cylinder bore. The piston assembly consists of the piston itself, its wrist pin (or piston pin), and the piston ring(s). An engine produces its power by burning the air-and-fuel mixture in the combustion chamber directly above the piston. Each time the spark plug fires, this air-and-fuel mixture ignites with an explosive force. The burning process heats the gases, causing them to expand rapidly, forcing the piston down the bore. That piston movement allows the engine to perform useful work. As you can imagine, the piston must withstand a lot of physical force as well as tremendous heat during an engine's operation. Therefore, as part of the rebuild procedure, you must inspect the entire piston assembly carefully for any damage.

Visual Examination

Start the piston assembly inspection with a visual examination of the piston itself. Check the piston for cracks or any other signs of surface damage, such as scratches or score marks. Pay particular attention to the front and the backsides of the piston. You should examine the piston in the areas of both the skirt and the rings; these areas are the most common sites of damage. Look for signs of surface damage on the piston's skirt. One of the most common types of piston damage is scoring on the piston's skirt. *Score marks* are deep, vertical scratches on the skirt. A similar type of damage is scuffing. Scuff marks are wide areas of wear on the piston. The scuff marks usually appear as shiny patches. Scuffing may or may not be accompanied by score marks and is normally present on the bottom of the piston as it is installed in the cylinder.

A variety of conditions may cause both scoring and scuffing. In most cases, excessive friction and heat create the marks. Under certain conditions, the temperature in a cylinder can approach the melting point, or weld point, of aluminum. A problem in an engine's cooling system or excess friction between the cylinder wall and the piston rings can cause these very high temperatures. Excessive friction is often caused by improper lubrication or from the piston fitting too tightly within the bore. Minor scuffing can be removed by using a fine emery cloth to clean the surface of the piston as illustrated in Figure 10-8.

If you find score marks or scuffing marks on a piston, try to determine the cause so that you can

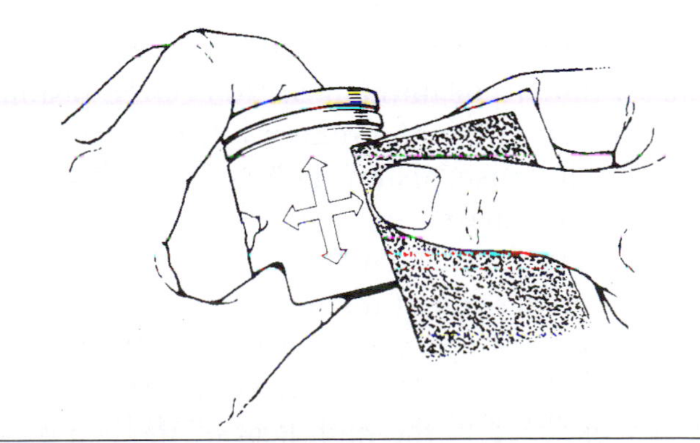

Figure 10-8 In this illustration, a technician is using a fine emery cloth to remove some light scuffing on a piston.

prevent the damage from recurring. This is one of the times you can take advantage of the notes and observations you made earlier in the disassembly process. During the disassembly, you should have noted any defective gasket surfaces. If you did note a possible source of damage during disassembly and later found marks on the piston, you'll already have important clues for use in the troubleshooting process. In such a situation, you may also want to talk to the customer or operator to find out whether the engine was overheating during operation.

Engine overheating, in addition to causing scuffing and scoring, usually produces a buildup of oil residue on the piston and the rings. Extreme heat breaks down the viscosity of oil and reduces its lubricating ability. Once oil breaks down, it begins to "bake" onto the engine components, forming a residue that resembles varnish. This residue can coat the piston rings and under extreme conditions can cause the rings to stick firmly to the piston. If this occurs, the rings will no longer be able to seal the combustion chamber properly. Therefore, always check to ensure that the rings are free to move on the piston and that both piston and rings are free of any buildup.

Although the skirt is the most frequently damaged site on a piston, you must also carefully examine the piston's crown. If you find any damage, try to determine the exact cause so you can prevent that damage from happening again. Any damage to the crown is usually the result of the fuel mixture burning improperly in the cylinder. If the fuel mixture ignites incorrectly or at the wrong time, a violent explosion can result. The concentrated heat created in such an explosion can burn a hole through the piston's crown. Also, the explosion itself can be powerful enough to knock a hole right through the top of the piston. We'll discuss common problems that occur in an engine that cause piston damage like this in a future chapter based on troubleshooting. As the final stage in your visual inspection of the piston, check the underside of the piston. It's in this area that the connecting rod is physically attached to the piston. Because the interior of the piston isn't directly exposed to the high heat of the combustion chamber and generally remains well lubricated, you'll usually find no signs of damage. However, you should still check to make sure there are no cracks and that the wrist pin fits properly into the piston.

Measuring the Piston

A typical piston appears to have a simple, cylindrical shape, similar to a can. However, looks can be deceiving. Pistons are not perfectly round, and they don't usually have perfectly straight sides when at room temperature. In fact, a typical piston is manufactured with a taper. That is, the diameter at the very bottom of the piston's skirt is larger than the diameter at the piston's head. Why does the piston have this taper? The reason is heat. You should also note that most all two-stroke pistons would have an arrow or marking that will point toward the exhaust port to ensure that the piston is installed correctly.

Most pistons are made of aluminum, and aluminum expands as the temperature rises. A piston, however, gets very hot at its head (where the combustion actually occurs) but remains relatively cool at the skirt (which is comparatively far away from the combustion chamber). Because the piston gets hotter at the top and also has more mass than at the bottom, its top expands more than its bottom. To compensate for this difference in expansion, a piston is made with a built-in taper. Then, as the piston gets hot, it expands and assumes a cylindrical shape.

A piston's shape is designed to compensate not only for the piston's expansion but also for the physical forces placed upon the piston. In the typical operation of an engine, the expanding gases force the piston both downward in the cylinder and against the cylinder walls. The piston's wrist pin links the piston to the connecting rod. Therefore, both the piston and the rod move together to rotate the crankshaft. Figure 10-9 shows the way the piston is connected to the crankshaft, causing a force to be directed toward one side of the piston on each up and down stroke. This side of the piston is called the *thrust face* and is in a plane that's parallel to a line running through the length of the wrist pin. The forces applied during a power stroke are greater at the thrust face than they are elsewhere on the piston's sides. Therefore, a piston is usually made so that the diameter measured at a right angle to the wrist

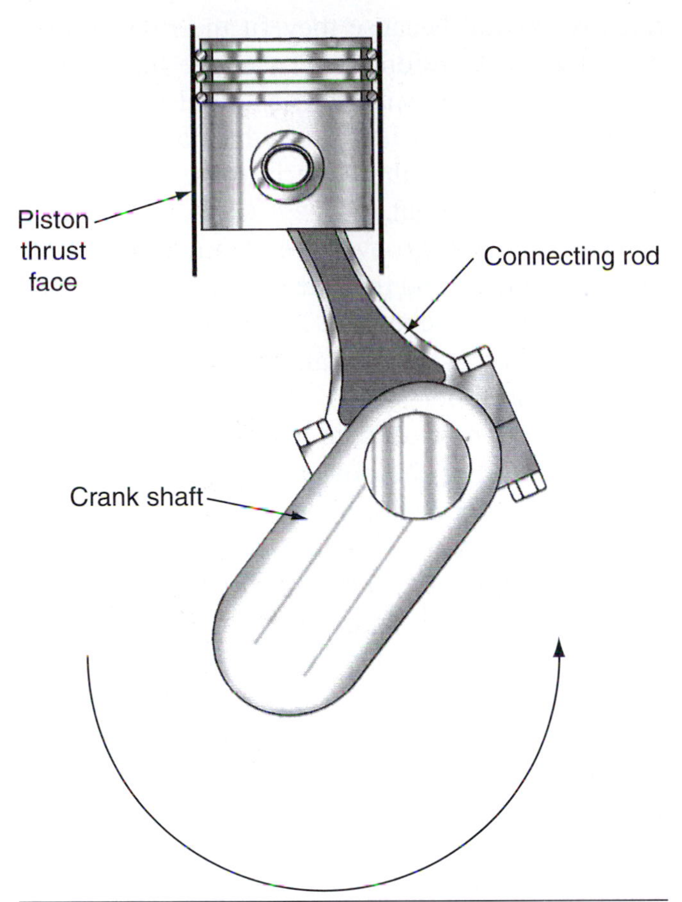

Figure 10-9 Pistons have two different thrust surfaces: the back side on the down stroke and the front side on the up stroke.

pin is slightly larger than the diameter measured along the wrist pin's length. That is, if you were to look at the head of a piston from above, the piston would appear to have a very slight egg shape. The wrist pin would divide the egg shape into its wide and narrow halves. The piston's thrust face would be at the wide half.

As we've already mentioned, the downward force applied to a piston on the engine's power stroke pushes the piston's thrust face against the cylinder wall. During the application of this force, the shape of a piston will become more round. Note that the egg shape of an actual piston is usually very slight and difficult to see with the eyes alone. However, if you measure the diameter at a right angle to the wrist pin and compare it to a diameter measured along the wrist pin's length, you should be able to note a difference between the two measurements.

Once the piston has been visually inspected, you can prepare the piston for measurement. Before measuring the piston remove the piston ring(s). To remove a piston ring, you must spread the ring open so that you can slide it out of its ring groove and off the piston. Most technicians spread piston rings open by hand (Figure 10-10).

Measure and record the piston outside diameter at an angle 90 degrees to the wrist pin bore as illustrated in Figure 10-11. Note that the micrometer is placed at the bottom of the piston skirt. Some manufacturers will have a specific distance at which to measure the piston. Be sure to check the appropriate service manual specifications and procedures when measuring a piston.

Once you've measured the piston's diameter, compare your measurement with the appropriate specification in the manufacturer's approved service manual. If the diameter of your piston is outside of specifications (generally this measurement would be smaller than the specification), the piston

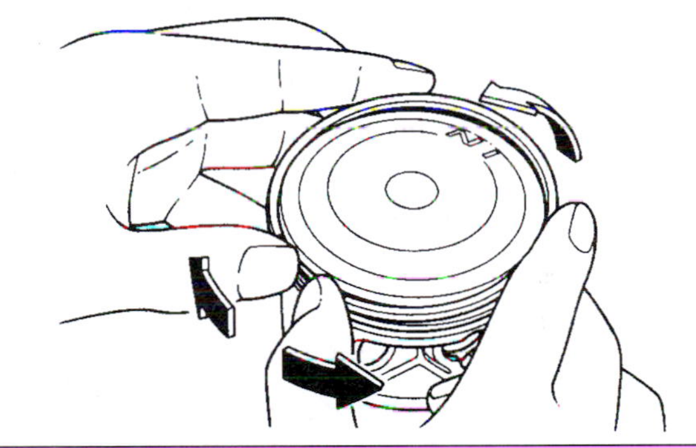

Figure 10-10 A technician is spreading and removing a piston ring in this illustration.

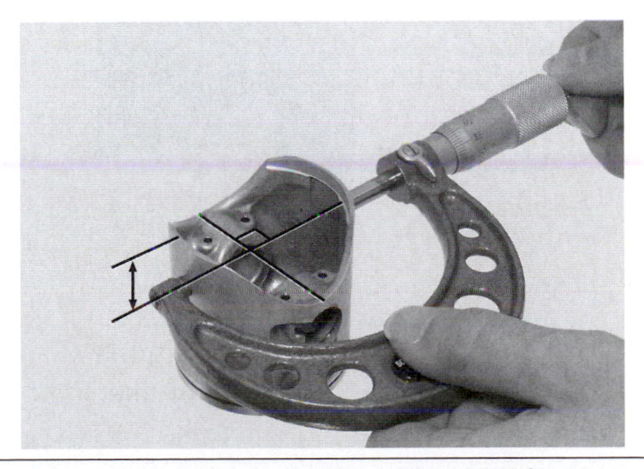

Figure 10-11 Measure the piston with an outside micrometer. Copyright by American Honda Motor Co., Inc. and reprinted with permission.

should be replaced. If the piston is within specifications and shows no signs of damage, you can reinstall it in the engine.

Always remember to replace the piston rings even if you reuse the piston. The **piston ring grooves** cut into the sides of the piston hold the piston rings in place. The **ring lands** are the uncut areas between the ring grooves. The ring grooves are actually slightly wider than the piston rings. As a result, the rings can move slightly, or float, within their grooves. Thus, the rings are able to actively conform to the cylinder walls while the engine is operating. The small amount of space between each piston ring and the inner side of its groove is called **piston ring side clearance**.

The combustion gases forcing themselves onto the piston get into the ring grooves and leave behind a residue. Therefore, to inspect the ring grooves for excessive wear, you must first clean the grooves thoroughly. When cleaning the grooves, remember that the piston is made of aluminum, a soft metal. Therefore, be careful not to dig into the piston and remove any metal, especially along the inner sides of the ring grooves. The best tool to use is an old piston ring. Made of a very tough material, old piston rings work well because they fit the ring grooves perfectly and therefore won't damage the sides of the grooves. If you want to use an old ring for this purpose, break it in half to produce a scraper-like edge. Then, insert the edge into the groove and scrape the residue out.

Once the ring grooves are cleaned, the piston can be wiped off and the side clearance for the piston rings can be checked. As mentioned before, this dimension is the clearance between the piston ring and the inner side of the ring groove. This small amount of clearance performs an important function. During the power stroke, the pressure produced by combustion pushes the piston down in the bore. Some of the expanding gases also force down along the side of the piston and behind the floating piston ring. The resulting pressure behind the piston ring forces the ring outward hard against the cylinder wall, thus helping to better seal the combustion chamber. By allowing the ring to seal better, the proper ring side clearance helps the engine produce more power.

Because a small amount of clearance should always be present, a piston ring will tip slightly under normal operating circumstances as illustrated

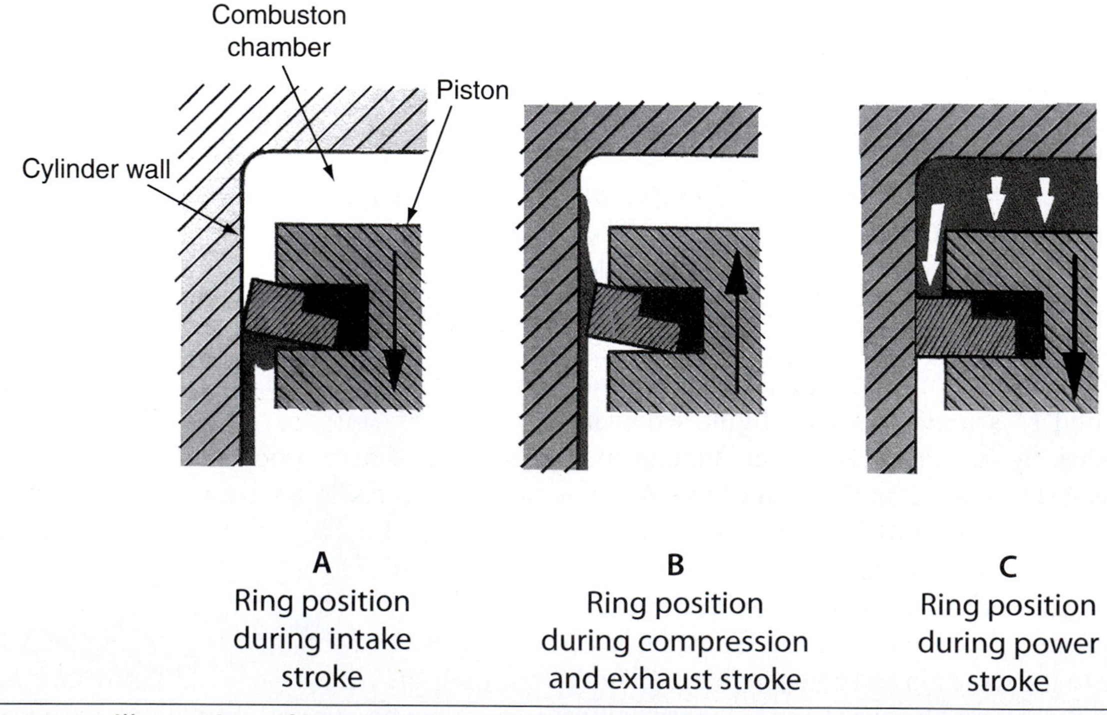

Figure 10-12 An illustration of how a piston ring moves within the piston ring groove is shown here. Figure A shows the ring position during the intake stroke. Figure B shows the ring position in the compression and exhaust stroke and Figure C shows the ring position during the power stroke.

in Figure 10-12. As the piston goes down in the cylinder during the intake stroke, the ring tips and scrapes excess oil off the cylinder wall. During the compression and exhaust strokes, the piston rises and the tipped ring glides over the oil film remaining on the cylinder wall. During the power stroke, the forces pushing down on the ring cause it to sit squarely, providing a better seal and therefore better power.

The proper ring groove clearance can be critical. If the clearance is too large, the ring will tip excessively as the piston moves up and down, reducing its ability to seal.

The excess movement of the ring on the piston may also cause the ring to break. If the clearance is too small, the ring may bind in its groove when the piston heats up and expands.

To measure the ring side clearance, a piston ring can be installed in its groove. A feeler gauge is then inserted between the ring and the bottom of the groove (Figure 10-13). After you've measured the piston ring side clearance, compare your measurement with the manufacturer's specification. Be sure to always refer to the appropriate motorcycle service manual to get the exact specification. In addition, each ring groove may be worn differently; therefore, you should check the side clearance in all of the piston's grooves.

Piston Ring Inspection and Measurement

The next step in the inspection process is to carefully check the piston rings for signs of unusual wear. As mentioned earlier, you should not re-install old piston rings into a cylinder; they should be replaced whenever an engine is taken apart. Normally, rings that are reused won't seat properly, resulting in poor engine performance. However, the condition of the old piston rings can provide clues to certain engine problems. For instance, small scratches found on the edge of the rings usually mean that dirt or other debris has been getting into the engine. This may likely indicate a faulty air-filtering system. Remember from Chapter 6 that two-stroke pistons have locating pins to prevent the piston ring from moving around on the piston.

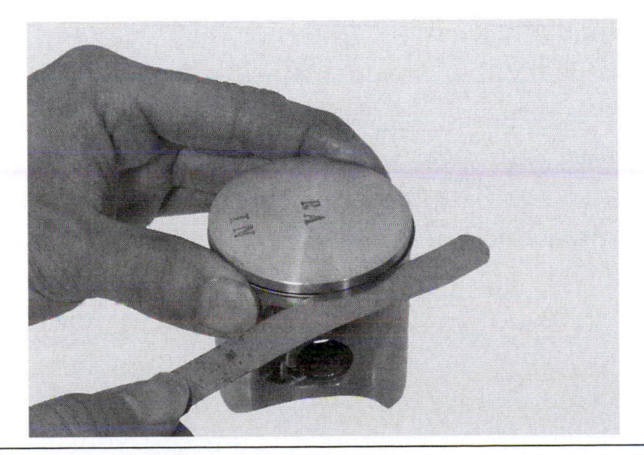

Figure 10-13 A technician is shown measuring the piston ring groove clearance with a feeler gauge. Copyright by American Honda Motor Co., Inc. and reprinted with permission.

You may recall that when new piston rings are installed in an engine, they must wear themselves into position against the cylinder walls to form a tight seal. Once this process of seating has occurred, the rings lose the ability to do so again. That is, if old rings are reinstalled in an engine, they won't be able to conform once again to the cylinder walls and make a tight seal. Without a tight seal, the combustion gases can leak past the rings. This reduces the amount of horsepower the engine can produce.

Before reassembly, you will be required to deglaze the cylinder. The importance of cylinder wall deglazing cannot be overemphasized. The proper cylinder finish will provide the quickest possible break-in and greatly reduce the possibilities of ring or piston scuffing during break-in. A glazed cylinder wall causes rings to "skate" on the highly polished finish and discourages the minute amount of wear, which is necessary to mate the piston rings with the bore of the cylinder.

A deglazed finish contains minute hills and valleys that carry a film of oil, which will prevent scuffing during break-in as well as produce the type of cylinder finish piston rings can mate to very rapidly. The finish produces a cross-hatched pattern that intersects at approximately a 45-degree angle. Probably the most critical part of the deglazing operation is the proper cleaning after deglazing. The residue of deglazing, tiny pieces of iron from the cylinder wall, if left in the engine, will rapidly destroy all moving parts. It is recommended that cylinders be cleaned thoroughly with warm soapy water. Clean until the bore can be wiped with a clean white cloth without soiling the cloth. Many professionals finish the cleaning process by wiping the bore with automatic transmission fluid due to its ability to pick up tiny pieces of metal. After clean up, lightly oil the area with the same oil that will be used to run the engine to prevent rust formation.

A worn piston ring is usually bright and shiny at the point where its edge contacts the cylinder wall. Worn rings also can be detected by performing a compression check on the engine before it's disassembled. A **compression check** is a simple test that measures the amount of pressure produced in the combustion chamber on the compression stroke. The compression is measured with a compression gauge that's inserted in the spark plug hole. If the piston rings are worn, the compression gauge displays a pressure reading that is lower than the manufacturer's specification. The reading is low because instead of being compressed, some of the air-and-fuel mixture is leaking down past the worn rings and into the crankcase. The compression gauge may also show an excessively high reading if there is excessive build-up in the combustion chamber area.

Measure the new piston ring(s) to ensure that the ring end gap is correct by placing the ring squarely into the cylinder (use the piston to do this effectively) and measuring the piston ring end gap with a feeler gauge (Figure 10-14). The **piston ring end gap** is the space between the ring at its opening when it is compressed in the cylinder.

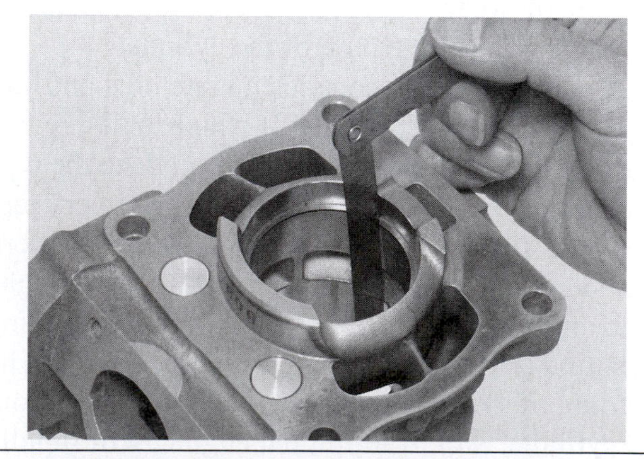

Figure 10-14 Piston ring end gap is the space between the piston ring at its opening when it is compressed in the cylinder. Measure this gap with a feeler gauge. Copyright by American Honda Motor Co., Inc. and reprinted with permission.

Measuring Piston-to-Cylinder Clearance

As explained earlier, a piston expands as its temperature rises. Since the metal of the piston typically expands more than the metal of the cylinder wall, some clearance must be allowed between these components when both are cold. This clearance is called the **piston-to-cylinder clearance** (or just piston clearance for short). The piston-to-cylinder clearance is a critical dimension. If the clearance is too small, the piston fits too tightly in the cylinder whenever the engine heats up, resulting in excessive friction that may cause the piston to seize in the bore. That is, the piston may actually melt onto the cylinder, therefore being unable to move up and down. Try rubbing your hands together lightly. There is little heat produced because there is very little friction. Now try quickly rubbing them together with much more force. You will almost immediately feel the heat being produced from the friction being created. If this occurs, the engine stops running, and the starter won't be able to rotate the engine at all. You may be able to free a seized piston after the engine cools down again; however, both the piston and the cylinder wall will probably be badly scored and damaged.

If the piston clearance is too large, the piston won't be held in place very well and tends to rock back and forth while the engine is running. This rocking motion creates a knocking noise and may eventually break the piston skirt. In addition, the piston ring's ability to seal the combustion chamber will be greatly reduced. You should always check the manufacturer's specification for the correct piston-to-cylinder measurement listed in the service manual because all engines are designed slightly different.

To determine the piston clearance in an engine, you'll need to measure the diameter of both the piston and the cylinder bore. Illustrations of these measurements can be seen back in Figures 10-5 and 10-11. Compare your measurements to the manufacturer's specifications. Then, subtract the outside diameter of the piston from the inside diameter of the cylinder bore. The result of your calculation is the actual piston clearance. Compare your calculated clearance to the manufacturer's specification. If the clearance is outside specifications, the piston and cylinder will have to be replaced or resized to make the clearance conform to specifications.

Measuring the Wrist Pin

The **wrist pin**, a cylinder-shaped piston assembly component, is used to link the connecting rod to the piston (Figure 10-15). The connecting rod's bearing for the wrist pin allows the end of the rod to rotate freely around the wrist pin as the piston travels up and down. The wrist pin must transfer each power stroke's downward physical force from the piston to the connecting rod. To ensure the wrist pin is strong enough to handle this task, the engine manufacturer usually makes the pin from high-quality steel, which is a very hard metal. Consequently, you won't see any wear on the pin itself. However, to guarantee that a wrist pin is not worn, measure the pin's diameter with an outside micrometer. Compare your measurement with the specification given in the service manual.

Inspecting the Connecting Rod

The connecting rod in a modern two-stroke engine is normally a one-piece unit and is used in conjunction with a multi-piece crankshaft. It takes special tools to disassemble and recondition this component and these will be discussed further in Chapter 11. The two-stroke one-piece rod will usually use needle bearings at the small end and a roller bearing at the large end. To inspect the needle bearing, place it along with the wrist pin into the connecting rod and move back and forth as shown in Figure 10-16. If it appears that there is too much clearance, remove the bearing and look for scoring or scuffing marks in the small end of the bearing. This would be an indicator of a lack of lubrication. Also, measure the small end of the connecting rod as shown in Figure 10-17 with an inside micrometer to ensure that it is within the correct manufacturer specification.

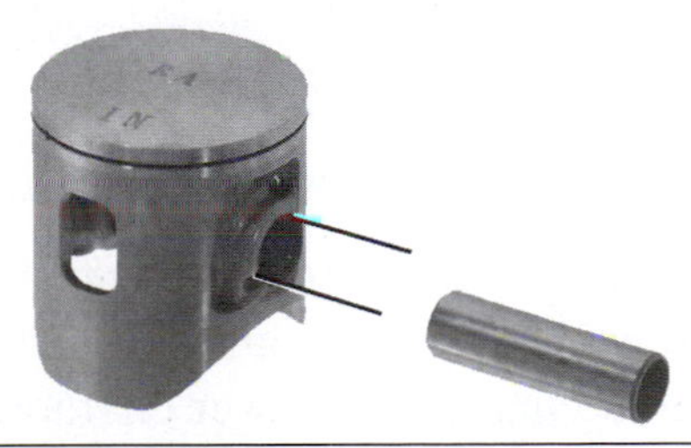

Figure 10-15 The wrist pin is a round, cylinder-shaped component that attaches the piston to the connecting rod. Copyright by American Honda Motor Co., Inc. and reprinted with permission.

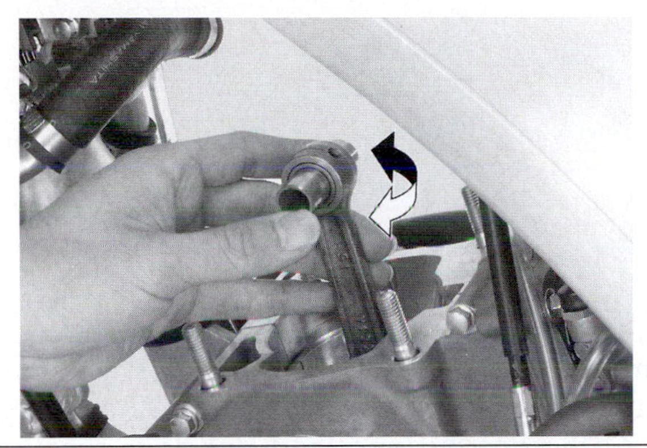

Figure 10-16 To inspect the needle bearing, place it along with the wrist pin into the connecting rod and move back and forth as shown here. Copyright by American Honda Motor Co., Inc. and reprinted with permission.

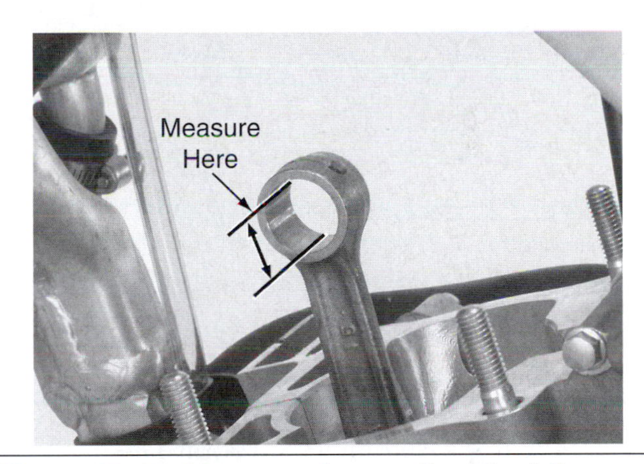

Figure 10-17 Measure the small end of the connecting rod with an inside micrometer and check your measurement with the manufacturer's specifications. Copyright by American Honda Motor Co., Inc. and reprinted with permission.

Inspecting the Reed Valves

Virtually every modern two-stroke motorcycle engine utilizes reed valves on the intake side of the induction system (Figure 10-18). Checking the reed valves is a simple process of inspecting the reed's valves, reed stopper, and the reed valve seat (Figure 10-19) for physical damage.

When looking at the reed valve assembly, there should be little to no air gap between the reed valve and the reed valve seat as illustrated in Figure 10-20.

TWO-STROKE POWER VALVES

Power valves in the two-stroke engines used on motorcycles have been produced by all motorcycle

Figure 10-18 Virtually all modern two-stroke engines utilize a reed valve-type induction system. Copyright by American Honda Motor Co., Inc. and reprinted with permission.

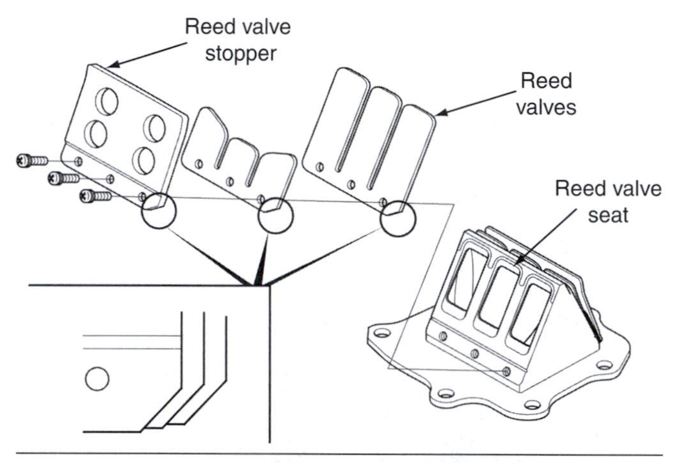

Figure 10-19 This illustration shows the parts of the reed valve assembly that require inspection. Copyright by American Honda Motor Co., Inc. and reprinted with permission.

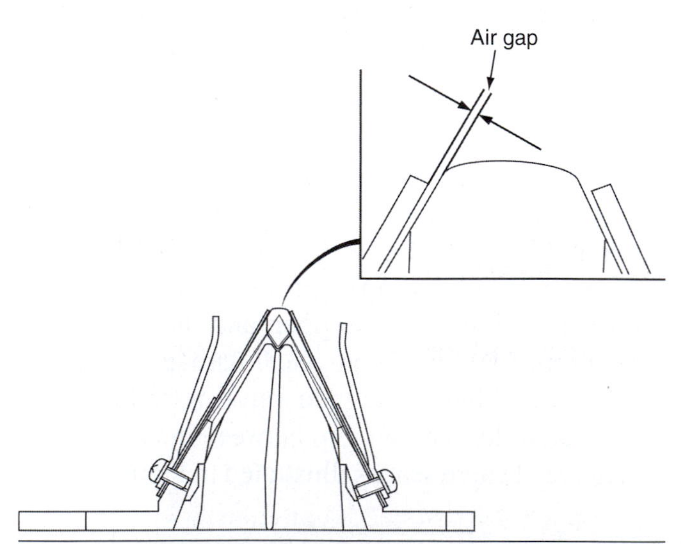

Figure 10-20 There should be little to no air gap between the reed valve and the reed valve seat. Copyright by American Honda Motor Co., Inc. and reprinted with permission.

manufacturers and vary widely in design. Two-stroke motorcycle manufacturers have produced variations of the power valve for their two-stroke engine's exhaust ports to help create a wider and more useful power band for greater engine performance. These power valve systems are designed to help control and improve the power of the two-stroke engine. The power valve allows for variations to the cylinder exhaust ports by changing the exhaust timing for optimum performance under a wide range of running conditions.

As we've discussed in Chapter 6, the two-stroke engine, by nature, has a very small useful power range (power band). In other words, it has a very small rpm range of useful power.

Reed valve engines became very popular in the mid-1970s to help broaden the two-stroke engine's power band and are still commonplace today. This was a great improvement over the standard piston-port engine design that was previously used. But the power band on the two-stroke engine was still quite narrow, especially when compared to the typical four-stroke engine design.

Although many different designs exist, power valves all have one common goal—to increase the usable power output of a two-stroke engine. Power valves accomplish this feat in one of two different ways:

- Increase exhaust system volume

By changing the volume of the two-stroke exhaust system (Figure 10-21), engineers can effectively change the point at which the engine makes the most power. For example, an exhaust system that has a large volume will have more usable low rpm power than an exhaust system that has a small volume. Therefore, if we can change the volume of the two-stroke exhaust system while it is running, we could have usable power at both low and higher engine rpms.

- Change the engine's exhaust port height

Long before the concept of power ports came about, performance tuners and engineers found that they could change the power characteristics of the engine by changing the height of the exhaust ports. It's known that the top of an exhaust port that is low in the cylinder as compared to the rest of the ports produces good power at low engine speeds.

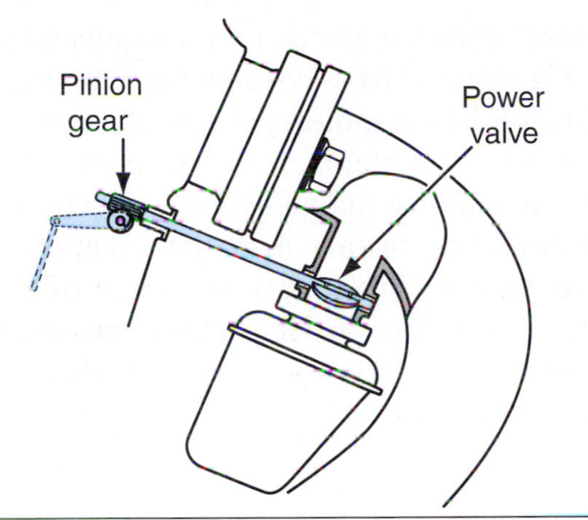

Figure 10-21 The power valve shown in this illustration increases the exhaust system volume to widen the engine's usable power band.

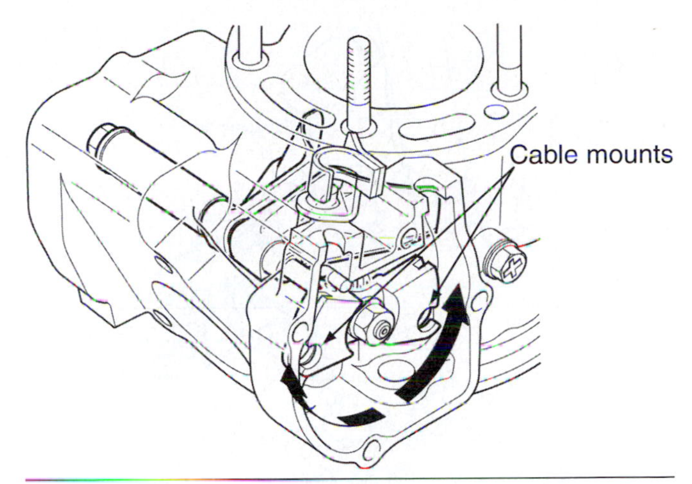

Figure 10-23 In this illustration, cables (not attached) are used to open and close the power valve. Copyright by American Honda Motor Co., Inc. and reprinted with permission.

On the other hand, an exhaust port that is higher in the cylinder produces good power output at high engine speed while suffering at low engine speed.

Now, if the height of the exhaust port on a two-stroke engine could be changed (Figure 10-22) while it was running, engineers thought that the rider could have the best of both worlds. After many years of research and development, power valves were introduced to the two-stroke engine with the first examples showing on GP road-racing engines in the late 1970s. Methods of opening and closing the valves are varied from centrifugally operated gear-driven systems to electronic servo-motor types that open and close the valve (Figure 10-23).

The two types of power valves mentioned have also been used together. Today, most manufactures use the method of changing the exhaust port height when using a power valve as it is the most compact of available technologies.

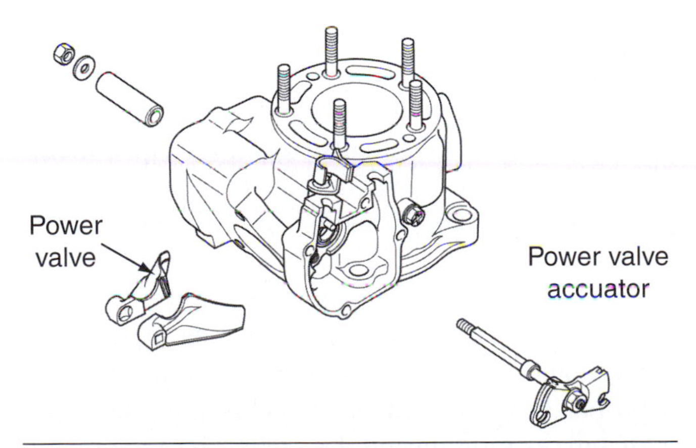

Figure 10-22 This style of power port changes the exhaust port height to widen the engine's usable power band. Copyright by American Honda Motor Co., Inc. and reprinted with permission.

How Power Valves Work

Power valves work by opening and closing valves in or near the two-stroke exhaust port at different engine rpms. Some engines use electronics to operate the power valve but most power valves work by using gears (Figure 10-24). The shafts move in and out or back and forth to either change the exhaust port height or change the volume of the exhaust system by allowing exhaust gases to flow into a separate chamber at certain engine speeds. By doing one or the other, or even both, the engine can have a much broader power band. Now, engine manufacturers can design very high horsepower two-stroke engines that are still very easy to ride at lower engine speeds, a task that was not possible before the use of the power valve.

Power Valve Maintenance

Since there are so many variations of power valves used in a two-stroke motorcycle and they have all evolved quickly over the past few years, we will not cover the cleaning and maintenance of them as they vary greatly in both design and complexity. Of course, the servicing of each manufacturer's power valve system will be covered in the individual service manual.

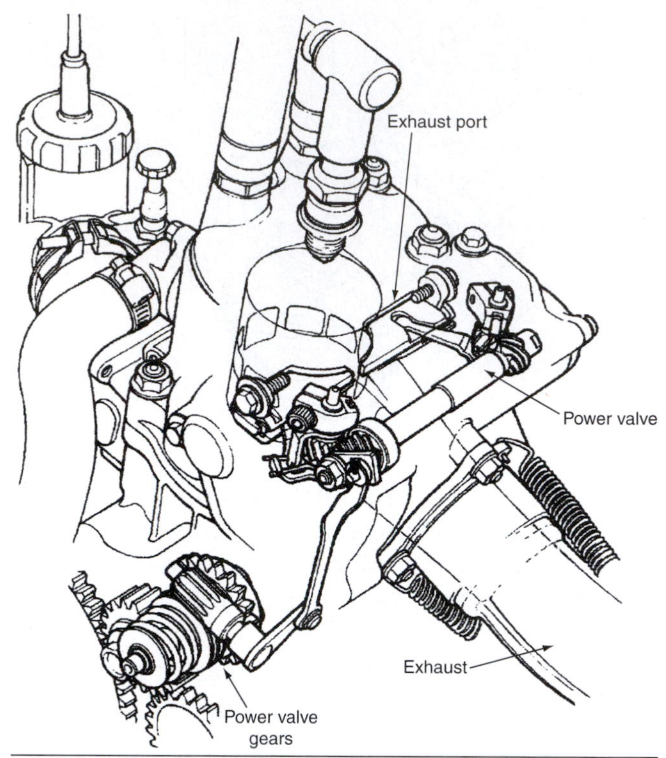

Figure 10-24 Centrifugal weights and gears are used to open and close this power valve system, which slides the exhaust valve higher or lower depending on engine rpm.

All power valve systems have a common maintenance procedure to follow—found in the appropriate manufacturer's service manual—which is to remove carbon deposits within each system. Remove all carbon deposits using a wire brush and a high-flash-point cleaning solvent. You should inspect all of the individual components of the power valve system for wear or any signs of damage. If damage is present, replace the power valve component.

Two-Stroke Motorcycle Power Valve Summary

As you should now understand, variable two-stroke exhaust power valve systems are all designed to do effectively the same basic thing: increase the two-stroke engine's power band. They just happen to have different designs to do this. Manufacturers have their own variations of power valves with many different designs as well as different looks. However, power valves are simply ways to vary the size of the two-stroke cylinder exhaust ports and exhaust systems. Variable exhaust port systems were first used on production-based motorcycles in the early 1980s and all of the systems have had many improvements to simplify maintenance and increase performance even further over the years as design and technology have moved forward. Today, it would be difficult to find a modern two-stroke motorcycle without some sort of power valve device being utilized.

STARTING THE REBUILT ENGINE

When you're certain that all components are reassembled and in place, all fasteners have been properly tightened, and the fluids have been properly added, it's time to start the engine. The engine should start within 5–10 kicks of the kick-start lever. If it doesn't start at this time, stop and verify that all electrical connectors are attached and then try again. Once started, let the engine idle or keep it as close to idle speed as possible. As the engine is warming up, check for any leaking fluids in and around the engine. Shut the engine off. Then wait until the engine has cooled to room temperature, and top off the coolant on a liquid-cooled engine.

Engine Break-In

Most manufacturers recommend that a new (or reconditioned) engine be properly broken in to make sure that all components are sealing well and that all components mesh together properly. During your assembly, use only the best possible materials and use original equipment-manufactured parts to assure the highest standards. Even though you're using the highest-quality components, it's still necessary to allow the parts to "break in" before subjecting the engine to maximum stress. The future reliability as well as the performance of the two-stroke engine depends on a proper break-in procedure. This includes extra care and restraint during the early life of the reconditioned engine. Follow the appropriate manufacturer's recommendations for the correct break-in procedure.

Summary

- Diagnosis is the process of determining what's wrong when something isn't working properly by checking the symptoms. Symptoms are the outward, or visible, signs of a malfunction.
- It is important to understand the necessity of proper engine component inspection, which includes visual inspections as well as measuring components for wear.
- There are many different types of motorcycle two-stroke engine power valves but all use three basic theories of operation:
- Change the exhaust port height
- Change the exhaust system volume
- A combination of the two above

Chapter 10 Review Questions

1. To measure the diameter of a piston, you should use a ________.
2. A piston ring groove clearance can be measured by using a ________.
3. An engine produces its power by burning the air-and-fuel mixture in the combustion chamber directly above the ________.
4. A worn piston ring will usually be bright and shiny at the point where its edge contacts the ________.
5. If the piston clearance is too small, the friction between the piston and the cylinder may cause the piston to ________ in the bore.
6. A ________ between the cylinder head and the cylinder helps create an airtight seal.
7. ________ are used to broaden a two-stroke engine's power band.
8. Wear within the cylinder is caused by the up-and-down motion of the ________ and ________.
9. An indication of worn piston rings can be seen before the engine is disassembled by conducting a ________ test.
10. The power valve systems used on motorcycles help control and improve the engine's power output at different engine ________.

CHAPTER

Two-Stroke Engine Lower-End Inspection

Learning Objectives

When students have completed the study of this chapter and its laboratory activities they should be able to:

- Recognize common lower-end engine failures in a two-stroke motorcycle
- Understand the importance of inspecting the various parts of the two-stroke lower end including the crankshaft and transmission for damage or wear
- Understand proper procedures for disassembling and inspecting the transmission in a typical two-stroke motorcycle engine
- Understand the importance of bench testing an engine prior to completing reassembly

INTRODUCTION

The process of disassembling the lower-end engine assembly is a process in which all of the engine parts located below the cylinder are removed, inspected, and replaced when necessary. As technicians it is important that we understand how to disassemble the engine right down to its heart: the crankshaft. It is one space important that you know how to inspect the engine components, do any necessary repair work, and reassemble the engine correctly.

Disassembly of the lower-end assembly will always require that the engine be removed from the chassis. Therefore, you should first be sure the malfunctioning component is located in the lower engine assembly. After all, you would not want to remove an engine from its chassis and disassemble it completely just to find out that the failed component did not require any major disassembly.

Common Lower-End Engine Failures

The following is a brief discussion of some areas in the lower-end assembly where malfunctions commonly occur.

Leaking Seals

Seals are used on transmission shafts and other rotating shafts within a motorcycle engine to prevent oil loss, keep contaminants from entering the engine and bearings, and at times to seal out atmospheric air when necessary.

While in many cases, replacement can be done while still in the chassis, a leaking oil seal sometimes requires that the engine's lower end be disassembled. Always confirm that the replacement of the faulty oil seal requires complete disassembly before disassembling the lower end. This is achieved by checking the appropriate service manual before you begin the work.

Worn Crankshaft Bearings

Crankshaft bearings are used to mount the crankshaft assembly into the crankcase. Bearing failure is indicated by a rough, growling sound. You may wish to use a mechanic's stethoscope (Figure 11-1) to help pinpoint the location of the bad bearing. A mechanic's stethoscope is very similar to a stethoscope that a doctor uses to listen to your heart and lungs during a physical examination. It picks up very faint sounds and can pinpoint the location of noises in an engine. Worn bearings may also allow excessive up-and-down movement of the crankshaft, or even prevent the engine from rotating.

Worn Connecting Rod Bearings

Connecting rod bearings are used to allow the connecting rod to rotate as the crankshaft assembly turns. The following are some symptoms of bad connecting rod bearings.

- The engine knocks
- The engine starts, but will not run freely
- The engine cannot be rotated (it's locked up)

Any of these symptoms will necessitate disassembly and repair of the lower end of the engine.

Transmission Problems

Another reason to disassemble the lower end of an engine may be because of transmission problems. Both the engine and transmission of two-stroke motorcycle engines are contained in a single casting (engine case). The procedure for separating the case is similar for most engines. Most two-stroke engines utilize a vertically split crankcase design that requires the removal of the top-end components.

This chapter includes the inspection of the most commonly found transmissions in the modern two-stroke engine—the constant-mesh transmission. Any suspected transmission malfunction should be investigated while you have the engine case opened.

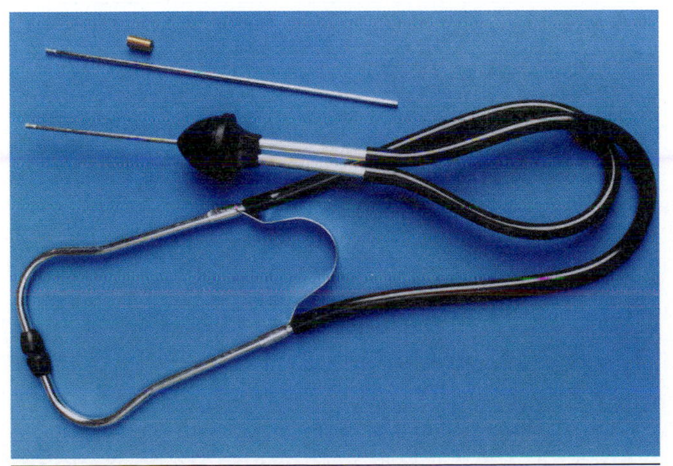

Figure 11-1 A mechanic's stethoscope is often used to listen for internal engine noises.

Repair Procedures

Being alert to other problems by inspecting for problems when you're performing repairs on an engine will help you to become a fully competent motorcycle technician. In this chapter, we'll list the necessary procedures to inspect and replace worn parts in single-cylinder, two-stroke liquid-cooled engines. These procedures also apply to other motorcycle two-stroke engines that you will work on as a technician.

As we've mentioned before, be sure that the motorcycle is clean before you begin any disassembly work. Use a water-soluble degreaser and use it according to the manufacturer's instructions. Remember that dirt or foreign particles can ruin your repairs if allowed to enter the working parts of the engine.

The disassembly of the lower end of a two-stroke engine requires that the top end be removed first. Therefore, we will assume that the top end of the engine has been removed, using the procedures found in the appropriate service manual.

The procedures in this chapter are general in nature. Their purpose is to familiarize you with the types of activities you'll encounter when inspecting the components of a typical two-stroke engine bottom end. Always refer to the appropriate motorcycle service manual for detailed information. The manufacturer's service manual contains all the information to do the job correctly, including detailed instructions for the specific make and model of motorcycle, special tools, and service tips. Above all, the service manual contains the appropriate safety information to ensure that your job be completed safely.

TWO-STROKE ENGINE REMOVAL AND DISASSEMBLY

As we discussed in Chapter 10, in most cases, the two-stroke engine top-end assembly can be removed and inspected without removing it from the chassis. However, to disassemble the two-stroke engine lower-end components, such as the crankshaft or transmission, you must first remove the engine from the chassis. While the procedure to remove a two-stroke motorcycle engine from its chassis is quite simple when a systematic approach is used, specific details are provided by the manufacturer. Therefore, the appropriate service manual should always be used as a reference.

As there are many different motorcycle manufacturers building two-stroke engines, we will not show you how to disassemble the bottom end of a two-stroke engine. We strongly recommend that you utilize the appropriate manufacturer's service manual and follow the procedures when disassembling an engine. Incidentally, for this chapter, just as in Chapter 10, we're using a liquid-cooled Honda CR125R motorcycle (Figure 11-2) for illustrative purposes. The procedures for the inspection of the bottom end of a two-stroke engine are very similar from one brand to another.

TWO-STROKE ENGINE LOWER-END INSPECTION

Once the lower end of the engine has been disassembled, it's time to inspect each component for damage and wear. Although you may be trying to locate a particular problem, you should carefully inspect all of the engine components while the engine is disassembled. Because a complete engine disassembly is a lengthy procedure that isn't done frequently, it's important to make sure that the job is done right and that there are no existing problems or soon-to-be problems as well. We'll learn how to inspect the lower end of a two-stroke engine by beginning with the engine crankcases.

Figure 11-2 A Honda CR125R engine will be used for illustrative purposes in this chapter. Copyright by American Honda Motor Co., Inc. and reprinted with permission.

Inspecting the Engine Crankcase

The engine crankcase halves (Figure 11-3) should be closely inspected for cracks, loose fitting bearings, and worn-out fastener anchoring points. If there are stripped bolt holes, they may be repaired by using a special thread repair kit that reconditions the hole (Figure 11-4). The Heli-Coil® and the Time-Sert® are two popular methods of thread repair and should be available from a motorcycle or automotive tool supplier.

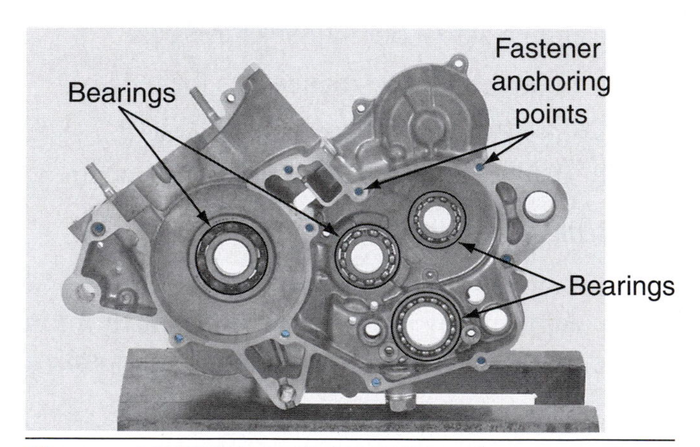

Figure 11-3 When inspecting the cases look closely for cracks, loose-fitting bearings and worn out fastener anchoring points. Copyright by American Honda Motor Co., Inc. and reprinted with permission.

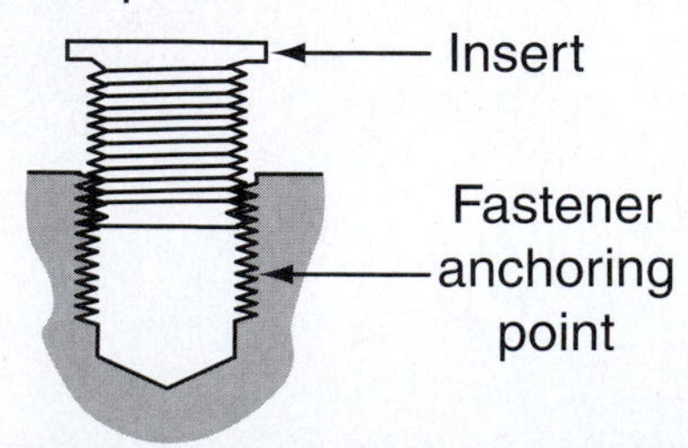

Figure 11-4 An insert is shown here that would assist in the repair of a stripped fastener anchoring point.

Inspecting and Replacing Engine Seals

Engine seals are designed to keep oil in an engine compartment or keep air out of the crankcases of a two-stroke bottom end. Seals are located on all two-stroke engine shafts that rotate and are exposed to the outside atmosphere and internal engine oil that must be separated from the engine's crankshaft (Figure 11-5).

While today's engine seals are very durable, it is important to inspect them to verify that they are in good condition. Inspect the oil-seal lip for wear and damage. Depending on its location, damage to the lip of the oil seal may result in leakage of air into the engine, transmission oil into the crankshaft compartment, or transmission oil from inside the engine to the ground. Inspect the seals very carefully. If a seal isn't in perfect condition, replace it. The rubber of the seal must be "live." That is, the lip of the seal must be soft and springy. Virtually all engine seals have a small coil spring that fits on the seal lip to apply pressure to the shaft (Figure 11-6). Be sure this spring is attached properly. If it is not in place, even a good seal will leak.

You should always install a new seal if you have any doubt about the condition of the old one. You should also install new seals whenever you replace any bearings that have a seal next to them.

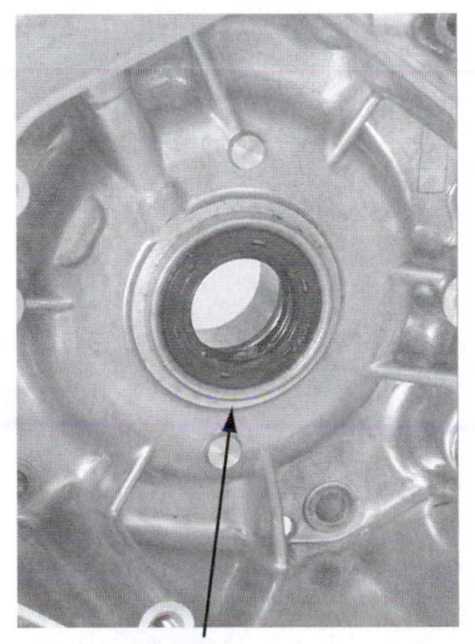

Figure 11-5 Engine seals are located on rotating shafts and are designed to keep oil in, keep oil out, or on a two-stroke engine in certain cases, keep air out of the crankcases. Copyright by American Honda Motor Co., Inc. and reprinted with permission.

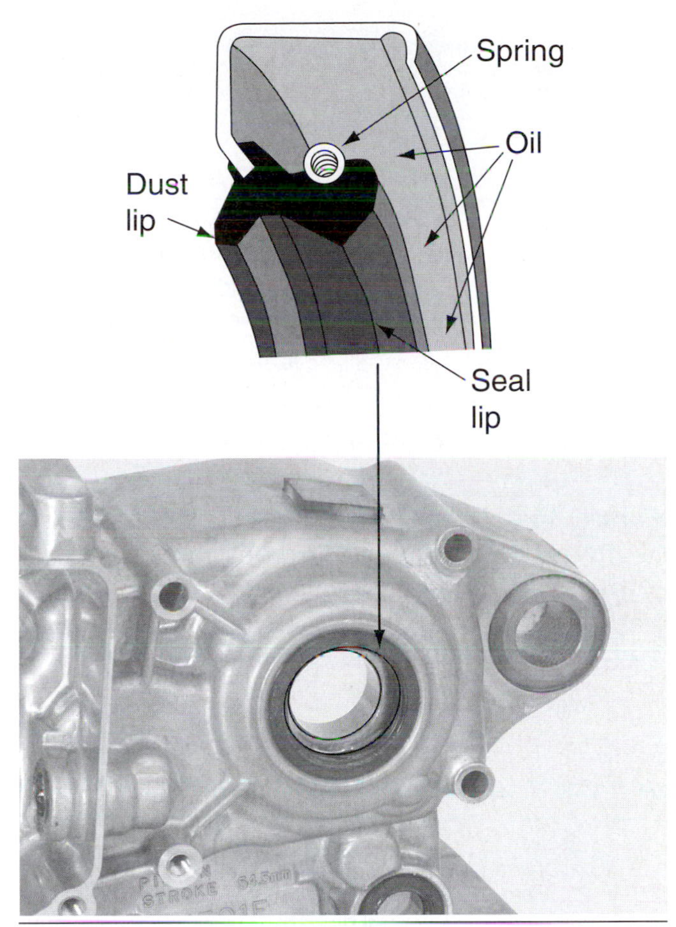

Figure 11-6 The parts of a seal are illustrated here. Copyright by American Honda Motor Co., Inc. and reprinted with permission.

Before you remove any seal, be sure to note how it is installed, as seals will only work correctly if installed in the proper direction. Once removed, you should never reuse a seal, as it will be damaged while being removed. If you remove a seal, replace it with a new one. Never try to reinstall an old seal. Since you should never reuse a removed seal, the removal process becomes pretty simple. They can be pried out of the cases by using a screwdriver or by using a special seal removal tool (Figure 11-7).

A new seal must be installed evenly in the hole. Use a seal installation tool to assist in placing a new seal in the cases correctly (Figure 11-8). Be sure to install the seal into the case properly; the seal manufacturer's identification number is usually on the side away from the bearing or shaft to be sealed. Check the appropriate service manual to be certain that you do not install the seal incorrectly.

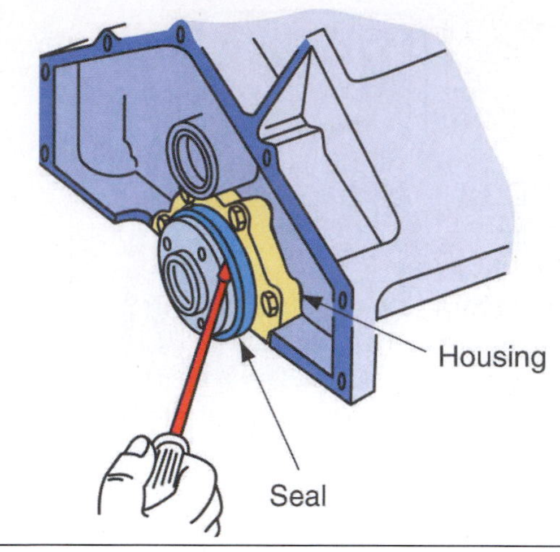

Figure 11-7 You can purchase special tools to remove seals or in many cases just pry it out with a screwdriver. Never reuse a seal once it has been removed.

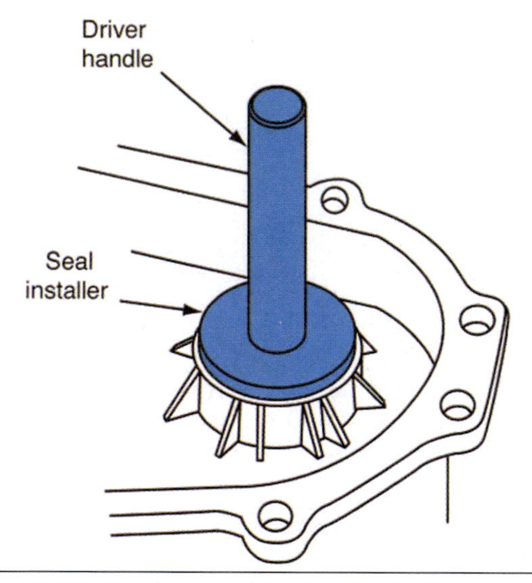

Figure 11-8 Seals need to be installed correctly and evenly by using a special seal installation driver that rests on the outside of the seal.

Some engine seals are located on the inside of the crankcase cavity, and can be replaced only when the crankcase halves are separated. In these cases, it is always a good idea to replace this type of seal whenever you have the crankcases separated.

Inspecting and Replacing Engine Bearings

The most common bearing that you'll find in a motorcycle two-stroke engine is the ball bearing. Remember that bearings usually make a low growling sound when they are failing. You can inspect a bearing by hand while it's still mounted in the crankcase or on its shaft (Figure 11-9) by rotating the bearing inner race (or outer race if the bearing is outside of the crankcase) and feeling for smooth operation. If a bearing has a rough feel or makes a noise when you rotate it, the bearing should be replaced.

Also, visually inspect the inner and outer races, the balls and ball cage of the bearing (Figure 11-10). If there is any sign of wear, chips, cracks, or damage, the bearing must be replaced.

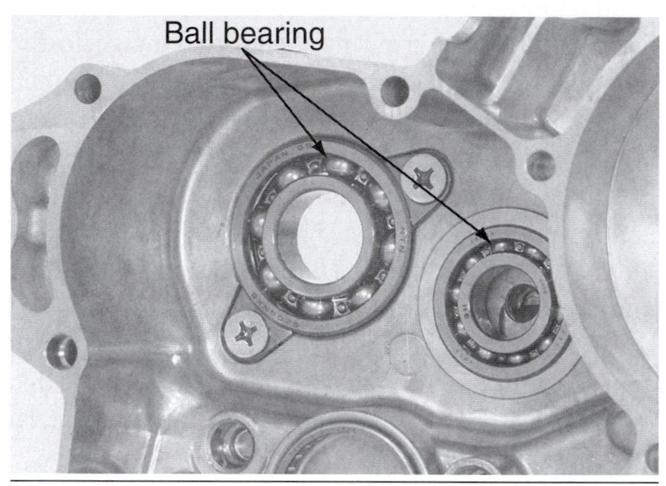

Figure 11-9 Ball bearings are the most common bearing found in a two-stroke engine. Copyright by American Honda Motor Co., Inc. and reprinted with permission.

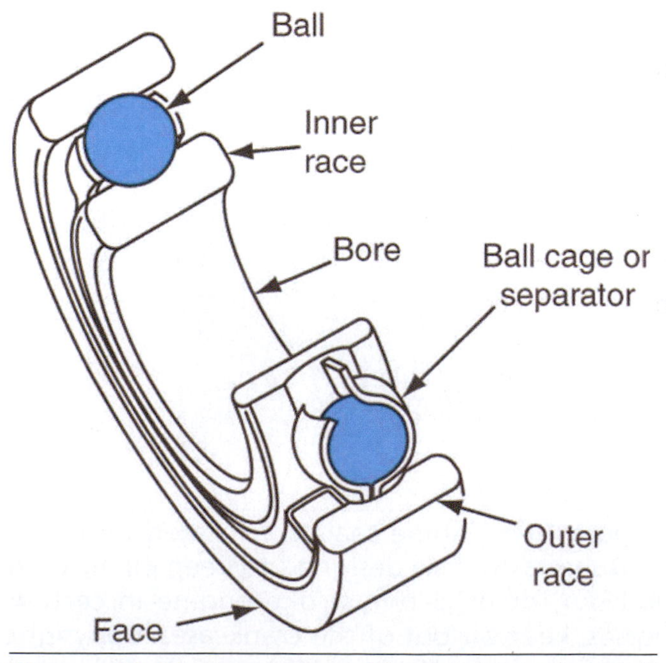

Figure 11-10 A cutaway illustration of a ball bearing is shown here.

To replace any two-stroke engine bearing, you must first remove the old bearing by driving it out of the crankcase or by using a special bearing puller (Figure 11-11) depending on its location.

Ball bearings are held in the case by a tight fit—which is called an interference fit. Before installing new bearings, you may want to put them in a freezer. When bearings are placed in a very cold environment they will shrink slightly, which will ease their installation. You can also heat the crankcases on a hot plate to expand the case metal before installing the bearings but doing so may allow any other bearing in the cases to fall out.

As just mentioned, placing cool bearings into a warm case makes for easy installation while ensuring a tight fit when both the bearing and case have returned to a normal temperature. This is because the cooled bearings expand as they warm up, and the warmed case shrinks as it cools down. A special tool, like the one shown in Figure 11-12, should be used to install the bearing evenly. Be sure to strike the bearing only on its outer cage or you will damage the bearing inner cage.

Inspecting the Clutch

The clutch is one of the most common two-stroke engine components to wear out or fail. Although lower-end engine disassembly isn't required to remove the clutch, the clutch must be removed to disassemble the lower end. Refer back to Chapter 9 for the procedure to inspect the clutch. As a review, thorough clutch inspection should include the following checks and measurements:

- Length of springs
- Thickness of friction and steel plates
- Warpage of clutch plates
- Smoothness of clutch basket and clutch center grooves
- Wear of clutch basket bearing or bushing
- Wear or cracks in clutch hub and primary driven-gear assembly

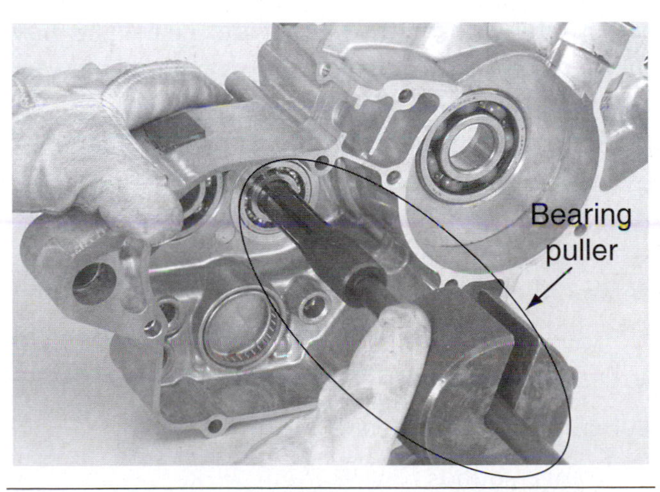

Figure 11-11 A bearing puller is shown here helping to remove a bearing from the crankcase half. Copyright by American Honda Motor Co., Inc. and reprinted with permission.

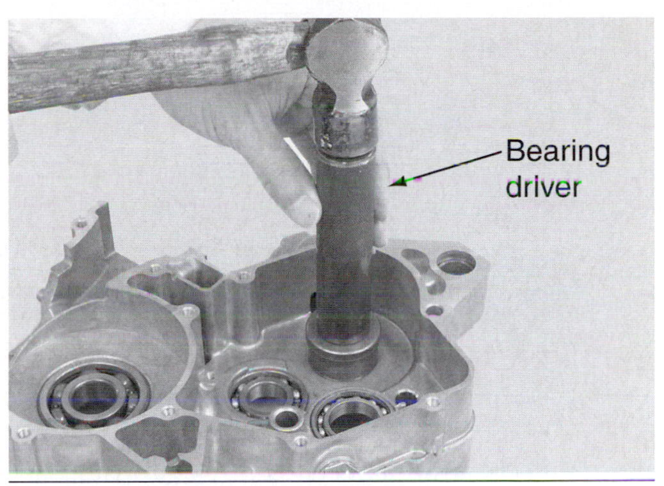

Figure 11-12 A bearing driver is shown here. Copyright by American Honda Motor Co., Inc. and reprinted with permission.

Inspecting the Transmission

Once the crankcases are split it is time to remove the transmission (Figure 11-13). Remove the transmission as a complete set as shown in Figure 11-14. Once removed, you can disassemble the transmission, inspect the transmission gears and other components, and then reassemble the transmission and return it to the crankcase.

Disassembling the Transmission

Now that the transmission has been removed, it is time to disassemble the gears from their respective shafts (Figure 11-15).

To remove the gears you will have to remove the retaining rings that hold some gears into place. To remove a retaining ring, expand it and pull it off using

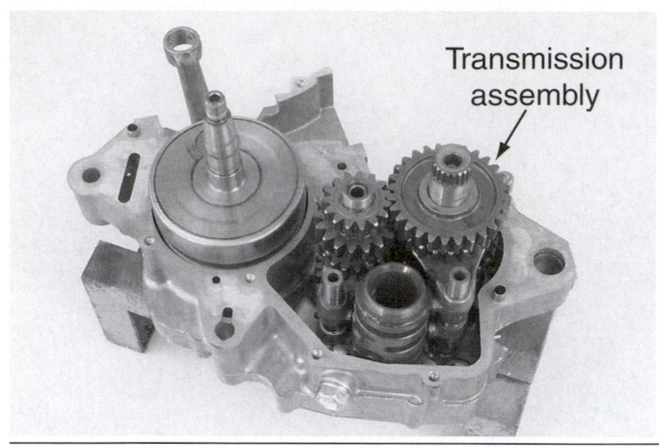

Figure 11-13 The gears of the transmission are shown in place within the crankcase. Copyright by American Honda Motor Co., Inc. and reprinted with permission.

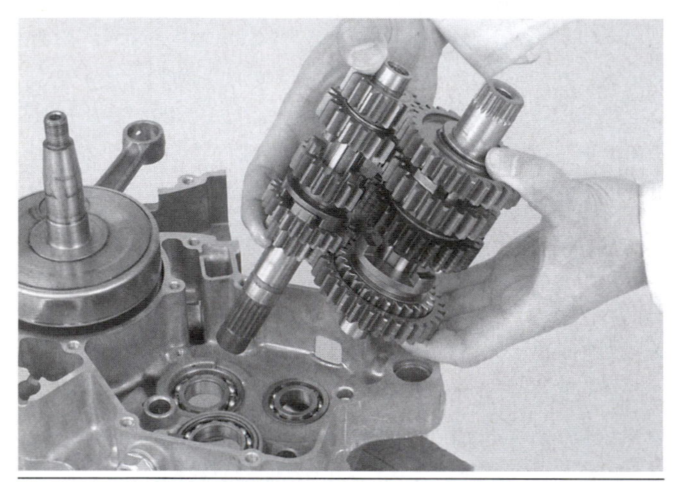

Figure 11-14 Remove the transmission as a complete set. Copyright by American Honda Motor Co., Inc. and reprinted with permission.

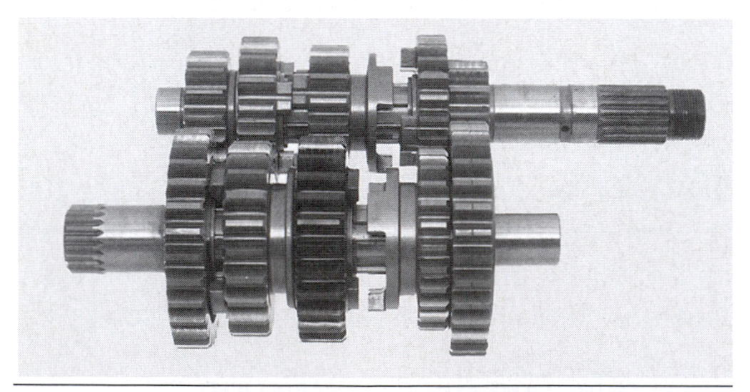

Figure 11-15 The transmission gears are shown on their shafts here. Copyright by American Honda Motor Co., Inc. and reprinted with permission.

Figure 11-16 Placing the transmission gears and washers on a long screwdriver will help keep them in the proper order. Copyright by American Honda Motor Co., Inc. and reprinted with permission.

the gear behind it. You should always replace any retaining rings that are removed because they can be weakened if spread too far out during removal.

When you remove the gears and other parts from the transmission shaft, it is a good idea to slip them onto a wire or long screwdriver to keep from losing them and to keep them in the proper order for replacement (Figure 11-16).

Transmission Power Flow and Symptoms

Now is a good time to review Chapter 9 on how transmissions operate and also to consider some of the problems that can occur in transmissions. Remember that modern two-stroke motorcycle engines use constant-mesh transmissions, which means that all gears rotate at all times. Power is developed at the engine crankshaft and is transmitted through the clutch to the transmission main shaft.

The main shaft transmits power through different sets of gears to the countershaft. Be sure to review the information about the gears and how they are moved to change the gear ratio between the main shaft and countershaft.

Some transmission problems can be diagnosed by the symptoms reported by the operator. Again, refer to Chapter 9 for a review of transmission symptoms and probable causes. The more common symptoms of transmission problems that you should be familiar with include the following:

- Difficulty shifting
- Inability to shift gears
- Strange sounds
- Jumping out of gear

Inspecting the Transmission Components

When inspecting the transmission, always carefully inspect each and every component, not just the pieces with obvious damage. For instance, if you find a chipped tooth on a gear, also check the shift fork and shift drum for damage as well.

Check the gears for damage or wear. Also inspect the gear dogs (Figure 11-17) and slots (Figure 11-18) for wear or damage. Inspect the inside of the gear for scoring and measure it for wear (Figure 11-19) using an inside micrometer (unless it's a splined gear). Check your measurements with the appropriate specification given in the service manual to verify that it is not out of specification. Inspect the gear bushings for wear or damage. Measure the inside and outside dimensions of the bushings to verify that they are within specifications.

Carefully inspect the main shaft and countershaft at their splined grooves (Figure 11-20). Check the shaft sliding surfaces for abnormal wear or damage.

The shift drum is the most critical component of the transmission that should be inspected but is often overlooked. The shift drum directs the shift forks back and forth in a timed succession to change from one gear to the next. The shift drum is a very precise component and the slightest damage could cause shifting problems. Check the shifting fork guide grooves very carefully for damage, such as a small chip or scoring (Figure 11-21). Also, inspect the shift drum bearing to ensure that the drum turns freely.

The next transmission component to be inspected is the shift fork. Although transmission problems aren't very common on most modern two-stroke motorcycles, when a problem does occur, the shift fork is usually damaged also. Visually check the shift fork for deformation or obvious wear. Use a Vernier caliper to measure the fork as shown in Figure 11-22 and compare the measurements to the factory specifications. It is also important to inspect visually the portion of the shift fork that rests in the shift drum for wear or damage (Figure 11-23).

Figure 11-17 Inspect the gear dogs for wear.

Figure 11-18 Inspect the gear slots for wear. Note that some wear is normal on gear slots. If the slots are rounded at the point of where the gear engages, it's time for a replacement.

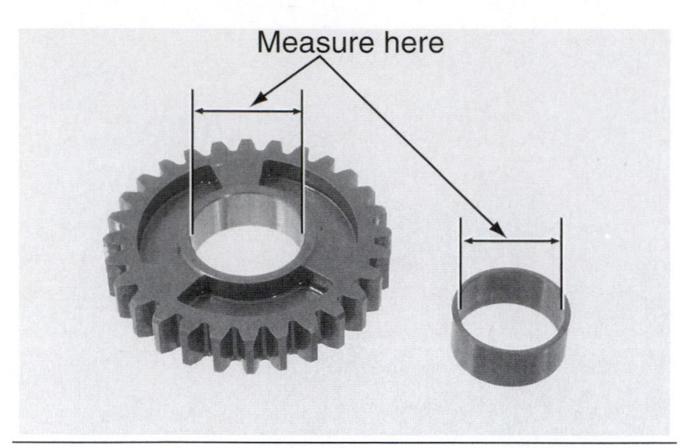

Figure 11-19 Inspect for scoring and measure the inside diameter of each gear. Copyright by American Honda Motor Co., Inc. and reprinted with permission.

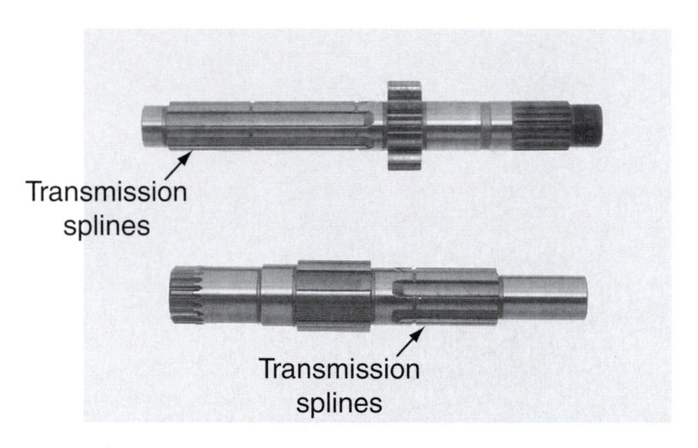

Figure 11-20 A transmission's mainshaft and countershaft are shown here. Inspect the splined grooves closely for wear or damage. Copyright by American Honda Motor Co., Inc. and reprinted with permission.

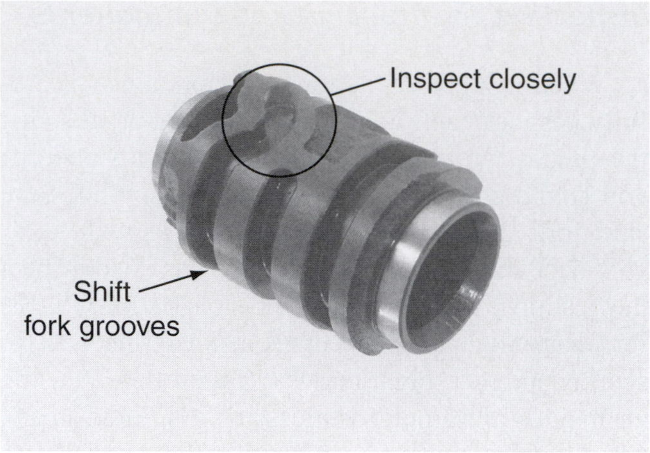

Figure 11-21 The most overlooked component of the transmission is shown here: the shift drum. Inspect the fork guide grooves very carefully for chipping or damage.

Assembling the Transmission

Before assembling the transmission, be sure that all parts are properly cleaned. Next, apply a light coating of oil to all sliding surfaces of each shaft before beginning the assembly process. This will ensure that there is adequate initial lubrication in the transmission.

Reassemble all of the transmission gears into the proper position on the appropriate shaft. Most service manuals contain an exploded view of the transmission (Figure 11-24) to assist with proper assembly of the gears, bushings, thrust washers, and retaining rings. When you install the thrust washers, be sure that the chamfered side faces away from the thrust load side of the gear. If a transmission retaining ring does not seat properly it will come out and

Figure 11-22 Measure the shift forks with a Vernier caliper.

Figure 11-23 Inspect the portion of the shift fork that rides in the shift drum for wear.

will likely cause serious transmission failure. Some gears use lock washer systems that have a splined washer and a lock washer that are engaged.

Be sure to align properly the oil holes in the bushings to allow appropriate oil flow to the gears, as shown in Figure 11-25.

Shift forks are marked to indicate their proper location. A fork marked with an "L" goes on the left side of the transmission and a fork marked "R" goes on the right side. A shift fork marked with a "C" would go in the center. Sometimes the center shift fork will have no markings on it (Figure 11-26).

After the transmission is properly assembled you can place it back into the crankcase as an assembly (Figure 11-27). You should then lubricate the transmission with engine oil while rotating the shafts, as illustrated in Figure 11-28.

Inspecting the Multi-Piece Crankshaft

Removing the crankshaft from a two-stroke engine often requires special tools that pull the crankcases away from the crankshaft (Figure 11-29) depending on the manufacturer's design.

You may also need to use a hydraulic press to remove the crankshaft from the crankcases (Figure 11-30).

Once removed, you will see that two-stroke engines use multi-piece crankshafts and can in most cases be rebuilt if worn or damaged. The most commonly replaced part of a two-stroke

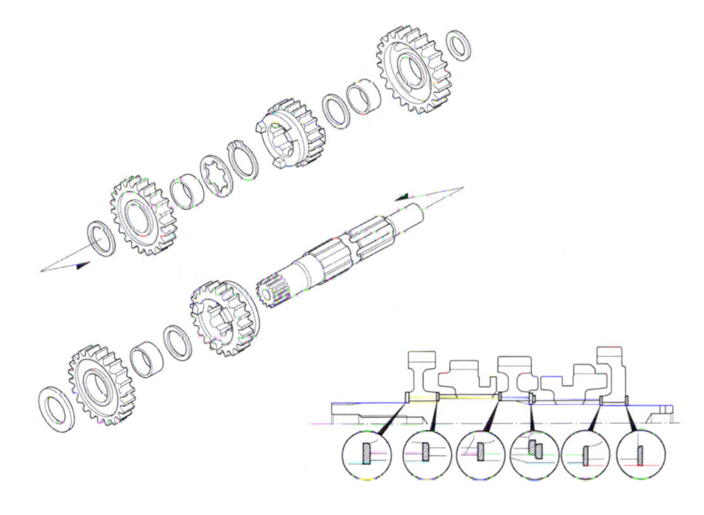

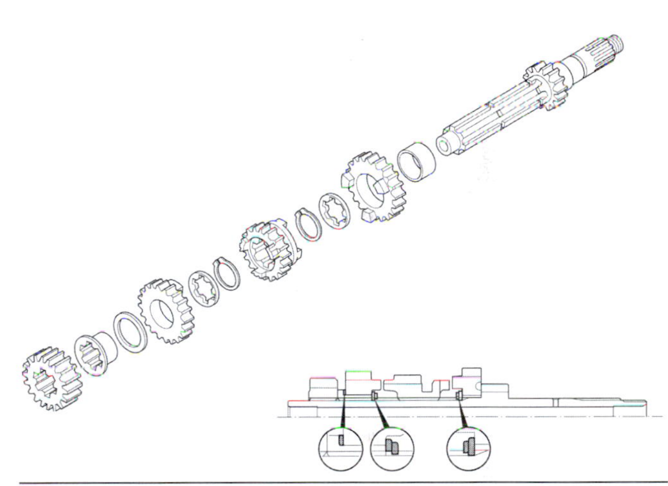

Figure 11-24 An exploded view of a transmission showing the correct placement of shims and clips is shown here. Copyright by American Honda Motor Co., Inc. and reprinted with permission.

Figure 11-25 Be sure to align any oil holes on the transmission gear bushings to ensure proper lubrication.

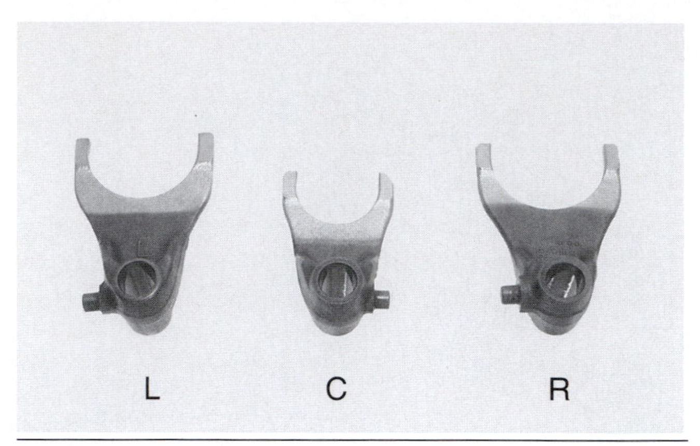

Figure 11-26 Shift forks are generally marked for correct installation. Copyright by American Honda Motor Co., Inc. and reprinted with permission.

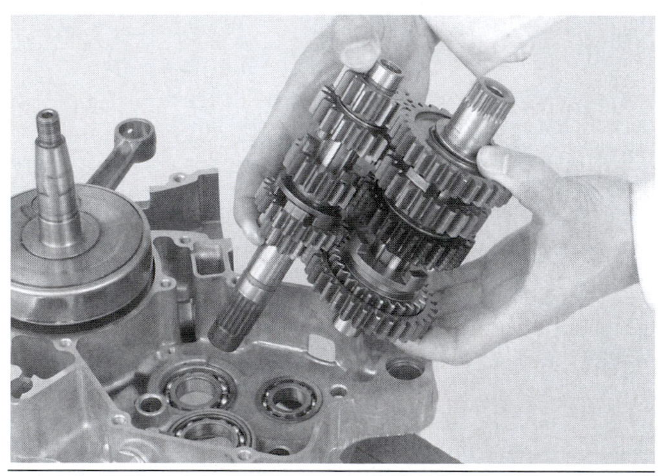

Figure 11-27 Install the transmission as a set back into the crankcase half. Copyright by American Honda Motor Co., Inc. and reprinted with permission.

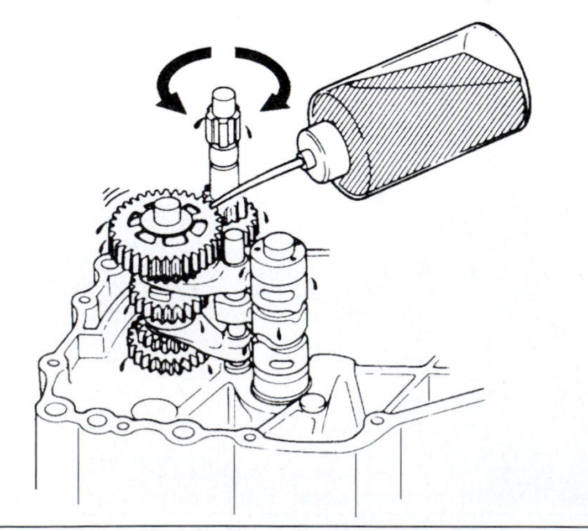

Figure 11-28 Lubricate the shafts with the same oil that is called for once the transmission is assembled. Copyright by American Honda Motor Co., Inc. and reprinted with permission.

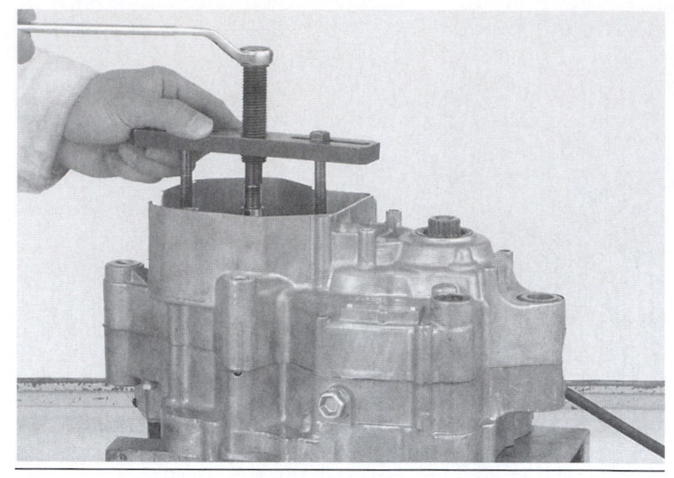

Figure 11-29 A special puller is often required to separate the crankcases from the crankshaft as shown here. Copyright by American Honda Motor Co., Inc. and reprinted with permission.

Figure 11-30 A hydraulic press is being used here to press the crankshaft out of the crankcase half. Copyright by American Honda Motor Co., Inc. and reprinted with permission.

crankshaft is the connecting rod and its bearing. The crankshaft connecting rod lower bearings are usually of the roller-type and are often replaced as a set with the lower crank pin and connecting rod.

Replacement of the connecting rod lower-bearing unit requires separation of the flywheels and is in most cases the type of job that is sent to a machine shop that specializes in such repair. Typically, a technician will visually inspect a crankshaft and check three measurements:

- Side clearance is required to ensure that there is enough space between the connecting rod and the crankshaft flywheels. This is a measurement that is done by using a feeler gauge (Figure 11-31) and the specification for this measurement can be found in the appropriate service manual.

- Radial clearance is the up-and-down motion of the connecting rod at the lower bearing. To check this measurement, the crankshaft will need to be placed on a pair of V-blocks that will hold it above the bench and then checked with a dial indicator (Figure 11-32).
- Runout is checked with the use of two dial indicators (Figure 11-33) and checks on the trueness of the crankshaft. This measurement is also done on V-blocks and assures the technician that the crankshaft is straight and not twisted, which will make the engine vibrate excessively.

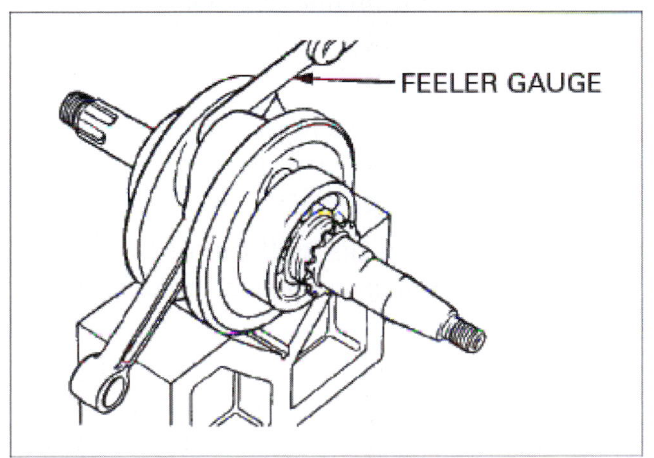

Figure 11-31 Measuring crankshaft side clearance with a feeler gauge. Copyright by American Honda Motor Co., Inc. and reprinted with permission.

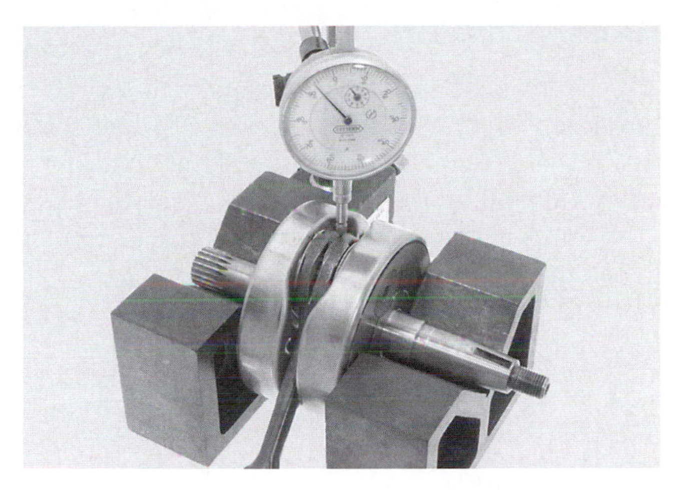

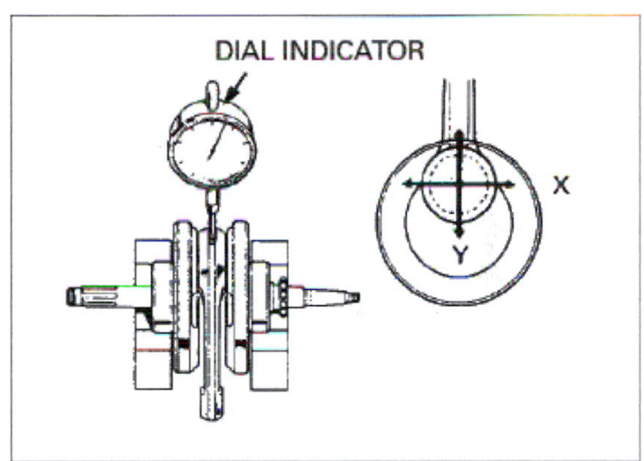

Figure 11-32 Measure for radial clearance with a dial gauge as shown here. Also measure in an X- and Y-axis as shown. Copyright by American Honda Motor Co., Inc. and reprinted with permission.

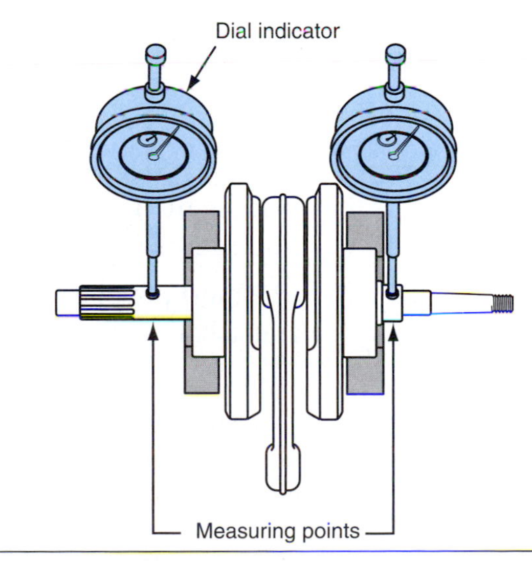

Figure 11-33 Measuring a crankshaft for trueness requires V-blocks to set the crankshaft on and two dial indicators. Copyright by American Honda Motor Co., Inc. and reprinted with permission.

Bench Testing

After you've completed the reassembly of the lower end of the engine, it is good practice to verify that all components move freely and properly. You would not want to discover that there is a problem after you have completely reassembled the engine and installed it into the chassis. This is known as bench testing.

As a check to ensure that the lower-end assembly is operating satisfactorily, turn the crankshaft to make sure it moves freely. Place the transmission into every gear to make sure that all gears mesh correctly and to verify there are no problems inside of the crankcases. When you're satisfied that all components have been assembled correctly and turn freely, you're ready to reinstall the top end onto the crankcases and finish the installation of the engine into the chassis. Keep in mind, always use the appropriate service manual when reassembling the engine as a guide to ensure that all specifications are correct.

Summary

- It is important to understand how to recognize common lower-end engine failures on a two-stroke motorcycle such as leaking engine seals, and transmission and crankshaft failures.
- It is important to inspect all of the components of the bottom end of a two-stroke lower end while the crankcases are apart to ensure that there will be no cause to have to disassemble the engine again after the initial repairs have been made.
- There are proper procedures for disassembling and inspecting the transmission in a typical two-stroke motorcycle engine that must be followed for a successful rebuild.
- By bench testing an engine prior to completing reassembly, you will know that all of the internals of the engine are working properly, removing any doubt that the crankcases were assembled incorrectly.

Chapter 11 Review Questions

1. An engine ________ should always be replaced with a new one if it's removed.
2. The shifting-fork groove surfaces should be measured by using a ________.
3. Ball bearings are held in the engine cases by an ________ fit.
4. A shift fork with a marking of "L" on it indicates that it's a ________ shift fork.
5. The ____ bearing is the most popular bearing used to support the crankshaft of a two-stroke engine.
6. Before assembling the transmission, apply ________ to all sliding surfaces of each shaft.
7. The two-stroke engine crankshaft connecting rod is normally a single-piece design. TRUE/FALSE
8. After you've completed the reassembly of the lower end of the engine, it is good practice to verify that all components move freely and properly. TRUE/FALSE
9. The ________ is the most critical component of the transmission that should be inspected but is often overlooked.
10. The ________ is one of the most common two-stroke engine components to wear out or fail.

CHAPTER

12 Four-Stroke Engine Top-End Inspection

Learning Objectives

When students have completed the study of this chapter and its laboratory activities they should be able to:

- Understand the importance of engine problem diagnosis
- Know the various components of a four-stroke engine top end
- Know the differences between the types of four-stroke engine top ends
- Understand the necessity of proper engine component inspection

Key Terms

Compression check
Detonation
Diagnostics
Leak-down test
Preignition
Prussian blue
Reamer
Scuffing or scoring
Stellite
Symptoms
Valve lapping

INTRODUCTION

Note: As in previous chapters, we will refer to both motorcycles and ATV engines as motorcycle engines only.

The vast majority of today's motorcycle engines are of the four-stroke design. The inspection of the components in a four-stroke motorcycle engine top end is a process by which the parts of the top end of an engine are removed, cleaned, inspected, and measured. The disassembly and inspection of four-stroke engines are tasks you'll perform often in your job as a motorcycle technician. An engine may be disassembled to make needed top-end repairs, or to complete the first step in a complete engine rebuild.

During a typical engine rebuild, the engine is restored to a like-new condition. This chapter introduces you to the procedures used to inspect the four-stroke engine top-end components. The procedures we describe apply to almost all four-stroke engines (regardless of model or manufacturer). We will explain how to visually inspect and measure the components of the engine top end. Throughout the discussion, we will point out the tools used in the process and provide you with some review information about the function of certain engine components. You may wish to review Chapter 6 as well to refresh your memory on each of the engine's individual components.

Repairs to the top-end assembly of many four-stroke motorcycle engines require that the engine be removed from the chassis. The service manual for the particular model you are working on will inform you if the engine can remain in the chassis or if it needs to be removed.

DIAGNOSTICS

If the motorcycle has an engine-related problem, before any components are disassembled the technician must diagnose the condition. **Diagnostics** is the process of determining what's wrong when something isn't working properly by checking the symptoms. **Symptoms** are the outward, or visible, signs of a malfunction. For example, a knock or a slipping clutch is a symptom. The actual cause might be a broken, worn, or malfunctioning part. Of course, these are only examples. To assist with diagnosing a problem you may need to use other testing equipment to help find the problem.

Often, complete diagnostics cannot be confirmed until the machine is actually disassembled. For example, if a four-stroke engine develops a rattle in the top end (the symptom), an experienced technician may recognize the sound and tentatively conclude that the cam chain tensioner is worn out. That would be the diagnosis. The technician would not be able to confirm the diagnosis until disassembling the engine and actually seeing that the cam chain tensioner and related parts are worn or damaged.

Correctly diagnosing problems is the most difficult and important part of a technician's job. Diagnosis is difficult because the technician often can't see the faulty part before disassembly. Correct diagnosis is important because the technician must not waste time disassembling and searching for parts that have not failed.

To diagnose problems, experienced technicians mentally divide a motorcycle engine into sections. For example, suppose the engine has a tapping sound. The technician first listens to hear if the tapping increases with the speed of the running engine while it is not moving. If it does, it can be determined that the noise is in relationship to the crankshaft or camshaft area of the engine. If the noise increases in speed once the motorcycle is in motion, the problem could be determined to be in the transmission area. By working in this manner, the technician inspects each part that's connected to the problem following a logical order of possible malfunctions. Troubleshooting will be covered in Chapter 20.

Before beginning any type of engine repair, be sure the motorcycle is clean. Dirt or other foreign particles cause damage to internal working mechanisms. Also, always use the correct and recommended tools.

REPAIR PROCEDURES

The procedures in this chapter are general in nature and their purpose is to familiarize you with the types of activities you'll encounter when working on a four-stroke top end. Always refer to the appropriate motorcycle service manual for complete disassembly information. The service manual contains all the information to do the job correctly, including detailed instructions for the specific model of motorcycle, special tools, and service tips. Above all, the service manual contains the appropriate safety information. To allow a broad understanding of the most common four-stroke top ends found on the modern motorcycle we will refer to a couple of different types of four-stroke engines throughout this chapter. The machines we will be referring to are a Honda CRF230F (Figure 12-1), which is an single-cylinder, air-cooled, single-overhead camshaft (SOHC) model and the Honda CBR1000RR (Figure 12-2), which is a four-cylinder, liquid-cooled, double-overhead camshaft (DOHC) model.

Figure 12-1 The Honda CRF230F off-road motorcycle. Copyright by American Honda Motor Co., Inc. and reprinted with permission.

Figure 12-2 The Honda CBR1000RR sport bike. Copyright by American Honda Motor Co., Inc. and reprinted with permission.

When disassembling an engine, be careful not to remove parts that don't need to be removed. Close inspection of the motorcycle, or reference to a service manual, can save many hours of time spent removing and replacing parts unnecessarily.

When rebuilding the top-end components of any engine, you need to be both patient and precise. Before you begin any repair job, take time to assemble the proper tools and the materials you'll need (micrometers, valve related tools, cleaning solvent, etc.). If an engine is to work properly, the measurements involving its piston and valves must fall within the manufacturer's specifications. Be patient and careful as you make the required measurements. Use the proper measuring instruments and record all your measurements accurately.

When you disassemble and rebuild an engine, examine the condition of each component. Check the parts before you clean them. For the most part, this should be a preliminary examination. The as-is condition of the parts can reveal a lot about the operation of an engine. After you've examined the parts and recorded your observations, clean the parts thoroughly and proceed with the rebuild by measuring and determining which parts need to be replaced and which can be reused.

FOUR-STROKE TOP-END DISASSEMBLY AND INSPECTION

The components of a typical SOHC air-cooled four-stroke engine top-end assembly are illustrated in Figure 12-3 and include the following components:

- Cylinder head-securing fasteners
- Cylinder head
- Cylinder head gasket
- Valve train components
- Cylinder studs
- Cylinder
- Cylinder base gasket
- Cylinder dowel pins
- Piston rings

- Piston
- Piston wrist pin
- Piston wrist pin retaining clips

The components of a typical DOHC liquid-cooled four-stroke engine top-end assembly are illustrated in Figure 12-4A.

You may notice that the cylinder on this engine is actually attached to the crankcases. To get at the pistons on this type of engine you will have to disassemble the crankcases, which will be covered in Chapter 13. Cylinders cast into the crankcases are seen often on today's multi-cylinder engines to make the engines more rigid and to assist the manufacturer to reduce costs.

The top-end assembly consists of the following parts:

- Cylinder head-securing fasteners
- Cylinder head
- Cylinder head gasket
- Valve train components
- Crankcase/cylinder
- Piston rings
- Piston
- Piston wrist pin
- Piston wrist pin retaining clips

As just mentioned, while the piston and related components are considered part of the top end of the four-stroke motorcycle engine, to get to them requires splitting the crankcases when the cylinder is cast into the crankcases (Figure 12-4B).

These exploded views of the top-end assemblies are what you would see at the beginning of most service manuals.

The differences between an air-cooled and a liquid-cooled engine were discussed in Chapter 7. You may wish to review this information before working on a four-stroke engine. In this chapter,

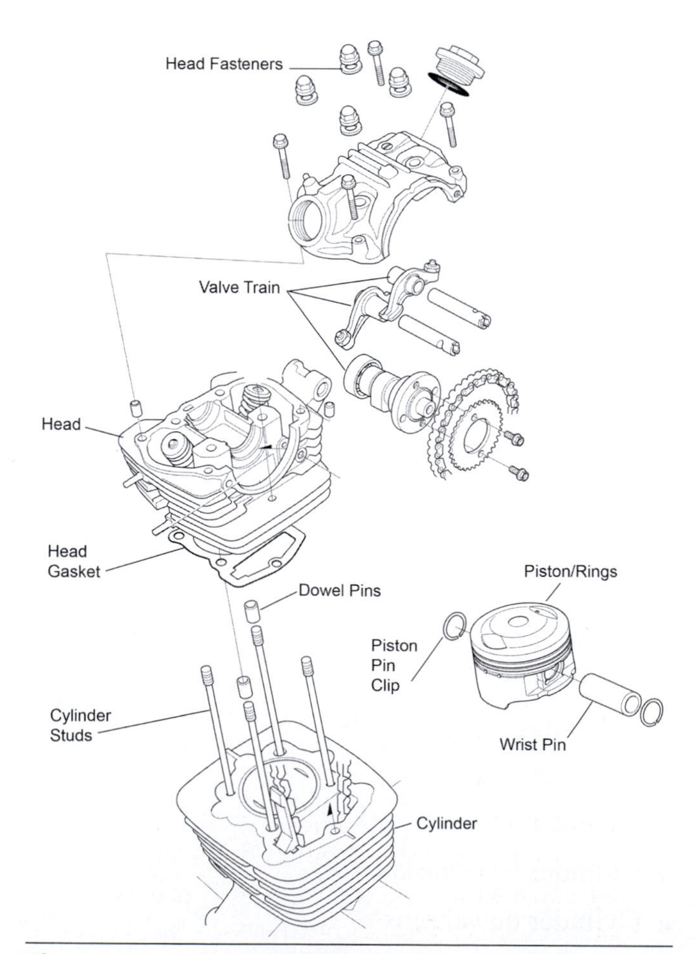

Figure 12-3 An exploded view of an SOHC air-cooled top end.

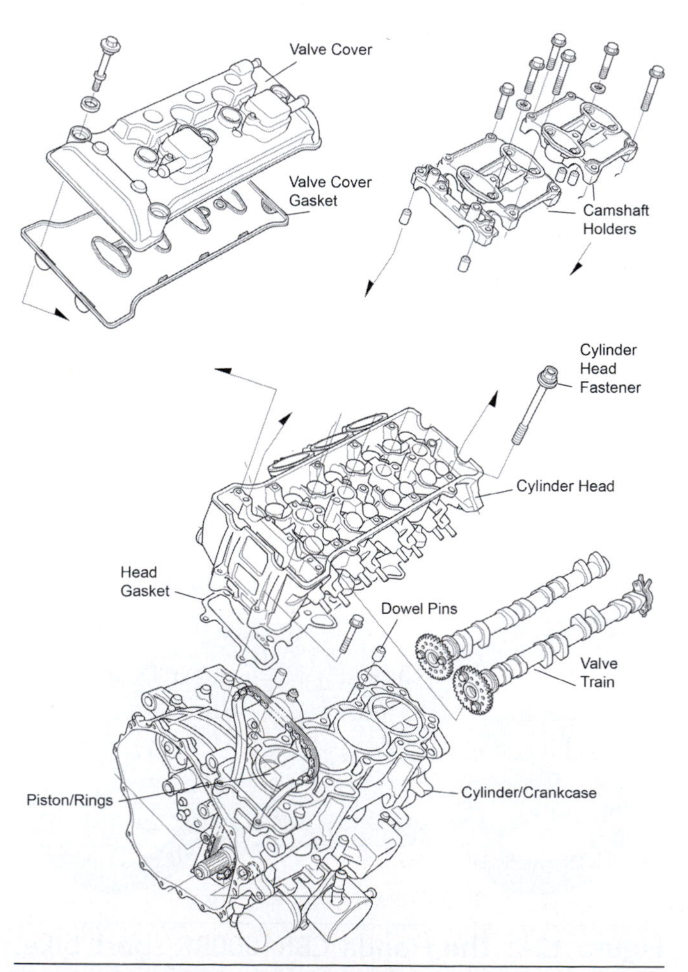

Figure 12-4A An exploded view of a DOHC liquid-cooled top end.

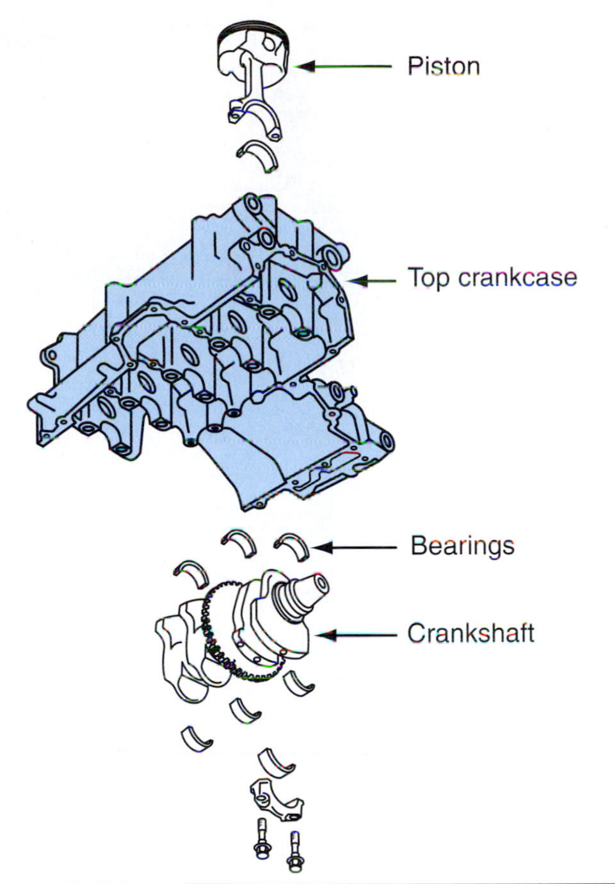

Figure 12-4B Some cylinders are built into crankcases as shown here.

you'll see images of both air- and liquid-cooled engines as both designs are commonly used in today's modern motorcycles. You will see that both air- and liquid-cooled four-stroke engine designs use similar approaches concerning disassembly and inspection procedures. You also may wish to review Chapter 6 for complete descriptions of the components we will be discussing in this chapter.

Four-Stroke Engine Removal

Removal of the engine from a motorcycle follows a pattern—that is, certain parts must be removed in a particular order. While we will not cover the removal or disassembly process of a four-stroke motorcycle engine here, it is important to understand that this pattern is similar between most makes and models of motorcycles. To ensure that the correct pattern is being followed, be sure to always use the manufacturer's service manual when removing the engine from the chassis.

Depending on the machine, an engine can weigh over 300 pounds. Therefore, it is advisable to use an assistant when taking the engine out of the chassis.

Disassembling the Top End of the Engine

As there are many different motorcycle manufacturers building four-stroke engines, we will not show you *how* to disassemble the top end of a four-stroke engine. We strongly recommend that you utilize the appropriate manufacturer's service manual and follow the procedures when disassembling an engine. The purpose of this chapter is to show you common procedures that you will see as you work on a four-stroke motorcycle engine.

FOUR-STROKE ENGINE TOP-END INSPECTION

There are many components in the four-stroke top end that require careful inspection and measurement. We'll begin our visual inspection of a four-stroke engine with the piston. Then we will move on to the piston rings, wrist pin, cylinder head, valve train, and camshaft.

Removing the Piston and Piston Rings

After you've exposed the piston and rings in a four-stroke motorcycle engine by removing the cylinder (Figure 12-5), you can remove the piston by taking the piston pin retaining clip out using a pair of needle-nose pliers (Figure 12-6), and then by sliding the piston pin out of the piston (Figure 12-7). Note that you should place a clean rag under the piston to prevent any loose pieces from falling into the bottom end of the engine. Also be sure to replace the retaining clips after removing them as they may not fit securely after they have been removed.

You'll notice that the typical four-stroke engine piston normally uses three piston rings (Figure 12-8). It consists of a three-piece oil ring on the bottom, one scraper ring in the middle, and a compression ring on the top.

Figure 12-5 A technician removing the cylinder is shown here. Copyright by American Honda Motor Co., Inc. and reprinted with permission.

Figure 12-6 Removing the piston clip. Note the rag under the piston to stop any outside debris from entering the bottom end. Copyright by American Honda Motor Co., Inc. and reprinted

Figure 12-7 The piston pin slides out of the piston. Copyright by American Honda Motor Co., Inc. and reprinted with permission.

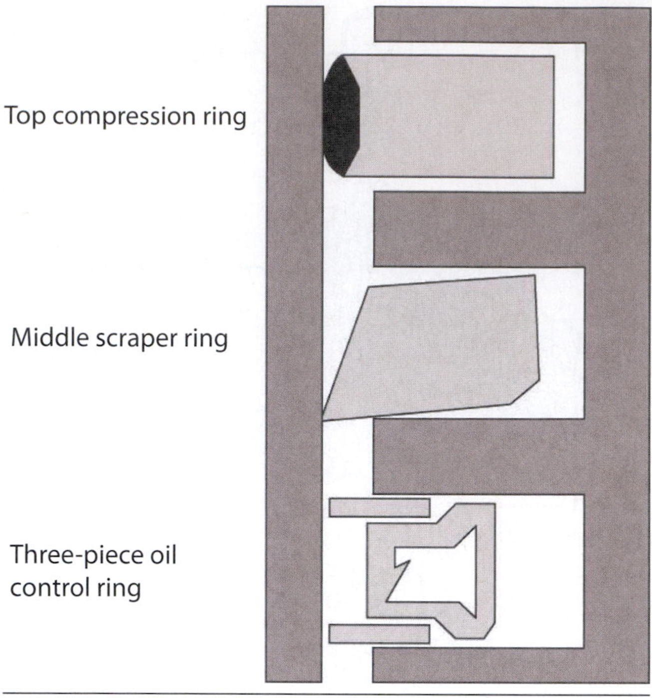

Figure 12-8 The three rings of a typical four-stroke piston are illustrated here.

Why three rings? The top ring is designed to seal most of the combustion pressure on the top of the piston. The middle ring is used as a compression ring as well but is used as a scraper to assist in removing excess oil from the cylinder wall when the piston is moving downward (Figure 12-9).

The bottom ring, also known as an oil control ring (Figure 12-10), is used to remove most of the oil from the cylinder walls through drain slots in the piston as it is moving downward.

The middle and bottom rings are designed to prevent oil from reaching the combustion chamber. If oil does get to the combustion chamber, the engine will smoke when running.

Remember, unlike a two-stroke engine, in the four-stroke cylinder fuel is burned without oil; instead, lubrication to the cylinder is supplied by a separate oil supply.

You should note that some modern performance-based four-stroke engines are now using just two piston rings (Figure 12-11): a compression ring and an oil control ring. Also, the pistons themselves are becoming shorter in size (Figure 12-12). These design tactics are used to reduce friction and reciprocating weight. These smaller pistons and two-ring systems originally came from Formula-One

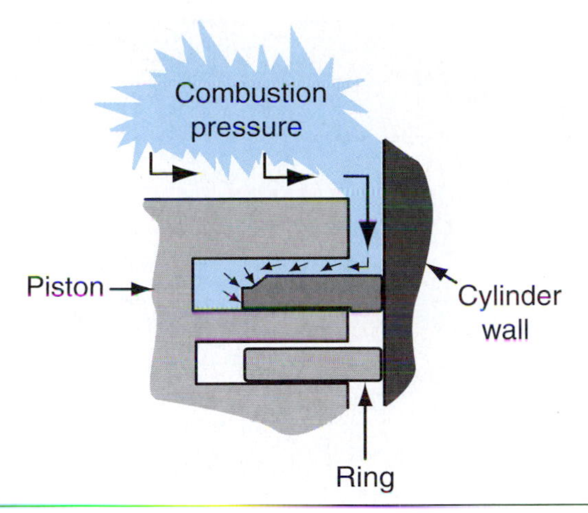

Figure 12-9 The top two piston rings are used to hold compression gases above the piston and the second ring is used to scrape oil from the cylinder on its way down the cylinder.

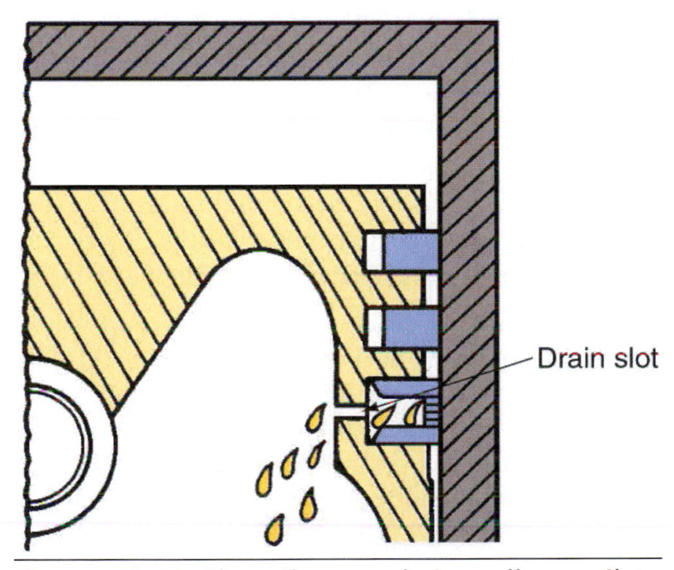

Figure 12-10 The oil control ring allows oil to flow back into the inside of the piston through the use of drain slots.

automotive and Moto-GP motorcycle technology. This is just one example of how technology used on the racetrack trickles down to production motorcycles.

Also found on four-stroke pistons are cutaways that are provided for valve head clearance (Figure 12-13). These cutaways, or pockets, are designed to allow clearance to prevent the piston from hitting and bending a valve as it opens and closes when the piston goes up and down.

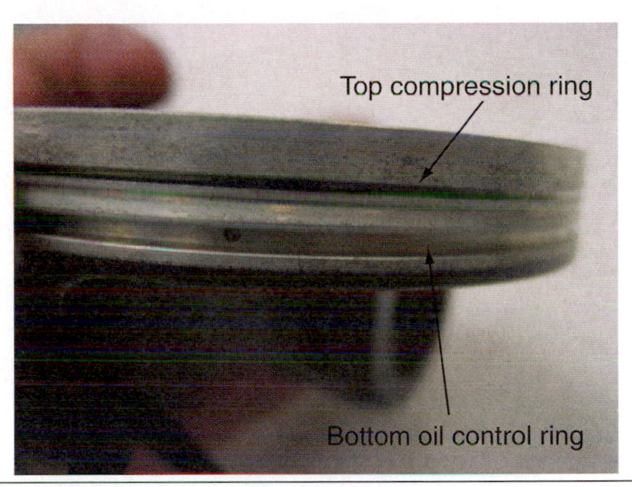

Figure 12-11 Some performance motorcycle engines now use just two piston rings, a top compression and bottom oil control ring.

Figure 12-12 The shorter piston on the right is becoming more commonplace in the motorcycle engine to reduce friction and reciprocating weight.

Inspecting the Piston

As you'll recall, the piston is the cylinder-shaped component that moves up and down the cylinder bore. The piston assembly consists of the piston itself, its wrist pin (or piston pin), and the piston rings. As you're now aware, an engine produces its power by burning the air-and-fuel mixture in the combustion chamber directly above the piston. Each time the spark plug fires, the air-and-fuel mixture ignites with a tremendous force. The burning process heats the gases, causing them to expand rapidly, forcing the piston down the bore. The piston movement is what allows the engine to perform useful work. The piston has to

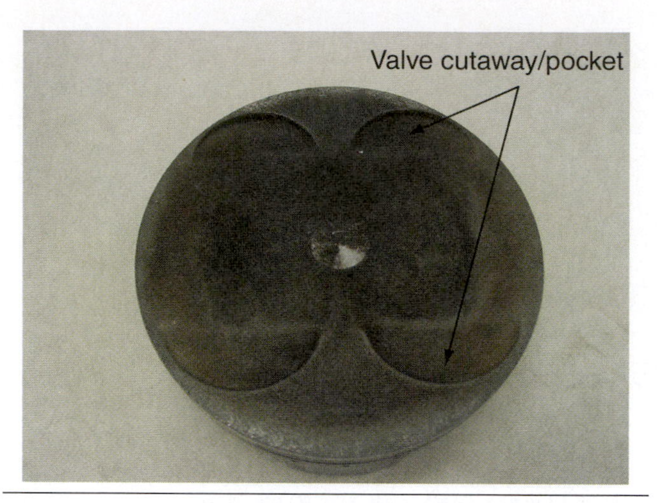

Figure 12-13 Cutaways are used to prevent piston-to-valve contact under normal engine use.

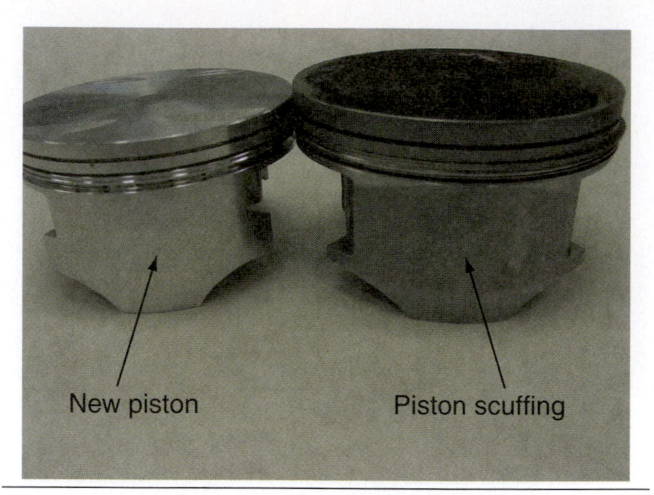

Figure 12-14 The piston on the left is new while the piston on the right shows signs of scuffing caused by dirt ingestion through the air-filtering system.

withstand a tremendous amount of physical force as well as extremely high temperatures during engine operation. Therefore, as part of the rebuild procedure, you must carefully inspect the entire piston assembly for damage.

Checking the Piston for Damage

Start your inspection of the assembly with a visual examination of the piston itself. Check the piston for cracks or any other signs of surface damage. Look closely in the areas of both the piston skirt and the rings; these areas are the most common sites of damage. Look for **scuffing** or **scoring** on the piston skirt. Scuff markings are wide areas of wear on the piston that usually appear as shiny patches (Figure 12-14). In most cases scuffing is caused by inadequate filtering of the air, which allows dirt to be ingested into the cylinder. Score marks are deep, vertical scratches that usually are caused by inadequate lubrication or overheating. Scuffing may or may not be accompanied by score marks.

Scoring and scuffing can be the result of a variety of conditions. In most cases, the marks are created by excessive friction and heat due to lack of lubrication or overloading the engine. Under certain extreme conditions, the temperature in a cylinder can approach the melting point, or weld point, of aluminum. These very high temperatures can be caused by a problem in the engine's cooling system or excess friction between the cylinder wall and the piston rings. Excessive friction is often due to improper lubrication.

If you find score marks or scuff marks on a piston, try to determine the cause so you can prevent the damage from recurring. This is one of the times you can take advantage of the notes and observations you made earlier in the disassembly process. During the disassembly, you should have checked to determine that the proper amount of oil was present in the engine or if the air filter was clean and installed correctly. If you noted a possible source of damage during disassembly and later found marks on the piston, you have important clues for use in the troubleshooting process. In such a situation, you may also want to talk to the customer or machine's operator to find out whether the engine was overheating during operation.

Oil Residue

Engine overheating, in addition to causing scuffing and scoring, usually produces a buildup of oil residue on the piston and the rings. Extreme heat breaks down the viscosity of oil and reduces its lubricating ability. When oil breaks down, it starts to bake onto the engine components, forming a residue that creates oil buildup and after time resembles varnish (Figure 12-15). This residue can coat the piston rings and eventually cause the rings to stick firmly to the piston. If this occurs, the rings are no longer able to seal the combustion chamber properly. Therefore, always check to ensure that the rings are free to move on the piston and that both the piston and rings are free of any buildup.

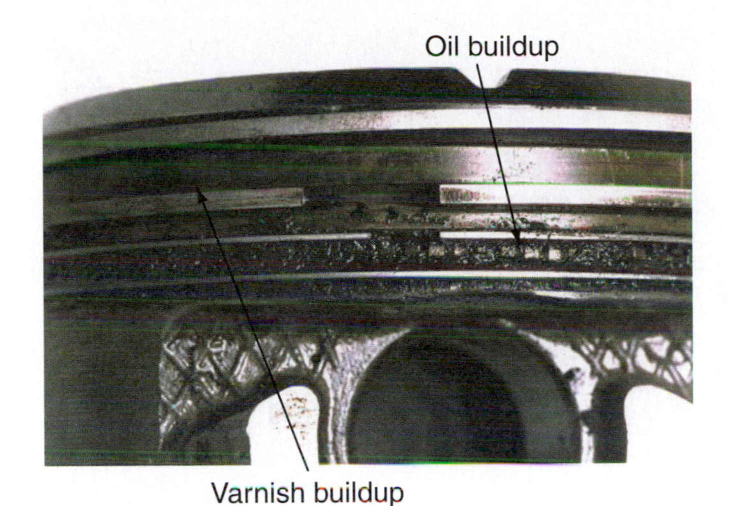

Figure 12-15 Oil build up on a piston can cause the rings to stick.

Examining the Piston Crown

Although the skirt is the most frequent wear site on a piston, you must also carefully examine the piston crown. If you find any damage, try to determine the exact cause so you can prevent the damage from happening again. Damage to the crown is usually the result of the fuel mixture burning improperly in the cylinder. If the fuel mixture ignites incorrectly, a violent explosion can result. The concentrated heat created in such an explosion can burn a hole right through the piston crown. Also, the explosion itself can be powerful enough to break right through the top of the piston. The following two terms describe different conditions that cause the fuel mixture to burn improperly:

- **Preignition** Preignition is the ignition of the fuel/air mixture in an engine before the spark plug actually fires. This may sound strange. How can the mixture ignite before there is a spark? The explanation is based on the fact that the burning air-and-fuel mixture produces a lot of heat. The lingering high temperature in the combustion chamber can cause small carbon deposits on the piston or in the combustion chamber to continue to burn, therefore causing the mixture to ignite without a spark (Figure 12-16). Preignition can also be caused by excessive compression of the air-and-fuel mixture. The carbon that sometimes builds up on the cylinder head and piston crown reduces the overall volume of the combustion chamber. The chamber's reduced volume results in an increase in the compression force exerted on the air-and-fuel mixture. This causes excess heat buildup.
- **Detonation.** Detonation is a condition in which, after the spark plug fires, some of the unburned air–fuel mixture in the combustion chamber explodes spontaneously, set off only by the heat and pressure of the air–fuel mixture that has already been ignited. In other words, the air-and-fuel mixture fails to burn smoothly. Instead, the mixture begins to burn normally in one area of the combustion chamber, then, as the pressure and heat in the chamber increases, the mixture ignites a second time in another area of the combustion chamber. Thus, two separate flames can burn at the same time in the chamber (Figure 12-17) even though they were initiated at different times (this occurs within a split second). This should not be confused with the use of two spark plugs in an engine, which would create normal combustion due to the fact that the combustion process was initiated when it was supposed to. When the two flames collide, a shock wave is created. The shock wave of detonation effectively hammers the top of the piston and the piston rings. Eventually, this hammering, or detonating, damages the piston and rings. The key point to understand is that detonation occurs *after* the initiation of normal combustion.

The most common cause of detonation is the use of gasoline with an octane rating that's too low for the engine. The recommended octane rating for the fuel in a particular engine can be found in the owner's manual and in the manufacturer's service manual for that engine. Detonation may also be caused by incorrect ignition timing. If the ignition timing is too far advanced, the spark occurs earlier than it should. In this case, the air-and-fuel mixture ignites and starts to burn when the piston is still rising on the compression stroke. This disruption of the normal burning pattern results in detonation.

Inspecting the Piston Rings

In most cases piston rings should be replaced when an engine is taken apart. Normally, rings that are reused won't seat-in properly, resulting in poor engine performance. You may recall that when new piston rings are installed in an engine, they must wear themselves into position against the cylinder

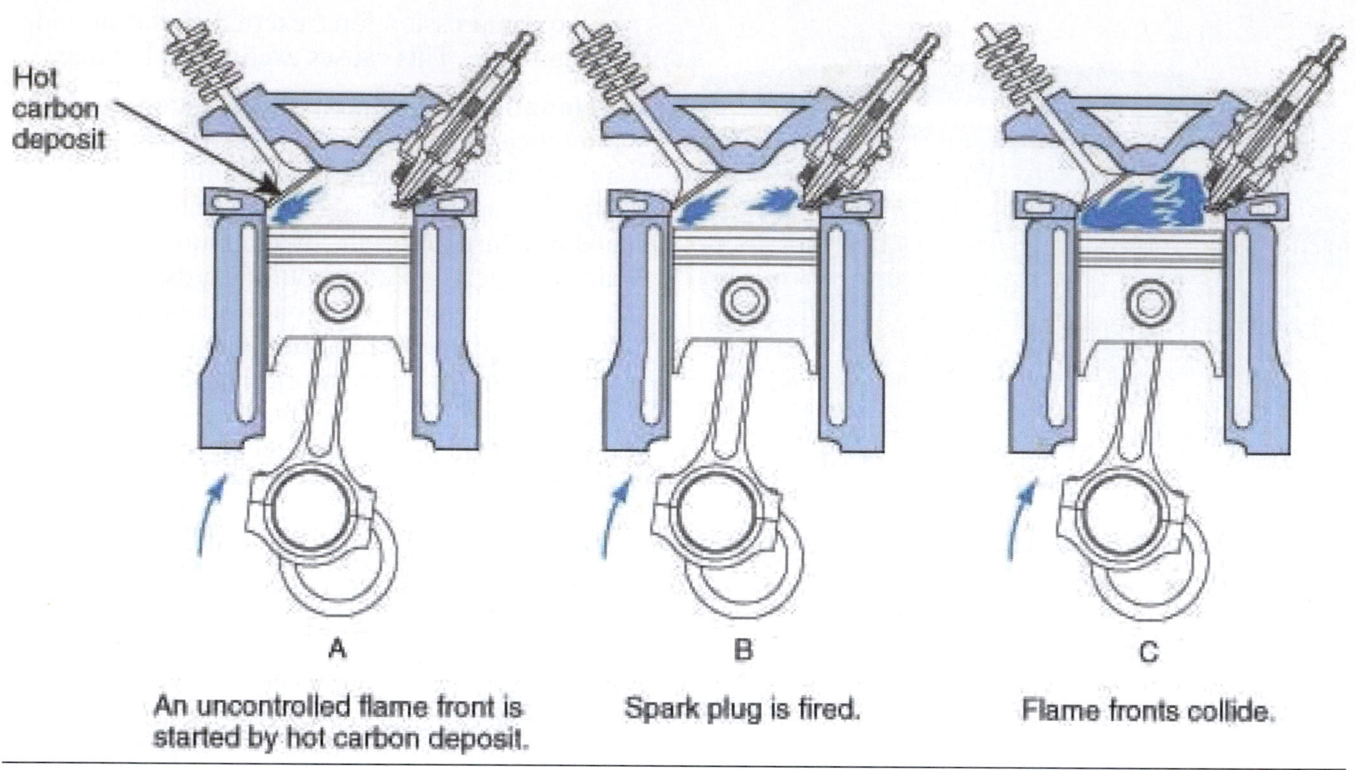

Figure 12-16 The three phases of preignition are illustrated here.

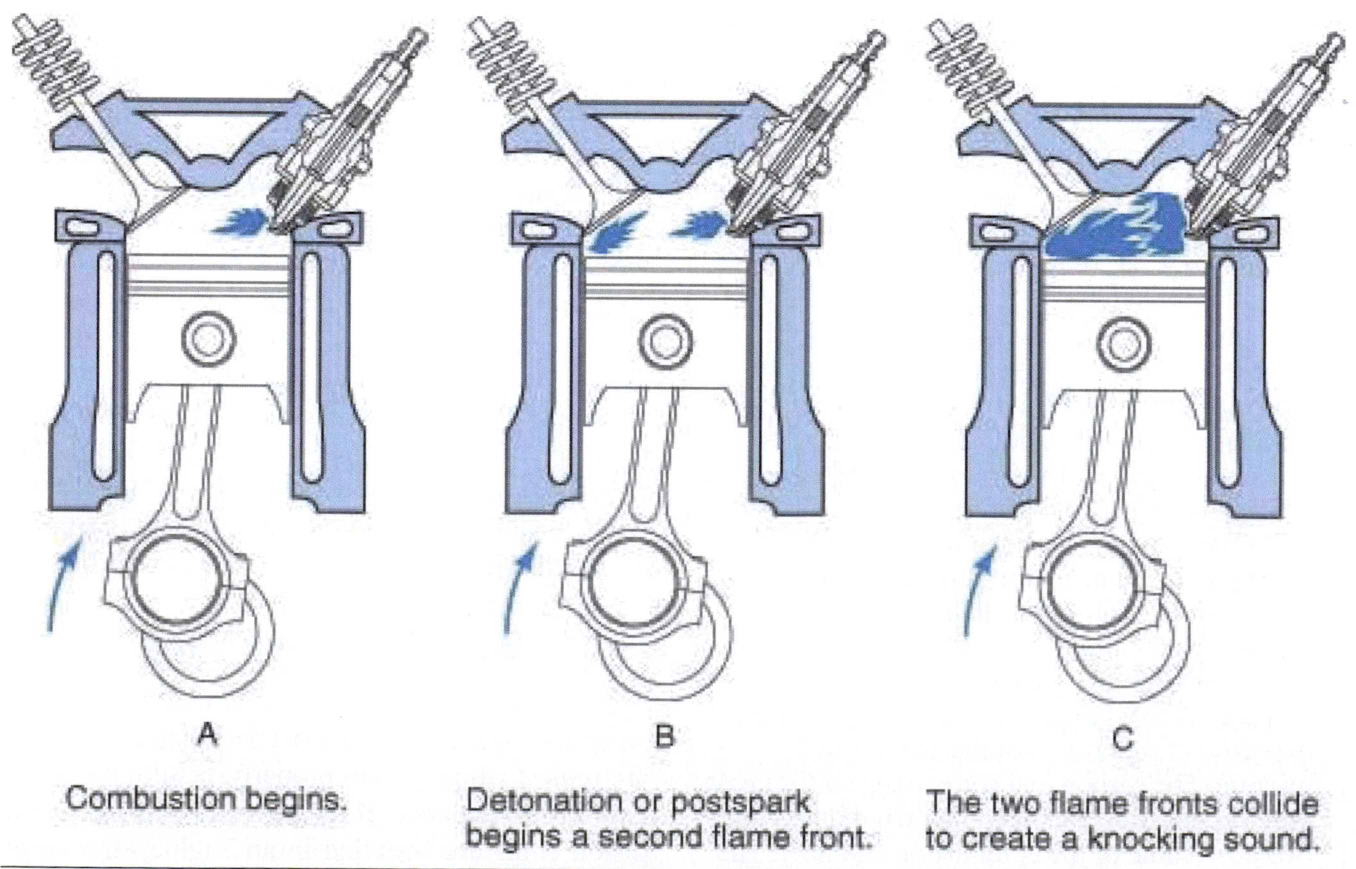

Figure 12-17 The phases of detonation are shown here.

walls to form a tight seal. Once this process of seating-in has occurred, the rings lose the ability to do so again. That is, if old rings are reinstalled in an engine, they won't be able to conform once again to the cylinder walls and make a tight seal. Without a tight seal, the combustion gases can leak past the rings. This reduces the amount of horsepower the engine can produce. In addition, oil from the crankcase seeps past the rings and into the combustion chamber. The engine thus consumes larger amounts of oil. Oil that enters the combustion chamber burns along with the air-and-fuel mixture. Any oil burning in the combustion chamber is revealed by excessive exhaust smoke as the engine runs.

Worn piston rings are usually bright and shiny at the point where the edge contacts the cylinder wall. Worn rings can also be detected by performing a **compression check** and an engine **leak-down test** on the engine *before it's disassembled.* A compression check is a simple test that measures the amount of pressure produced in the combustion chamber on the compression stroke (Figure 12-18A). The compression is measured with a special gauge inserted in the spark plug hole. If the piston rings are worn, the gauge displays a pressure reading that's much lower than the manufacturer's specification. The reading is low because, instead of being compressed, some of the air-and-fuel mixture is leaking past the worn rings and into the crankcase. To ensure a correct compression reading be sure to hold the throttle open fully while turning the engine over. This allows the most available amount of air into the combustion chamber, which in turn will give the highest possible reading on the compression gauge.

An engine leak-down test (Figure 12-18B) is a more comprehensive engine diagnostic test than a compression test as this type of test allows you to measure the percentage of air that leaks past the piston rings and valves. Engine leak-down testers are available from general tool sources.

A leak-down test provides a clear indication of whether the combustion chamber is sealing properly. The test involves pressurizing the combustion chamber and measuring the rate at which the air is lost past the rings and valves (or head gasket).

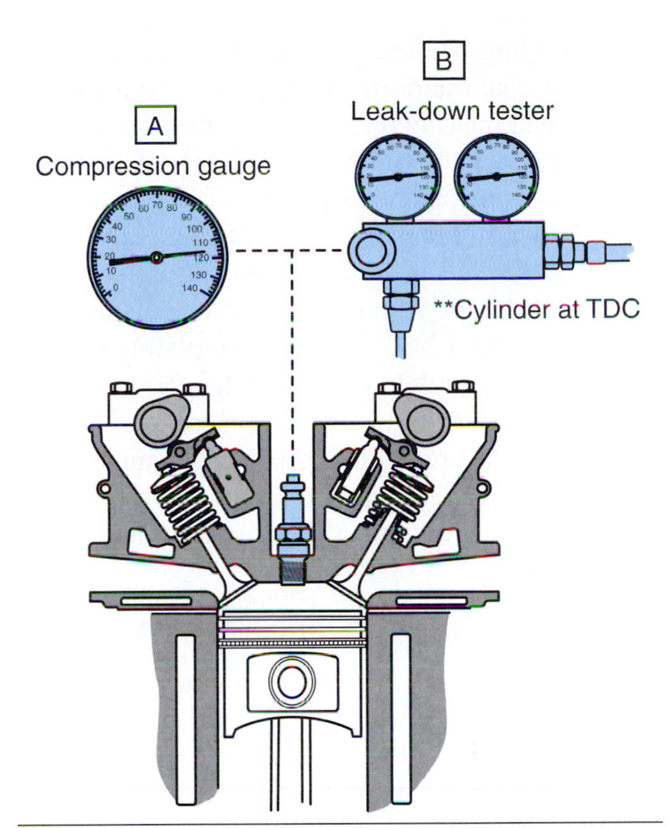

Figure 12-18 (A) A compression test can help to diagnose a problem before disassembling a four-stroke motorcycle. (B) An engine leak-down tester gives a better understanding of where the problem is in a four-stroke engine.

For instance, if the supply of air pressure is 100 psi, and the cylinder is able to maintain a pressure of 90 psi, the cylinder is said to have 10 percent leakage based on the supply flow rate. But more important than a determination of whether the engine needs repair, is to find out more precisely where the problem actually is coming from. The directions for performing this type of test will be provided by the tool manufacturer.

Once installed, simply listen to the air-box, exhaust pipe end, and crankcase filler cap to determine whether the intake valve(s), exhaust valve(s), or piston rings are leaking. Squirting a little soapy water around the cylinder and head-mating area will tell you if the head gasket is leaking to the outside atmosphere.

Measuring the Piston Rings

Even though in most cases it is advised to replace the piston rings after an engine has been taken apart you may wish to measure the piston rings that are already in the engine that you are disassembling. Before you can measure the rings you must take them off of the piston. To remove a ring, spread the ring open so you can slide it out of its ring groove and off the piston (Figure 12-19).

Piston rings are measured by fitting them into the cylinder and then checking the end gap with a feeler gauge. This is done by inserting a piston ring into the cylinder squarely, using the piston as a guide (Figure 12-20). After the piston ring is inserted, you can then check for the end gap using a feeler gauge (Figure 12-21). Each piston ring should be measured at the top, middle, and bottom of the cylinder. The specification for the proper ring end gap is given in the appropriate service manual.

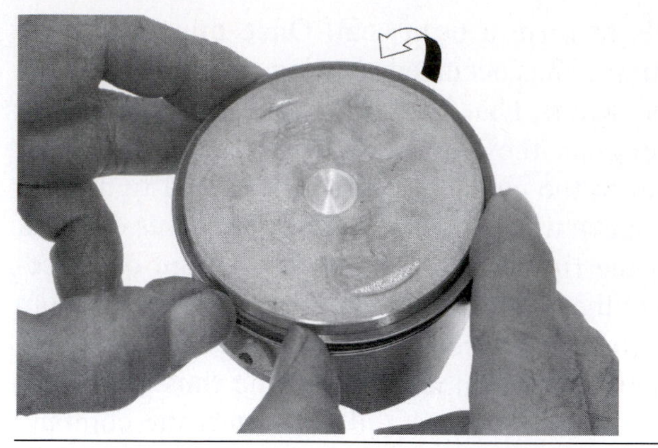

Figure 12-19 A technician removing a piston ring. Copyright by American Honda Motor Co., Inc. and reprinted with permission.

Inspecting the Piston Ring Grooves

The ring grooves, cut into the sides of the piston, hold the piston rings in place. The ring lands are the uncut areas between the ring grooves. The ring grooves are actually slightly wider than the piston rings. As a result, the rings can move slightly, or float, within their grooves. The rings are able to actively conform to the cylinder walls while the engine is operating. The small amount of space between each piston ring and the bottom side of its groove is called the piston ring side clearance.

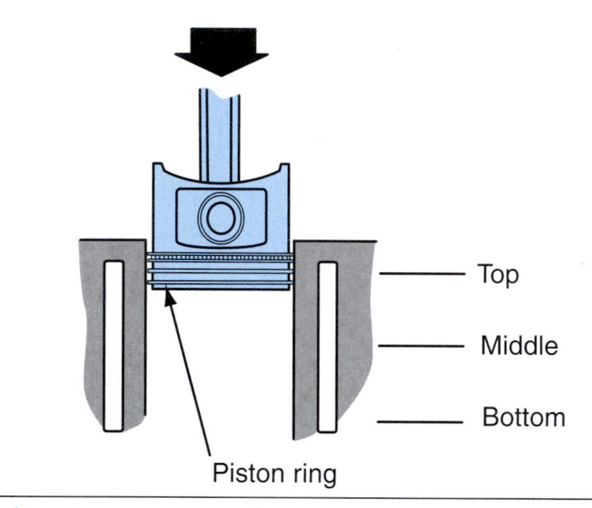

Figure 12-20 Use the piston to place the rings correctly into the cylinder prior to measuring the ring end gap.

The combustion gases forcing themselves onto the piston get down into the ring grooves and leave behind a residue. Therefore, to inspect the ring grooves for excessive wear, you must first clean the grooves thoroughly. When cleaning the grooves, remember that the piston is made of aluminum, a soft metal. Be careful not to dig into the piston and remove any metal, especially along the inner sides of the ring grooves. The most common tool used to clean the piston ring grooves is an old piston ring. Made of a very tough material, old piston rings work well because they fit the ring grooves perfectly and, therefore, won't damage the sides of the grooves. If you use an old ring for this purpose, break it in half to produce a scraper-like edge. Then, insert the edge into the groove and scrape the residue out.

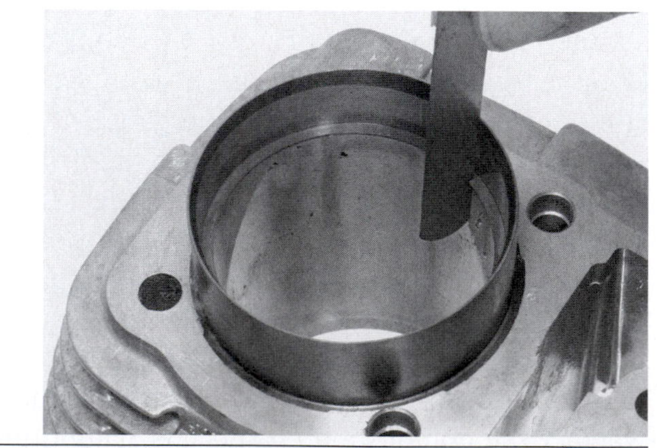

Figure 12-21 Checking the ring end gap with a feeler gauge. Copyright by American Honda Motor Co., Inc. and reprinted with permission.

After the ring grooves are cleaned, the piston can be wiped off and the side clearance for the piston rings can be checked. As mentioned earlier, this dimension is the clearance between the piston ring and the inner side of the ring groove (Figure 12-22). This small amount of clearance performs an important function. During the power stroke, the pressure produced by combustion pushes the piston down the bore. Some of the expanding gases are also forced down the side of the piston and behind the floating piston ring. The resulting pressure behind the piston ring forces the ring outward, hard against the cylinder wall, thus helping to better seal the combustion chamber. By allowing the ring to seal better, the proper ring side clearance helps the engine produce more power.

Note that because a small amount of clearance should always be present, a ring tips slightly under normal operating conditions. As the piston goes down the cylinder during the intake stroke, the ring tips and scrapes excess oil off the cylinder wall.

During the compression and exhaust strokes, the piston rises and the tipped ring glides over the oil film remaining on the cylinder wall.

During the power stroke, forces pushing down on the ring cause it to sit squarely, providing a better seal and, therefore, better power. The proper clearance between the piston ring groove and the piston ring can be critical. If the clearance is too large, the ring tips excessively as the piston moves up and down, reducing its ability to seal. The excess movement of the ring on the piston may also cause the ring to break. If the clearance is too small, the ring binds in its groove when the piston heats up and expands.

After you've measured the piston ring side clearance, compare your measurement with the manufacturer's specification. In addition, each ring groove may be worn differently. Therefore, check the side clearance in all of the piston's ring grooves.

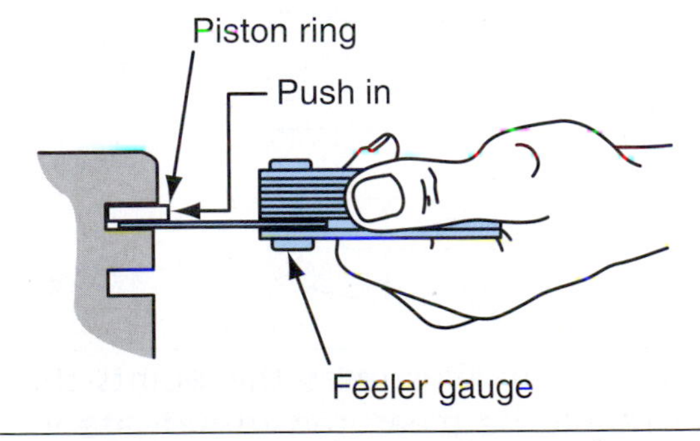

Figure 12-22 Checking piston ring groove clearance.

Measuring the Piston

After the piston and rings have been visually inspected, you can prepare the piston for measurement.

A typical piston appears to have a simple shape, like a can. However, looks can be deceiving. As mentioned previously, pistons are manufactured with a taper. That is, the diameter at the very bottom of the piston's skirt is larger than the diameter at the piston's crown. This is due to the varying amounts of material and the different rates of expansion of the material with heat. Once up to operating temperature, a piston's shape becomes the same from top to bottom.

The appropriate manufacturer's service manual shows where to measure the diameter of the piston. Figure 12-23 provides an example.

The piston must be measured with a micrometer. It's a good idea to keep track of the piston's actual diameter because you can use that measurement when calculating the clearance between the piston and the cylinder walls. Once you've measured the diameter of the piston, compare your measurement to the appropriate specification or specification range. If the diameter of the piston is outside the specification given by the manufacturer, the piston should be replaced. If the piston is within specifications and shows no signs of damage, you can reinstall it in the engine.

Measuring the Cylinder

Now we'll measure the cylinder to determine the amount of wear that has occurred on the cylinder walls. Movement of the piston and rings within the cylinder contributes to cylinder wear. The areas of wear are the locations in which the rings travel, as well as the areas in which the piston skirt contacts the cylinder walls. Cylinder wear is also caused by the piston rocking on the wrist pin, due to the piston tipping slightly during its travel. Piston rocking can create a noise known as piston slap. Under these conditions, the cylinder bore wears more on the front and back than on the sides. By front and back, we mean at a 90-degree angle to the wrist pin.

Cylinder measurements are taken from front to back and side to side. These areas are called

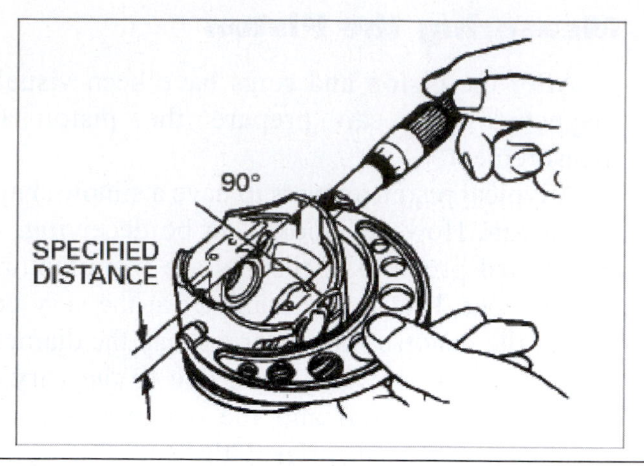

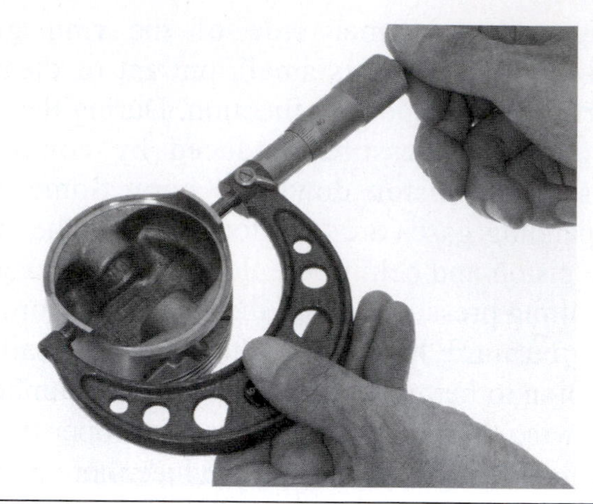

Figure 12-23 Engine manufacturers will specify where to take piston measurements. Copyright by American Honda Motor Co., Inc. and reprinted with permission.

the X- and Y-axis of measurement. To measure the cylinder, a cylinder bore gauge is used (Figure 12-24).

Insert the cylinder bore gauge at a point near the top of the cylinder and rock it back and forth slightly to find the smallest diameter. Move the dial gauge to "0" and use this as your baseline reading. The gauge is then moved to a point near the center of the cylinder and a reading is taken there as well. Finally, the gauge is positioned at the bottom of the cylinder and another reading is taken. This is done for both the X- and Y-axis measurements to determine the cylinder's trueness (Figure 12-25). The readings are compared and the difference indicates the amount of wear.

The difference between measurements taken on the same axis is known as cylinder taper. The difference between the two axes is called out-of-round. The taper and out-of-round must not exceed factory specifications for the engine on which you're working. Each model has its own specifications. If the measurements exceed allowable limits, in many cases the cylinder may be bored or recut to a new size and fitted with a new and larger piston and ring set. Not all cylinders can be bored; check the appropriate service manual to determine if the cylinder is capable of being bored. Boring a cylinder is a job that requires the use of special machine tools and is normally done by a specialist. Boring can be done at most machine shops. Some dealerships do their own boring of cylinders as well.

Measuring the Piston-to-Cylinder Clearance

A piston expands as its temperature rises. Because the metal of the piston typically expands more than the metal of the cylinder wall, some clearance must be allowed between these components

Figure 12-24 Using a cylinder bore gauge. Copyright by American Honda Motor Co., Inc. and reprinted with permission.

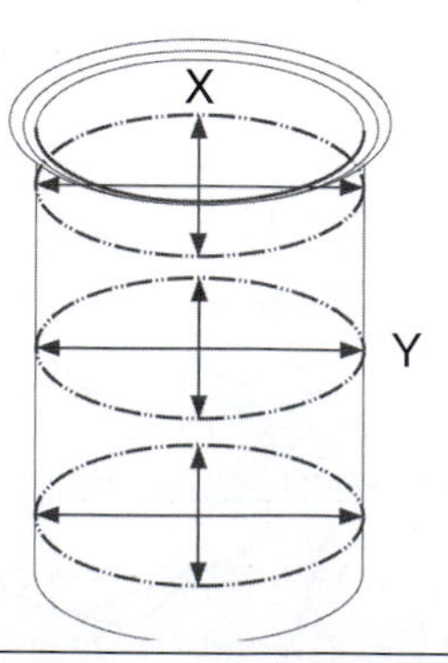

Figure 12-25 This illustrates the points that a cylinder should be measured to indicate wear. Copyright by American Honda Motor Co., Inc. and reprinted with permission.

when both are cold. This clearance is called the piston-to-cylinder clearance, or piston clearance, for short. The proper piston clearance for an engine is given in the appropriate service manual.

If the piston clearance is too small, the piston fits too tightly in the cylinder when the engine heats up, resulting in excessive friction. Friction between the piston and the cylinder can be so great that the piston seizes in the bore. That is, the piston may wedge itself so tightly into the cylinder that it cannot move up or down.

You may be able to free a seized piston after the engine cools down again; however, both the piston and the cylinder wall will probably be badly scored and damaged. If the piston clearance is too large, the piston isn't held in place and tends to rock back and forth while the engine is running. This rocking motion creates a knocking noise and may eventually break the piston skirt. In addition, the ability of the piston rings to seal the combustion chamber is greatly reduced.

Determining Piston Clearance To determine the piston clearance in an engine, you'll need to measure the diameter of both the piston and the cylinder bore. Compare your measurements to the manufacturer's specifications. Then, subtract the outside diameter of the piston from the inside diameter of the cylinder bore. The result of your calculation is the actual piston clearance. Finally, compare your calculated clearance to the manufacturer's specification. If the clearance is outside specifications, the piston and cylinder must be resized to make the clearance conform to specification. This method is the most accurate way to measure the piston-to-cylinder clearance.

Inspecting the Wrist Pin

The wrist pin is a cylinder-shaped component of the piston assembly (Figure 12-26). It's used to link the connecting rod to the piston. The connecting rod's bearing surface for the wrist pin allows the end of the rod to rotate freely around the pin as the piston travels up and down. The wrist pin must transfer each power stroke's downward physical force from the piston to the connecting rod.

To ensure the wrist pin is strong enough to handlethe task, the engine manufacturer usually makes the pin of very high-quality steel. For this reason,

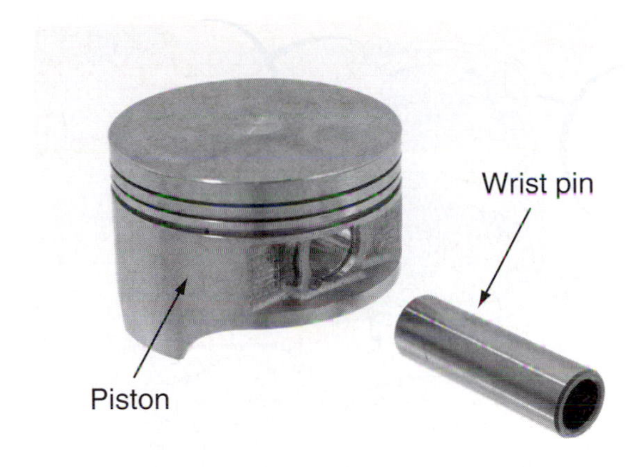

Figure 12-26 The piston wrist pin connects the piston to the connecting rod. Copyright by American Honda Motor Co., Inc. and reprinted with permission.

you normally won't see much wear on the pin itself unless there is a lack of lubrication in which case you will be able to feel and see scoring marks on the pin. To verify that a wrist pin is not worn, measure the pin's diameter with an outside micrometer and compare your measurement with the specification given in the service manual (Figure 12-27).

Piston Ring Installation

When installing rings on a four-stroke piston it is important to know that they must be installed with any ring markings facing upward (toward the piston crown) and their end gaps set at the manufacturer's recommended angle apart from each other (generally 120 degrees) as seen in Figure 12-28.

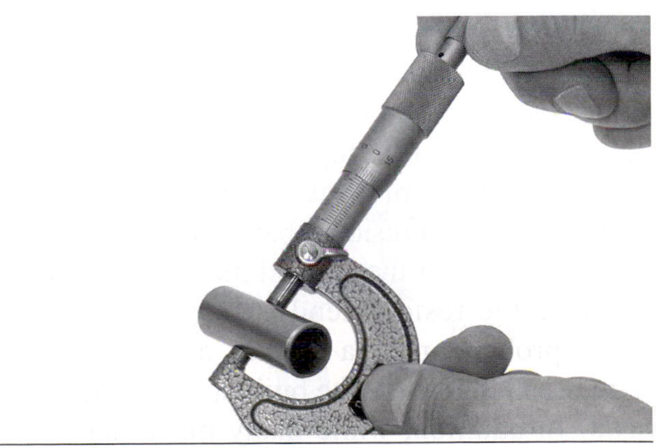

Figure 12-27 Measuring a wrist pin with an outside micrometer. Copyright by American Honda Motor Co., Inc. and reprinted with permission.

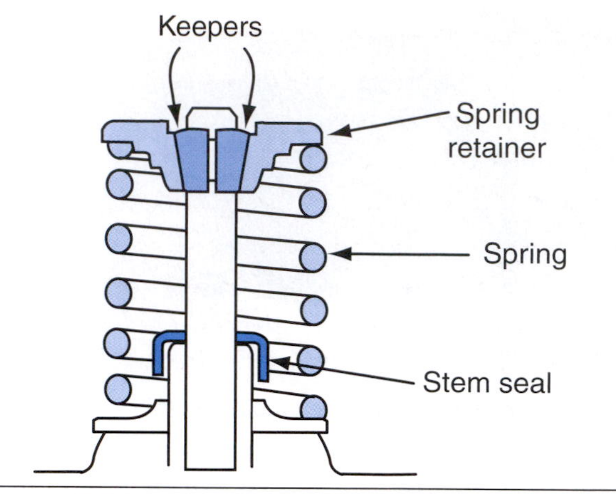

Figure 12-32 A typical valve spring and retaining devices are shown here.

Figure 12-33 Most engines use two valve springs per valve, one inner (right) and one outer (left). Copyright by American Honda Motor Co., Inc. and reprinted with permission.

seen in Figure 12-33) but in many of today's motorcycle engines you will find a single spring per valve as well (Figure 12-34). In addition, you will commonly find a valve seal under the valve spring, which is used to keep oil from entering the combustion chamber.

Although the basic purpose of a valve spring is to close the valve while also ensuring that the valve train stays in contact with the cam lobe, it must perform this feat under grueling conditions that vary tremendously several times every minute. The expected rpm range, camshaft profile, and cylinder head design just begin the list of criteria for engineers to use when choosing the right valve spring for an engine.

Manufacturers use different numbers of valves for various engine applications including two-valve engines (one intake and one exhaust valve), three-valve engines (two intake and one exhaust), four-valve engines (two intake and two exhaust), and five-valve engines (three intake and two exhaust). These different valve arrangements are utilized to obtain the optimal flow of intake and exhaust gases for the intended design. The most common valve arrangement is the four-valve engine design (Figure 12-35).

There are five critical areas of typical four-stroke engine valve. The valve areas you'll need to pay close attention to are shown in Figure 12-36.

Valves must operate under a variety of extreme conditions. As a result, certain areas of the valve assembly often show signs of wear or physical damage. Because they're located in the combustion chamber, valves can reach temperatures of

Figure 12-34 Some performance machines use only one valve spring per valve. Copyright by American Honda Motor Co., Inc. and reprinted with permission.

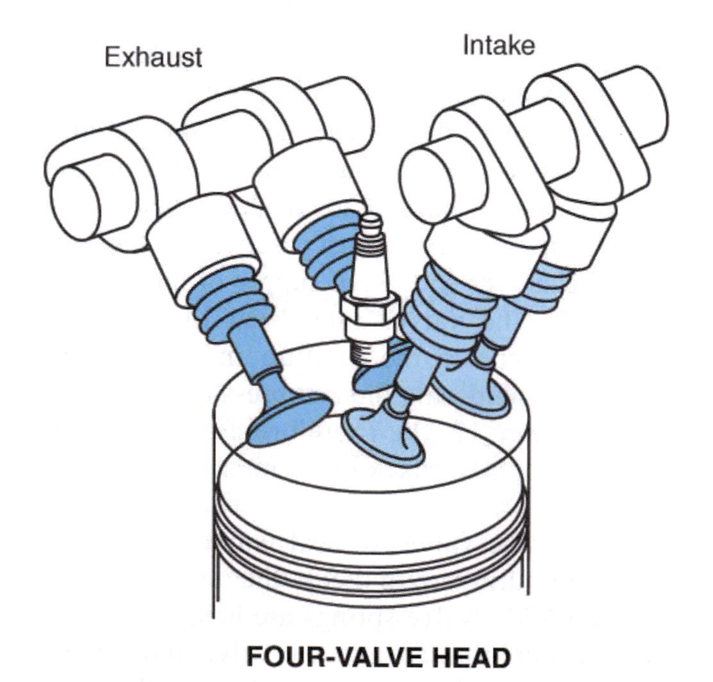

Figure 12-35 An illustration of a typical four-valve design.

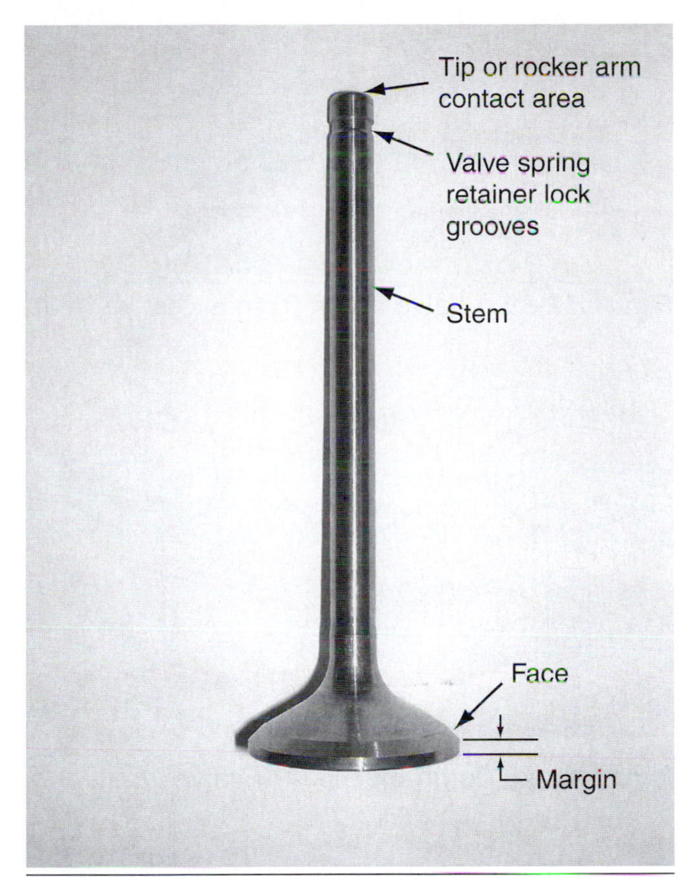

Figure 12-36 The five critical areas to look at on a valve are shown here.

well over 1,000 degrees Fahrenheit under normal operating conditions! Heat tends to wear away the exposed surfaces, particularly the valve heads. Heat, however, isn't the only problem valves face. In addition, friction between the valve stems and the valve guides produces wear. Keep in mind that each valve must open and close for every power stroke in a four-stroke engine. Because many motorcycle engines operate at 10,000 rpm or more, valves must open and close above 5,000 times per minute under these circumstances. When a valve opens and closes this fast, friction builds up between the valve stem and the valve guide. This eventually leads to wear in the stem, the guide, or both. The rapid movement of the valve also tends to hammer on the valve seat. This hammering action distorts the valve face and the cylinder head valve seat, eventually allowing combustion gases to leak past the valve, even when the valve is closed. For these reasons, all components of the valve assembly must be thoroughly inspected, at times replaced, or reconditioned as part of any top-end engine inspection or rebuild.

Valve Inspection

Before a thorough visual inspection can be performed, the valves must be removed and cleaned. Removal of a valve requires the use of a spring compression tool to compress the valve spring to allow you to remove the valve keepers (Figure 12-37).

Anything that comes in contact with a valve (for instance oil, gas, and carbon) tends to get baked onto the valve surface. This can sometimes build up to create large carbon deposits on the valves (Figure 12-38).

Ordinary cleaning solvents may not have the ability to remove all of the carbon on the valve. In most cases, the best way to clean a valve is to use a wire brush. This ensures the removal of any buildup. An ordinary handheld wire brush could be used, but a wire brush on a bench grinder will

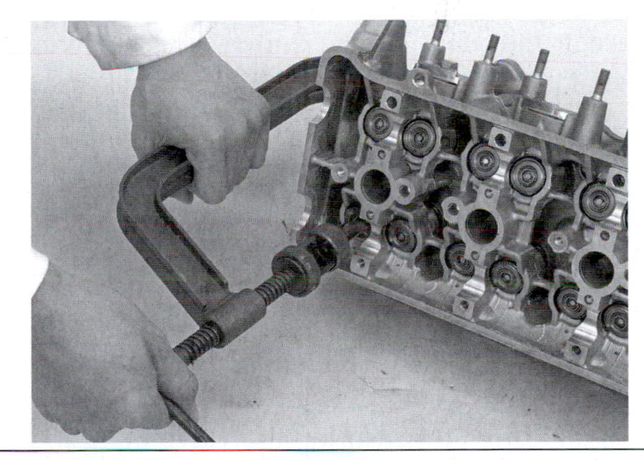

Figure 12-37 A typical valve spring compressor is shown here. Copyright by American Honda Motor Co., Inc. and reprinted with permission.

Figure 12-38 An example of a valve with excessive carbon buildup.

make the job much easier. After you have cleaned the valves with the wire brush, wash them in cleaning solvent to remove any leftover dirt particles.

Now you can begin the actual visual inspection of the valve. Visually inspect both the intake and exhaust valves. This includes inspecting for signs of physical damage and determining if each valve is sized within specifications. Valves can be damaged as the result of several conditions. Most often, though, valves are damaged by excessive heat (Figure 12-39).

If a valve becomes overheated, its edges can melt or its head can crack. If the damage is severe enough, pieces of the valve can actually break off. Common types of valve damage are illustrated in Figure 12-40. If you notice any of these types of damage, you must replace the valve.

Inspect the valve margin to detect any signs of distortion (Figure 12-41). The margin is the area between the valve's head and the line where the valve face begins. The valve margin is usually measured with a small ruler or with a Vernier caliper and is very important as it helps the valve to withstand the heat in the combustion chamber. If the valve margin is too small, the valve will possibly crack or burn through. When you are checking valve margins, always remember to verify the manufacturer's specifications for the motorcycle engine on which you're working.

In addition to measuring the valve margin, you should measure the valve stem, which is the part extending down from the valve's head (Figure 12-42). An outside micrometer is used to measure the valve stem's diameter. It's a good idea to measure the valve stem at the top, middle, and bottom. The diameters at all three of these locations should match the manufacturer's specification for the stem's diameter.

The valve springs must also be measured for their free length to ensure there's enough seat pressure placed on the valve itself. This measurement is done using a Vernier caliper, as illustrated in Figure 12-43.

Pay close attention to the valve spring coils. In most cases they are wound in a progressive fashion meaning that the coils are closer together on one end as compared to the other. This allows for a lower rate of pressure to start the opening of the

Figure 12-39 A valve burnt from excessive heat.

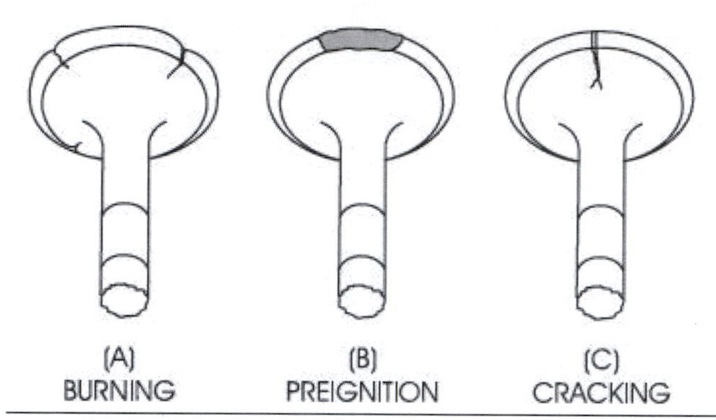

Figure 12-40 Common types of valve wear.

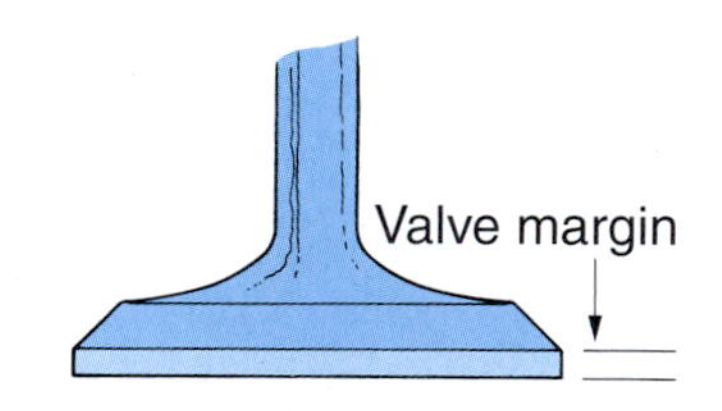

Figure 12-41 The valve margin is a critical part of the valve and is measured with a Vernier caliper or small machinist ruler.

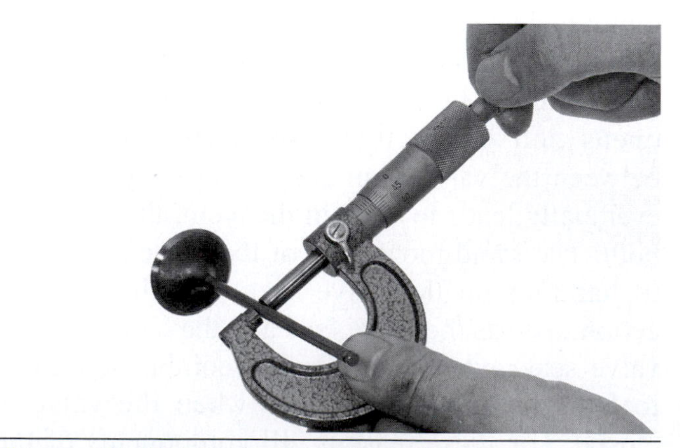

Figure 12-42 Measuring the valve stem with an outside micrometer. Copyright by American Honda Motor Co., Inc. and reprinted with permission.

valve while increasing pressure as the valve is opened to its full open position. In most cases, the valve spring coils that are closer together will be installed on the surface of the cylinder head (Figure 12-44).

Refacing Valves

If you inspect the valves and find that they're in good condition, they can be reused in the engine. However, all valves will eventually experience wear and distortion from use. Valve resurfacing is not common on motorcycle engines as most all modern motorcycle exhaust valves and many intake valves utilize a coating called **stellite** on the valve face that is used to harden them to increase longevity. Stellite is a very hard alloy of cobalt and chromium with cobalt as the principal ingredient. It is used to make cutting tools and for surfaces subject to heavy wear and is also very resistant to corrosion. Stellite alloys are so hard that they are very difficult to machine, and therefore, valves that use it are very expensive. Typically a valve using stellite alloy will have only a very thin coating of it on the face as well as on the tip of the valve. Stellite alloys also tend to have extremely high melting points due to the cobalt and chromium content that makes the material very useful for exhaust valves on today's high-performance engines. Because of use, refacing a valve with stellite should be carefully considered as a valve that has the hard stellite coating ground off will wear considerably faster than replacing the valve.

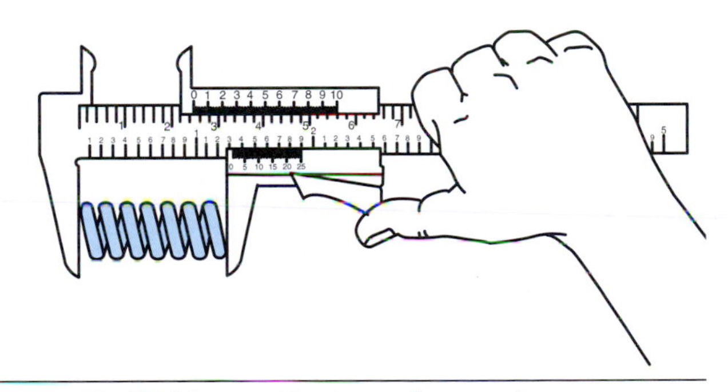

Figure 12-43 Measuring a valve spring.

Figure 12-44 The valve spring should be installed into the head with the compressed coils facing the head. Copyright by American Honda Motor Co., Inc. and reprinted with permission.

The process of reconditioning the face of the valve is commonly called valve grinding, or refacing. A machine like the one in Figure 12-45 is used for this process.

It cannot be mentioned enough that most modern engines come with stellite-coated valves and, therefore, shouldn't be resurfaced. If these valves show excess wear, they should be replaced.

Inspecting Valve Guides

Now that the valves have been inspected and deemed acceptable, we can move on to the valve guides. These components of the valve assembly are used to properly position the valves in the engine and to guide the valves as they move up and down. The valve guide's job is difficult. Not only must it keep the valve in position, but it must also allow the valve to move freely up and down and dissipate extreme heat (Figure 12-46). The extreme heat around the valve guide makes it difficult for oil to properly lubricate the guide. If insufficient oil is available for lubrication, excessive friction rapidly wears down the guide. If more oil is present than is needed, the excess oil becomes baked onto the valve stem. The baked-on oil can build up and block the valve openings.

Valve guide wear is a phrase referring to the amount of clearance between the valve stem and the valve guide. Keep in mind that the valve opens

Figure 12-45 A typical valve re-facing machine is shown.

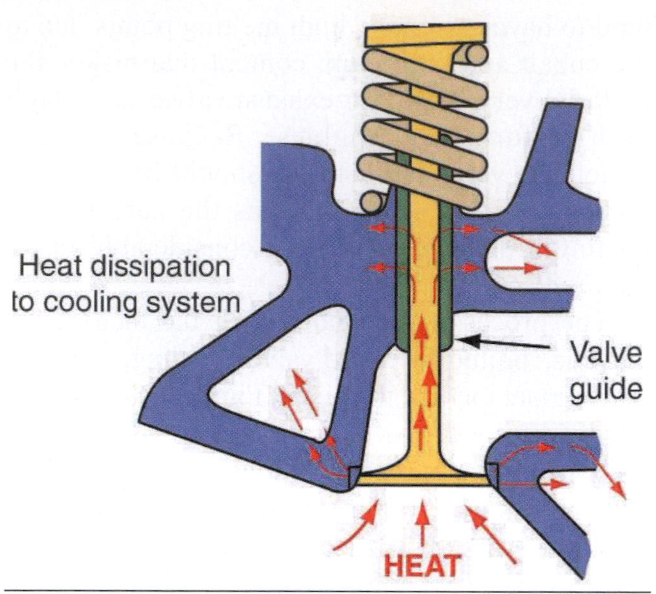

Figure 12-46 The valve guide helps to dissipate heat from the valve.

and closes thousands of times each minute. This leads to wear in the valve guide. As the valve guide wears, the valve begins to move slightly side to side as it opens and closes. This side-to-side movement, if excessive, can cause the valve to seat improperly and thus fail to completely seal the cylinder and in extreme cases break a valve (Figure 12-47). For this reason, the guides must be checked and replaced if they are found to be worn beyond the manufacturer's specifications.

Valve guide wear is determined by comparing the measurement of the inside diameter of the guide to the outside diameter of the valve stem. Because the inside diameter of the guide is quite small, a small-bore gauge is required to measure that dimension (Figure 12-48). As discussed earlier, a typical outside micrometer can be used to measure the valve stem's diameter. The stem diameter is then subtracted from the guide diameter to find the clearance between the stem and guide. Check the guide inside diameter in three locations: top, middle, and bottom of the guide as the guide will tend to wear more on the top and bottom (Figure 12-49). Finally, the calculated clearance is compared with the manufacturer's specifications given in the service manual.

Replacing Valve Guides

Generally, valve guides are made of soft metals, such as bronze or cast iron, to reduce the amount

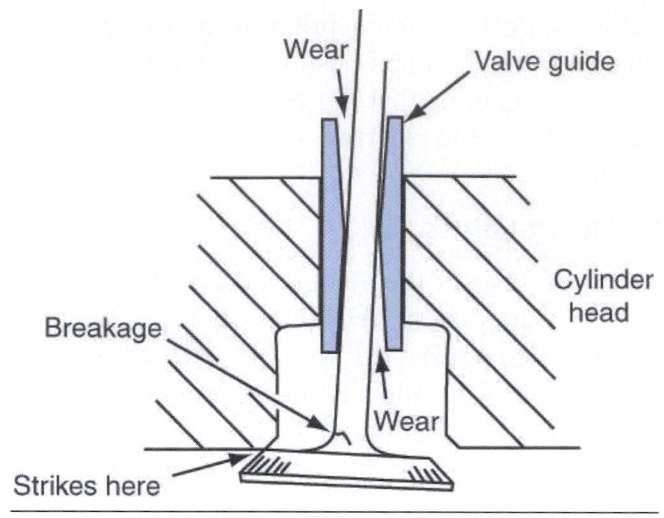

Figure 12-47 An exaggerated illustration of a worn valve guide.

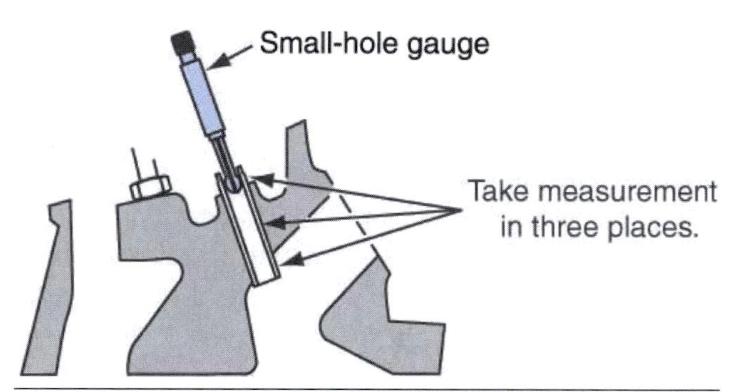

Figure 12-48 Measure the guide inside diameter in three locations using a small-hole gauge.

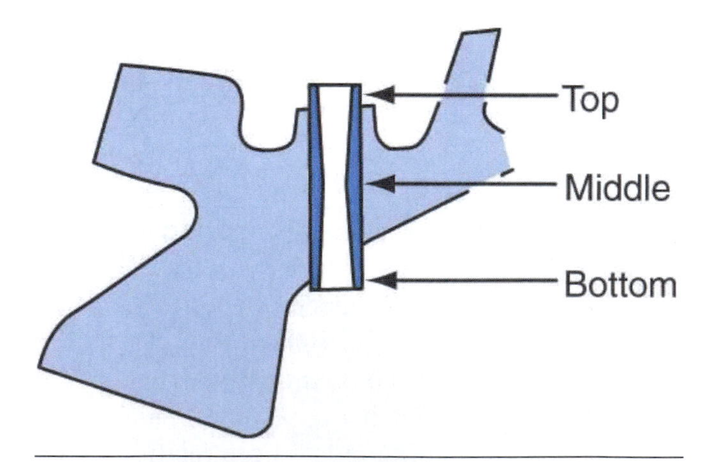

Figure 12-49 Most wear on a valve guide occurs at the top and the bottom of the guide.

of friction created by the moving valve stem. Worn valve guides can be removed with a driver and a ball peen hammer and a special driver tool (Figure 12-50). These tools are available from the motorcycle manufacturer or a specialty tool maker.

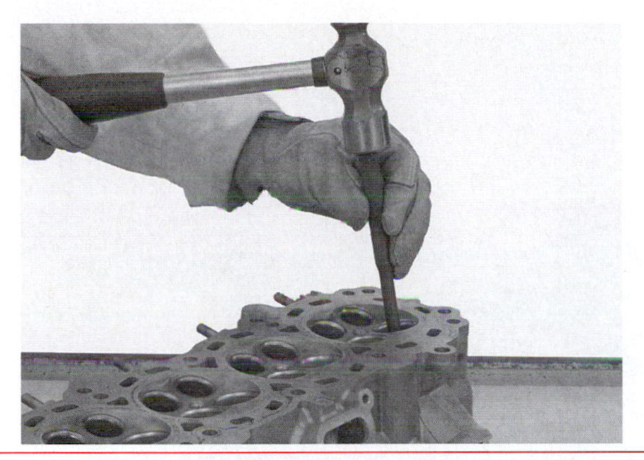

Figure 12-50 Driving the guide out from the combustion chamber side of the cylinder head. Copyright by American Honda Motor Co., Inc. and reprinted with permission.

Figure 12-51 Installing a new guide from the valve spring side of the cylinder head. Copyright by American Honda Motor Co., Inc. and reprinted with permission.

When driving out the old valve guide, be sure that the cylinder head is supported so it won't move. A few small blocks of wood under the cylinder head provide the proper support. After you've obtained the proper size driver, place it on the valve guide from the combustion side of the cylinder head. Then, use a ball peen hammer or press to knock or push out the guide.

The new valve guide must fit very tightly in the guide bore. Remember that when metals are chilled, they contract (shrink), and when heated, they expand (swell). Therefore, to make it easier to insert the valve guide, place the guide in a freezer for about an hour. The cold temperature causes the valve guide to shrink. Also, you may want to heat the cylinder head on a hot plate to allow the head to expand. When the guide has cooled off and the head has heated up, insert the guide into the cylinder head. A special driver is used again with a ball peen hammer to install the new guide into the cylinder head (Figure 12-51), but this time from the top of the cylinder head instead of from the combustion chamber side as when the guide is removed. Because the cold guide has shrunk somewhat and the cylinder head has been heated, the guide should fit into the guide bore relatively easily. Be sure to follow the manufacturer's instructions when installing a new valve guide as there are different procedures required from one manufacturer to another.

After you've replaced the guides, you'll need to ream out the newly inserted valve guide to meet specifications. A new guide has a slightly smaller inside diameter than is necessary. After the guide is installed, its inside diameter must be enlarged to be slightly larger than the valve stem. The hole in the guide is enlarged using a **reamer**. A reamer is a long, round cutting tool with cutting edges along its length. The tool operates much like a drill bit. Unlike a drill bit, however, a reamer doesn't cut on its end; it can't be used to actually drill a hole in a piece of metal. The cutting surfaces of a reamer are along its sides. The tool is used to remove material only along the inside surface of an already existing hole.

To ream out a valve guide, the reamer is inserted into the hole and turned clockwise until it penetrates the entire length of the guide (Figure 12-52). Because of the typical design of the reamer's cutting edges, the tool should always be turned in the clockwise direction (turning it in the opposite direction will dull the edges). Even when you are backing the reamer out of the valve guide, you should continue turning it in the clockwise direction.

The appropriate inside diameter of a valve guide depends on the size of the valve stem. The service manual for the engine specifies the proper diameter. This allows sufficient clearance for the stem to move through the guide as the valve opens and closes. After a valve guide has been reamed, any metal particles should be removed with compressed air. Then the area should be washed with solvent. After a valve guide is replaced you must re-cut or recondition the valve seats.

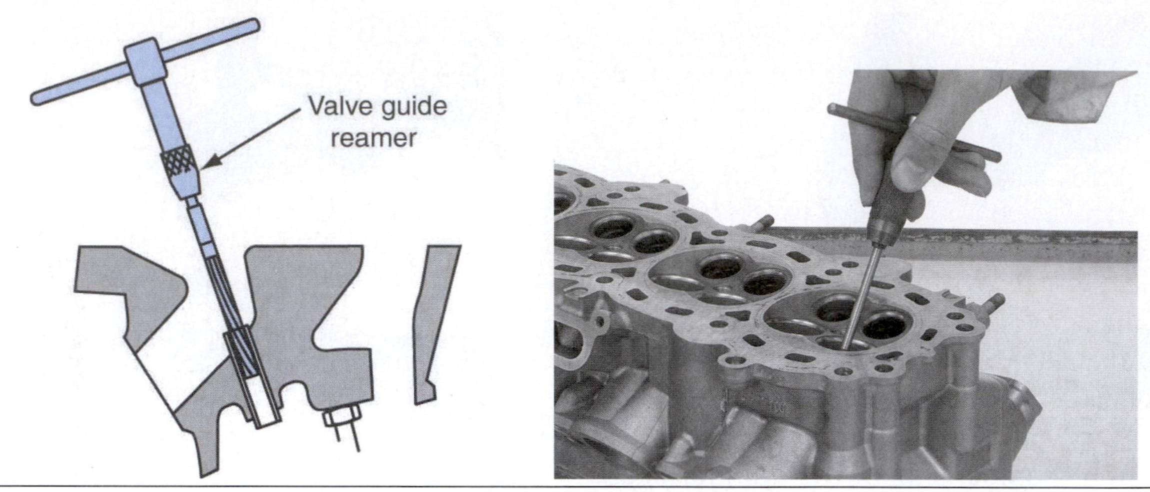

Figure 12-52 Using a valve guide reamer. Copyright by American Honda Motor Co., Inc. and reprinted with permission.

Reconditioning Valve Seats

A valve seat is the part of the cylinder head that mates with the valve face (Figure 12-53A). Usually, a worn valve seat can be reconditioned to get it back into shape (Figure 12-53B). The seal formed by the valve seat's precise fit with the valve face prevents leakage from the cylinder when a valve is closed. Due to the seat's location near the combustion chamber, a valve seat, like the valve itself, must be able to withstand high temperatures. A valve seat must also be able to conduct the heat from the closed valve and dissipate it to the engine's cooling system. If a valve seat did not help dissipate the valve's heat, the valve would get so hot it would simply begin to melt.

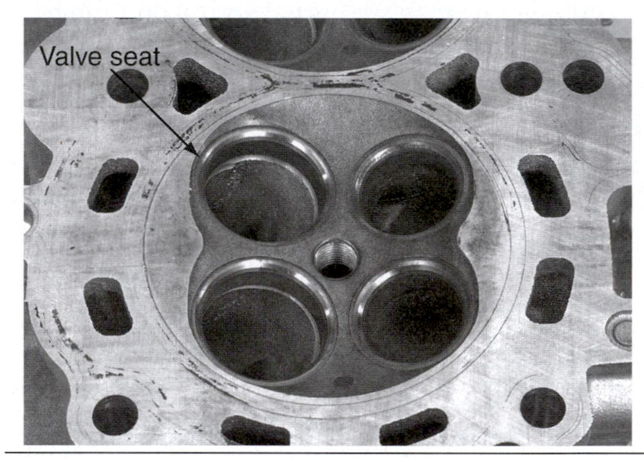

Figure 12-53A The valve seat is part of the cylinder head. Copyright by American Honda Motor Co., Inc. and reprinted with permission.

Melting of a valve is often referred to as burning (Figure 12-39). Most often, the exhaust valves are the valves that burn. Exhaust valves get very hot from passing exhaust gases. An intake valve doesn't get quite as hot because the incoming air-and-fuel mixture tends to slightly cool it. When a valve closes, the valve face fits closely into the valve seat. Accordingly, the heat from the valve head is passed into the valve seat. From there, the heat can be dissipated to the engine's cooling system to the air in an air-cooled engine, or to the coolant in a liquid-cooled engine. The valve seat's ability to dissipate heat is just as important as its ability to provide a proper seal.

Figure 12-53B A worn valve seat will not seal the combustion chamber properly.

In most engines, valve seats are made of a very hard steel alloy. In an aluminum cylinder head, valve seats are usually in the form of inserts pressed into place by the manufacturer. Because of the extreme heat under which the valve seats operate, the seats, like the valves, become distorted and worn out over time. When the valve face or the valve seat becomes distorted, the sealing surfaces no longer match up; therefore, the valve doesn't seal completely when it's closed. For this reason, valve seats normally need to be refinished during the rebuilding process. The refinishing process, called valve seat refacing, restores the valve seat to a perfectly round shape with a smooth sealing surface. Furthermore, the valve seat is beveled to match the angle of the valve face. By refinishing the valve seat, you ensure that the valve forms a proper seal when closed.

While there are various types of cutting tools available, the most common hand tool used to refinish a valve seat is pictured in Figure 12-54. The appropriate angle for a valve seat is found in the service manual for the engine. The equipment for valve seat cutting is normally available from specialty tool manufacturers.

The valve seat reconditioning tool uses a pilot to position the cutting device and to ensure that the tool remains centered properly. The pilot is simply a round piece of metal that fits tightly into the valve guide. The cutting device has a hole in its center that fits over the end of the pilot. The pilot can thus hold the tool centered in the valve guide. Because a pilot must be inserted into the valve guide to refinish the seat, the valve guide must be in proper condition before the seat is refinished.

After the pilot is inserted into the valve guide, the cutting device is placed over the pilot and into contact with the valve seat (Figure 12-55). The cutting tool is then rotated. The rotation of the tool removes metal from the valve seat and refinishes its surface.

There are normally three cuts made when reconditioning a valve seat (Figure 12-56). The angle that is cut first is considered to be personal preference by many. The first cut cleans and reconditions the actual valve seat area and this cut is normally 45 degrees. The second and third cuts are for the area below and above the valve seat. Normally angles are cut between 60 and 75 degrees for the area below the seat and between 15 and 30 degrees for the area above the seat. Some machinists prefer to cut the 45-degree angle last instead of first. You should note that the seat cut is usually about one degree different than the valve itself. This causes an interference that allows for correct mating of the surfaces once the engine is reassembled and started (Figure 12-57).

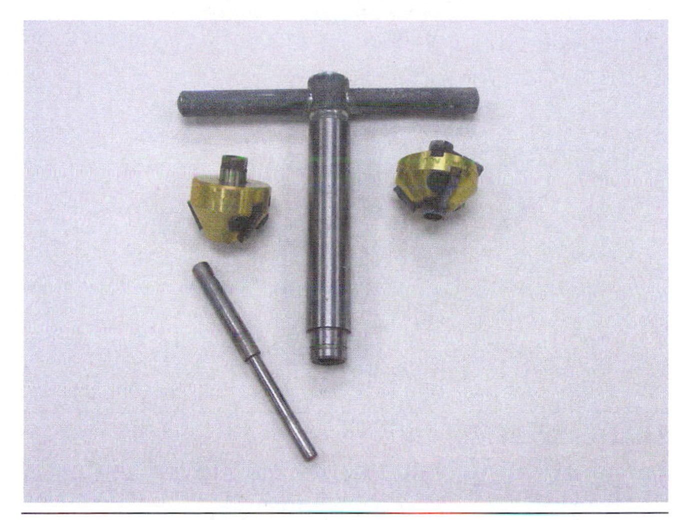

Figure 12-54 A typical valve seat cutting tool with its various components is shown here.

Figure 12-55 A valve seat cutting tool in use.

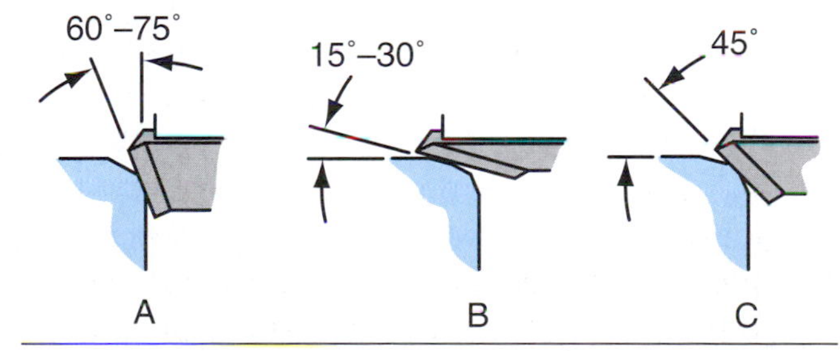

Figure 12-56 Typically, three cuts are made to a valve seat.

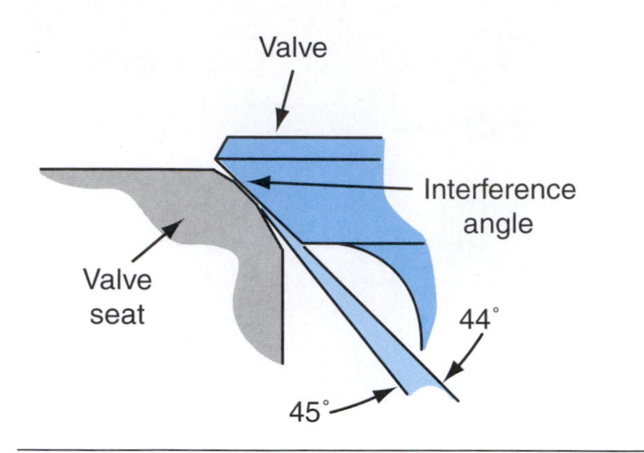

Figure 12-57 An interference fit is made when reconditioning valve seats to allow for better sealing of the valve and valve seat.

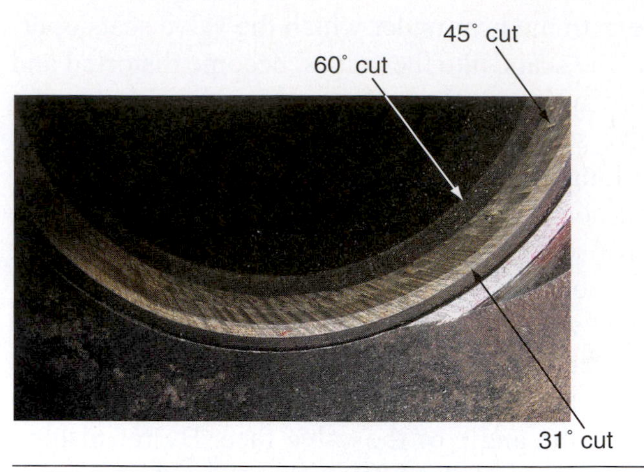

Figure 12-58 You can clearly see the three cuts of the valve seat in this image.

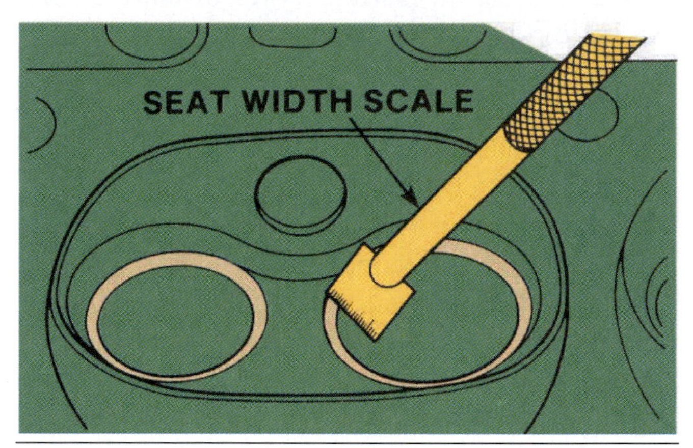

Figure 12-59 A seat width scale is illustrated here.

Metal is removed from the valve seat until a smooth, uniform surface appears. At first, the tool may cut in only a few spots due to the distortion of the valve seat, but as the tool continues rotating, metal continues to be removed until the tool cuts evenly at all points all the way around the valve seat.

The finished valve seat should look like the one illustrated in Figure 12-58. Note the different angles at the three different locations in the figure. The middle bevel (45 degrees) is the actual seat that makes contact with the valve face. The width of the valve seat must now be measured and checked against specifications. A seat width scale (Figure 12-59), machinist's ruler, or a Vernier caliper can be used to measure this dimension. Making the seat too narrow prevents the valve from transferring enough heat and will quickly wear or burn the valve. Check the manufacturer's service manual for exact specifications.

The valve seat can be cut incorrectly in a variety of ways as seen in Figure 12-60. After cutting the seat and measuring its width, you may find the valve seat is too wide or too narrow or too high or too low. The width can be decreased by partially re-cutting the seat with either the top cutter or bottom cutter.

So, how do we know if we should cut material off of the top or bottom of the seat? You can decide based on where the valve face makes contact with its seat as it should be centered on the valve face (Figure 12-61). At the same time, you can check on whether the valve face makes contact all the way around its seat. It's very important that you finish the job with a proper and complete seal between the face and the seat.

To decide which end of the seat's width you should cut, you can use a technique that involves the application of a blue dye called **Prussian blue**. This is a special blue dye that identifies the contact between the valve face and the valve seat and can be purchased from most automotive supply stores. First, remove the cutting tool and pilot from the valve guide. Then, place a coating of the blue dye on the valve face. Insert the valve into the guide and press it in until it firmly contacts the valve seat. When the face is in contact, apply a slight amount of pressure and rotate the valve one-quarter turn in the seat. Then, remove the valve and look carefully at the valve face. The dye mark left on the valve seat indicates exactly where the valve face contacts the seat.

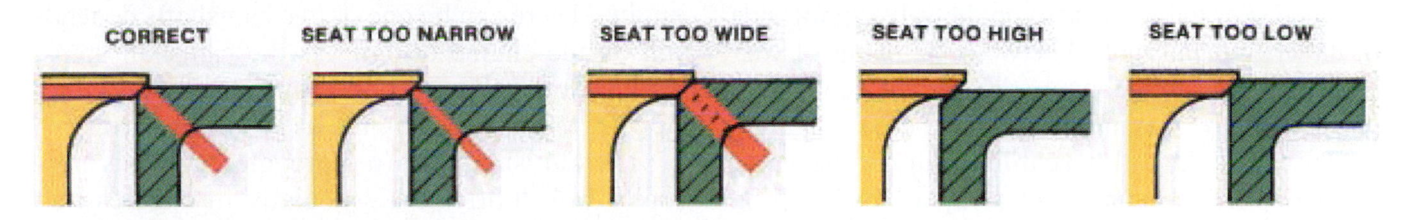

Figure 12-60 Different cuts to the valve seat can cause the valve to seat incorrectly.

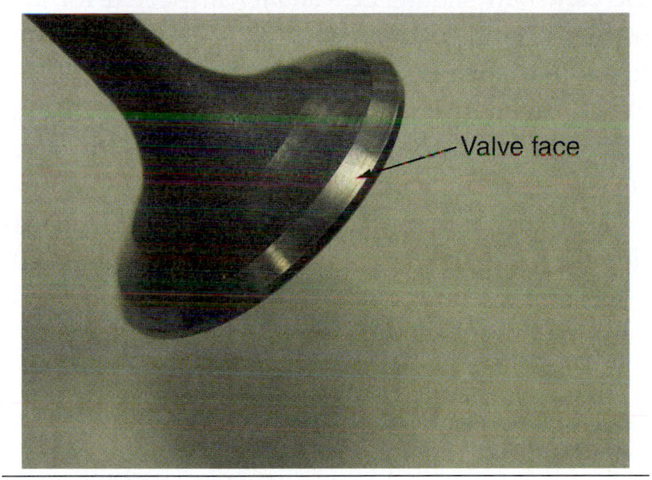

Figure 12-61 The valve seat should be centered with the face of the valve.

If the contact area appears to be closer to the bottom of the valve seat, you should narrow the seat from the top, thus helping to center the contact area. Similarly, if the contact area is closer to the top of the valve seat, you can center the contact area by narrowing the seat from the bottom.

Valve Lapping

Valve lapping is the process of mating the valve and the seat together to ensure a complete fit between the valve and the valve seat. Valve lapping produces the closest possible fit between the valve face and the valve seat. As a general rule, valves should be lapped to the seats any time the valves have been removed from the engine, even if they appear to be in good shape and you don't plan on reconditioning the seats. Some manufacturers recommend that brand new valves should be lapped before installing them into an engine to ensure a perfect seal. There are certain conditions when valves should not be lapped such as those made of certain materials like titanium. In these cases the manufacturer's service manual will point out not to lap these valves.

Valves are lapped using a grinding paste, or lapping compound, a substance that feels a lot like ordinary toothpaste, but contains fine, abrasive grains. When the compound is rubbed onto metal, the abrasive grains smooth the metal's surface. The paste is used with all four-stroke engines. Lapping compound is a common product that can usually be purchased from a local auto parts store. The compound is available in versions with grains of varying abrasives. Usually, a fine-grain compound is used on motorcycle engines.

To begin lapping the valves, apply a thin coating of lapping compound to the face of a valve. When you've covered the contact area, insert the valve into the valve guide and push it down until it makes contact with the valve seat. When installed, each valve rotates within its own seat. Remember that the abrasive lapping compound is between the valve face and the valve seat. Therefore, when the valve is rotated, the abrasives in the compound wear away the surfaces slightly, thereby mating them to one another.

A valve lapping stick is a tool sometimes used to rotate the valves. The lapping stick consists simply of a round wooden or plastic shaft with a suction cup on the end. The suction cup is attached to the head of the valve. To help the suction cup stick better, many technicians moisten the cup slightly before attaching it. After you've attached the lapping stick, you can rotate the valve back and forth by spinning the shaft of the tool between the palms of your hands (Figure 12-62). While rolling the shaft back and forth, apply a moderate amount of downward pressure. This helps the lapping compound to mate the valve and the seat together. Many of today's engines use valves that are too small in diameter to use a lapping stick. In these cases, a small piece of rubber hose can be installed over the valve stem on the valve spring side of the engine and rubbed between the palms of your hands (akin to the lapping stick except now gently lifting the valve against the cylinder head).

To check the valve seating, remove the valve from the engine and clean away all the lapping compound using solvent and a clean cloth. When the valve is clean, apply a thin coat of Prussian blue dye to the valve face. Then, insert the valve back into the valve guide. Apply a slight downward pressure with your thumb, and rotate the valve slightly. Remove the valve and observe the valve seat. If the blue dye is evenly distributed around the seat, the valve has been properly lapped. If the dye is distributed unevenly around the seat, more lapping compound should be applied and the valve should be re-lapped or you should consider reconditioning the valve seat.

After all of the valves have been appropriately lapped, the valves and their seats should be thoroughly cleaned with solvent and then with soap and water. This removes any leftover lapping compound. Remember that lapping compound is abrasive; if it's allowed to get into the working engine, it may do serious harm to the bearings and other vital engine parts. You can test the seal of the valves by inserting them and pouring a liquid (like cleaning solvent) into the port and observe that there is no leakage past the valve.

Inspecting the Camshaft

The camshaft is the component that controls the opening and closing of the valves in a four-stroke engine. There can be one or two camshafts depending on the engine design. As the camshaft spins, the cam lobes (Figure 12-63) open the valves either directly or indirectly by the use of rocker arms.

The camshaft in a motorcycle engine is normally well lubricated, especially in engines that use a high-pressure lubrication system. When you're rebuilding a motorcycle four-stroke engine, the camshaft should be visually inspected for any signs of damage. Specifically, look for any cam lobes that appear to have surface damage. Also, check the camshaft's ends that are supported in bearings. Look for any signs of scoring or other surface damage.

Motorcycle manufacturers provide a specification for the diameters of each part of the camshaft. Measure those areas using a micrometer and check your measurements against the specifications.

Inspecting the Camshaft Drive System

There are several types of camshaft drive systems with the most common being belt, gear, and chain. The belt and chain drive systems utilize a tensioning device that varies from manufacturer to manufacturer. The chain camshaft drive system is most popular and uses sprockets on both the camshaft and crankshaft (Figure 12-64). You should pay close attention to the tensioning system on all drive systems, as they can be prone to wear as the tension on the chain varies greatly due to the constantly changing engine rpm.

Figure 12-62 Using a valve lapping tool. Copyright by American Honda Motor Co., Inc. and reprinted with permission.

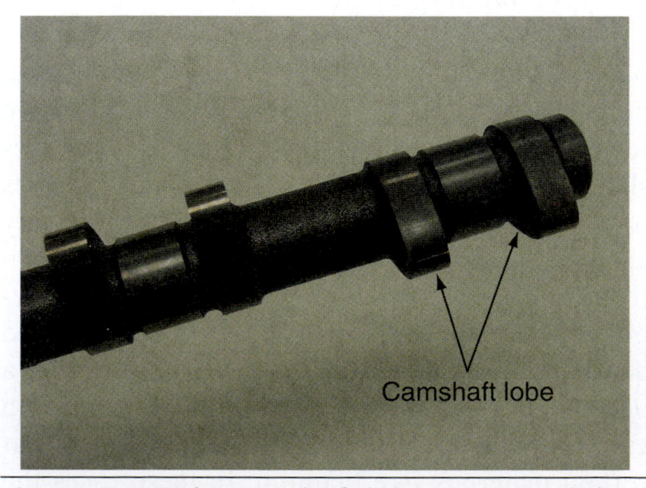

Figure 12-63 The camshaft lobes open and close the valves.

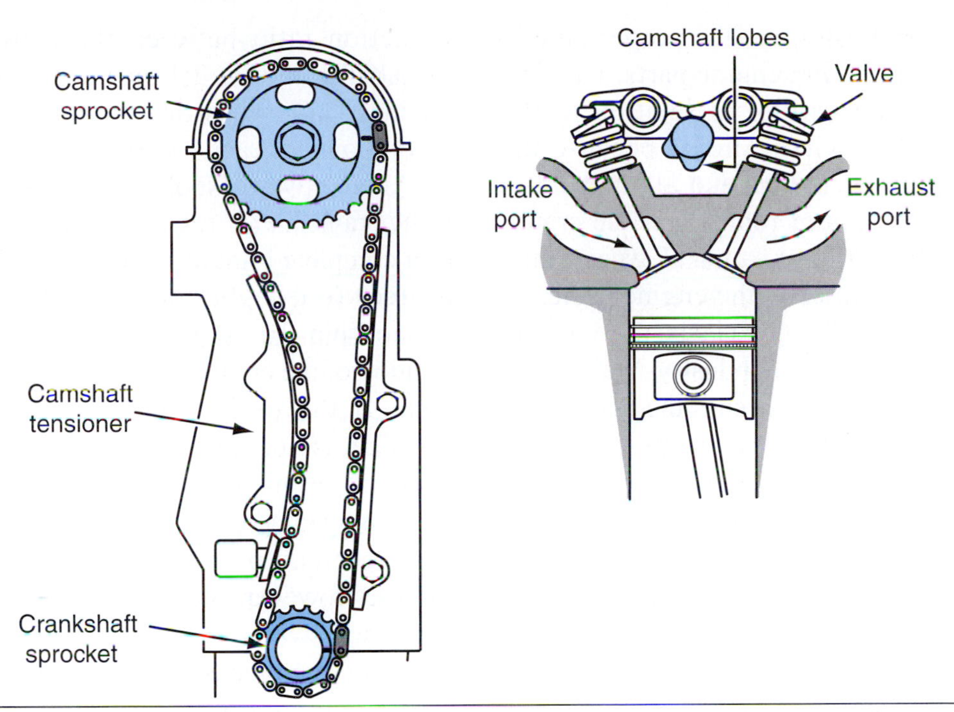

Figure 12-64 The chain camshaft drive system is the most popular valve train drive system used on four-stroke motorcycle engines.

FOUR-STROKE TOP-END REASSEMBLY

While it is important to use the manufacturer's service manual for any work on a motorcycle engine, here are a few common tips to use during the reassembly process of any four-stroke engine.

Installing the Piston

Place a clean rag over the base of the cylinder to prevent foreign objects from falling into the crankcase. Install the piston pin through the piston, making sure the piston directional arrow is facing in the correct direction. Always install new piston pin retaining rings.

Installing the Cylinder

Before installing the cylinder, clean all surfaces to remove old gasket material. Lightly oil and compress the piston rings. Lower the cylinder over the piston. Make sure you pull the cam chain out of the crankcase cavity to prevent the chain from being caught between the cylinder and the crankcase. Push the cylinder down tightly to the crankcase joint. Install the cam chain tensioner guide into the cylinder, as well as the cylinder head alignment pins. To complete the installation procedure for the cylinder, install the cylinder head gasket onto the top of the cylinder.

Installing the Valves

As mentioned before in this chapter, valve springs must be installed so that the narrow pitch of the spring sits on the surface of the head. Always replace any valve stem oil seals when the valves are removed to prevent any chance of oil seal-related problems. Good valve springs provide adequate seat pressure to allow a tight fit between the valve face and the valve seat to seal the combustion chamber. The proper valve springs also prevent the valve from bouncing on its return to the seat (especially at high engine speeds), thus losing cylinder pressure. Valves are held in place with valve keepers, which should be installed using the manufacturer's instructions. To ensure that the valve keeper is installed correctly it is a common practice to gently tap on the valve after installation to ensure that the keepers are in place (Figure 12-65).

Installing the Camshaft

You should be aware that all four-stroke engines must be adjusted, or timed, so that each rotating part is in the proper position at the proper time in relation to the other moving engine parts. The timing of the camshaft rotation, in relation to the crankshaft rotation, is vital. This is because the camshaft must open the valves and allow them to close at specific degrees of rotation of the crankshaft. The induction and exhaust of gases must take place at specific times in the engine cycle.

The overhead camshaft engine is the most widely used engine in the motorcycle industry. We'll focus on this type of engine as we discuss the installation of the camshaft. We will discuss common and general procedures for both the single-overhead cam (SOHC) and double-overhead cam (DOHC) designs.

Camshaft and Crankshaft Rotation Relationships

The timing for the valves to open is determined by the position of the camshaft. This position is indicated by degrees of crankshaft rotation. You should already know that there are 360 degrees in a circle. A quarter of a turn is, therefore, equivalent to 90 degrees; half a turn is equivalent to 180 degrees; a three-quarter turn is equivalent to 270 degrees. The valves must open and close at specified degrees of crankshaft rotation. The gear reduction ratio between the crankshaft and the camshaft is always 2:1 on a four-stroke motorcycle engine. This means that the crankshaft makes two revolutions for each single revolution of a camshaft.

Figure 12-65 Tapping on the valve after assembly helps to ensure that the valve keepers are in place. Copyright by American Honda Motor Co., Inc. and reprinted with permission.

Another way of looking at this relationship is that if the crankshaft is rotated 90 degrees (or one-fourth of a complete rotation), the camshaft must rotate 45 degrees (one-eighth of a complete rotation). This relationship must remain constant during a complete rotation of the engine or certain failure will be imminent. If the timing between the crankshaft and camshaft is not correct, valves will hit onto the piston causing engine failure.

Timing of the camshaft position in relation to the crankshaft position is done by aligning certain marks. However, there are various ways and various places where these marks may appear. One of the principal pieces of information you must have for every four-stroke engine is the exact location of the timing marks. Normally the marks appear on the alternator rotor and the cam sprocket. These marks might be lines, dots, letters, or a combination of these.

The alternator rotor is keyed to the end of the crankshaft. As a result of its positioning, the rotor can be used to indicate the position of the piston. It's also used in connection with timing the ignition system, which will be discussed in a later chapter. When you align the rotor with the correct spot as found in the appropriate service manual (Figure 12-66), on the engine crankcase the crankshaft will be at top-dead center (TDC), which will normally be indicated with a "T" mark. This is the mark used in the majority of engines but not on all. Be sure to check the manufacturer's service manual for the correct cam timing marks and procedures. There are also other marks on the rotor, which are used for ignition timing.

When you know the position of the crankshaft, you must get the correct corresponding position for the camshaft. Find the timing marks on the camshaft sprocket (Figure 12-67). The

Figure 12-66 The crankshaft rotor will usually have a mark for TDC to assist with timing the camshaft to the crankshaft. Copyright by American Honda Motor Co., Inc. and reprinted with permission.

Figure 12-67 Timing marks on an SOHC camshaft sprocket. Copyright by American Honda Motor Co., Inc. and reprinted with permission.

marks on the camshaft sprocket are usually aligned with a mark or a part of the cylinder head (Figure 12-68). As indicated earlier, there are other methods of timing. In a gear-driven overhead cam where you do not have a sprocket and chain, one tooth on each gear may be marked for timing alignment.

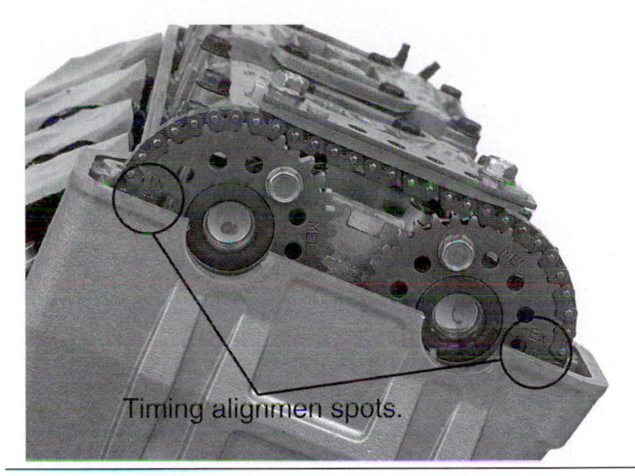

Figure 12-68 Note the line alignment with the cylinder head. Copyright by American Honda Motor Co., Inc. and reprinted with permission.

Checking and Adjusting Valve Clearance

It's important to check the clearance between camshaft and valve stem tip after any work is done on the valves. Since the space between the valve and camshaft may have changed by the reconditioning and lapping procedures, the clearances should be measured and compared with the manufacturer's specifications. If a valve clearance is incorrect, it needs to be adjusted. There are different adjustment procedures for different motorcycles that will be discussed in Chapter 19 covering maintenance procedures.

There are different adjustment methods used for different valve arrangements. Valve clearance is normally checked when the cylinder is at top-dead center (TDC). At the TDC position on the compression stroke, both the intake and exhaust valves for the cylinder should be completely closed. As a general rule, the valve clearance is greater on the exhaust valve than on the intake valve because the exhaust valve gets hotter, producing a greater expansion rate of metal. Valves are adjusted when the engine is at room temperature and specifications for valve clearance vary with each model of motorcycle.

Summary

- While a complete diagnostics may not be able to be confirmed until the machine is actually disassembled, when there is an engine-related problem, the technician must diagnose the condition. Diagnostics is the process of determining what's wrong when something isn't working properly by checking for symptoms. Symptoms are the outward, or visible, signs of a malfunction.
- The key components in a typical four-stroke engine top end include:
  - Cylinder head
  - Cylinder head gasket
  - Valve train components
  - Cylinder
  - Cylinder base gasket
  - Piston rings
  - Piston
- The two primary types of four-stroke engine top ends are the single-overhead cam (SOHC) and the double-overhead cam (DOHC).
- There are many components in the four-stroke top end that require careful inspection and measurement. A visual inspection is critical and proper measuring with the appropriate tools is critical to successfully completing a four-stroke engine top-end rebuild.

Chapter 12 Review Questions

1. Pistons used on most motorcycle four-stroke engines have _________ piston rings.
2. The difference between measurements taken on the same axis (X or Y) of a cylinder is called _________.
3. The process used to mate the valve and the seat together is called _________.
4. The _________ must be installed so the narrow pitch sits on the surface of the head.
5. The _________ valve usually has more clearance between it and the camshaft.
6. Valves are adjusted at _________ temperature.
7. The gear ratio between the crankshaft and the camshaft is _________ .
8. Camshafts are timed in relation to the _________ .
9. Cutaways are installed on the top of many four-stroke pistons to allow for _________ .
10. The valve seat is part of the
    a. valve spring
    b. cylinder head
    c. piston
    d. camshaft
11. The marks on the piston rings should always be installed facing
    a. toward the piston crown
    b. each other
    c. downward
    d. away from the piston crown
12. To measure piston ring end gap, use a _________ .
13. The most common cylinder head problem is damage to the _________ .
14. Refacing valves that are coated with stellite is
    a. a common practice in the motorcycle industry
    b. not recommended
    c. only occasionally needed
    d. better than replacing them

CHAPTER 13 Four-Stroke Engine Lower-End Inspection

Learning Objectives

When students have completed the study of this chapter and its laboratory activities they should be able to:

- Recognize common lower-end engine failures in a four-stroke motorcycle
- Identify the various components in a four-stroke engine lower-end assembly
- Understand the importance of inspecting the various components of the four-stroke lower end including the crankshaft and transmission for damage or wear
- Understand proper procedures for disassembling and inspecting the transmission in a typical four-stroke motorcycle engine
- Understand the proper procedures to measure the bearing surfaces of a single-piece crankshaft
- Recognize the importance of bench testing an engine prior to completing reassembly
- Understand the importance of properly breaking in a four-stroke motorcycle engine

Key Terms

Babbitt bearing
Connecting rod bearings
Crankshaft bearings
Engine seal
Oil pump
Plain bearing
Plastigage
Rotor-type pump
Trochiod-type pump

INTRODUCTION

While the lower-end components of a modern four-stroke motorcycle engine are very reliable they will occasionally need repairs. For this reason, it's important to understand how to disassemble the engine right down to its heart—the crankshaft. It is important to know how to inspect the various lower-end engine components, do any necessary repair work, and reassemble the engine correctly.

You may need to disassemble an engine to make needed repairs, or you may perform a disassembly as the second major step in a complete engine rebuild.

In Chapter 12 we used a Honda CRF230F (Figure 13-1) and a Honda CBR1000RR (Figure 13-2) as primary examples for a four-stroke motorcycle engine top-end inspection. Therefore, we will continue to use these machines for our discussion on bottom-end inspections due to the fact that both of these models do a good job of representing the majority of the types of engines you will see in the motorcycle industry. Of course, just as in Chapter 12, our explanations will be general in nature as other four-stroke motorcycles will vary somewhat, but the basic principles of inspection remain the same.

Most repairs of the motorcycle engine lower-end assembly will require that the engine be removed from the chassis. Therefore, be sure the malfunctioning component is located in the lower-end assembly before you remove the engine from its chassis and disassemble it. After all, you wouldn't want to go to all that effort only to find out that repairing the failure didn't require all that work. For instance, a slipping clutch would be located in the lower end of the engine but to do the needed repairs, you would not have to remove the engine from the frame of the motorcycle.

Figure 13-1 A Honda CRF230F. Copyright by American Honda Motor Co., Inc. and reprinted with permission.

Figure 13-2 A Honda CBR1000RR. Copyright by American Honda Motor Co., Inc. and reprinted with permission.

REPAIR PROCEDURES

The procedures in this chapter are general in nature and not intended to be used for actual disassembly and repair. Their purpose is to familiarize you with the types of activities you'll encounter when working on a four-stroke motorcycle bottom end. Always refer to the appropriate motorcycle service manual for accurate specifications and proper repair procedures. The service manual contains all the information to do the job correctly, including detailed instructions for the specific make and model of motorcycle-required special tools and service tips. Above all, the service manual contains the appropriate safety information.

This chapter will cover the two most popular styles of motorcycle engines: a single-cylinder air-cooled engine and the multi-cylinder four-stroke engine. The majority of the procedures discussed will apply to all motorcycle four-stroke engines that you will work on.

This chapter will not, however, cover step-by-step procedures to disassemble an engine, as these steps vary from one motorcycle to another. This includes the removal and installation of the engine into the frame. You will want to rely on the appropriate service manual for these types of procedures. Keep in mind that some lower-end engine repairs may not require the removal of the top-end components so be sure that you are aware of what exactly will have to be removed from the engine before starting any repairs. Identifying these areas of repair will be covered later in this chapter.

As mentioned in previous chapters, be sure that the motorcycle is clean before you begin any disassembly work. A water-soluble degreaser should be used according to the manufacturer's instructions. Remember that dirt or foreign particles can cause serious problems if allowed to get into the engine's internal working parts.

COMMON LOWER-END ENGINE FAILURES

In the lower-end engine assembly, malfunctions most commonly occur with the engine seals, crankshaft and connecting rod bearings, and transmissions. The following brief descriptions explain these potential trouble areas. Being alert to the possibility of problems in one section of an engine, while you are performing repairs on another section, will help you become a better motorcycle technician.

Leaking Engine Seals

An **engine seal** is designed to prevent oil from getting out of a four-stroke motorcycle engine and leaking between the crankcases and a moving part like a transmission shaft (Figure 13-3).

A leaking engine oil seal will cause oil to leak out of the engine. Sometimes this will be a small spot of oil under the engine while the motorcycle is sitting and other times the seal may not leak unless the engine is running. In most cases, a leaking seal will not require that the engine's lower end be disassembled. But depending on the type of seal you may need to disassemble the crankcases to replace it. Always verify that the replacement of the faulty oil seal requires complete disassembly before tearing down the lower end. You can do this by referring to the appropriate service manual before you begin working on the engine.

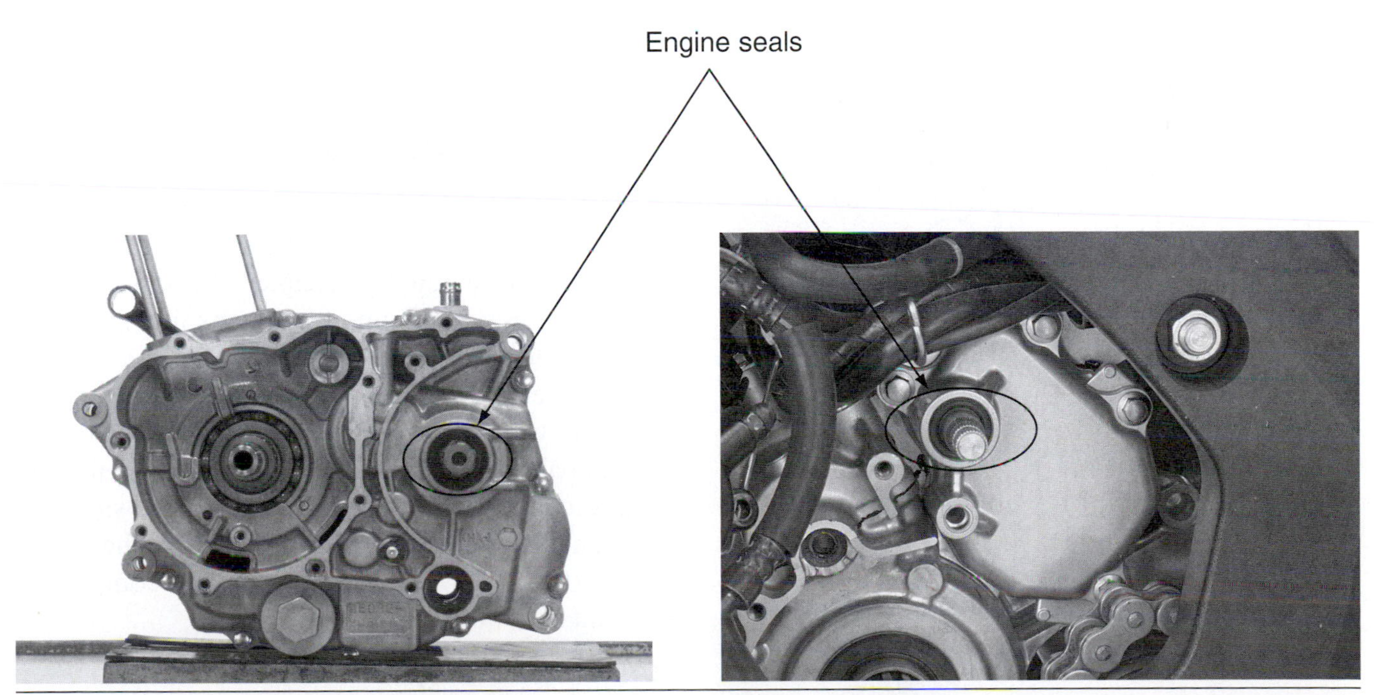

Figure 13-3 Engine seals are designed to prevent oil from getting out of a four-stroke motorcycle engine and leaking between the crankcases and a moving part like a transmission shaft. Copyright by American Honda Motor Co., Inc. and reprinted with permission.

Keep in mind that an oil leak may not always be due to a faulty seal. The cause of the leaking may be due to a scored or bent shaft.

Worn Crankshaft Bearings

Crankshaft bearings are used to connect the crankshaft assembly to the crankcase and require constant lubrication. Two types of bearings can be found on the four-stroke motorcycle engine depending on its design. A ball bearing is used on multi-piece crankshafts and a **Babbitt bearing**, also known as a plain bearing, will be found on motorcycle engines using a single-piece crankshaft. An illustration of these types of crankshafts can be seen in Figure 13-4. A bearing that is failing usually makes a rough growling sound. You may want to use a mechanic's stethoscope (Figure 13-5) to help pinpoint the location of a bad bearing. Bad bearings may cause excessive up-and-down movement of the crankshaft, or worse, prevent the crankshaft from rotating.

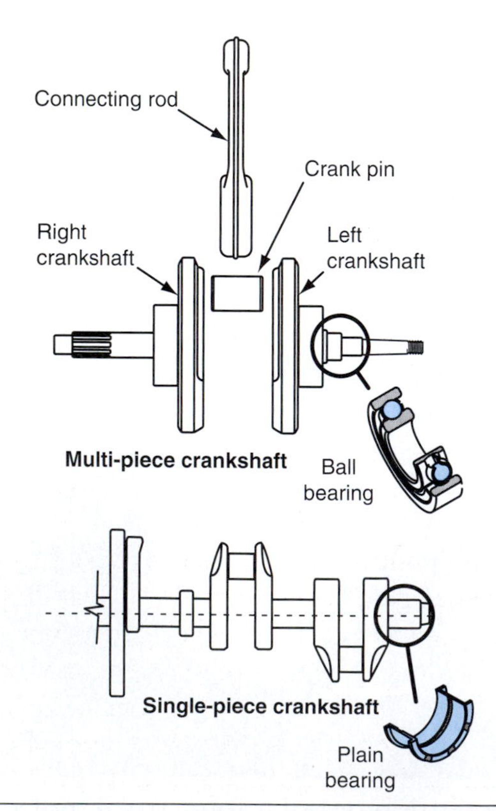

Figure 13-4 The different types of crankshafts found on a four-stroke engine are illustrated here.

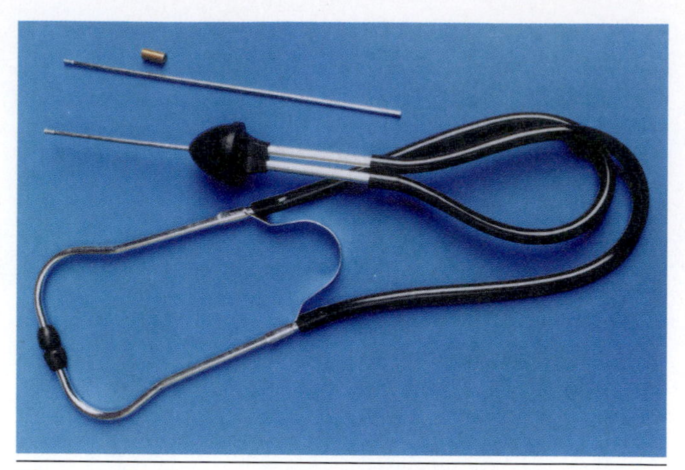

Figure 13-5 A mechanic's stethoscope can be used to find internal engine noises.

Worn Connecting Rod Bearings

Connecting rod bearings also come in two varieties on the typical four-stroke motorcycle engine. A roller type on the multi-piece crankshaft (Figure 13-4) and another plain bearing on the single-piece (Figure 13-6) allow the connecting rod(s) to pivot freely when the crankshaft turns.

Symptoms of a failed connecting rod bearing are knocks, vibrations, or an engine that cannot be turned over (seized up). Any of these symptoms necessitate the disassembly, inspection, and repair of the lower end of the engine.

Transmission Problems

Chapter 9 discussed how transmissions function in motorcycle engines; therefore, it may be a good time to review that chapter. We will compare the symptoms of common problems in a transmission with what we suspect could be wrong. Because each part in the transmission does a certain job, when a failure occurs you can usually tell which part is at fault by the symptoms. The following is a description of some common transmission malfunctions, hints on how to recognize them, and suggestions for the almost certain component at fault.

Difficult Shifting

When it is difficult to shift between gears while the motorcycle is moving and it's been determined that the clutch is not at fault, the cause is usually a

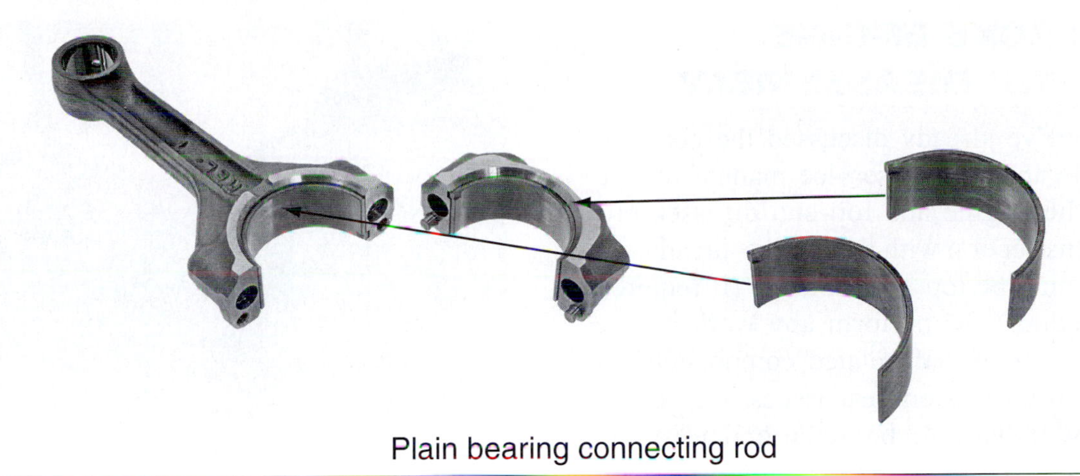

Figure 13-6 Babbitt-style bearings, better known as plain bearings, are used on the single-piece crankshaft. Copyright by American Honda Motor Co., Inc. and reprinted with permission.

bent or burnt shift fork. A damaged shift fork no longer fits properly in the grooves of the gear. This problem requires replacement of the shift fork. Replacement of the sliding gear to which the shift fork is attached is also likely necessary as well as the gear to which the sliding gear is engaged. Shift forks usually fail when put under extreme loads, such as shifting without using the clutch or forcing the transmission into gear while it's still under a load. If a shift fork is bent or burnt, it is visually noticeable by gouging or a discoloration on the shift fork ears. Difficult shifting can also be caused by a seized gear on a transmission shaft, which is generally caused by a lack of proper lubrication.

Inability to Shift Gears

Sometimes you'll find a machine that shifts into one gear, but not into the next. This problem is often caused by the shift return spring, which returns the shifting lever to its original position. Repair is usually possible by replacing the spring and, in most cases, won't require the complete disassembly of the engine. The spring is usually attached to the shift shaft assembly and is located behind or near the clutch assembly.

Strange Sounds

A low growling sound when the motorcycle is in gear and rolling usually indicates a bearing failure. When a bearing failure occurs, it may cause a transmission shaft to move slightly out of position. When this occurs, the gears don't mesh together properly and produce a low growling noise. In these cases, not only the bearing needs replacement, but many times the gears may also need to be replaced. You can distinguish this sound as it will only occur when the motorcycle is moving as compared to a crankshaft bearing that will be noisy anytime the engine is running, whether moving or not.

Clunking Noises

Another characteristic sound that indicates a transmission problem is an excessive clunking sound when the engine is in a particular gear while under a load. Normally this indicates broken teeth on one or more gears. This warrants a complete inspection of all transmission components, as the broken teeth normally damage other parts in the transmission.

Jumping Out of Gear

Jumping out of gear is usually caused by worn transmission gear dogs and/or slots. Gear dogs and slots are used to lock the two gears together. When dogs and slots become excessively rounded, the gears tend to slip out of the holes when the engine rpm increases. Thus, the engine "jumps" out of gear. The gears, as well as the shift forks, need replacement in this case. The shift forks become damaged due to the excessive pressure they encounter.

FOUR-STROKE ENGINE LOWER-END DISASSEMBLY

Because we've already discussed the need for the use of the appropriate service manual for the removal of the engine and top-end disassembly, we'll begin this section with the engine already out of the frame and the top end removed (if required for the procedure). To perform any work on the crankshaft and crankshaft-related components, as well as most transmission gear issues, the engine crankcases will have to be separated to allow access to these components. Be careful when separating the crankcase halves as there are often many different sized fasteners holding them together. *Never* use a hammer to attempt to separate the crankcase halves! Most crankcases require the use of a sealant when reassembling if a gasket is not required.

Figure 13-7 A vertical spilt-type crankcase requires the removal of the top end to separate the cases. Copyright by American Honda Motor Co., Inc. and reprinted with permission.

Separating the Crankcases

There are two types of crankcases found on the four-stroke motorcycle engine: the vertical crankcase, which is found on most single-cylinder engines and the horizontal crankcase, which is found on most multi-cylinder motorcycle engines.

Vertical Crankcase Separation

With the vertical crankcase engine design, the cylinder and cylinder head are attached to both sides of the crankcase assembly (Figure 13-7). Therefore, it is mandatory that the top end of the engine be removed before the engine crankcase halves can be separated (Figure 13-8).

Figure 13-8 A technician is separating a vertical crankcase here. Copyright by American Honda Motor Co., Inc. and reprinted with permission.

Horizontal Crankcase Separation

With the horizontal crankcase engine design, you are not required to remove the top-end components unless you're going to work with the crankshaft assembly (Figure 13-9). This is because the cylinder and cylinder head are attached to the top crankcase half. If, for example, the transmission requires repair, it's not necessary to remove the top end of the engine in most cases. If you do not have to remove the top-end components, you will save yourself a considerable amount of rebuild time.

You should take note that in some four-stroke motorcycle engines the transmission can also be removed without separating the crankcases. This type of transmission is known as a cassette-type transmission. A cassette transmission is removed out of the side of the crankcase as a complete unit (Figure 13-10) and is normally found in sport bikes used in racing applications where the changing of the actual gear ratios would be allowed for use at different race tracks.

Figure 13-9 The top end may be left installed for many horizontal-type four-stroke bottom-end engine repairs. Copyright by American Honda Motor Co., Inc. and reprinted with permission.

FOUR-STROKE ENGINE LOWER-END INSPECTION

After all the components have been removed from the separated crankcases, it's time to inspect each one individually to check for damage or wear. Depending on the reason for disassembly, you might think that you should look only for a particular problem while the engine is completely apart. This is far from being correct.

While the engine is disassembled, you should carefully inspect all components. Because complete disassembly of the engine isn't common practice, it's important not only to ensure that the job is done right the first time, but also to ensure that no existing problems or soon-to-be problems are present. Let us start this discussion by inspecting the components that are attached to the crankcases.

Inspecting Engine Crankcases

The engine crankcases should be closely inspected for cracks, loose-fitting bearings, and worn-out fastener-anchoring points. If there are stripped anchoring points, they may be repaired using a special tool and process that places inserts into the hole (Figure 13-11) making the stripped fastener point useful again. These tools are available at your local hardware store or tool supplier.

Inspecting Engine Seals

Engine seals are located on all shafts that rotate and are exposed to the outside atmosphere on a four-stroke engine (Figure 13-12). Inspect all seals to verify that they're in good condition. Ensure that

Cassette-type transmission removal

Figure 13-10 A cassette-type transmission allows removal of the transmission gears without separating the crankcases. Copyright by American Honda Motor Co., Inc. and reprinted with permission.

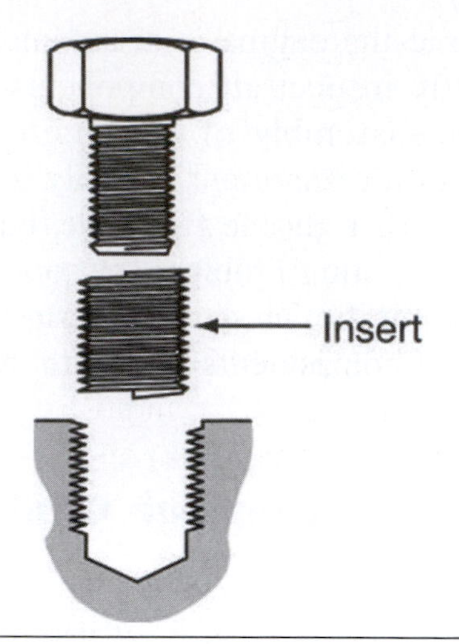

Figure 13-11 Inserts can be used to repair crankcases with stripped anchoring points.

they fit on the shafts properly. Inspect the seal lips for tears or rough surfaces. The rubber must be *live*. That is, the lip of the seal must be soft and springy. Most seals have a small coil spring that fits on the outer side of the seal lip. Be sure this spring is in place.

Remove seals by sliding them off of the shaft, tapping them out, or prying them out of the crankcase half. If you remove a seal, you should replace it with a new one. New seals are tapped into place with a mallet or a special seal installation tool (Figure 13-13). Be sure the new seal is installed evenly in the hole. Incorrect fitting of a seal will cause oil leaks. Be sure to install the seal into the case properly. Normally, the manufacturer's identification number is on the side away from the bearing to be sealed.

Inspecting Engine Bearings

The most popular bearing found in a motorcycle or ATV engine is the ball bearing. You can inspect the bearing race by hand while it's still mounted on a shaft (Figure 13-14) or if it remains in the crankcase half (Figure 13-15). Rotate the inner race by hand and inspect it for any abnormal noise or lack of smooth operation. Visually inspect the race and balls. If they show signs of wear, chips, cracks, or damage to the hard bearing surface, you must replace them.

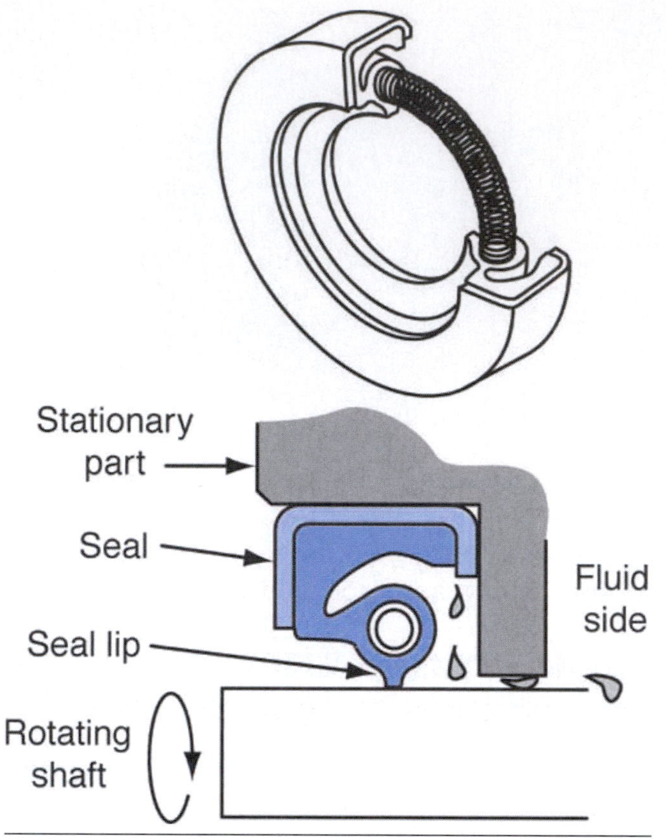

Figure 13-12 Seals keep oil in and debris out of the engine.

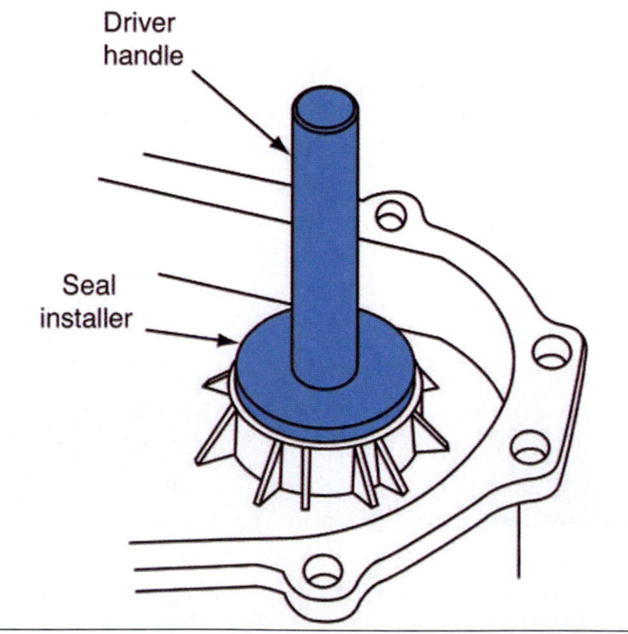

Figure 13-13 A typical seal installation tool is shown here.

Figure 13-14 Feeling for roughness on a bearing while it is mounted on a shaft.

Figure 13-15 A bearing being checked for roughness while in the crankcase. Copyright by American Honda Motor Co., Inc. and reprinted with permission.

Replace the bearing if there's any doubt that it's not in good condition. To replace most bearings, first remove the old bearings with the manufacturer's recommended special tools (Figure 13-16). Be sure to remove any clips that may be holding the bearing in place as you could break the crankcase half if any retaining clips are not removed.

Most ball bearings are held in the case by interference or a very tight fit. Cooling the bearing and heating the crankcase half may be recommended by the manufacturer. Placing cool bearings into a warm case ensures easy installation, as well as a snug fit when both have returned to normal temperature. This is because the cooled bearings expand as they warm up, and the warmed case shrinks as it cools down. Use the manufacturer's recommended special tools when replacing bearings to ensure that you do not damage the new bearing during installation (Figure 13-17).

The other popular bearing found in the four-stroke engine is the Babbitt or plain bearing (Figure 13-18). This bearing is used in most horizontal crankcase engines as a crankshaft journal support bearing, or as a multi-piece connecting rod bearing. Plain bearings are coded with colors or numbers to describe the size of the bearing to

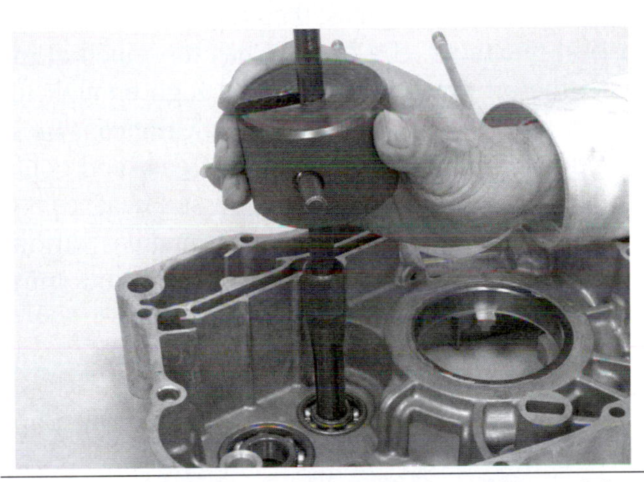

Figure 13-16 An internal bearing puller with a slide hammer is sometimes required to remove a bearing as shown here. Copyright by American Honda Motor Co., Inc. and reprinted with permission.

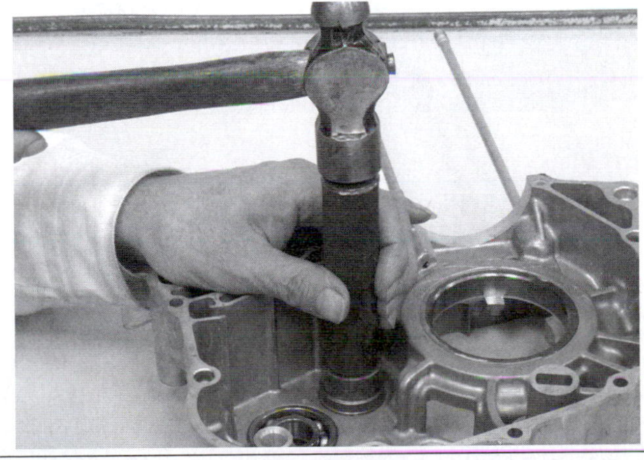

Figure 13-17 A bearing driver is a tool that sets on the outside bearing race to prevent damage to the bearing during installation. Copyright by American Honda Motor Co., Inc. and reprinted with permission.

ensure that the correct bearing is being used. Plain bearings are also used in some vertical crankcase engines that use a single-piece crankshaft.

Plain bearings are used with single-piece crankshafts, two-piece connecting rods, and also (through precision-machined surfaces) with some camshafts. They utilize high oil pressure from the running engine for lubrication and are measured for the amount of oil clearance they have, using a fine plastic string called **plastigage**. Plastigage is a plastic measuring clearance material that is compressed between bearing surfaces, then compared to a scale to find thickness. If a bearing has too much clearance, the oil pressure will not be high enough for proper lubrication and if too little clearance is present, the component will not spin freely and could actually seize from excess friction and heat.

Plastigage comes in long pieces and has various diameters. Before using plastigage, you must thoroughly clean and dry the surface to be measured (Figure 13-19A). Place a piece of the plastigage string on each bearing's inside surface (Figure 13-19B). Then install the opposite bearing surface and torque to the factory specification. Next, take the pieces back apart carefully and measure the amount of oil clearance that the bearing has (Figure 13-19C). You should also inspect the plain bearing visually for signs of wear.

Inspecting the Oil Pump

The **oil pump** used in the majority of four-stroke motorcycles is the **Trochoid-type pump**, which is better known as a **rotor-type pump**. The rotor pump consists of a pair of rotors: an inner and an outer (Figure 13-20) and is located in the lower end of the engine. The inner rotor is shaft driven, while the outer rotor is moved by the inner rotor

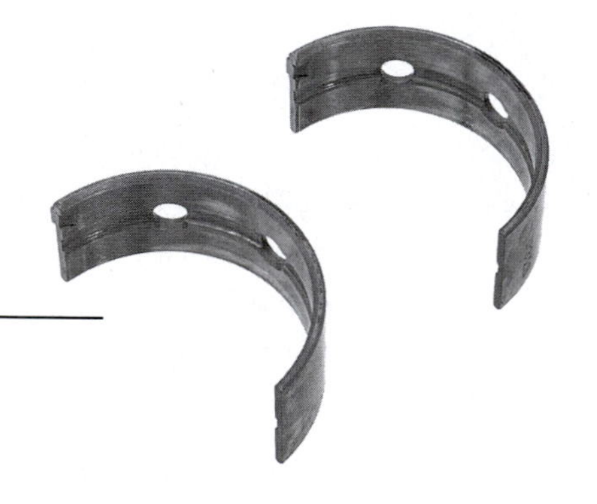

Figure 13-18 Plain bearings are widely used on today's four-stroke engines with horizontal-type crankcases. Copyright by American Honda Motor Co., Inc. and reprinted with permission.

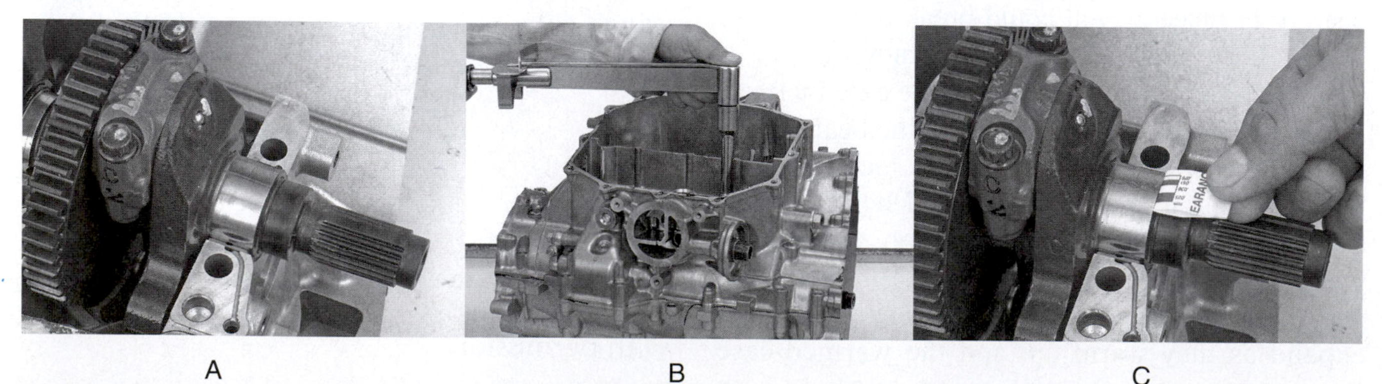

Figure 13-19 The procedure of using plastigage to verify oil clearance is shown here. Copyright by American Honda Motor Co., Inc. and reprinted with permission.

and is free to turn in the housing. The lobes on the rotors squeeze oil through passages in the pump body. As the inner rotor rotates, oil is constantly picked up from the inlet side, transferred, and pumped through the outlet side. Oil pressure is created when the oil is squeezed between the inner and outer rotors. The rotor-type oil pump design is capable of creating both high volume and high pressure.

To inspect this type of pump, disassemble and clean the parts of the pump and set the inner and outer rotors into the pump body properly. Measure the body clearance and tip clearance using a feeler gauge (Figure 13-21). If all measurements are within the manufacturer's specification, reassemble the oil pump in the reverse order of disassembly.

Inspecting the Clutch

Although lower-end engine disassembly isn't required to remove the clutch, the clutch must be removed to disassemble the lower end. Because the clutch is among the most common engine components to wear out or fail, it would be good practice to review Chapter 9 where the inspection and repair of the common multi-plate wet clutch is covered. We will cover a couple of key notes for the inspection of a typical clutch.

The most popular choice for measuring the thickness of the clutch friction and steel plates is the Vernier caliper (Figure 13-22). All of the plates must be measured individually. The clutch plates should also be checked for warpage. Lay each plate on a flat surface and attempt to run a feeler gauge under the edge (Figure 13-23). If the feeler gauge slips in anywhere under the edge, the plate is warped and must be replaced.

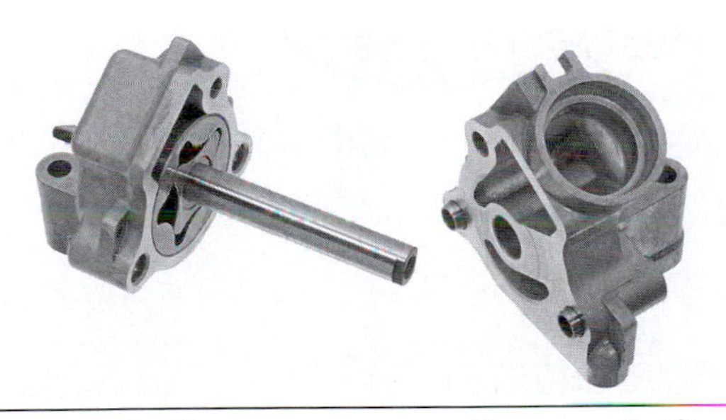

Figure 13-20 The rotor-type oil pump is the most commonly found oil pump on a four-stroke motorcycle engine. Copyright by American Honda Motor Co., Inc. and reprinted with permission.

Figure 13-21 Measuring oil pump clearance with a feeler gauge. Copyright by American Honda Motor Co., Inc. and reprinted with permission.

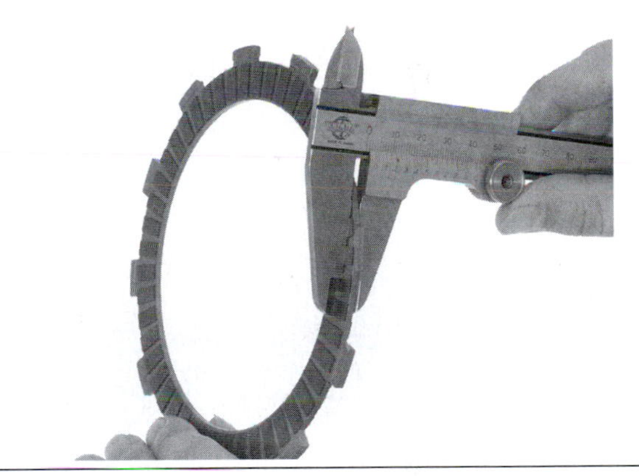

Figure 13-22 A Vernier caliper is used to measure the friction disk of a clutch. Copyright by American Honda Motor Co., Inc. and reprinted with permission.

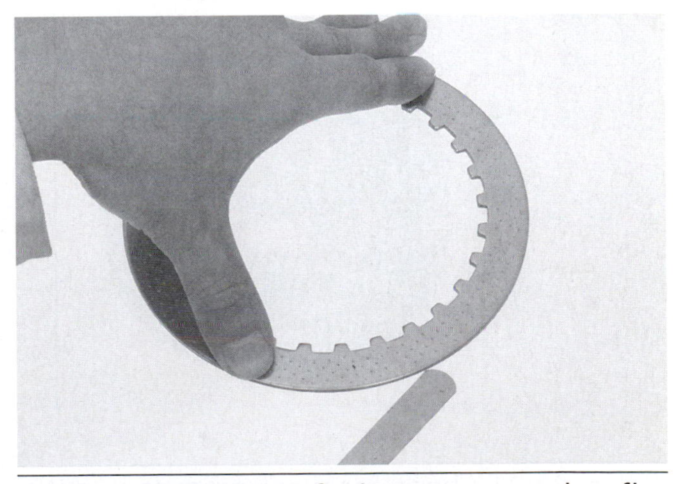

Figure 13-23 Use a feeler gauge and a flat plate to check that the clutch plate is not warped. Copyright by American Honda Motor Co., Inc. and reprinted with permission.

Inspect the clutch basket and clutch center grooves. Ensure that the grooves are smooth so the prongs of both the driving and driven plates fit properly (Figure 13-24). Inspect the clutch basket bearing and/or bushing for excessive wear. The bearing should be replaced if the clutch housing has excessive wobble when placed on the shaft or if the bushing is scored from heat or lack of lubrication.

Inspecting a Single-Piece Crankshaft

A single-piece crankshaft can be inspected and measured after the connecting rods have been removed. One measurement can be to check it to ensure it is not bent by placing it in a crankshaft jig and rotating it (Figure 13-25). Note that there are holes in the crankshaft for pressurized oil to flow to the bearings. Before disassembly, mark the location of the connecting rods and their big-end caps, to ensure that the rods are reassembled in their original position. Remove the connecting rod big-end cap nuts, and take off the rod and cap with the bearing inserts attached (Figure 13-26). The connecting rods and crankshaft will normally have bearing size and weight code letters and/or numbers stamped in them for information purposes (Figure 13-27). This information is provided to assist with proper replacement bearing selection.

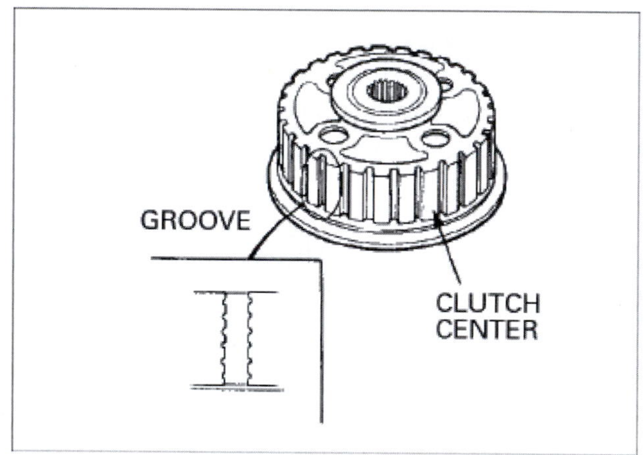

Clutch center

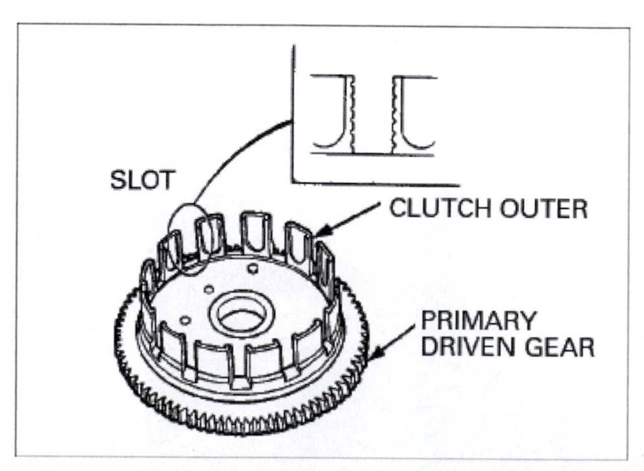

Clutch outer

Figure 13-24 Check for wear grooves in both the clutch basket and clutch center. Copyright by American Honda Motor Co., Inc. and reprinted with permission.

Inspect the crankshaft and connecting rod journals closely for signs of scoring or roughness. The surfaces must be very smooth. You should also measure the bearing surfaces with a micrometer and check your measurements with the factory specification given in the appropriate service manual (Figure 13-28). In addition, check for crankshaft run-out (Figure 13-25). Note that there are

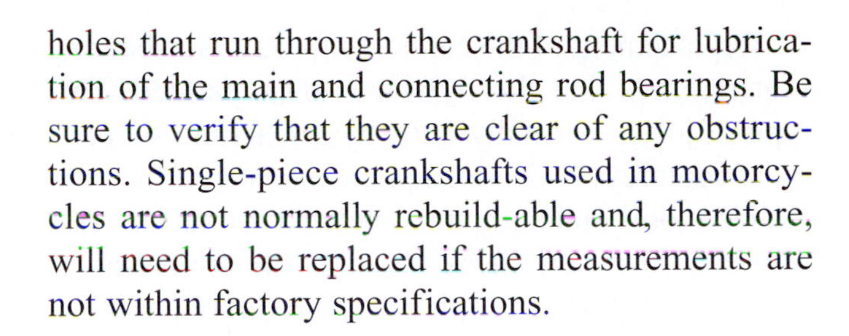

holes that run through the crankshaft for lubrication of the main and connecting rod bearings. Be sure to verify that they are clear of any obstructions. Single-piece crankshafts used in motorcycles are not normally rebuild-able and, therefore, will need to be replaced if the measurements are not within factory specifications.

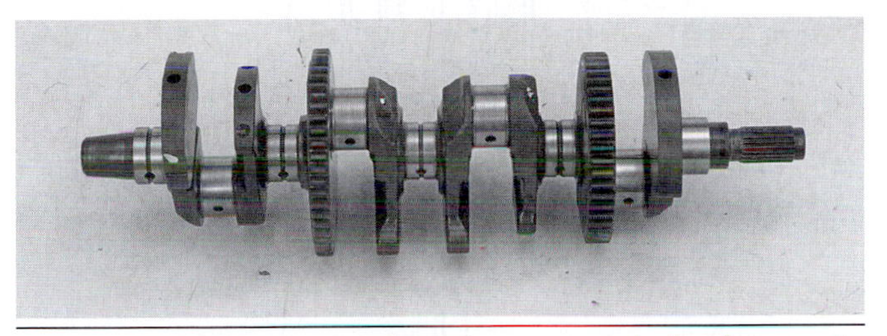

Figure 13-25 A single-piece crankshaft ready for measurement. Note the holes in the bearing surface areas for the flow of oil to the bearings.

Figure 13-26 A separated connecting rod is shown here. Look closely to see the bearing inserts. Copyright by American Honda Motor Co., Inc. and reprinted with permission.

Inspecting a Multi-Piece Crankshaft

Multi-piece crankshafts are normally rebuildable. Multi-piece crankshaft connecting rod lower bearings are usually of the roller type and are replaced as one unit. The unit consists of the connecting rod, rollers, cage, and crank pin.

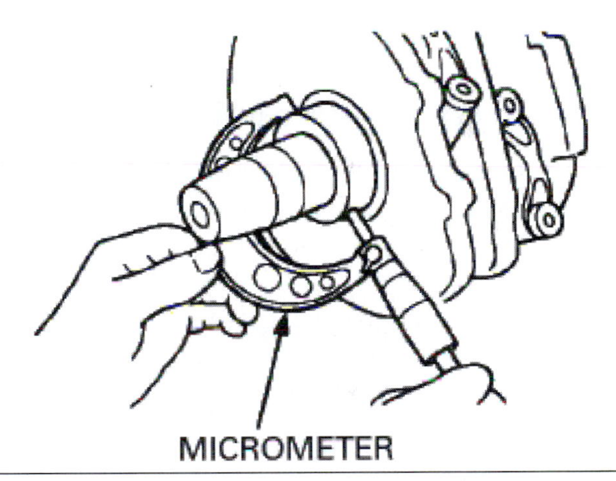

Figure 13-28 Use a micrometer to measure the bearing journals. Copyright by American Honda Motor Co., Inc. and reprinted with permission.

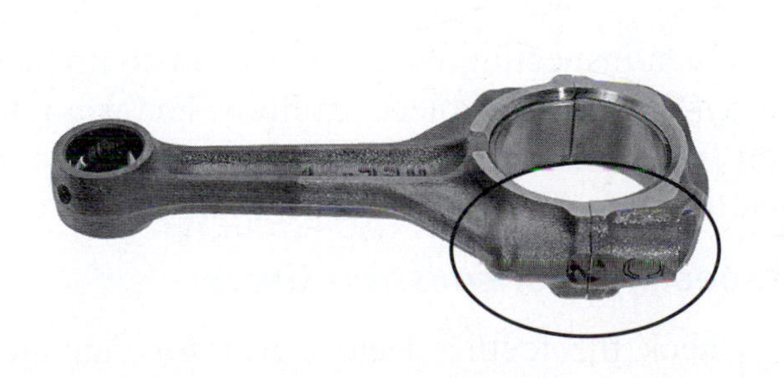

Figure 13-27 The manufacturer places markings on the crankshaft and connecting rod to help with bearing selection. Copyright by American Honda Motor Co., Inc. and reprinted with permission.

Replacement of the connecting rod lower-bearing unit requires that the flywheels be separated. This type of job is normally done by a machine shop, as most motorcycle dealerships do not carry the specialized tools and equipment needed for such a task. You can, however, check to see if the crankshaft is aligned properly by placing the crankshaft on V-blocks and measure the ends of the crankshaft with a dial indicator (Figure 13-29). A symptom of an out-of-line crankshaft would be excessive engine vibration when the engine is running at any speed.

Most manufacturers will also give a specification for the connecting rod side clearance that is measured with a feeler gauge (Figure 13-30), as well as radial clearance (Figure 13-31) that is checked with a dial indicator placed on the connecting rod.

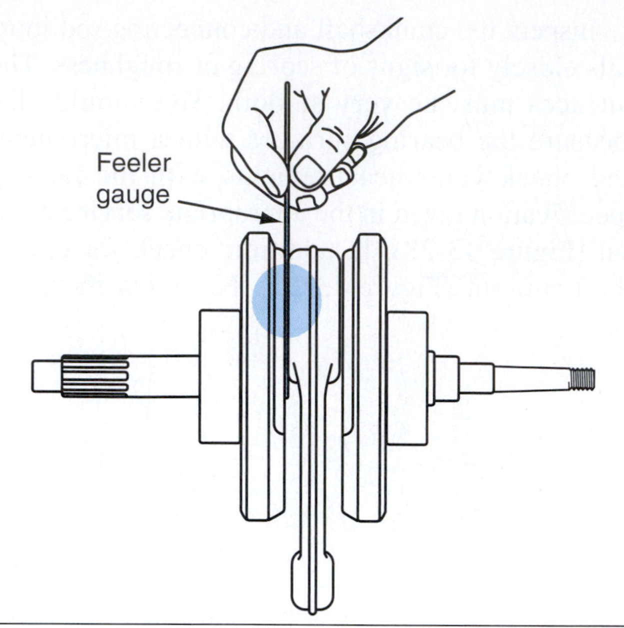

Figure 13-30 Checking a crankshaft connecting rod for side play.

Transmission Removal

The transmission gear cluster on a vertically split four-stroke engine is removed as a complete set (Figure 13-32) out of one side of the crankcase half.

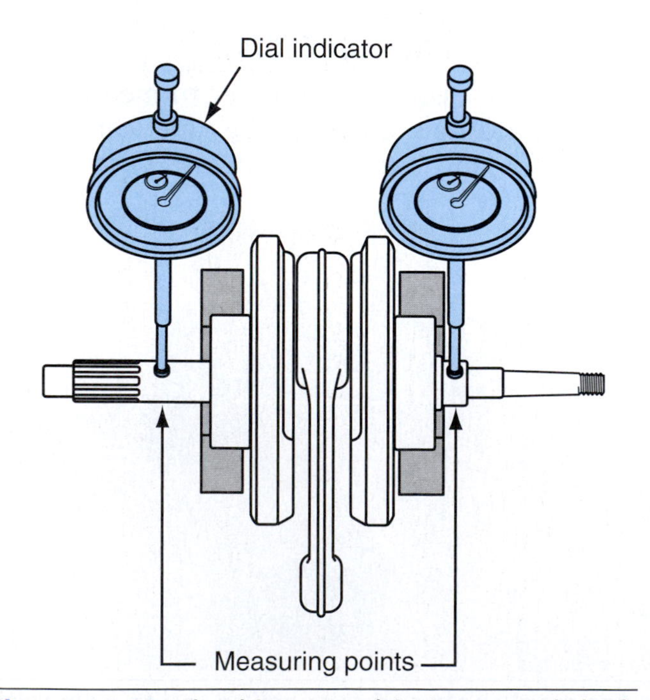

Figure 13-29 Checking a multi-piece crankshaft for trueness.

The transmission gears are removed from the top half of the crankcase set on a horizontally split engine (Figure 13-33) one shaft at a time.

When taking gears, bushings, spacers, and retaining clips off a transmission shaft, keep track of the disassembled parts (Figure 13-34). To remove a retaining ring, expand it and pull it off using the gear behind it. Always replace retaining rings that have been removed from a transmission shaft. After removal, they are generally spread too far out to be useful again.

Inspecting the Transmission

When inspecting the transmission components, always inspect each piece carefully. For example, if you find a bent or burnt shift fork, check the gear and the shift drum for damage as well.

Inspecting Transmission Gears

Check the teeth on each gear for damage or excessive wear. Inspect all the gear dogs or slots for wear or damage as well (Figure 13-35). Measure the inside diameter of the gear (unless it's a splined gear) with a telescoping gauge and outside micrometer

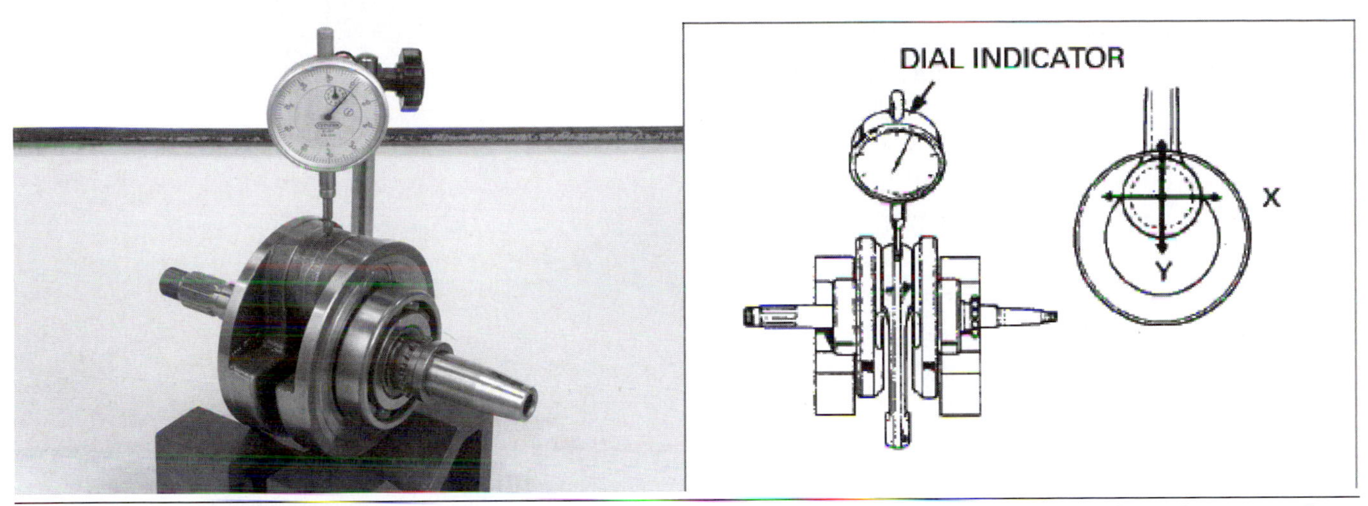

Figure 13-31 Checking for radial clearance on a multi-piece crankshaft. Copyright by American Honda Motor Co., Inc. and reprinted with permission.

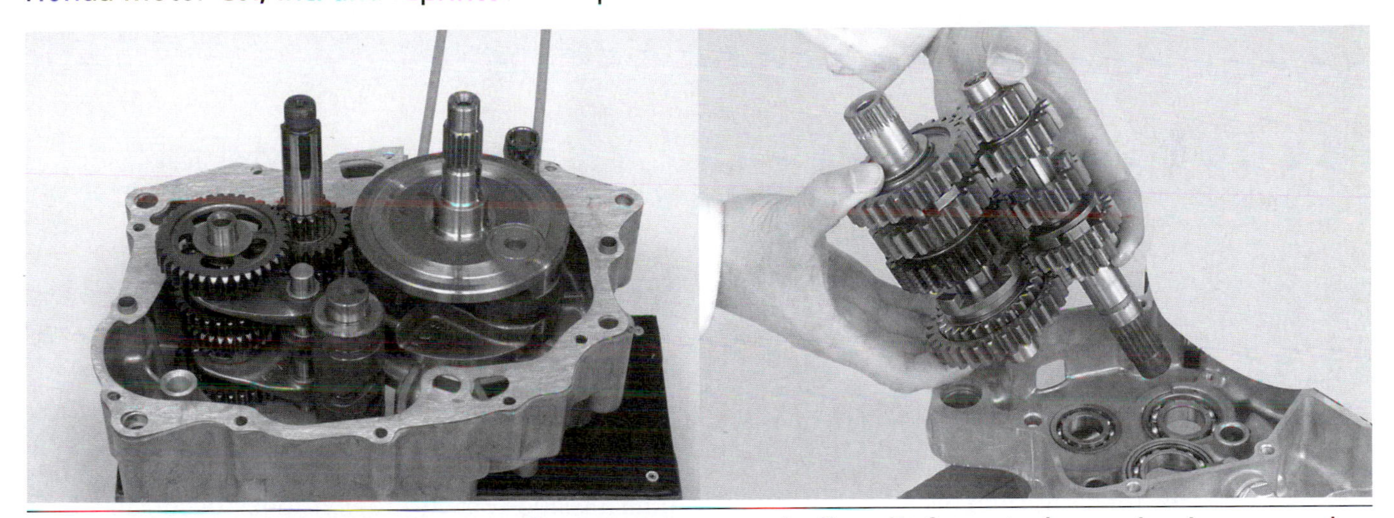

Figure 13-32 The transmission gear cluster set on a vertically split four-stroke engine is removed as a complete unit out of one side of the crankcase half. Copyright by American Honda Motor Co., Inc. and reprinted with permission.

Figure 13-33 Horizontal-type crankcases allow each transmission shaft to be removed separately. Copyright by American Honda Motor Co., Inc. and reprinted with permission.

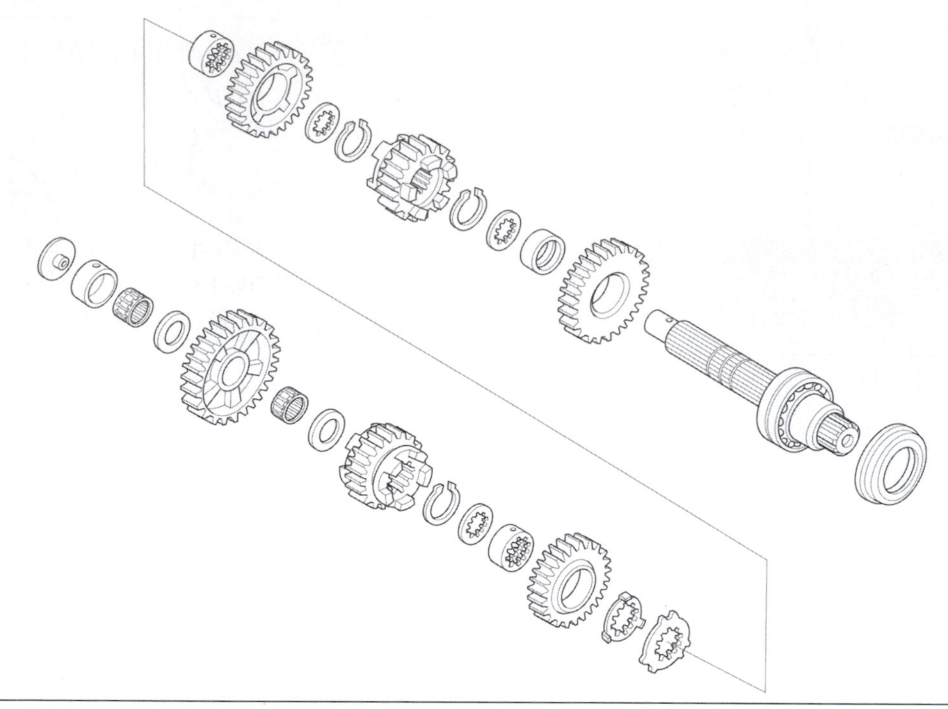

Figure 13-34 To allow for a more efficient assembly, keep the parts of a transmission in order. Copyright by American Honda Motor Co., Inc. and reprinted with permission.

and compare the measurement with the specification in the service manual (Figure 13-36).

Inspecting Transmission Shafts

Carefully inspect the main shaft and counter shaft along their splined grooves, as well as their sliding surfaces for abnormal wear or damage. Measure the shafts for proper dimensions, and compare to the specified sizes given in the service manual for the motorcycle that you are working on (Figure 13-37).

Inspecting Shift Forks

Although transmission problems aren't very common with most modern motorcycles, when a transmission-related problem does occur, the shift fork is most likely to be damaged. Check the shift fork for deformation, abnormal wear, size, and

Figure 13-35 Checking dogs and slots for wear is a typical procedure when inspecting the gears of a transmission.

Figure 13-36 Measure the inside diameter of the gear with a telescoping gauge and outside micrometer. Copyright by American Honda Motor Co., Inc. and reprinted with permission.

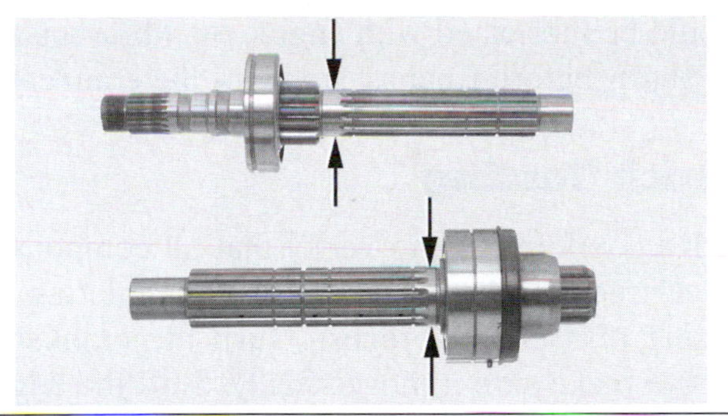

Figure 13-37 Measure the transmission shafts where the manufacturer indicates as shown here. Copyright by American Honda Motor Co., Inc. and reprinted with permission.

straightness. Measure the fork at the locations shown in Figure 13-38 and compare them to the factory specifications. A commonly missed inspection point on a shift fork is the area where the fork slides in the shift drum. Be sure to inspect this area closely for burn marks or chips (Figure 13-39).

Inspecting the Shift Drum

The shift drum is a critical component of the transmission that's often overlooked. Check the shifting fork guide grooves for damage such as a small chip or scoring (Figure 13-40). Also inspect the bearing in which the shift drum rotates. Many shifting problems relate to a faulty shift drum but are overlooked due to a lack of attention to detail.

Assembling the Transmission

Before assembling the transmission, be sure that all parts are cleaned. Apply a very small amount of molybdenum disulfide grease (Molylube) to all sliding surfaces of each shaft before beginning the assembly process. This ensures adequate initial lubrication in the transmission. Reassemble all the transmission gears in the proper position on the appropriate shaft. All service manuals have an exploded view of the transmission that assists with the assembly of the transmission gears, bushings, thrust washers, and retaining rings (Figure 13-41).

Bushings have oil holes that allow oil to flow to the gears. They must be properly aligned (Figure 13-42). When you install the thrust washers, be sure that the chamfered side faces away from the thrust load side of the gear. Also align the retaining snap ring with one of the grooves of the spline (Figure 13-43). If the retaining ring rotates easily in the groove, replace it as it will not properly seat and may come out, causing serious transmission failure at the most inappropriate time.

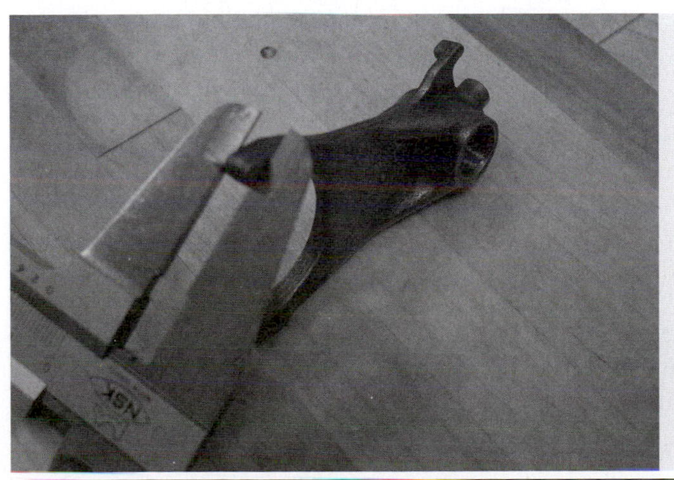

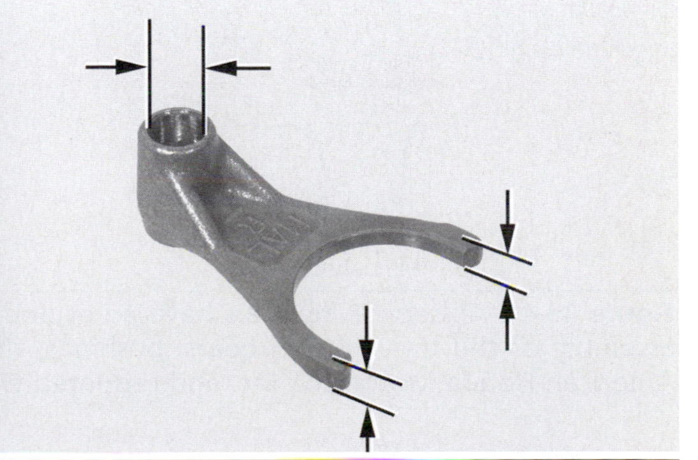

Figure 13-38 Measurement points of a common shift fork.

Figure 13-39 A commonly missed inspection point on a shift fork is the area where the fork slides in the shift drum.

Figure 13-40 Check the shift drum very closely for any chips or cracks. Copyright by American Honda Motor Co., Inc. and reprinted with permission.

There will be one shift fork for every two gears in a motorcycle transmission. Shift forks are marked for their proper location. A fork marked with an "L" is placed on the left side of the transmission. A fork marked with a "C" is the center shift fork; the fork marked with an "R" is the right side shift fork (Figure 13-44).

After the transmission shafts are properly assembled, they must be placed into the crankcase as an assembly in a vertical crankcase engine design (Figure 13-45). In a horizontal crankcase engine design, set the gear shafts back into their appropriate position in the crankcase individually (Figure 13-46). All transmission gears and shafts should be lubricated with engine oil while rotating the shafts before final assembly of the crankcases.

Bench Testing

It's good practice to verify that all components move freely and properly before completing the engine reassembly process. Turn the crankshaft over to make sure it moves freely. Shift the engine into every gear to ensure there are no apparent problems inside the engine. Nothing is more frustrating than finding a problem in the crankcase after you have completely reassembled the engine and installed it into the chassis.

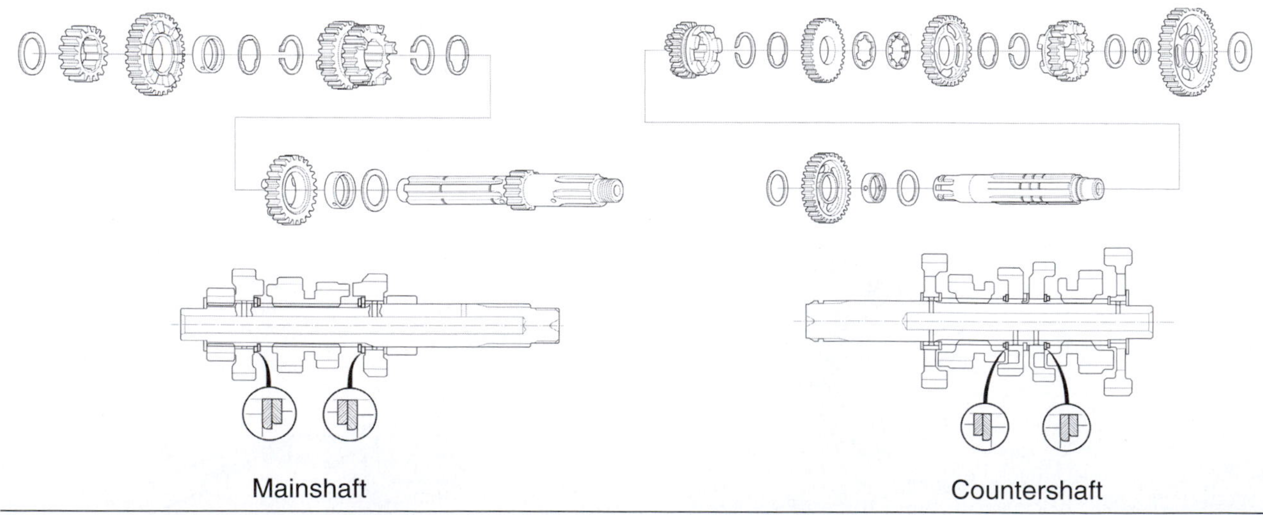

Figure 13-41 All service manuals have an exploded view of the transmission that assists with the assembly of the transmission gears, bushings, thrust washers, and retaining rings. Copyright by American Honda Motor Co., Inc. and reprinted with permission.

FOUR-STROKE LOWER-END REASSEMBLY

Before you begin reassembling the lower end of any engine, be sure to thoroughly clean every part with a cleaning solvent. When you're prepared to begin the assembly process, stay organized and

Figure 13-42 Be sure to align the oil holes of the transmission shaft bushings correctly. Copyright by American Honda Motor Co., Inc. and reprinted with permission.

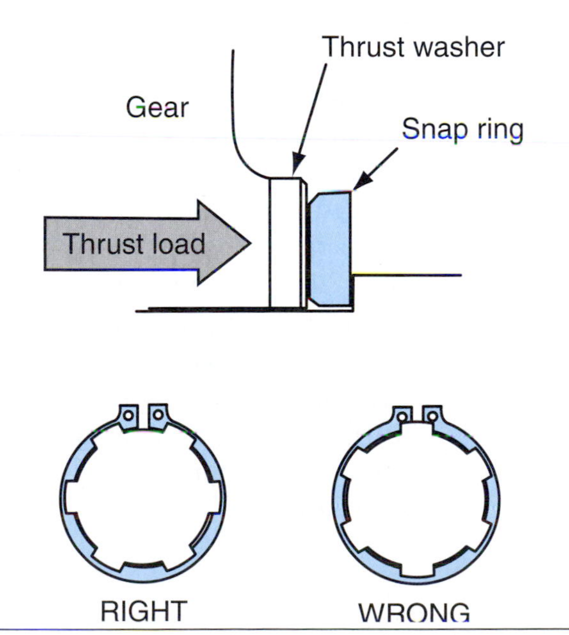

Figure 13-43 Align the retaining snap ring with one of the grooves of the spline as illustrated here.

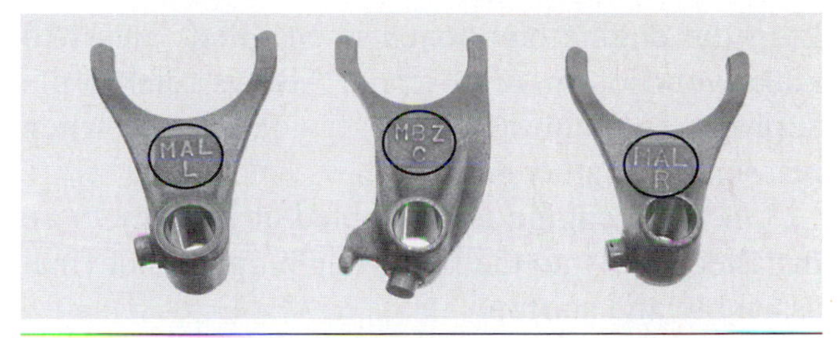

Figure 13-44 Shift forks are marked for proper placement as shown here. Copyright by American Honda Motor Co., Inc. and reprinted with permission.

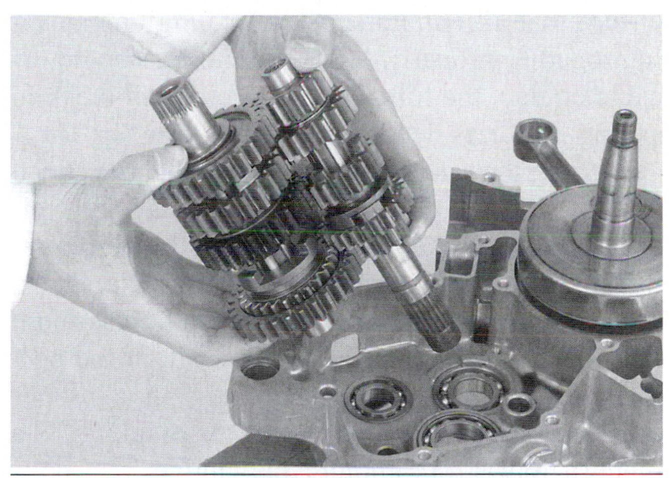

Figure 13-45 Installing the transmission as a set is required in the vertical split crankcase design. Copyright by American Honda Motor Co., Inc. and reprinted with permission.

Figure 13-46 In a horizontal crankcase engine design, the gear shafts are placed back into position individually. Copyright by American Honda Motor Co., Inc. and reprinted with permission.

keep the engine components separated. This will make your job more efficient. Always refer to the appropriate manufacturer's service manual when reassembling an engine.

Once the engine is assembled, it is ready to be installed back into the frame and prepared for final assembly and startup.

Engine Break-In

Most manufacturers recommend that a new (or reconditioned) engine be properly broken-in to ensure that all components are sealing and meshing together properly. With the four-stroke motorcycle engine, the time needed to properly break in an engine varies depending on its use.

During the assembly process, it's recommended that you use the best possible materials and original equipment-manufactured parts. Nonetheless, it's still necessary to allow the parts to break in before subjecting the engine to constant maximum stress. The future reliability, as well as the performance of the engine, depends on a proper break-in procedure. The reason for breaking in a modern four-stroke motorcycle engine is to optimize the compression seal between the combustion chamber and the underside of the piston, which in turn will minimize pressure loss in the cylinder(s).

The piston ring seal is really what the break-in process is all about. Piston rings do not seal the combustion pressure by spring tension. Ring tension is necessary only to "scrape" the oil as the piston goes back down the bore of the cylinder. A typical piston ring exerts less than 10 pounds of spring tension against the cylinder wall. That level of low spring tension cannot seal 3,000-plus pounds of pressure created from the combustion chamber gases.

The true sealing of the piston ring to the cylinder wall comes from the actual combustion gas pressure. The gases created by the rapid, ignition and burning of the fuel/air mixture in the

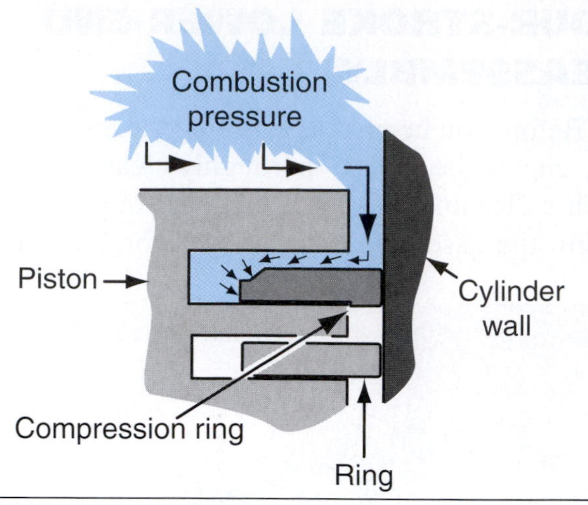

Figure 13-47 The gases created by the rapid ignition and burning of the fuel/air mixture in the combustion chamber gets behind the ring to force it outward against the cylinder wall.

combustion chamber takes the path of least resistance, which means it passes over the top of the ring, and gets behind the ring to force it outward against the cylinder wall (Figure 13-47).

The problem is that new piston rings are far from a perfect match for the cylinder and must be worn in properly and as quickly as possible in order for the ring to completely seal all the way around the bore of the cylinder. If the gas pressure is strong enough and the piston ring is mated correctly with the cylinder, then the entire ring will make contact with the cylinder surface, thus minimizing the combustion gas loss and allowing for maximum power in the engine. For street or typical off-road use, many engine builders recommend that a motorcycle be ridden as the owner would normally ride, paying close attention not to sustain high engine rpms for long periods of time for the first 100 miles or two hours. These numbers are general and it is recommended that you follow the manufacturer's instructions. Break-in is up to you, but the loss in power from improper ring sealing can be as high as 10 percent.

Summary

Common lower-end engine failures in a four-stroke motorcycle include the following:

- **Leaking Engine Seals**
  An engine seal is designed to prevent oil from getting out of a four-stroke motorcycle engine and leaking between the crankcases and a moving part like a transmission shaft.
- **Worn Crankshaft Bearings**
  A bearing that is failing usually makes a rough growling sound. You may want to use a mechanic's stethoscope to help pinpoint the location of a bad bearing.
- **Worn Connecting Rod Bearings**
  Symptoms of a failed connecting rod bearing are knocks, vibrations, or an engine that cannot be turned over (seized up). Any of these symptoms necessitate the disassembly, inspection, and repair of the lower end of the engine.
- **Difficult Shifting**
  When it is difficult to shift between gears while the motorcycle is moving and it has been determined that the clutch is not at fault, the cause is usually a bent or burnt shift fork.
- **Inability to Shift Gears**
  When you find a machine that shifts into one gear but not into the next the problem is often caused by a faulty shift return spring, which returns the shifting lever to its original position.
- **Strange Sounds from the Transmission**
  A low growling sound when the motorcycle is in gear and rolling usually indicates a bearing failure. When a bearing failure occurs, it may cause a transmission shaft to move slightly out of position. When this occurs, the gears do not mesh properly and produce a low growling noise.
- **Clunking Noises**
  Another characteristic sound that indicates a transmission problem is an excessive clunking sound when the engine is in a particular gear while under a load. More often than not, this indicates broken teeth on one or more gears.
- **Jumping Out of Gear**
  When a transmission jumps out of gear, it is usually caused by worn transmission gear dogs and/or slots. Gear dogs and slots are used to lock the two gears together.

There are various components found in a four-stroke engine lower-end assembly. It is important to know the way that these components operate, as they must all move in harmony with one another to allow the engine to run effectively.

- As a technician, you must understand the importance of inspecting the various components of the four-stroke lower end, including the crankshaft and transmission, for damage or wear using the correct and appropriate tools and equipment.
- There are basic common procedures for disassembling and inspecting the transmission in a typical four-stroke motorcycle engine but each manufacturer has their own specific details on the proper procedures.
- The use of Plastigage is required to verify the correct measurement of the bearing surfaces of a single-piece crankshaft.
- Without utilizing bench testing during the reassembly process of an engine, the technician risks the consequence that a component was incorrectly installed and the engine will not function properly.
- Breaking in a newly rebuilt engine correctly is critical in the longevity of the engine. Improper break-in can reduce power by up to 10 percent.

Chapter 13 Review Questions

1. Ball bearings are held in place by
   a. an interference fit
   b. the bearing and the shaft
   c. a 6 mm bolt
   d. spacers
2. When any transmission problem is detected, a related component that's likely to be damaged is/are the
   a. gear dogs
   b. shift fork
   c. gear bushings
   d. counter shaft
3. Which of the following tools should you use to measure clutch springs?
   a. Inside micrometer
   b. Machinist rule
   c. Vernier caliper
   d. Outside micrometer
4. A component in a four-stroke engine that normally has letters and numbers stamped on it indicating size and weight information, is the
   a. crankshaft
   b. shift fork
   c. shift drum
   d. clutch hub
5. When comparing differences in vertical and horizontal crankcase design, which of the following statements is true?
   a. Horizontal crankcase designs require removal of the top-end components before the crankcase can be disassembled.
   b. A special pulling tool is needed to remove the crankshaft from both vertical and horizontal crankshaft designs.
   c. Separating the engine cases exposes the crankshaft but not the transmission gears on a vertical crankcase.
   d. Vertical crankcase designs require removal of the top-end components before the crankcase can be disassembled.
6. When comparing a single-piece crankshaft to a multi-piece crankshaft, it can be said that
   a. single-piece crankshafts are normally rebuild-able; multi-piece crankshafts are not
   b. on multi-piece crankshafts, connecting-rod lower bearings are replaced with the connecting rod
   c. on single-piece crankshafts, connecting rods don't need to be removed prior to inspection
   d. measurements of both single-piece and multi-piece crankshafts are made using a Vernier caliper
7. Which of the following tools would you use to measure a shift fork?
   a. Verneir caliper
   b. Machinist rule
   c. Hydrometer
   d. Outside micrometer
8. The transmission component that's often overlooked due to a lack of attention to detail is the
   a. shift fork
   b. main shaft
   c. countershaft
   d. shift drum
9. A low growling sound heard when the motorcycle or ATV is in gear and rolling, normally indicates a faulty ________.
10. The ________ is installed into the vertically split crankcase as a complete unit.

CHAPTER

14 Electrical Fundamentals

Learning Objectives

When students have completed the study of this chapter and its laboratory activities they should be able to:

- Understand the importance of proper safety procedures when working with electrical systems
- Explain the two basic theories of electricity
- List the types of electrical circuits
- Explain the terms voltage, current, and resistance
- Calculate voltage, current, and resistance using Ohm's law
- Describe how to use a multi-meter to measure voltage, resistance, and current
- Understand the term "schematic" and understand how to read a simple wiring diagram

Key Terms

AC
Alternator
Ammeter
Amp
Amperes
Amp hour
Armature
Battery
Bench test
Block diagrams
Capacitor
Circuit
Circuit breaker
Coil
Conductor
Continuity
Current
DC
Dielectric
Diode
Dynamic
Electrical potential
Electricity
Electrolysis
Electrolyte
Electromagnet
Electromotive force (EMF)
Electron
Field coil
Flux lines
Free electron
Fuse
Ground
Grounded circuit
Ignition
Insulator
Lines of force
Load
Magnetic induction

Magnetism
No-load test
Normally closed solenoid
Normally open solenoid
NPN
Ohm
Ohmmeter
Open circuit
Parallel circuit
Permeability
PNP
Polarity
Pole
Rectifier
Rectifier
Regulator
Reluctance
Reluctor
Resistance
Rotor
Schematic
SCR
Selenium
Series circuit
Series/parallel circuit
Short circuit
Silicon
Sine wave
Solder
Static
Stator
Switch
Thermo-switch
Thyristor
Unloaded
Valence electrons
Volt
Voltage
Voltage Drop
Voltmeter
Watt
Wire gauge
Wiring diagram
Zener diode

INTRODUCTION

This chapter is the first of three that will concentrate on the subject of electricity. The text will show you the basics of electricity, where electricity comes from, and how we measure electricity. In the following chapters, we'll discuss charging systems, ignition systems, and other electrical circuits that will be found when working on a motorcycle.

Since electricity can't normally be seen, many technicians in the motorcycle industry know little about it and are somewhat afraid of it. Electricity isn't a difficult subject area to learn as long as you understand the basics of how electricity works. We can assure you that you don't have to be an engineer with a background in the theory of electrical systems in order to competently service the electrical systems on modern motorcycles. However, for you to understand why something is not functioning properly, you must first know how it works.

The technician who understands how electrical systems produce, conduct, store, and use electrical energy will find it easier to locate and correct problems in these systems. Therefore, in this chapter we'll discuss the fundamentals of electricity, including the terms used in this field, and cover the primary tool used to measure various electrical components. You may be wondering, what exactly is electricity? While there are many answers to that question, we will state that for our purposes of working on motorcycle electrical systems, electricity is a natural force produced by the movement of electrons.

In this chapter we're going to give you a general understanding of electricity and how it works. While you should understand the basic theory and facts presented here, you won't be expected to become an electrical genius and memorize complex electrical formulas and theories. In fact, we will try very hard to make learning about electricity and electrical systems as easy as possible and use the simplest terms we can to help you to understand this interesting subject area of motorcycle repair.

The electrical system is perhaps the most important support system of a motorcycle. Without electricity, the engine on a motorcycle would not run. Electricity provides the needed spark for combustion as well as the power required for electric starting, lights, instrumentation, and many other accessories found on a typical motorcycle. Advances in the field of electronics have brought new technologies to the motorcycle industry that allow them to be much more efficient. Today's technician needs to understand basic electrical circuits before they can diagnose and service these new technologies. Although test equipment used on the modern motorcycle assists in simplifying working on the modern motorcycle, understanding electrical principles, component operation, circuit design, and testing procedures are all necessary to be a successful modern motorcycle technician.

SAFETY PRECAUTIONS WITH ELECTRICITY

Electrical devices and circuits can be dangerous. Safe practices are necessary to prevent shock, fires, explosions, mechanical damage, and injuries resulting from the careless or improper use of tools.

Perhaps the greatest hazard is electrical shock. Electricity affects the body by overriding brain impulses and contracting muscles. Therefore, a current through the human body in excess of 10 milliamperes can paralyze the victim and make it impossible to let go of a "live" conductor.

Your skin can have approximately one thousand times more resistance to the flow of electricity when dry, which would be in the vicinity of several hundred thousand ohms. When moist or cut, the skin's resistance may become as low as several hundred ohms. In this circumstance, even so-called safe voltages as low as 30 or 40 volts might produce a fatal shock. Naturally, the danger of harmful or fatal shock increases directly as the voltage increases. You should be very cautious, even with low voltages. Never assume a circuit is dead, even though the switch is in the OFF position.

General Electrical Safety Rules

Safe electrical practices will protect you and those around you. Study and know the following rules:

- Do not work when you're tired or taking medicine that makes you drowsy.
- Do not work in poorly lighted areas.
- Do not work in damp or wet areas.
- Use approved tools, equipment, and protective devices.
- Do not work if you or your clothes are wet.
- Remove all rings, bracelets, and similar metal items.
- Never assume that a circuit is off. Check it with a device or piece of equipment that you are sure is operating properly.
- Don't tamper with safety devices. Never defeat an interlock switch. Verify that all interlocks operate properly.
- Keep your tools and equipment in good condition. Use the correct tool for the job.
- Verify that capacitors have discharged. Some capacitors may store a lethal charge for a long time.
- Don't remove equipment grounds. Verify that all grounds are intact.
- Do not use adapters that defeat ground connections.
- Use only an approved fire extinguisher. Water can conduct electric current and increase the hazards and damage. Carbon dioxide (CO_2) and certain halogenated extinguishers are preferred for most electrical fires. Foam types may also be used in some cases.
- Follow directions when using solvents and other chemicals. They may explode, ignite, or damage electrical circuits.
- Certain electronic components affect the safe performance of equipment. Always use the correct replacement parts.
- Use protective clothing and safety glasses.
- Don't attempt to work on complex equipment or circuits without proper training. There may be many hidden dangers.
- Some of the best safety information for electrical and electronic equipment is the literature prepared by the manufacturer. Find it, read it, and use it!
- When possible, keep one hand in your pocket while working with electricity. This reduces the possibility of your body providing an electrical path through the heart.

Any of the above rules could be expanded. As your study progresses, you'll learn many of the details concerning proper procedures. Learn them well because they're the most important information available. Remember, always practice safety; your life depends on it.

BASIC PRINCIPLES OF ELECTRICITY

As mentioned previously, perhaps the main reason that most people find it difficult to understand electricity is that they cannot actually see it. By actually knowing what electricity is and what it is *not*, one can easily understand it. One thing is for sure—electricity is *not* magic. It is something that either takes place or can take place in everything around us. It not only provides power for the lights and refrigerator in your house, but it is also the basis for the communication between your brain and eyes as you read this sentence.

Although electricity cannot be seen, its effects of can be seen, felt, heard, and even smelled. One of the most common displays of electricity is a lightning bolt. Lightning is electricity—albeit a large amount of electricity. In fact the power of lightning is incredible. Using the power in much smaller amounts to perform some work is the basis for a motorcycle's electrical system. Electricity cannot be seen because it results from the movement of extremely small objects that move at close to the speed of light, which is 186,000 miles per second.

The typical motorcycle electrical system has many different paths through which electricity can flow. The four major electrical systems are:

- Starting systems, which are used on many motorcycles to rotate the engine to start it
- Ignition systems, which provide high-energy sparks to ignite the fuel-and-air mixture inside the engine's combustion chamber
- Lighting systems, which are used to power the lights as well as operate other electrical equipment on the machine
- Charging systems, which are used to produce the electricity to recharge the battery that is used to store electricity

Besides these major systems, there are also many other electrical subsystems.

Let us take a closer look at electricity by first looking at some basic electrical theories, a basic electricity storage container, and a simple circuit.

Electrical Theories

There are two basic theories of electricity that you should be aware of and understand their similarities as well as their differences.

Conventional Theory of Electricity

The conventional theory of electricity states: "Electric current flows from the positive terminal of the voltage source, through the circuit, to the negative terminal." Simply stated, this theory supports that electricity flows from positive to negative.

It is not known with certainty as to who came up with the conventional theory of electricity but in the mid-1700s, Ben Franklin studied it and is credited with the first use of the terms "battery," "positive," and "negative" to describe the terminals of a battery. Many of the characteristics of electricity that we take for granted today were not known in Franklin's time. It was thought that electric current flowed from the positive terminal of the voltage source, through the circuit, to the negative terminal.

You should know that the conventional theory of electricity is used in the motorcycle industry.

The Electron Theory

One hundred and fifty years after the conventional theory was considered, a group of scientists developed the electron theory of electricity. Like most new ideas, the electron theory of electricity was not accepted at first. In many industries, including ours, it is still not in wide use. However, the electron theory is the basis of modern electronics.

The electron theory of electricity states: "Electricity is the flow of electrons from the negative terminal of a source through a conductor, completing the circuit back to the positive terminal of the source." Simply stated, this theory supports that electricity flows from negative to positive.

Scientists now generally accept the electron theory concerning the nature of electricity.

Atoms

All matter is composed of molecules, and each molecule contains two or more atoms. Atoms, in turn, are made up of neutrons, protons, and electrons (Figure 14-1). It is the arrangement of these particles that makes materials such as liquids, solids, and gases differ from one another.

The core of the atom, called the nucleus, contains protons and neutrons. Protons have a positive electrical charge and neutrons are neutral, meaning that they have no electrical charge. Electrons have a negative charge and rotate around the nucleus of the atom. Atoms normally have an equal number of protons and electrons, and therefore, an equal number of positive and negative electrical charges. These charges cancel each other out, resulting in an atom with no positive or negative electrical charge. When an atom has more protons than electrons, it's positively charged. When an atom has more electrons than protons, it's negatively charged (Figure 14-2).

Atoms, which make up different kinds of material, have different numbers of electrons and protons. For example, a carbon atom has only 12 protons and 12 electrons, while a uranium atom has 234 protons and 234 electrons.

In some materials, the electrons are tightly bound in orbit around the nucleus of the atom and aren't free to travel to other atoms. This condition exists in materials that are poor conductors of electricity. In other materials, the orbits of the electrons are relatively large and the electrons are able to travel to other atoms. Such materials are good electrical conductors (Figure 14-3). The orbits of the electrons in copper are large and the electrons can move relatively easily so copper conducts electricity well. For this reason, most electrical circuits use copper wire as the conductor through which the current flows.

When an atom is positively charged or negatively charged, the condition (lack of or excess number of electrons) will cause a flow of electrons from one atom to another. The idea of removing electrons from an atom may seem strange. However, we remove electrons from atoms all the time without realizing it. For example, if you shuffle across a carpet and then touch a metal surface, what usually happens?

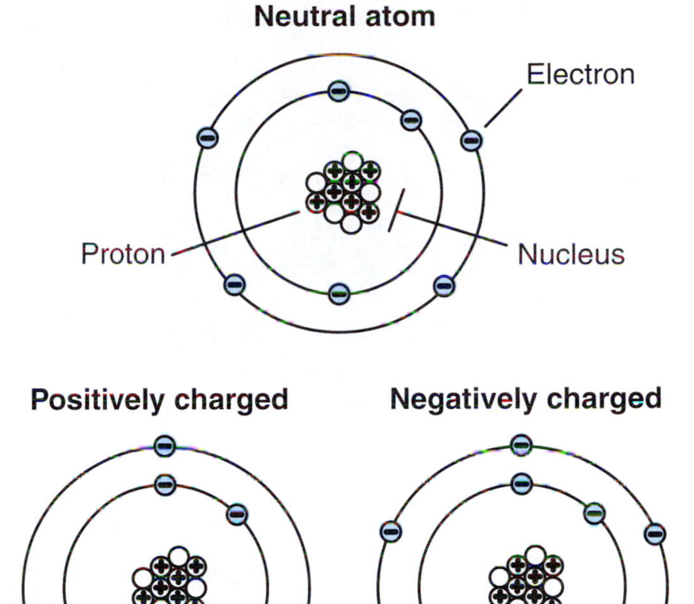

Figure 14-2 A neutral atom has an equal number of protons and electrons. A positively charged atom has more protons than electrons. A negatively charged atom has more electrons than protons.

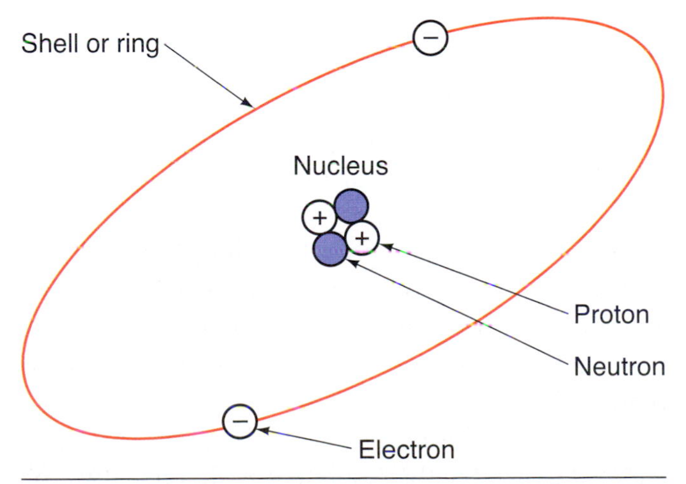

Figure 14-1 Atoms are composed of electrons, protons, and neutrons.

Copper (conductor)

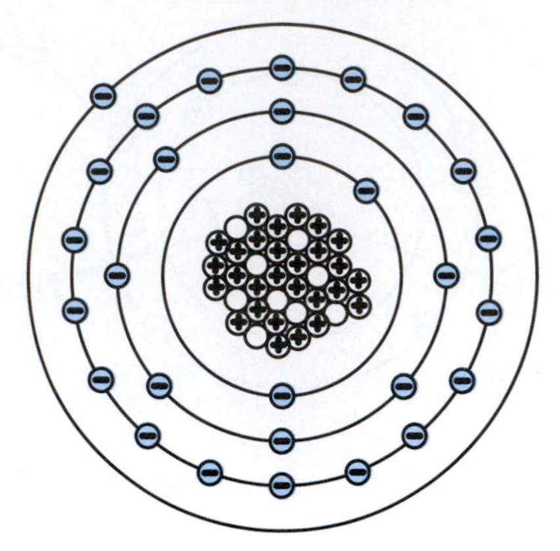

The outer valance ring contains a single electron

Silicon (semiconductor)

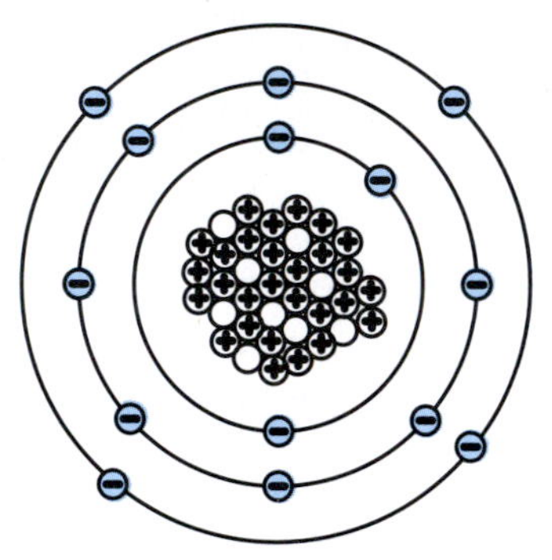

The outer valance ring contains four electrons

Oxygen (insulator)

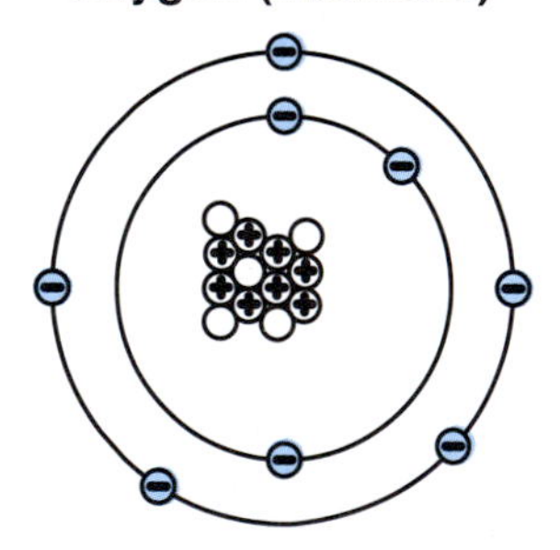

The outer valance ring contains five electrons

Figure 14-3 Hydrogen and oxygen are poor conductors, silicon is a semiconductor, and copper is a good conductor.

You probably receive a small shock, and maybe even see a spark. This occurs because as you scuffed your socks along the carpet, you actually rubbed free electrons off the carpet. Free electrons are also called valence electrons. Your body held onto these electrons, and you became negatively charged. When you touched the metal surface, the free electrons from your body transferred to the metal, restoring your body to a neutral charge. The discharge of electrons from you to the metal caused the small shock that you felt.

Thus, you can see that it's not impossible to get electrons moving from one place to another. However, it's easier to get electrons moving in some materials than in others. The structure of an individual atom will determine how easily an electron can be removed from it. For example, in Figure 14-3, you saw that the structure of the hydrogen atom makes it very difficult to remove an electron from its orbit. So, it's very difficult to produce a flow of electricity in hydrogen. However, in a copper atom, the outermost electron can easily be dislodged from its orbit. Therefore, it is very easy to get a flow of electricity moving in copper.

Any substance in which electrons can move freely is called an electrical conductor. Copper, silver, gold, and other metals are good electrical conductors. In fact, silver and gold are better electrical conductors than copper, but because silver and gold are so expensive, they generally are not used to make electrical wires in motorcycles. Materials in which the electrons are tightly bonded to the nucleus are called insulators. Plastic, nylon, ceramic, and other similar materials are very resistant to the flow of electricity and are classified as insulators.

The Simple Battery

We will now discuss the battery, as it is a storage device for electricity used on motorcycles. We will discuss the very basics of batteries here as they are discussed in more detail in Chapter 15.

Electricity is an invisible form of energy that can be transformed into magnetism, light, heat, or chemical energy. Because we know how to control electrical energy, we can use it to perform many jobs.

Note that batteries have two different ends. The end of the battery that's labeled with a negative, or minus sign (–) is called the negative terminal. The opposite end of the battery that's labeled with a positive, or plus sign (+), is called the positive terminal. The negative terminal of the battery has a negative charge, as it contains too many electrons. The positive terminal of the battery has a positive charge, as it contains too few electrons.

The negative and positive charges in a battery are produced by a simple chemical reaction. The battery terminals, or electrodes, are two strips of lead. Each electrode is made from a different type of lead. When the strips of metal are placed into an electrolyte solution, a chemical reaction occurs. Electrolyte is a chemical compound that, when molten or dissolved in certain solvents (usually water), will conduct an electric current. Because of this reaction, a negative charge forms on one electrode and a positive charge forms on the other electrode.

You've probably heard the phrase "opposites attract." This phrase holds true with electricity. Opposing electrical charges (positive and negative) strongly attract each other and try to balance each other out. Because of this attraction, whenever too many electrons are in one place, the electrons will try to move to a place where there are fewer electrons. This is the basic operating principle of a battery. The negative terminal of a battery has a high concentration of electrons, while the positive terminal has very few electrons. So, the electrons at the negative battery terminal will be drawn toward the positive battery terminal. But to actually move from the negative terminal to the positive terminal, the electrons need a path to follow. We can create a path for the electrons by connecting a conductor and a load between the battery terminals. By attaching these pieces, we actually build a circuit.

In the simple circuit shown in Figure 14-4, electrons flow from the negative battery terminal to the positive terminal through the conductors and light bulb that are attached to them. Note that the flow of electricity produced by the battery will continue as long as the chemical reaction in the battery keeps up. After some time, the chemical reaction in the battery will stop and the battery will stop functioning. At that point the battery will need to be recharged or replaced. This is why motorcycles have charging systems.

Capacitors

Capacitors are also used to store electricity. In an electrical circuit using a capacitor, the capacitor charge usually equals the circuit voltage. If the circuit voltage falls or the circuit is open, the capacitor will release its charge. In a circuit where short bursts of voltage are needed, capacitors are very useful.

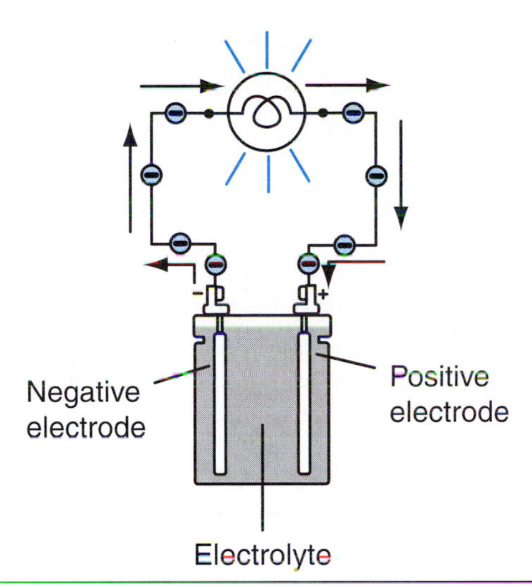

Figure 14-4 In this illustration of a battery, a chemical reaction takes place between the electrodes and the electrolyte solution. This chemical reaction produces an electrical charge on each of the electrodes.

Circuits

There are three requirements to complete a simple circuit—a power source, a conductor, and a load (Figure 14-5). A power source is simply a source of electrical power. The power source in a motorcycle electrical circuit is generally a battery. The conductors are the wires that carry the electricity. In order to use the electrons to perform useful work, we've connected a light bulb to the circuit, which is our load. A load is any device, such as a light bulb, that we want to run with electricity.

We may want to choose to allow a circuit to be closed or open. In a closed circuit, when the switch is turned on electrical power from the power source flows through an unbroken path to the load, flows through the load, and then returns back to the power source. In other words, when we turn the switch on, electrons from the negative battery terminal travel to the positive battery terminal. This flow of electrons through a circuit is called electric current. In contrast, in an open circuit, the switch

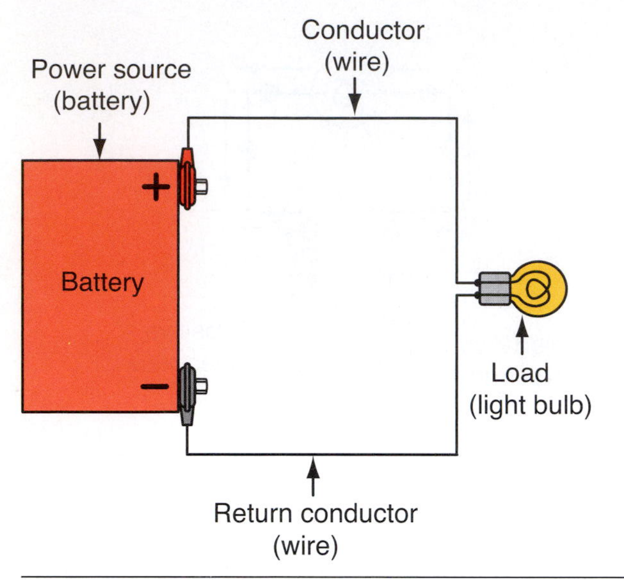

Figure 14-5 There are three requirements to complete a circuit—a power source, a conductor, and a load.

is turned off, which breaks the path of the circuit so power doesn't reach the load. A simple flashlight circuit is shown in Figure 14-6. The power source in this circuit is a battery. The conductors are copper wires. The load is a light bulb. In Figure 14-6A, the switch is open (which would be considered in the OFF position). The electrical circuit is therefore open, and power can't flow through the wires to reach the bulb. In Figure 14-6B, the switch is closed (in the ON position). This circuit is complete and electricity flows through the wires to reach the bulb, causing the filament to heat up and glow. The simple circuit that we have just described is known as a series circuit.

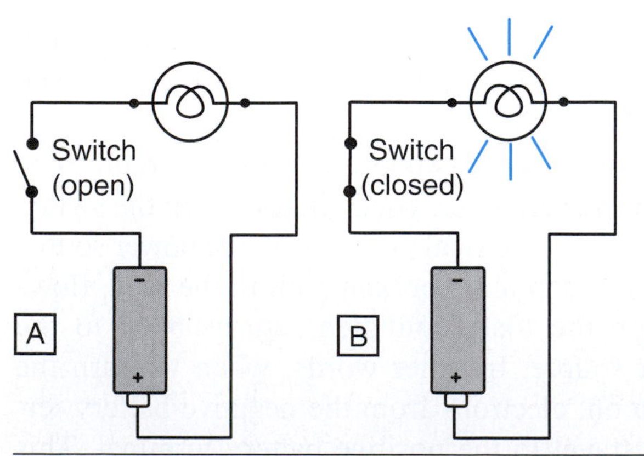

Figure 14-6 This figure shows a simple circuit. In (A), the circuit is open. In (B), the circuit is closed.

You should note that all electrical systems could be broken down to simple circuits very similar to the circuit that we have just discussed!

Electron Flow in a Circuit

Let's take a closer look at how electrons flow in an electrical circuit. Figure 14-7 shows a simple series circuit in which a copper wire is attached to a battery. One section of the copper wire is enlarged so that you can see how electrons flow through it.

In the figure, the circuit is closed, and the electrons from the negative battery terminal are drawn to the positive terminal. Remember that the outermost electron in each copper atom is easily dislodged from its orbit. An electron is drawn from the negative battery terminal into the copper conductor wire. This electron then collides with a free electron in a copper atom, bumping a free electron and taking its place. The displaced free electron moves to a neighboring copper atom, bumps a free electron out of the copper atom's orbit, and takes its place. As this chain reaction continues, free electrons bump their neighbor out of orbit, taking

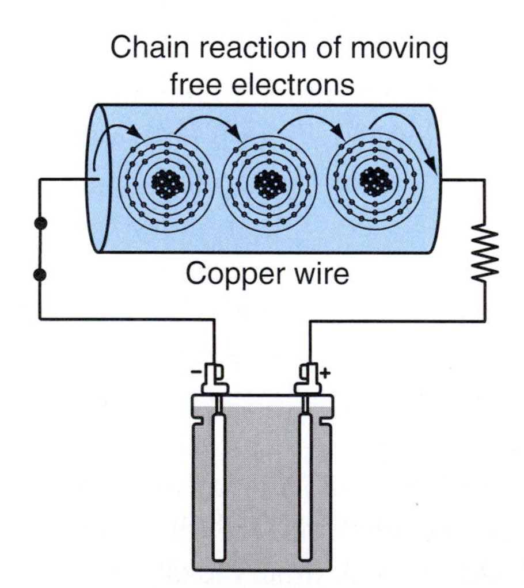

Figure 14-7 A section of the conductor in this illustration has been enlarged so that you can see how electrons flow through the wire. A free electron from the battery enters the wire. The free electron then creates a chain reaction within the wire where free electrons bump other electrons from the outer shell of the atoms. Remaining free electrons are drawn to the positive side of the battery, completing the circuit.

their place. This chain reaction of moving electrons is called electric current.

In reality, of course, atoms are much too small to see, so we can't follow the movement of just one electron through a wire. Many millions of copper atoms make up a single strand of wire. When a circuit is closed, millions of electrons move through the wire at the same time, at a high rate of speed; consequently, the more electrons moving through a circuit, the higher the current in the circuit.

Types of Circuits

As we discussed earlier, there are different types of circuits used in electrical systems (Figure 14-8). We've already talked about one simple electrical circuit—the **series circuit**. A series circuit is a circuit that has only one path back to its source. In a series circuit, if one light bulb burns out, the whole circuit shuts down because there's no path for the electricity to continue to flow. A **parallel circuit** is a circuit that has more than one path back to the source of power. In a parallel circuit, if a light bulb burns out, it won't have any effect on the other bulbs because they each have a separate return path to the source of power. A popular type of circuit that you'll find on motorcycles is a combination of the series and the parallel circuits, called the **series/parallel circuit.** The series/parallel circuit contains a load in series and a parallel load in the same circuit.

Unwanted Circuit Conditions

There are different electrical circuit conditions that have an adverse effect on electrical systems. These circuit conditions are opens, shorts, and grounds.

As you already know, an **open circuit** is a circuit that has an incomplete path for current to flow. An example of an *unwanted* open circuit is a broken wire or a blown light bulb.

A **short circuit** is a circuit that has developed a path to the source of power before it reaches the load in the circuit. A short circuit will blow fuses in the circuit as well as damage wires and components in the electrical system.

A **grounded circuit** is a circuit that allows the power to flow back to the source after the load, but before the means of control.

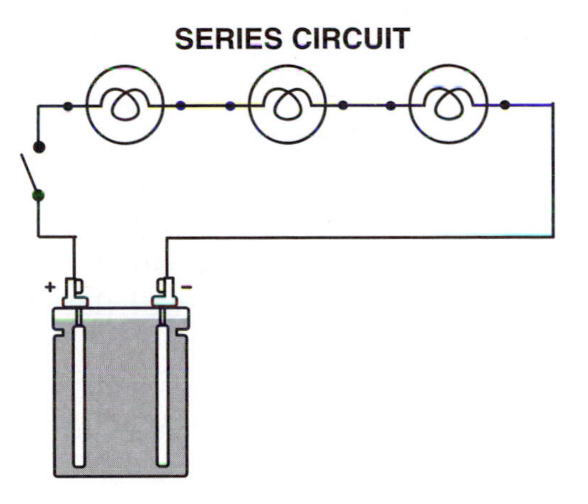

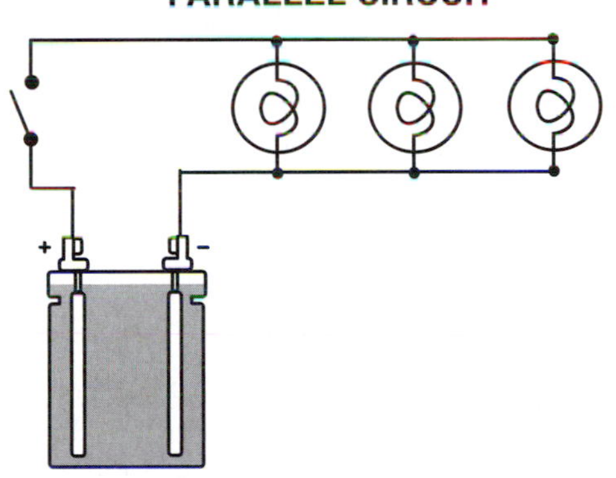

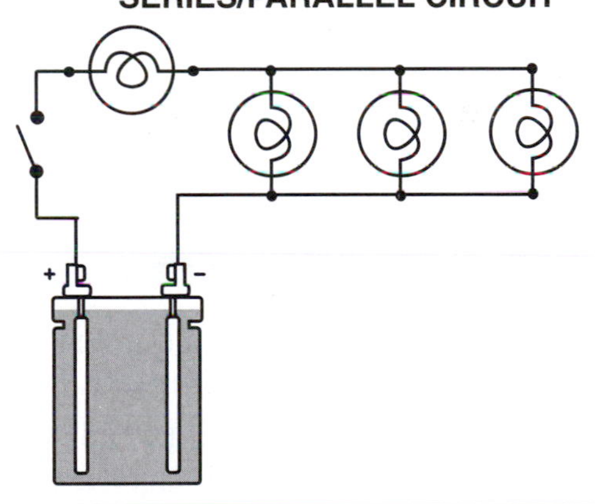

Figure 14-8 Series, parallel, and series/parallel circuits.

Properties of Electricity

Confused yet? If so, don't worry, we understand. Lets take a look at the properties of electricity in a way that may make all of this easier to comprehend. Electric current flowing through a wire can be compared to water flowing through a pipe (Figure 14-9). The laws governing electric circuits are easily explained by this analogy.

Voltage

Water pressure is measured in pounds per square inch, while electrical pressure is measured in volts. When the two water tanks in our illustration, A and B, are connected, water flows from tank A to tank B.

This flow is the result of a height difference between the two tanks. Water will flow through the pipe from the full tank to the empty tank until the water level is even in both tanks. The pressure from the weight of water in the full tank will naturally cause the water to flow. A valve could be installed to open or close the water passage and another could be used to slow the flow or increase it as desired.

Similarly, electrical current will flow through a wire due to electrical "pressure" created by a battery or alternator. The electrical pressure is measured as "voltage (V)."

Resistance

Water will have a lower rate of flow through a smaller or longer pipe due to increased resistance (Figure 14-9). Similarly, electrical current will have a lower rate of flow through a smaller or longer conductor such as wire. Partially closing off the flow from the pipe decreases water flow. Similarly, a load or resistance such as a light bulb in an electrical circuit decreases voltage.

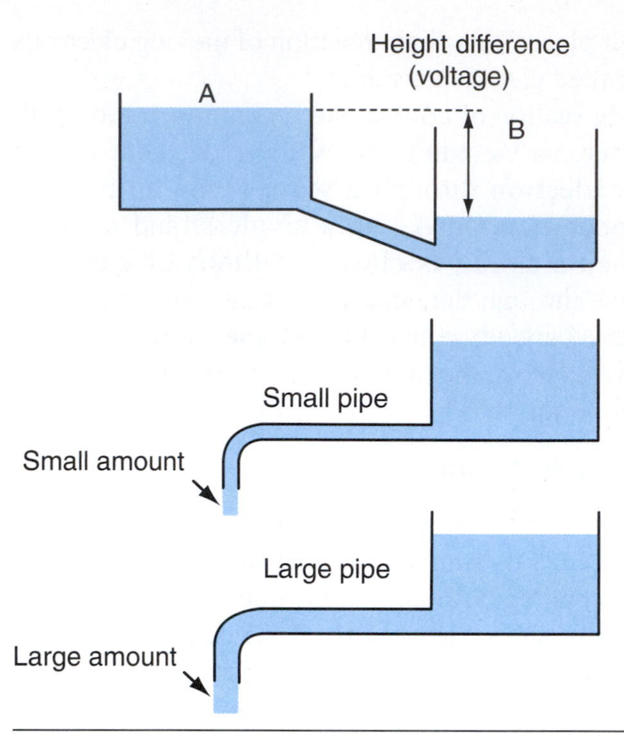

Figure 14-9 Electric current flowing through a wire can be compared to water flowing through a pipe. When the two water tanks are connected, water will flow from tank A to tank B. This flow is the result of a pressure difference that is similar to the pressure difference found in voltage. When we add a small pipe to a water tank less water flows as compared to a larger pipe. This relationship is comparative to electrical resistance.

UNITS OF ELECTRICITY

We just briefed you on a couple of the key electrical units of measurement, voltage, and resistance. Throughout this chapter, you'll learn terms that are used in connection with electrical systems and will learn some basic formulas. There are three basic units of measurement used to measure electricity: current, voltage, and resistance.

Each unit of measurement is named after a famous experimenter in electricity: The **amp** after Frenchman Andre M. Ampere, the **volt** after Italian Alessandro Volta, and the **ohm** after German Georg Simon Ohm.

Current (Amperes)

As you've already learned, when a complete conducting path is present between two opposing electrical charges, electrons will flow between the two points. **Current** is the rate of flow of electrons through a conductor. Current is measured in units called amperes, or amps, which is often abbreviated with the letter "A." For instance, the quantity 3 amperes would be abbreviated 3A. In other electrical-related work, electrical drawings, diagrams, and mathematical formulas, the letter "I" is used to represent current. Small amounts of current can be measured in milliamperes, which is abbreviated mA. One milliampere of current is equal to one-thousandth of an ampere, or 0.001A of current.

Voltage (Volts)

Now, let's look at the electrical quantity called voltage. Remember that in a battery, one terminal has a negative charge and the other terminal has a positive charge. Whenever a positive charge and a negative charge are positioned close to each other, a force is produced between the two charges. This force is called **electrical potential**. Electrical potential is simply the difference in electrical charge between the two opposing terminals. Electrical potential can also be thought of as the amount of electrical pressure in an electrical system. The bigger the difference between the two opposing charges, the greater the electrical potential will be. Voltage is a measurement of the amount of electrical potential in a circuit. Voltage is measured in units called volts, which is often abbreviated with the letter "V." For instance, the quantity 12 volts would be abbreviated 12V. In electrical diagrams and mathematical formulas, the letter "E" usually represents voltage.

Resistance (Ohms)

The last electrical quantity we'll look at is called resistance. Resistance is a force of opposition that works against the flow of electrical current in a circuit. You've already seen that current flows easily through copper wires in a circuit. However, frayed wires, corroded connections, and other obstructions will reduce the flow of electrons through a circuit. That is, the circuit will resist the flow of current through it. When a lot of resistance is present in a circuit, more voltage is needed to increase the flow of electrons moving through the circuit. Resistance is measured in units called ohms, which is often abbreviated with the Greek letter omega, represented by the symbol Ω. In electrical diagrams and mathematical formulas, the letter "R" is usually used to represent resistance. Motorcycle service manuals often provide electrical specifications in ohms. A service manual may tell you, for example, that the resistance you should be able to measure between the leads on a charging system stator should be .2Ω. Note that we will discuss charging systems, their components, specifications, and how to measure circuit quantities in more detail in an upcoming chapter.

Ohm's Law

Now, it seems reasonable that the letter "R" is used for resistance but you must be wondering why the abbreviation for voltage is the letter "E" and current is "I," right? Well, the letter "I" for current represents the "Intensity of electron flow" and the letter "E" is used to represent "Electromotive force."

The values of resistance, current, and voltage have a very important relationship in a circuit. The amount of current flowing through a completed circuit is directly proportional to the voltage applied to the conductor. This relationship between resistance, current, and voltage is known as Ohm's law. Ohm's law states that a resistance of one ohm (1Ω) permits a current flow of one ampere (1A) in a circuit that has a source voltage of one volt (1V). This rule can be applied to any voltage, resistance value, or amperage. If you have any two of the values, the third can be determined by using Ohms law.

This relationship that's summarized by Ohm's law is expressed with the mathematical formula:

Voltage (E) = Current (I) × Resistance (R)

Two other useful variations of the Ohm's law equation are:

Current (I) = Voltage (E) divided by Resistance (R)

Resistance (R) = Voltage (E) divided by Current (I)

To help you understand how to use these equations, use the illustration in Figure 14-10 and let's look at the following example. As you can see in the illustration, if you cover the unknown quantity with your finger, you can determine the unknown by using the remaining pieces of the formula.

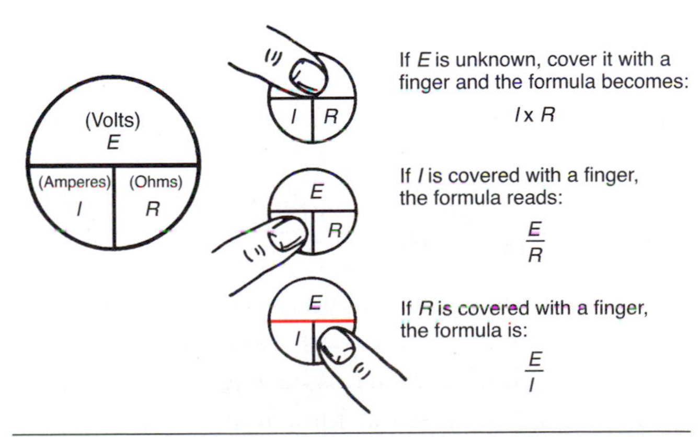

Figure 14-10 This illustration shows how Ohms law works. By covering the unknown quantity with your finger you can determine the unknown by using the remaining pieces of the formula.

If you have a circuit that draws 3 amps of current from a 12-volt battery, how much resistance is in the circuit? To solve this equation, simply install the known measurements in the formula as follows:

Resistance (R) = 12 (E) divided by 3 (I)

The answer is 4 ohms.

Ohm's law is a very useful formula that you should know. The Ohm's law formula is frequently used to analyze circuits and troubleshoot problem areas. By using these three given variations of the Ohm's law formula, it is easy to find the proper voltage, resistance, and current values for any circuit.

You should note that as the resistance in a circuit increases, the current decreases. Conversely, if the resistance in a circuit decreases, the current increases. All circuits are designed to carry a particular amount of current. In fact, most circuits are protected by fuses that are rated at an amperage value that's just slightly higher than the current value of the circuit. Thus, if a problem develops in a circuit, the circuit will draw too much current from the battery and the fuse's elements will melt (the fuse will blow), creating an opening in the circuit. This design prevents any further damage to the electrical circuit from occurring.

Electrical Power

When consumers buy electricity, they buy power. The unit of power for electricity is the watt. A simple formula for relating watts to voltage and current is:

Power = voltage × current, or watts = volts × amps

Thus, if there are 5 amps going through a resistor due to a voltage of 200 volts, the power consumed by the resistor is:

Power = voltage × current = 200 volts × 5 amps = 1,000 watts

These measurements, plus an understanding of the nature of electricity, are essential to anyone working with electricity. The user of the multimeter should have some knowledge of the operation and mechanics of the particular circuits and/or the device being tested.

Current, Voltage, and Resistance Relationships

Let us use the water analogy illustration once again to help better understand the relationship among current, voltage, and resistance in an electrical circuit and compare it to an electrical circuit example (Figure 14-11).

- The water pipes form a path for the water to follow. The water pipes are similar to the conductors in the adjacent electrical system.
- The water valve turns the flow of water on and off. The water valve is similar to the switch in the electrical system.
- The water wheel is being operated by the flow of water. The water wheel is similar to the light bulb (the load) in the electrical circuit.
- The water reservoir (the water source) is similar to the battery (the power source) in the electrical circuit.
- The flow of water is similar to the flow of electrons. The amount of flow would be the current.
- The water pump is the pushing force that causes the water to flow into the pipes, just as voltage does in the electrical circuit.

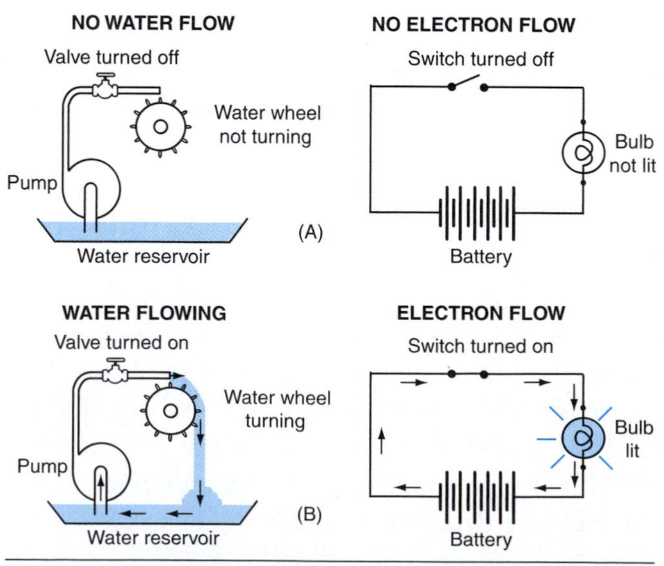

Figure 14-11 The basic principles of electricity can be easily visualized when you compare an electrical circuit to a water system.

In the illustration in Figure 14-11A, both the water system and the electrical circuit are turned off. Both the water valve and the electric switch are in the off position, so no water or current flows. The water wheel doesn't turn and the light bulb doesn't illuminate. In the illustration in Figure 14-11B, in the water system, the water valve is turned on. Water is pumped out of the reservoir and into the pipes; the water flows through the pipes, turns the water wheel, and then returns to the reservoir. In the electrical circuit, the switch is also turned on. Electric current flows out of the battery through the wires, lights the bulb, and returns to the battery.

In this example, you can think of resistance as being like a blockage or a clog in the water pipe. If some debris were stuck in the pipe, the flow of water through the pipe would be reduced. Similarly, excessive resistance in an electrical circuit reduces the flow of current through the circuit.

AC/DC Current and Voltage

There are two different types of electrical current—direct current and alternating current. It is important that you understand the differences between these two types of current.

Direct Current

Direct current (DC) is the flow of electrons in one direction only. A DC voltage reading is non-varying and on a motorcycle is usually produced by a battery. For example, if we were to graph a DC voltage of 12 volts over a period of time, the graph would appear as shown in Figure 14-12. Whatever the voltage value, a DC voltage remains constant and unchanging over time.

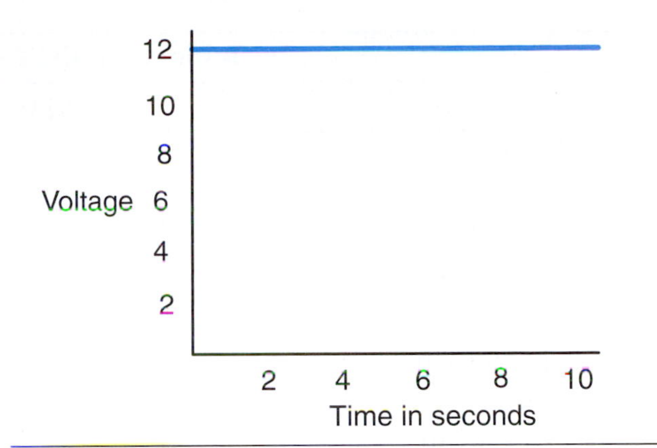

Figure 14-12 A DC voltage level remains constant over time.

Alternating Current

In contrast to direct current, alternating current (AC) is the flow of electrons first in one direction, and then in the opposite direction. Alternating current reverses direction continually and is produced by an AC voltage source. Alternating current is the type of current found in household electrical systems and wall outlets. Motorcycles also produce AC in various ways (We'll discuss alternating current in more detail in Chapter 15). As you see in Figure 14-13, the alternating current starts at zero, then rises to a maximum positive value. At the maximum positive point, the current reverses direction and falls back to zero and continues to drop until it reaches the maximum negative value. The current then reverses direction again and rises back to zero. One complete transition of the current from zero to the positive peak, down to the negative peak, and back up to zero is called a cycle. These alternating current cycles repeat continuously as long as the current flows.

As related to motorcycles, there are three key items needed to produce AC voltage: a magnetic field, a conductor, and motion.

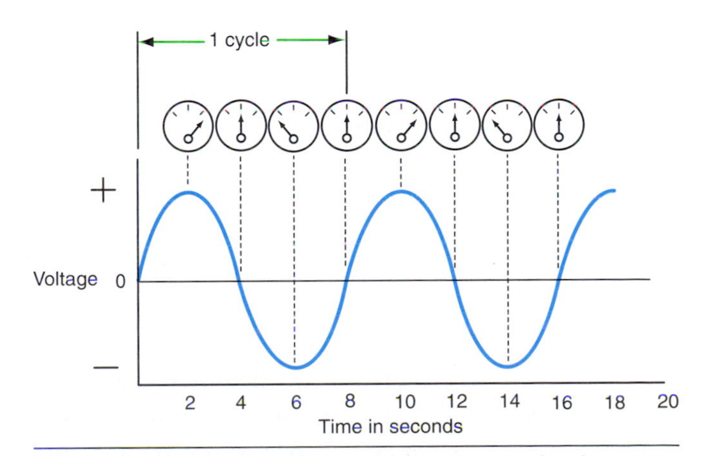

Figure 14-13 An AC voltage level changes over time.

Table 14-1: ELECTRICAL QUANTITIES		
Unit	**Abbreviation**	**Value**
Ampere	A	1 ampere
Milliampere	mA	0.001 ampere
Volt	V	1 volt
Kilovolt	KV	1,000 volts
Millivolt	mV	0.001 volt
Megavolt	MV	1,000,000 volts
Ohm	Ω	1 ohm
Megohm	MΩ	1,000,000 ohms
Kilohm	KΩ	1,000 ohms

Motorcycles that do not have a battery or a DC storage device use AC voltage and current for their lighting and ignition systems. Machines that use a battery use the DC voltage produced by the battery to power the starter, lights, horn, and other accessories. These machines also use AC voltage to keep the charging system working properly to keep the battery's voltage level correct.

Common Electrical Quantities

Table 14-1 has common electrical quantities, their abbreviations, and their values. You should become familiar with these abbreviations, as you will see them in different areas of a service manual when working with the electrical systems on motorcycles.

ELECTRICAL METERS AND MEASUREMENTS

Although we can observe the effects of electricity, such as a glowing light bulb, we can't see the flow of electrons that we call electricity. We can, however, use various meters to observe the action of electric current in a circuit. When you begin working with electrical meters, you'll notice that there are two basic types of readouts, or displays, that meters use to present the information: analog and digital. The function of these two types of meters is the same—displaying electrical information, but the ways in which the each display data is different.

Analog Electrical Meter

The display of an analog electrical meter has a movable pointer and a scale (Figure 14-14). The meter is usually enclosed in a case and has terminals (test leads), which connect to jacks on the front of the case. In most cases, a red jack indicates a positive terminal and a black jack is negative.

Most analog meters have scales from zero up to some maximum number. Some meters may have zero centered in the middle of the scale with numbers on the right and left. Most analog meters have what's called a mechanical zero adjustment. This means that by turning a screwdriver inserted into a small screw on the front of the meter, the pointer can be adjusted so that it's exactly over the zero on the scale. (During this adjustment the meter shouldn't

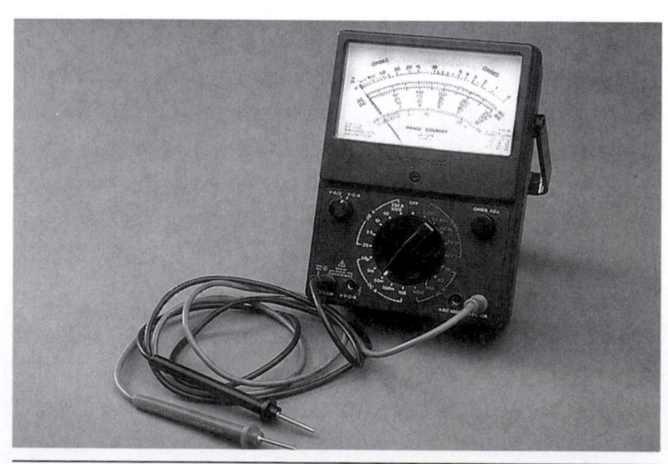

Figure 14-14 An analog electrical meter.

be connected to any circuit.) The amount of movement of the pointer is called pointer deflection.

Digital Electrical Meter

The display of a digital electrical meter has a numeric readout (Figure 14-15). Like the analog meter, the digital meter is also enclosed in a case and has positive and negative terminals (test leads) that connect to jacks on the front of the case.

Electrical Meter Uses

As you use electrical meters, you'll find that there are three primary types: voltmeter, ammeter, and ohmmeter. Most meters come with a variety of these, which describes a multi-meter. Keep in mind that the display of these meters may be either analog or digital.

Voltmeter

Voltmeters are used to measure the voltage, or potential difference, between two points (Figure 14-16). A voltmeter can be used to check voltage at any point in a circuit. Remember that voltage is like pressure and exists between two points; it doesn't flow like current. Therefore, a voltmeter isn't connected in series, but must be connected across a circuit, or in parallel (Figure 14-17). You can measure volts in both AC and DC.

Figure 14-15 A digital electrical meter.

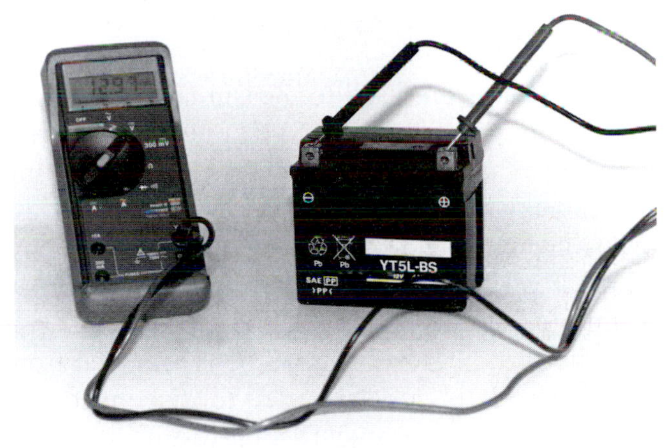

Figure 14-16 Voltmeters are used to measure the voltage between two points.

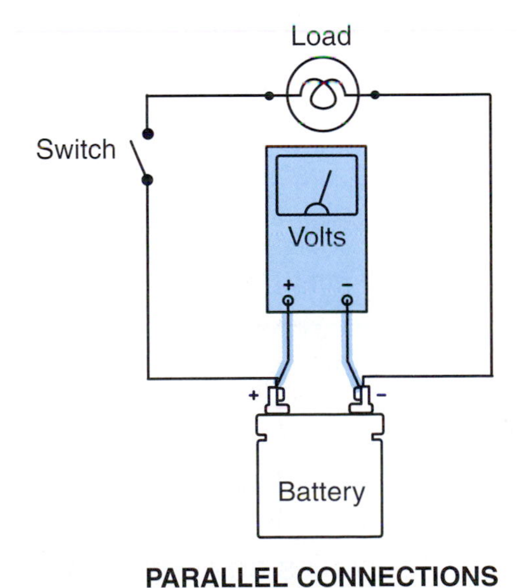

Figure 14-17 A voltmeter is connected in parallel to check the voltage of the battery.

Ammeter

Ammeters are used to measure the current flow through a circuit. As we discussed before, current is measured in amperes (or amps). The scale of an ammeter shows the number of amps in a particular circuit. Unlike a voltmeter, ammeters are always connected in series in a circuit (Figure 14-18). An ammeter must be connected in series because the entire current must flow through both the circuit and the ammeter. Like voltmeters, there are both AC and DC ammeters.

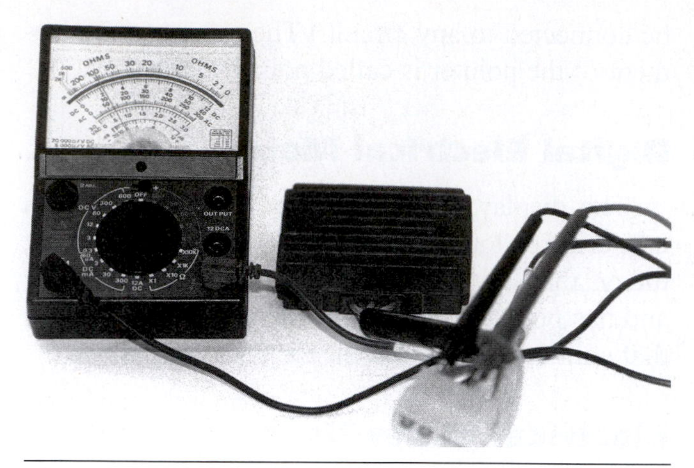

Figure 14-19 Ohmmeters are used to measure the resistance of a circuit or component by applying a known voltage to the circuit and measuring the resulting current.

Ohmmeter

Ohmmeters are used to measure the resistance of a circuit or component by applying a known voltage to the circuit and measuring the resulting current (Figure 14-19). Ohmmeters usually have a built-in power supply such as a 9 volt battery, which supplies the voltage to test the part. Thus, when connecting an ohmmeter to a circuit, you must be certain the power source is removed from the circuit. It is good practice to disconnect the battery of the motorcycle when testing using an ohmmeter.

Multi-Meters

A multi-meter is a meter that combines the testing capabilities of a voltmeter, ammeter, and ohmmeter into one meter. Multi-meters are the most popular item among electrical testing equipment in the motorcycle industry. The multi-meter may be referred to as a volt/ohmmeter (VOM) or if it is a digital meter with ammeter testing capabilities, it may be referred to as a digital volt/ohm/amp meter (DVOA). The DVOA is also called a digital/multimeter (DMM).

The meter illustrated in Figure 14-20 is an example of a DMM. A dial on the front of the multi-meter is used to select what you want to measure—voltage, current, or resistance. Many

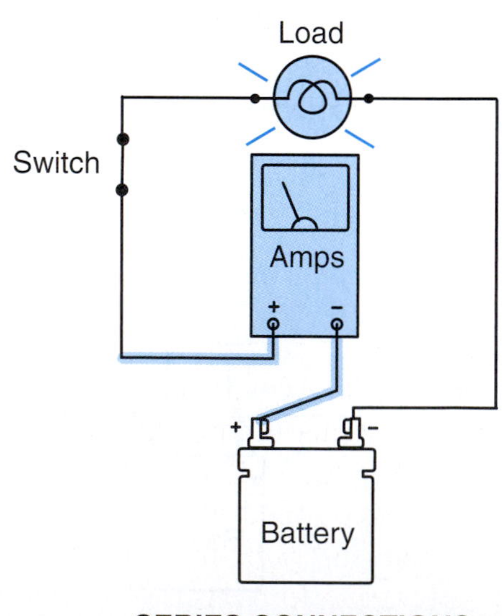

Figure 14-18 An ammeter is connected in series to check the current of the draw of the light bulb in this illustration.

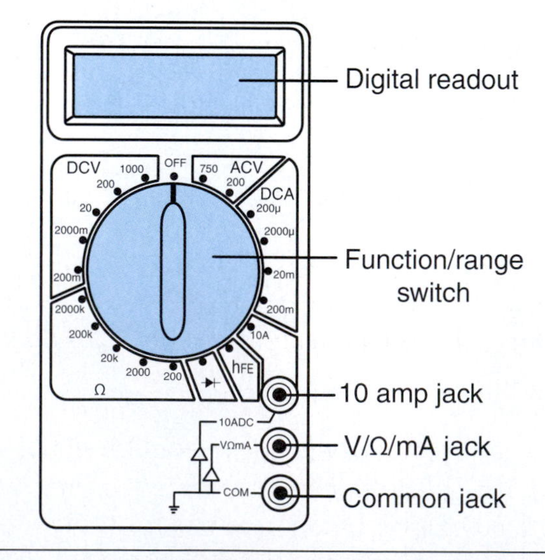

Figure 14-20 A multi-meter is a meter that combines the testing capabilities of a voltmeter, ammeter, and ohmmeter into one meter and is the most popular electrical testing tool found in a technician's toolbox.

multi-meters will also have range selectors that can be set for the quantity being measured. For example, in our illustration, you can see multiple DCV ranges: 200 millivolts, 2,000 millivolts, 20 volts, 200 volts, and 1,000 volts. When the selector switch is in the 20V position, the meter will only measure from 0 to 20 volts, which is the most common range tested on motorcycles.

You will also note the jacks that are used to hook up the leads for different testing applications. Having the test leads in the correct jack is crucial when doing electrical tests.

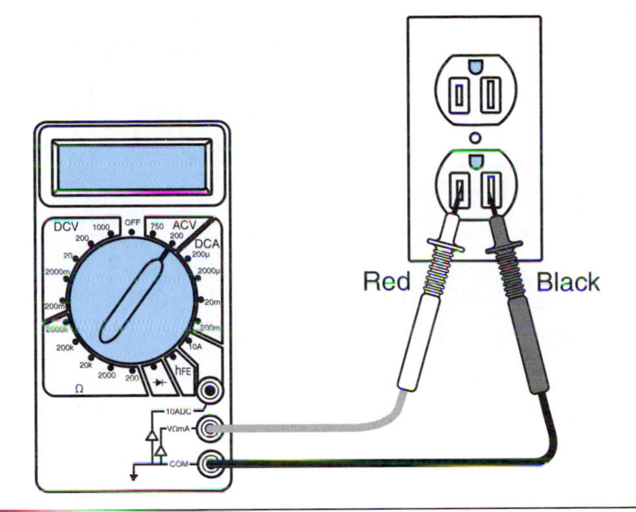

Figure 14-21 Measuring AC voltage from a home electrical outlet.

Multi-Meter Operation

Since they are the most popular in the motorcycle industry, we'll cover the basic information you need to know about operating the multi-meter primary functions. First, let us discuss some basic information.

Measuring AC

The connection of the positive and negative terminals of an AC voltmeter or ammeter doesn't matter because, as you'll recall, AC current is constantly reversing itself (Figure 14-21). Therefore, the positive and negative terminals are constantly alternating from positive to negative and negative to positive.

When measuring AC voltage, the following requirements must be met:

- Whenever you're checking for any kind of voltage (AC or DC), the meter must be connected in parallel to the circuit.
- We briefly mentioned earlier that to produce AC voltage, you must have three items: a magnet, a conductor, and motion. To get an AC voltage reading on a motorcycle, there must be engine motion. That is, the crankshaft of the motorcycle you are testing must be turning.

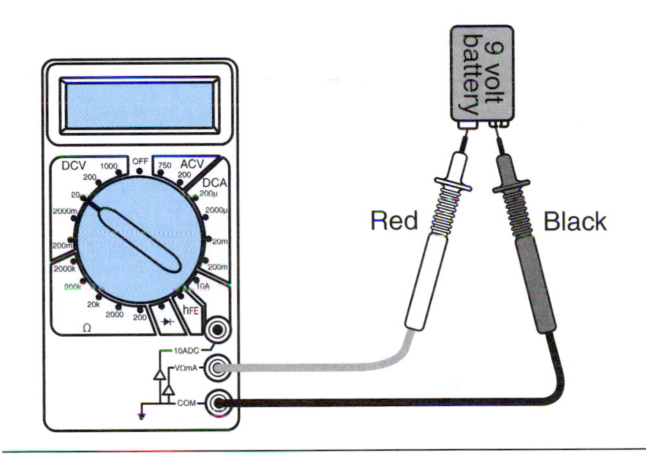

Figure 14-22 Measuring DC voltage across a 9 volt battery. Note that the leads must be connected correctly to provide an accurate reading.

Measuring DC

The connection of the positive and negative terminals of a DC voltmeter or ammeter is important.

When measuring a DC circuit, you must be sure to connect the meter so that the negative terminal of the meter is connected in the circuit toward the negative terminal of the battery (Figure 14-22). Likewise, the positive terminal of the meter must be connected in the circuit toward the positive terminal of the battery. If the meter is improperly connected, it will read exactly opposite of the actual measurement!

When working on motorcycles, the only current readings that you'll take will be for DC current. Be sure to hook the meter up in series when checking for DC current (Figure 14-23). Also be sure to hook up the meter leads correctly. One way to verify that the leads are correctly hooked up is to turn the power on after the meter is attached and look at the reading while the engine isn't running. The meter must read a negative number if the key is on and the engine is off. If the meter is reading a positive number, simply switch the meter leads and check again.

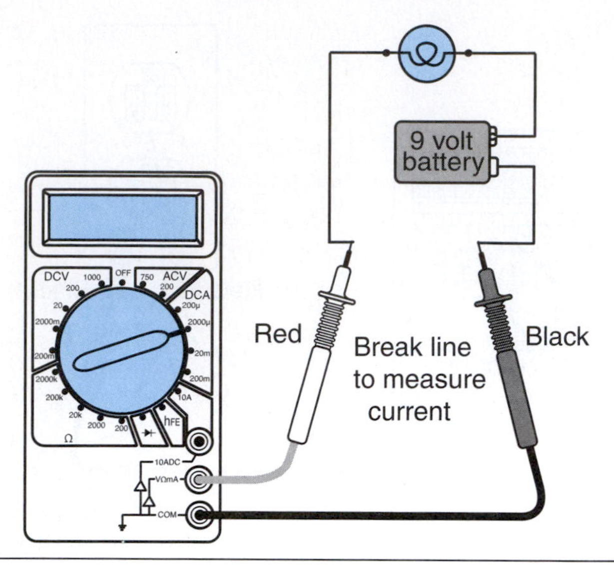

Figure 14-23 An ammeter must be connected in series to the circuit to allow the current to flow through the meter.

WARNING *Never electric start a motorcycle engine while the ammeter is hooked up in series to the battery. The meter is not designed to handle the large amount of amperage that the starter motor requires to turn the engine over, which will almost certainly damage the meter.*

A type of ammeter commonly used in connection with motorcycle electrical troubleshooting is a 20-0-20 DC ammeter. This meter has a scale with a zero in the center and the number 20 on the far right and left of the scale. Thus, 20 amps is the full-scale value, or maximum amperage, that can be measured with this instrument. The pointer rests over the zero when no amps are being measured. When current is being measured, the position of the pointer to the right or left of the zero not only tells you the rate of current flow, but also whether current is flowing into or out of the battery. If the needle points to the right side of the scale, it indicates a flow of direct current into the battery. In other words, the battery is being charged. If the needle points to the left side of the scale, it indicates that current is flowing from the battery. In other words, the battery is being discharged. Let us look at a couple of examples.

On a motorcycle, AC current flows from an alternator to the battery where it's then stored (Figure 14-24). Before it reaches the battery, it flows through a component called a **rectifier**, which changes the current from AC to DC. Therefore, to check the current, you would connect a DC ammeter between the rectifier and the battery. In this example, our ammeter is reading a current of about 17 amps. This tells us that the battery is being charged.

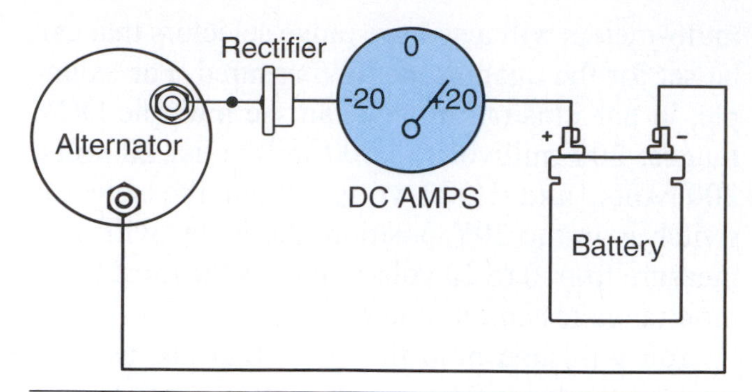

Figure 14-24 This ammeter is measuring a current flow into the battery at a rate of about 17 amps.

If an ammeter were connected between the lights and the battery on our motorcycle with the key on and engine off), the ammeter would read a current of negative amps (Figure 14-25). In this scenario, the battery is being discharged.

Measuring Resistance

As we discussed earlier, whenever you're using an ohmmeter, it's very important to first disconnect the component being tested from the rest of the electrical system (Figure 14-26). In other words, isolate and de-energize the component. If you don't isolate the component from the rest of the electrical system, you risk damaging the meter. You may also receive a false resistance reading, as

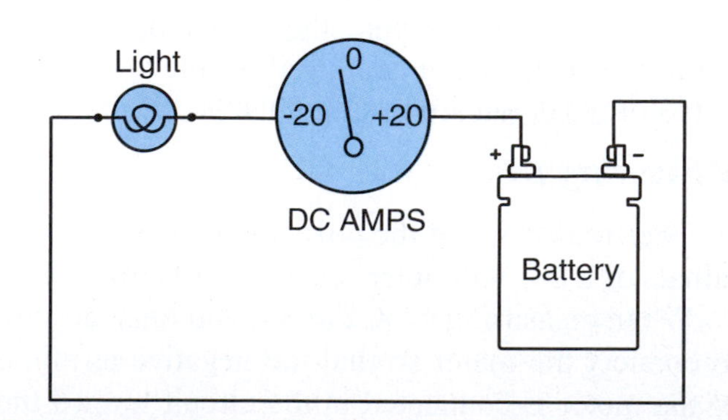

Figure 14-25 This ammeter is measuring a current discharge of about 3 amps from the battery to the lights.

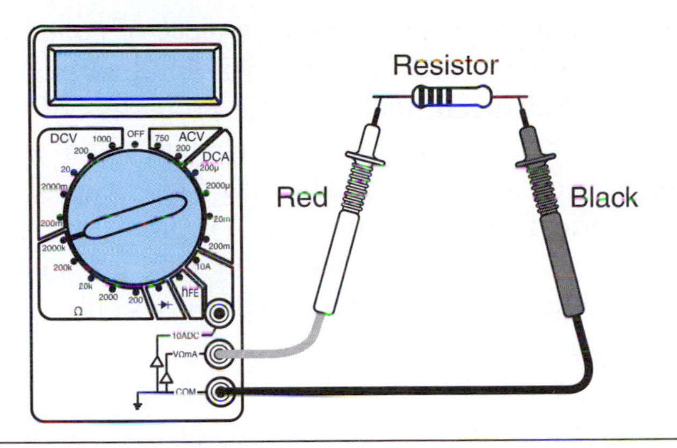

Figure 14-26 Resistance is measured with the component disconnected from the circuit.

there may be other resistance in the circuit that you are about to test.

Voltage Drop

As previously discussed, voltage is the electrical force or push of electricity. Voltage drop is the consumption of the available voltage from the battery as it crosses some form of resistance such as a light bulb or a switch. A voltage drop measurement allows you to identify if all of the available voltage is being used up by the component it is intended to power.

When the electrical current reaches a load such as a light bulb, current is forced through the filament of the bulb. Each connection in a circuit will offer a very slight resistance to electron flow. Normal voltage drop for a connection in a circuit is 0.1 (one tenth) volt or less.

In the motorcycle industry, voltage drop should not exceed the limits set for the following components:

- Wires or electrical cables: 0.2 VDC
- Switches: 0.3 VDC
- Electrical connections: 0.1 VDC
- Electrical grounds: 0.1 VDC

As you should recall, current remains the same throughout any circuit; what is lost or used up is the voltage. Therefore, if you measure 12 volts before the load and 0 volts after the load, all is working well. However, if the full 12 volts is not powering the load, there is unwanted resistance somewhere in the circuit. Now it must be determined if the unwanted resistance is before or after the load. You can quickly isolate the area of the unwanted resistance by dividing the circuit in two: positive and negative.

The purpose of any circuit is to allow an electrical load to operate as designed. No matter what the load is—a light bulb, clock, starter motor, or electronic control module—it must have power to it and it must have a ground to make the circuit complete.

The supplied voltage will be consumed by the load as it operates. This provides a clue that will make troubleshooting electrical problems much simpler. We know that voltage is going into the load and being consumed within the load. Therefore, in a basic circuit we should have full source voltage from the battery to the load, and because the load consumes the voltage as it operates, we should have 0 volts coming out of the load. Remember, you should expect about 12 volts (depending on the total source voltage on hand) on the positive side and 0 volts on the negative side of the load (Figure 14-27).

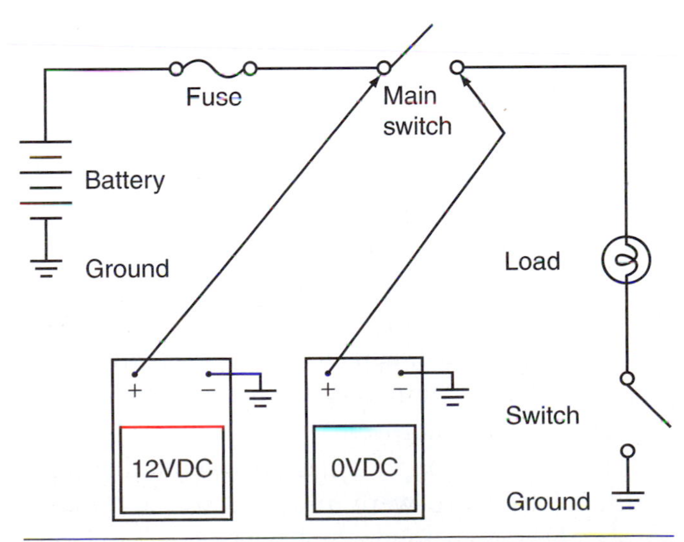

Figure 14-27 Testing the positive side of an electrical component in a circuit should show the source voltage, while testing on the negative side of the circuit should read zero or near zero volts. This is known as voltage drop.

Precautions When Using a Multi-Meter

You can destroy a multi-meter if you use it improperly. Here are some basic steps you should know about how to operate a multi-meter.

1. Determine what you want to measure (voltage, current, or resistance).
2. Set the meter up to the proper unit of measurement (volts, amps, or ohms).
3. Attach the test leads to the meter.
4. Select the quantity you want to measure by turning the dial.
5. Holding the two test leads, touch the probes to two points in a circuit.
6. Read the resulting information on the meter's display.

Note that this is a basic description of the operation of a typical multi-meter. The actual operation may be somewhat different than described on these pages so be sure to read the instruction manual provided with the meter.

As a general note, when you're using meters, you need to ensure that they're properly connected to the circuit being tested. Improper connection can result in an incorrect measurement and, in some cases, can damage the meter. When using an analog meter it's also important that you learn to read each meter scale properly.

MAGNETISM

Magnetism, like electricity, is a force we can't see. However, like electricity, we can observe its effects. The exact explanation of magnetism isn't completely understood, and most of this field is well beyond the scope of this textbook. However, it's important for you to understand some basic information about magnets so that you can better understand how alternators and generators produce electricity in a motorcycle.

Many years ago, scientists discovered that fragments of iron ore were attracted to each other. Researchers also found that when a magnetized iron bar was suspended in the air, one end would always point north. This was called the north pole of the magnet. The opposite end of the bar became the south pole of the magnet.

It was also found that when a piece of nonmagnetized metal, such as steel, was rubbed over a magnetized metal, the magnetic properties of the metal were transferred to the steel. The area affected by a magnet is called the field of force or magnetic field (Figure 14-28A). Note that the "lines" of force, or flux lines, as they are sometimes called, are for illustrative purposes only—we cannot actually see the lines.

As with electricity, one important property of magnets you should know is that opposites attract. When opposite poles of a magnet or magnets are placed near each other, they attract each other (Figure 14-28B). Conversely, when two like poles are placed together, they repel each other (Figure 14-28C). This is because the lines of force are going in opposite directions. Another property of magnets is that when a nonmagnetic substance

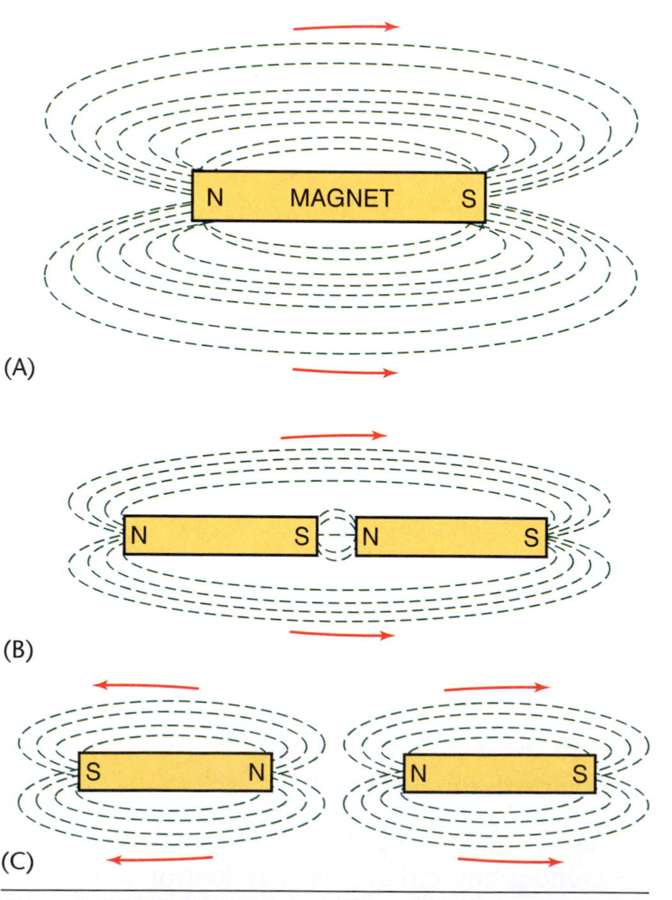

Figure 14-28 The area affected by a magnet is called the field of force or magnetic field (A). When opposite poles of a magnet or magnets are placed near each other, they'll attract each other (B). When two like poles are placed together, they repel each other (C).

(such as a piece of wood) is placed in a magnetic field, the lines of force aren't deflected. Magnetic forces pass through nonmagnetic materials.

Types of Magnets

Let's explore the three different types of magnets.

Magnetite

A natural magnet, called magnetite, comes in rock form. Magnetite is a weak magnet and isn't used in any motorcycle components.

Permanent Magnet

A man-made material, permanent magnets are very strong and long lasting. Permanent magnets are commonly found in different parts of motorcycles.

Electromagnet

An electromagnet, which is also man-made, is another commonly found magnet on motorcycles. An electromagnet consists of a coil wound around a soft iron or steel core. The core becomes strongly magnetized when current flows through the core and becomes almost completely demagnetized when the current is interrupted; hence, the term electromagnet, as it combines electric current with magnetic properties.

Magnetic Forces

Electromagnetism

The concept of electromagnetism is very important to the operation of electrical systems used in motorcycles. Electromagnetism is the magnetic effect produced when an electric current flows through a conductor (wire). When the current flows through the wire, the wire becomes surrounded by a magnetic field as seen in Figure 14-29. The magnetic field is strongest in the space immediately surrounding the conductor.

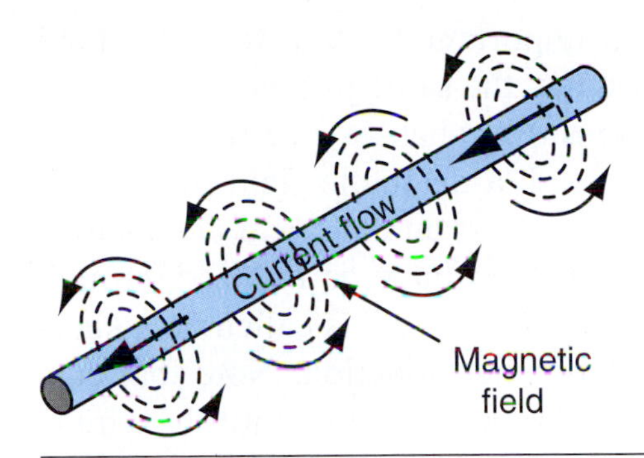

Figure 14-29 When current flows through wire, a magnetic field is produced.

Magnetic Coils The force of electromagnetism has many interesting and highly useful applications. If an insulated piece of conductor wire is looped around an iron bar to form a coil, the resulting device is called a magnetic coil (Figure 14-30). When current flows through a magnetic coil, each separate loop of wire develops its own small magnetic field. The small magnetic fields around each separate loop of wire then combine to form a larger and stronger magnetic field around the entire coil. The coil develops a north pole and a south pole. The magnetic field at the center of a magnetic coil is stronger than the fields above or below the coil.

Magnetic Induction

When a conductor (wire) is moved through a magnetic field so that it passes across the lines of force, an electromotive force (EMF), or potential voltage, is induced in the wire. If the wire is part of a complete electrical circuit, current will flow through the wire. This important fact is the basis for the various kinds of AC-producing generators used in motorcycles. This kind of generator may also be called an alternator. We'll explain electrical generating systems in greater detail in Chapter 15,

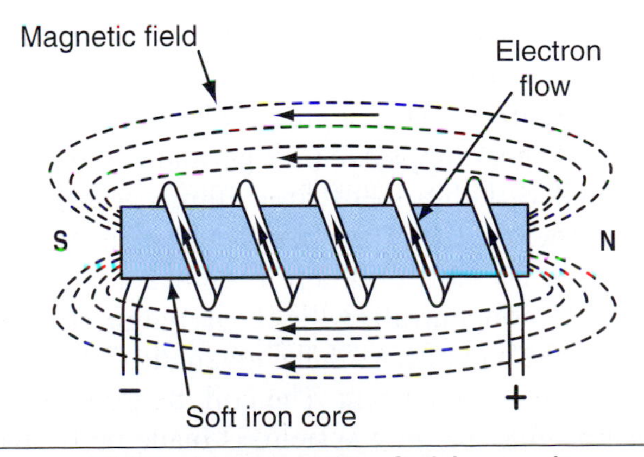

Figure 14-30 A magnetic field can become highly concentrated when an iron core is installed in a coil of wire.

but for now it's important for you to understand that each is based on the same principle.

The principle is that when an electrical conductor is passed through a magnetic field (or a magnetic field is moved past an electrical conductor), an electric current, or voltage, is induced through the conductor wire. This effect is called the generator action of magnetic induction. Note that current will not flow through the wire until the wire is connected in a complete circuit.

Mutual Induction

The final electromagnetic property we'll look at is called mutual induction. If two conductors are placed close together and current is applied to one of the conductors, a voltage will be induced in the other conductor. This occurs because when two conductors are physically close to each other, the energy in the "live" conductor will stimulate the other conductor to become energized too. This effect is called mutual induction, and it can be used to operate ignition coils. Note that if the conductors are moved apart from each other, the effect of mutual induction is not as great. If the conductors are moved far enough apart, the energy of the "live" conductor will not be strong enough to influence the second conductor, and the mutual induction effect will stop.

In Chapter 15, we will show you how the principle of mutual induction is used to help operate a motorcycle's ignition system.

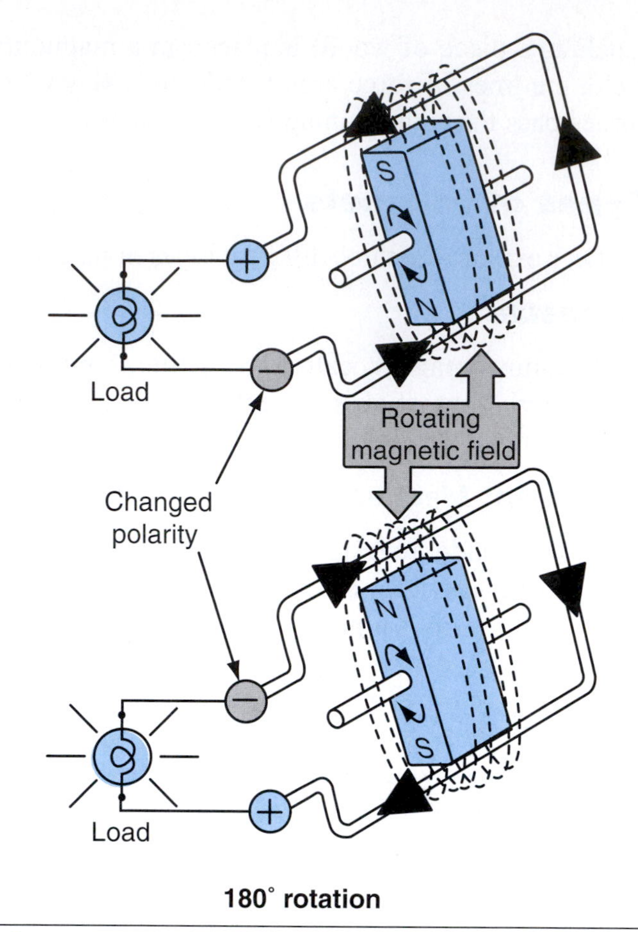

Figure 14-31 In this AC generator, as the magnet rotates, the induced current reverses.

AC Generator Operation

The generator action of magnetic induction is the basic property that is used to operate a motorcycle charging system. We'll explain this property using Figure 14-31 as a guide, which shows a very basic AC charging system. A permanent magnet is suspended within a soft iron frame, which completes the circuit for the permanent magnet's lines of force. The soft iron core becomes a temporary magnet, concentrating lines of magnetic force around the coil of wire in the magnetic field to produce an electric current. The coil, better known as a stator in a charging system, is made up of many loops of conductor wire. As the magnet rotates, the magnetic polarity of the soft iron frame is reversed. With each 180 degrees of rotation, the magnetic lines of force around the soft iron frame collapse and then reestablish themselves in the opposite direction. Each time the lines of force collapse and rebuild, the coil of wire within the magnetic field cuts them and an electric current is produced.

The voltage and current produced by the simple generator shown would be quite low. But if we wound many loops of wire into a coil and rotated the coil in the magnetic field, a much larger voltage and current would be produced. This is the arrangement in a real AC generator. The amount of voltage and current produced by a generator is based on the following three things:

- The number of turns in the coil and the diameter of the wire
- The strength of the magnetic field
- The speed at which the wire coil passes by the magnets

All motorcycles and ATVs that contain batteries use the AC generator action of magnetic induction to charge their batteries. In such machines, generators or alternators charge the batteries, and the energy from the batteries is then used to power the various electrical systems.

Solenoids

Some electromagnets have special movable cores. This type of electromagnet is called a solenoid or relay. Inside the solenoid coil, there is a movable piece of metal called a plunger. When a solenoid coil is energized by a flow of current, the resulting magnetic field moves the plunger in the coil. When the flow of current stops, a spring forces the plunger back into its original position. Solenoids are used in electric starter systems, as well as in many safety devices on motorcycles. Solenoids are designed in one of two ways:

- A **normally open solenoid** is a solenoid that doesn't allow current flow unless the solenoid is activated. This type of solenoid is found in electric starting systems and allows a high current flow after a very small current flow activates the solenoid.
- A **normally closed solenoid** is a solenoid that allows current to flow unless the solenoid is activated. This type of solenoid will be found in safety devices such as kickstand safety devices and creates an open circuit after a very small current flow activates the solenoid.

Electromagnetism in Motors

You've just learned that when a conductor moves through a magnetic field, a voltage is produced in the conductor. Now, suppose that a current-carrying conductor is placed in a magnetic field. What happens? Well, the interaction between the magnetic field and the moving electrons in the conductor causes a physical force to be applied to the conductor. If the conductor is free to move, this physical force will cause the conductor to move for as long as the conductor current and the magnetic field are maintained. This property is called the motor action of electromagnetic induction.

The motor action of electromagnetic induction is shown in Figure 14-32. In this figure, a conductor

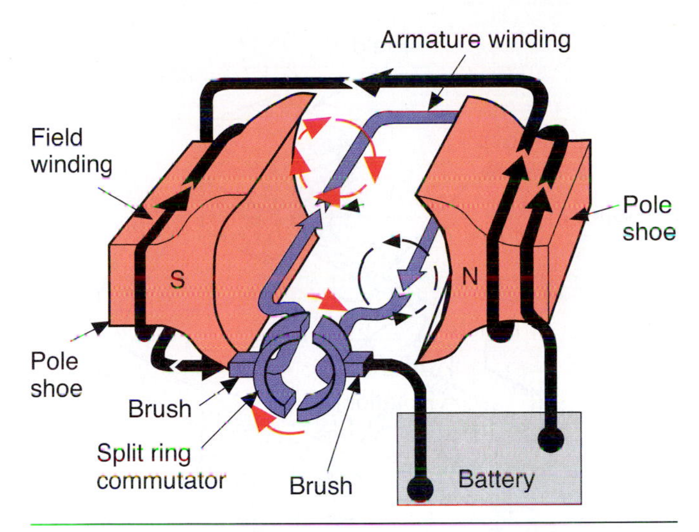

Figure 14-32 When the field windings are energized, the field magnets produce a magnetic field in the motor. When current flows through the armature windings, magnetic fields are produced around the windings. The interaction of these fields causes the armature to rotate.

(wire) is connected to a battery to form a complete circuit. Current is already flowing in the conductor when it's placed in a magnetic field between two magnets. The reaction between the magnetic field and the moving electrons in the conductor causes the conductor to rotate as shown by the arrow in the figure. The motor action of electromagnetic induction is the basic property that's used to operate electric starter motors.

An illustration of the basic parts of an electric starter motor is shown in Figure 14-33. In a starter motor, the armature is a rotating component that's mounted on a shaft and positioned between the motor's field magnets. Loops of conductor wire, called armature windings, are connected to the armature's commutator. Note that for simplicity, only one winding is shown in the figure. The brushes are electrical contacts that slide over the surface of the commutator as the armature rotates. The brushes are connected to a battery. Electrical wires, called field windings, are wound around the field magnets. When current flows into these wires, the field magnets become electromagnets and produce a powerful magnetic field inside the motor. When current is applied to the brushes, the current moves through the brushes and into the commutator and armature windings. The current flowing through the armature

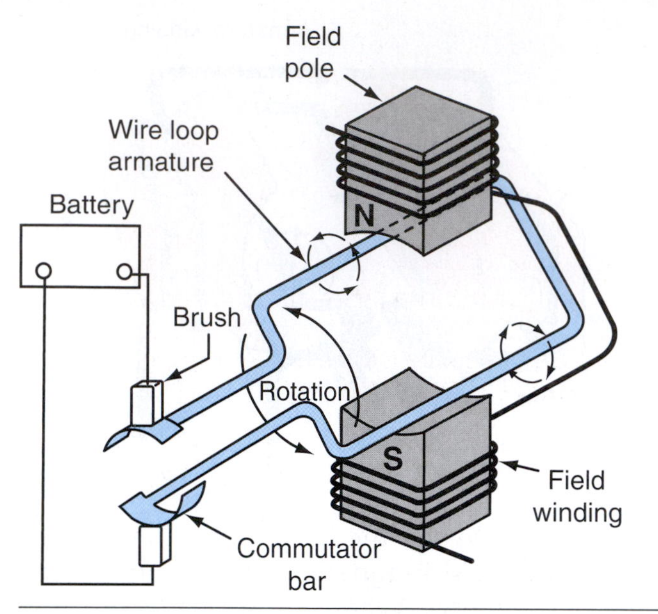

Figure 14-33 The basic parts of an electric motor.

windings produces magnetic fields around the windings. The interaction of all these powerful magnetic forces causes the armature to spin.

Many motorcycles use electric starter motors. The output shaft of the electric motor in such a system is generally connected to gears that engage the crankshaft. The spinning motion of the electric motor's armature is transferred through these gears to the crankshaft of the motorcycle engine. Some people may use the word "motor" when talking about either the electric starter motor or the motorcycle engine. Do not confuse the starter motor with an engine!

ELECTRONIC DEVICES

Now you should have a basic understanding of electrical and magnetic principles. Next, let's take a brief look at some electronic devices. Let's start by reviewing a few terms. You'll remember that a conductor, such as copper wire, is a material that allows electrical current to flow through it easily. An insulator, such as plastic or nylon, is a material that resists the flow of electricity through it. There are other materials called semiconductors, which as the name implies, allow some flow of electricity through them.

A semiconductor is a substance whose electrical conductivity is between that of a conductor and an insulator. A semiconductor's electrical conductivity also increases as its temperature increases. Silicon, germanium, and selenium are common semiconductor materials that are used to make electronic components.

Semiconductor devices are manufactured in laboratories under very special conditions. The semiconductor materials are specially processed and combined to form electronic devices such as diodes and transistors. Because of the way semiconductor materials are processed during manufacturing, the finished diodes and transistors are capable of controlling the flow of electrons. Because of these special manufacturing processes, the conducting and insulating properties of semiconductor materials can be used to perform useful work in a circuit.

Electronic Components

Electronic devices contain components that are used to control the flow of electrons in a circuit. Many different electronic components are used in circuits, but we'll look just at the most common ones used in motorcycle and ATV electrical systems. These devices are the diode, Zener diode, transistor, and silicon-controlled rectifier (SCR).

Diode

A diode is a simple electronic device that has two terminals and acts like a one-way valve. The two terminals are called an anode and a cathode. The anode is the positive (+) terminal, whereas the cathode is the negative (–) terminal.

When a positive voltage is applied to the anode end of a diode, electric current moves through the diode and exits at the cathode end. In this situation, the diode acts like a conductor. When a positive voltage is applied to the cathode end of a diode, the diode resists the flow. Current won't flow through the diode. In this situation, the diode acts like an insulator. Diodes allow current to flow through them in one direction only. The electrical symbol for a diode, shown in Figure 14-34, illustrates this principle.

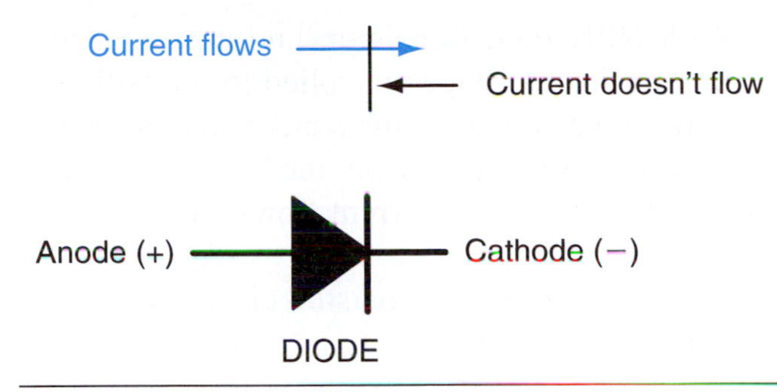

Figure 14-34 This is the electrical symbol for a diode. A diode will allow electrical flow in only one direction.

Zener Diode

Like a regular diode, a Zener diode allows current to flow in one direction. However, the Zener diode will also allow current to flow in the opposite direction if the voltage exceeds a predetermined value called the trigger or breakdown voltage (Figure 14-35). At the predetermined voltage, the diode becomes conductive in the opposite direction. This characteristic makes the Zener diode useful for voltage regulation in motorcycle charging systems.

Silicon-Controlled Rectifier (Thyristor)

A silicon-controlled rectifier (SCR) is another type of semiconductor component. SCR's are used as switching devices in electronic circuits. An SCR, is often called a thyristor, has three terminals: the anode, the cathode, and the gate. The construction of an SCR is similar to that of a diode, except that an SCR has an additional terminal—the gate (Figure 14-36).

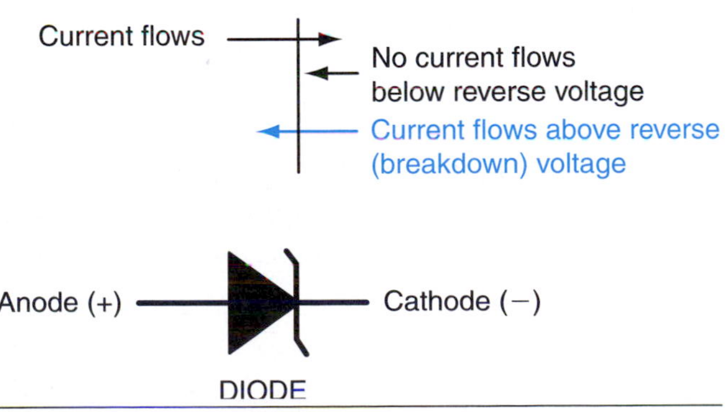

Figure 14-35 A Zener diode allows current to flow in the reverse direction when a predetermined voltage is sent in the opposite direction.

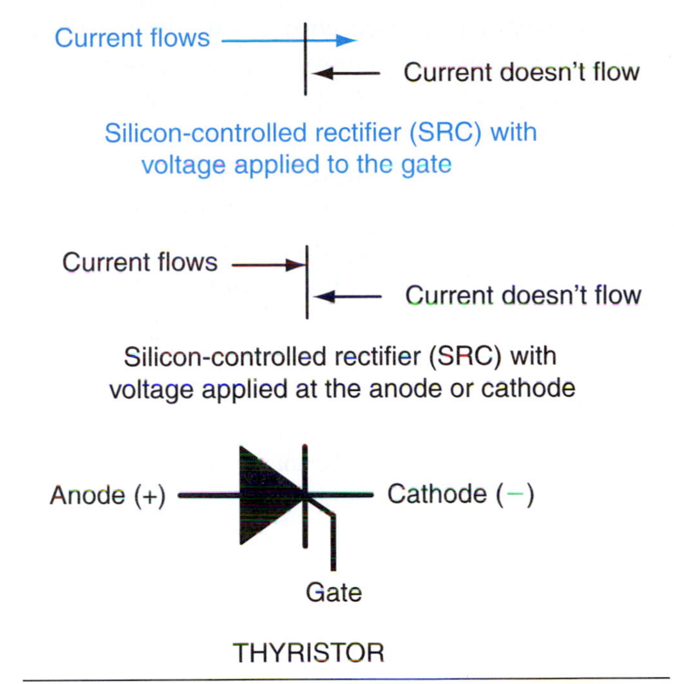

Figure 14-36 An SCR, or thyristor, is a diode with a gate that allows current to flow only when a predetermined voltage is applied to the gate.

Unlike a diode, an SCR will block current in both directions. If you apply a voltage across an SCR, current won't flow. If a small amount of voltage is applied to the gate of an SCR however, current will flow through the SCR in the forward direction. Current will continue to flow until the voltage is removed from the gate. Thus, an SCR can be switched on and off by applying a voltage to the gate. In Chapter 15, we will look at how these electronic components function in electronic ignition systems and charging system circuits.

Transistor

A transistor is another type of electronic device that's widely used in motorcycle electrical systems. A transistor is a semiconductor device. Transistors are used to control the flow of current in a circuit and they function like relays. They switch a current on and off when they receive a small current. The key difference between a relay and a transistor is that a transistor has no moving parts.

A transistor has three wire terminals: the base, the collector, and the emitter. There are two types of transistors—PNP and NPN (Figure 14-37).

With PNP-type transistors, when positive voltage is applied to the emitter and the base is at a lower voltage level, a small amount of current flows from the emitter to the base. At that time a large amount of current flows from the emitter to the collector.

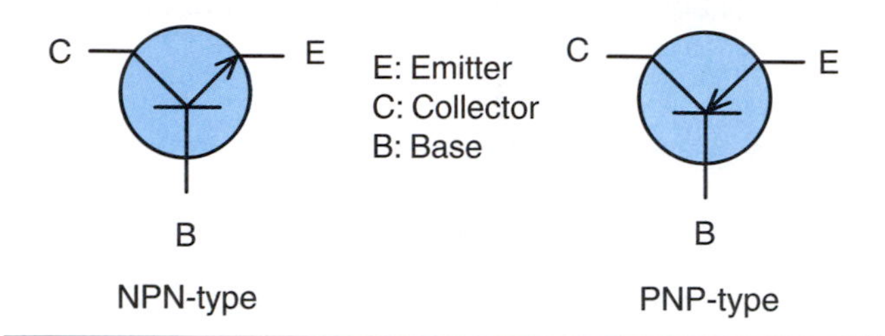

Figure 14-37 The electrical symbols of the PNP and NPN transistor types are shown here.

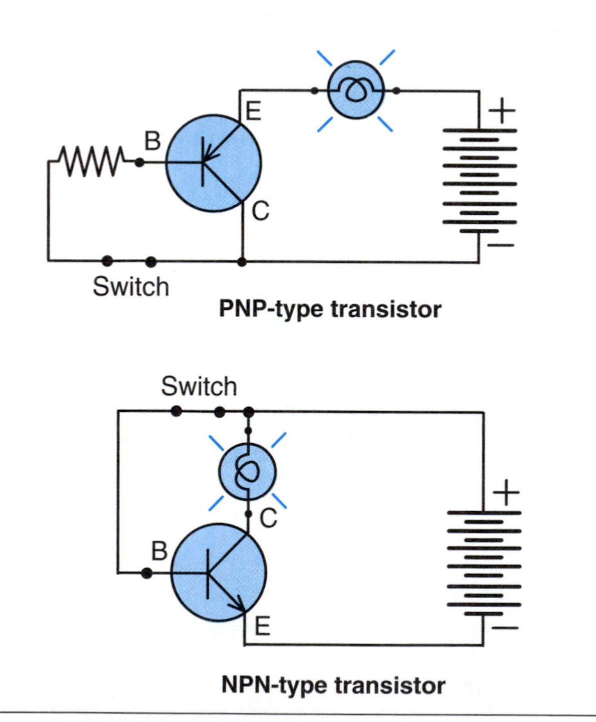

Figure 14-38 The circuitry of the PNP and NPN transistors and their power flow are shown here.

With NPN-type transistors, no current flows when a positive voltage is applied to the collector and a negative voltage to the emitter. When a small amount of current flows from the base to the emitter, a large amount of current flows from the collector to the emitter.

An illustration of transistor circuitry can be seen in Figure 14-38.

ELECTRICAL SCHEMATICS AND SYMBOLS

Before you can begin working on a motorcycle electrical system you must know how to read the electrical schematics (also known as wiring diagrams). A schematic is like a road map but instead of connecting cities and towns through the numerous roads and highways, it connects the wires in the electrical system to the various components that require electricity to function properly. Schematics use symbols to describe the various components in the system being looked at. The symbols shown in Figure 14-39 are some of the most common type of electrical symbols used on motorcycle wiring diagrams.

All manufacturers use schematics to assist you with tracing a wire to a component. While not standardized, all manufacturers use certain colors to help you know right away if the wire you are looking at is a wire that corresponds with the component you are trying to test or trace. For instance, some manufacturers use the color green for ground and black for switched-on power, while another manufacturer may use those same colors for entirely different reasons! It is because of this possibility that you must rely on the manufacturer's service manual to ensure that you are on the right path. While some manufacturers illustrate their schematics in full color, most use simple

SYMBOLS USED IN WIRING DIAGRAMS			
	Positive		Temperature switch
	Negative		Diode
	Ground		Zenner diode
	Fuse		Motor
	Circuit breaker		Splice
	Condenser		Not connected
	OHMS		Volt meter
	Fixed value resistor		Ammeter
	Variable resistor		Eyelet terminal
	Fusible link		Thermal element
	Coil		Multiple connectors
	Open contacts		Digital readout
	Closed contacts		Single filament bulb
	Closed switch		Dual filament bulb
	Open switch (SPST)		Light-emitting diode
	Double-throw switch (SPDT)		Thermistor
	Momentary contact switch		PNP bipolar transistor
	Pressure switch		NPN bipolar transistor
	Relay		Battery

Figure 14-39 Some of the most common electrical symbols found on a motorcycle schematic or wiring diagram are shown here.

black-and-white diagrams in their service manuals (Figure 14-40).

To assist a technician with diagnosis of an electrical complaint, most manufacturers create a more precise schematic of the various subsystems within the electrical system of the motorcycle. These are known as **block diagrams**. With a block diagram you can separate the desired subsystem like the ignition system (Figure 14-41), or charging system (Figure 14-42), from the rest of the machine. This aids in the technician's ability to focus on the system in question.

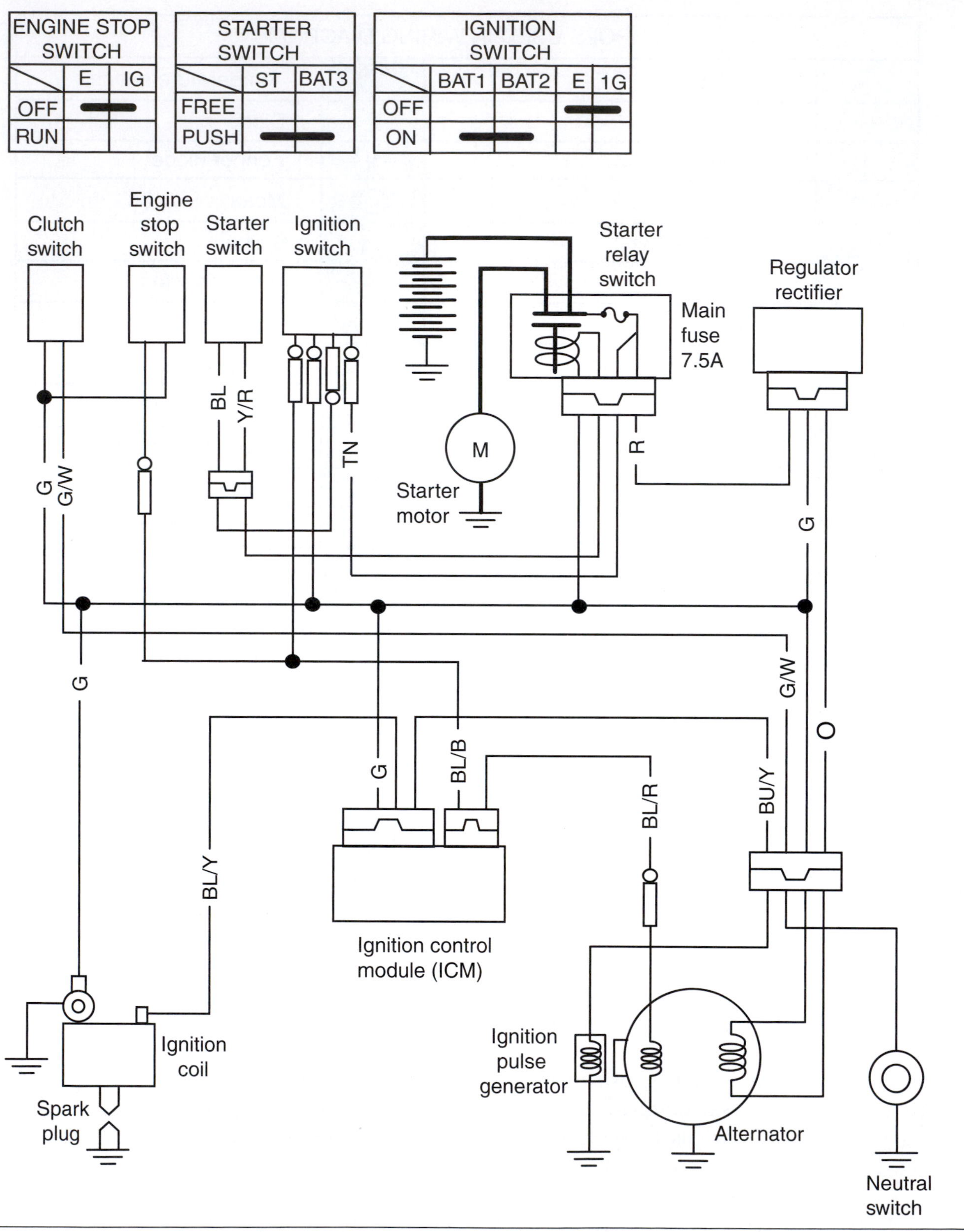

Figure 14-40 A complete schematic for an off-road motorcycle is shown here. Copyright by American Honda Motor Co., Inc. and reprinted with permission.

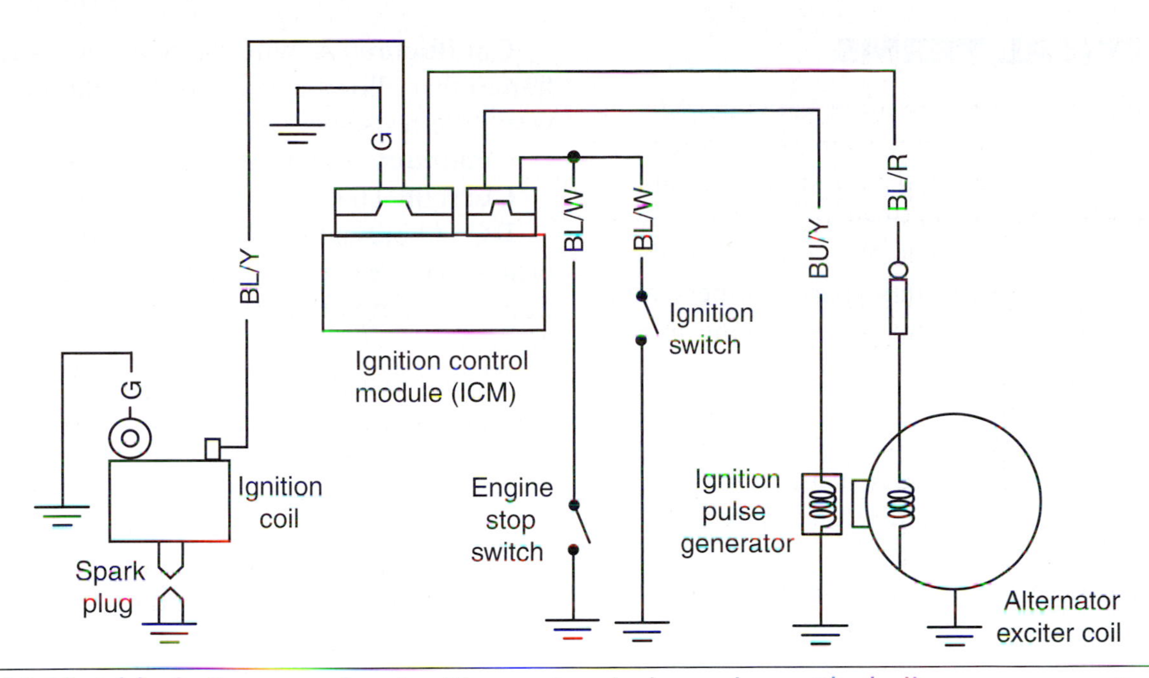

Figure 14-41 A block diagram of an ignition system is shown here. Block diagrams separate the system you are looking at from the rest of the electrical system. This aids in diagnosing problems within the system in question. Copyright by American Honda Motor Co., Inc. and reprinted with permission.

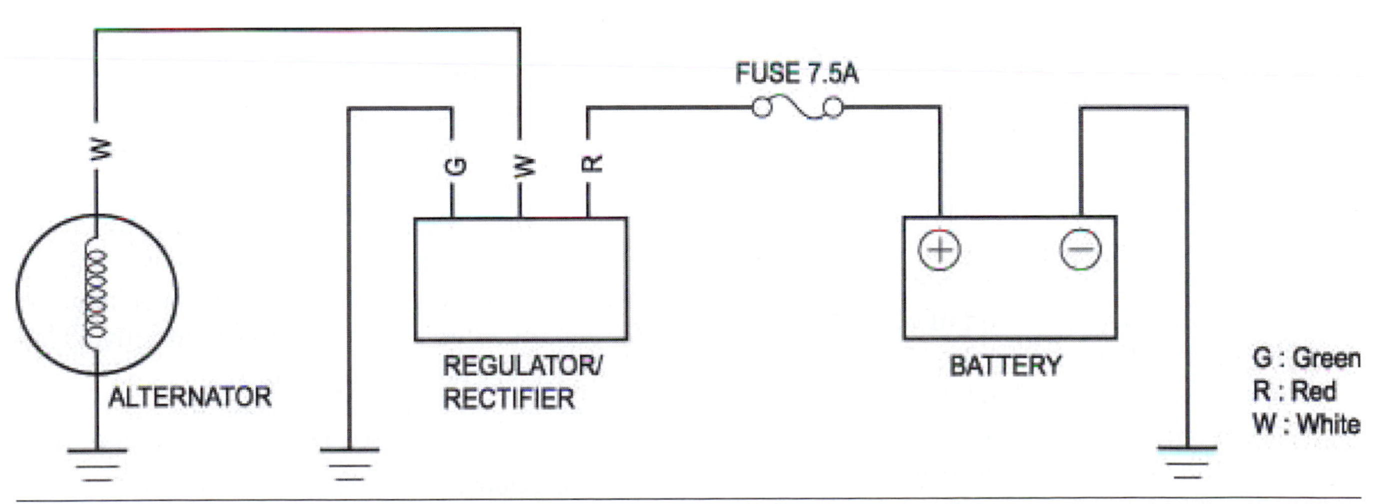

Figure 14-42 A block diagram of a charging system is shown here. Copyright by American Honda Motor Co., Inc. and reprinted with permission.

ELECTRICAL TERMS

The following terms are often found in material related to electrical repairs. Your job as a technician will be much easier if you know and understand these terms. For ease of reference, they're listed in alphabetical order. While many of these terms were discussed in this chapter, others were not but are important terms to understand just as well.

AC: Abbreviation for "alternating current," which is electricity that reverses direction and polarity while flowing through a circuit. Examples: 110 volts AC in a household reverses direction and polarity 60 times per second (60 Hz).

Alternator: An AC "generator" that uses magnetic induction to produce electricity. A revolving magnet and stationary stator windings are used. The current produced is AC.

Amperes: Commonly called "amps," which are electrical units of current flow through a circuit (similar to gallons per minute of water through a hose).

Amp hour: Discharge rate of battery in amperes times hours.

Armature: A group of rotating conductors that pass through a magnetic field. The current produced is usually DC after passing through a commutator device. Armatures are also used in electric motors.

Battery: A chemical device used to store electrical power. Within the battery a chemical reaction takes place that produces a voltage potential between the positive and negative terminals.

Bench test: Isolated component inspection.

Capacitor (condenser): A component, which in a discharge state, has a deficiency of electrons and will absorb a small amount of current and hold it until discharged again.

Circuit: Composed of three items: a power supply, load, and completed path.

Circuit breaker: Heat-activated switch that interrupts current when overloaded. A circuit breaker can be reset and replaces the function of a fuse.

Coil: A conductor looped into a coil-type configuration, which when current is passed through, will produce a magnetic field.

Conductor: A wire or material (such as a frame) that allows current to flow through it with very little resistance.

Continuity: Having a continuous electrical path.

Current: The flow of electrons in a circuit.

DC: Abbreviation for "direct current," which means that the current will only flow in one direction—from positive to negative (conventional theory).

Dielectric: A nonconductor of electricity. Dielectric materials do not allow electricity to flow through them. (See *insulator*.)

Diode: A semiconductor often used in a rectifier on motorcycles. A diode has the characteristic of allowing current to pass through in only one direction. Thus it's used to change AC to DC current.

Dynamic: Spinning or rotating in motion, refers to making a test when the component is in use.

Electricity: The flow of electrons through a conductor.

Electrolysis: The movement of electrons through an electrolyte solution. A battery charges and discharges through electrolysis. Electroplating (chroming) is an example where electrolysis is used to move and deposit metals from one electrode to another. In cooling systems, contaminated (tap water) coolant becomes an electrolyte, allowing electrolysis and the deposition of metal oxide scale on cooling system components.

Electrolyte: The sulfuric acid and distilled water solution that batteries are filled with at setup.

Electromagnet: A coil of wire that is wound around a soft iron core, that acts as a magnet when current is passed through it.

Electromotive force (EMF): The pressure of electrons in a circuit (also known as voltage). Created by difference in potential between positive and negative terminals of power supply. Also called pushing force of electricity.

Electron: The revolving part or moving portion of an atom. The electrons moving from atom to atom are electrical current.

Field coil: The field coil is an electromagnet. The flux lines may be used for generating electricity, for electric motor operation, or for operating a solenoid/relay.

Flux lines: All the magnetic lines of force from a magnet. (See *lines of force*.)

Free electron: An electron in an atom's outer orbit, which is held only loosely within the atom. Free electrons can move between atoms.

Fuse: A short metal strip that's protected by a glass or plastic case, which is designed to melt when current exceeds the rated value.

Ground: A common conductor used to complete electrical circuits (negative side). The ground portion of motorcycle electrical systems is often the frame.

Ignition: The spark produced by the high-tension coil by which the spark plug "ignites" the air-and-fuel mixture.

Insulator: A material that does not conduct electricity and therefore prevents the passage of electricity. All electrical wires are protected by plastic or special rubber insulation. (See *dielectric*.)

Lines of force: Refers to a magnetic field whose lines run from its north pole to its south pole. (See *flux lines*.)

Load: Anything that uses electrical power such as a light bulb, coil, or spark plug.

Magnetic induction: When a conductor is moved through a magnetic field, electricity will be "induced" into the conductor when the flux field cuts through the conductor.

Magnetism: The characteristic of some (ferrous) metals to align their molecules. The alignment of the object's molecules will cause the object to act as a magnet. Every magnet has both a north and south pole. Like polarities repel, opposites attract. Around every magnet, there's a magnetic field, which contains lines of force.

No-load test: A dynamic test with the component insulated or disconnected from its main system.

NPN: A transistor in which the emitter and collector layers are N-type and the base layer is P-type (negative, positive, negative).

Permeability: Ability of material to "absorb" magnetic flux (can be temporary or permanent) (See *reluctance*).

PNP: Transistor in which the emitter and collector layers are P-type and the base layer is N-type (positive, negative, positive).

Polarity: In magnets, polarity is north and south; in electricity, polarity is positive and negative.

Pole: The north "pole" or south "pole" of a magnet. Also refers to the lugs (iron cores) of a stator around which the AC generator's wires are wound.

Rectifier: Changes AC to DC. Usually a group of four or six diodes comprises a "bridge rectifier." (See *diode*.)

Regulator: Used to limit the output of a generator or alternator.

Reluctance: Resistance to magnetism. (See *permeability*.)

Reluctor: Magnetic field interrupter used as a signal generator in ignition systems.

Resistance: The opposition offered to the flow of current in a circuit.

Rotor: The revolving magnets or electromagnets that form the magnetic field in an alternator or ignition signal generator.

Schematic: A wiring diagram showing the electrical components and circuitry in detail.

SCR: An abbreviation for silicon-controlled rectifier, which is an electronically controlled switch. (See *thyristor*.)

Selenium: Similar to silicon materials in characteristics; it's also used as a rectifier on older motorcycle models.

Silicon: A material used in the construction of semiconductors. Because of its characteristics, the material allows current flow only under certain prescribed conditions.

Sine wave: A graphic depiction of the form of alternating current usually taken from an oscilloscope.

Solder: Tin/lead alloy with rosin core used to form lower-resistance connections of electrical components or wires.

Static: Stationary. Usually a test made of a stationary component rather than a bench test.

Stator: A stationary conductor (usually several coils of wire). When magnetic flux cuts the stator windings, a voltage potential is induced in the windings.

Switch: A device that opens or closes an electrical circuit.

Thermo-switch: A bimetallic switch, which when heated, opens or closes a circuit.

Thyristor: An electronically controlled switch that opens when signaled at the gate and closes after current flow falls. (See *thyristor*.)

Unloaded: (See *no-load test*.)

Valence electrons: The electrons contained in the outermost electron shell of an atom. Valence electrons are also known as free electrons.

Voltage: (See *electromotive force*.)

Watt: The unit of electric power; W = E I (Wattage = Voltage Current).

Wire gauge: Wire diameter. Usually specified by an AWG (American Wire Gauge) number. The smaller the number, the larger the wire diameter.

Wiring diagram: Similar to a schematic, but less detail. A wiring diagram usually shows components in block form rather than illustrating their internal circuitry.

Zener diode: Similar to a standard diode, but allows current flow in the reverse direction when the breakdown voltage is reached.

Summary

- Electrical devices and circuits can be dangerous. Safe practices are necessary to prevent shock, fires, explosions, mechanical damage, and injuries resulting from the careless or improper use of tools.
- There are two theories of electricity. The conventional theory of electricity states that "Electric current flows from the positive terminal of the voltage source through the circuit to the negative terminal." Simply stated, this theory supports that electricity flows from positive to negative. The electron theory of electricity states: "Electricity is the flow of electrons from the negative terminal of a source through a conductor, completing the circuit back to the positive terminal of the source." Simply stated, this theory supports that electricity flows from negative to positive.
- There are three types of electrical circuits. A series circuit is a circuit that has only one path back to its source. A parallel circuit is a circuit that has more than one path back to the source of power. The series/parallel circuit contains a load in series and a parallel load, both within the same circuit.
- Voltage resistance and current are all common terms used with electricity. Voltage is a measurement of the amount of electrical potential in a circuit and is measured in volts. Resistance is a force of opposition that works against the flow of electrical current in a circuit and is measured in ohms. Current is the rate of flow of electrons through a conductor and is measured in amps.
- The values of resistance, current, and voltage have a very important relationship in a circuit. The amount of current flowing through a completed circuit is directly proportional to the voltage applied to the conductor. This relationship between resistance, current, and voltage is known as Ohm's law. Ohm's law states that a resistance of one ohm (1Ω) permits a current flow of one ampere (1A) in a circuit that has a source voltage of one volt (1V).
  - The relationship that's summarized by Ohm's law is expressed with the mathematical formula: Voltage (E) = Current (I) × Resistance (R)
  - Two other useful variations of the Ohm's law equation are:
    - Current (I) = Voltage (E) divided by Resistance (R)
    - Resistance (R) = Voltage (E) divided by Current (I)
- Multi-meters are the most popular electrical testing tool used in the motorcycle industry and integrates three popular measuring devices into one handy-to-use tool. Voltmeters are used to measure the voltage, or potential difference, between two points. Ammeters are used to measure the current flow through a circuit. Ohmmeters are used to measure the resistance of a circuit or component by applying a known voltage to the circuit and measuring the resulting current.
- Before you can begin working on a motorcycle electrical system you must know how to read

the electrical schematics (also known as wiring diagrams). A schematic is like a road map but instead of connecting cities and towns through numerous roads and highways, it connects the wires in the electrical system to the various components that require electricity to function properly.

Chapter 14 Review Questions

1. A circuit that has more than one path to the power source is a
   a. series circuit
   b. open circuit
   c. parallel circuit
   d. short circuit
2. When a conductor is moved through a magnetic field, a voltage will be induced in the conductor. However, current will not flow through the conductor unless it's
   a. formed into a coil
   b. connected in a complete circuit
   c. moved very quickly through the magnetic field
   d. made of copper
3. What is the voltage of a circuit having a current flow of 3 amperes and a resistance of 4 ohms?
   a. 0.12V
   b. 12V
   c. 7V
   d. 36V
4. A circuit that has an incomplete path for current to flow is called
   a. a grounded circuit
   b. an open circuit
   c. a short circuit
   d. a closed circuit
5. Electromotive force (EMF) is measured in
   a. watts
   b. ohms
   c. amps
   d. volt
6. An ohmmeter is always connected to an electrical circuit
   a. in series
   b. to measure voltage
   c. after it has been isolated from the rest of the electrical system
   d. after the power has been turned on
7. The terminals of a diode are the
   a. gate, anode, and cathode
   b. primary and secondary
   c. emitter, collector, and base
   d. anode and cathode
8. What would be the resistance of a 12-volt light bulb with 5 amps flowing through it?
   a. 2.4 ohms
   b. 2.4 volts
   c. .416 ohms
   d. .417 volts
9. An ammeter is always connected to an electrical circuit
   a. after it has been isolated from the rest of the electrical system
   b. in parallel
   c. in series
   d. after the power has been turned on
10. The conventional current flow theory can best be described as
    a. electron flow from the negative terminal to the positive terminal
    b. electron flow from the cathode terminal to the anode terminal
    c. electron flow from AC to DC and then back to AC
    d. Electron flow from the positive terminal to the negative terminal

11. Electrical current flow is measured in
    a. ohms
    b. amps
    c. watts
    d. volts
12. The abbreviation “AC” stands for ________.
13. The abbreviation “DC” stands for ________.
14. With a ________ you can separate the desired subsystem like the ignition system, from the rest of the schematic.
15. Which theory of electricity is used in the motorcycle industry?
    a. electron
    b. electro-conventional
    c. conventional
    d. non-conventional

CHAPTER 15 Motorcycle Charging Systems and DC Circuits

Learning Objectives

When students have completed the study of this chapter and its laboratory activities they should be able to:

- Explain why motorcycles have charging systems
- Describe the theory behind a basic charging system
- Visually identify the different types of charging systems found on motorcycles from a schematic
- Describe how alternators generate AC power
- Describe how a charging system changes alternating current into direct current
- Read block diagrams for various DC electrical system circuits

Key Terms

Alternator
Battery
Distilled water
Excited-field Alternators
Permanent-magnet Alternators
Rectifier
Rotor
Schematic
Stator
Switches
Voltage regulator

INTRODUCTION

This chapter is the second of three chapters devoted to motorcycle electrical systems. In Chapter 14, you learned about the basics of electricity, where electricity comes from, and how to measure it. In this chapter, you'll learn how to apply this electrical theory to understand motorcycle charging and other related electrical systems. We'll begin by describing the basics of a charging system. Next, we'll take a closer look at each of the components in the charging system and how they operate. After we've discussed each of the components, we'll review the overall operation of the charging system and explain how to maintain and troubleshoot them. We will then look at some of the other electrical circuits found on motorcycles.

A charging system is necessary in any motorcycle that uses a battery to assist with powering electrical components. As we mentioned in Chapter 14 on basic electrical fundamentals, the first step to understanding electrical systems and how they work is to start with the basics.

To understand how a charging system works, you must first understand what a charging system does. Then, you need to know the components that make up the charging system and how they work. Many of the images in this chapter will be line drawings of the discussed component. This is to allow you to best understand how these components are illustrated in the manufacturer's service manual. We will also show you actual images of the components on an actual motorcycle when discussing the components.

CHARGING SYSTEMS

The purpose of a charging system is to replenish the current in a battery as it is used when the motorcycle is operating. An **alternator** provides the electrical power source for the charging system. The alternator provides an alternating current (AC) output but a battery requires direct current (DC).

In order to convert the AC output of the alternator to DC, which is needed by the battery, a **rectifier** is used. A rectifier is a device that permits electrical current to flow in one direction only. The rectifier converts the AC (which you'll remember alternately flows in one direction, then in the other direction) into DC, which flows in only one direction. This process is known as *rectification*. The voltage from the charging system to the battery is maintained within a predetermined range by a **voltage regulator**. A voltage regulator is a device used to maintain constant voltage levels on the electric system. By controlling the output of the charging system, the voltage regulator prevents undercharging or overcharging the battery.

From the smallest 50cc scooter to large 1,800cc touring motorcycles, all charging systems have the same common basic components (Figure 15-1). These components may have different designs and sizes in various motorcycle applications but still provide the same functions in the charging system.

The Alternator

Depending on the manufacturer, there are many different terms used for an alternator including generator, dynamo, and magneto. An alternator is a device powered by an engine that is designed to create electrical energy to charge batteries and power on-board devices. The alternator is driven by the rotation of the engine's crankshaft and may be directly attached to the crankshaft or it may be driven by a gear and placed elsewhere on the motorcycle's engine. An alternator only produces electrical output when the engine is rotating. Remember from Chapter 14 that there are three key items needed to produce AC voltage: a magnetic field, a conductor (coils of wire), and motion (Figure 15-2). The output from the alternator varies with the speed of the engine. The faster the engine's crankshaft turns, the more AC voltage is produced.

An alternator has two main components—the **rotor** and the **stator**. The rotor has a series of magnets

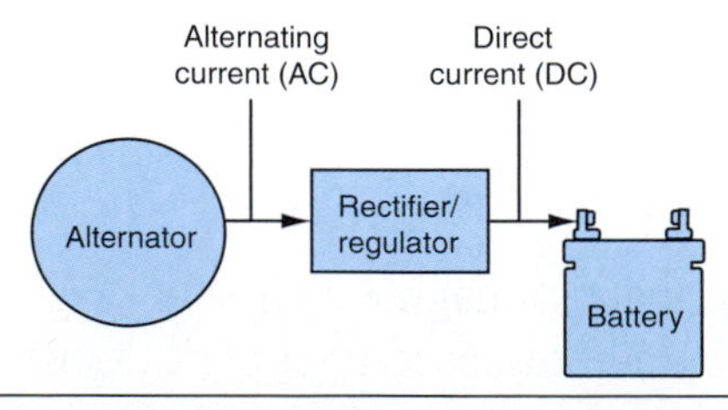

Figure 15-1 All charging systems have the same basic components. A simple diagram of a charging system is shown here.

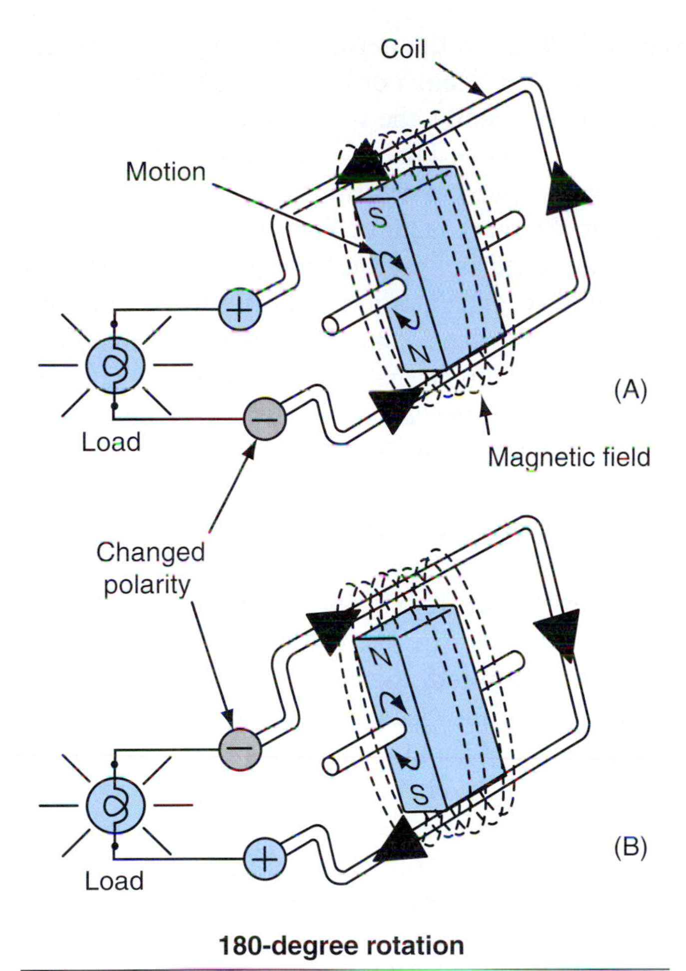

Figure 15-2 The three key items needed to produce AC voltage: a magnetic field, a conductor (coils of wire), and motion. In (A), current is flowing in one direction while in (B) the current flows in the opposite direction creating alternating current.

and rotates either inside or outside the stationary windings of the stator. The stator consists of sets of coils, which are used to produce power for the motorcycle's electrical circuits and to charge the battery.

From the viewpoint of design and construction, alternators used in the motorcycle industry can be divided into two general types: **permanent-magnet alternators** and **excited-field electromagnet alternators**.

Permanent-Magnet Alternators

The permanent-magnet alternator (Figure 15-3) is the most commonly used type of AC-generating system found on motorcycles. You will find this type of alternator mounted directly on the crankshaft (Figure 15-4). Permanent magnets are incorporated into the flywheel on the rotor (Figure 15-5). With this design, the flywheel is fitted onto the crankshaft and is held in place with a woodruff key (Figure 15-6) and a mounting bolt. The stator may be fitted on the crankcases or on the outside cover (Figure 15-7).

Excited-Field Electromagnet Alternators

Excited-field electromagnet alternators don't use a permanent magnet. Instead, they have a field coil, which is energized with DC current. The field coil becomes a powerful magnet when DC power is supplied. Power is generated as the rotor spins past the stator. Excited-field electromagnets are located in different areas of the engine besides the

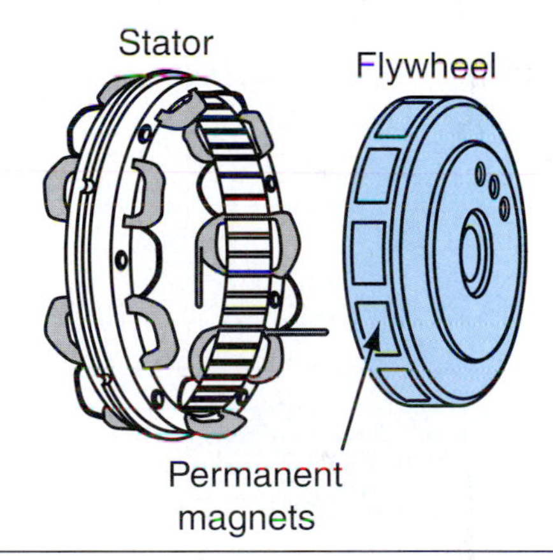

Figure 15-3 Permanent-magnet systems are the most popular types of charging systems found on motorcycles.

Figure 15-4 Permanent-magnet charging systems are primarily located on the crankshaft of the engine under the alternator cover. Copyright by American Honda Motor Co., Inc. and reprinted with permission.

Figure 15-5 Permanent magnets are attached to the rotor. Copyright by American Honda Motor Co., Inc. and reprinted with permission.

Figure 15-6 Alternator rotors are usually held in place by a woodruff key as shown here. Copyright by American Honda Motor Co., Inc. and reprinted with permission.

Figure 15-7 The stator is mounted onto the outer case on this motorcycle. Copyright by American Honda Motor Co., Inc. and reprinted with permission.

crankshaft as on the permanent-magnet alternator to help keep them cool. If not attached to the crankshaft directly, the rotor's speed can be multiplied by gears or chains. This is done to increase the rotor's speed of rotation if it is deemed necessary by the manufacturer's engineers.

The excited-field alternator is potentially the most powerful AC generator available because of the high amount of magnetism that it can create. In most cases the excited-field alternator is fully self-contained on today's motorcycles just as you would see an alternator on an automobile. There are two types of excited-field electromagnet alternators: brush and brushless.

Brush-Type Excited Field The brush-type excited-field coil has the field coil placed within the rotor (Figure 15-8). Current flows through carbon brushes to the field-coil slip rings. When current is applied, the rotor is induced electromagnetically and becomes a very strong magnet. The rotation of the magnetized core acts on the stator coils to produce AC current.

Brushless-Type Excited Field The brushless-type excited-field coil (Figure 15-9) eliminates the maintenance factor of the excited-field coil design by placing the rotor around the inner field coil. When the field coil is energized, the magnetic field

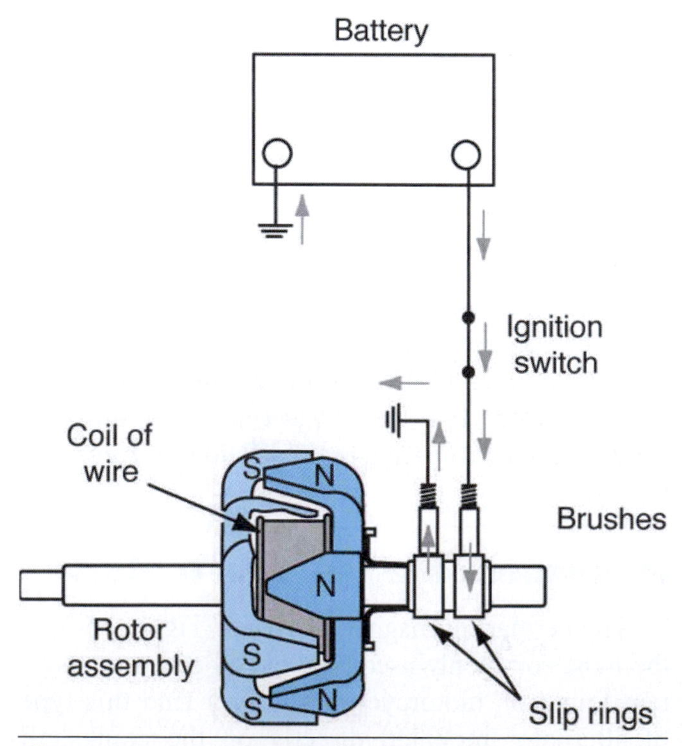

Figure 15-8 A simple diagram of a brush-type excited-field alternator is illustrated here.

magnetizes the rotor core. The rotation of the magnetized core acts on the stator coils placed on the outside of the rotor to produce AC current.

The Rectifier and Voltage Regulator

A rectifier is required to convert the AC from the alternator into DC that is used by the battery. The rectifier uses a diode or a group of diodes to convert the AC into DC by allowing current flow in one direction only and therefore preventing the AC voltage to continue. Voltage regulators are used to allow current to flow into the battery to charge it when needed and also to stop the current flow to the battery to prevent it from being overcharged. Virtually all voltage regulators use thyristors (SCRs) and Zener diodes, which provide a current limiting function to control battery charging.

Although they are two separate components, voltage regulators and rectifiers are normally integrated as a single unit to reduce the cost of production (Figure 15-10).

A simplified schematic of a charging system is shown in Figure 15-11. Note the dotted lines surrounding the regulator/rectifier. Inside the lines you can see diodes and SCRs. The diodes make up the rectifier while the SCRs are used for the regulator. To ease the understanding of these critical components of a motorcycle's charging system, we will discuss the rectifier and regulator separately even though they are normally combined as one unit.

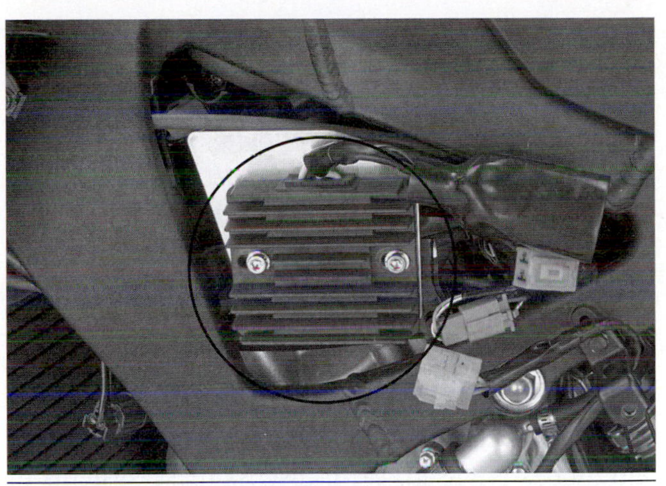

Figure 15-10 A voltage regulator/rectifier is shown here. The cooling fins help to remove the heat produced from the regulator when it sends current back to ground. Copyright by American Honda Motor Co., Inc. and reprinted with permission.

Rectifiers

The purpose of a rectifier is to change the AC that's produced by the alternator into DC to charge the battery. A rectifier consists of as few as one or as many as six diodes, depending on the charging system's needs as determined by the manufacturer's engineers. Remember that the diode serves as a one-way electrical device that allows current to flow in one direction but not in the other.

The basic principle behind a rectifier's function is that it allows current to pass through in only one

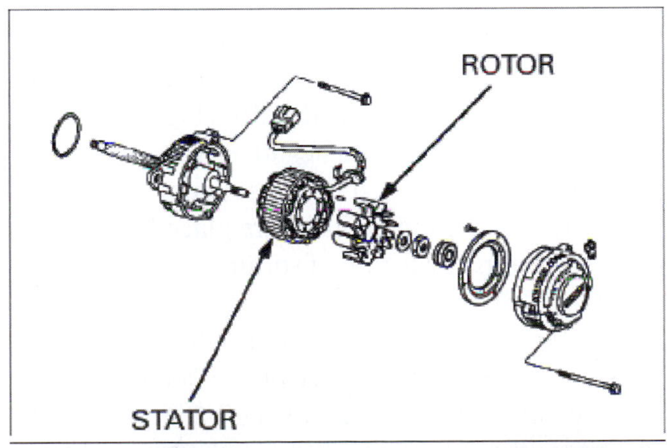

Figure 15-9 A brushless-type-excited-field alternator is shown here. Copyright by American Honda Motor Co., Inc. and reprinted with permission.

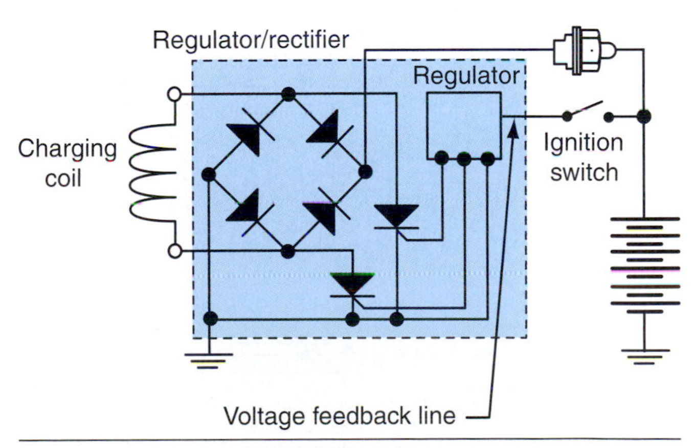

Figure 15-11 Although two separate components, the regulator/rectifier are normally assembled as one unit as the dotted lines indicate in this illustration.

direction—like a one-way electricity gate. Because alternating current is continually reversing direction, the rectifier must change it to direct current so that it can be utilized by the battery, which stores electrical energy in DC. A single diode rectifier wired in series into a circuit will block half of the AC current flowing into it and allow the other half of the current to flow to the battery as seen in Figure 15-12. This is known as half-wave rectification and is used on some smaller motorcycles.

In order to allow more current produced by the alternator to reach the battery, additional stator coils and diodes are used. We'll discuss these systems in more detail later in this chapter. When rectifiers no longer work properly, they must be replaced as a unit as they cannot be repaired. We will also cover how to test rectifiers later in this chapter.

Voltage Regulators

The purpose of the voltage regulator is to control the voltage to prevent undercharging or overcharging the battery. There are two types of voltage regulators: mechanical and electronic. The mechanical voltage regulator was widely used on motorcycles until the mid-1970s, but now is all but extinct. Since that is the case, we will concentrate on electronic voltage regulators.

Electronic voltage regulators are used on virtually all motorcycles today that have a need for voltage regulation. Electronic regulators contain no moving parts and never need to be adjusted. Generally speaking, most electronic regulators, or current limiters as they are sometimes called, have a solid-state, transistorized arrangement of electronic devices, such as thyristors and Zener diodes. Voltage regulators are difficult to test as they require a voltage input to turn the component on and off. As mentioned previously, we cannot test a diode with an ohmmeter when there is power applied to it. Therefore, most manufacturers set a predetermined voltage to check to verify that the voltage regulator is working properly. Because electronic voltage regulators are sealed units (in the same manner as a rectifier), they cannot be repaired. If they fail, the unit must be replaced.

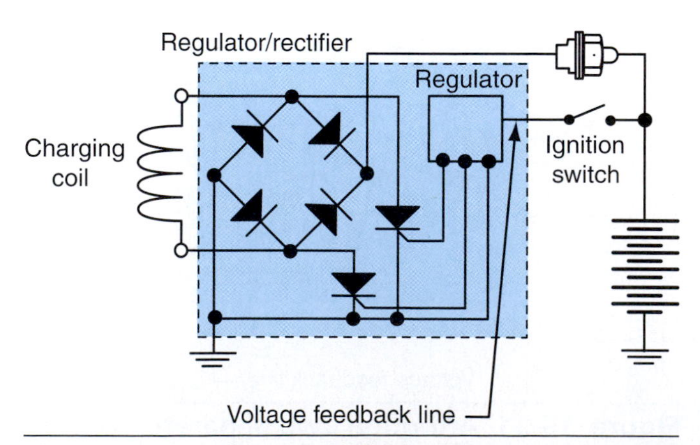

Figure 15-12 This simplified schematic shows a half-wave rectifier that blocks one-half of the AC waveform.

A disadvantage of having both the regulator and the rectifier assembled as one complete unit is that if either the regulator or the rectifier portion fails, the entire unit must be replaced.

The Battery

Technically speaking, a **battery** is an electrochemical device that converts chemical energy to electrical energy used to store electrical power to supply uninterrupted energy for an electrical system. For example, street motorcycles need lights that operate when the engine isn't running. They get it from the battery. Accessories such as clocks and alarms require a battery, as does the electric starter. As the battery is discharged from use, the charging system charges the battery as required. Most modern motorcycles use 12-volt electrical systems and therefore utilize 12-volt batteries.

Lead Acid Batteries

Motorcycle batteries are also known as lead acid batteries. A conventional wet-cell motorcycle battery consists of a series of cells (Figure 15-13). Each cell has positive and negative metal plates and is capable of storing just over 2 volts of electricity.

The dissimilar metal plates are placed in an electrolyte solution consisting of sulfuric acid and water. These are then insulated from each other with a permeable, non-conductive material, which allows the transfer of ions. The transfer of ions occurs during the discharge and recharge of the battery. The battery produces electricity from a chemical change that takes place between the positive and negative plates in the electrolyte solution (Figure 15-14).

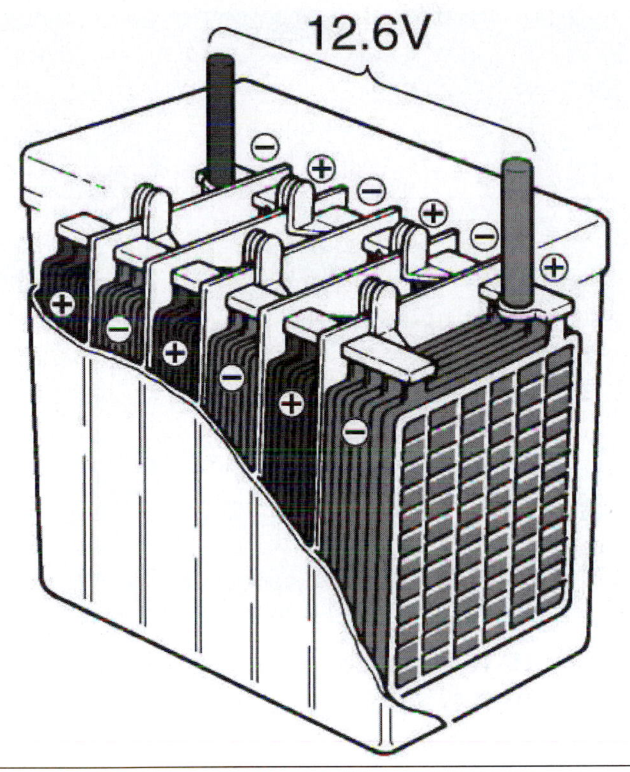

Figure 15-13 Motorcycle batteries consist of a series of cells that contain positive and negative plates.

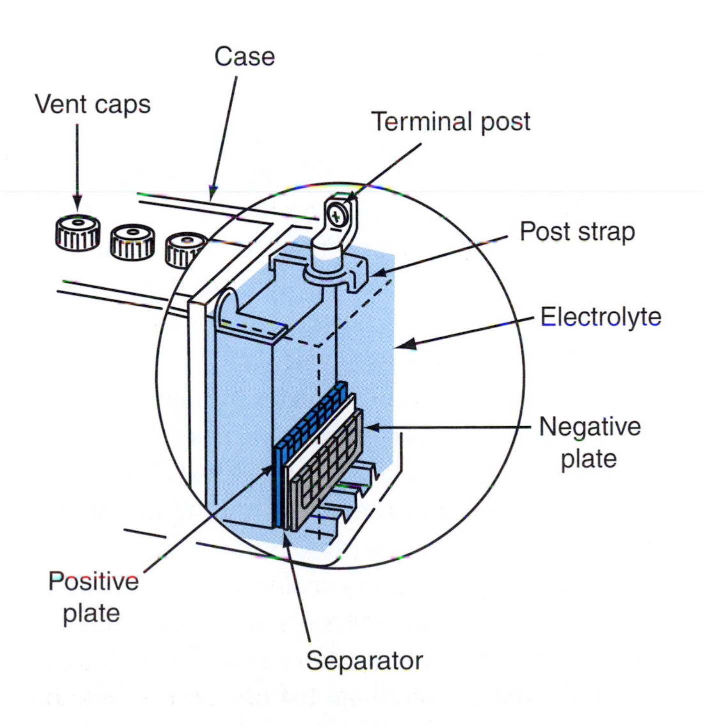

Figure 15-14 When the plates of a battery are placed in an electrolyte solution electricity is due to a chemical reaction.

Also occurring is the change in specific gravity or density (weight) of the electrolyte (Figure 15-15). During the discharge cycle, sulfuric acid is drawn from the electrolyte into the pores of the plates. This reduces the specific gravity of the electrolyte and increases the concentration of water. During recharge, this action is reversed and the sulfuric acid is driven from the plates, back into the electrolyte, increasing the specific gravity.

A key piece of information to remember is that during the discharge cycle, lead sulfate is being formed on the battery plates (Figure 15-16). This is known as sulfating. Although sulfation is a normal function with all batteries during the discharge cycle, the battery must be recharged to drive out the sulfuric acid into the electrolyte. Without this recharge, the lead sulfate will continue to develop and become difficult, and eventually impossible, to break down during a normal recharge cycle. There are two basic

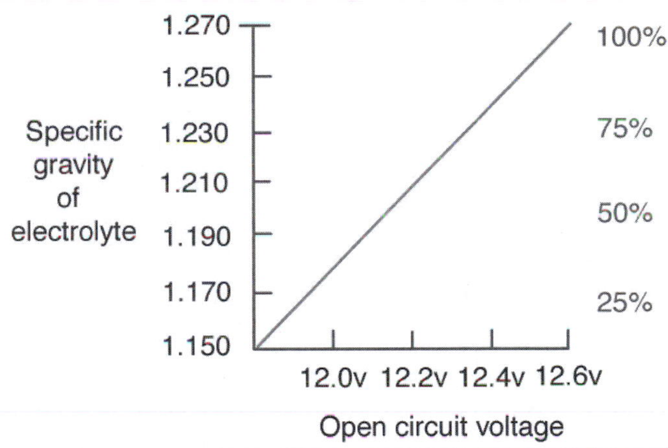

Figure 15-15 Electrolyte is measured by its specific gravity or weight as compared to water.

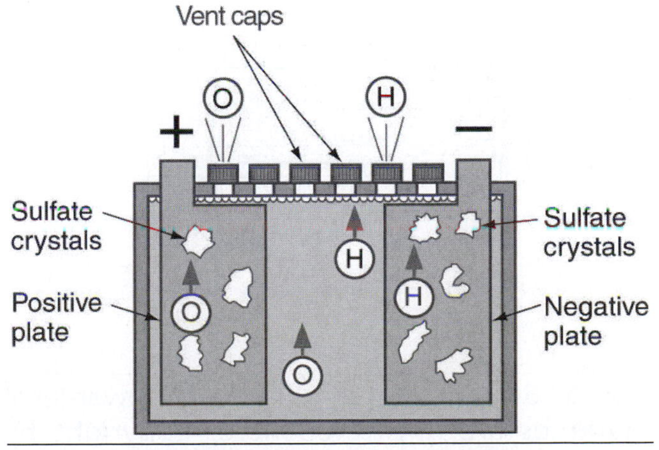

Figure 15-16 Sulfation is a normal function of any battery during its discharge cycle.

types of batteries used in a motorcycle electrical system: conventional and maintenance-free.

Conventional Batteries Not commonly used today on new motorcycles, the conventional battery design has a cap over each of the battery cells. When you charge a lead-acid battery, electrolysis breaks the water down into its components: hydrogen and oxygen gas. Conventional batteries have a vent, usually routed into a tube, to remove the gases produced during the normal battery charge cycle. When current is supplied to the battery, gas is emitted from the plates and electrolyte temperature increases. This increase in heat causes a loss of water from the battery electrolyte over time. The loss of water and increased heat drastically reduces the life of the battery and, if left uncorrected, will damage the battery beyond repair. With the conventional-type battery, the water must be replaced as the level drops.

Because a battery is constantly subjected to charging and discharging cycles, the water in the electrolyte is slowly boiled off during normal use with a conventional battery. Because of this, the electrolyte level should be checked on a regular basis. When the electrolyte level becomes low, add distilled water until the electrolyte reaches the upper level line on the battery case (Figure 15-17).

The volume of acid lost in this process is so small that acid replenishment is never required during the service life of a battery. However, the vent tube must be routed so that it does not discharge near the chain or other critical parts that are susceptible to acid damage. Use only **distilled water** when refilling the battery (Figure 15-18). Distilled water is water that has virtually all of its impurities removed through distillation. Distillation involves boiling the water and re-condensing the steam into a clean container, leaving the contaminants behind. Tap water contains chlorine, iron, and other elements, which will contaminate the electrolyte and reduce its effectiveness.

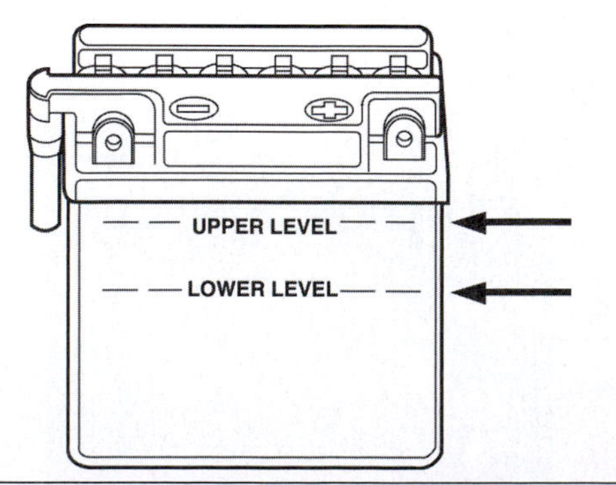

Figure 15-17 A battery's upper- and lower-level markings are illustrated here. Copyright by American Honda Motor Co., Inc. and reprinted with permission.

Figure 15-18 Distilled water can be purchased at any grocery store.

When the water is evaporated to the point where the plates become exposed, the sulfation process drastically accelerates. This damages the battery and shortens the battery life. Sulfation can occur not only when the electrolyte level is low, but also when the battery is discharged for long periods.

Remember, as the electrolyte level goes down when the water in the battery evaporates, replenish the battery with distilled water only.

Maintenance-Free Batteries Maintenance-free (MF) batteries are very similar in design to conventional batteries. The difference is that the positive and negative lead plates in the maintenance-free battery use Absorbed Glass Mat (AGM) Technology, which totally seals all of the acid in the special plates and separators. Therefore, you do not need to add water to a maintenance-free battery.

Unlike the conventional battery, MF do not have a vent to allow for the escape of excess gases. Instead, they use a safety valve that's designed to open when extreme gas pressures are produced. The safety valve closes and seals the battery when the

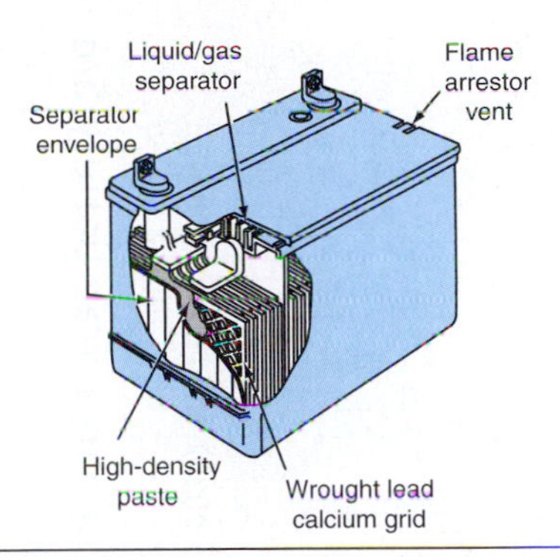

Figure 15-19 The parts of a maintenance-free battery are shown in this cutaway drawing.

internal pressure returns to normal. Figure 15-19 shows a cutaway illustration of a typical maintenance-free battery and its internal components.

The majority of MF batteries require that the included special acid pack (Figure 15-20) be installed prior to initial set up (Figure 15-21). Some maintenance-free batteries come prefilled directly from the manufacturer (Figure 15-22).

CHARGING SYSTEM OPERATION

Now that you know about the basic components of a motorcycle charging system, let's see how these components work together to charge a battery. Later in this chapter, you'll learn about various charging circuit tests. Let's look at an example of a complete **schematic** (Figure 15-23). A schematic is a diagram that shows, by means of graphics and symbols, the electrical connections and functions of a specific circuit arrangement.

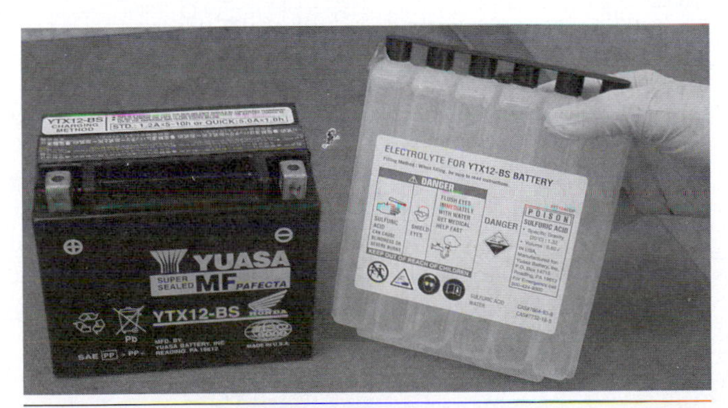

Figure 15-20 Most maintenance-free batteries come with their own acid pack that must be installed prior to initial use.

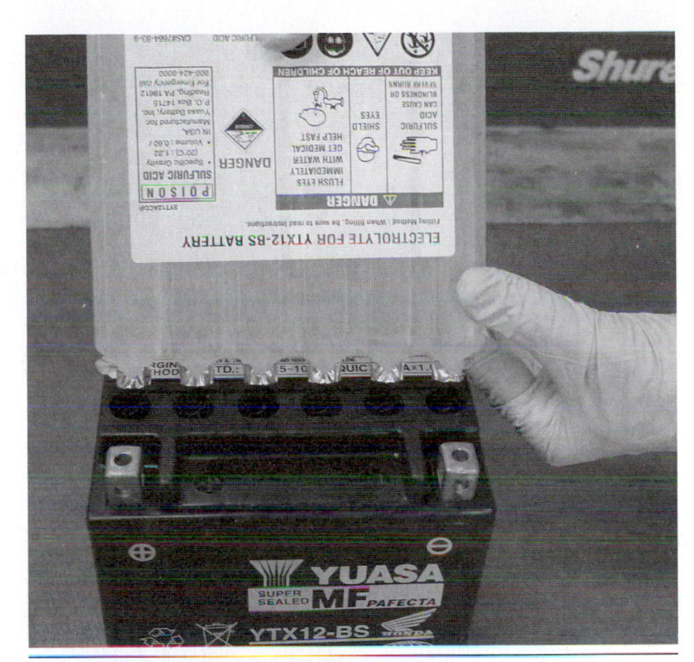

Figure 15-21 The acid pack is installed using the manufacturer's instructions.

Note that all of the wires are color-coded and that a color-code chart is shown in the lower right corner of the schematic. Observe at the bottom of the schematic that the connections are shown for each position of the switches.

The schematic may seem quite complex at first, but if you break down the individual electrical system that you're working on within a schematic, it becomes much easier to read and understand. You can break down a schematic by drawing a "block diagram" of the system as seen in Figure 15-24. Use this figure to follow along as we talk about the operation of a basic charging system.

Figure 15-22 Some maintenance-free batteries come prefilled with acid from the battery manufacturer.

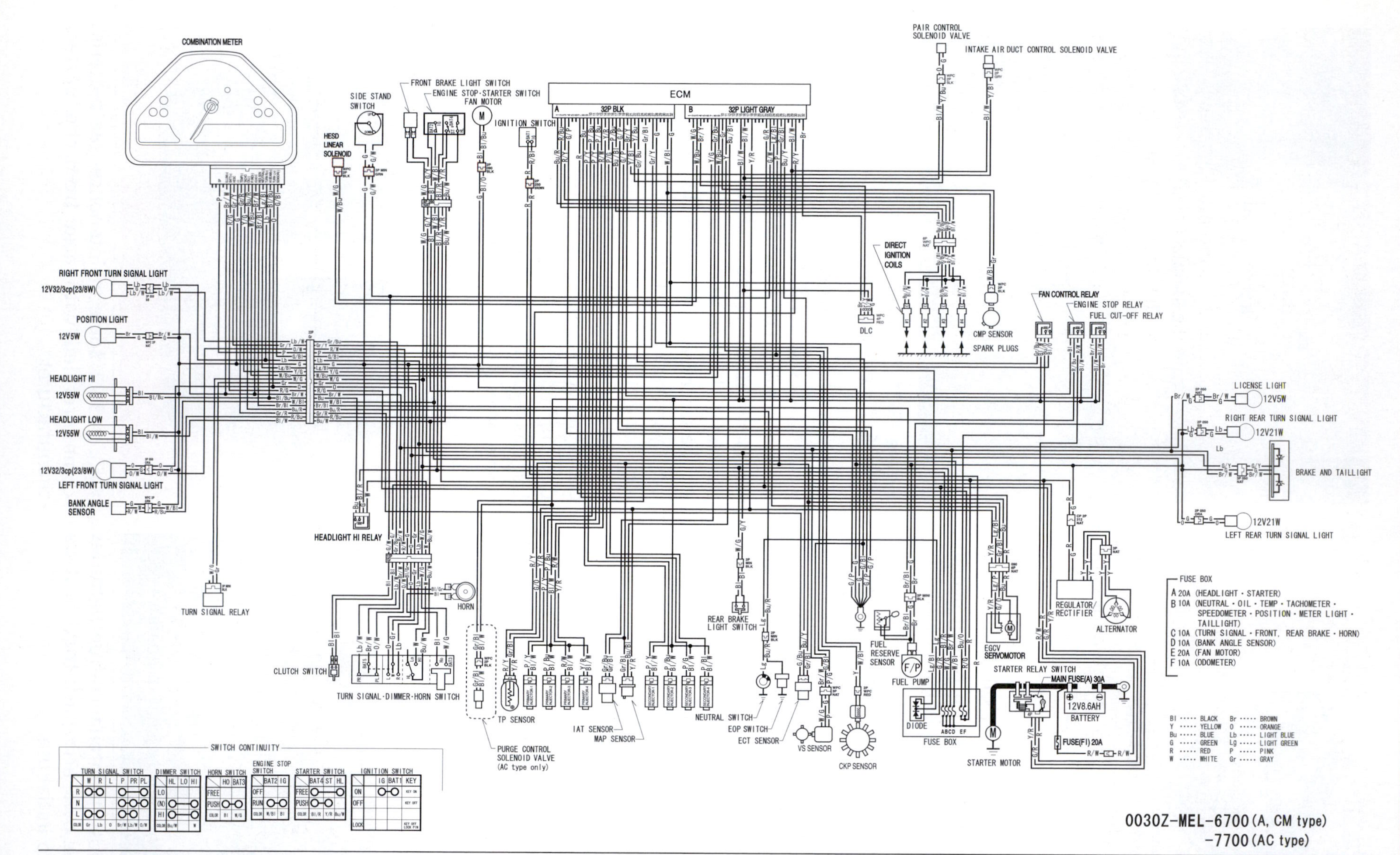

Figure 15-23 A complete schematic of a motorcycle electrical system is shown here. The parts of a maintenance-free battery are shown in this cutaway drawing. Copyright by American Honda Motor Co., Inc. and reprinted with permission.

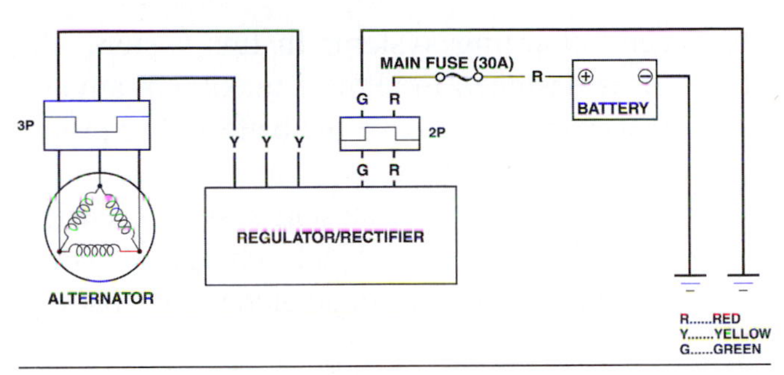

Figure 15-24 A block diagram of a charging system from the schematic in Figure 15-23 is shown here. Copyright by American Honda Motor Co., Inc. and reprinted with permission.

Find the alternator on the block diagram. The alternator produces electricity in the form of alternating current when the engine is running. Now locate the alternator on the actual schematic. As we refer to each of the charging system components on the block diagram, locate them on the actual schematic as well to see how the block diagram was derived from the schematic.

Notice the three stator leads connected to the alternator. Current flows through these leads, which are color-coded and labeled Y (yellow) on the schematic. These wires carry AC current from the alternator to the regulator/rectifier as the alternator rotor spins past the stator windings. The AC current enters the rectifier and is changed to direct current, which leaves the rectifier through the R (red) wire. This red wire provides direct current to the battery for charging. The current from the red wire also sends a signal to the regulator portion of the regulator/rectifier. The rectifier and regulator are connected to the common ground by the G (green) wire. When the voltage reaches a predetermined level, the voltage regulator adjusts the excess rectified DC current to ground to prevent the battery from being overcharged.

As you can see, by connecting individual components to work together as a system, direct current is provided for charging the battery and supplying power to the lighting system. The alternating current is changed into direct current by the rectifier. The amount of charging current is controlled by the voltage regulator. Each component in the charging system must be kept in good working condition to allow the charging system to continue to function properly. This includes keeping all wiring connections clean and tight-fitting to prevent excessive resistance.

TYPES OF CHARGING SYSTEMS

Now that you understand how a basic charging system operates and can identify the individual components in a charging system, we'll move on to a discussion of the various types of charging systems that are found on motorcycles and ATVs. We'll begin with the simplest charging system and then learn about the more complex systems. Remember, all charging systems operate in the same basic fashion; they just have different ways of producing AC current. One way that charging systems differ is related to the number of charging coils at the input. The charging system is also based on the needs of the electrical system. More electrical components require a larger output charging system.

Half-Wave Charging System

The half-wave charging system is the simplest charging system. This charging system uses only one "grounded" charging coil, and only one-half of the AC output is actually used. As shown in Figure 15-25, the alternator has two pairs of magnets and it produces two cycles of AC for each complete rotation (360 degrees) of the rotor (flywheel).

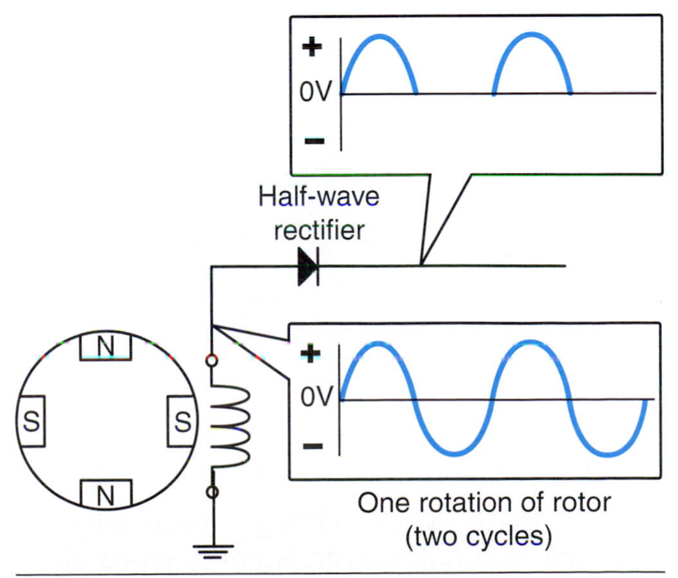

Figure 15-25 The half-wave charging system is the simplest design as it only has one diode.

A single diode is used to rectify the AC output into DC to charge the battery. When the AC flows through the diode, the negative voltage wave of the AC is cut off and the positive voltage wave is passed to charge the battery.

This type of charging system has a low output, and its small size is best suited for very small machines with small electrical loads. Because of its low output potential, this charging system is not used very much anymore.

These low output systems regulate the DC voltage by the use of a relatively simple half-wave regulating system as shown in Figure 15-26. In this system, the charging current from the alternator is rectified by diode D1 and charges the battery. When the AC voltage increases as the engine rpm increases, the AC wave rectified by diode D2 goes through the Zener diode (ZD) and allows the gate of the SCR to open. The SCR intentionally shorts the AC input from the alternator to ground. Therefore, the half-wave charging system with a voltage regulator is either fully charging the battery or not charging the battery at all! This is the major disadvantage of the half-wave charging system and the main reason that it's not used much any longer. You may also notice that this particular system uses AC to power the headlight system. When the headlight is not on the excess power is sent back to ground through a resistor.

One other type of voltage regulation found on some half-wave charging systems is known as the "balanced" charging system. In this system, the alternator is designed to allow a maximum amount of AC that won't overcharge the battery. The proper AC level is maintained independent of engine speed. Therefore, the balanced charging system needs no voltage regulator. The complete charging system consists of an alternator, a single diode, and a battery.

Full-Wave Charging System

A full-wave charging system also uses one charging coil similar to the half-wave system, but instead uses the full output potential of the charging coil instead of sending one half of the coil's output to ground. Full-wave charging systems are used on some medium-sized motorcycles. When comparing this system to the half-wave charging system, you will notice that it is more efficient by using the most alternator potential for charging the battery.

The full-wave charging system uses four diodes to rectify the AC from the alternator into DC (Figure 15-27). When the AC input voltage is positive, current flows from the alternator through diode D1 to the battery, through diode D2, and back to the alternator as shown by the blue arrows. When the AC input voltage reverses direction, current flows from the alternator through diode D3 to the battery, through diode D4, and back to the alternator as shown by the black arrows. Operating in this fashion, the AC output of the alternator is converted into a full-wave DC waveform.

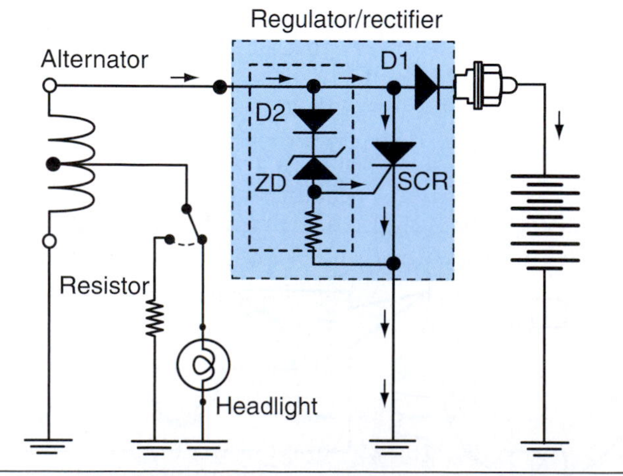

Figure 15-26 When voltage regulation is required in a half-wave system it sends all of the current to ground, meaning that it is either fully charging or not charging at all.

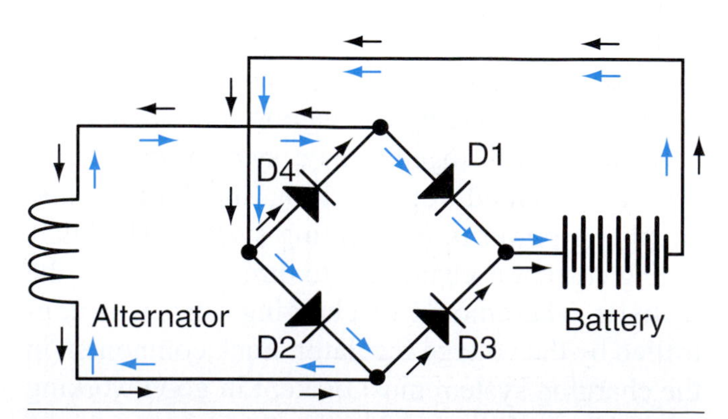

Figure 15-27 By using four diodes, both the upper and lower halves of the alternating current waveform are used and we have a full-wave rectification system.

The voltage regulation system used on a full-wave charging system normally has a voltage feedback line (Figure 15-28). The feedback line tells the voltage regulator when the battery no longer needs charging. The regulator then opens the gates on the SCRs. The SCRs intentionally short the AC input from the alternator to ground, and cuts off one-half of the current to the battery, and has the capability to cutting all of the current to the battery if needed.

3-Phase Permanent-Magnet Charging System

The 3-phase permanent-magnet charging system is the most widely used system on motorcycles because of the large charging potential to the electrical system. This system uses permanent magnets like the charging systems previously mentioned; however, the 3-phase system uses three charging coils instead of one (Figure 15-29). The alternator for this type of charging system typically has the rotor mounted on the crankshaft and the stator mounted on the outside cover of the engine (Figure 15-30).

The rectifier in the 3-phase system consists of six diodes and is connected to the alternator. The voltage regulation system is the same as the full-wave system except that it has the ability to change the charging system from 3-phase into a full-wave or a half-wave system as the battery approaches a full charge. This is done by independently controlling the gates to the three SCRs, which short the alternator output to ground (Figure 15-31). The waveform created by a 3-phase charging system (Figure 15-32) more closely approximates a pure DC output because of the three AC waves that are produced in a single revolution of the alternator's rotor.

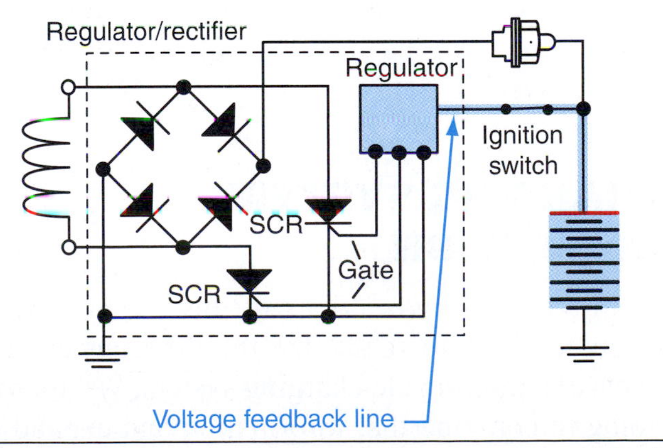

Figure 15-28 A full-wave charging system block diagram with a regulator is shown here.

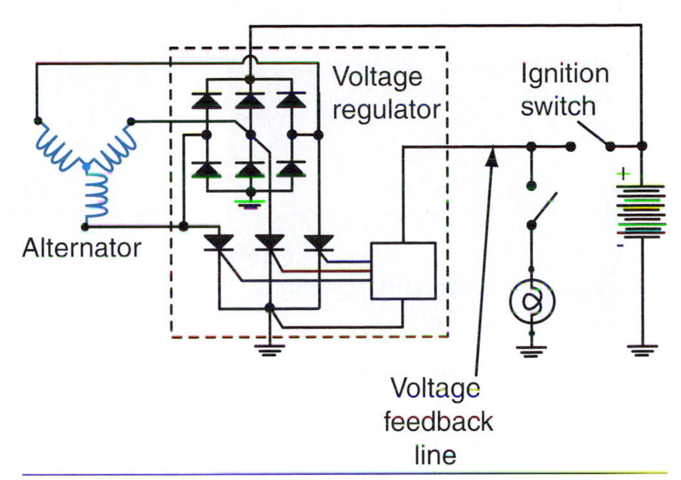

Figure 15-29 A 3-phase permanent-charging system has three coils of wire at the alternator instead of one as on the half-wave and full-wave systems.

3-Phase Excited-Field Electromagnet Charging System

The 3-phase excited-field electromagnet charging system is used on many larger motorcycles and on motorcycles that have the alternator in a location that's not directly mounted on the crankshaft (Figure 15-33). The 3-phase electromagnet system differs from the 3-phase permanent-magnet system primarily because it uses an electromagnet instead of a permanent magnet in the alternator to produce the input AC. Also, in most cases it is completely self-contained (Figure 15-34), meaning that all charging system components are within one unit. As mentioned previously, there are two types of excited-field systems: brush and brushless.

Voltage regulation on the 3-phase electromagnet occurs by changing the field coil current to create a stronger or weaker electromagnet. By having more current pass through the field coil, a stronger electromagnetic field is created. Conversely, having less current pass through the field coil produces a weaker electromagnetic field.

The voltage regulator monitors the voltage at the battery and controls the base of the transistor. When the regulator turns the transistor on, the battery feeds current through the ignition switch,

Figure 15-30 The alternator for the 3-phase permanent-magnet charging system typically has the rotor mounted on the crankshaft (A and B) and the stator mounted on the outside cover of the engine (C). Copyright by American Honda Motor Co., Inc. and reprinted with permission.

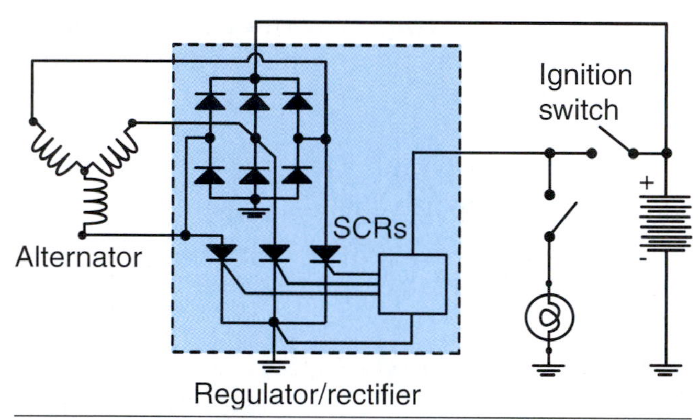

Figure 15-31 The regulator/rectifier has six diodes and three SCRs.

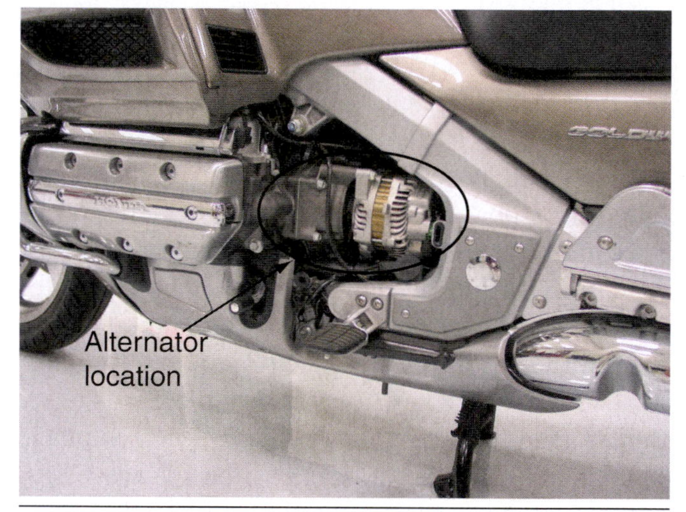

Figure 15-33 An excited-field charging system is not normally located on the crankshaft as seen here.

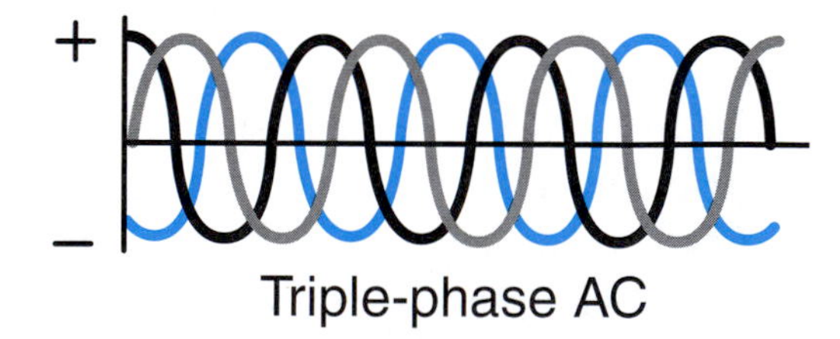

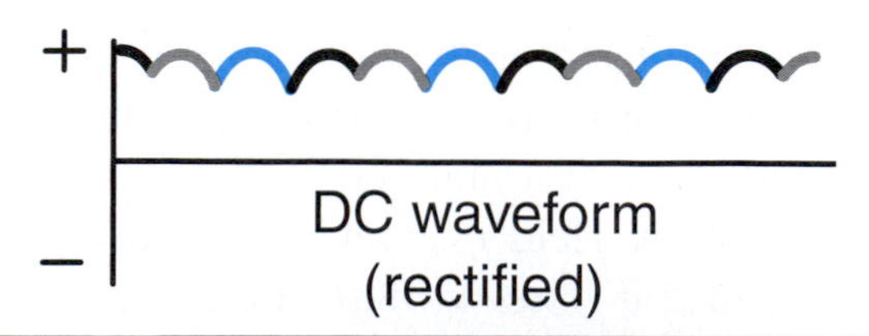

Figure 15-32 The waveform created by a 3-phase charging system more closely approximates a pure DC output because of the three AC waves that are produced in a single revolution of the alternator's rotor.

field coil, and transistor to ground. The field coils magnetize the rotor and the alternator generates AC current as the engine rotates.

The charging system waveform created by the 3-phase excited-field electromagnet charging system is the same as for the 3-phase permanent-magnet system.

CHARGING SYSTEM INSPECTION

In the previous sections of this chapter, we've discussed how to recognize the different components of a motorcycle charging system. We are now going to combine that information and expand on it to help you understand how to inspect for problems found within a charging system.

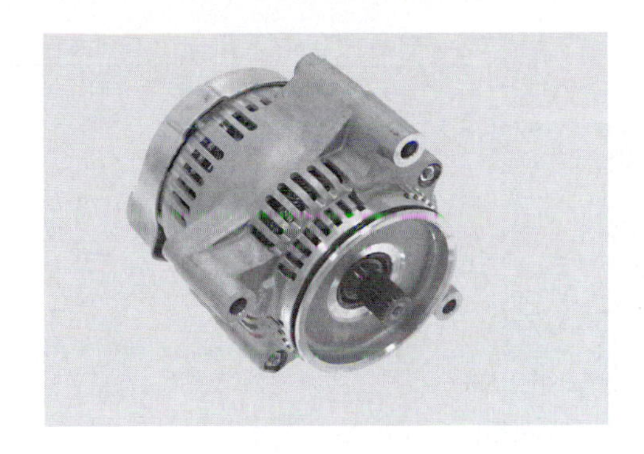

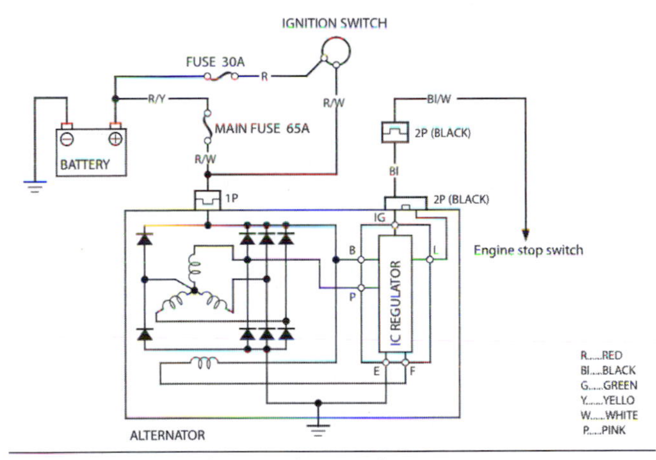

Figure 15-34 The 3-phase electromagnet system differs from the 3-phase permanent-magnet system primarily because it uses an electromagnet instead of a permanent magnet in the alternator to produce the input AC. Also, in most cases it is completely self-contained. Copyright by American Honda Motor Co., Inc. and reprinted with permission.

Figure 15-35 A typical soldering gun is shown here.

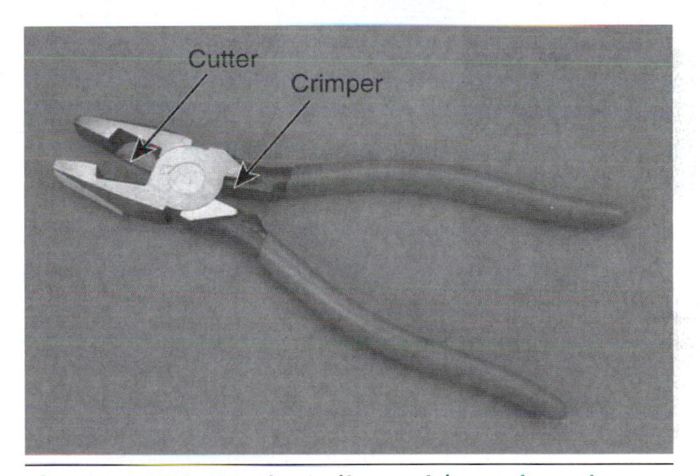

Figure 15-36 Cutting pliers with a wire crimper.

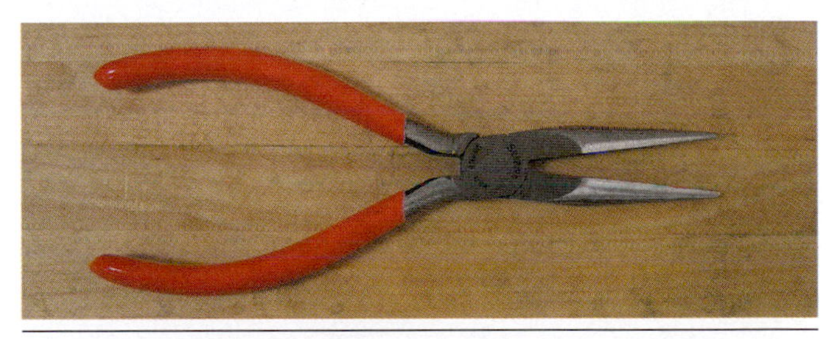

Figure 15-37 Needle-nose pliers.

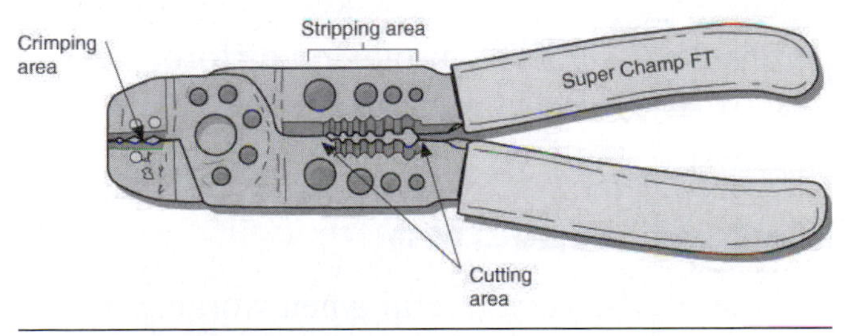

Figure 15-38 Wire striper/crimping pliers.

Hand Tools for Electrical Work

The basic hand tools you will need for most electrical repairs are as follows:

- Soldering gun (Figure 15-35)
- Cutting pliers (Figure 15-36)
- Needle-nose pliers (Figure 15-37)
- Wire stripper/crimping pliers (Figure 15-38)
- Test light (Figure 15-39)
- Multi-meter (Figure 15-40)
- Hydrometer (Figure 15-41)

There are also other electrical system specialty tools that normally are manufacturer specific.

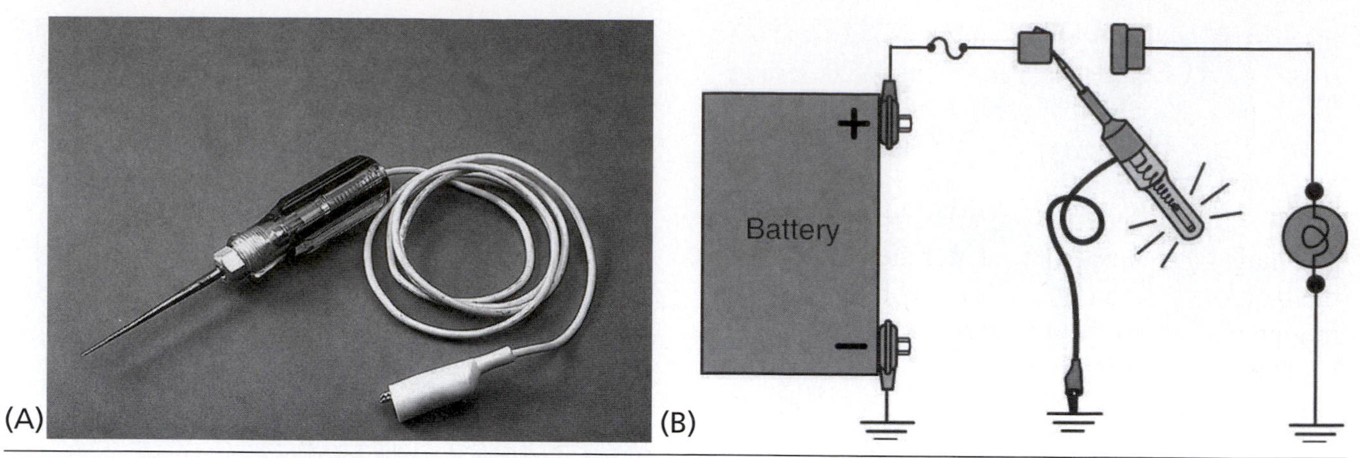

Figure 15-39 A typical test light is shown in (A) and in use in (B).

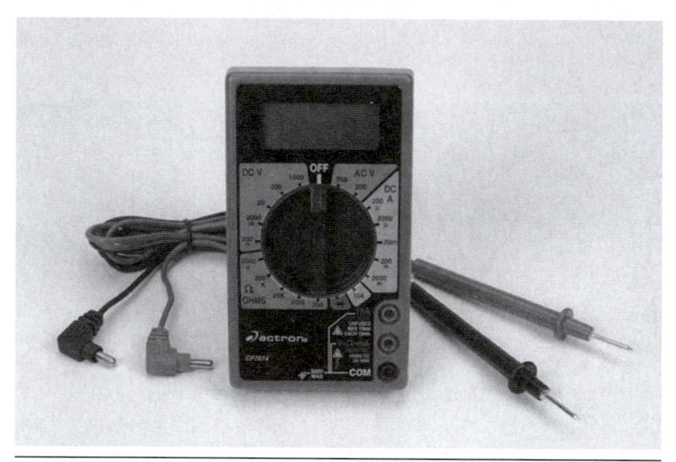

Figure 15-40 A typical digital multi-meter.

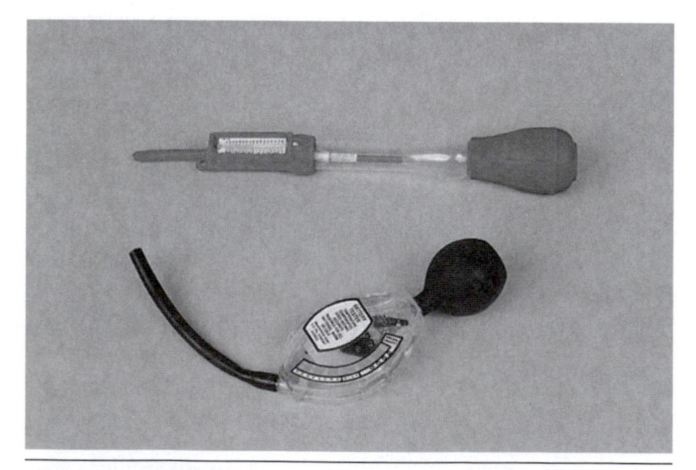

Figure 15-41 Two examples of different types of a hydrometer.

Battery Inspection

First off, be very careful when working with a battery. Always wear safety glasses when working around batteries. Battery acid will destroy clothing, paint, etc., and could also cause severe burns if it gets on your skin. If you accidentally spill some battery acid, the spill should be washed quickly using water and baking soda, which will help to neutralize the acid.

Before doing any testing, you should always visually inspect the battery. If there are cracks in the casing, broken terminals, or other signs of severe damage, such as heavy white lead sulfate on the internal plates, the battery should be replaced.

Next, check the battery cables and ensure they have good contact with the battery terminals. If the cables or terminals are corroded or loose, be sure to clean and tighten the connection. A bad battery connection can cause very high resistance, which will interfere with the flow of electrical current. This can cause many different problems in the electrical system. Clean the battery cables and terminals with a wire brush. A smear of dielectric grease on the cables and terminals will help prevent corrosion.

The electrolyte in a battery is very caustic. The condition of a battery is determined by the specific gravity of the electrolyte. When working with a conventional battery you can use a hydrometer (Figure 15-41), which is available from most automotive parts stores. When a battery is new or fully charged, you should get a specific gravity reading of 1.280 to 1.320 (depending on air temperature). As the battery is discharged, this reading will decrease.

A battery provides direct current for operating the motorcycle. One way of knowing the amount of current that can be drawn out of a battery is to know its ampere-hour capacity. For example, a 12-amp/hour battery will discharge fully if one amp of current is drawn out continuously for a 12-hour period.

Motorcycle batteries are rated in ampere-hours. The larger the ampere-hour number, the stronger the battery. The voltage does not change in relation to ampere-hours of the battery.

The battery amp/hour rating is not always stated within the battery model number. To determine the amp/hour rating on most batteries you must look on the battery cover. Some batteries will state the amp/hour rating right on the battery (Figure 15-42).

Most other maintenance-free battery amp/hour ratings will have to be determined using some good-old-fashioned math! To determine what the amp/hour rating is on a particular battery, simply look at the standard charge rating (Figure 15-43), shown as STD imprinted on the battery and multiply that number by ten.

Batteries that are weak or have been out of service for a long period of time can be recharged using a battery charger. Note that there are two types of battery chargers available.

Motorcycle manufacturers generally do not recommend the use of automotive-style battery chargers (Figure 15-44). Most automotive chargers are of the constant voltage, variable current type and are slow to charge a small motorcycle-style battery. Automotive chargers do not shut off; therefore, a battery can actually overcharge and become damaged left attached to the incorrect battery charger for a long period of time.

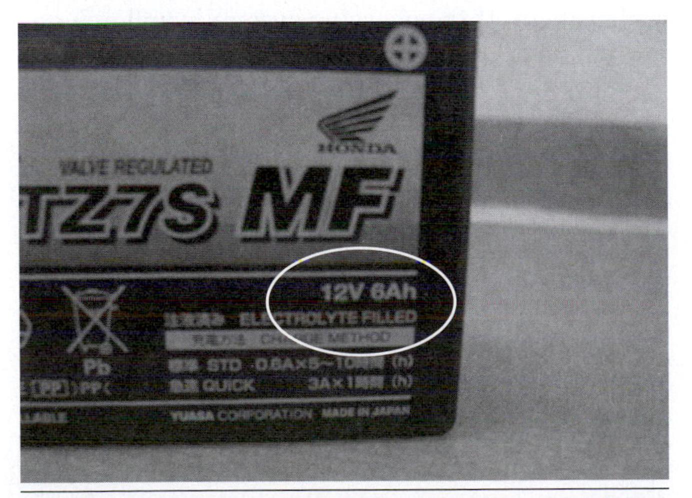

Figure 15-42 On some batteries, the amp/hour rating may be easily determined by looking at the battery cover.

Instead, most manufacturers recommend the more sophisticated constant current battery charger-maintainer (Figure 15-45) that has advanced features to recover, monitor, and maintain 12-volt batteries. A battery charger-maintainer knows when to shut off and will maintain the battery properly without fear of overcharging if left on the charger for a long period of time.

In the past, and even in some cases today, a battery would be tested for its ability to hold a charge by using a battery load tester (Figure 15-46). A load tester tests the battery under a heavy electrical load condition (such as using the electric starter) while it is out of the motorcycle.

The motorcycle industry is beginning to utilize a tool that has been used in the automotive industry for some time now to measure the health of a battery (Figure 15-47). A battery conductive analyzer describes the battery's ability to conduct current. It measures the plate surface available in a battery for chemical reaction. Measuring conductance provides a very reliable indication of the battery's condition and is correlated to battery capacity. This type of tester can be used to detect cell defects, shorts, normal aging, and open circuits in a battery, all of which can cause the battery to fail.

A fully charged battery will have a high conductance reading, up to 110 percent of its internal rating. As a battery ages, the plate surface will sulfate or shed active material, which will lower its capacity and conductance.

The conductance tester will display the service condition of the battery. It will indicate if the battery is good, needs to be recharged and tested again, has failed, or will soon fail. In addition to giving the state of charge, this type of tester will show the state of health (Figure 15-48).

Once sulfation develops to an advanced state, permanent capacity loss or total failure of the battery will occur. Besides the sulfation concerns, many other detrimental actions are taking place inside the battery while in a discharged condition. The corrosive effect on the lead plates and connections within the battery is greatly increased due to the reduced specific gravity of the electrolyte. The corrosion of the plates will typically result in a

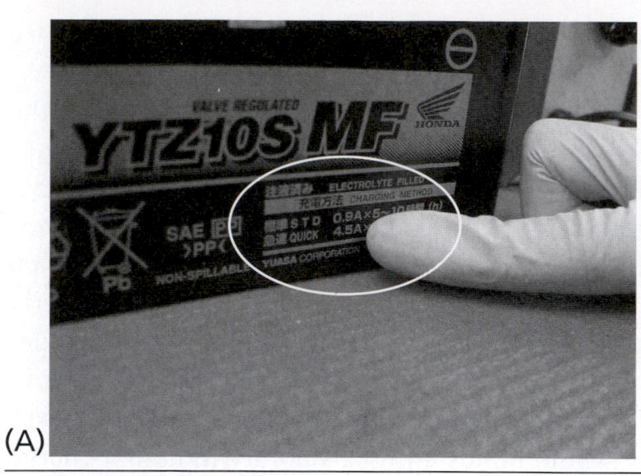

Figure 15-43 Some battery amp/hour ratings must be determined by locating the STD number. Multiply the STD number by ten to determine the amp/hour rating. The STD rating may be shown on the front (A) of the battery or on the top (B) of the battery.

gradual reduction in performance followed by battery failure. The corrosion associated with the inter cell connectors and the connecting welds will, in many instances, result in a sudden battery failure. The corroded connector may have sufficient integrity to support low drain accessories such as lights and instruments, but lack the necessary strength to provide the high discharge current required to start the vehicle. This corrosive effect can also dissolve the lead into solution, which in turn may compromise the plate insulators and result in micro shorts. Another condition that frequently occurs in a discharged battery is freezing because the solution in a discharged battery has lower acid content and therefore higher water content.

Figure 15-44 Automotive-type constant voltage battery chargers are not usually recommended to charge a motorcycle battery.

Rectifier/Regulator Inspection

As we've mentioned, electronic voltage regulators and rectifiers have no internal moving parts and must be replaced if found to be defective. The main symptoms of a faulty voltage regulator are the following:

- The battery discharges
- The battery becomes overcharged
- The lights in the electrical system burn out quickly

In most cases, to inspect a voltage regulator, you simply run the engine at the manufacturer's recommended engine speed and check for DC voltage at the battery. If the system is overcharging, the regulator is at fault and will require replacement. If the charging system is undercharging and all other charging system components have been proven to be in proper working order, the regulator is probably at fault.

Rectifiers are relatively easy to test. An ohmmeter is used to test rectifiers as shown in Figure 15-49. Simply connect the ohmmeter to the ends of each of the diodes and check the resistance in both directions. The resistance should be low in one direction and very high in the opposite direction. The specification will be given in the appropriate service manual. A general guideline for testing most diodes is to have 5–40 ohms of resistance in the forward bias direction (where current is allowed to pass) and infinite resistance in the reverse bias direction (where current is not allowed to pass).

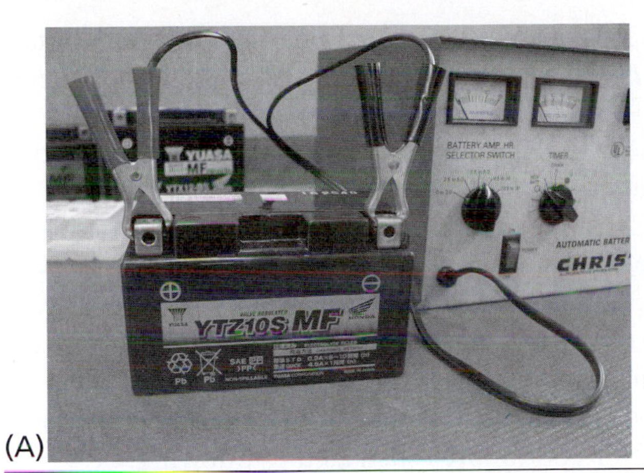

Figure 15-45 Most motorcycle battery manufacturers recommend a constant current battery charger over the constant voltage automotive type. These battery chargers can be purchased to charge one or two batteries (A) or up to ten at once (B).

To test the 3-phase six-diode rectifier in Figure 15-50, attach the black probe of the ohmmeter to the ground side of the rectifier (E) and the red probe to P1, P2, and P3. Record your measurements. Then swap the meter leads and take the three resistance readings again. You have now measured the groundside of the rectifier.

You can next test the battery side of the rectifier by attaching the meter probes to the battery (B) side of the rectifier and test the diodes in the same manner. When you have completed testing, you should have twelve readings consisting of forward- and reverse-bias measurements for each of the six diodes.

Alternator Inspection and Testing

The function of the alternator is to produce electricity that can be used to charge the battery or supply current for other electrical system components such as the lights, horn, and so on. As you learned before, the principal parts of an alternator are the rotor and the stator. Permanent magnets or

Figure 15-46 Load testers are used to test a battery's ability to withstand high amounts of electrical load.

Figure 15-47 A battery conductive analyzer describes the battery's ability to conduct current. This type of test is a more accurate way to determine a battery's internal condition.

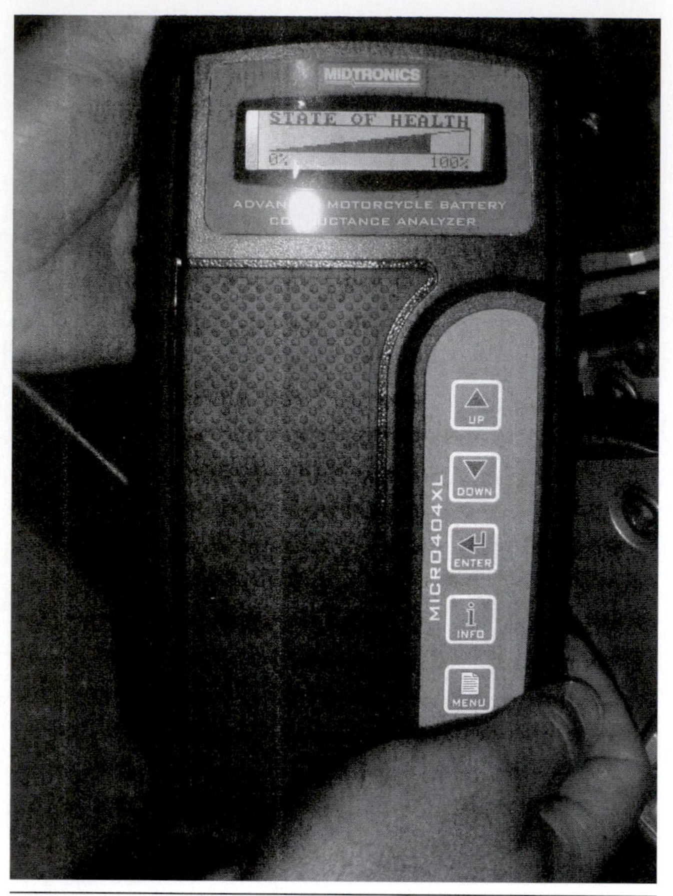

Figure 15-48 A battery conductive analyzer has the capability to test the state of health of a battery.

electromagnets are attached to the rotor and rotate when the crankshaft rotates. Rotors are positioned so that the magnets closely pass the stator coils. As the magnets pass the coils, electricity is induced in the coils.

Problems seldom occur with alternators because they have few moving parts. Servicing is generally not required except on brush-type excited-field alternators. If a problem does occur on an alternator, the problem may be due to a faulty stator coil. You should be aware of possible stator coil failures such as the following:

- **Open stator wire** If the stator wire is open, there is no AC output, therefore the battery will not charge. In this case, an ohmmeter will indicate no continuity (infinite resistance) between the terminals being tested.
- **Shorted circuit** Diagnosis of a shorted circuit is a bit more difficult. The symptom may simply be poor AC output performance or low AC output when the engine is hot. Vibration or shock can be the cause of such problems.

Rotors

Rotors must not contact the stator coils as they turn. Therefore, care should be taken to ensure that the rotor is correctly positioned on the crankshaft. If the rotor is loose or crooked, it will wobble and damage the stator coils.

Rotors do not require service and rarely fail; however, care should be taken to prevent the loss of magnetism on permanent-magnet charging

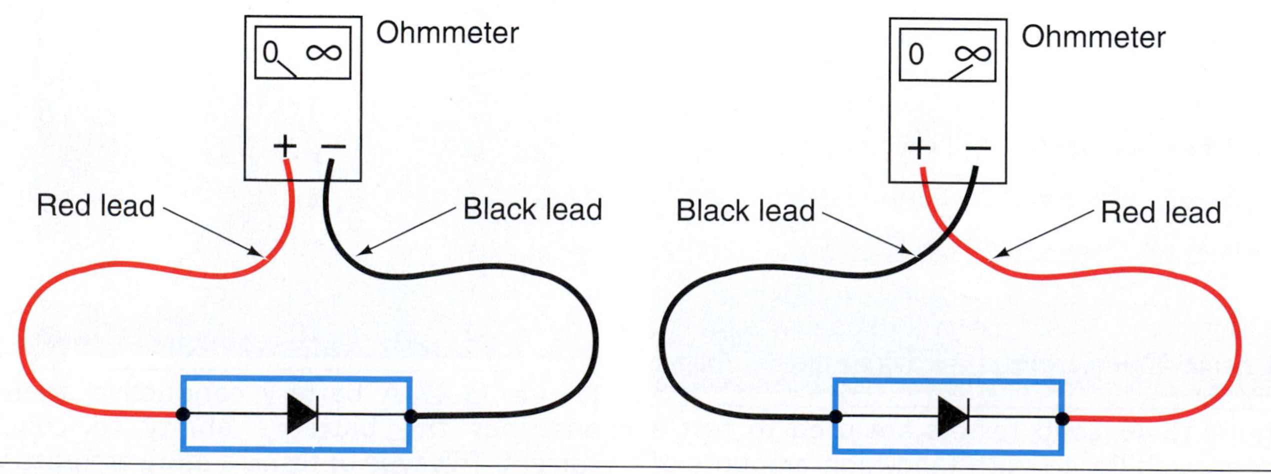

Figure 15-49 Testing a diode with an ohmmeter.

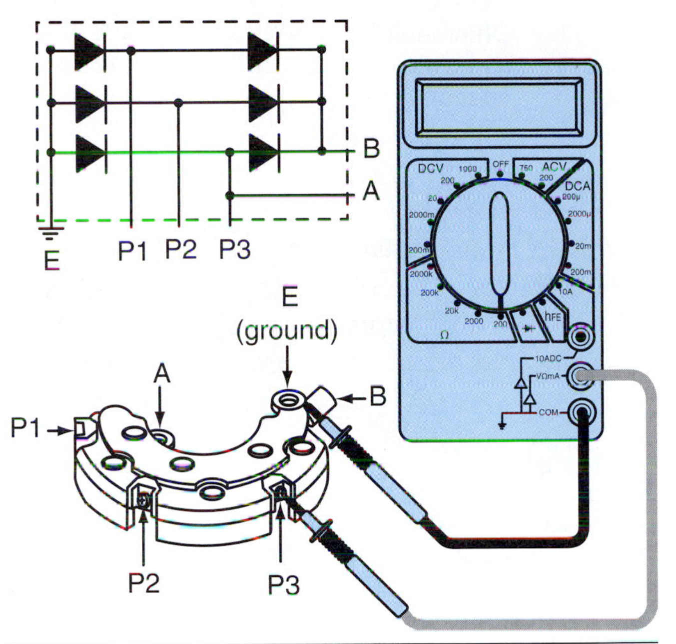

Figure 15-50 Testing a six-diode rectifier requires twelve individual tests.

systems, which can be caused in several ways, such as the following:

- Dropping the rotor
- Hitting the rotor with a hammer (for example, to remove it)
- Allowing the rotor to come into contact with another magnetic field

Heat and aging are other factors that can cause rotors to lose their magnetism. If the magnets on the rotor are weakened for any reason, you should replace the rotor. Keep in mind that there are no specifications given to tell you how much magnetism a magnet should have; therefore, you should be aware that this could be a cause of a weak charging system if all other tests indicate that the charging system is working properly but the battery does not come to a full charge.

Electromagnet Excited-Field Inspection

Electromagnetic excited-field coils can be inspected by measuring the field coil resistance. To measure the coil resistance, set an ohmmeter to the R1 range. Connect the test leads of the ohmmeter to the slip rings as shown in Figure 15-51. If the meter doesn't read as specified in the service manual, replace the rotor. Rotor coil resistance should normally be about 4 ohms. In addition, check to make certain that there is no short to ground at the excited field by testing the slip ring at the end of the rotor shaft (Figure 15-52).

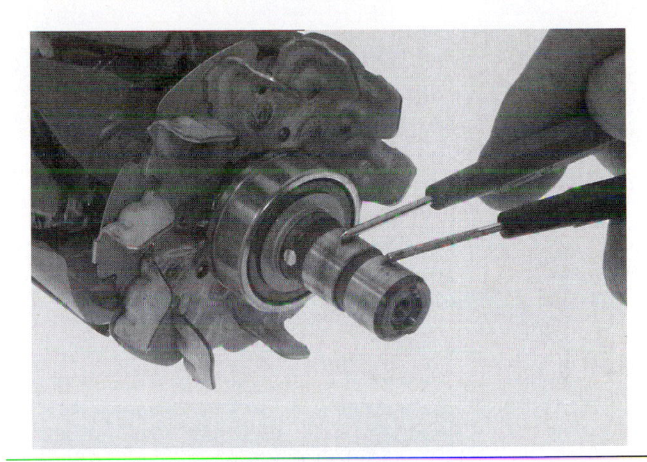

Figure 15-51 Testing the field coil resistance by measuring the resistance through the slip rings is shown here. Copyright by American Honda Motor Co., Inc. and reprinted with permission.

Stator Inspection

The stator can be tested measuring the coil resistance. Set an ohmmeter to the R1 range and connect the test leads between each of the sets of stator coil wires (Figure 15-53). The stator coil resistance should be less than one ohm for each coil.

Next, verify that there are no shorts to ground by setting the ohmmeter to the highest range. Measure the resistance between the stator coil core and each coil winding. The reading should be infinity. If there is a short, the stator is defective and must be replaced.

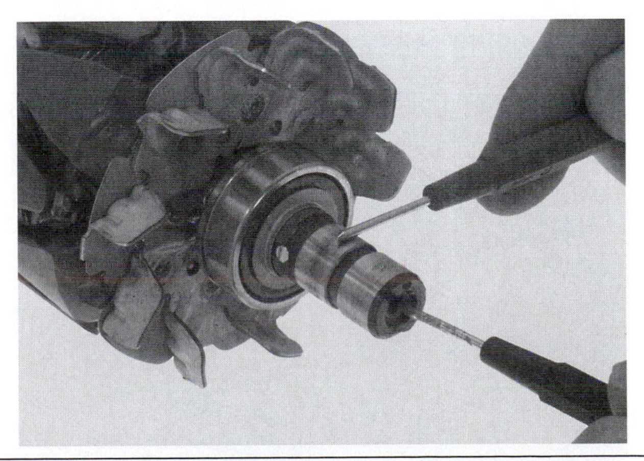

Figure 15-52 Checking for a short to ground. Copyright by American Honda Motor Co., Inc. and reprinted with permission.

(A)

(B)

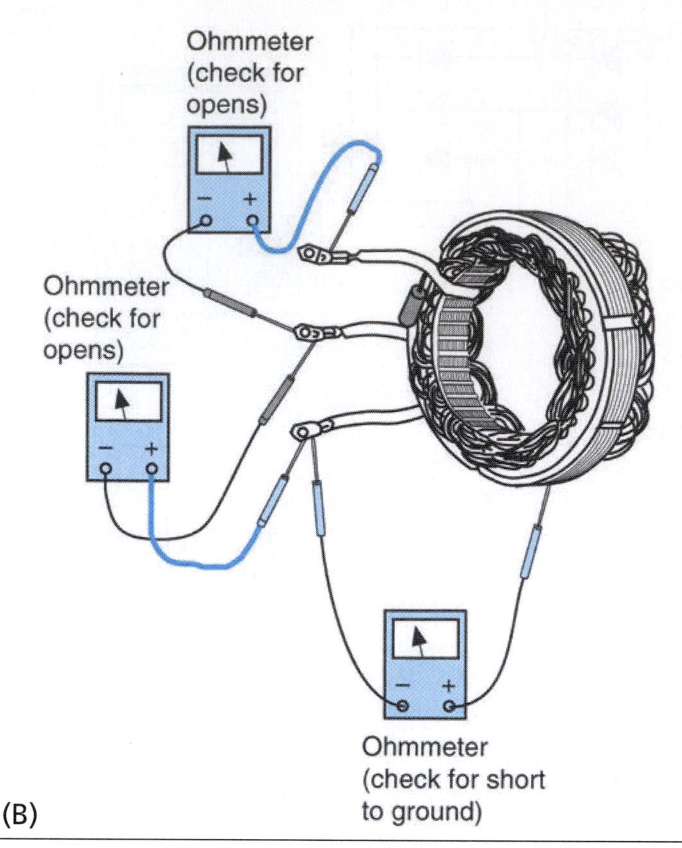

Figure 15-53 Checking for stator resistance on an actual alternator (A) and an illustration of the main tests to verify that a stator is good (B). Copyright by American Honda Motor Co., Inc. and reprinted with permission.

Another test of the stator can be made with the engine running. Set your multi-meter to measure AC voltage and connect the meter to the output of the alternator (Figure 15-54). Increase the engine speed to 3,000–4,000 rpm. The voltage should be at least 20 volts AC at each connection point with the stator disconnected from the rest of the electrical system.

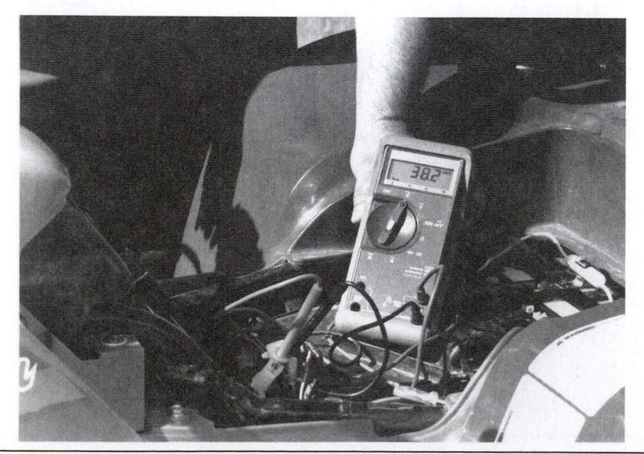

Figure 15-54 Testing for AC voltage at the alternator while it is disconnected from the rest of the electrical system.

When checking the continuity of a stator, remember to isolate the stator from the rest of the electrical system by disconnecting it from the rest of the motorcycle.

DC ELECTRICAL CIRCUITS

When electric current leaves the battery in a motorcycle electrical system, some of the current travels to various electrical components. In this section, we're going to discuss the various DC electrical circuits found on motorcycles. We'll give a brief description of each circuit and show how it can be thought of as a separate electrical subsystem. Each of the systems that we are about to discuss have the following four things in common:

- Each is powered by the battery
- Each is operated by a switching device
- Each must complete its circuit to operate
- Each has a load device to create resistance in the system (lights, horn, etc.)

Switches

Switches are components designed to open and close an electrical circuit. If a switch is defective, it must be replaced. You can check a switch using an ohmmeter. The ohmmeter should indicate continuity (a complete connection or "0" ohms when the switch is in the "On" position and should indicate infinite resistance when the switch is in the "Off" position (Figure 15-55).

Some switches have more switch positions and leads than the simple On/Off type of switch. These switches should have a switch matrix shown in the motorcycle or ATV electrical schematic similar to that shown in Figure 15-56. A switch matrix shows you the switch positions and the switch leads that should have continuity for each position.

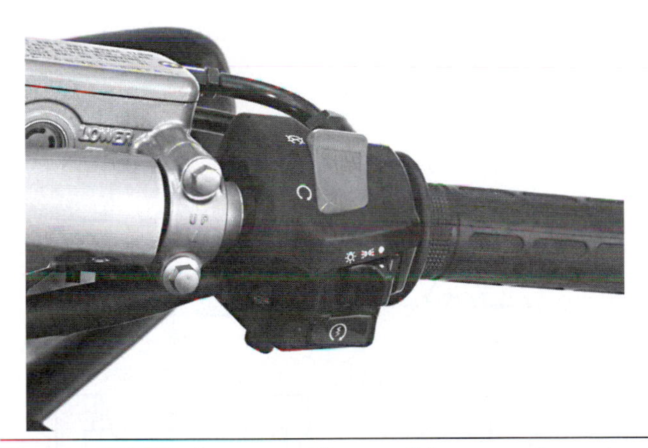

Figure 15-55 A properly working engine stop switch will show a complete connection in one position and an open connection in the other. Copyright by American Honda Motor Co., Inc. and reprinted with permission.

	BAT1	IG1	BAT2	ACC	IG2	KEY
ACC			O—	—O		KEY ON
ON	O—	—O	O—	—O—	—O	KEY ON
OFF						KEY OFF
LOCK						KEY OFF LOCK PIN
COLOR	R	R/Bl	R	R/W	Bu/O	

Figure 15-56 Many switches have multiple uses for different positions. Copyright by American Honda Motor Co., Inc. and reprinted with permission.

Headlight Circuits

The headlight is turned on whenever the ignition switch is in the "On" position, but is momentarily turned off as the electric starter motor is activated. The purpose of this is to allow maximum current flow from the battery to the starting circuit. The headlight is momentarily turned off during starting by the starter switch. When the starter button is pressed, the starter switch opens a set of contacts in the lighting circuit (Figure 15-57).

Turn Signal/Hazard Relay Circuits

A typical turn signal and hazard relay wiring diagram is illustrated in Figure 15-58. The diagram illustrates the system as it would be with the turn signal and hazard switches in the "Off" position.

When the turn signal switch is turned on (either left or right), power flows from the battery through the turn signal relay, through the L (left) or R (right) switch contacts, and through the left or right indicator and signal lights to ground.

When the hazard switch is activated, power flows from the battery through the hazard switch contacts, and through both the left and right indicator and signal lights to ground. The hazard relay and turn signal relay are special circuits that open and close to make the lights flash.

Brake-Light Circuits

A typical brake-light circuit is shown in Figure 15-59. Note that there are two separate brake-light switches—one for the front brake and

ENGINE STOP SWITCH

	IG	BAT
OFF		
RUN	O—	—O
COLOR	W	W/Bl

STARTER SWITCH

	ST	IG	BAT4	HL
FREE			O—	—O
PUSH	O—	—O		
COLOR	Y/R	Bl	Bl/R	Bu/W

Figure 15-57 In this illustration, the engine stop switch is connected when in the "On" position and the (electric) starter switch creates an open to the headlight (HL) when it is being pushed. This allows maximum current to be available for the starter motor to turn the engine over. Copyright by American Honda Motor Co., Inc. and reprinted with permission.

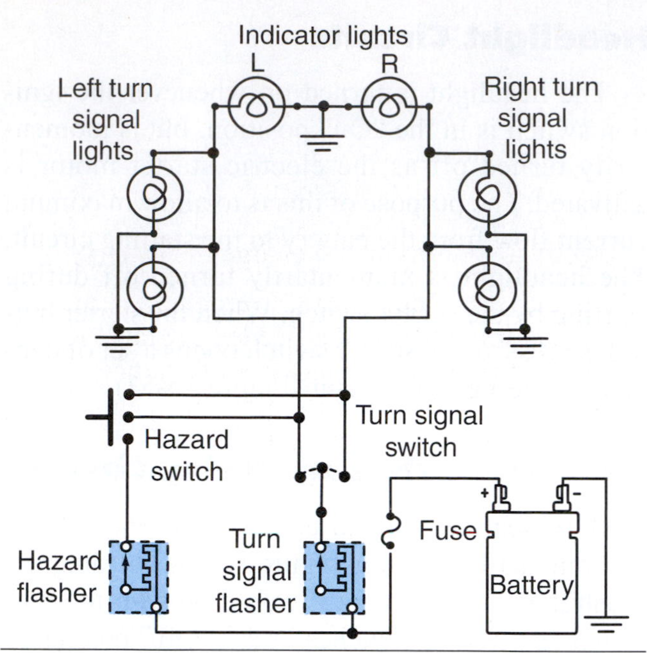

Figure 15-58 A typical turn signal and hazard lighting system block diagram.

one for the rear brake. As either brake switch is activated (by applying the brakes) the circuit is completed and the brake light is illuminated. The ignition switch in the brake-light circuit keeps the brake light from operating except when the ignition switch is turned on.

Horn Circuits

Horn circuits are almost identical in design to the brake circuit just mentioned except that there's only one switching device (the horn button). A horn circuit may have its switching device located on the groundside of the horn instead of the positive side as the brake switch diagram illustrates.

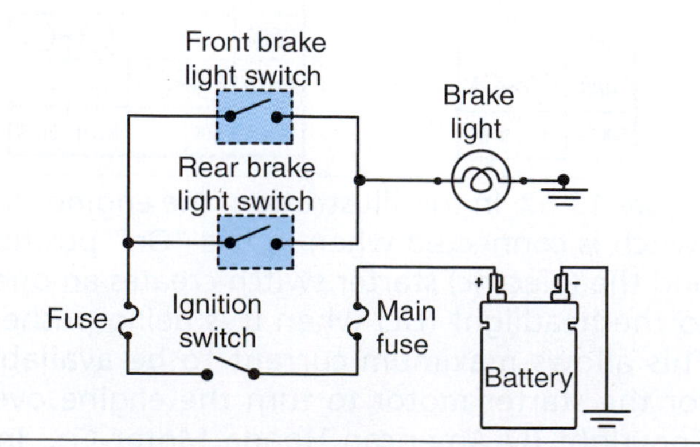

Figure 15-59 A typical brake-light circuit.

Neutral-Light Circuits

When the ignition switch is turned on and the transmission is in neutral, the neutral indicator light turns on (Figure 15-60). The neutral switch is normally operated by the shift drum. When the transmission is in any gear other than neutral, the path to ground is broken and the neutral light is turned off.

Warning Lights

A typical oil pressure warning-light system used on a four-stroke engine is shown in Figure 15-61. If the engine oil pressure falls below a specified amount, the oil pressure switch senses it and turns on the warning light. When the oil pressure is too low, the switch provides a ground to turn on the indicator. When the oil pressure is satisfactory, the oil pressure switch removes the ground connection and the light turns off.

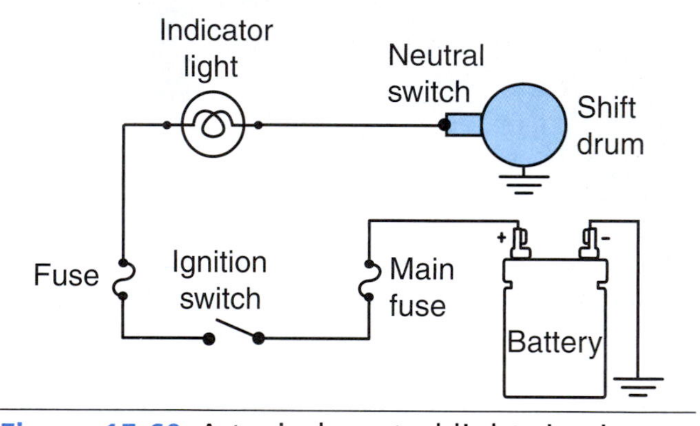

Figure 15-60 A typical neutral light circuit.

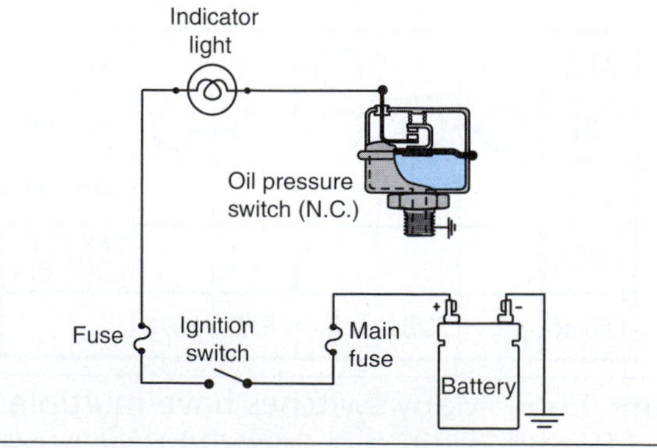

Figure 15-61 A typical oil pressure warning-light circuit.

Summary

- The purpose of a charging system is to replenish the voltage in a battery as it is used when the motorcycle is running.
- The alternator generates AC voltage. The rectifier changes the AC into DC. The battery stores the DC voltage and the voltage regulator controls the voltage being sent to the battery.
- There are three basic types of charging systems used on motorcycles: half wave, full wave and 3-phase. Once you understand the basics of these charging systems, it is easy to distinguish them on a schematic.
- There are three key items needed to produce AC voltage: a magnetic field, a conductor (coils of wire), and motion. An alternator is used to generate electricity on a motorcycle and has two main components: the rotor and the stator. The rotor has a series of magnets and rotates either inside or outside the stationary windings of the stator. The stator consists of sets of coils, which are used to produce power for the motorcycle's electrical circuits and to charge the battery.
- A rectifier is required to convert the AC from the alternator into DC that is needed to charge the battery. Depending on the type of charging system being used, the rectifier uses a diode or a group of diodes to convert the AC into DC by allowing AC current to flow in only one direction.
- It is important to understand that block diagrams are often used to separate an electrical subsystem from the rest of the motorcycle. This is done by drawing out the specific components in the system in question by using the complete schematic as a starting point.

Chapter 15 Review Questions

1. What is used to generate AC current in a motorcycle or ATV?
   a. Alternator
   b. Rectifier
   c. Battery
   d. Thyristor
2. Resistance in an isolated component can be quickly checked using
   a. an AC voltmeter
   b. an ammeter
   c. a DC voltmeter
   d. an ohmmeter
3. In a permanent-magnet alternator, the magnets are attached to or are part of the
   a. countershaft
   b. crankshaft body
   c. rotor
   d. stator
4. What device is used to control the rate of charging to prevent undercharging or overcharging the battery?
   a. Diode
   b. Voltage regulator
   c. Resistor
   d. Rectifier
5. A 12-volt motorcycle or ATV conventional wet-cell battery has how many cells?
   a. 2
   b. 3
   c. 6
   d. 9
6. What device is used to convert AC current into DC current?
   a. Voltage regulator
   b. Resistor
   c. Rectifier
   d. Battery

7. A charging system that uses the complete AC waveform from a single charging coil is called a _______ charging system.
   a. half-wave
   b. 3-phase permanent-magnet
   c. full-wave
   d. 3-phase electromagnet
8. Which of the following is potentially the most powerful type of alternator?
   a. Excited-field electromagnet
   b. Alnico inner-rotor
   c. Permanent magnet
   d. Current limiter
9. What can be used on a sulfuric acid spill to help neutralize it?
   a. Distilled water
   b. Motor oil
   c. Baking soda
   d. Sand
10. The neutral light indicator is lit by a switch normally operated by the
    a. countershaft
    b. shift drum
    c. gearshift lever
    d. sprocket
11. White deposits on battery plates are known as
    a. chalking
    b. sedimentation
    c. cracking
    d. sulfation
12. When AC is changed to DC, it's known as
    a. rectification
    b. mutual induction
    c. electromagnetism
    d. chemical reaction
13. The two main components of an alternator are the
    a. rotor and flywheel
    b. regulator and rectifier
    c. rotor and stator
    d. motor and starter
14. Another term to describe a voltage regulator is
    a. current limiter
    b. rectifier
    c. inverter
    d. thyristor
15. What device is used to detect a short between a stator coil winding and the core?
    a. Ammeter
    b. Ohmmeter
    c. Converter
    d. Voltmeter

CHAPTER

16 Ignition and Electric Starting Systems

Learning Objectives

When students have completed the study of this chapter and its laboratory activities they should be able to:

- Describe the three major functions of an ignition system
- Describe the common components found in all types of ignition systems
- Name the two major electrical circuits used in an ignition system and their common components
- Name the different types of ignition systems
- Describe how the electric starter systems used on motorcycles operate

Key Terms

Breaker points
DC motor operating principle
Electric starting systems
Hall-effect sensor
Ignition coil
Ignition system
Ignition timing
Magnetic pulse generator
Preignition
RPM
Solenoid
Spark plugs
Starter clutch

INTRODUCTION

In Chapters 14 and 15, you learned the basics of electricity and how charging systems operate. You also learned about battery-powered electrical circuits found on motorcycles.

In this chapter, you'll learn about the different types of ignition systems. First, we'll explain basic ignition system operation and identify the main components in an ignition system. Then, we'll look at the different types of ignition systems and learn about ignition system timing. Finally, we will discuss the electric starting systems found on motorcycles.

Do you remember the stages of operation in both a two-stroke and four-stroke engine? In each cylinder of the engine, the piston rises during the compression stage to compress the air-and-fuel mixture in the combustion chamber. Just before the piston reaches top-dead center, a spark plug fires in the cylinder and ignites the compressed air-and-fuel mixture. The ignition of the air-and-fuel mixture forces the piston down in the cylinder, producing the power stage. The power produced by the ignition of the air-and-fuel mixture turns the crankshaft, which in turn keeps the piston moving and the engine running.

One of the requirements for an efficient engine is the correct amount of heat shock, delivered at the right time. This requirement is the responsibility of the ignition system. The ignition system supplies properly timed, high-voltage surges to the spark plug(s). These voltage surges cause combustion inside the cylinder.

The ignition system must create a spark, or current flow, across each pair of spark plug electrodes (Figure 16-1) at the proper instant, under all engine operating conditions. This may sound relatively simple, but when one considers the number of spark plug firings required and the extreme variation in engine operating conditions, it is easy to understand why ignition systems are so complex.

If a four-cylinder engine is running at 8,000 revolutions per minute (rpm), the ignition system must supply either 8,000 or 16,000 sparks per minute. Many ignition systems fire the spark plugs in multiple cylinders once per revolution. When this occurs it is known as a wasted spark as only one cylinder requires a spark. Most of today's ignition systems use a digital-type ignition system that does not use a wasted spark. Spark plugs must fire at the correct time and generate the correct amount of heat. If the ignition system fails to do these things, fuel economy, engine performance, and emission levels will be adversely affected.

MOTORCYCLE IGNITION SYSTEMS

The sole purpose of an **ignition system** is to provide a spark that will ignite the air-and-fuel mixture in the combustion chamber. The spark must be timed to occur at a precise point relative to the position of the piston as it reaches top-dead center (TDC) on the engine's compression stroke. The difference between one ignition system type and another is how the spark is activated. Many of today's ignition systems are used in unison with electronic fuel injection systems as well.

For each cylinder in an engine, the ignition system has three main functions. First, it must generate an electrical spark that has enough heat to ignite the air/fuel mixture in the combustion chamber. Second, it must maintain that spark long enough to allow for the combustion of all the air

Figure 16-1 The sole purpose of an ignition system is to provide a spark that will ignite the air-and-fuel mixture in the combustion chamber of an engine.

and fuel in the cylinder. Last, it must deliver the spark to each cylinder so combustion can begin at the right time during the compression stroke of each cylinder.

When the combustion process is completed, a very high pressure is exerted against the top of the piston. This pressure pushes the piston down on its power stroke and is the force that gives the engine power. For an engine to produce the maximum amount of power it can, the maximum pressure from combustion should be present when the piston is at 10 to 23 degrees after top-dead center (ATDC). Because combustion of the air/fuel mixture within a cylinder takes a short period of time, usually measured in thousandths of a second (milliseconds), the combustion process must begin before the piston is on its power stroke. Therefore, the delivery of the spark must be timed to arrive at some point before the piston reaches top-dead center.

Determining how much before TDC the spark should begin gets complicated because of the fact that as the speed of the piston moving from its compression stroke to its power stroke increases, the time needed for combustion stays about the same. This means the spark should be delivered earlier as the engine's speed increases (Figure 16-2). However, as the engine has to provide more power to do more work, the load on the crankshaft tends to slow down the acceleration of the piston and the spark should be somewhat delayed.

Figuring out when the spark should begin gets more complicated due to the fact that the rate of combustion varies according to certain factors. Higher compression pressures tend to speed up combustion. Higher octane gasolines ignite less easily and require more burning time. Increased vaporization and turbulence tend to decrease combustion times. Other factors, including intake air temperature, humidity, and barometric pressure, also affect combustion. Because of all of these complications, delivering the spark at the right time is a difficult task.

How does an ignition system produce a spark, time it perfectly, and keep making sparks over and over again? Let's find out.

Ignition Timing

Ignition timing in an internal combustion engine is the process of setting the time that a spark will occur in the combustion chamber during the power stroke relative to piston position and crankshaft velocity. Ignition timing reference marks are normally located on the engine's crankshaft flywheel to indicate the position of the piston. Motorcycle manufacturers specify initial or base ignition timing.

When the marks are aligned at TDC, the indicated piston is at top-dead center of the engine's stroke. Additional marks indicate the proper number of degrees of crankshaft rotation before TDC

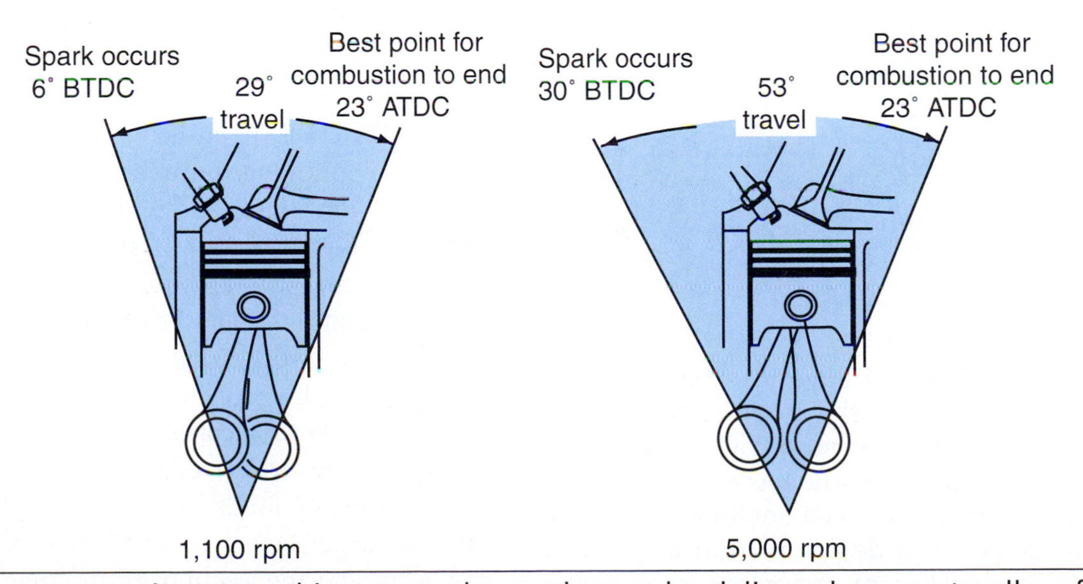

Figure 16-2 As an engine's speed increases, the spark must be delivered sooner to allow for complete combustion of the air-and-fuel mixture.

(BTDC) or after TDC (ATDC). In a majority of engines, the initial timing is specified at a point between TDC and 15 degrees BTDC depending on the manufacturer's predetermined specification.

If optimum engine performance is to be maintained, the ignition timing of the engine must change as the operating conditions of the engine change. These conditions affect the speed of the engine and the load on the engine. All ignition timing changes are made in response to these primary factors.

Ignition Timing Advance

Since motorcycle engines run at various engine speeds and ignition timing needs to be varied for these engine speeds, it is necessary to advance and/or retard the ignition. On motorcycles, two different methods are used to advance the ignition.

Older motorcycle ignition systems are equipped with centrifugal advance mechanisms (Figure 16-3) that advance and retard ignition timing in response to engine speed. Centrifugal advance uses a set of pivoted weights and springs connected to the shaft with the point cam (crankshaft or camshaft) attached to it. When engine speed increases, the weights move outward shifting the plate where the triggering device is mounted. This shifting of the plate causes the triggering device its signal earlier causing an advance in the ignition timing.

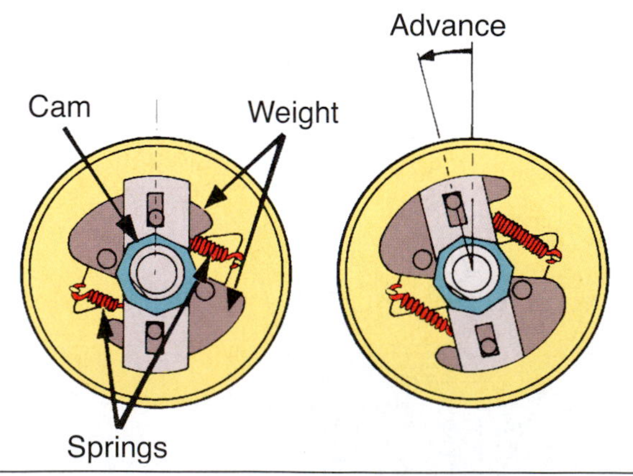

Figure 16-3 Centrifugal advancer uses a set of pivoted weights and springs connected to the shaft holding the triggering device attached to it (left image). When engine speed increases, the weights move outward shifting the plate where the triggering device is mounted (right image). This shifting of the plate causes the triggering device its signal earlier causing an advance in the ignition timing.

Most all modern-day motorcycles use an electronic advance system to control the ignition. Electronic advance systems require no mechanical parts and therefore have no parts to wear out. The overall design eliminates the need for maintenance. Electronic advance systems use multiple sensors to determine the correct timing advancement for any given condition. Electronic advance systems offer a greater variety of timing choices for different engine running conditions instead of basically only two with a centrifugal advance system.

Engine RPM

Engine **rpm** refers to the revolutions per minute that an engine is operating. At higher rpms, the crankshaft turns through more degrees in a given period of time. If combustion is to be completed by a particular number of degrees ATDC, ignition timing must occur sooner or be advanced.

However, air/fuel mixture turbulence (swirling) increases with rpm. This causes the mixture inside the cylinder to turn faster. Increased turbulence requires that ignition must occur slightly later or be slightly retarded.

These two factors must be balanced for best engine performance. Therefore, while the ignition timing must be advanced as engine speed increases, the amount of advance must be decreased some to compensate for the increased turbulence.

Engine Load

The load on an engine is related to the work it must do. Riding up hills or pulling extra weight increases engine load. Under load there is resistance on the crankshaft, therefore, the pistons have a harder time moving through their strokes.

Under light loads and with the throttle partially opened, a high vacuum exists in the intake manifold. The amount of air/fuel mixture drawn into the manifold and cylinders is small. On compression, this thin mixture produces less combustion pressure and combustion time is slow. To complete combustion by the desired degrees ATDC, ignition timing must be advanced.

Under heavy loads, when the throttle is opened fully, a larger mass of air/fuel mixture can be drawn in, and the vacuum in the manifold is low. High combustion pressure and rapid burning results. In such a case, the ignition timing must be retarded to prevent complete burning from occurring before the desired degrees ATDC.

Firing Order

Up to this point, the primary focus of this discussion has been ignition timing as it relates to any one cylinder. However, the function of the ignition system extends beyond timing the arrival of a spark to a single cylinder. It must perform this task for each cylinder of the engine in a specific sequence.

Each cylinder of an engine must produce power once in every 720 degrees of crankshaft rotation. Each cylinder must have a power stroke at its own appropriate time during the rotation. To make this possible, the pistons and connecting rods are arranged in a precise fashion called the engine's firing order. The firing order is arranged to reduce rocking and imbalance problems. Because the potential for this rocking is determined by the design and construction of the engine, the firing order varies from engine to engine. Vehicle manufacturers simplify cylinder identification by numbering each cylinder. Regardless of the particular firing order used, the number one cylinder always starts the firing order, with the rest of the cylinders following in a fixed sequence.

The ignition system must be able to monitor the rotation of the crankshaft and the relative position of each piston to determine which piston is on its compression stroke. It must also be able to deliver a high-voltage surge to each cylinder at the proper time during its compression stroke. How the ignition system does these things depends on the design of the system.

BASIC IGNITION SYSTEM COMPONENTS

Figure 16-4 shows a simplified drawing of a basic ignition system. The main components of the system are as follows:

- Power source
- Ignition switch
- Ignition coil
- Spark plug
- Triggering switch
- Stop switch

All ignition systems contain these components. The difference is how the components function. Beginning with the power source, an in-depth analysis of each of these components will be discussed.

Power Sources

In motorcycle ignition systems, there are just two different power source options. These power sources are the battery (DC) or an AC generator (AC).

In a battery ignition system a battery is connected to the ignition coil. A triggering switch device is used to alternately turn the DC voltage on and off for its operation.

AC generator power sources are far more common than battery systems for off-road motorcycles. The AC-powered ignition system uses the principles of magnetism to produce a voltage. In Chapter 15, we discussed generators and magnetic induction. Remember that when a conductor wire is moved through a magnetic field, a voltage is induced in the conductor. It's also true that if a magnet is moved near a conductor, a voltage is induced in the conductor. If this conductor wire is connected to a complete circuit, current will flow in the circuit.

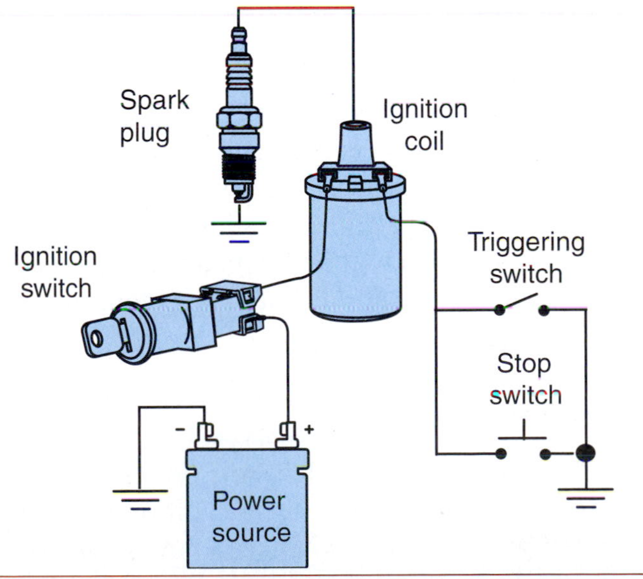

Figure 16-4 This simplified drawing shows the basic components of an ignition system.

In an AC ignition system, permanent magnets are installed in the engine's flywheel. As the flywheel turns, the moving magnets cause a voltage to be induced to the ignition coil.

We'll look at the design and operation of both the AC-powered system and the battery system in detail a little later in this chapter. For now, just keep in mind that the power source for a motorcycle or ATV ignition system can be provided by either AC power or a battery.

Battery Power Source System Advantages

Battery-type ignition systems have some advantages over an AC ignition system. First, the battery that powers an ignition system can also be used to run other devices, such as lights, accessories, and electric starter systems. In contrast, the majority of AC-powered ignition systems supply power only to fire the spark plug. Because a battery can be used to run an electric starter system, machines that contain battery systems can be started with a simple push of an electric starter button. AC-powered ignition systems are generally activated by manually starting the engine with a kick-start or pull-start device. Therefore, street-type motorcycles generally use battery systems, while smaller, off-road motorcycles generally use AC-powered systems.

AC Generator Power Source System Advantages

The AC-powered ignition system has certain advantages over the battery as a power source. First, when a motorcycle uses an AC generator, no onboard battery is needed. Batteries can be heavy and inconvenient on machines such as small dirt bikes and racing machines. In addition, no separate charging system is required with an AC generator, while batteries require a charging system to keep them working.

Ignition Switch

The ignition switch allows the power source to provide electrical power to the ignition system. Generally, a key-type switch also powers all components that utilize a power source such as lights and accessories.

Ignition Coil

An **ignition coil** provides the spark to the engine through the spark plug and is essentially a transformer that consists of two wire windings that are wound around an iron core (Figure 16-5). The first winding is called the primary winding, and the second winding is called the secondary winding. The secondary winding has many more turns of wire than the primary winding.

In an ignition coil, one end of the coil's primary winding is always connected to a power source. Depending on the type of ignition system, the power source may be a battery (DC) or a rotor with a permanent magnet (AC). Either type of power source can be used to apply a voltage to the primary winding of the coil.

When current passes through the primary winding of the coil, a magnetic field is created around the iron core. When the current is switched on, the magnetic field expands around the iron core. As the magnetic field expands, the magnetic lines of flux cut through the wires of the secondary winding and induce a voltage in the secondary winding. If the current in the primary winding is switched off, a voltage is again induced into the secondary winding by the magnetic lines of flux as they collapse and again cut through the secondary winding. The current induced into the secondary winding flows in opposite directions when the current in the primary is turned on and turned off. This is because the magnetic lines of force around the iron core cut through the secondary winding in opposite directions as the magnetic field expands and collapses.

Because the secondary winding of the coil has many more wire coils than the primary, the voltage produced in the secondary winding is much higher than the original voltage applied to the primary winding. In a typical motorcycle ignition system, the power source supplies about 12 volts to the primary winding of the ignition coil. From this 12-volt input, the ignition coil produces 20,000 to 60,000 volts or even more at the secondary coil.

The secondary winding of the coil is always connected to the spark plug through the spark plug wire. The spark plug wire is a heavily insulated wire that contains the high voltage and keeps it from arcing to ground until it reaches the spark plug.

When the magnetic field in the ignition coil expands or collapses, the high voltage in the secondary winding is applied to the spark plug and causes a spark to jump across the spark plug gap. The spark ignites the air-and-fuel mixture, allowing the motorcycle engine to run.

It's important to remember that the high voltage in the secondary winding of the coil is produced each time the primary current is turned on or off. In a collapsing-field ignition system, the high voltage from the secondary winding is used when the current to the primary winding is switched off. In a rising-field ignition system, the high voltage from the secondary winding is used when the current to the primary winding is switched on. This means that all ignition systems need some type of a device that will keep turning the current from the power source on and off. The high voltage is carried through a highly insulated wire from the coil to the spark plug.

Spark Plugs

Spark plugs provide the crucial air gap across which the high voltage from the coil causes an arc or spark. The main parts of a spark plug are a steel shell; a ceramic core or insulator, which acts as a heat conductor; and a pair of electrodes, one insulated in the core and the other grounded on the shell. The shell holds the ceramic core and electrodes in a gas-tight assembly and has threads for plug installation in the engine (Figure 16-6). The insulator material is made of ceramic materials to provide for increased durability and strength. The shell may be coated with corrosion resistance material and/or materials that prevent the threads from seizing to the cylinder head. Most of today's spark plugs have a resistor (normally about 5K ohms) between the top terminal and the center electrode. Some spark plugs use a semiconductor material to provide for this resistance. The resistor reduces RFI, which can interfere with, or damage, radios, computers, and other electronic accessories. If an engine was originally equipped with resistor plugs, resistor plugs should be installed when the originals are replaced.

A terminal post on top of the center electrode is the connecting point for the spark plug cable. Current flows through the center of the plug and arcs from the tip of the center electrode to the ground electrode. The center electrode is surounded by the ceramic insulator and is sealed to the insulator with copper and glass

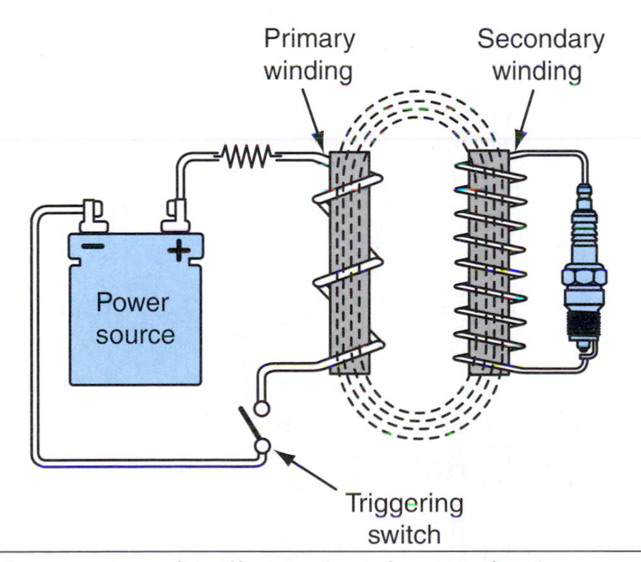

Figure 16-5 This illustration shows a basic transformer. When a voltage is applied to the primary winding, a voltage is induced into the secondary winding that is many times greater than the voltage in the primary winding.

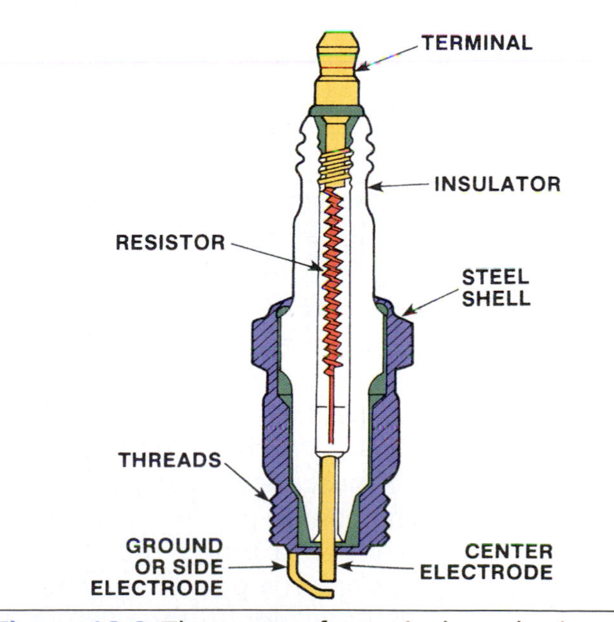

Figure 16-6 The parts of a typical spark plug.

seals. These seals prevent combustion gases from leaking out of the cylinder. Ribs on the insulator increase the distance between the terminal and the shell to help prevent electric arcing on the outside of the insulator. The steel spark plug shell is crimped over the insulation, and a ground electrode, on the lower end of the shell, is positioned directly below the center electrode. There is an air gap between these two electrodes.

A key point to remember with spark plugs is that they come in various sizes and designs to accommodate different engine designs.

Spark Plug Reach

One important design characteristic of spark plugs is the reach (Figure 16-7). This refers to the length of the shell from the contact surface at the seat to the bottom of the shell, including both threaded and non-threaded sections. Reach is crucial because the plug's air gap must be properly placed in the combustion chamber to produce the correct amount of heat. When a plug's reach is too short, its electrodes are in a pocket and the arc is not able to adequately ignite the mixture. If the reach is too long, the exposed plug threads can hit the piston or get so hot they will ignite the air/fuel mixture at the wrong time and cause preignition. **Preignition** is a term used to describe abnormal combustion, which is caused by something other than the heat of the spark.

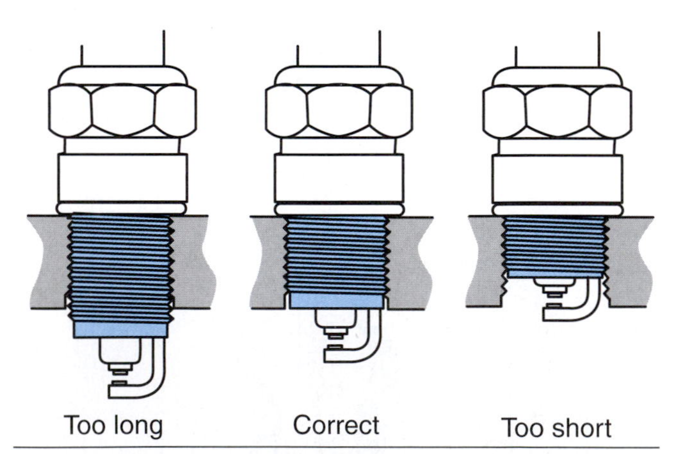

Figure 16-7 Spark plug reach is crucial because the plug's air gap must be properly placed in the combustion chamber to produce the correct amount of heat. When a plug's reach is too short, its electrodes are in a pocket and the arc is not able to adequately ignite the mixture. If the reach is too long, the exposed plug threads can hit the piston or get so hot they will ignite the air/fuel mixture at the wrong time.

Heat Range

When the engine is running, most of the plug's heat is concentrated on the center electrode. Heat is quickly dissipated from the ground electrode because it is attached to the shell, which is threaded into the cylinder head. On liquid-cooled engines coolant circulating in the head absorbs the heat and moves it through the cooling system. On air-cooled engines the heat is absorbed through the cylinder head. The heat path for the center electrode is through the insulator into the shell and then to the cylinder head. The heat range of a spark plug is determined by the length of the insulator before it contacts the shell. In a cold-range sparkplug, there is a short distance for the heat to travel up the insulator to the shell. The short heat path means the electrode and insulator will maintain little heat between firings (Figure 16-8).

In a hot spark plug, the heat travels farther up the insulator before it reaches the shell. This provides a longer heat path and the plug retains more heat. A spark plug needs to retain enough heat to clean itself between firings, but not so much that it

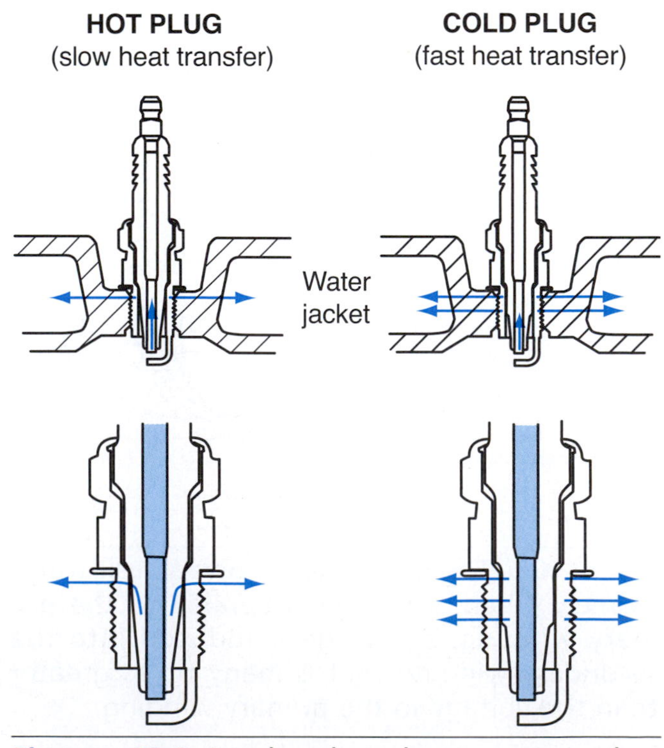

Figure 16-8 Spark plug heat range: hot versus cold.

damages itself or causes premature ignition of the air/fuel mixture in the cylinder.

The heat range is indicated by a code imprinted on the side of the spark plug, usually on the porcelain insulator.

Spark Plug Gap

The correct spark plug air gap (Figure 16-9) is essential for achieving optimum engine performance and long plug life. A gap that is too wide requires higher voltage to jump the gap. If the required voltage is greater than what is available, the result is misfiring. Misfiring results from the inability of the ignition to jump the gap or maintain the spark. A gap that is narrow requires lower voltages and can lead to rough idle and prematurely burned electrodes, due to higher current flow.

Electrodes

The materials used in the construction of a spark plug's electrodes determine the longevity, power, and efficiency of the plug. The construction and shape of the tips of the electrodes are also important.

The electrodes of a standard spark plug are made with copper and some use a copper-nickel alloy. Copper is a good electrical conductor and offers some resistance to corrosion.

Platinum electrodes are used to extend the life of a plug (Figure 16-10). Platinum has a much higher melting point than copper and is highly resistant to corrosion. Although platinum is an extremely durable material, it is an expensive precious metal; therefore, platinum spark plugs cost more than copper plugs. Also, platinum is not as good a conductor as copper. Spark plugs are available with only the center electrode made of platinum (called single-platinum) and with the center and ground electrodes made of platinum (called double-platinum). Some platinum plugs have a very small center electrode combined with a sharp, pointed, ground electrode designed for better performance.

Until recently, platinum was considered the best material to use for electrodes, because of its durability. However, iridium is six times harder, eight times stronger, and has a melting point 1,200 degrees higher than platinum. Iridium is a precious, silver-white metal and one of the densest materials found on earth. A few spark plugs use an iridium alloy as the primary metal complemented by rhodium to increase oxidation and wear resistance. This iridium alloy is so durable that it allows for an extremely small center electrode. A typical copper/nickel plug has a 2.5mm diameter center electrode and a platinum plug has a 1.1mm diameter. An iridium plug can have a diameter as small as 0.4mm (Figure 16-11), which means the firing voltage requirements are decreased. Iridium is also used as an alloying material for platinum.

Electrode Designs Spark plugs are available with many different shapes and numbers of electrodes. When trying to ascertain the advantages of each design, remember the spark is caused by electrons moving across an air gap. The electrons will

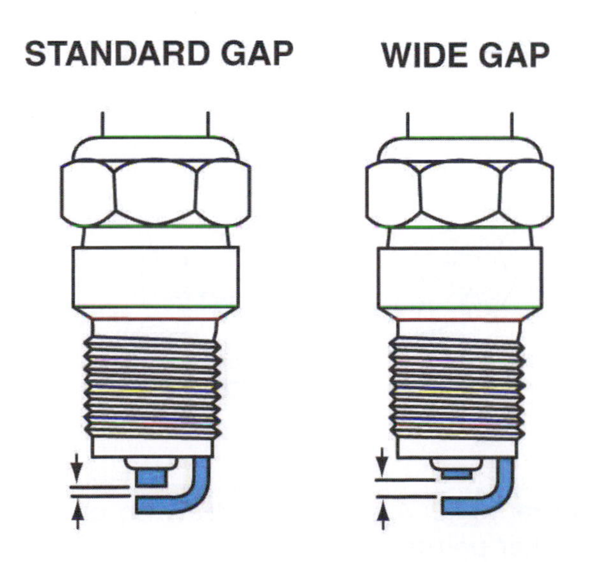

Figure 16-9 Examples of spark plug gap.

Figure 16-10 A platinum-tipped spark plug is shown here.

Figure 16-11 The spark plug here has a small diameter iridium center electrode and a grooved ground electrode.

always flow in the direction of the least electrical resistance.

The shape of the ground electrode may also be altered. A flat, conventional electrode tends to crush the spark, and the overall volume of the flame front is smaller. A tapered ground electrode increases flame front expansion and reduces the heat lost to the electrode. Many motorcycle spark plug ground electrodes have a U-groove machined into the side that faces the center electrode. The U-groove allows the flame front to fill the gap formed by the U. This ball of fire develops a larger and hotter flame front, leading to a more complete combustion.

Triggering Switch Devices

Different types of ignition systems use different types of switching devices. There are two basic types of trigger switching devices used in motorcycle ignition systems. Older ignition systems use a set of electrical contacts called breaker points and a condenser to do the switching. While no longer used by any major manufacturer of motorcycles, there are millions of motorcycles still in use that use points and condensers. All modern motorcycle systems use electronic components to do the switching. Either way, the result on the ignition coil and the spark plug is the same.

Breaker Points and Condenser

Breaker points are mechanical contacts that are used to stop and start the flow of current through the ignition coil. The points are usually made of tungsten, a very hard metal that has a high resistance to heat. One breaker point is stationary (fixed), and the other point is movable and insulated from the stationary point. The movable contact is mounted on a spring-loaded arm, which holds the points together. A simplified drawing of a set of breaker points is shown in Figure 16-12.

When the two breaker points touch, the ignition circuit is complete and the primary winding of the ignition coil is energized. When the end of the spring-loaded movable breaker point is pressed, its contact end moves apart from the stationary breaker point. This opens the circuit and the flow of current stops. Each time the breaker points move apart, the spark plug fires. This action is shown in Figure 16-13.

The movable breaker point is moved to the open position by a turning cam with a single or multiple lobes. Depending on the engine design, the cam may be located on the crankshaft or on the end of the camshaft. Each lobe on the cam forces the movable breaker point away from the stationary point, and the spark plug fires. The spring mounted under the movable point holds the movable breaker point against the cam.

Another important component of a breaker points system is the condenser (or capacitor). Remember that each time the breaker points touch, current flows through them. Unless this current

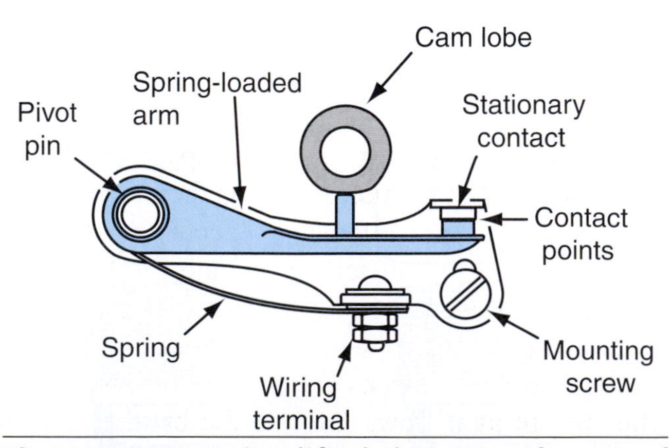

Figure 16-12 A simplified drawing of a set of breaker points.

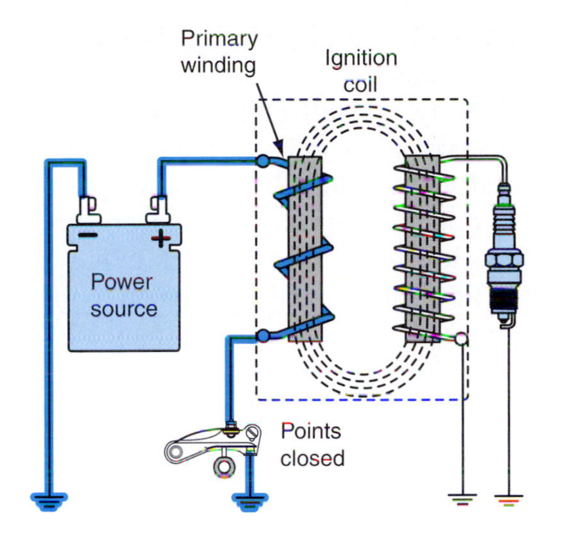

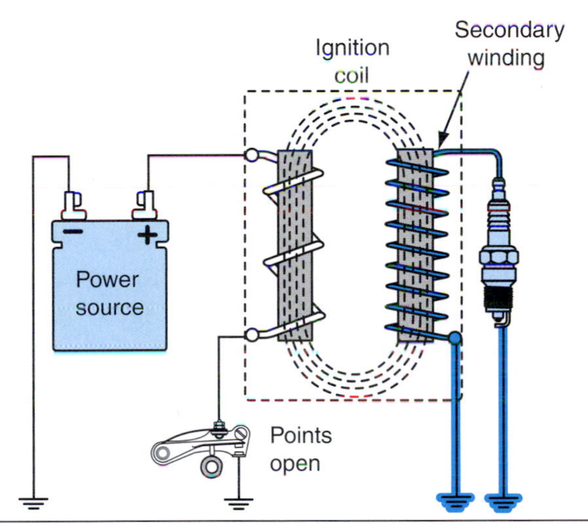

Figure 16-13 This figure illustrates the action of the breaker points in a simple ignition circuit. When the points are closed, current flows through the ignition coil primary winding. When the points open, the circuit is broken. The magnetic field in the coil collapses, which induces a voltage into the coil secondary winding and fires the spark plug.

flow is controlled in some way, a spark or arc will occur across the breaker points as they move apart. If this sparking is allowed to occur, the breaker points will burn and fail to operate properly. The points would also absorb the electrical energy and reduce the output voltage of the ignition coil.

For these reasons, a condenser is used to control the current as it flows through the breaker points. A condenser absorbs current and stores it like a miniature battery. In an ignition circuit, the condenser is connected across—or parallel to—the breaker points. As the breaker points begin to separate, the condenser absorbs the current created by the collapsing magnetic field around the primary winding of the coil so that it can't jump between the points and make a spark. When the opening of the points breaks the circuit, the condenser releases its charge back into the primary circuit.

The breaker-points-and-condenser switching system is used in both AC and battery-powered ignition systems. An illustration of a breaker-points system is shown in Figure 16-14. Note the location of the breaker points and condenser in the circuit.

Electronic Trigger Devices

When an electronic ignition system is used on a motorcycle, a sensor is used to monitor the position of the crankshaft and control; the flow of current to the primary side of the ignition coil. These sensors primarily include magnetic pulse generators and Hall-effect sensors. An electronic switch eliminates the need for breaker points and a condenser.

Magnetic Pulse Generator A **magnetic pulse generator** is a generator of single or multiple voltage pulses that uses a magnet to trigger instead of mechanical points and will be located on the engine's crankshaft or camshaft in most cases. It consists of two parts: a timing disc (also known as

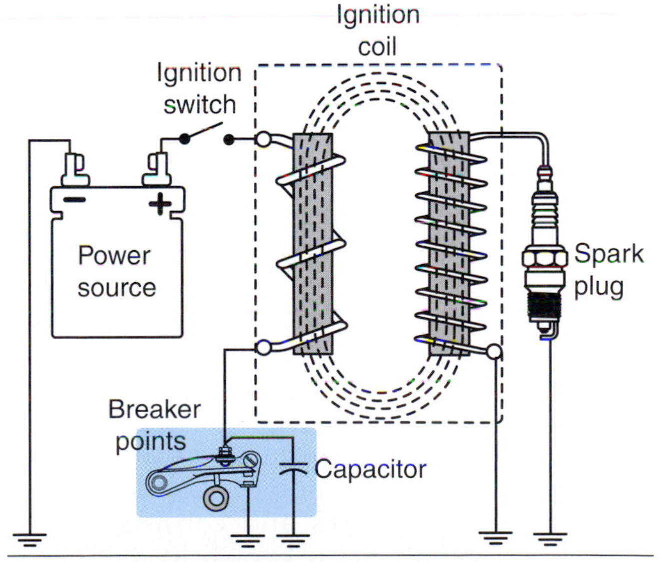

Figure 16-14 This illustration shows a typical battery-powered breaker point system.

a reluctor) and a pickup coil (Figure 16-15). The pickup coil consists of a length of wire wrapped around a permanent magnet. The magnetic pulse generator operates on basic electromagnetic principles. Remember that voltage can only be induced when a conductor moves through a magnetic field. The magnetic field is provided by the pickup unit and the reluctor provides the movement through the magnetic field needed to induce voltage when the crankshaft or camshaft is turned. As the reluctor teeth approach the pickup coil a voltage is induced and used just as the opening and closing of the contact points control the voltage to the primary side of the ignition coil. A specific, manufacturer-determined air gap is required to ensure the proper signal is being produced.

Hall-Effect Sensor The **Hall-effect sensor** or switch is a device in which an output voltage is generated in response to the intensity of a magnetic field applied to a wire and is the most common engine position sensor used on a motorcycle using any type of electronic ignition system. There are several reasons for this. Unlike the magnetic pulse generator, the Hall-effect sensor produces an accurate voltage signal throughout the entire rpm range of the engine. Furthermore, a Hall-effect switch produces a square wave pattern that is more

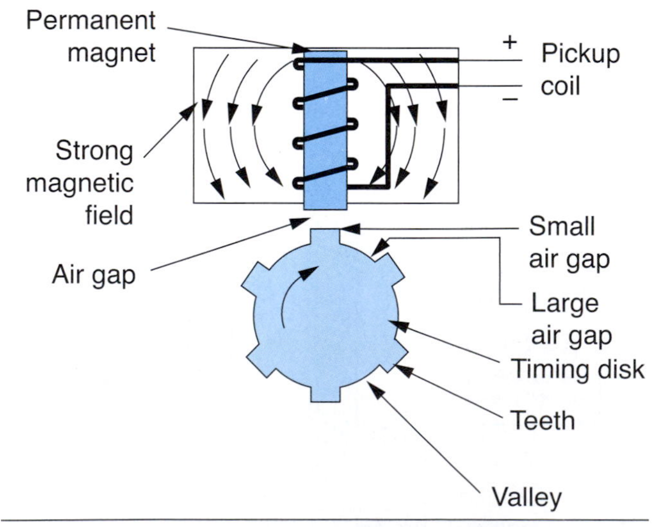

Figure 16-15 A magnetic pulse generator will be located on the engine's crankshaft or camshaft in most cases and consists of two parts: a timing disc (also known as a reluctor) and a pickup coil.

compatible with the digital signals required by onboard computers.

Functionally, a Hall-effect switch performs the same tasks as a magnetic pulse generator. But the Hall-effect switch's method of generating voltage is quite unique. It is based on the Hall-effect principle, which states: If a current is allowed to flow through a thin conducting material and that material is exposed to a magnetic field, voltage is produced. In essence, a Hall-effect switch is either on or off. It also uses a reluctor that is used to switch the power on and off as it passes by the sensor.

Stop Switch

Once an engine is started, it will keep running until it runs out of fuel or is put under a heavy enough load to cause it to stall. The stop switch provides a convenient means to stop the engine.

Different types of stop switches are found on different types of ignition systems. On some motorcycles, the stop switch interrupts the flow of electricity to the spark plug by giving the electrical current an easier path to ground. This type of switch consists of a button that grounds the ignition system, which is seen back in Figure 16-4.

In other engines, the stop switch is designed to prevent the flow of electricity through the primary winding of the ignition coil. This type of stop switch is connected in series with the primary side of the ignition coil. When you turn the switch to the OFF position, the ignition circuit is opened and the engine will stop.

TYPES OF IGNITION SYSTEMS

Now that you understand how a basic ignition system operates, let's take a closer look at the construction of some different types of ignition systems. The two general types of ignition systems used in motorcycle applications are the:

- Breaker point ignition system
- Electronic ignition system

There are two types of breaker point systems. The magneto breaker point ignition systems are usually found on older machines where electricity is needed only to power the spark plug—not a starter system or lights. The battery-and-points ignition system is found on older (pre-1980s) street motorcycles that have electric starter systems and lights.

Electronic ignition systems of one type or another are found on virtually all, modern-day motorcycles.

As you read through the following information on these ignition systems, remember that all systems contain the same basic components. The magneto system and the battery system are very similar except that they use different power sources. Electronic ignition systems use electronic components to perform the switching function, but their power source can be either a battery or an AC generator. Finally, all ignition systems have a switch device to turn the ignition system ON/OFF.

Magneto Ignition Systems

In older motorcycles and ATVs without any lights or a battery, the AC source may have the sole function of operating the ignition system. In other models that include lighting systems, one AC generator coil may be used for lighting while the other is used for the ignition. All magneto ignition systems operate without a battery, or are independent of the battery if one is used for the operation of other electrical functions.

The magneto ignition system uses permanent magnets, which are installed in the engine's flywheel or rotor. Magnetos are classified as being one of three types:

- High tension
- Low tension
- Energy transfer

High-Tension Magneto Ignition System

High-tension magneto ignition systems (Figure 16-16) haven't been used on motorcycles for many years, but they were once used quite often. With this ignition system design, the ignition coil (magneto primary and secondary windings) is mounted in a stationary position near the flywheel. When the flywheel turns, the magnets induce a voltage in the primary winding of the ignition coil.

The position of the magnets on the flywheel is very important. To generate the voltage at the exact time needed, the magnets in the flywheel must be properly aligned. This means that the flywheel must be located in exactly the proper position on the crankshaft. The flywheel is held in position on the crankshaft by a flywheel key. The flywheel key is inserted into matching slots that are cut into the crankshaft and flywheel.

In order for the high-tension magneto system to work, the ignition coil must be mounted in a stationary position close to the flywheel. The gap between the edge of the flywheel and the iron core of the ignition coil is an important specification in an ignition system. The engine manufacturer will specify the proper width for this gap in thousandths of an inch or hundredths of a millimeter. This is one of the specifications that must be checked when you are servicing a high-tension magneto ignition system.

Now, let's take a closer look at the operation of a high-tension magneto system. Figure 16-17 illustrates a simplified drawing of a high-tension magneto system in operation. The drawing shows only the outer edge of the flywheel. The center of the

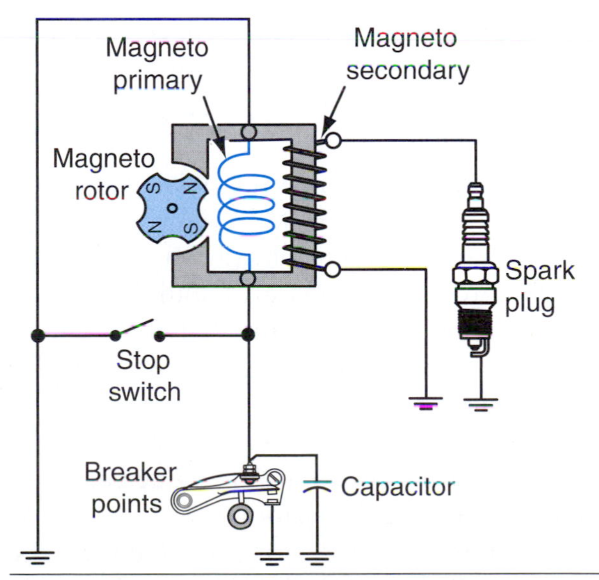

Figure 16-16 High-tension magneto system wiring diagram.

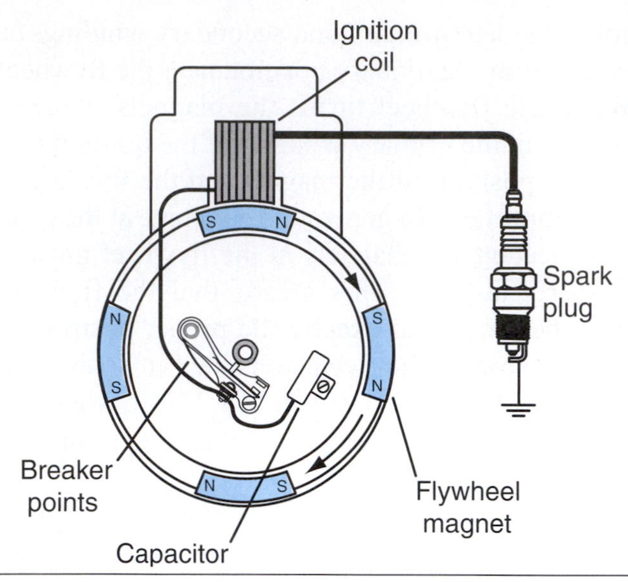

Figure 16-17 This is a simplified drawing of a high-tension magneto ignition system. A permanent magnet is mounted near the edge of the flywheel. As the flywheel turns, the magnet passes near the ignition coil and induces a voltage in the primary winding.

flywheel is cut away so that you can see the breaker points, which are located underneath the flywheel.

Remember that the ignition coil is a transformer that contains a primary winding and a secondary winding of conductor wire. In a typical high-tension magneto ignition coil, the primary winding consists of about 150 turns of fairly heavy copper wire and the secondary winding consists of about 20,000 turns of very fine copper wire. This difference in the windings is what causes the voltage to be multiplied from the primary to the secondary in a transformer.

As the flywheel turns, the permanent magnets mounted near the edge of the flywheel move past the ignition coil. This movement magnetizes the soft iron core (coil armature) and induces a current in the primary winding of the ignition coil. The magnetic field produced by the primary winding induces a voltage in the secondary winding. However, the buildup and collapse of the magnetic field isn't fast enough to induce the voltage necessary to fire the spark plug.

The primary winding is connected to the breaker points. When the breaker points are closed, a complete circuit is formed and a current flows through the primary winding to produce a magnetic field. The cam is timed to open the breaker points just as the magnetic field in the primary winding begins to collapse. This interrupts the current flow in the primary circuit, causing the magnetic field around the primary winding to rapidly collapse. At the same time, the condenser, which protects the breaker points from burning, releases its charge back through the primary winding to hasten the collapse of the magnetic field. This action helps to increase the voltage induced in the secondary winding.

The high voltage induced in the secondary winding causes a current to flow through the spark plug wire and arc across the spark plug gap. After the high voltage in the secondary winding is released as a spark, the flywheel continues to turn until the magnet positions itself by the ignition coil again, and the process repeats itself.

Low-Tension Magneto Ignition System

The operation of the low-tension system is very similar to that of the high-tension magneto system that was just described. The main difference between the low-tension magneto ignition system and the high-tension system is that the low-tension system uses a separate ignition coil. The breaker points in both the high- and low-tension magneto ignition system are connected in series with the primary circuit. When the breaker points are closed in the low-tension magneto system, the primary circuit is completed (Figure 16-18). As the magneto rotor turns, alternating current is generated in the magneto windings and flows through the ignition coil primary winding. The primary

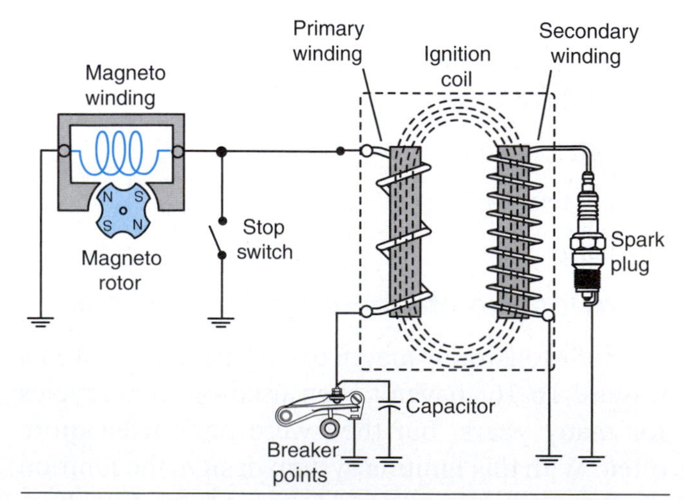

Figure 16-18 This is a simplified wiring diagram of a low-tension magneto ignition system.

winding in the ignition coil produces a magnetic field in the ignition coil; however, the buildup and collapse of the field isn't fast enough to induce the voltage required to fire the spark plug.

Energy-Transfer Ignition System

The energy-transfer ignition system (Figure 16-19) is the most popular type of magneto ignition system found on motorcycles. The primary difference between the energy-transfer system and the magneto systems previously discussed is that the breaker points are connected in parallel with the primary circuit instead of in series. By having the points wired in parallel, the primary winding in the ignition coil induces voltage into the secondary windings by using a rapid buildup of a magnetic field instead of a rapid collapse of the field.

Battery-and-Points Ignition Systems

Now, let's look at a battery-and-points ignition system. Remember that battery ignition systems were used in older street-type motorcycles. In a battery-and-points ignition system, a battery is used to provide power to the ignition coil instead of a magneto; however, the remainder of the system is similar to the magneto systems we've discussed. The battery-and-points system (Figure 16-20) uses the same type of breaker points, condenser, and spark plug as magneto-type ignition systems.

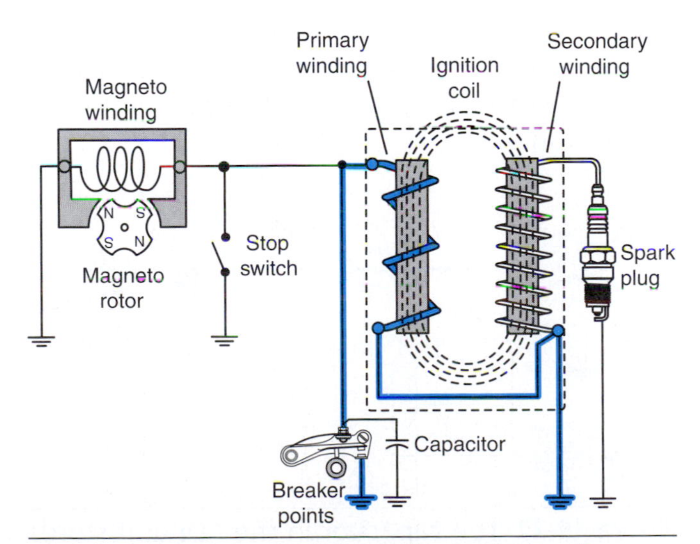

Figure 16-19 This is a simplified wiring diagram of an energy-transfer ignition system.

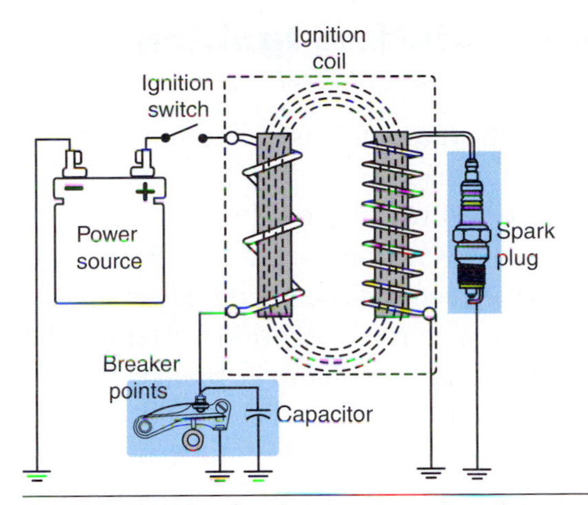

Figure 16-20 The battery-and-points system uses the same type of breaker points, condenser, and spark plug as magneto-type ignition systems, with the primary difference being the source of electrical power.

The battery used in this type of system is the lead-acid storage battery discussed previously in Chapter 15. Besides providing electricity to power the ignition coil, the battery may also be used to power lights, horns, electric starter systems, and other accessory circuits.

The battery-and-points ignition system uses breaker points to trigger the ignition. The battery provides the voltage to energize the primary winding of the ignition coil. The voltage to the ignition coil is controlled by a key-operated ignition switch. When the ignition switch is turned on, power from the battery passes through the ignition switch and through the primary winding of the ignition coil. The opposite end of the primary winding is connected to the breaker points and condenser. The breaker points, the secondary winding, and the spark plug operate in exactly the same manner as in the high- and low-tension magneto systems. The contact points are opened by the breaker-point cam at the proper time. As the points open, the primary magnetic field rapidly collapses, causing a high voltage to be induced into the secondary winding. The only difference in the battery system is that the battery energizes the primary winding of the ignition coil with DC current, instead of the AC current used in the magneto systems.

When the ignition switch is turned off, the switch contacts open, and the flow of power is stopped from the battery to the primary winding of the ignition coil. As a result, the engine stops running.

Electronic Pointless Ignition Systems

Breaker-points-and-condenser ignition systems have been used for many years. You'll still occasionally see these types of ignition systems on older motorcycles. However, points-and-condenser ignition systems have been replaced in all newer motorcycle engines by electronic ignition systems. The reason for this is that mechanical breaker points eventually wear out and fail. The result is poor engine performance at first and, ultimately, total ignition failure. Electronic ignition systems use permanent magnets, electronic sensors, diodes, transistors, and SCRs in place of mechanical switching components, so they last for a very long time.

Except for the breaker points and condenser, electronic ignition systems use the same basic components that we've discussed. In place of the breaker points and condenser, the electronic ignition system uses an electronic ignition control module (ECM or ICM). This module is a sealed, non-repairable unit that's normally mounted on a bracket on the chassis. The unit is frequently black in color, which has led to the term "black box" often being used for the module.

Other than the rotor and its magnets, electronic ignition systems have no moving parts, so the performance of the system won't decrease through operation. Electronic ignition control modules are very resistant to moisture, oil, and dirt. They're very reliable, don't require adjustments, and have very long life spans. An electronic ignition system provides easy starting and smooth, consistent power during the operation of the motorcycle.

Although there are many variations, there are three basic types of electronic ignition configurations that we will discuss:

- Capacitor discharge ignition (CDI)
- Transistorized ignition
- Digitally controlled transistorized ignition

The electronic ignition system most often used on off-road motorcycles is the capacitor discharge ignition system. The basic components of a CDI system may be configured in several different ways. Although various CDI systems may have different arrangements of wiring and parts, all CDI systems operate in much the same way.

Figure 16-21 shows how the components of a CDI system are arranged for a typical off-road motorcycle. Note that the CDI system contains two coils (windings) that are triggered by magnets in the flywheel or AC generator. The larger coil is the charging or exciter coil and the smaller coil is called the trigger coil. The trigger coil controls the timing of the ignition spark.

As the flywheel rotates past the exciter coil, the alternating current produced by the exciter winding is rectified (changed to DC) by the diode in the CDI unit. The capacitor in the CDI unit stores this energy until it's needed to fire the spark plug (Figure 16-22). As the flywheel magnet rotates past the trigger coil, a low-voltage signal is produced, which activates the electronic switch (SCR) in the CDI unit (Figure 16-23). This completes the primary circuit to allow the energy stored by the capacitor to pass through the primary winding of the ignition coil. The transformer action of the

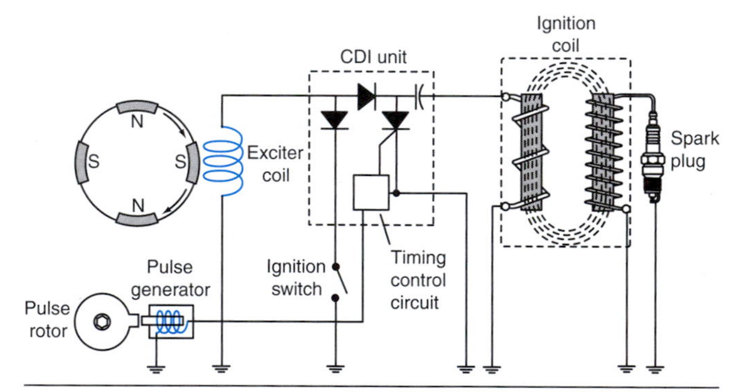

Figure 16-21 This is a simplified wiring diagram of a typical CDI system. Copyright by American Honda Motor Co., Inc. and reprinted with permission.

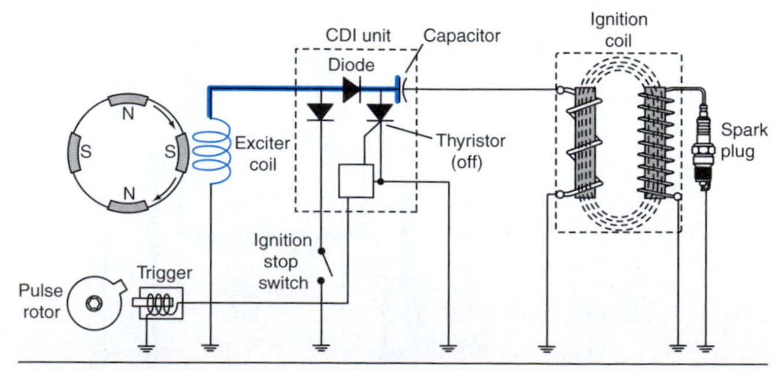

Figure 16-22 The capacitor in the CDI unit stores the energy until it's needed to fire the spark plug.

ignition coil causes a high voltage to be induced in the secondary winding of the ignition coil to fire the spark plug (Figure 16-24).

Another type of a CDI system is found on many ATVs and also on some motorcycles. It uses DC current from a battery as its source of voltage. A voltage booster is placed in the CDI unit instead of the AC generator and an exciter coil (Figure 16-25) amplifies the battery voltage to over 200 volts. This type of CDI system uses the same components we have discussed and operates in the same fashion.

Transistorized Ignition Systems

The transistorized ignition system (Figure 16-26) operates by controlling the flow of electricity to the primary coil of the ignition. With this type of ignition system, transistors are contained within the ICM and are used to supply electricity to the primary coil. When the voltage level in the primary winding reaches a certain level, the second transistor turns off the first transistor. This causes the magnetic field around the primary coil to collapse and create the high voltage across the secondary coil. The high voltage is then discharged across the spark plug.

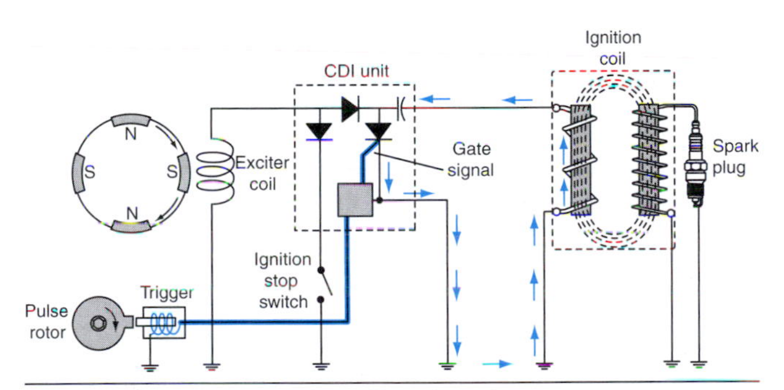

Figure 16-23 As the flywheel magnet rotates past the trigger coil, a low-voltage signal is produced, which activates the electronic switch (SCR) in the CDI unit. This completes the primary circuit to allow the energy stored by the capacitor to pass through the primary winding of the ignition coil.

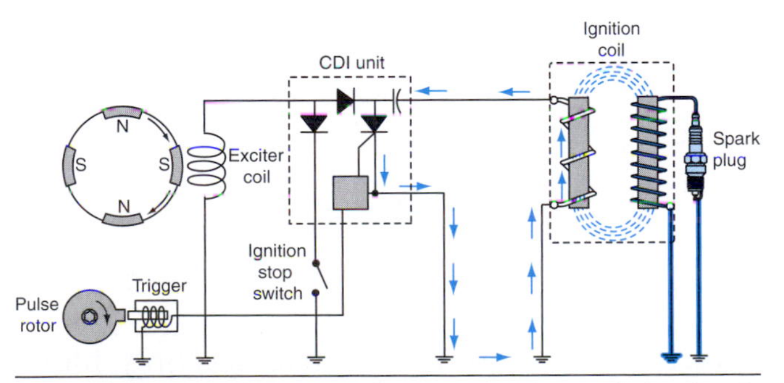

Figure 16-24 The transformer action of the ignition coil causes a high voltage to be induced in the secondary winding of the ignition coil to fire the spark plug.

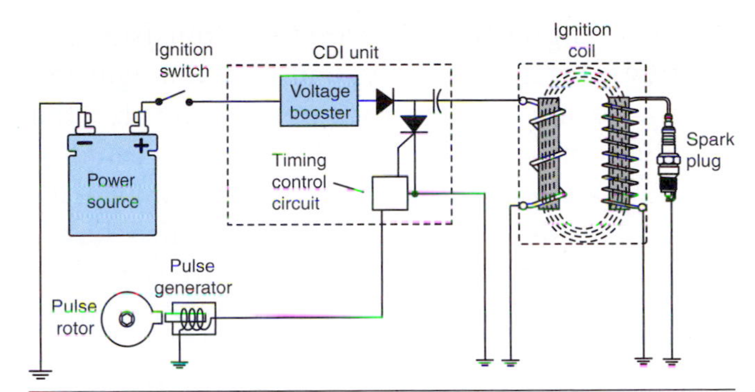

Figure 16-25 A simplified DC CDI system is shown here. Copyright by American Honda Motor Co., Inc. and reprinted with permission.

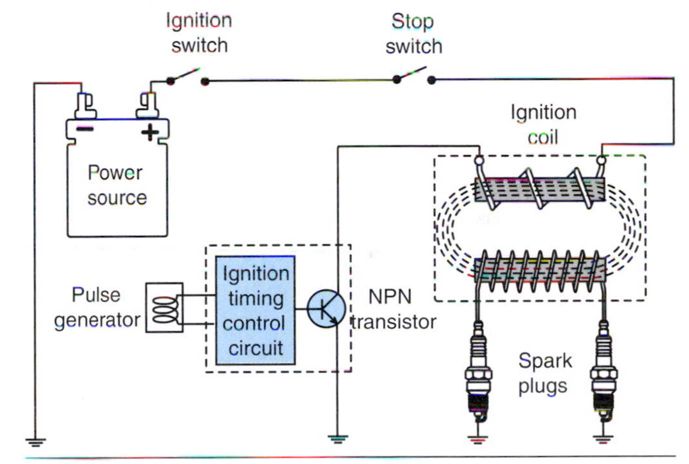

Figure 16-26 A transistorized ignition system is illustrated here.

Digitally Controlled Transistorized Ignition Systems

The digitally controlled transistorized ignition system is a type of transistorized pointless ignition that's found in most street motorcycle engine applications today. The electronic components of a digitally controlled ignition system are contained in one unit that can be mounted directly to the motorcycle chassis. In this type of system, a transistor and a microcomputer are used to perform the trigger switching function.

The digitally controlled transistorized ignition system digitally controls the ignition timing using a microcomputer inside the ignition control module (Figure 16-27). The microcomputer calculates the ideal ignition timing at all engine speeds. The microcomputer also has a fail-safe mechanism, which cuts off power to the ignition coil in case the ignition timing becomes abnormal.

The generator rotor has projections, known as reluctors, that rotate past the ignition pulse generator, producing electronic pulses. The pulses are sent to the ignition control module. The engine rpm and crankshaft position of the cylinder are detected by the relative positions of the projections that are located on the rotor.

The ICM consists of a power distributor, a signal receiver, and a microcomputer. The power distributor distributes battery voltage to the ICM when the ignition switch is turned to the ON position and the engine stop switch is in the RUN position. The signal receiver uses the electronic pulse from the ignition pulse generator and converts the pulse signal to a digital signal. The digital signal is sent to the microcomputer, which has a memory and an arithmetic unit. The microcomputer memory stores predetermined characteristics of the timing for different engine speeds and crankshaft positions. The memory then determines when to turn the transistor on and off to achieve the correct spark plug firing time.

When the transistor is turned on, the primary winding of the ignition coil is fully energized. The computer turns the transistor off when it's time to fire the spark plug. This collapses the magnetic field and induces a high voltage in the ignition coil secondary winding to fire the spark plug.

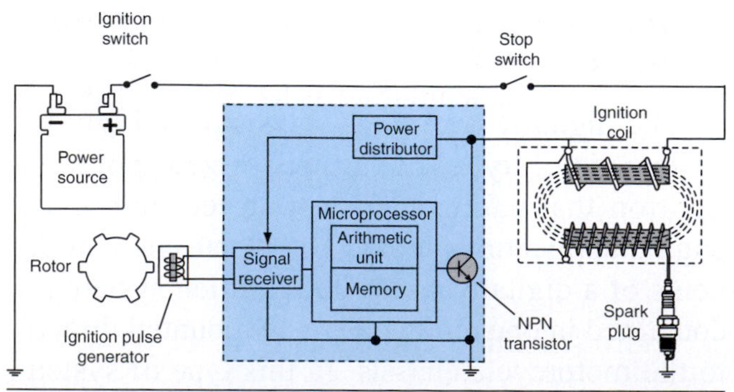

Figure 16-27 This is a simplified wiring diagram of a digitally controlled transistorized ignition system.

Visually, both the standard TPI and the digital TPI look very similar. The primary visual difference between these two popular ignition systems is the ignition pulse generator rotor. When used on a standard TPI, the pulse generator rotor will have only one reluctor to signal the pulse generator. On the digital TPI system there are several reluctors to inform the microcomputer of the engine's rpm and crankshaft position.

ELECTRIC STARTER SYSTEMS

The **electric starting systems** found on motorcycles use a direct-current (DC) motor to transform the battery's electrical energy into mechanical energy to turn the engine crankshaft quickly enough to start the engine. The amount of current required for a starting system is very high. Therefore, a starter solenoid (also known as an electromagnetic switch) and heavy gauge electrical leads are used to make the connection between the battery and starter motor. When the starter motor electrical circuit is completed, it engages a starter drive clutch that directly or indirectly engages the engine crankshaft. Reduction gears between the starter motor and starter clutch are used to multiply the starter motor's torque output. There are numerous safety features found in electric starting circuits to ensure that the engine cannot be started under certain circumstances (Figure 16-28).

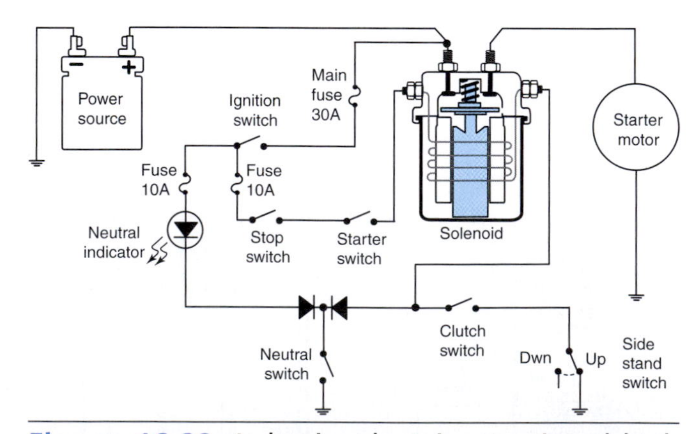

Figure 16-28 A basic electric starting block wiring diagram is illustrated here. There are numerous safety features found in electric starting circuits to ensure that the engine cannot be started under certain circumstances.

DC Motor Operating Principle

The electric starter motor uses the **DC motor operating principle**. As we've discussed in Chapters 14 and 15, when an electric current flows through a wire, magnetic lines of force encircle the wire. If the current-carrying wire is placed between the north and south poles of a magnet, a reaction occurs between the magnetic field encircling the wire and the magnetic field between the magnets.

If the directions of the magnetic fields are as indicated in Figure 16-29, the magnetic lines of force will reinforce each other below the wire, where they run in the same direction. Conversely, the lines of force will tend to cancel each other out above the wire, where they run in opposite directions. This causes the wire to be forced upward. The current-carrying wire is always pushed away from the side having the stronger magnetic field. If the electrical current through the wire were reversed, just the opposite reaction would occur and the wire would be forced downward.

If a loop of current-carrying wire is located between the north and south poles of a magnet as seen in Figure 16-30, the direction of the current flow (and therefore the direction of the magnetic field encircling the wire) in the loop at A is opposite to the direction of current flow (and magnetic field) in the other side of the loop at B. Therefore, side A of the loop is forced upward while side B is forced downward. This causes the black-colored loop in the figure to rotate in a clockwise direction until it stands perpendicular to the lines of magnetic force between the magnetic poles.

If both the blue and white wires in Figure 16-30 are fixed so that they rotate together,

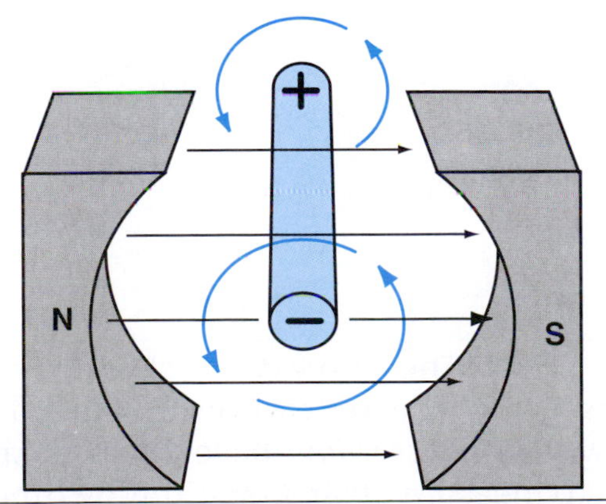

Figure 16-29 A current-carrying conductor placed in a magnetic field will cause motion.

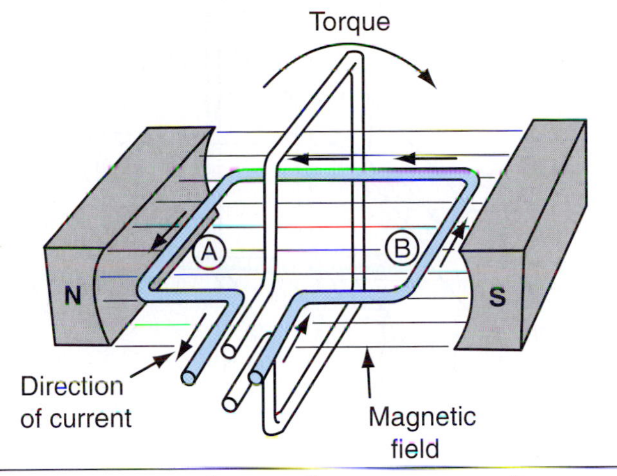

Figure 16-30 A loop of current-carrying conductor placed in a magnetic field will cause a rotary motion.

the white-colored wire would be in the vertical position when the blue wire is horizontal. Now, if we pass a current through the white wire as we did for the blue wire when it was in the horizontal position, the white wire will be forced to turn in the same (clockwise) direction. This continues the rotary motion of the wires. As the white wire is turned to the vertical position, the blue wire is returned to the horizontal position. However, to make the motor continue to rotate in the same direction, the current in the blue wire must now be reversed. The reversal of current flow is accomplished by a commutator-and-brush arrangement as shown in Figure 16-31. The battery is connected to carbon brushes, which slide against commutator segments. Each commutator segment is connected to one end of a wire loop. The commutator segments rotate with the wire loops. As the segments turn, each brush slides from one commutator segment to the next. The direction of current flowing through each wire loop is reversed when the brushes contact opposite commutator segments, allowing the loop to continue rotating as long as there is battery current being sent to the brushes.

The DC motor we've described has been greatly simplified to illustrate basic DC motor principles. In an actual DC motor, many loops of wire, called armature windings, are used to make the DC motor run more smoothly and develop more power. In addition, many starter motors use four electromagnets rather than the two permanent magnets shown in this simple illustration.

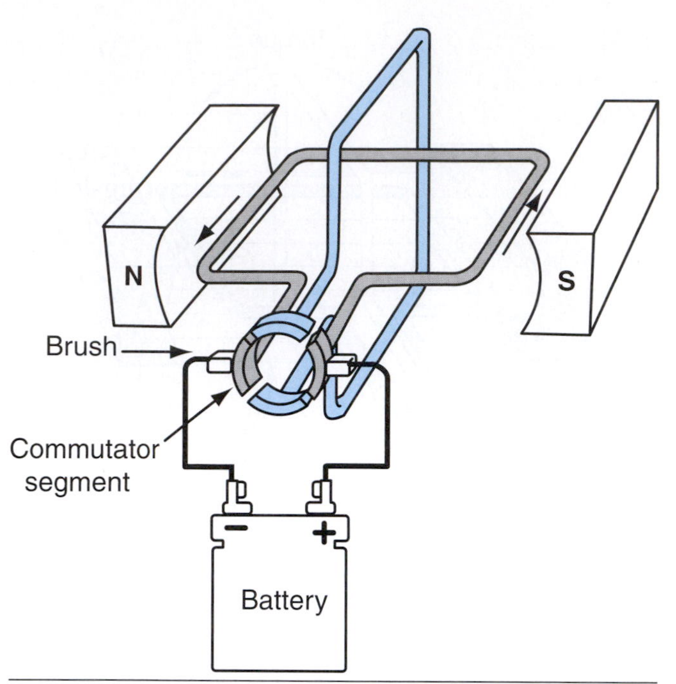

Figure 16-31 This illustration shows how the commutator and brushes operate in a DC motor.

Starter Motor Construction

A cutaway view of a typical starter motor is shown in Figure 16-32. The motor contains coils of wire wound around a laminated-iron armature core. At one end of the armature there are copper commutator segments that directly correspond to the number of armature coils of wire. Each of the commutator segments is insulated from the others. The armature coils are spaced so that, for any position of the armature, there will be coils near the poles of the field magnets. This makes the torque both continuous and strong. Electromagnets are used in many starter motors instead of permanent magnets because they can be made to provide a stronger magnetic field than a permanent magnet.

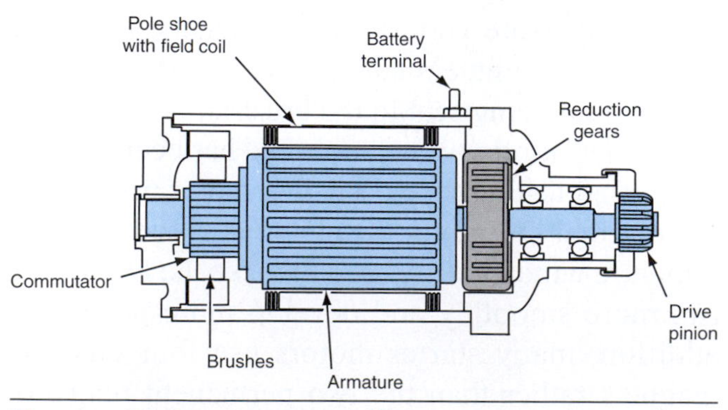

Figure 16-32 Cutaway illustration view of an electric starter motor.

The brushes are pieces of carbon, which have a long service life and cause minimum commutator wear. Springs are used to hold the brushes firmly against the commutator (Figure 16-33). The brushes and commutator connect the field coil windings with the armature windings in series. Therefore, any increase in current will strengthen the magnetism of both the field and armature. DC motors produce high starting torque, which is necessary in a starter motor. The brushes have a specification for length as they can wear with time (Figure 16-34).

The armature shaft is connected to a gear reduction system, which multiplies the motor's torque. This enables the starter to turn the engine over rapidly under compression. The gear reduction system may be contained in the engine crankcase or built into the starter motor housing, depending on the model.

Figure 16-33 The end cap of this starter motor is removed to show the brushes, brush springs, and commutator. Copyright by American Honda Motor Co., Inc. and reprinted with permission.

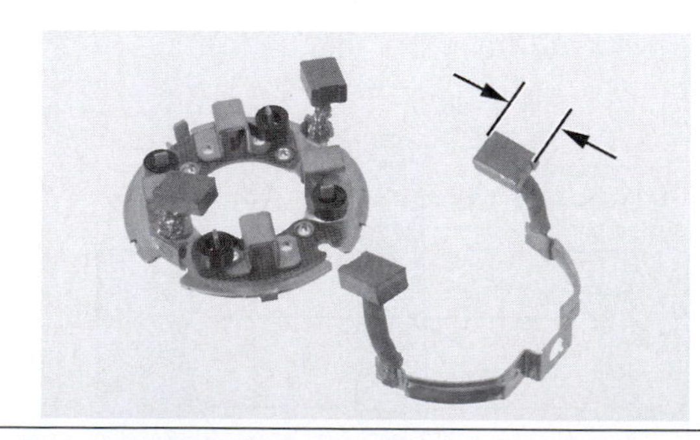

Figure 16-34 The contact brushes on starter motors are made of carbon and wear with time. Manufacturers will have a length specification to measure wear. Copyright by American Honda Motor Co., Inc. and reprinted with permission.

Starter Solenoids

A starter motor can draw in excess of 120 amperes of current when cranking the engine. Heavy electrical cable and a heavy-duty switch are required to properly handle this high current flow. It would seem obvious that it wouldn't be practical to run heavy cables up to the handlebar and install a large, heavy-duty switch there. Instead, a small push-button switch on the handlebar activates an electromagnetic starter **solenoid** switch, as shown in Figure 16-35. A solenoid is an electromagnetic coil that mechanically opens and closes a circuit when electric current is run through it. The starter solenoid connects the battery to the starter motor. You will normally find the solenoid mounted on the motorcycle frame, near the battery.

When the main switch is turned on and the starter button is pressed, the starter solenoid primary circuit is completed. DC current flows from the battery through an electromagnet in the solenoid. The electromagnet pushes the plunger into contact with the starter switch terminals, completing the circuit between the battery and the starter motor.

Starter Clutches

The **starter clutch** is a one-way clutch mechanism that allows the starter motor to engage only while the starter motor is operating to start the engine. Starter clutches are also known as sprag clutches (Figure 16-36).

When the engine starts, the engine's increased speed automatically disengages the starter motor. Figure 16-37 shows an illustration of a starter clutch. This particular starter clutch would be installed on the crankshaft and is chain driven.

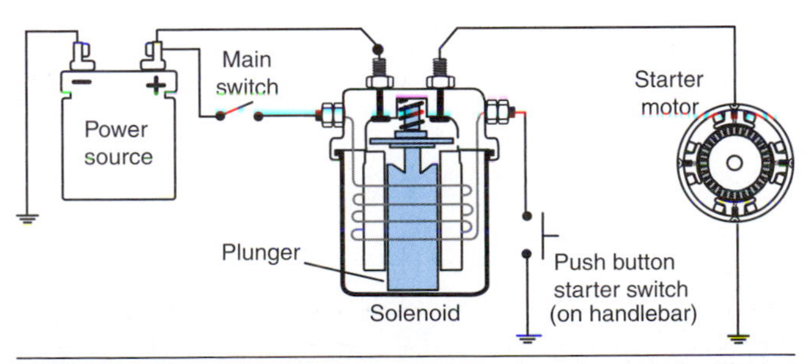

Figure 16-35 This illustration shows a simplified wiring diagram of an electromagnetic starter switch/solenoid.

The starter clutch housing is attached to the engine crankshaft. Starter engagement is achieved by locking the gears of the starter motor to the starter clutch housing, and disengagement is achieved by unlocking these parts. Spring-loaded rollers in the clutch housing perform the locking and unlocking functions.

The rollers ride on ramps within the starter clutch housing. When extended, the rollers wedge the hub tightly against the clutch housing. When the rollers are retracted, the sprocket hub and clutch housing are no longer locked together.

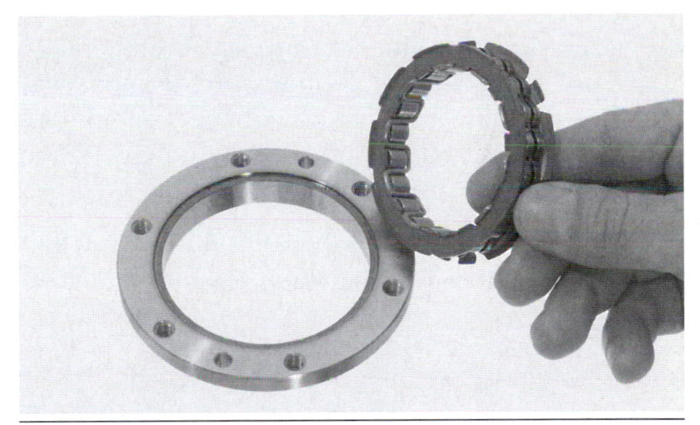

Figure 16-36 An example of a one-way starter clutch is shown here. Copyright by American Honda Motor Co., Inc. and reprinted with permission.

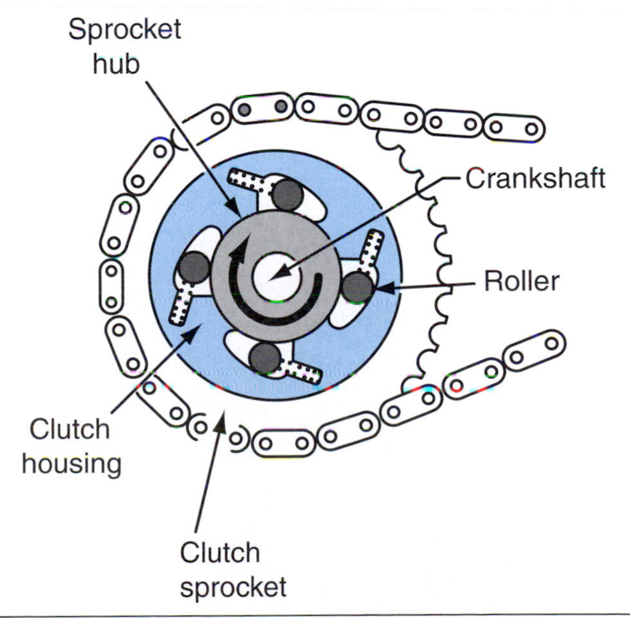

Figure 16-37 A typical chain-driven starter clutch is illustrated here.

Summary

- For each cylinder in an engine, the ignition system has three main functions. First, it must generate an electrical spark that has enough heat to ignite the air/fuel mixture in the combustion chamber. Second, it must maintain that spark long enough to allow for the combustion of all the air and fuel in the cylinder. Last, it must deliver the spark to each cylinder so combustion can begin at the right time during the compression stroke of each cylinder.
- The main components of an ignition system are the power source, ignition switch, ignition coil, spark plug, triggering switch, and stop switch.
- Ignition systems all use a primary circuit and secondary circuit. The primary circuit induces a relatively small voltage through the principle of magnetism to create a high output voltage to the spark plug.
- There are two general types of ignition systems: breaker point and electronic ignition. There are four types of breaker-point systems: low-tension magneto, high-tension magneto, energy transfer, and battery point. There are three basic types of electronic ignition systems: capacitor discharge, transistorized, and digitally controlled transistorized.
- The electric starting systems found on motorcycles use a direct current (DC) motor to transform the battery's electrical energy into mechanical energy to turn the engine crankshaft quickly enough to start the engine. The amount of current required for a starting system is very high. Therefore, a starter solenoid and heavy-gauge electrical leads are used to make the connection between the battery and starter motor. When the starter motor electrical circuit is completed, it engages a starter drive clutch that directly or indirectly engages the engine crankshaft. Reduction gears between the starter motor and starter clutch are used to multiply the starter motor's torque output.

Chapter 16 Review Questions

1. The ________________ winding of an ignition coil uses relatively few turns of heavy copper wire in relation to the ________________ winding.
2. A capacitor discharge ignition system has fewer moving parts than an energy-transfer ignition system. True/False
3. The power source in a motorcycle ignition system is connected directly to the secondary winding of the ignition coil. True/False
4. Electric starters are used to transform the battery's electrical energy into ________________ energy to turn over the engine.
5. In a CDI system, which one of the following components stores the energy to fire the spark plug?
   a. Capacitor
   b. Diode
   c. Charging coil
   d. SCR
6. The function of the condenser in a breaker points ignition system is to
   a. advance the engine timing at high rpm.
   b. induce a voltage in the primary coil.
   c. delay the opening of the points.
   d. prevent arcing across the points.
7. The length of the metal threads at the end of a spark plug is called the
   a. ground.
   b. reach
   c. insulator.
   d. shell.
8. Which of the following is an ignition component that has no means of adjustment?
   a. Contact breaker points
   b. Mechanical advancer
   c. Electronic advancer
   d. Voltage regulator
9. Which of the following statements about the ignition coil in a motorcycle ignition system is correct?
   a. The coil will be found only in an ignition system that's powered by a magneto.
   b. The coil's iron core is called the primary.
   c. The secondary winding is connected to the spark plug wire.
   d. The secondary winding contains fewer coils of wire than the primary winding.
10. An electric starter system uses a ________________ to carry the high current from the battery to the starter.
    a. condenser
    b. magneto
    c. capacitor
    d. solenoid

CHAPTER

17 Frames and Suspension

Learning Objectives

When students have completed the study of this chapter and its laboratory activities they should be able to:

- Identify the different frame designs used by motorcycle manufacturers
- Explain how front suspension components function
- Identify rear suspension systems used by motorcycle manufacturers
- Explain what is meant by the term "rising rate rear suspension"
- Explain how an oil damper functions
- Explain the different coil spring designs

Key Terms

Cartridge
Dampers
De Carbon shock
Fork bushings
Fork oil seal
Fork springs
Frame
Inner fork tubes
Offset
Outer fork tubes
Rake
Spring adjusters
Spring rate
Standard damping rod
Steering stem
Suspension
Swing arm
Trail
Viscosity
Wheelbase

INTRODUCTION

This chapter deals with two important, interrelated motorcycle systems: the frame and the suspension. As a knowledgeable motorcycle technician, you must thoroughly understand these systems. Correctly aligned and serviced, these systems provide comfort and safety to the rider.

In this chapter, you will learn to identify some common frame designs as well as learn how modifying the frame design changes the way a motorcycle handles.

Steering and wheel alignment work with the suspension system to give the rider a smooth, comfortable ride over rough road surfaces and terrain without transmitting a great amount of shock to the rider. The suspension connects the main body of the motorcycle with the wheels and, if it's working properly, provides as smooth a ride as possible. This chapter will also explore the various parts that comprise the suspension system.

To develop faster and lighter motorcycles, manufacturers have made significant changes in frame geometry, shock-absorber capabilities, and swing-arm arrangements as well as frame and suspension designs. This chapter will cover basic frame and suspension theory and design.

MOTORCYCLE FRAMES

The **frame** is the skeleton of the motorcycle in which virtually everything pertaining to the motorcycle is attached. It must be straight to provide the proper wheel alignment and steering. It must provide a secure mounting for the steering. The frame provides a rigid structure to mount the suspension and engine. The frame must be strong enough to support the weight of the rider, the engine, and all of the other motorcycle components attached to it.

Frame Design

Manufacturers use many frame configurations, depending on how the motorcycle will be used. Stress and vibration from both the suspension and the engine act against the frame. These forces are major factors when a manufacturer is designing a frame whether it's for street use, track use, or off-road use. The frame must also be as lightweight as possible for ease of handling.

Frame design also depends on such factors as engine displacement, intended use of the motorcycle, cost, visual appeal, and materials. Most frames are made of steel. But aluminum is used in frames for sport-type motorcycles and for off-road, motocross motorcycles. Generally speaking, frames made of aluminum alloys are lighter than steel frames. Aluminum frames are bulkier and more expensive to produce than steel frames.

Manufacturers design frames using a variety of tubing in different shapes. Round tubing has the same strength in all directions, but rectangular and square tubing has different strength ratings in different directions (Figure 17-1). Thin-wall, rectangular aluminum tubing can be strengthened by

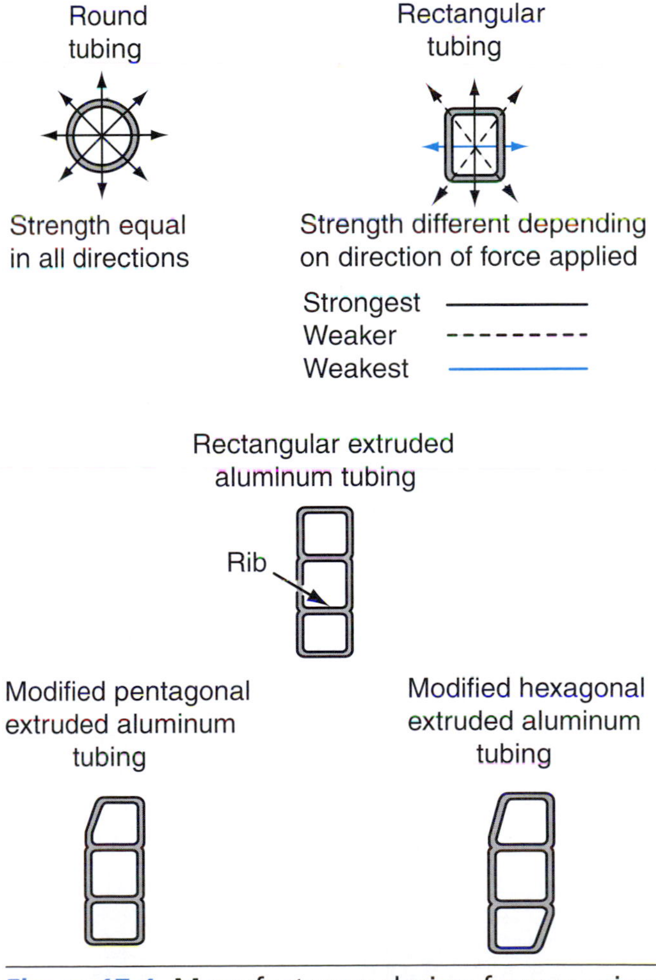

Figure 17-1 Manufacturers design frames using many different types of tubing. This image illustrates the different strength factors of different tubing.

utilizing internal ribs and extrusion production (Figure 17-2). Manufacturers also use a special modified pentagonal or hexagonal extruded aluminum tubing with internal strengthening ribs in order to improve the frame member's strength-to-weight ratio, its rigidity in one or more specific directions, and, in some cases, to allow a more compact and unobstructed riding position. Different castings and forgings combine to form the optimal frame design for a given model's needs. Many motorcycle frames are made almost entirely of round steel tubing of various outside diameters and thicknesses. Frames can also be made from square tubing.

Most frames include some casting or pressed sections in order to form strong, compact joining areas for major attachment points (Figure 17-3).

The various material types, forms, and dimensions used in frame design are linked directly to the experience gained from various types of testing including racing. For most manufacturers, as new knowledge is gained through competition, it is combined with input from non-competition testing and utilized in the construction of new generation production motorcycles.

Single-Cradle Frame

This frame design has one down tube and one main backbone pipe (Figure 17-4). The structural material of the frame surrounds the engine. This frame is widely used for lightweight, off-road, as well as some midrange, on-road motorcycles.

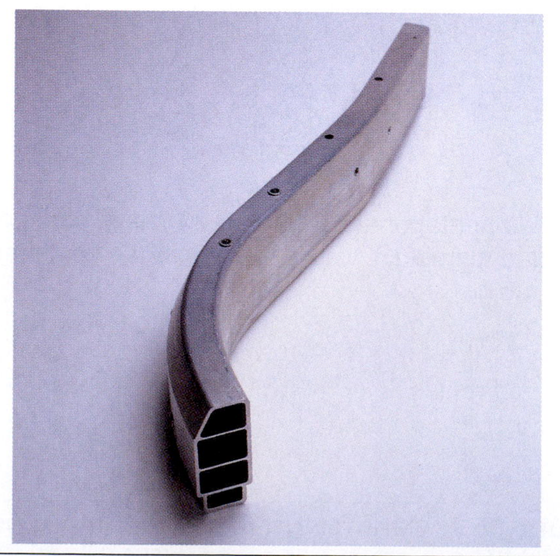

Figure 17-2 Internal ribs are used with thin-wall rectangular tubing. Copyright by American Honda Motor Co., Inc. and reprinted with permission.

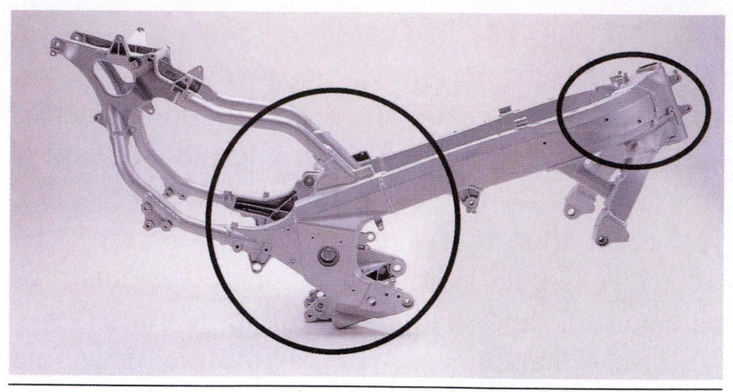

Figure 17-3 In this image, castings are circled. Castings are included to form a strong and compact joining area for the major attachment points of the frame. Copyright by American Honda Motor Co., Inc. and reprinted with permission.

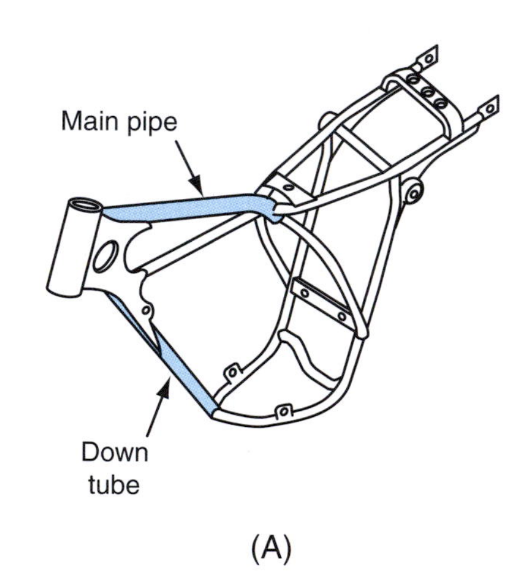

(A)

(B)

Figure 17-4 Illustration (A) shows a single-cradle frame. Copyright by American Honda Motor Co., Inc. and reprinted with permission. Illustration (B) shows a motorcycle that uses a single-cradle-type frame. Courtesy of American Suzuki Motor Corp.

Double-Cradle Frame

This frame is similar to the single-cradle frame (Figure 17-5). The major difference is that the double-cradle frame has two down tubes and two main pipes. The double main pipes and down tubes increase frame strength. In some cases, a section of the down tube can be removed on some models to help the technician remove the engine when such a task is necessary. Larger displacement, on-road motorcycles commonly use this design.

Backbone Frame

The backbone frame is not always considered to be the most desirable design for stiffness with its single, wide, main beam from which the engine is suspended (Figure 17-6). This frame design does, however, allow for a lot of flexibility for designers, since it is virtually concealed inside the finished product. The engine appears to hang in mid air. The backbone frame is simple and inexpensive to produce, and is used mainly on naked and off-road motorcycles.

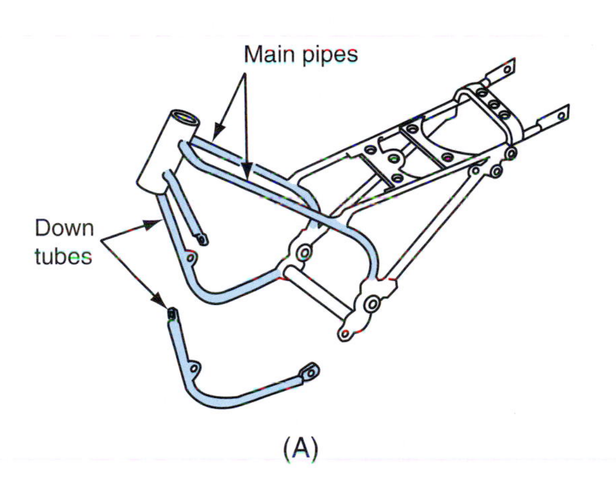

(A)

(B)

Figure 17-5 Illustration (A) shows a double-cradle frame. Copyright by American Honda Motor Co., Inc. and reprinted with permission. Illustration (B) shows a motorcycle that uses a double-cradle-type frame. Courtesy of American Suzuki Motor Corp.

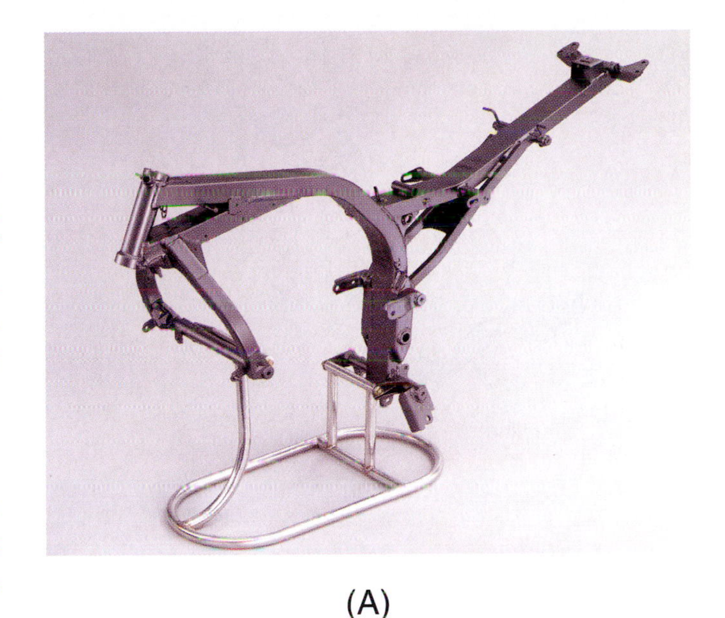

(A)

(B)

Figure 17-6 Illustration (A) shows a backbone frame. Illustration (B) shows a motorcycle that uses an backbone-type frame. Copyright by American Honda Motor Co., Inc. and reprinted with permission.

Underbone Frame

The underbone frame design is made of steel tubing and pressed plates (Figure 17-7) and is used mostly on scooters and other small-engine

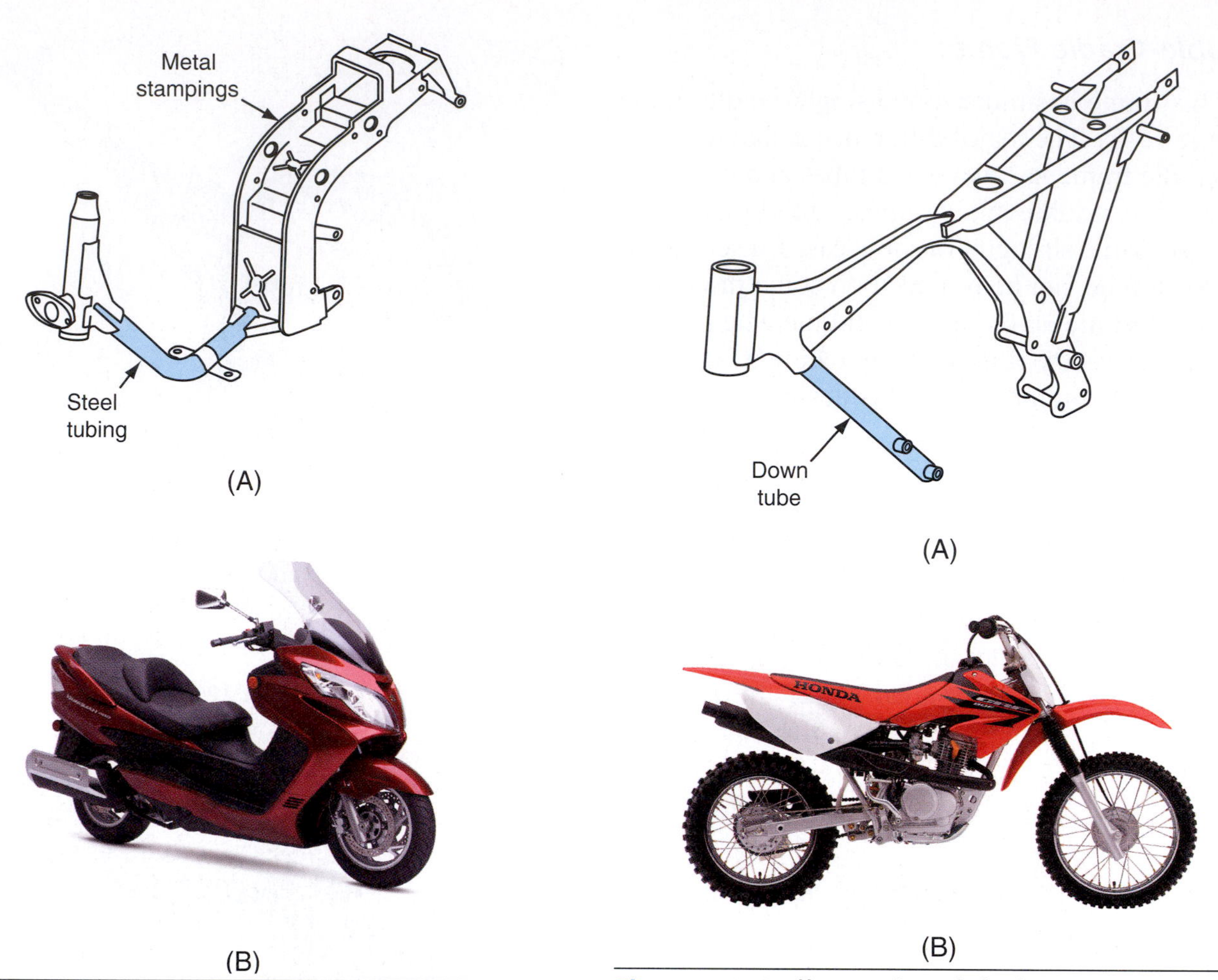

Figure 17-7 Illustration (A) shows an underbone frame. Copyright by American Honda Motor Co., Inc. and reprinted with permission. Illustration (B) shows a scooter that uses an underbone-type frame. Courtesy of American Suzuki Motor Corp.

Figure 17-8 Illustration (A) shows a diamond-type frame. Illustration (B) shows a motorcycle that uses a diamond-type frame. Copyright by American Honda Motor Co., Inc. and reprinted with permission.

displacement motorcycles. This frame design allows motorcycle manufacturers to produce motorcycles at a low cost while still offering many design options.

Diamond Frame

Similar to the backbone frame, the lower section of the down tube isn't connected to any other frame tubes (Figure 17-8). The diamond design uses the engine as a part of the frame's structure. When it's secured, the engine generates the extra needed frame strength. In these situations, the engine is known to be a stressed member of the frame. Used primarily on small- to mid-sized motorcycles, this frame design is desirable because of its simple structural design, serviceability, and light weight.

Perimeter Frame

The perimeter frame was designed to minimize twisting of the steering neck by having the upper frame tubes at a wide angle to the neck (Figure 17-9). The upper frame tubes also wrap around the perimeter of the engine, and joins the swing arm to the steering head in as short a distance as possible, further increasing the stiffness of the overall frame. Manufacturers mainly construct this frame with aluminum. Some, but not all, perimeter frames contain a removable subframe to improve service access (Figure 17-9B). This type of frame was originally designed for road-racing machines but quickly was adapted for sport-type, on-road motorcycles and now it is found on many off-road bikes (Figure 17-10) as well.

(A)

(B)

Figure 17-9 Illustration (A) shows a perimeter-type frame. Illustration (B) shows a motorcycle that uses a perimeter-type frame. Note that the motorcycle has a removable subframe. Copyright by American Honda Motor Co., Inc. and reprinted with permission.

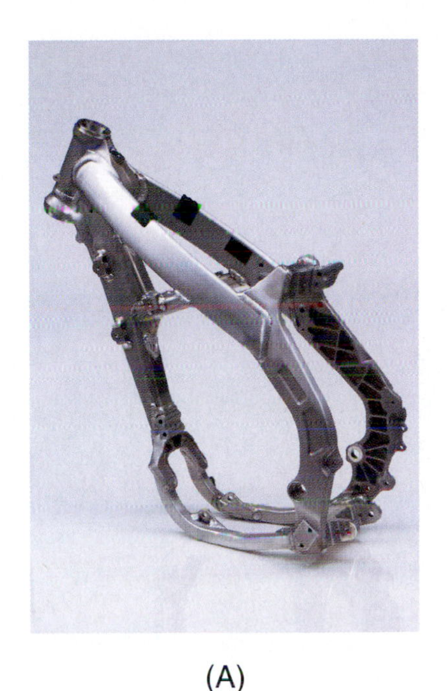

(A)

(B)

Figure 17-10 Illustration (A) shows an off-road motorcycle perimeter-type frame. Copyright by American Honda Motor Co., Inc. and reprinted with permission. Illustration (B) shows a moto-cross bike that uses a perimeter-type frame. Note that the motorcycle has a removable subframe. Courtesy of American Suzuki Motor Corp.

Pivotless Frame

Very similar to the perimeter frame design, the pivotless frame features a large triple box design (Figure 17-11). The swing arm for this frame attaches directly to the engine and provides more torsional rigidity. This design allows for the simplest, lightest frame and is less affected by the swing arm-imposed loads during hard cornering and acceleration.

Trellis Frame

The trellis frame design rivals the aluminum perimeter frame for rigidity and weight and can be either pivotless (Figure 17-12), or it may have the swing arm pivot in place depending on the manufacturer's intention. Well known as a favorite of Italian and European manufacturers it has proved to be a great success in racing and competition.

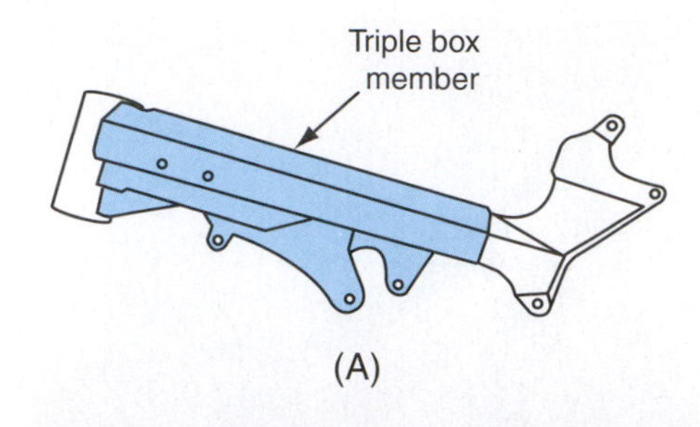

(A)

(B)

Figure 17-11 Illustration (A) shows a pivotless-type frame. Illustration (B) shows a motorcycle that uses a pivotless-type frame. Note that on a pivotless frame, the swing arm and rear wheel are attached directly to the engine. Copyright by American Honda Motor Co., Inc. and reprinted with permission.

Figure 17-12 A trellis frame design is pictured here. Courtesy of Ducati.

This frame design is now seen on some Japanese motorcycles as well. The trellis frame uses the same principles as the perimeter frame, and connects the steering head and swing arm as directly as possible. The frame is made up of a large number of short steel (or aluminum) tubes welded together to form a trellis. The trellis frame is not only easy to manufacture, but is extremely strong as well.

Frame Modifications

You may encounter a motorcycle with a modified frame. A considerable part of modifying the motorcycle's frame consists of changing the angle of the steering head-pipe joint where it attaches to the top frame rail and the front-frame down tube. Changing the angle of these tube joints changes the angle of the steering-stem axis in relation to the ground. This angle is called the steering castor, or more commonly known as the **rake**. Changing the angle of the head-tube joint also changes the steering trail and wheelbase. **Trail** is a word used to identify the distance along the ground between an imaginary vertical line through the center of the front axle to the ground and another line through the center of the steering stem to the ground. **Wheelbase** is the distance from the center of the front wheel axle to the center of the rear wheel axle (Figure 17-13). The **offset** is the distance between the center of the front axle and the steering axis.

Changing both the rake and the trail directly affects both steering ease and the stability of the motorcycle. Often, motorcycles that have been chopped also have the length of the front-fork tubes increased. This is usually done in conjunction with changing the head-pipe angle. This style is used to create a "chopper" style of motorcycle (Figure 17-14).

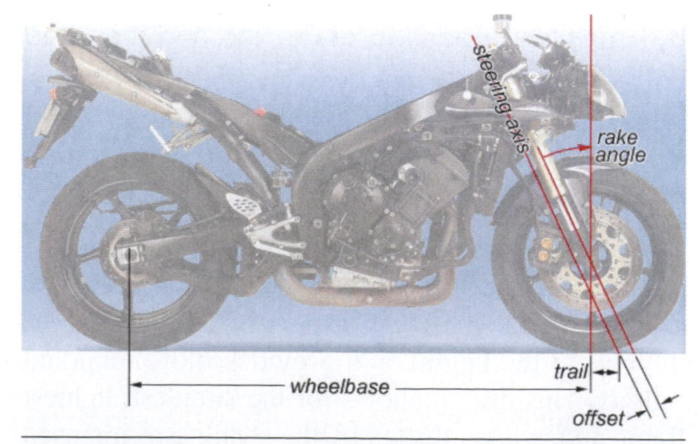

Figure 17-13 Motorcycle rake, trail, offset, and wheelbase is illustrated here. Copyright © Chris Longhurst 1994-2007. All rights reserved.

Figure 17-14 A chopper-style motorcycle has a high rake angle.

When you increase the rake, you increase the trail. This change lengthens the wheelbase and makes the motorcycle more difficult to turn. Therefore, choppers are more suited for riding on straight highways. However, it's possible to change the rake and trail too much. When this happens, the steering is very slow and requires a lot of pressure to turn the wheel until the wheel reaches a 15 degree to 25 degree angle from the frame line. Then, the wheel tends to turn sharply to a 90 degree angle, and it requires a lot of pressure on the handlebars to prevent it from completing the turn to a 90 degree angle. This makes steering at slow speeds much more difficult.

Motorcycles that have the setting reversed—small rake, smaller trail–are much easier to turn; the pressure on the handlebars can be very light. This reversed rake and trail condition affects the bike's stability. The motorcycle is more difficult to hold in a straight line at higher speeds and therefore may require a steering damper to assist at high speeds (Figure 17-15). The triple clamp holds the steering damper, if one is included.

Thus, most manufacturers install the steering head pipe at an angle that allows the rake to be just large enough for the designed purpose of the motorcycle (Figure 17-16). Any modification to the frame head angle, length of fork legs, or size of wheels and tires affects—to some extent—both the rake and trail.

Most manufacturers make more than one size of motorcycle, each designed for a basic purpose (dirt bikes, street bikes, etc.), along with some bikes designed for multiple uses. Motorcycle manufacturers don't recommend making any frame-design changes; but if frame work is to be done, remember that it should be done by an expert who knows metal, welding, and frame-design factors. The frame can be easily ruined by an inexperienced technician. In addition, a dangerous situation may be created if the frame is not correctly aligned.

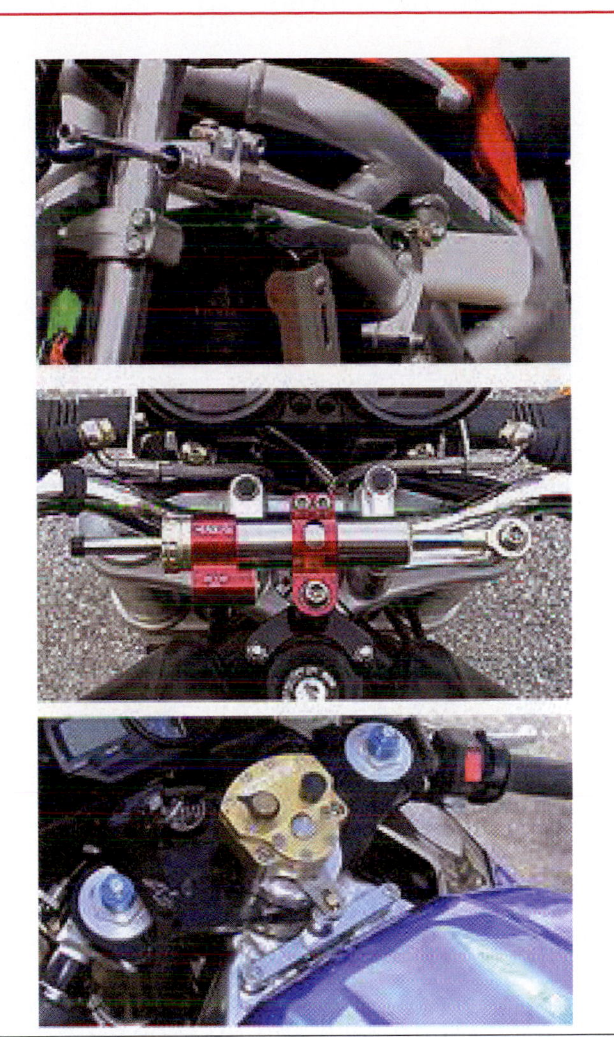

Figure 17-15 Different types of steering dampers are shown here. All have the same effect of increasing the stability of the motorcycle.

MOTORCYCLE SUSPENSION SYSTEMS

A **suspension** is a mechanical system of springs or shock absorbers connecting the wheels and axles to the chassis of a motorcycle. Suspension

(A)

(B)

Figure 17-16 Different types of motorcycles use different rake angles. (A) shows a sport bike with a very short rake and (B) shows a custom cruiser with a tall rake. Courtesy of American Suzuki Motor Corp.

systems are designed for comfort and safety. A well-designed suspension will absorb irregular surfaces and ensure that the tires of the motorcycle stay in contact with the road while transmitting a minimum amount of shock through the frame. Suspension system designs vary according to the model, intended look, and how the vehicle will be used: touring, sport, moto-cross, dual-purpose, and so on. A sport or moto-cross bike will have a stiffer suspension for better control and cornering ability as compared to a touring bike or cruiser. The main parts of the suspension system are as follows:

- Front fork
- Swing arm
- Rear damper(s)

Front Suspension

The front suspension supports the front wheel and allows it to pivot from side to side for steering by means of a triple clamp, steering head, and stem. Over the years, a number of front suspension designs have been introduced on motorcycles. Two designs utilize what is known as a link type suspension. Both of these systems have their axles supported by links. One design uses a leading link where the links pivot toward the front of the bike (Figure 17-17) in front of the lower portion of the front end. The trailing link design has its "trail" as the lower portion of the front end (Figure 17-18).

Another type of front suspension, introduced by BMW, is called the telelever (Figure 17-19). The

Figure 17-17 A leading link-type front suspension is shown here. Copyright by American Honda Motor Co., Inc. and reprinted with permission.

Figure 17-18 A trailing link-type front suspension is shown here. Copyright by American Honda Motor Co., Inc. and reprinted with permission.

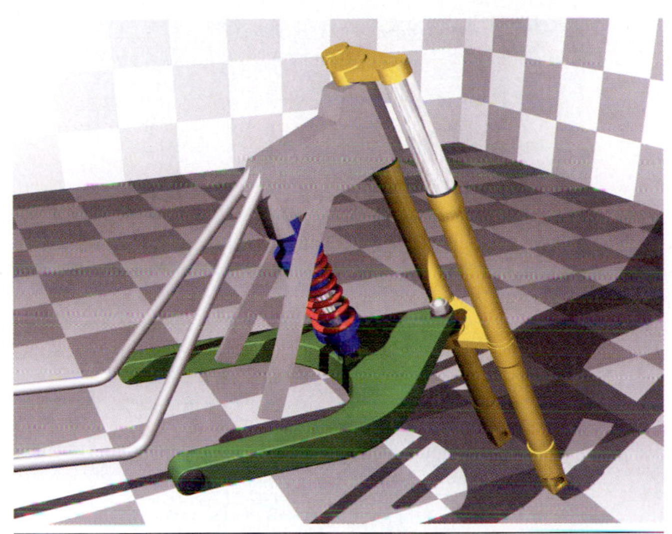

Figure 17-19 The telelever-type front suspension was introduced by BMW. Copyright © Chris Longhurst 1994-2007. All rights reserved.

Figure 17-20 The conventional telescopic-type front suspension system is the most popular system found on motorcycles. Copyright by American Honda Motor Co., Inc. and reprinted with permission.

telelever uses a linkage and single hydraulic shock with external spring. The telelever, according to its maker, "separates suspension and steering." It is most commonly praised for virtually eliminating braking dive. It is standard on many BMW motorcycles.

Most motorcycle front forks incorporate telescopic hydraulic shock absorbers to absorb the vertical shock of the front wheel when hitting bumps, thus providing a smooth ride. The most popular telescopic motorcycle front suspension system is the conventional type (Figure 17-20), which utilizes a pair of upper fork tubes and lower fork sliders that move into one another. Inside each of the telescopic forks are springs held in place with caps and a damping system (Figure 17-21). Seals are used to prevent oil from escaping from the fork.

The telescopic front suspension cushions the shock of the front wheel hitting bumps in road surfaces. As the wheel hits a bump, the sliders are pushed upward over the inner fork tubes and compress the springs. The oil in the outer fork tube flows through the holes or valves into the inner-fork tube (Figure 17-22). Since the transfer of oil into the inner fork tube takes up space, trapped air is compressed and the increased air pressure increases oil flow resistance. This has a damping effect on the shock and limits the upward movement

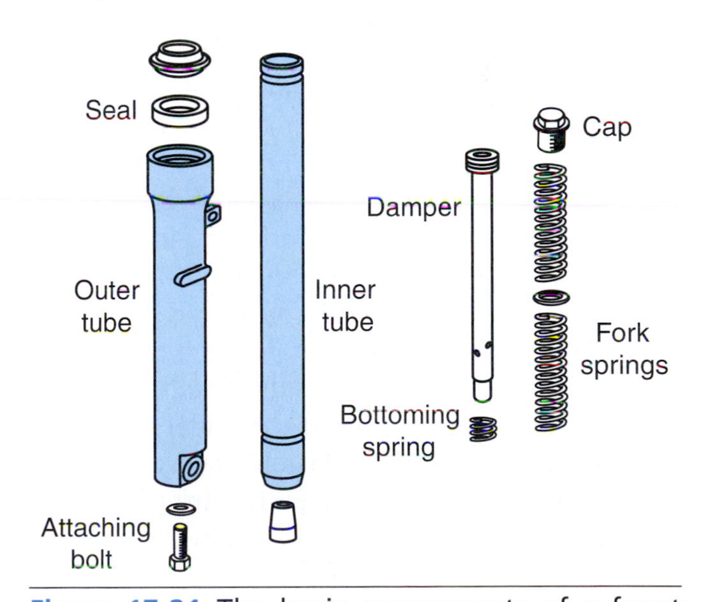

Figure 17-21 The basic components of a front telescopic suspension system are shown here.

of the outer fork tubes. As the shock load is relieved, the springs push the slider back to the extended position, and oil is pulled back through the holes (valves) into the vacuum created by the

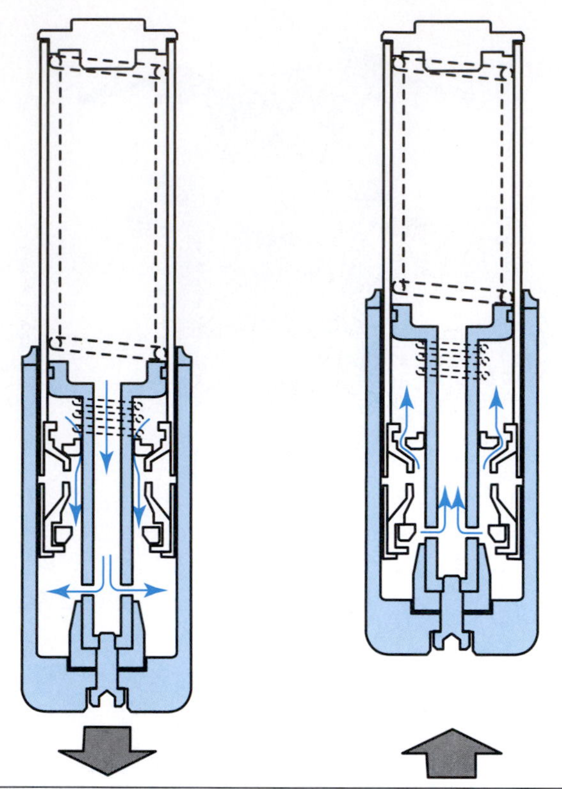

Figure 17-22 The oil flow of a basic telescopic front suspension system is shown here.

Figure 17-23 An inverted (also known as upside-down) front suspension is shown here. Note that the sliding tubes are mounted to the front wheel. Courtesy of American Suzuki Motor Corp.

extension of the sliders. The flow of oil is restricted by the size of the holes or valves, and a damping effect is obtained with each movement of the sliders.

Inverted ("upside-down") forks (Figure 17-23) are mounted with the outer fork tube in the triple clamps and the inner slider mounted to the axle. This is done to increase rigidity. The rigidly mounted outer tube houses large fork bushings and large slider tubes, adding to the bike's increased stability. These designs benefit the rider by providing increased handling precision and are found on both off-road and on-road motorcycles.

Figure 17-24 The motorcycle steering stem is highlighted in this illustration.

Front Suspension Components

There are various components that comprise the front suspension. Let us take a closer look at these components.

Steering Stem The **steering stem** (Figure 17-24) supports the right and left inner fork tubes by clamping around them. The stem has an axle or steering shaft in the center, mounted through bearings into the frame head (Figure 17-25). The bearings allow the stem to turn left and right. The crown attaches around the stem axle using a clamp and/or securing nut. The crown also attaches to the top of each inner fork tube by bolts that provide a solid, secure attachment of the inner fork tubes. The crown also serves as a mounting brace for the handlebars.

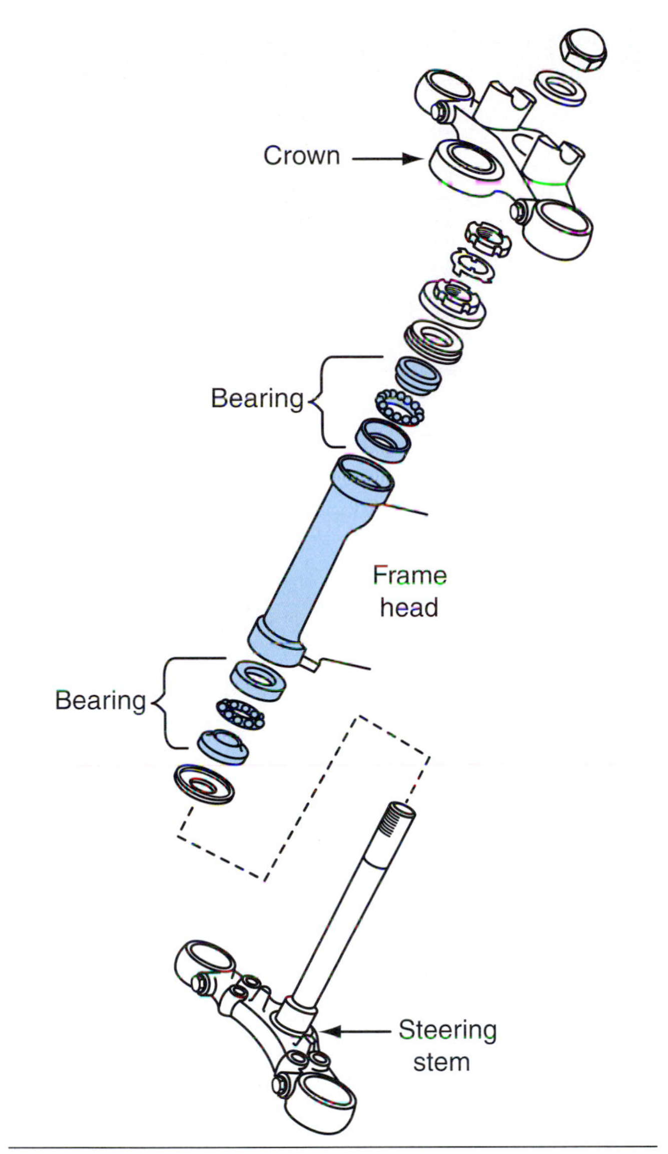

Figure 17-25 The components of a typical steering stem are shown here.

There are three types of bearings used in the steering stem on motorcycles (Figure 17-26): loose ball, caged ball, and tapered roller. Each type of bearing has its own advantages. Worn or improperly adjusted steering stem bearings can cause many different symptoms, such as front-end shimmy, difficult steering, and unusual noises and feel.

Inner Fork Tubes The **inner fork tubes** serve as a guide and mount for the outer fork tubes. The outside of the inner fork tube is machined to provide a smooth surface for the bushing and oil seal to slide over. On the inside, the fork tube serves as a cylinder for the damper rod to slide up and down in. Figure 17-27 shows an exploded view of a front fork.

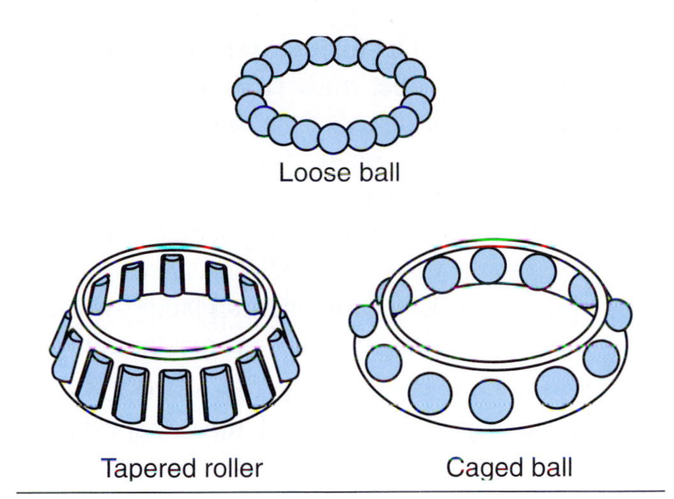

Figure 17-26 The three types of steering head bearings are shown here: loose ball, tapered roller, and caged ball.

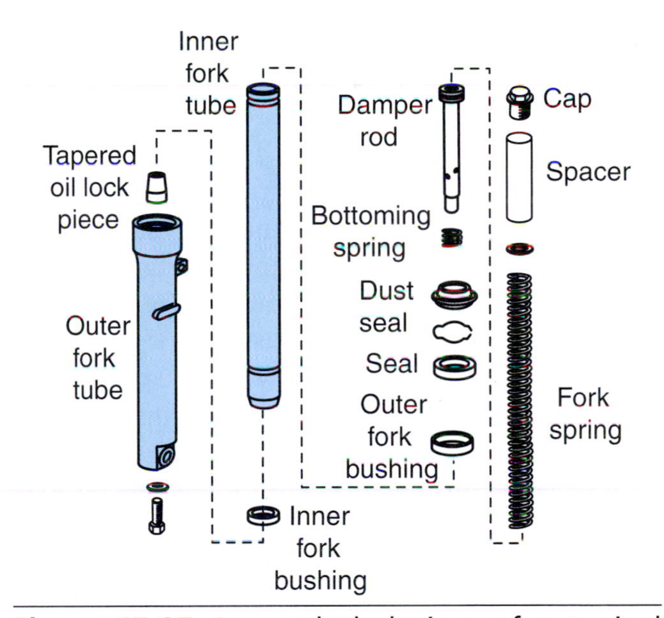

Figure 17-27 An exploded view of a typical front fork.

Outer Fork Tubes The **outer fork tubes** attach to the wheel axle and move up and down over the inner fork tubes. The outer fork tubes hold the fork oil and the damping system.

Fork Bushings Most all fork tubes have **fork bushings** to prevent lateral movement of the fork during its up-and-down movement. These bushings may wear with time and should be inspected whenever the forks are apart. The appropriate service manual will give the specifications for bushing inspection and measurement.

Fork Oil Seal The **fork oil seal**, held in the top of the outer fork tube, must fit snugly around the inner-fork tube because its function is to prevent oil leakage. In addition, many motorcycles have dust seals to help protect the oil seals. Removal and replacement of a fork seal usually requires a special tool. Refer to the manufacturer's service manual for this information. Use care when replacing seals, as they must not be damaged during installation.

Fork Springs **Fork springs** extend the forks and allow movement of the outer fork tubes. Fork springs fit inside the inner fork tube and press against the tube cap nut at one end and the outer fork tube at the other end. Many forks will have fork spring spacers in place to take up space in long travel suspension systems. We will get into more detail on springs later in this chapter.

Hydraulic Damping Units Various front suspensions hold hydraulic damping units in place. Two examples of typical systems follow.

Looking back at Figure 17-22 you will see a **standard damping-rod** fork design. As the outer tube moves upward on the compression stroke, oil flows from the outer fork tube lower chamber through an orifice in the damper rod and into the fork tube, while the oil remaining in the outer fork tube lower chamber pushes past the valve into the outer tube upper chamber. The resistance in the fork created by the oil flow is designed to absorb shock on the compression stroke of the fork. As the fork gets close to full compression, a tapered oil lock piece hydraulically locks the fork before it has metal-to-metal contact with any portion of the fork.

On the fork rebound stroke, the oil in the damper rod flows through an orifice in the top of the damping rod into the outer fork tube lower chamber. The result is resistance that serves as a damping force to control the spring from rebounding too quickly. A rebound stopper spring is utilized to prevent metal-to-metal contact during the rebound stroke.

The **cartridge** fork design (Figure 17-28) includes a hollow piston and rod that fit into a damping rod inside the fork to form a cartridge. Most of the oil in the fork stays within the cartridge, which maximizes the fork's damping effectiveness. In most cases, there are adjusters (Figure 17-29) that allow the rider to change the compression and rebound oil flow, making the suspension softer or stiffer, thereby providing adjustable damping.

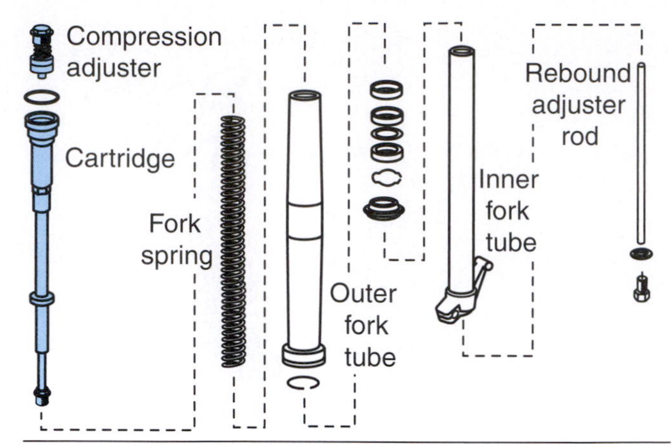

Figure 17-28 A cartridge-type fork exploded view is illustrated here.

Figure 17-29 Most cartridge-type forks have damping adjusters as shown here. Courtesy of American Suzuki Motor Corp.

Swing Arm

A **swing arm** (sometimes called a rear fork) is the main component of the rear suspension of most modern motorcycles and ATVs. It is used to hold the rear axle in pace while pivoting vertically to allow the suspension to absorb bumps. The swing arm is mounted to the frame by a pivot bolt, which prevents side-to-side movement while at the same time allowing up-and-down movement of the rear wheel (Figure 17-30).

The swing arm serves as the mounting bracket for the rear wheel assembly and the rear shock absorber(s). The most common type of swing arm is the conventional (Figure 17-31) swing arm that

Figure 17-30 The swing arm is attached to the frame and allows the rear wheel to move up and down. Courtesy of American Suzuki Motor Corp.

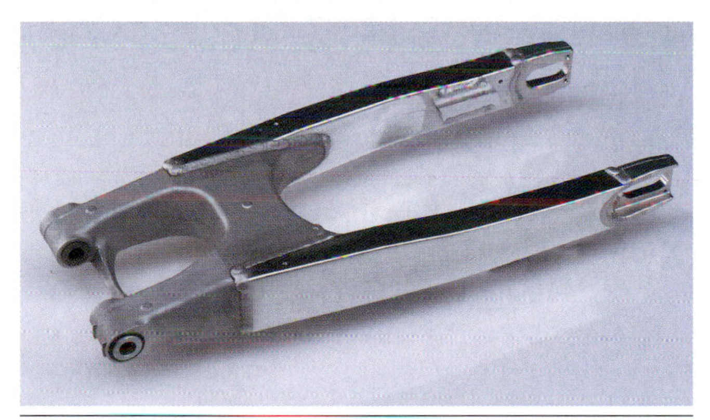

Figure 17-31 A conventional-type swing arm is pictured here. Courtesy of American Suzuki Motor Corp.

allows for the rear wheel axle to slide through both sides of the unit.

This type of swing arm may be made of steel or aluminum and in many cases is heavily braced (Figure 17-32) to accept extreme load pressures from aggressive riding.

Another type of swing arm is the single-sided design (Figure 17-33), which first became popular in racing applications due to the ability to easily remove and reinstall the rear wheel quickly.

Rear Suspension Systems

Motorcycle suspension-system technology, especially in rear suspensions, has advanced considerably over the past decade. Twin-shock rear suspensions are still widely used, but single-shock models have become the industry standard (Figure 17-34). Why a single-shock system?

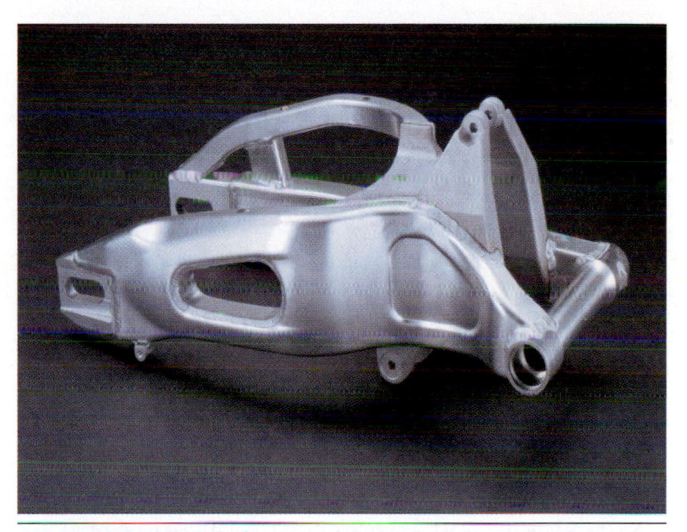

Figure 17-32 Swing arms may come with bracing to help increase the rigidity of the rear suspension system. Copyright by American Honda Motor Co., Inc. and reprinted with permission.

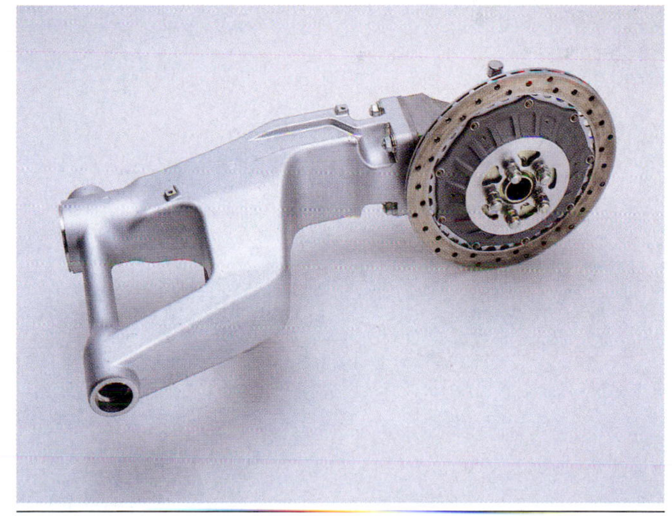

Figure 17-33 A single-sided swing arm is pictured here. Copyright by American Honda Motor Co., Inc. and reprinted with permission.

In a twin-shock system (Figure 17-35), the suspension units are typically attached very close to the rear axle. This means that as the suspension compresses and expands, the shock absorber pistons are traveling in a stroke that is nearly the same as the full deflection of the swing arm. Hitting a large bump might deflect the rear axle upward by 100mm and back, resulting in the same 100mm stroke in the shocks.

If this is done often (ridden on a bumpy road or track) the shock absorber piston begins to agitate the damper oil so much and so frequently that the

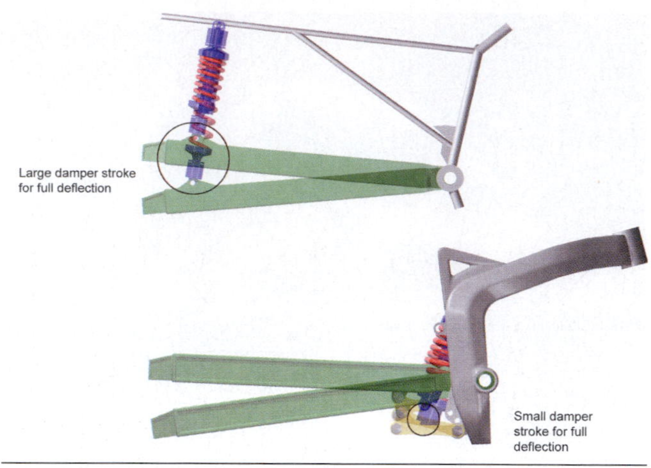

Figure 17-34 A comparison of a twin-shock and single-shock rear suspension is illustrated here

Figure 17-35 A twin-shock rear suspension system is illustrated here.

oil begins to heat up and foam; consequently, losing the ability to perform as it should. This is known as shock absorber fade. When this occurs, the shock acts like a spring with no damping qualities, sort of like a pogo stick! In the early 1970s, engineers placed the twin shock absorbers closer to the pivot point but were limited with the technology of the day.

In the mid-1970s, engineers began to mount a single shock absorber system toward the front of the swing arm (Figure 17-36). The swing arm might still have a lot of travel at the axle, but basic

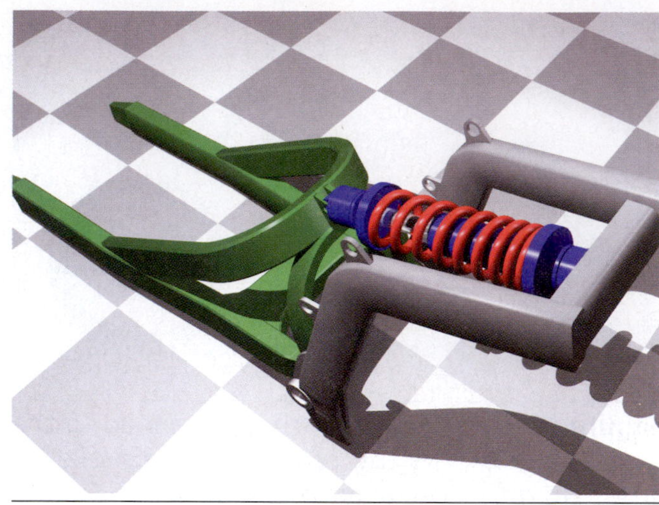

Figure 17-36 The first single rear shock systems were mounted as illustrated here.

geometry shows you that when mounted closer to the pivot point, the deflection of the shock is much less. This translates into shorter shock absorber movements, which translates to less opportunity for the damper oil to foam.

As technology evolved the single-shock rear suspension evolved to utilize complex link single-shock systems, where a series of levers reduce the shock absorber travel even further than a single shock alone (Figure 17-37). All link-type rear suspension systems operate on the principle of *rising rate.* When a rider goes over small bumps and at low speeds, the first portion of the rear suspension travel is engineered to ensure a smooth, comfortable ride. Then, as the rider hits larger bumps or as the speed of the machine increases, the suspension becomes progressively stiffer to resist bottoming. This is known as a rising rate suspension.

REAR-DAMPER DESIGN AND THEORY OF OPERATION

Springs are used on motorcycles to absorb impacts and bumps. However, when impact is made if only a spring is used as suspension, inertia causes the motorcycle to bounce up and down and continue to do so (Figure 17-38).

Dampers are used to convert the kinetic (bouncing) energy (Figure 17-39) into heat by utilizing friction and oil flow resistance in the

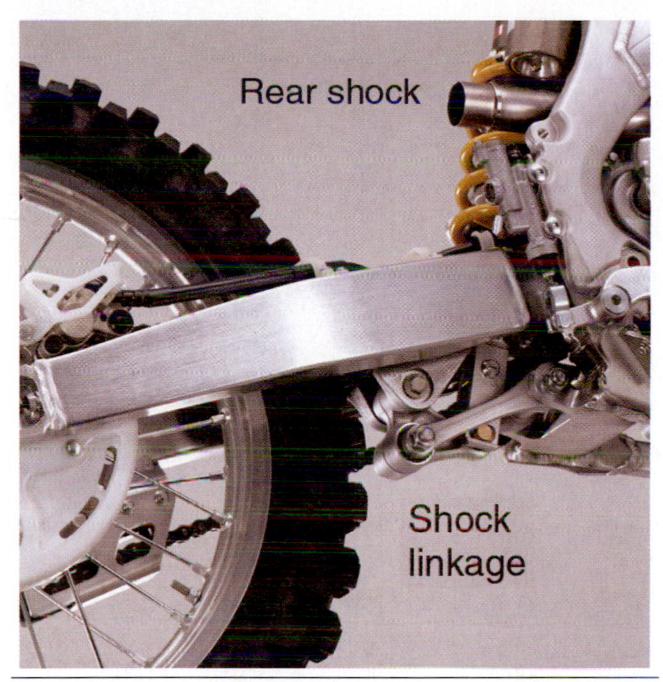

Figure 17-37 Current single-shock rear suspension systems utilize linkage systems to allow for a rising rate, which changes the stiffness of the ride depending on the terrain.

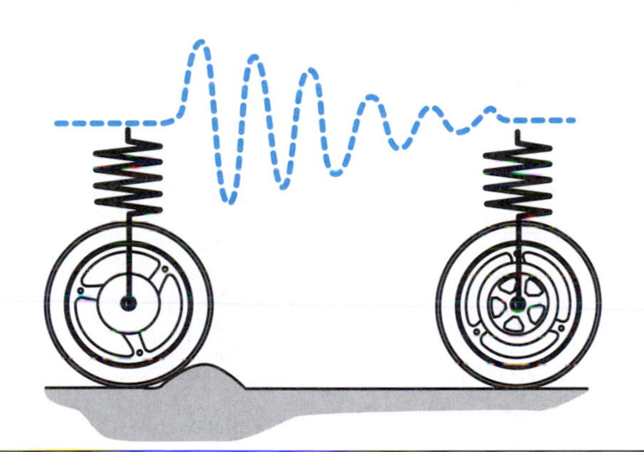

Figure 17-38 If only springs are used for suspension, the machine would bounce continually after hitting a bump.

damper; thus to stabilize the movement of the spring. This stabilization is called damping force.

To keep the rear wheel on the ground while traversing different types of terrain, a combination of an oil damper and spring are used on motorcycles (Figure 17-40) to keep the rider comfortable and keep rear wheel traction at a maximum. Dampers are also known as shock absorbers. Most shock absorbers are designed to control the unwanted rebound effects (rebound damping) of the spring. Typically, most shocks have very little resistance during the compression of the shock (compression damping) as most of the resistance is handled by the selected spring.

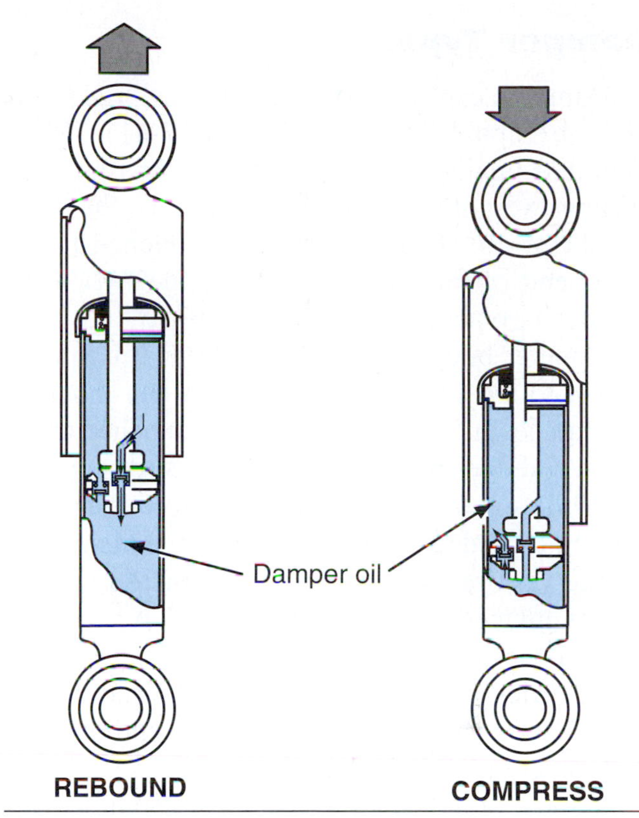

Figure 17-39 Dampers are used to convert kinetic (bouncing) energy into heat by utilizing friction and oil flow resistance in the damper to control the movement of a spring.

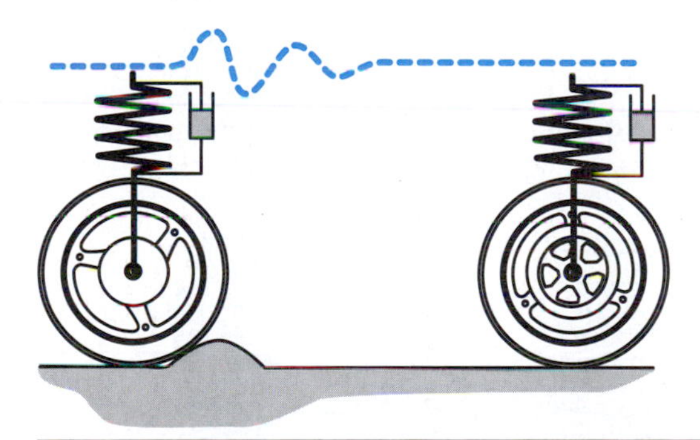

Figure 17-40 Dampers are used on motorcycles to prevent the suspension from bouncing after hitting a bump.

Damper Types

Dampers can be designed either as "right side up" or "upside down" (Figure 17-41). When mounted upside down, the damper has less un-sprung weight. Un-sprung weight is the total weight of all of the components attached to the lower end of the damper at the frame. When the damper is placed in the "right-side-up" position, the damper body is added as additional unsprung weight to the suspension, whereas when mounted "upside down," the damper body is attached to the frame and does not move with the suspension.

While some very lightweight motorcycles use a friction-type damper (Figure 17-42), which relies on the friction of a non-metallic piston against grease placed inside the damper cylinder to counter the rebound effect of the spring, most all motorcycles use oil damper units. The friction-type dampers are mostly seen on the front end of small motorcycles.

Some motorcycles use a single damping design that provides damping only on the rebound stroke. This type of shock relies primarily on spring resistance for compression damping but the best choice of damper has the ability to offer resistance for both compression and rebound.

The most popular rear dampers on motorcycles today are the **de Carbon shock** design. This damper uses nitrogen gas in a separate chamber to keep the oil from foaming. As you should remember, foaming of the oil causes the damper to lose its damping qualities. A "piggyback" remote-reservoir de Carbon shock absorber is pictured in Figure 17-43. This type of damper is very popular on modern motorcycles and allows an increase in oil capacity while providing a separate nitrogen chamber. The damper may also be produced with a remote reservoir (Figure 17-44) that can be mounted separately on the frame.

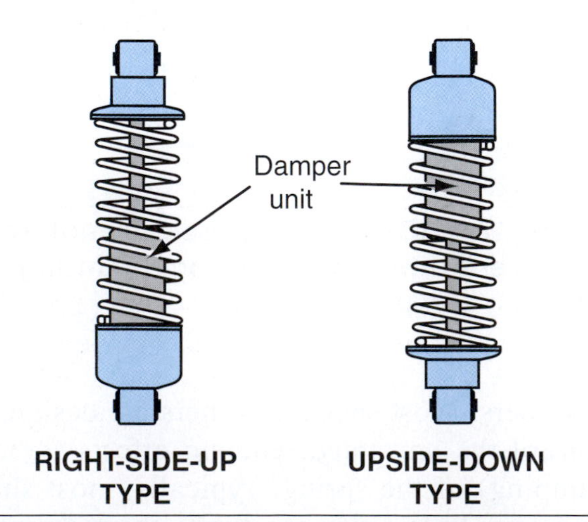

Figure 17-41 Dampers can be designed one of two ways as illustrated here.

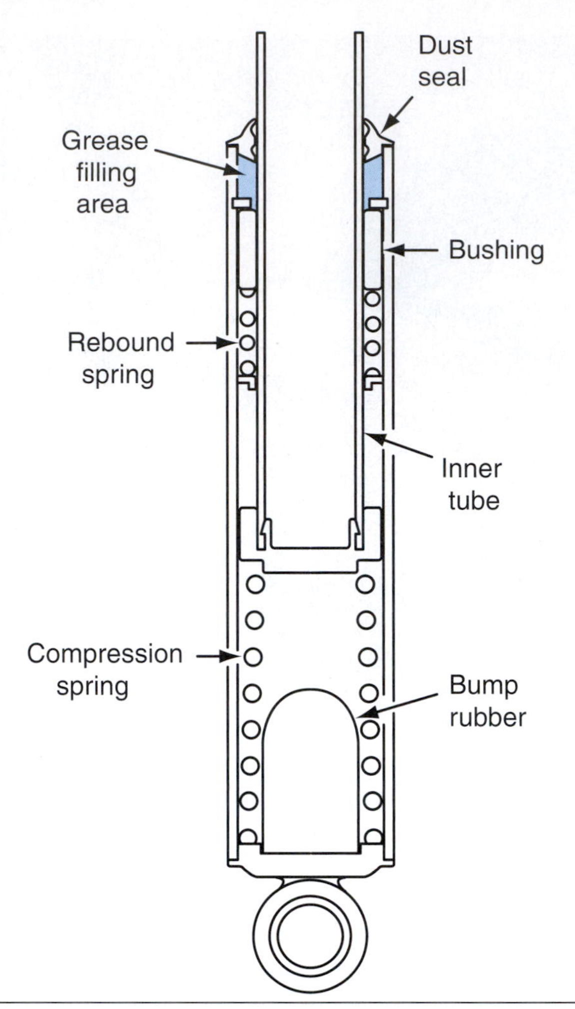

Figure 17-42 A friction-type damper uses grease in the damper cylinder to counter the rebound effect of the spring.

The oil in the de Carbon shock is normally controlled by spring washers that are forced to move when the oil pushes them open as the shock is moved back and forth (Figure 17-45). Damping characteristics can be changed by moving these shims into different positions. This type of work is recommended for experienced shock-absorber professionals only. Figure 17-46 shows the shock further disassembled so we can see the shims as they are removed in order.

Figure 17-43 A piggyback-type rear damper unit is shown here. Courtesy of American Suzuki Motor Corp.

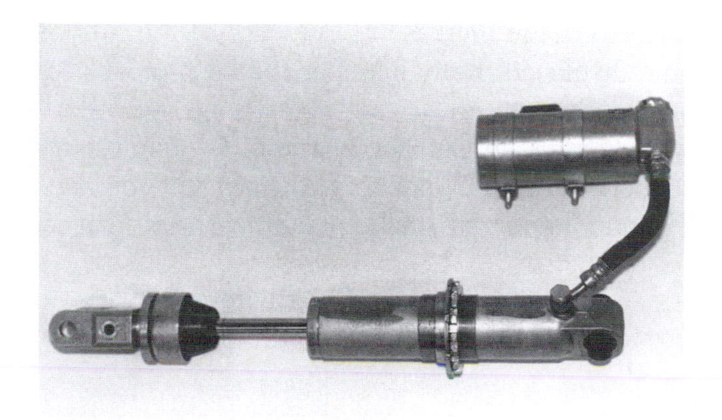

Figure 17-44 A separate remote reservoir de Carbon rear damper unit is shown here.

Figure 17-45 Spring washers control the oil flow in a de Carbon-type rear shock.

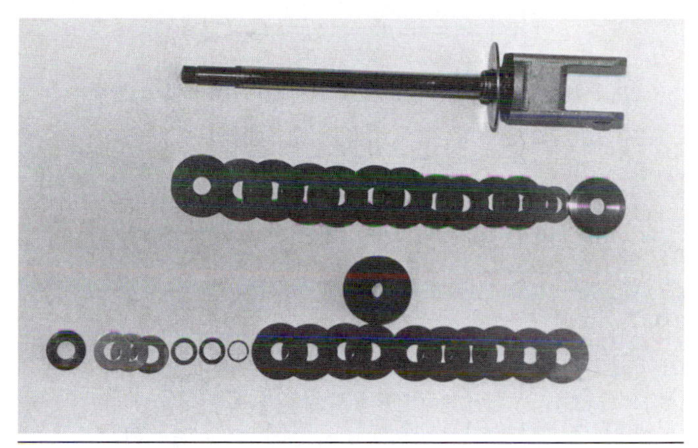

Figure 17-46 An experienced suspension technician can change the spring washers in a rear damper to change the damping characteristics.

Damper Operation

While there are different types of oil dampers their operation is very similar.

Damping is provided when we utilize the resistance of oil passing through a hole. The resistance to flow increases in the following manner:

- Higher oil viscosity
- A smaller hole
- Faster oil flow

Using the image in Figure 17-47, you can see that when we have a small hole at the end of a cylinder, the resistance to push the piston is high and the fluid pressure is high. When a large hole is utilized, the resistance to move the piston is lower and the fluid flow pressure is lower as well.

Viscosity is the "thickness" of a liquid or its ability to flow. Oil viscosity will generally decrease as its temperature increases. When oil passes through holes and slits, the friction produces heat. Quality oil-type dampers use special oil that provides very little viscosity change in respect to a change in temperature. It is important to understand that it is critical to use the correct oil when rebuilding any suspension component. The manufacturer's service manual will provide the recommended oil. Let's next discuss the process of compression and rebound damping.

Compression Damping

During the compression stroke (Figure 17-48), the damper rod is pushed and the damper rod piston

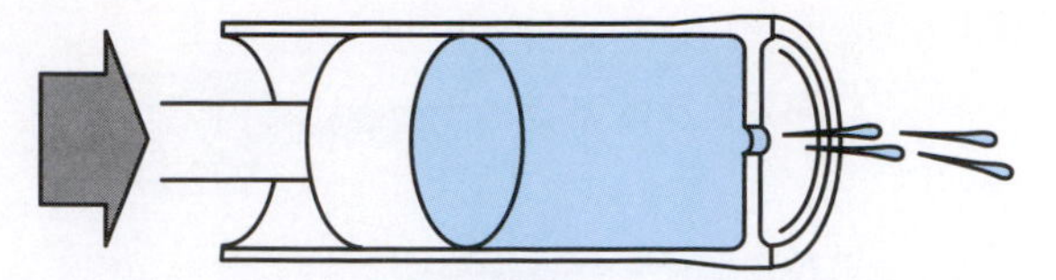

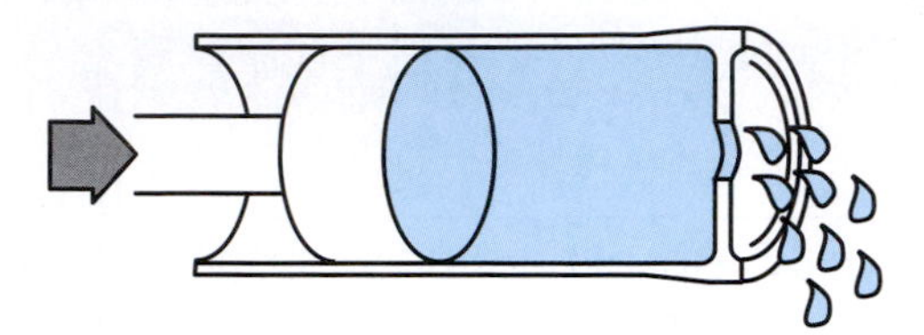

Figure 17-47 This illustration shows the different characteristics of a damper with different-sized holes for the oil to flow from. Copyright by American Honda Motor Co., Inc. and reprinted with permission.

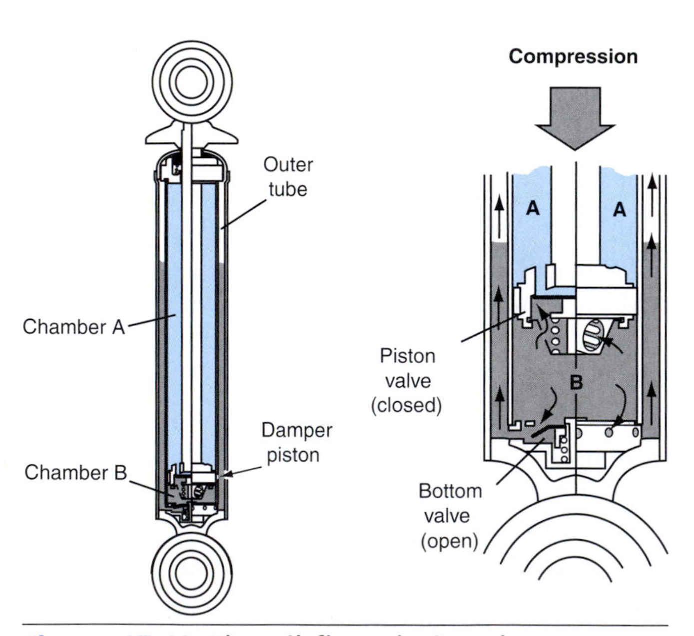

Figure 17-48 The oil flow during the compression stroke is illustrated here.

moves downward. This causes the pressure in chamber A to decrease and the pressure in chamber B to increase. The damper rod piston has a valve that opens to allow the oil in chamber B to flow with little restriction to chamber A. As the damper rod moves down the cylinder the volume of oil displaced by the piston flows to the outer tube through the hole in the bottom compression valve.

Rebound Damping

During the rebound stroke (Figure 17-49), the damper rod piston moves upward. During this movement, the pressure in chamber A increases, and the pressure in chamber B decreases. The compression valve in the damper rod piston closes and the oil in chamber A flows to chamber B through the rebound valve in the piston. When the oil passes through the rebound valve in the piston, damping is provided. During the rebound stroke, the bottom valve opens and allows the oil to flow from the outer tube unrestricted.

Air Chambers

Dampers that use oil have a small air chamber inside the damper casing (Figure 17-50). Why? Because oil is a liquid and a liquid cannot be compressed where air can be compressed easily. Therefore, the increase or decrease of the oil volume displaced by the damper rod as it moves can be absorbed.

If there is no air, the volume of the damper rod cannot be accommodated, and the rod will not move. The air chamber is small enough to ensure that the

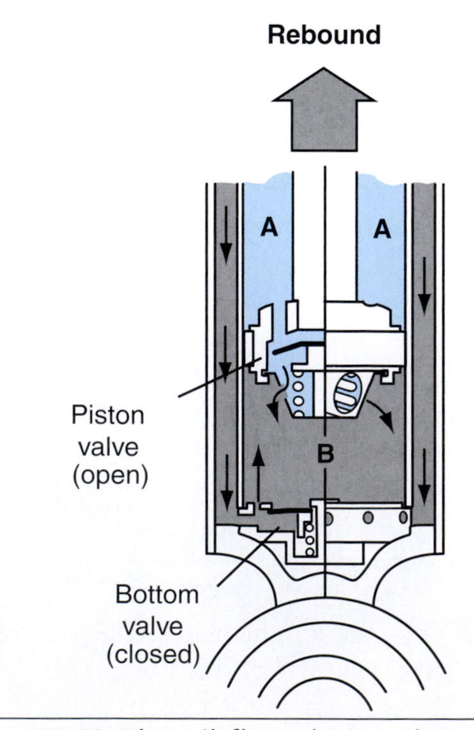

Figure 17-49 The oil flow during the compression stroke is illustrated here.

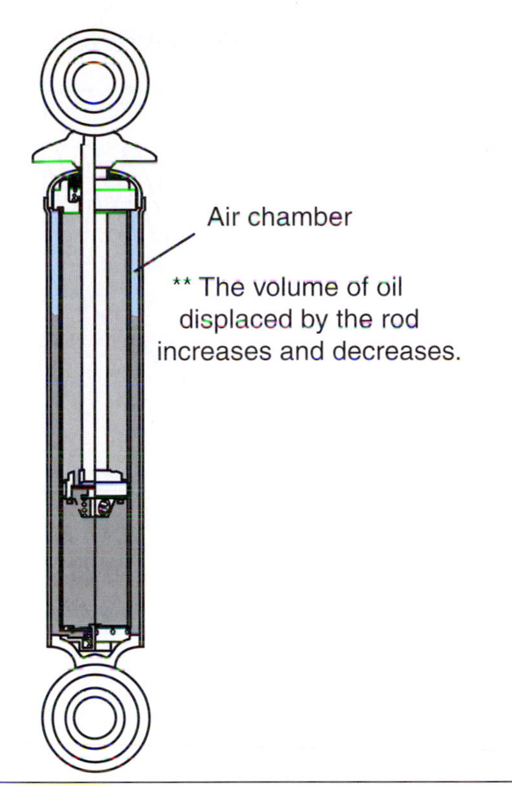

Figure 17-50 An air chamber must be used to accommodate the damper rod as it moves up and down.

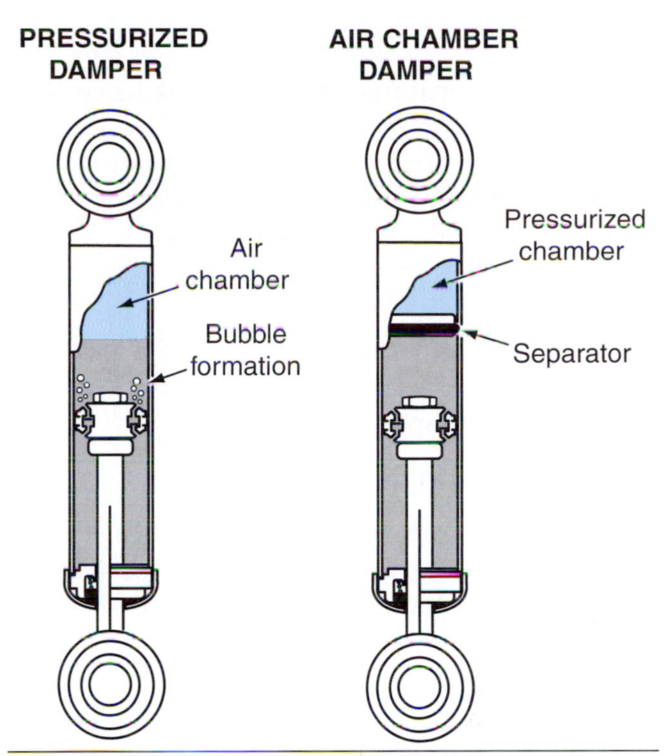

Figure 17-51 The difference between a damper with an air chamber and a pressurized chamber. With a pressurized damper unit, there is less chance for the oil to foam as the damper heats up from use.

piston will not be exposed to air but large enough to ensure movement when fully compressed.

Pressurized Dampers

When the temperature increases or when pressure decreases in a damper, air is dissolved in the oil and tends to foam and produce bubbles. In oil dampers, high pressure and low pressure are produced across the piston during the compression and rebound strokes; damping is produced as oil passes through small holes, creating heat from the friction. During the rebound stroke when pressure in the air chamber is low and the damping force is high, the flow rate of the oil that passes through the rebound valve is fast and the temperature rises due to the pressure resistance, which causes foaming. Foaming causes a reduction of the damping effect.

When pressurized dampers are used (Figure 17-51), the air chamber is pressurized with nitrogen to prevent the formation of bubbles. To prevent the nitrogen gas in the chamber from mixing with the oil, a floating piston or a rubber diaphragm is used to separate the nitrogen gas and the oil.

Damper Springs

Motorcycles use steel coil damper springs attached to the outside of the damper and have specific performance characteristics that are determined by the spring wire diameter, coil diameter, pitch (Figure 17-52), and material quality. The spring absorbs the impacts and bumps as the motorcycle is ridden. There are three different types of coil springs used on motorcycle suspension systems.

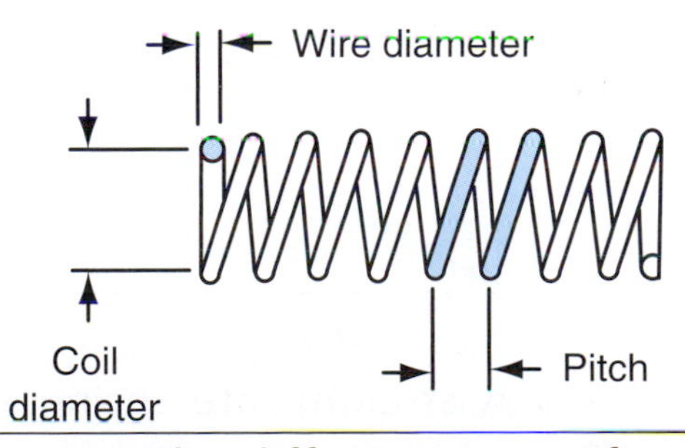

Figure 17-52 The different areas of a spring that have an effect on the spring performance characteristics.

Spring Rate

The **spring rate** is the amount of force (lbs) required to compress the spring by 1 inch (Figure 17-53). For example, if 100 lbs of force is required to compress a spring by 1 inch, the spring rate for that spring is 100 lbs/in.

Straight-Rate Springs

A straight-rate spring (Figure 17-54) allows for a reactive force that increases proportionally with the stroke of the spring.

Dual-Rate Springs

A dual-rate spring is created by changing the pitch of the spring windings in a phased manner. The spring with both A and B pitches, as shown in Figure 17-55, starts with a spring rate of A for the first half of the spring stroke. As the stroke increases, rate A compresses and then allows rate B to react.

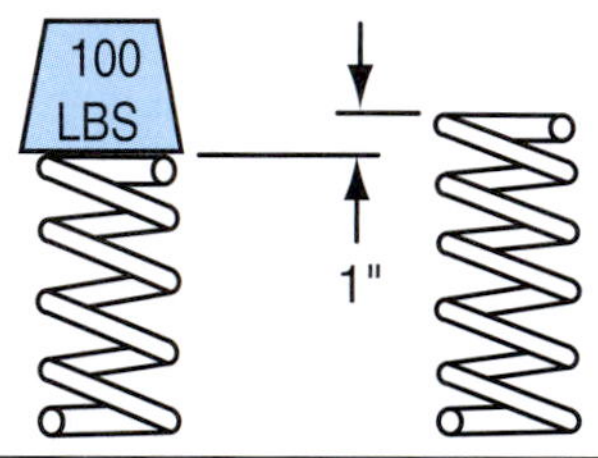

Figure 17-53 An example of how spring rate is measured is illustrated here.

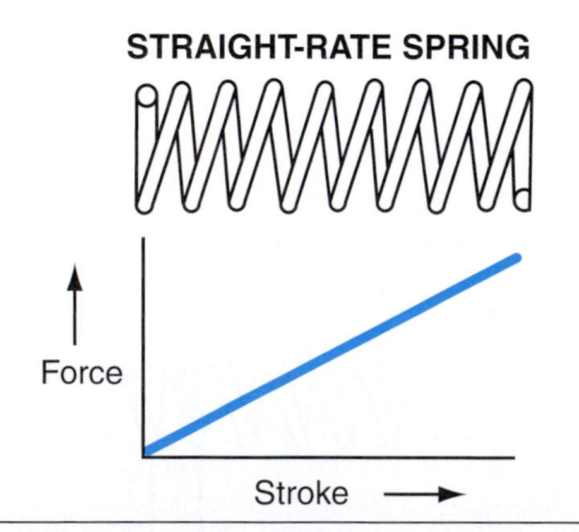

Figure 17-54 A straight-rate spring is illustrated here.

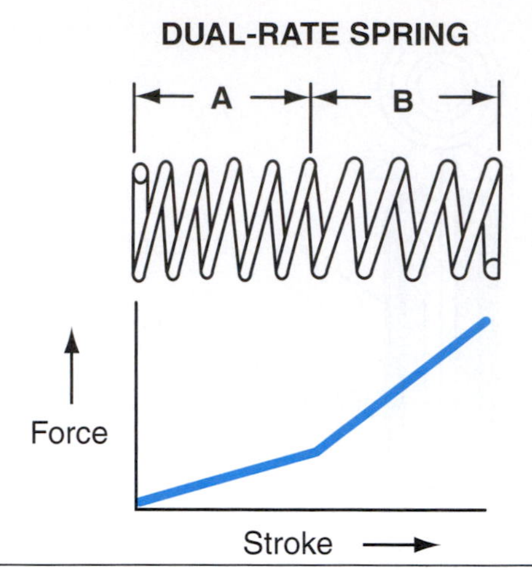

Figure 17-55 A dual-rate spring is illustrated here.

Progressive-Rate Springs

Changing the distance between each coil throughout the length of the spring (Figure 17-56) makes the spring rate change in a linear manner in relationship with the stroke. At the beginning of the stroke, the spring has a lower rate and absorbs small bumps of the road surface. As the stroke increases, the reaction forces an increase to absorb large amounts of impact.

Spring Adjusters

Spring adjusters are utilized to adjust the set length of the damper spring. Changing the set length changes the initial spring rate (Figure 17-57). Spring adjusters ensure riding comfort and tire grip for a lightweight or a heavy load condition. In the case of a need to raise the initial spring rate, the set spring length is reduced from the standard position and makes the spring more rigid. The basic concept for spring adjusters allows the rider to make the initial spring rate more rigid if the spring set length is reduced. On the other hand, the initial spring rate will become less rigid if the spring set length is increased. Refer to the shop manual for details on the method for adjustment. The two primary methods of adjusting the spring length is by cam adjuster or a screw and locknut (Figure 17-58).

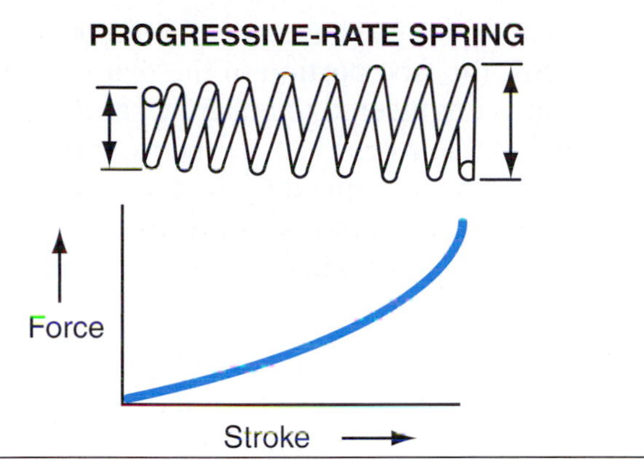

Figure 17-56 A progressive-rate spring is illustrated here.

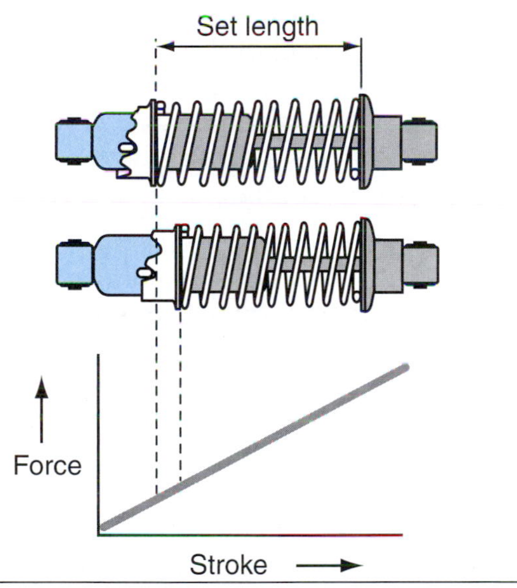

Figure 17-57 The set length of a spring determines the initial spring rate pre-load.

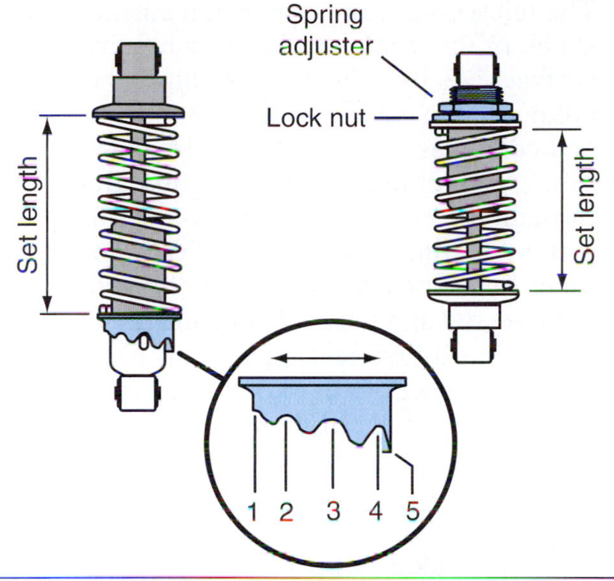

Figure 17-58 The cam type and screw and lock-nut type are the most popular types of spring adjusters found on motorcycles.

Summary

- Manufacturers use many frame configurations, depending on how the motorcycle will be used. Frame design also depends on such factors as engine displacement, intended use of the motorcycle, cost, visual appeal, and materials. Frames are designed with a variety of tubing in different shapes. There are eight different basic frame types: single cradle, double cradle, backbone, underbone, diamond, perimeter, pivotless, and trellis.
- The front suspension supports the front wheel and allows it to pivot from side to side for steering. Most motorcycle front forks incorporate telescopic hydraulic shock absorbers to absorb the vertical shock of the front wheel when hitting bumps. The most popular telescopic motorcycle front suspension system is the conventional type, which utilizes a pair of upper fork tubes and lower fork sliders that move into one another.

The telescopic front suspension cushions the shock of the front wheel hitting bumps in road surfaces. As the wheel hits a bump, the sliders are pushed upward over the inner fork tubes and compress the springs that are placed in the tubes. The oil in the outer fork tube flows through the holes or valves into the inner fork tube. Since the transfer of oil into the inner fork tube takes up space, trapped air is compressed and the increased air pressure increases oil flow resistance. This has a damping effect on the shock and limits the upward movement of the outer fork tubes. As the shock load is relieved, the springs push the slider back to the extended position, and oil is pulled back through the holes (valves) into the vacuum created by the extension of the sliders. The flow of oil is restricted by the size of the holes or valves, and a damping effect is obtained with each movement of the sliders.

- Motorcycle rear suspensions are made in two basic designs: twin-shock and single-shock systems. With a twin-shock design, the suspension units are typically attached very close to the rear axle. This means that as the suspension compresses and expands, the shock absorber pistons are traveling in a stroke that is nearly the same as the full deflection of the swing arm. Single-shock rear suspension systems came to motorcycles in the mid-1970s. As technology evolved the single-shock rear suspension evolved to utilize complex link single-shock systems, where a series of levers reduce the shock absorber travel even further than a single shock alone.
- When a rider goes over small bumps and at low speeds, the first portion of the rear suspension travel is engineered to ensure a smooth, comfortable ride. Then, as the rider hits larger bumps or as the speed of the machine increases, the suspension becomes progressively stiffer to resist bottoming. This is known as a rising rate suspension.
- To keep the rear wheel on the ground while traversing different types of terrain, a combination of an oil damper and spring are used on motorcycles to keep the rider comfortable and keep rear wheel traction at a maximum. Dampers are also known as shock absorbers. Most shock absorbers are designed to control the unwanted rebound effects (rebound damping) of the spring. Typically, most shocks have very little resistance during the compression of the shock (compression damping), as most of the resistance is handled by the selected spring.
- Motorcycles use metal coil springs usually made of steel (but now titanium can also be seen on some performance motorcycles) attached to the outside of the damper. They have specific performance characteristics that are determined by the spring wire diameter, coil diameter, pitch, and material quality. Springs are designed to absorb the impacts and bumps as the motorcycle is ridden. There are three different types of coil springs used on motorcycle suspension systems: single, dual, and progressive rate.

Chapter 17 Review Questions

1. The frame design that joins the swing arm to the steering head in as short a distance as possible is the ________ frame.
   a. backbone
   b. perimeter
   c. diamond
   d. underbone
2. What is the most commonly used front fork design used on motorcycles?
   a. Leading link
   b. Telescopic
   c. Trellis
   d. Friction
3. The ________ is the component that allows rear wheel up-and-down movement.
   a. steering stem
   b. pivot shaft
   c. shock
   d. bushing
4. Hydraulic damping in a telescopic suspension system is obtained by the transference of ________ trapped between the inner and outer fork tubes through small holes drilled in the inner fork tube.
5. The ________ is the component that supports the right and left inner fork tubes by clamping around them.
6. Which type of bearing is NOT used in the steering stem on motorcycles?
   a. Loose ball
   b. Caged ball
   c. Straight roller
   d. Tapered roller
7. ________ forks are mounted with the outer fork tube in the triple clamps and the inner slider mounted to the axle.
8. When you increase the ________ on a motorcycle the wheelbase is lengthened and the motorcycle will be more difficult to turn.
9. The frame design that has the swing arm attached directly to the engine is the
   a. diamond
   b. backbone
   c. pivotless
   d. perimeter
10. When mounted upside down, a damper has less ________ weight.
11. The ________ serves as the mounting bracket for the rear-wheel assembly and the rear shock absorber(s).

CHAPTER 18 Brakes, Wheels, and Tires

Learning Objectives

When students have completed the study of this chapter and its laboratory activities they should be able to:

- Identify the different brake systems and brake system components used on motorcycles
- Understand basic antilock braking system function
- Understand basic integrated braking system function
- Identify the differences between tube-type and tubeless-type tires
- Identify the different types of tire construction used on motorcycles

Key Terms

Antilock braking systems (ABS)
Belted bias ply tires
Bias ply tire
Brake caliper
Brake fluid
Brake lever
Brake rotor
Braking systems
Cast wheels
Disc brake
Drum brake
Hydraulic brake lines
Integrated braking systems (IBS)
Master cylinder
Organic brake pads
Radial ply tires
Sintered brake pads
Spin balancing
Spoke wheels
Static balancing
Tire
Tire balancing
Truing

INTRODUCTION

This chapter will be an exploration into the areas of brake systems, wheels, and tires used on motorcycles. You'll begin by learning about the different types of brakes. We'll describe how each type of brake operates and identify its components. Next, we'll look at the wheels commonly found on motorcycles. In this discussion, we'll cover both the spoke and nonspoke types of wheels used on motorcycles. In addition, you'll learn about repairing spoke-type wheels. Our tire information includes both tube-type and tubeless-type tires. We'll provide information for tire repair and balancing.

Two very important areas of safety for motorcycles are braking systems and tires. Better brakes and tires are constantly being developed to keep up with vehicles that are being designed to be more powerful and easier to handle over a wide variety of conditions.

As a motorcycle technician, you'll often be called upon to adjust, repair, and replace tires, wheels, and brakes. These jobs can be very difficult for a novice technician; however, a few simple procedures, such as those you will learn in this chapter, can do a lot to help speed up these jobs and avoid problems.

BRAKING SYSTEMS

The **braking systems** used on motorcycles, like virtually any type of braking system, are designed to reduce the machine's kinetic energy by transforming it into heat energy known as friction heat. The brakes on motorcycles are energy conversion devices that convert the energy of motion (kinetic energy) into heat energy.

Motorcycle braking is accomplished by the friction (resistance to movement) produced when a brake lining is forced against a rotating drum or rotor. Frictions created by the braking system slow, and eventually stop, the wheels from rotating.

The types of brakes used on motorcycles fall into two general categories:

- Drum brakes
- Disc brakes

A **drum brake** is a brake in which the friction is caused by a set of shoes or pads that press against the inner surface of a rotating drum. A **disc brake** relies on friction being applied to both sides of a spinning disk by brake pads. These types of brakes can be actuated by either hydraulic (fluid pressure) or mechanical (cable or linkage) mechanisms.

Mechanical Drum Brakes

First, let's look at the mechanical drum brake (Figure 18-1). While sometimes found on the front as well, a mechanical drum brake, when used, is typically found on the rear wheel. With this type of brake, a backing plate is connected to the swing arm and holds the two brake shoes. The wheel and brake drum rotate around the brake shoes. When the rider applies the brake, a cam pushes the two semicircular shoes outward. The circle formed by the two shoes expands. When the shoes expand, they press against the rotating drum, thereby limiting its free rotation.

It is very important that brake systems quickly dissipate the heat that is generated by the friction of the braking action to ensure that the stopping force is consistent. Because drum brakes contain all of the heat-generating components within the wheel hub (Figure 18-2), it is important that engineers design these components using materials that conduct heat rapidly. It is also important that the brake be designed of the proper size for the anticipated requirements of the motorcycle.

Figure 18-1 This motorcycle utilizes a mechanical drum brake on the rear wheel. Courtesy of American Suzuki Motor Corp.

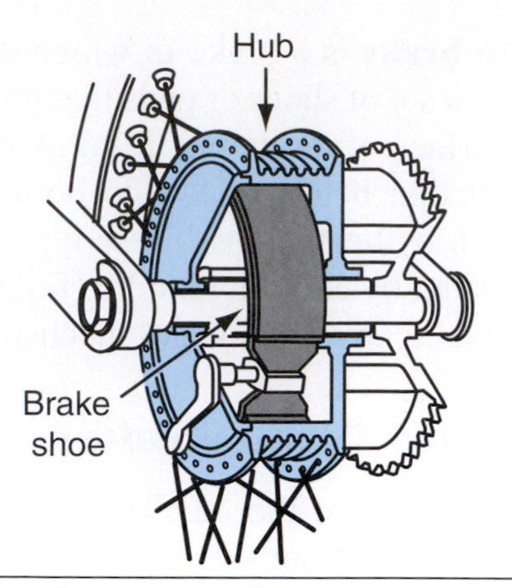

Figure 18-2 A cutaway illustration of a mechanical drum brake. Note the cooling fins on the outside of the hub. These are used to assist in removing the heat caused by the friction created by applying the brakes.

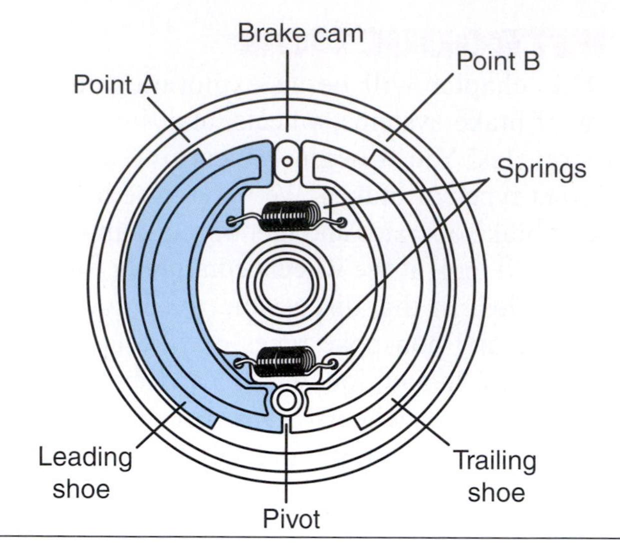

Figure 18-3 A single-leading shoe brake cutaway illustration.

In order to enhance heat conductivity while providing excellent wear properties on the inner surface of the brake drum, the drum itself is generally made of cast iron. In most cases, the other portions of the drum and wheel hub are made of aluminum alloy with cooling fins cast into the outer circumference for heat conductivity and dispersion. To help to speed conductivity, the cast iron drum is pressed into the aluminum hub and cannot be removed.

Types of Mechanical Drum Brakes

Mechanical drum brakes may use either one or two cams to expand the shoes. In the one-cam arrangement, sometimes called a single-leading shoe and single-trailing shoe brake system, springs hold one end of the shoe to the cam. The other end is anchored to a round peg or pivot. Turning the elliptical cam spreads the shoes. Figure 18-3 illustrates the single-leading shoe arrangement.

The leading shoe makes contact with the drum at Point A. Wheel rotation draws the shoe tighter into the drum. The trailing shoe makes contact at Point B, but the motion of the wheel tends to push the shoe away from the drum. In other words, the rotation of the drum helps the leading shoe develop braking power, whereas the opposite is true in the case of the trailing shoe.

Figure 18-4 illustrates the double-leading shoe brake system. In this system, a cam is used at both ends of the shoes. Therefore, expansion is equal at each end of each shoe. The double-leading shoe system requires less pressure to operate because the shoes expand equally around the drum, and the rotation of the drum pulls the leading edge of each brake shoe tighter against the drum.

Hydraulic Drum Brakes

The hydraulic drum brake is very similar to the mechanical drum brake except for its method of actuation (Figure 18-5) and is most likely to be found on an ATV. The hydraulic drum brake uses hydraulic

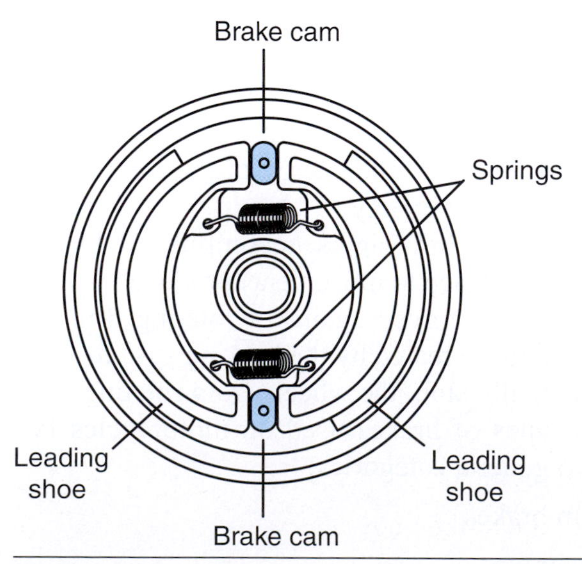

Figure 18-4 A double-leading shoe brake cutaway illustration.

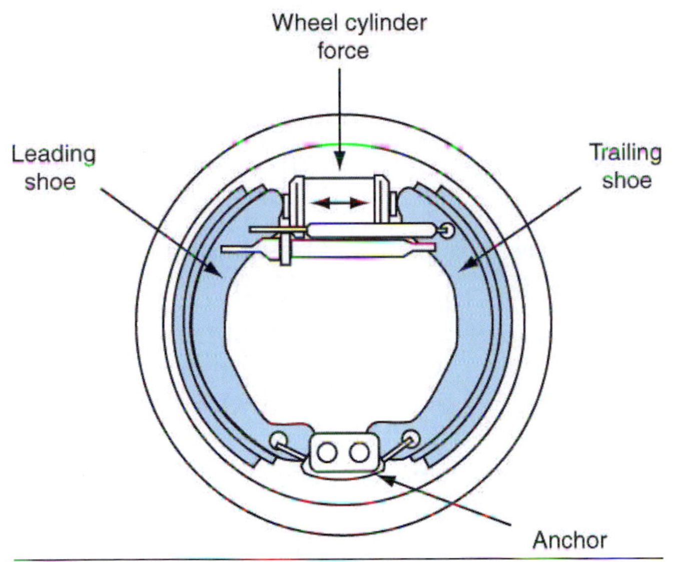

Figure 18-5 Hydraulic drum brakes use an inner wheel cylinder to apply hydraulic pressure to move the brake shoes outward.

brake fluid to transfer the pressure from the brake pedal or lever to the actual brake mechanism.

The most common arrangement of hydraulic drum brakes consists of a brake lever or pedal, a master cylinder, hydraulic lines, a "slave cylinder," and the brake linings and drum. We get into more details on the actuation of a hydraulic brake system when we discuss hydraulic disc brakes.

Hydraulic Disc Brakes

The other category of brake systems is the disc type. Disc brakes are commonly used on both front and rear wheels (Figure 18-6). Disc brakes use two pads that apply pressure on the sides of a rotor to stop wheel rotation. The clamping action is normally caused by hydraulic pressure moving the pads together. As pressure is applied to the brake pads, they squeeze the rotor, preventing its turning.

Hydraulic Disc Brake Components

The major components of a hydraulic disc brake are the following:

- Brake lever
- Master cylinder
- Hydraulic brake lines
- Brake fluid
- Brake caliper

Figure 18-6 This off-road motorcycle uses disc brakes front and rear. Courtesy of American Suzuki Motor Corp.

- Brake pads
- Brake rotors

Let us take a closer look at each component.

Brake Lever The **brake lever** is a lever mounted to the handlebar (or a pedal on the frame) used for activating the brake and is connected to the master cylinder (Figure 18-7). As the rider moves the lever, a piston in the cylinder moves and forces fluid to flow from the master cylinder to the caliper assembly. Here the pressure of moving fluid causes the pads to grip the rotor, thereby stopping wheel rotation.

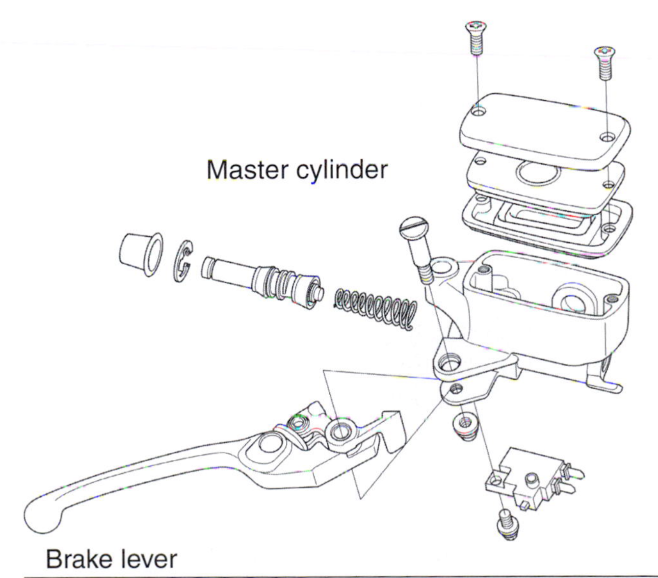

Figure 18-7 An exploded view illustration of a brake lever and master cylinder. Copyright by American Honda Motor Co., Inc. and reprinted with permission.

When the brake lever is released, a spring moves the lever and master cylinder piston back to their starting positions. This relieves the fluid pressure and the pads move away from the rotor, freeing the wheel.

Master Cylinder The **master cylinder** is the part of the brake system that stores the brake fluid and when the lever or pedal is applied, forces the fluid to the brake caliper. It consists of a piston, piston return spring, cylinder, and a reservoir (Figure 18-8). The master cylinder is used to apply pressure to the brake system. As the brake lever is applied, the master cylinder piston moves against the fluid and builds pressure in the system. The pressure is exerted against the brake caliper piston, which pushes the brake pads against the rotor. When the lever is released, the spring within the master cylinder returns the piston to its original position. A diaphragm is used in the reservoir and expands as the brake fluid level drops from pad wear.

There is a significant rise in hydraulic pressure between the master cylinder and the caliper due to the difference in the size of the pistons (Figure 18-9). The leverage ratio of the lever to the master cylinder piston also increases the braking action.

Hydraulic Brake Lines Hydraulic brake lines connect the master cylinder and the brake caliper. When the rider applies pressure to the brake lever, the master cylinder converts the force applied on the brake lever to hydraulic pressure. This hydraulic pressure passes through the hydraulic brake lines from the master cylinder to the caliper assembly. The increased pressure causes the brake-caliper piston to force the brake pad against the rotor, which stops the wheel. Brake lines are designed to resist flexing as the pressure builds in the brake system.

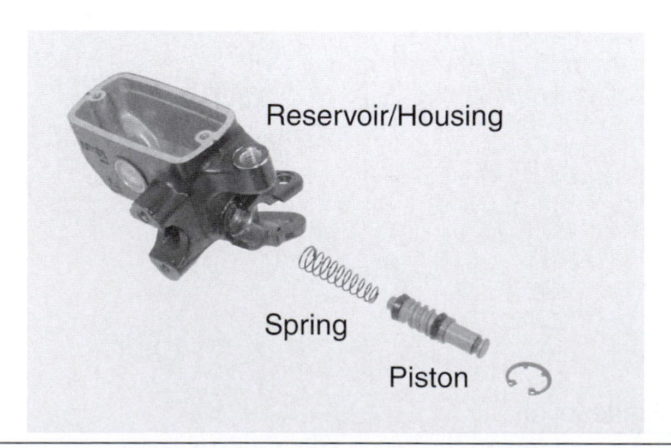

Figure 18-8 The internal components of an actual master cylinder are pictured here. Copyright by American Honda Motor Co., Inc. and reprinted with permission.

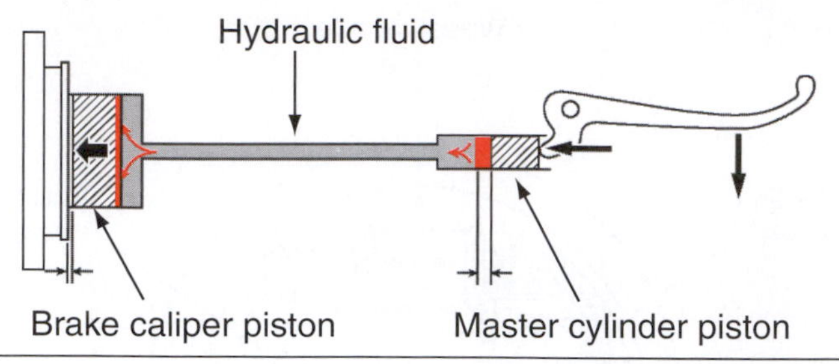

Figure 18-9 The different sizes of the master cylinder and caliper pistons create an increase in braking action. Copyright by American Honda Motor Co., Inc. and reprinted with permission.

Brake Fluid Hydraulic brake systems commonly use one of four types of **brake fluid**. Brake fluid is a type of hydraulic fluid used in brake applications in motorcycles. Brake fluids are rated by the U.S. Department of Transportation (DOT). These brake fluids are designated as DOT 3, DOT 4 (Figure 18-10), DOT 5, and DOT 5.1. Brake fluids are divided into two distinct groups: glycol based and silicone based. The most important thing to remember about the two types of brake fluid is that neither is compatible with the other!

DOT 5 brake fluid is a silicone-based fluid. Water or moisture won't mix with DOT 5 brake fluid. DOT 5 can't be mixed with any other type of

Figure 18-10 Typical containers of DOT 3 and 4 brake fluid.

brake fluid. A disadvantage that DOT 5 fluid has in relation to any other type is that it emulsifies (foams) very easily; therefore, be careful not to shake the container when using it. DOT 5 also has additives that counteract the fluid's tendency to make the seals swell. It's important to understand that these brake additives impregnate the seals in the brake system. If you change from one type of fluid to another, you must replace the seals and all rubber parts to prevent brake-system failure.

DOT 3, 4, and 5.1 brake fluids are glycol-based fluids and are hygroscopic. This means that they absorb moisture from the atmosphere. There are many ways for moisture to enter a brake system. Condensation from regular use, washing the vehicle, and humidity are the most common, with little hope of prevention.

This is very important to know because over time, moisture contaminates the brake fluid, which in turn lowers the fluid's boiling point and may lead to corrosion in the master cylinder and brake caliper.

The three glycol-based brake fluids are compatible although you should shy away from replacing a higher-number fluid with a lower number as the boiling point for the higher-number fluid is greater than a lower number (we will discuss boiling points next).

Also, note that glycol-based brake fluids are harmful to plastic and painted parts, so you must use extreme care when draining or adding these brake fluids to the master cylinder.

The primary difference between the three glycol-based brake fluids is their boiling points.

Before we get started on the minimum boiling points, we should discuss briefly how the standard is set for boiling points by the U.S. Department of Transportation for brake fluid.

- Dry boiling point is the temperature that the brake fluid will boil at with no measurable amount of water mixed in with it.
- Wet boiling point is the temperature that the brake fluid will boil once it has absorbed a prescribed amount of water measured by its weight.

According to the U.S. Department of Transportation standards, DOT 3 brake fluid must have a minimum dry boiling point of 401 degrees Fahrenheit and a wet boiling point of 284 degrees Fahrenheit. DOT 4 brake fluid's minimum dry boiling point must be 446 degrees Fahrenheit and a wet boiling point of 311 degrees Fahrenheit. DOT 5.1 has a minimum dry boiling point of 518 degrees Fahrenheit and a wet boiling point of 375 degrees Fahrenheit.

In addition, DOT 5.1 brake fluid has a lighter viscosity rating than DOT 3 or 4 as it was initially designed for vehicles with antilock braking systems (ABS). These systems generally operate better with lighter viscosity fluids.

By the way, DOT 5 (silicone-based) brake fluid has a minimum dry boiling point of 500 degrees Fahrenheit and a wet boiling point of 356 degrees Fahrenheit.

No matter which type of brake fluid you use it important to use only fresh brake fluid from a sealed container to avoid contaminating the brake system.

Caliper Assembly The **brake caliper** (Figure 18-11) is the part of a brake system that, when applied by the rider, clamps the brake disc to slow down the motorcycle. Pistons inside the calipers are in direct contact with brake pads, which press against the brake rotor. The brake fluid pushes the pistons outward, which in turn pushes the pads toward the brake rotor. As the pads press

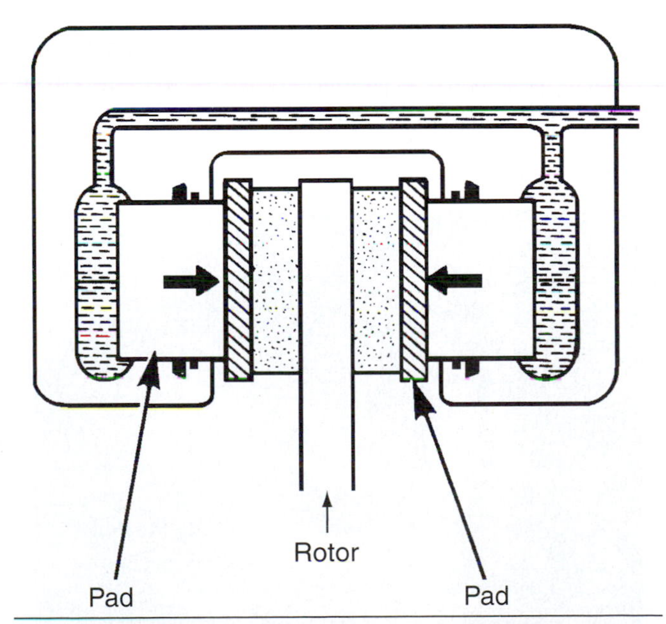

Figure 18-11 An illustration of a simple brake caliper is illustrated here. Copyright by American Honda Motor Co., Inc. and reprinted with permission.

against the rotor, the friction slows the rotation of the wheel. When the brakes are released, the pressure against the caliper piston is relieved and then retracts with the aid of the caliper seal to allow the wheel to turn freely (Figure 18-12).

Two types of hydraulic calipers are used on motorcycles today: single-push calipers and opposite-piston calipers.

With a single-push caliper design (Figure 18-13) both pads press against the brake rotor through a reaction of a sliding caliper yoke. This system may have one piston or two pistons (Figure 18-14).

The opposite-piston caliper design (Figure 18-15) is similar to a single-push caliper except that this design has two or more pushing pistons to apply braking pressure across from each other. Some designs have two, four, or even six pistons doing the same job as the single-piston caliper, but

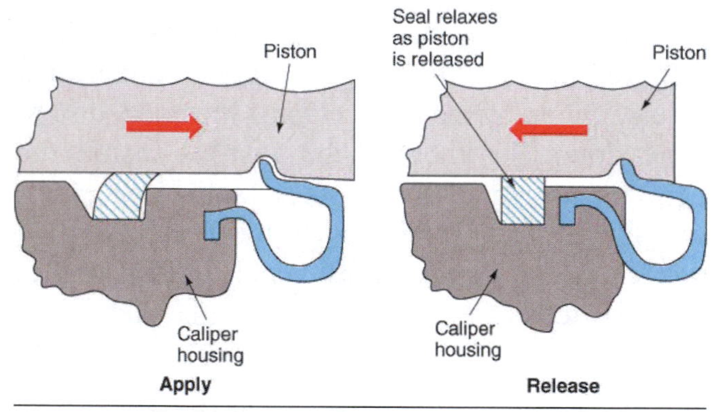

Figure 18-12 The piston caliper seal assists in moving the caliper piston away from the rotor when pressure is released.

Figure 18-13 A single-piston push caliper is shown here.

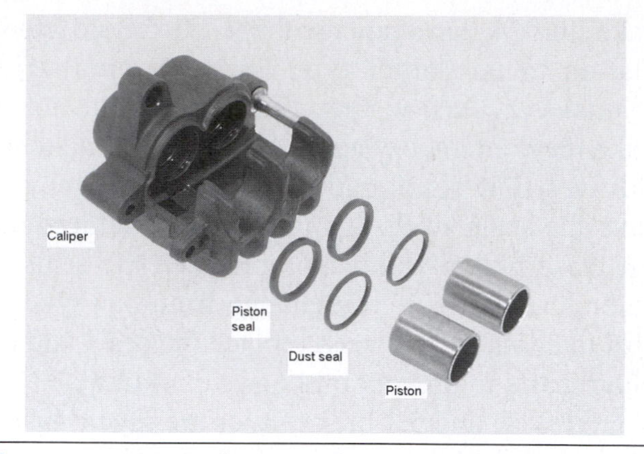

Figure 18-14 A disassembled two-piston, push-type caliper is pictured here. Copyright by American Honda Motor Co., Inc. and reprinted with permission.

Figure 18-15 An opposite-piston caliper is pictured here.

applying up to six times the force of a single-piston caliper system! With the opposite-caliper design, it is necessary to split the caliper (Figure 18-16) when rebuilding it, as the pistons face each other and would not be able to be removed if the caliper were not created in two pieces.

Brake Pads There are two basic types of brake pads used on motorcycles: organic and sintered.

- **Organic brake pads** (Figure 18-17) are constructed by mixing non-asbestos fibers, such as glass, rubber, carbon, and Kevlar®, with filler materials and high-temperature resins. The resins act like a thermo-set plastic, which holds the components together like glue. Organic-type brake pads are the most popular type of brake pad on the market and offer lower

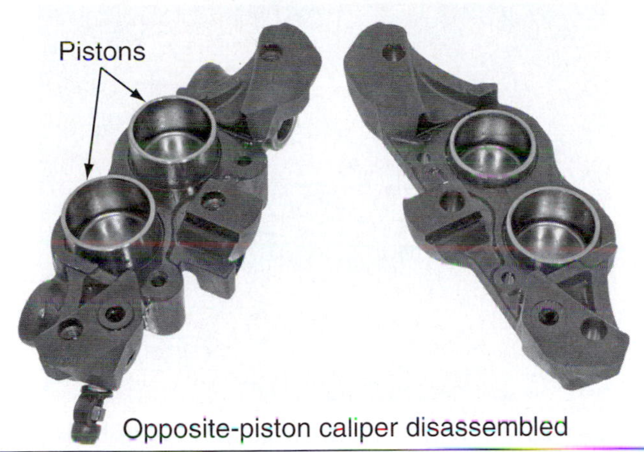

Figure 18-16 Opposite-piston calipers come apart into two pieces to allow rebuilding. Copyright by American Honda Motor Co., Inc. and reprinted with permission.

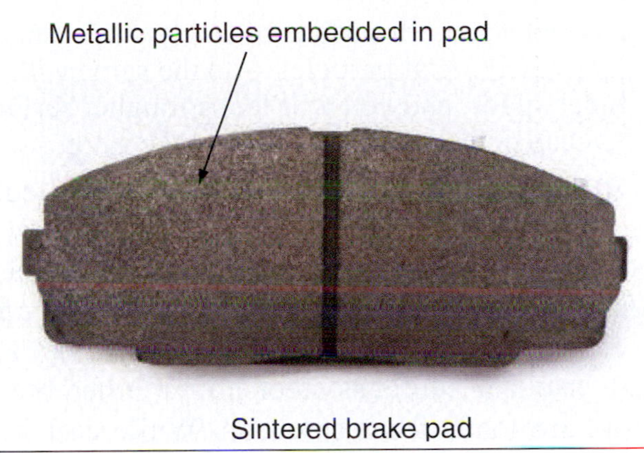

Figure 18-18 Sintered brake pads have a high content of metallic content. An example is shown here.

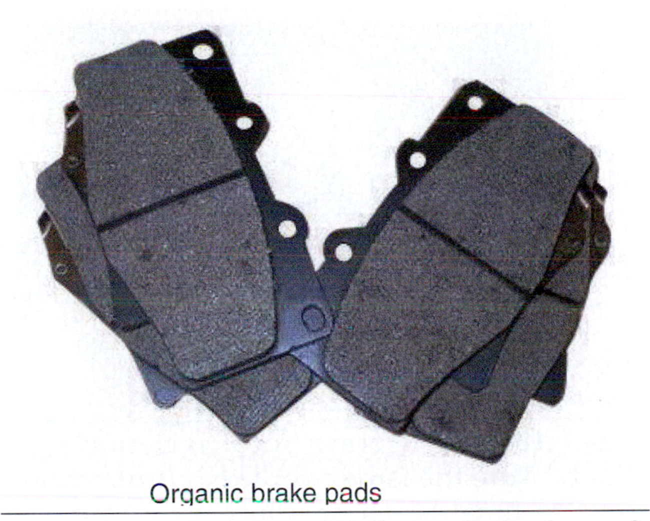

Figure 18-17 An example of organic brake pads is shown here.

rotor wear and a softer feel. Therefore, they are easier to control. They generally are made of softer compounds that create less noise but will wear fast and create more brake dust over time. They require more time to "bed in," which means fit with the rotor and are prone to glazing when overheated under hard use.

- **Sintered brake pads** (Figure 18-18) have a very high metal content and heat up quickly, which offers more friction earlier. By converting kinetic energy more quickly to heat, they are very effective in stopping a motorcycle by offering higher friction levels and maintaining these levels even in wet conditions. Sintered brake pads have a low tendency to fade when used in extreme conditions and have a shorter bed-in period. These pads have an additional defined bite point, meaning that they have a more sensitive feel when being used, sometimes considered to be "touchy." Since they run at a higher operating temperature, they can affect the brake rotor surface and are not suitable for all types of brake rotors or calipers. Finally, motorcycles that use sintered brake pads require more frequent brake fluid inspections as the heat transfer from the pads can affect the longevity of the fluid.

Generally, it is recommend that brake pads be replaced with original equipment pads from the manufacturer, but there are a large number of after-market manufacturers that offer replacement pads for motorcycles. Key things to remember when considering brake pad replacement are the following:

- Continued and controlled quality performance along the heating cycle of the pad from cold to very hot and across the various different riding conditions that the rider may encounter.
- Look for a thermal barrier between the brake pad and the caliper piston, preferably a ceramic-based material.
- Look for an advertised long life span under normal usage while delivering excellent braking qualities when required.
- When using sintered pads look for fine and evenly distributed metal particles. You can run your fingers over premium sintered brake pads

and feel a smooth surface of finely cut, evenly distributed metal particles—do the same with budget pads, and you will feel a rougher surface that can actually damage your brake rotor.

Brake Rotors A **brake rotor** (also called a disc) is attached directly to the wheel of the motorcycle and is used to slow down or stop the wheel from rotating. To help protect them from rust, most original equipment motorcycle brake rotors are made from a stainless-steel alloy. Other brake rotors are made from cast iron. While cast iron transfers heat better than stainless steel it is prone to rusting when exposed to water and therefore not often used on motorcycles. Because of the limited materials that can be used to make lightweight rotors, thermal distortion becomes a problem if the rotor is made too thin, therefore, the most common problem found with a motorcycle rotor is warpage. Rotors are measured for warpage by using a dial indicator by attaching the dial gauge and turning the wheel slowly (Figure 18-19).

Rotors are commonly drilled with holes or grooves to help remove dust or water from the rotor surface while the brakes are being applied. Rotors can be mounted solid to the wheel (Figure 18-20) or they may be mounted in a way to allow them to "float" (Figure 18-21), which helps to cool the rotor quickly as well as offers better braking under severe conditions.

Hydraulic Disc Brake Precautions

When working with hydraulic disc brakes, observe the following precautions:

Figure 18-19 A dial indicator is used to verify that the rotor is not warped as seen in this picture. Copyright by American Honda Motor Co., Inc. and reprinted with permission.

Figure 18-20 Most modern rotors have holes or grooves to help remove debris and water when the brakes are applied. This rotor is mounted solid to the wheel. Note the directional arrow. Copyright by American Honda Motor Co., Inc. and reprinted with permission.

Figure 18-21 This brake rotor is designed to "float." Note the large rivet-like button mounting attachments. These allow a slight movement in the rotor that allows full pad contact even when parts are not perfectly aligned. Copyright by American Honda Motor Co., Inc. and reprinted with permission.

- Never reuse brake fluid and do not use fluid from a container that has been left open. The fluid is hygroscopic (it absorbs moisture from the air).
- Do not mix two types of fluid (glycol-based with silicone-based) for use in brake systems.
- Do not leave the reservoir cap off for any length of time, because the fluid will absorb moisture from the air and the system may collect other contamination, such as dust and dirt.
- DOT 3 and DOT 4 brake fluid will damage painted surfaces. Wipe up any spilled fluid immediately.

- Don't use gasoline, motor oil, or any other mineral oils near disc-brake parts; these oils cause deterioration of rubber brake parts. If oil spills on any brake parts, it is difficult to wash off and will eventually react and break down the rubber.
- If any of the brake-line fittings or the bleeder valves are loosened at any time or for any reason, the air must be bled from the brake.

Hydraulic Brake Service

Hydraulic brake service includes the inspection and replacement of the brake pads as well as replacement and bleeding of the brake fluid. Follow the manufacturer's instructions carefully when doing any of these services as safety is of great concern when it comes to the brakes on a motorcycle. The most common tool used to bleed a brake system is a vacuum brake bleeder (Figure 18-22), which allows you to use vacuum to pull the brake fluid from the master cylinder to the caliper.

Antilock Brake Systems (ABS)

Antilock braking systems (ABS) prevent the wheels from locking up under excessively hard braking or when attempting to stop on slippery surfaces such as wet roads. This system controls braking torque during heavy over-operation of the brakes (such as a panic stop to avoid an accident) and helps ensure optimum tire-to-road surface traction during hard braking. ABS was originally offered on large touring motorcycles (Figure 18-23) but is now offered to virtually every type of street motorcycle including sport bikes (Figure 18-24).

While there are various types of antilock braking systems offered for motorcycles, they are all designed to perform the same basic function: prevent the wheels from locking up under extreme braking pressure.

The antilock braking system is activated only when the motorcycle rider applies the front or rear brakes hard enough to lock up the wheel. A common type of ABS operates through an ABS control

Figure 18-23 ABS for motorcycles was initially introduced on large touring-type motorcycles. Copyright by American Honda Motor Co., Inc. and reprinted with permission.

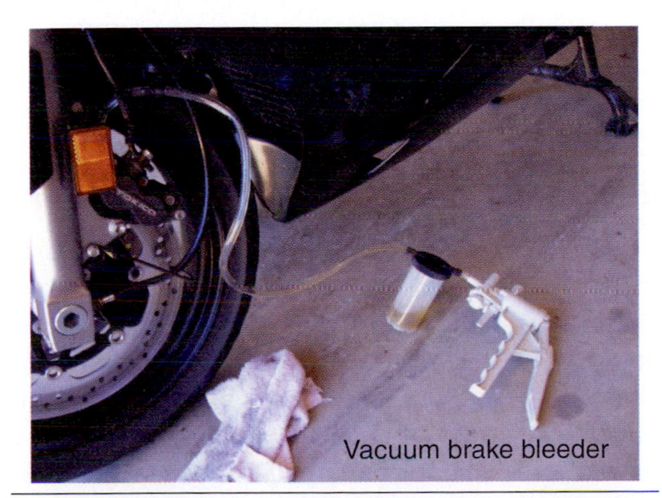

Figure 18-22 A vacuum-type brake bleeder is a very popular tool when bleeding brakes on a hydraulic brake system.

Figure 18-24 ABS can be found on all types of motorcycles today including sport bikes. Courtesy of American Suzuki Motor Corp.

unit, which in itself is a computer that measures wheel speed by using pulsers placed on the brake rotors and wheel speed sensors mounted on the chassis (Figure 18-25). The sensor information is processed through the control unit and the information is sent to modulators for both the front and rear wheel to control brake-fluid pressure accordingly when appropriate. When activated, the ABS varies the braking torque to allow the front and rear wheels a limited amount of controlled slipping. When the computer senses a lock up, it eases the brake-line pressure to allow the wheel to keep rotating while at the same time bringing the motorcycle to a safe stop.

When activated the rider can feel the ABS working through the levers as braking pressure is applied. The braking torque is altered by the use of a three-stage cut-off valve and controlled by pistons as illustrated in Figure 18-26. Stage one in the illustration shows the operation when the ABS is activated. The control valve is moved down and cuts flow to both the cut-off valve and orifice valve, which eliminates the flow of brake fluid to the brake caliper. During stage two, the control piston moves slightly upward to open the cut-off valve but leaves the orifice valve closed. This allows only a small amount of hydraulic pressure to flow from the master cylinder to the brake caliper. At stage three, the piston is returned to its normal position, which opens both the cut-off valve and the orifice valve, allowing full braking pressure to flow from the master cylinder to the brake caliper. When activated, the ABS cut-off valve control pistons move many times per second.

So, the question always arises: Does ABS stop a motorcycle faster than a non-ABS equipped machine? With a highly skilled rider at the controls, a non-ABS motorcycle can be stopped quicker on dry pavement, but in wet or gravel-covered road surfaces the rider can stay in better control with an ABS equipped motorcycle than without ABS. The true functional development of ABS was to keep the rider in control more so than to stop the motorcycle faster than a machine without ABS.

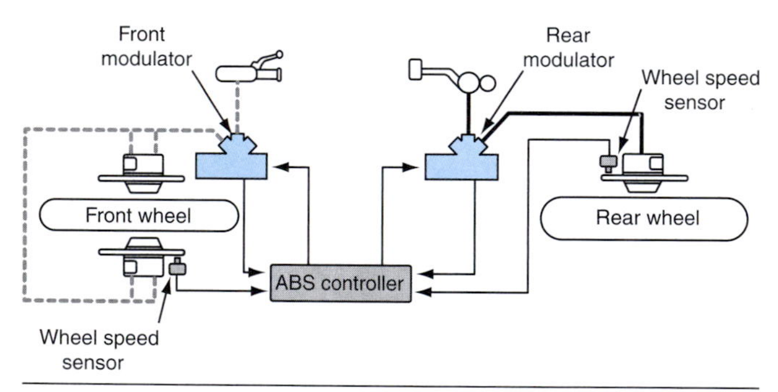

Figure 18-25 A basic ABS design is illustrated here.

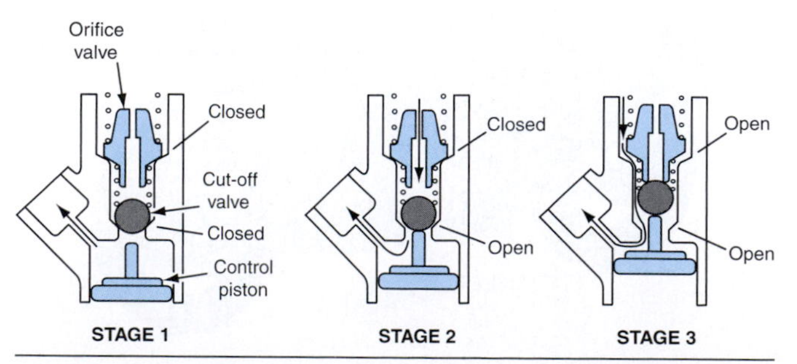

Figure 18-26 Typically, a three-stage control valve is used to vary the hydraulic pressure when ABS is activated.

Integrated Braking Systems

Integrated braking systems (IBS) also known as Linked Brake Systems (LBS) and Combined Brake Systems (CBS) offer the rider the use of both front and rear brakes when stepping on the rear brake pedal, squeezing the front brake lever, or a combination of both. It should be noted that IBS may be combined with ABS for an even wider range of braking and steering control. While IBS are virtually maintenance free, it is very important to follow the service manual procedures when it comes to bleeding the hydraulic system on motorcycles with this system.

IBS Components

The idea behind integrated braking systems is to attempt to provide an optimal balance of front and rear braking forces whenever the brake lever and/or the brake pedal is used. As mentioned above, most of these systems use hydraulics and no electronic controls. Instead of using hydraulic brakes some smaller motorcycles (generally scooters) use mechanical brakes integrated by the use of cables. We will discuss the components typically found in an IBS.

Brake Caliper Typically, an IBS uses three piston calipers (Figure 18-27) that control two independent hydraulic systems.

Generally speaking, you will find that when applying the rear brakes, the center piston of the rear caliper and center piston of the front caliper are operated directly by the brake pedal. For the front brakes, the two outer pistons of the front calipers are controlled by the front brake lever and the outer piston of the rear caliper is controlled by a secondary master cylinder that is activated by the centrifugal force of the brake application of the front brakes (Figure 18-28). This arrangement delivers a broad, yet easily controlled range of braking force, depending on the individual or both of the two (lever and pedal) brakes being engaged.

Delay Valve A delay valve (Figure 18-29) may be positioned between the rear brake master cylinder and the center piston of the front calipers on a three-disc brake system (two front calipers and one rear). The delay valve engages only one front caliper at first (Figure 18-30A), which prevents the front end from diving under rear braking. As the rear brake pedal pressure increases, the delay valve introduces pressure to the remaining front caliper to a predetermined pressure (Figure 18-30B). The resulting feel to the rider is braking on the rear wheel with very little front-end dive that is commonly felt when the front brakes are suddenly applied.

Secondary Master Cylinder Many integrated brake systems use the centrifugal force exerted on the front caliper when engaged to actuate a secondary master cylinder (Figure 18-31). This applies a

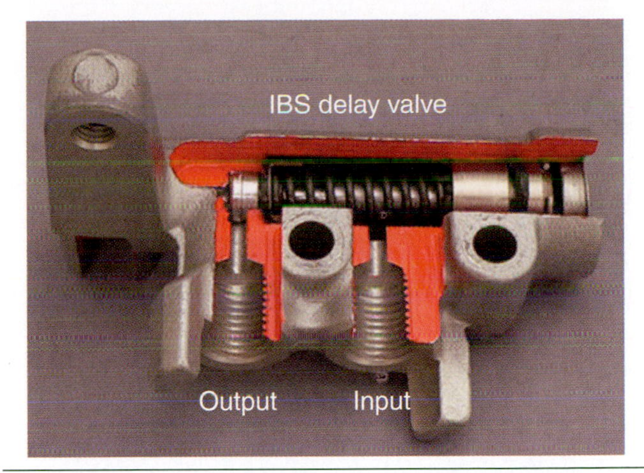

Figure 18-29 A cutaway view of a delay valve is shown here.

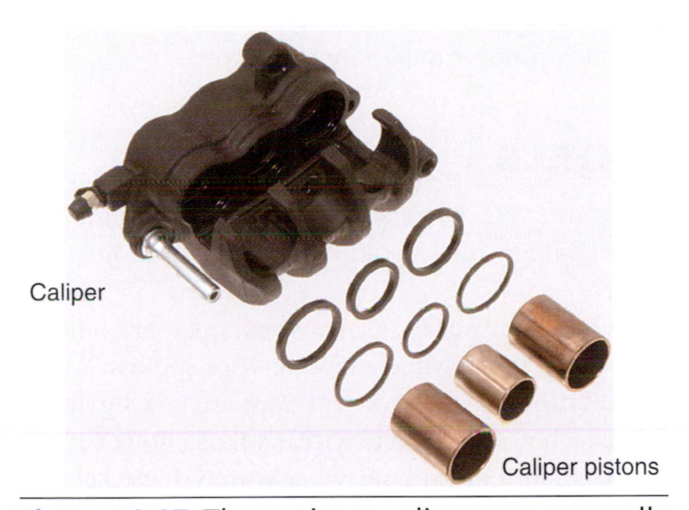

Figure 18-27 Three piston calipers are normally used with an IBS. The center piston is controlled by a separate hydraulic circuit than the outer pistons. Copyright by American Honda Motor Co., Inc. and reprinted with permission.

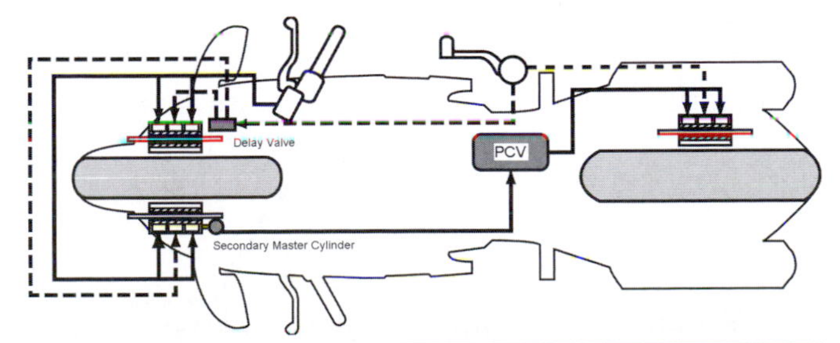

Figure 18-28 A line drawing of a typical IBS is illustrated here. The dotted lines activate the rear brake while the solid lines are for the front brake.

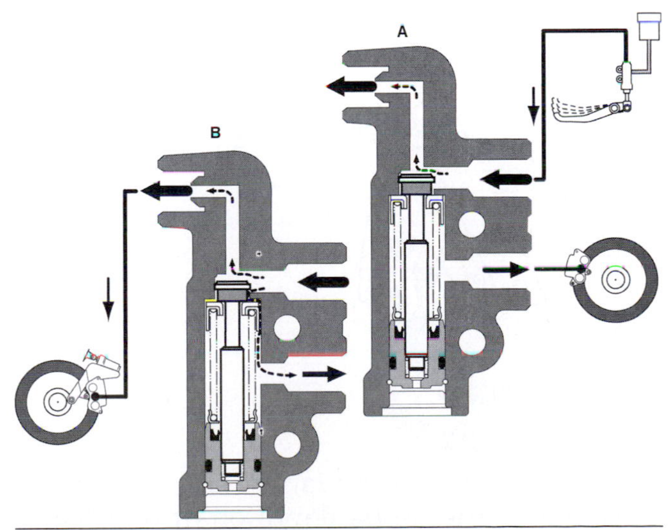

Figure 18-30 The function of the delay valve is illustrated here. Copyright by American Honda Motor Co., Inc. and reprinted with permission.

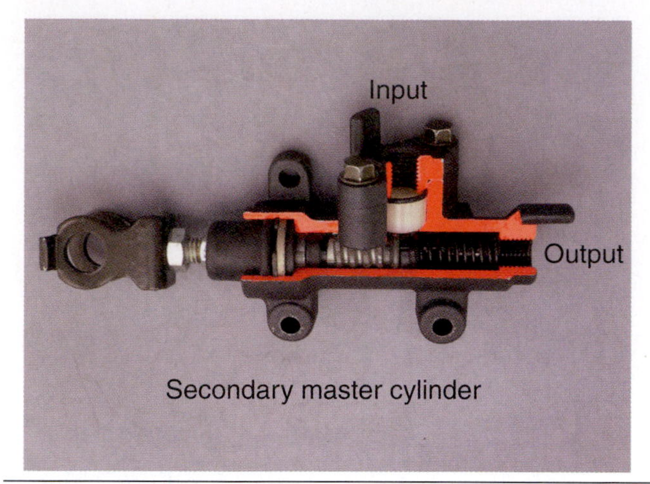

Figure 18-31 A secondary master cylinder cutaway is shown here.

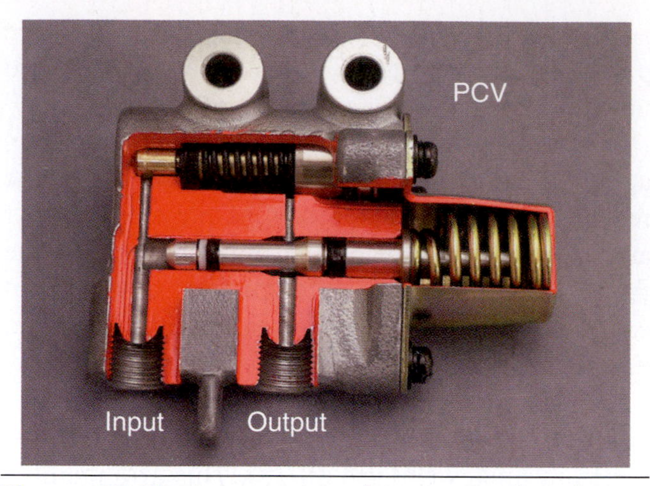

Figure 18-33 A cutaway of a proportional control valve (PCV) is shown here.

corresponding amount of pressure to the rear brake caliper through a proportional control valve (PCV) (Figure 18-32).

Proportional Control Valve (PCV) The PCV (Figure 18-33) is located between the secondary master cylinder and the rear caliper piston that it is designed to operate. It regulates pressure in stages of operation.

Initially, the PCV's output pressure increases in direct proportion to the increasing input pressure originating from the secondary master cylinder. As input pressure continues to increase, a cutout piston activates and causes the output pressure to hold. Increasing the brake input pressure forces a decompression piston down, which expands a subchamber that draws pressure off the output side of the PCV. Figure 18-34 illustrates a graphic display of how a proportional valve operates.

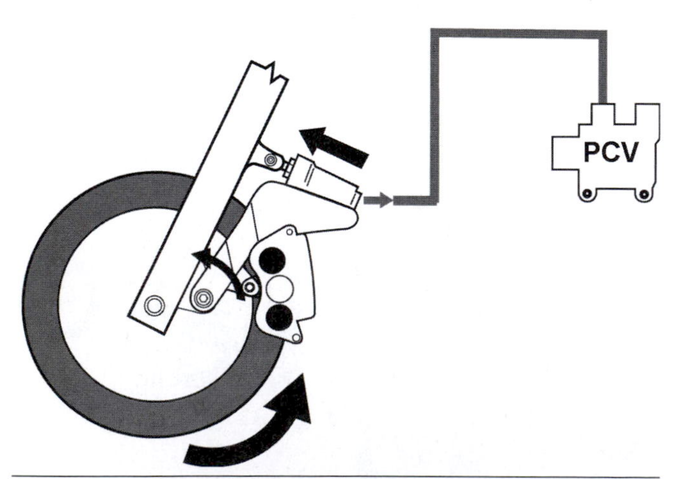

Figure 18-32 The secondary master cylinder is activated by the forward motion of the caliper when the front brake is applied. Fluid pressure goes to the proportional control valve (PCV) and then applies pressure to the rear brakes.

WHEELS

Motorcycle wheels can be separated into two basic categories: spoke wheels and nonspoke wheels.

When we refer to spoke wheels, we are talking only about those wheels having wire spokes. When we refer to non-spoke wheels, we are talking about wheels that don't have wire spokes; however, as you'll soon learn, some wheels may have several rigid metal supports between the hub and rim of

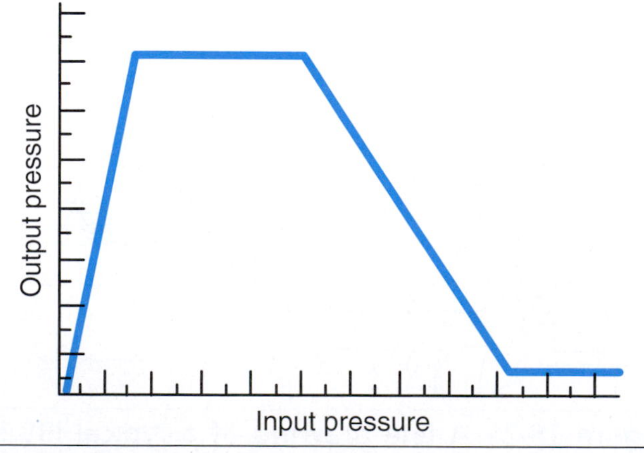

Figure 18-34 This image illustrates the characteristics of the operation of a PCV.

the wheel. These supports are sometimes referred to as spokes, although we will not consider them spoke wheels.

All wheels are designed to support the weight of the motorcycle and rider and to assist in providing the riding, braking, and steering forces. Motorcycle wheel designs are light, but strong, to ensure a safe and comfortable ride.

Spoke Wheels

Spoke-type are made with a steel or aluminum outer rim combined with strong wire spokes (Figure 18-35). Most spoke wheel assemblies are laced with a cross pattern. In a cross pattern, one spoke crosses other spokes that are going in the opposite direction from the same side of the hub and rim (Figure 18-36). Different cross patterns are determined by the number of times that one spoke crosses other spokes. The higher the cross-pattern number, the more vertical and radial strength the wheel assembly will have.

Maintaining Spoke Wheels

Spoke-type wheels require maintenance to ensure that the spokes are tight and to keep the wheels true (round). The replacement of a motorcycle wire-spoke wheel rim can be a difficult job. The job becomes much more complex because the spokes are connected to the hub flange and run in different directions. This arrangement enables the spokes to:

- Support the weight of the motorcycle
- Accept the force of acceleration
- Accept the force of braking

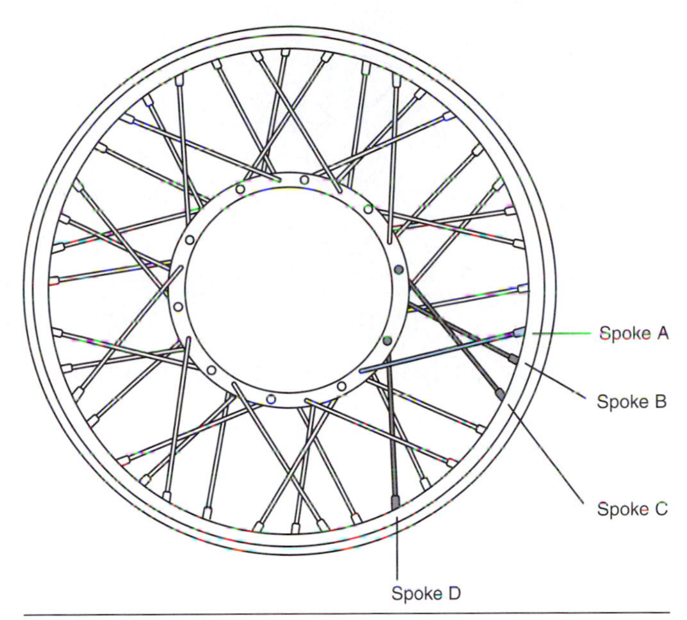

Figure 18-36 Note that spoke A crosses over three other spokes.

Figure 18-35 This motorcycle comes equipped with spoke-type wheels. Courtesy of American Suzuki Motor Corp.

Figure 18-37 shows how the different spokes support the stresses between the wheel hub and rim. The vertical spokes shown in Figure 18-37A mainly support the weight of the motorcycle and the shock received when the motorcycle hits a bump. The spokes shown in Figure 18-37B are angled to the rear of the motorcycle and take the majority of stress from acceleration. The spokes shown in Figure 18-37C are angled to the front and take the majority of stress during braking.

Spokes must be maintained to prevent the spokes from coming loose. Loose spokes allow the rim to warp and increase the possibility of breaking the spokes. Loose spokes can also cause excessive wear to the hub and spoke nipple. Broken spokes should be replaced immediately. Broken spokes increase the load on the other spokes, which eventually causes the other spokes to break as well, eventually causing wheel failure.

Spokes must be kept tight. Because each spoke will stretch a certain amount, the spokes should be tightened regularly—and evenly. Spokes that are

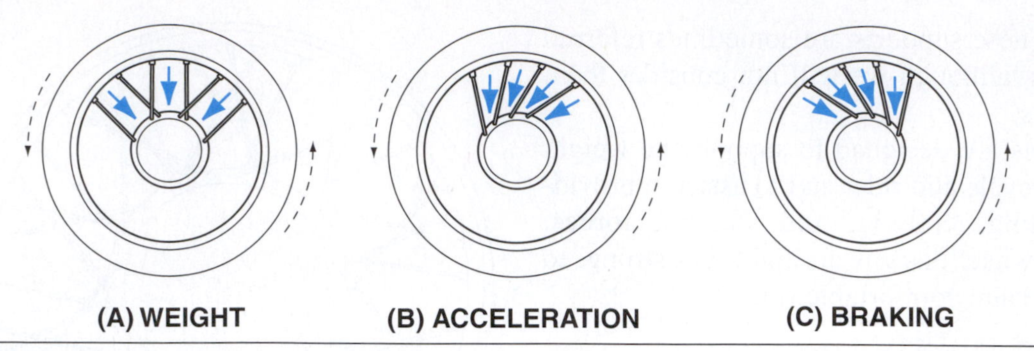

Figure 18-37 Each spoke has a job to support the different stress points between the wheel and the hub.

too tight or too loose will cause the wheel to wobble. Spokes that are too loose or too tight can cause breakage of the spoke, nipple, or the hub.

Installing a new set of spokes (also known as lacing or relacing) into a hub and rim can be very difficult for the beginner because of the pattern that must be followed when inserting the spokes. However, if done correctly, installing spokes is quite easy and you'll never bend a spoke when installing it. There are two things you must ensure before you begin installing spokes into a rim. First, be certain you have a rim with the correct number of holes. Usually motorcycles use a 36- or 40-spoke wheel. The rim must have the same number of holes as the hub. Second, be sure the angles of the rim holes for the nipples are correct. Each nipple will just fit through the rim. If the angle is too much or not enough, it will cause the spoke to bend when tightened. Each manufacturer has its own recommendations pertaining to relacing a wheel so it is highly recommended that you refer to the appropriate service manual prior to attempting to do any repairs on a spoke-type wheel.

Truing Spoke Wheels

Truing a wheel consists of tightening each spoke so the wheel rolls straight. A wheel-truing jig can be a valuable aid because it holds the wheel as you tighten the spokes. This type of jig can be purchased from motorcycle tool supply companies and also from some wholesale parts distributors. To check the trueness of a wheel, dial indicators are used to check both radial (up-and-down) and lateral (side-to-side) trueness (Figure 18-38).

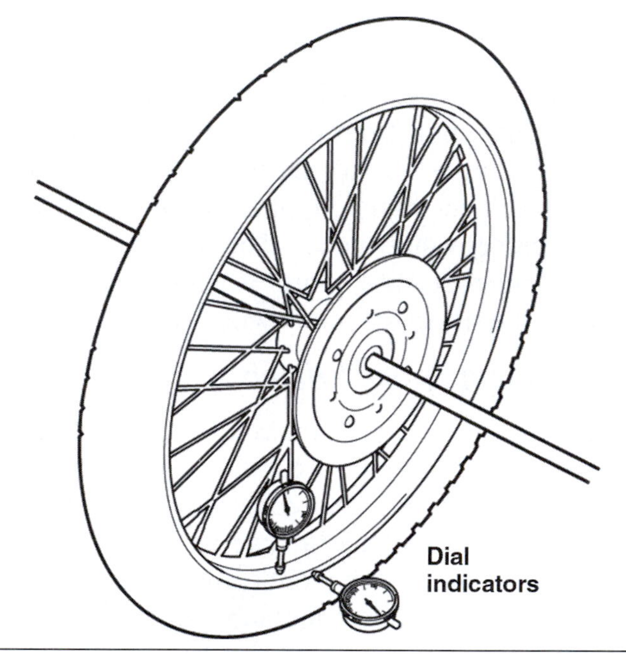

Figure 18-38 Dial indicators are used to help true a spoke-type rim.

Nonspoke Wheels

Nonspoke wheels used on modern motorcycles are generally cast from aluminum or magnesium (Figure 18-39). **Cast wheels** were first used on racing motorcycles to reduce weight and to maintain greater precision. The weight reduction was provided partly because the cast wheel had a smaller hub than was required by the spoke-type wheel. The smaller hub also simplified the mounting of disc brakes to the wheel. Today, cast wheels are designed to be used with either a tube-type or a tubeless-type tire and will have a marking on the rim stating if it is designed for one particular tire use only.

Figure 18-39 This motorcycle uses cast nonspoke wheels. Courtesy of American Suzuki Motor Corp.

Nonspoke wheels can also be spun from a solid block of aluminum (Figure 18-40).

Nonspoke wheels require very little maintenance other than checking for wheel bearing wear and occasionally checking for cracks if the motorcycle has been ridden over extremely rough roads.

Axles, Wheel Bearings, and Wheel-Bearing Seals

The wheel axle holds the wheel securely to the chassis. Whenever the wheel assembly's axle is removed, you should always check the axle's run-out using V-blocks and a dial indicator, as illustrated in Figure 18-41. Check the model-specific service manual for the actual service-limit run-out.

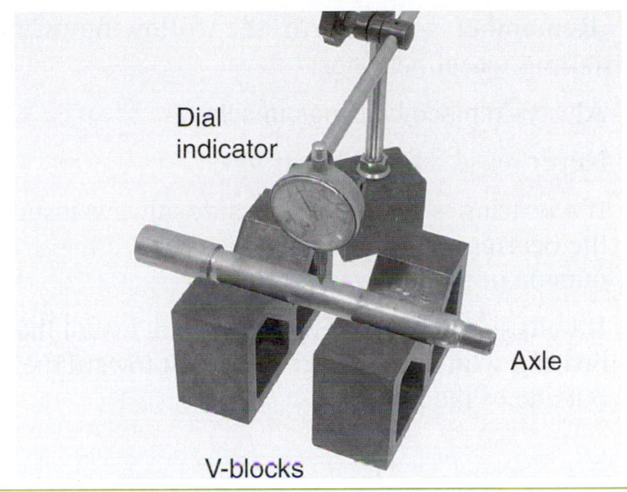

Figure 18-41 To check for axle run-out, a dial indicator and V-blocks are used. Copyright by American Honda Motor Co., Inc. and reprinted with permission.

Wheel bearings should also be checked when the wheel has been removed or when the manufacturer has stated a recommended replacement time. Replace the bearings if they show any sign of roughness or discoloration. Before replacing a wheel bearing, you must first remove the wheel-bearing dust seal. It's good practice to always replace the seal when it's been removed. Remove the bearing from the hub using a drift or manufacturer's special tools and hammer (Figure 18-42).

While tapered bearings are sometimes used on wheels, the most common wheel bearing found on motorcycles is the caged ball bearing. Always replace wheel bearings in sets.

Figure 18-40 A one-piece wheel cut from a block of aluminum is shown on the rear while a cast nonspoke wheel is on the front. Courtesy of American Suzuki Motor Corp.

Figure 18-42 A special tool to assist with wheel bearing removal is shown here. Copyright by American Honda Motor Co., Inc. and reprinted with permission.

Remember to perform the following when installing wheel bearings:

- Always replace bearings in sets.
- Never reuse removed bearings.
- If a bearing is sealed on one side, always install the bearing with its sealing face toward the outside of the hub.
- If both sides of a bearing are sealed, install the bearing with its stamped size mark toward the outside of the hub.

MOTORCYCLE TIRES

Just as on an automobile, the primary purpose of a motorcycle **tire** is to provide traction and carry the weight of the motorcycle. Tires must also withstand different types of thrust over varying speeds and surface conditions. Tires help to absorb road shock just as the suspension does. Tires must perform under a variety of conditions such as wet, dry, hot, and cold pavement as well as gravel. The motorcycle may be traveling straight up or leaned over. As a tire rolls on the road, friction is created between the tire and the surface. This friction gives a tire its traction. While good traction is needed, it must actually be limited as too much traction will limit the ability of the motorcycle to roll, therefore causing a waste of engine power and fuel.

There are two basic types of tires used on motorcycles: tube type and tubeless. Generally speaking, if a motorcycle has spoke wheels it uses a tube-type tire, while those using nonspoke wheels are generally for tubeless-type tires.

Tube-Type Tires

Tube-type tires are designed to use an inner tube to hold air in the tire. This air-filled tube is inside the tire's casing, as illustrated in Figure 18-43. If the tire gets penetrated with a nail or other sharp object, air will leak out instantly and the tire will have to be removed so the tube can be replaced. You should never repair a tube; always replace it with a new one. When fitting a new tube to a tire, be sure the tube-size and tire-size markings are the same.

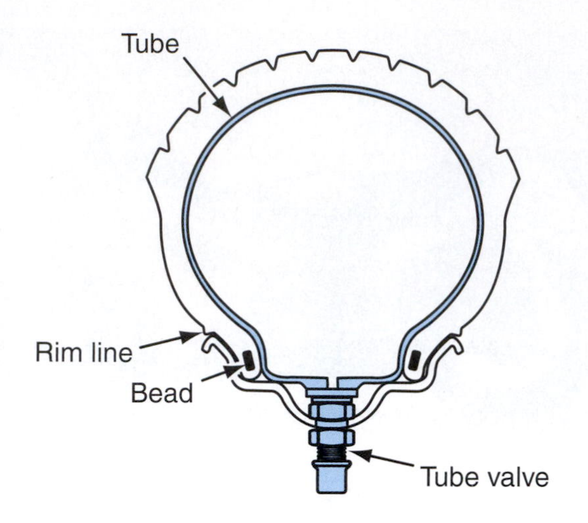

Figure 18-43 The components of a tube-type tire are illustrated here.

Tubeless Tires

Tubeless tires share most of the design features of a tube-type tire, except the tubeless tire includes an inner liner, which prevents air from filtering through the tire (Figure 18-44). This inner liner acts in place of the inner tube used in a tube-type tire. A snap-in rim valve allows for the input of air into the tire and rim.

Tire Size

Whenever you replace a tire, you must use a replacement that is the correct size. The correct tire

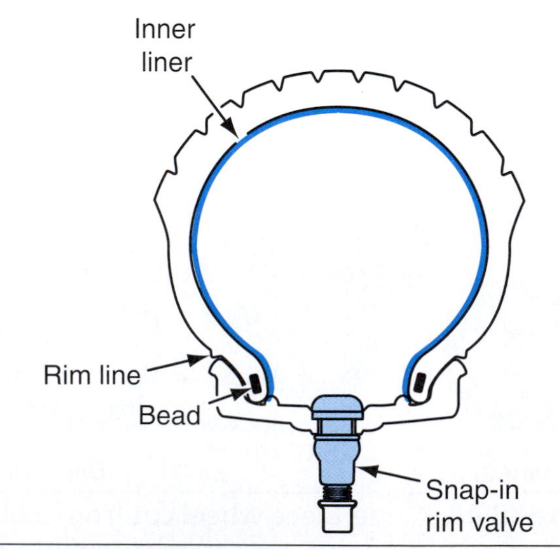

Figure 18-44 The components of a tubeless-type tire are illustrated here.

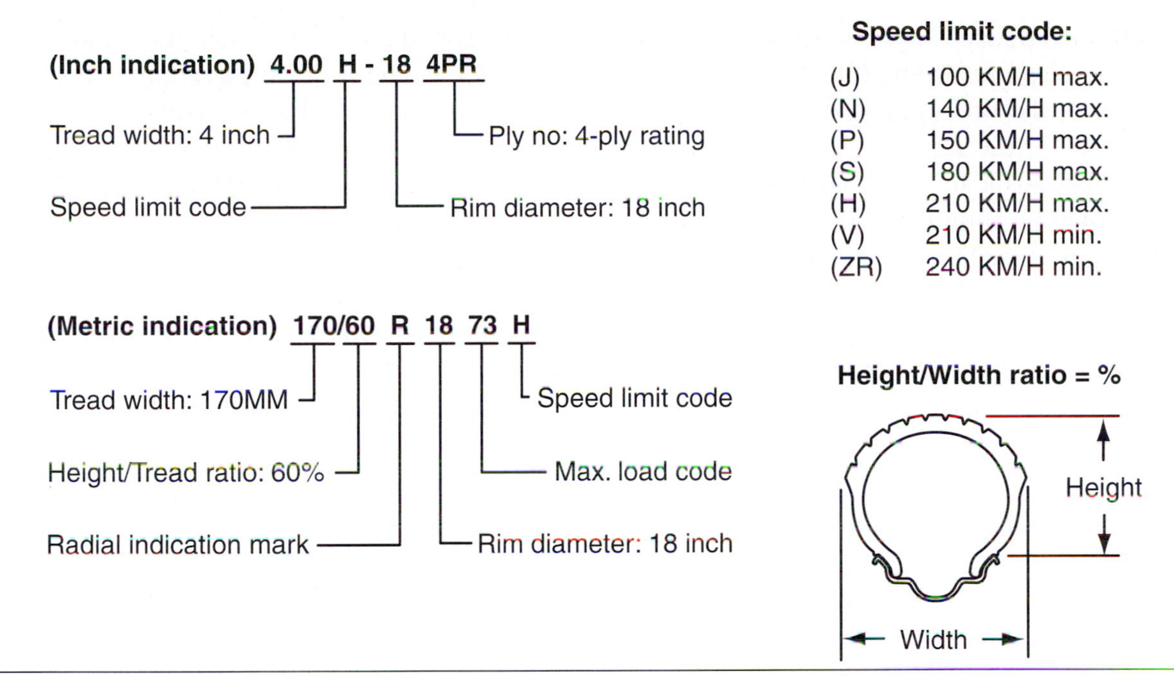

Figure 18-45 This chart helps to explain how to read tire sizing codes.

sizes are provided by each motorcycle manufacturer. Changing to a different-sized tire can cause problems. Using an oversized tire can cause the tire to make contact with the fender or forks. Also remember that the motorcycle's geometry is a critical aspect of the design. Any changes in tire size can dangerously affect handling and the stability of the motorcycle.

Tire code numbers and letters are printed on the sidewall of the tire. These code numbers and letters provide an inch or metric indication of the tire size and speed rating. Figure 18-45 shows how to interpret the tire code. When replacing a tire, refer to the technical data published by the manufacturer. Consider the tire size, motorcycle use, tire design, load-and-speed ratio, and tread pattern before choosing a tire.

Tire Construction

There are three basic types of construction on today's motorcycle tires: bias ply, belted bias ply, and radial ply (Figure 18-46). The main difference between radials and bias ply tires lies in their construction. All tires are reinforced with cords of steel or synthetic materials such as nylon or Aramid.

While radial tires perform better than bias tires, some older motorcycles can't be fitted with radial tires because of differences in rim profiles. It is recommended to stick with the type of tire that the manufacturer suggests as the motorcycle was designed with a specific type of tire.

The different ways in which radial and bias tires react to side loads can lead to unpredictable behavior while cornering. If they are mixed, the reactions can be multiplied and could cause a dangerous situation on a motorcycle.

Bias Ply Tires

The oldest design currently in use is the **bias ply tire** (Figure 18-47). It has a body of fabric plies that

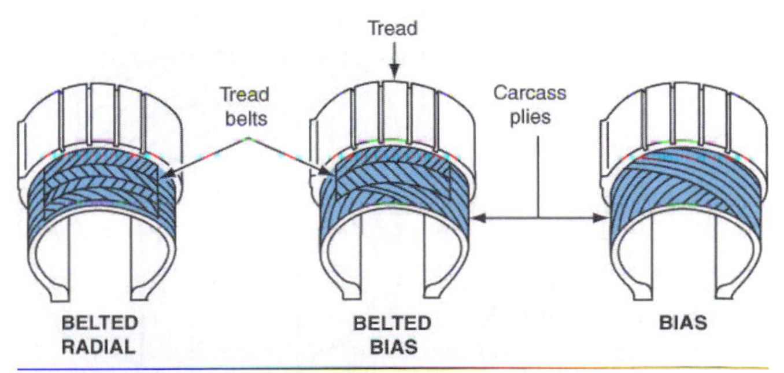

Figure 18-46 The three main types of tire construction used on motorcycles are shown here.

run alternately at opposite angles to form a crisscross design. The angle varies from 30 to 38 degrees with the centerline of the tire and has an effect on high-speed stability, ride harshness, and handling. Generally speaking, the lower the cord angle, the better the high-speed stability, but also the harsher the ride. Bias ply tires usually are available in 2- or 4-ply.

Belted Bias Ply Tires

Belted bias ply tires are similar to bias ply tires, except that two or more belts run the circumference of the tire under the tread (Figure 18-48). This construction gives strength to the sidewall and greater stability to the tread. The belts reduce tread motion during contact with the road, thus improving tread life. Plies and belts of various combinations of rayon, nylon, polyester, fiberglass, and steel are used with belted bias construction. Belted bias ply tires generally cost more than conventional bias ply tires, but last up to 40 percent longer.

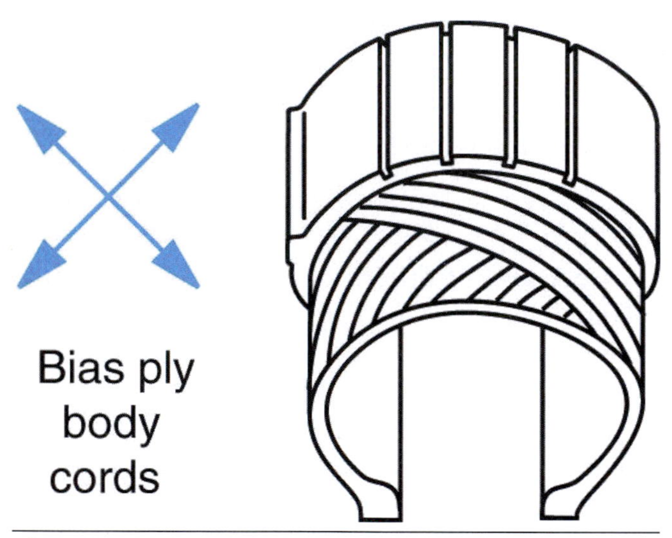

Figure 18-47 A cross-section illustration of a bias ply tire.

Radial Ply Tires

Radial ply tires have body cords that extend from bead to bead at an angle of about 90 degrees—"radial" to the circumferential centerline of the tire—plus two or more layers of relatively inflexible belts under the tread (Figure 18-49). The construction of various combinations of rayon, nylon, fiberglass, and steel gives greater strength to the tread area and flexibility to the sidewall. The belts restrict tread motion during contact with the road, thus improving tread life and traction. Radial ply tires also offer greater fuel economy, increased skid resistance, and more positive braking.

Tire Balancing

Tire balancing can be defined as the proper distribution of weight around a tire and wheel assembly to counteract centrifugal forces acting upon the heavy areas in order to maintain a true running wheel perpendicular to its rotating axis. Tire balance is important because it affects the wear of the tire and provides handling and a smooth ride. To balance a wheel, you must determine where the wheel-and-tire assembly is the heaviest and then place a small weight opposite

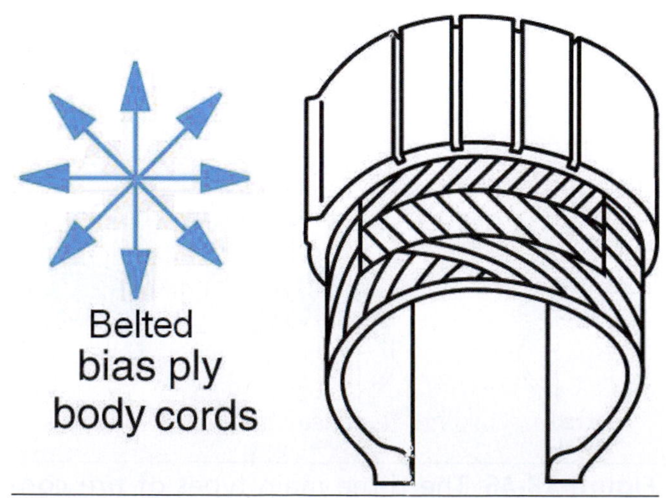

Figure 18-48 A cross-section illustration of a belted bias ply tire.

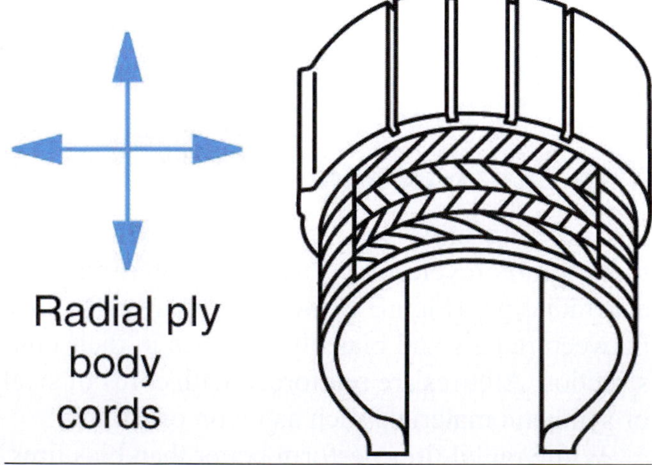

Figure 18-49 A cross-section illustration of a radial ply tire.

the heavy portion of the wheel. The weights are placed either on the spokes or on the rim of the wheel.

There are several ways to balance a motorcycle tire. The two most common ways are **static balancing** and **spin balancing**. Spin balancing requires the use of a special machine. If a wheel-balancing machine isn't available, the wheel can be balanced on the motorcycle. To balance a wheel on the motorcycle, remove the front-wheel assembly and remove the wheel's backing-plate assembly or hydraulic-caliper assembly. Reinstall the wheel on the motorcycle without the backing plate or hydraulic caliper. Spin the wheel slowly and wait for it to stop. The wheel will stop with the heaviest point facing toward the ground. Mark the top of the wheel with a piece of chalk to indicate the lightest part of the tire and wheel assembly. This is illustrated in Figure 18-50, which shows a static wheel-balancing stand.

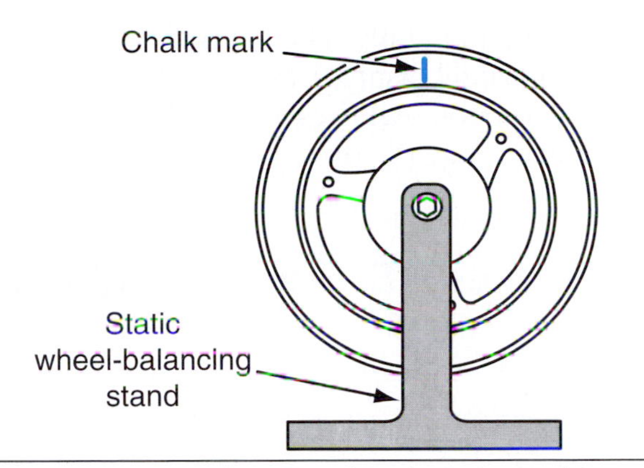

Figure 18-50 A static wheel balancer is shown here. When the wheel is at rest, the lightest point of the wheel will be at the top. A chalk mark is used to place a wheel weight at that point.

To balance the wheel, add a wheel weight to the light side of the wheel as indicated by the chalk mark. Spin the wheel again several times, each time noting where the wheel stops. When the wheel doesn't come to rest in the same spot each time it's rotated, the wheel is balanced. This job takes patience but once you've mastered it, the job will only take a few minutes. It is extremely important to remember to balance wheels whenever the tire has been separated from the rim.

Summary

- The types of brakes used on motorcycles fall into two general categories: drum brakes and disc brakes. Brakes can be actuated by either hydraulic (fluid pressure) or mechanical (cable or linkage) mechanisms.
- Antilock braking systems (ABS) are designed to prevent the wheels from locking up under excessively hard braking or when attempting to stop on slippery surfaces such as wet roads. This system controls braking torque during heavy over-operation of the brakes (such as a panic stop to avoid an accident) and helps ensure optimum tire-to-road surface traction during hard braking. While there are various types of antilock braking systems offered for motorcycles they are all designed to perform the same basic function: prevent the wheels from locking up under extreme braking pressure. The antilock braking system is activated only when the motorcycle rider applies the front or rear brakes hard enough to lock up the wheel.
- Integrated braking systems (IBS) offer the motorcycle rider the use of both front and rear brakes when stepping on the rear brake pedal, squeezing the front brake lever or a combination of both. The idea behind integrated braking systems is to attempt to provide an optimal balance of front and rear braking forces whenever the brake lever and/or the brake pedal is used.
- Tube-type tires are designed to use an inner tube to hold air in the tire. This air-filled tube is inside the tire's casing. If the tire gets penetrated with a nail or other sharp object, air will leak out instantly and the tire will have to be removed so the tube can be replaced. Tubeless tires share most of the design features of a tube-type tire, except the tubeless tire includes an inner liner, which prevents air from filtering through the tire. This inner liner replaces the inner tube used in a tube-type tire.
- There are three basic types of construction on today's motorcycle tires: bias ply, belted bias

ply, and radial ply. The main difference between radials and biasply tires lies in their construction. All tires are reinforced with cords of steel or synthetic materials such as nylon or Aramid. While radial tires perform better than bias tires, some older motorcycles cannot be fitted with radial tires because of differences in rim profiles.

Chapter 18 Review Questions

1. Which brake fluid will not harm plastic surfaces?
   a. DOT 2
   b. DOT 3
   c. DOT 4
   d. DOT 5
2. The two basic wheel categories are ________ and ________.
3. The two most common ways to balance a motorcycle tire are spin balancing and
   a. air balancing.
   b. static balancing.
   c. fast balancing.
   d. water balancing.
4. The oldest tire design currently in use is the
   a. bias ply tire.
   b. belted bias ply tire.
   c. radial tire.
   d. bias radial tire.
5. Why is it important to only use brake fluid that has come from a sealed container?
   a. It's more cost effective.
   b. So it doesn't lose color.
   c. It's stronger.
   d. To avoid contamination.
6. DOT 3 and DOT 4 brake fluid will damage painted surfaces. True or False.
7. In its most basic form, how does an antilock braking system operate?
   a. It controls brake fluid return.
   b. It controls brake fluid exit.
   c. It controls brake fluid pressure.
   d. It controls brake fluid color.
8. ________ brakes were first used on racing machines to reduce weight and to maintain greater precision.
9. What is the most common type of wheel bearing?
   a. Caged ball
   b. Tapered needle
   c. Needle
   d. Plain roller
10. IBS and ABS are systems that perform exactly the same way. True or False.
11. What does a tubeless tire use in its design features that prevents air from filtering through the tire?
    a. Inner liner
    b. Inner plate
    c. Inner tube
    d. Inner strap
12. What type of brake arrangement uses a cam at both ends of the shoe?
    a. Single leading shoe
    b. Hydraulic
    c. Double leading shoe
    d. Trailing shoe
13. A ________ is used to check the side-to-side run-out on a spoke wheel.
14. Pistons inside the brake caliper are in direct contact with ________, which press against the brake rotor.
15. IBS may be combined with ABS. True or False.

CHAPTER

19 Motorcycle Maintenance and Emission Controls

Learning Objectives

When students have completed the study of this chapter and its laboratory activities they should be able to:

- Understand the importance of performing various engine and chassis maintenance procedures
- Describe the correct procedures for storing a motorcycle
- Identify the different emission-control systems used on street-legal motorcycles

Key Terms

Air filter
Battery acid
Battery conductive analyzer
Battery load tester
Belt-driven final drive
Cable maintenance
California Air Resources Board (CARB)
Carbon dioxide
Carburetor synchronization
Chain-driven final drive
Compression test
Conventional batteries
Cupping
Electronic ignition systems
Emission-control system
Environmental Protection Agency (EPA)
Hydrocarbons
Leak-down tests
Maintenance-free batteries
Maintenance intervals
Mechanical seal
Oil changes
Oil filters
Oxides of nitrogen
Power valves
Scheduled maintenance
Shaft drives
Spark plug
Timing light
Tire air pressure
Tune-up
Vacuum gauges
Valve clearance
Water

INTRODUCTION

In this chapter, you'll learn the importance of scheduled maintenance intervals as well as engine and chassis maintenance procedures. Included in this chapter is an explanation of motorcycle emission-control systems and their function.

What are the benefits to having motorcycles maintained and serviced on a regular basis? Motorcycle technicians are frequently asked this question. Peak performance from a motorcycle requires that each part be in good working condition and correctly adjusted. An experienced technician knows that if a part isn't functioning correctly, it can affect the performance of other related parts and the performance of the entire machine. For example, a spark plug that doesn't fire when it should affects the power output of the engine. Spark plug failures can be caused by an electrical system malfunction or by fuel system problems. Scheduled servicing ensures that marginal parts and out-of-tolerance adjustments are routinely corrected.

Many service shops use the words "tune-up" when referring to particular service procedures on motorcycles. You will find the word **tune-up** means different things to different people. For instance, some motorcycle owners assume that certain items will be serviced during a routine tune-up, when in fact a tune-up may not include that level of service. A fairly common customer misconception is that a tune-up automatically includes carburetor overhaul service. Some shops may include this in a tune-up, but most do not. Another example of tune-up variations is that some service shops will adjust the valves in a four-stroke engine during a tune-up while other service shops do not include this level of work in a tune-up. Most shops consider this level of repair to be a separate service. The differences in what different shops provide during a tune-up are one of the main reasons that you'll find a wide range of prices for them.

With today's modern motorcycles the focus should be pointed more toward **scheduled maintenance** instead of yearly "tune-ups." This is due to the higher quality of materials used on today's motorcycles as well as the technology that is available preventing the need for an annual service.

There's an important point that we want to emphasize. You, as a motorcycle technician, should encourage owners and riders to adopt a policy of routine periodic service inspections for their vehicles. Scheduled quality maintenance is an important element in a program to ensure trouble-free operation of the vehicle, as is a complete pre-ride inspection each time the vehicle is used.

MAINTENANCE INTERVALS

All manufacturers recommend that motorcycles be serviced at specific mileage or time intervals. The suggested **maintenance intervals** are listed in the owner's manual to help remind owners, and in the service manual to help technicians set up a realistic and appropriate maintenance schedule. To set an example of what was previously mentioned, many motorcycle owners have a vehicle "tune-up" every spring, even though it may not be necessary. For example, if a motorcycle owner had a full scheduled maintenance completed last spring with 8,000 miles on the machine and rode only 1,000 miles over the summer, the motorcycle will not need a full maintenance this spring.

To ensure maximum performance, a technician should check parts for wear and perform needed adjustments during specific service interval inspections. This allows technicians to alert owners if minor or even major repair work is required. If caught in time this process can potentially eliminate many serious mechanical failures, benefiting both the owner and the technician.

In a service environment, it will be important for you to convey to vehicle owners the results and findings of a routine service inspection in a way that generates confidence in your ability. If, for example, during the inspection, you find a defective part, you should be able to provide the owner with an accurate assessment of the problem and an estimate of what it will cost to correct.

The following two tables are examples of typical engine and chassis maintenance intervals as they would appear in a service or owner's manual. Table 19-1 pertains to an ATV that has no method of recording mileage so time frames are utilized, whereas Table 19-2 refers to a motorcycle that comes equipped with a speedometer. You should note that the maintenance schedules you'll see in many manuals are based upon average riding conditions. Machines that are subjected to severe usage would require more frequent servicing than what is listed in the tables found in the service or owner's manual. This will be mentioned in each manual.

As you can see by these suggested maintenance intervals, different adjustments are recommended at different times. Always keep in mind that if a machine is being used under harsh conditions, such as extreme heat or dusty conditions, you will need to perform certain maintenance procedures more frequently.

Table 19-1: PERIODIC MAINTENANCE FOUR-STROKE ATV

Item	Interval		
	Initial 1 month	Every 3 months	Every 6 months
Cylinder Head & Exhaust Pipe Nuts	Tighten	Tighten	
Valve Clearance	Inspect, Adjust, Clean	Inspect, Adjust, Clean	
Spark Plug (Replace every 12 months)		Inspect, Adjust, Clean	
Air Cleaner		Clean	
Engine Oil & Oil Filter	Replace	Replace	
Clutch	Inspect, Adjust, Clean		Inspect, Adjust, Clean
Carburetor	Inspect, Adjust, Clean		Inspect, Adjust, Clean
Fuel Line (Replace every 4 years)	Inspect, Adjust, Clean		
Spark Arrester			Clean
Drive Chain	Inspect, Adjust, Clean	Inspect, Adjust, Clean	
Sprockets	Inspect, Adjust, Clean	Inspect, Adjust, Clean	
Brakes	Inspect, Adjust, Clean	Inspect, Adjust, Clean	
Brake Fluid (Replace every 2 years)	Inspect, Adjust, Clean	Inspect, Adjust, Clean	
Brake Hose (Replace every 4 years)		Inspect, Adjust, Clean	
Tires	Always inspect before riding	Always inspect before riding	Always inspect before riding
Steering	Inspect, Adjust, Clean	Inspect, Adjust, Clean	
Chassis Nuts & Bolts	Tighten	Tighten	
General Lubrication		Lubricate	

Table 19-2: PERIODIC MAINTENANCE
FOUR-STROKE MULTI-CYLINDER STREET MOTORCYCLE

Item	Interval (miles)					
	600	4,000	8,000	12,000	16,000	20,000
Fuel Line		IAC		IAC		IAC
Throttle Operation		IAC		IAC		IAC
Carburetor Choke		IAC		IAC		IAC
Air Cleaner			Replace			Replace
Spark Plug	IAC	Replace	IAC	Replace	IAC	Replace
Valve Clearance				IAC		
Engine Oil		Replace		Replace		Replace
Engine Oil Filter		Replace		Replace		Replace
Carburetor Synchronization		IAC		IAC		IAC
Carburetor Idle Speed	IAC	IAC	IAC	IAC	IAC	IAC
Radiator Coolant		IAC		IAC		Replace
Cooling System		IAC		IAC		IAC
Secondary Air Supply System		IAC		IAC		IAC
Evaporative Emission-Control System			IAC		IAC	
Brake Shoe/ Pad Wear	IAC	IAC	IAC	IAC	IAC	IAC
Brake System		IAC		IAC		IAC
Brake Light Switch		IAC		IAC		IAC
Headlight Aim		IAC		IAC		IAC
Clutch System	IAC	IAC	IAC	IAC	IAC	IAC
Side Strand		IAC		IAC		IAC
Suspension		IAC		IAC		IAC
Nuts, Bolts, Fasteners		IAC		IAC		IAC
Wheels/Tires		IAC		IAC		IAC
Steering-Head Bearings		IAC		IAC		IAC

Note: IAC means Inspect, Adjust, and Clean

Maintenance Process Order

Many of the adjustments and repairs related to motorcycle and ATV maintenance can be completed in any order the technician chooses. However, you will find that a standard routine, applicable to servicing every type of vehicle, will help you do the job more efficiently.

All technicians will adopt and practice a routine or system that works best for them. We suggest that you adopt a structured approach that fits your unique requirements. The important thing is to get into a fixed routine so that your service technique is not "hit or miss," but instead structured and complete.

MOTORCYCLE ENGINE MAINTENANCE

As a motorcycle technician, you'll need to perform engine service and vehicle maintenance including cleaning, adjusting, and replacing parts. This will ensure that parts meet the manufacturer's specifications. The parts and systems listed below are subject to dirt, wear, and/or vibration, and should be carefully inspected (and replaced or adjusted as necessary) when you're doing an engine service and vehicle maintenance. In addition, certain tests may need to be performed, such as compression and leak-down tests, to ensure that the engine internals are operating properly.

- Oil and oil filter replacement
- Cooling system and coolant inspection
- Valve inspection/adjustment (four-stroke engines)
- Power valve (two-stroke engines) inspection/adjustment
- Clutch inspection/adjustment
- Spark plug inspection/replacement
- Battery inspection
- Ignition system testing
- Fuel system inspection/adjustment
- Air filter inspection/cleaning

Good Maintenance Starts with a Clean Machine

The first step in performing quality maintenance on any motorcycle is to start with a clean vehicle. It's very important that dirt or other foreign material not contaminate the internal working parts of the engine. Use soap and water or a commercially available degreaser to clean the exterior of the machine. Clean motorcycles are also easier to work on and look better when you have cleaned them after you finish the maintenance work. Always wipe down the external parts of the motorcycle after you have worked on it to remove any fingerprints that may have been left behind. You can be assured that customers will notice and appreciate the fact that they receive their motorcycle back after a service has been performed cleaner than when it came into the service shop.

Oil and Oil Filter Inspection and Replacement

Engine and transmission **oil changes** are among the most frequent services done at any motorcycle repair facility. Both two-stroke and four-stroke engine machines fall under this category.

Two-Stroke

The only maintenance required with the two-stroke upper engine lubrication system is to be certain that there's a correct ratio of oil to fuel with the premix-type lubrication system. For an oil-injection lubrication system, you must verify that there is an adequate supply of two-stroke oil in the oil reservoir.

The two-stroke transmission and clutch have a separate oil drain plug that's normally found on the bottom of the transmission (Figure 19-1). An oil-level check bolt is often used to verify the proper oil level in many two-stroke transmissions. It's necessary to drain and replace this oil on a regular basis as described in the appropriate owner's or service manual. Neglecting to service the transmission can result in premature transmission and bearing failures.

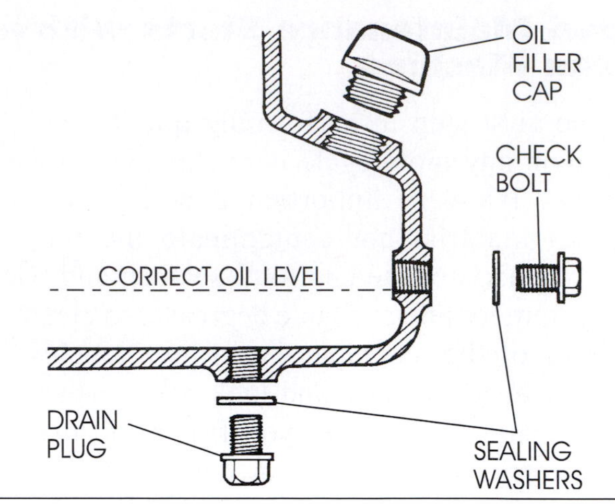

Figure 19-1 Two-stroke engines will often have an oil-level check bolt to ensure the correct amount of oil is in the crankcase. Oil will drip out of the check bolt hole when the level is correct. The drain plug will typically be on the bottom of the crankcase.

Four-Stroke

Most four-stroke motorcycle engines use the same oil to lubricate all of the engine's internal components (Figure 19-2). These engines usually have only one drain plug for removing the oil from the engine's crankcase (Figure 19-3). This is not always the case, so be sure to verify before changing the oil and filter by reviewing the manufacturer's service manual. For example, Harley Davidson, Moto-Guzzi, and BMW, as well as some Japanese models such as the Honda CRF450R, have more than one cavity to store oil for the engine.

In most cases, there's also an oil filter that should be cleaned or replaced whenever the engine oil is changed. Paying close attention to the oil filter is just as important as changing the engine oil. The filter traps and contains most of the dirt and contaminants that the engine has released into the engine oil. If the engine oil is replaced without replacing the oil filter, the engine will filter new oil through a dirty filter, which may contaminate the oil and cause it to lose its effectiveness much sooner than it would if a new filter were installed.

Four-Stroke Oil Filters The three most common types of motorcycle **oil filters** are the cartridge-type paper filter (Figure 19-4), the element-type paper filter (Figure 19-5), and the centrifugal-type oil filter (Figure 19-6). Both paper filter types have an oil pressure bypass valve to allow oil to pass through the filter even if the filter itself is so dirty that oil cannot pass through it. The theory behind the use of a bypass valve is that dirty oil is better than no oil at all!

The cartridge-type filter is removed and installed by using an oil filter wrench. The wrench is designed to grasp the outside of this type of filter as shown in Figure 19-7.

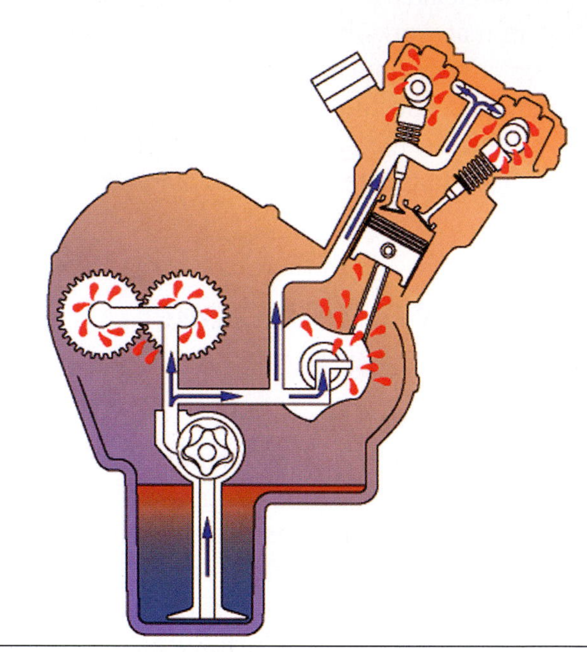

Figure 19-2 The oil in the crankcase of a four-stroke engine will typically lubricate all of the engine's internal components as shown here. Copyright by American Honda Motor Co., Inc. and reprinted with permission.

Figure 19-3 The drain plug for a four-stroke engine is normally at the bottom of the engine crankcases as pictured here. Copyright by American Honda Motor Co., Inc. and reprinted with permission.

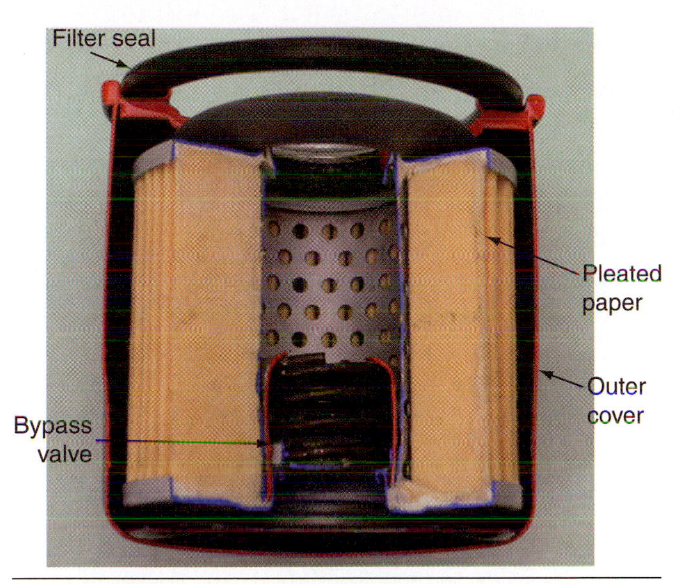

Figure 19-4 A cutaway view of a cartridge-type oil filter.

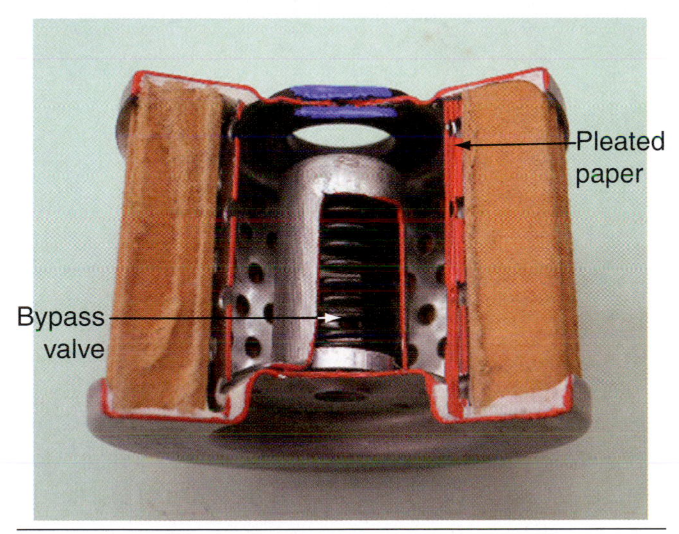

Figure 19-5 A cutaway view of an element-type paper oil filter.

The element-type filter is removed by first removing the filter cover (Figure 19-8) and then removing the oil filter from the filter cover correctly. Reference the service manual for the specific model you are working on for the proper assembly procedure. It's critically important that you correctly install the new oil filter, springs, washers, and seals (Figure 19-9). If installed incorrectly, this type of oil filter will not provide proper engine oil filtering. In a worst-case scenario, the filter could prevent the oil from reaching the engine components, which would cause major engine damage.

Figure 19-6 A centrifugal-type oil filter will be mounted on the crankshaft. To clean this type of filter, you will typically have to remove the clutch cover. Copyright by American Honda Motor Co., Inc. and reprinted with permission.

Figure 19-7 A typical oil filter wrench can be seen here removing a cartridge-type filter. Copyright by American Honda Motor Co., Inc. and reprinted with permission.

Figure 19-8 An element-type paper filter is held in place by a filter cover as seen here.

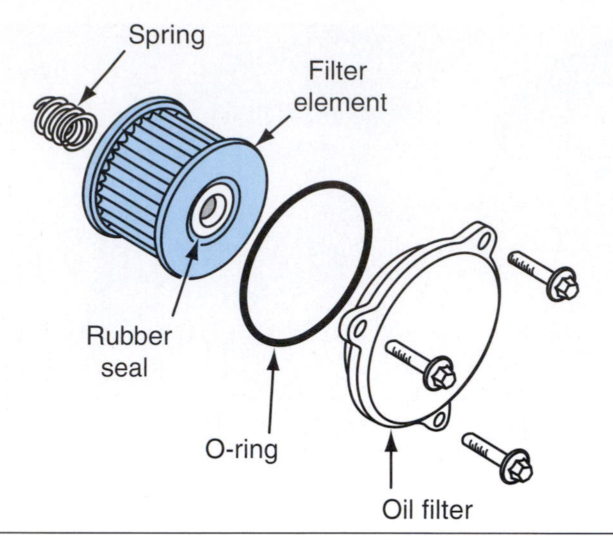

Figure 19-9 It is important that the element-type paper filter be installed correctly and in the proper order to function properly.

Many small engine motorcycles use a centrifugal oil filter that is attached to the engine's crankshaft and uses centrifugal force to keep the oil clean. Particles heavier than the oil, like tiny pieces of metal, are separated by centrifugal force and collect on the outer side of the chamber, causing sludge to form. To clean this type of filter you will likely have to remove an engine side cover (normally the clutch cover) and take the filter apart. You clean this type of filter by removing the built-up sludge (Figure 19-10) and then reassemble the filter unit.

Figure 19-10 Cleaning the inside of a centrifugal oil filter. Copyright by American Honda Motor Co., Inc. and reprinted with permission.

Checking the Oil Level

When you're checking the oil level of a motorcycle engine, be sure the motorcycle is in an upright position, on level ground, and not parked on the side stand. If the engine oil is checked while the motorcycle is on the side stand, the oil may appear to be low or high because of the angle of the motorcycle. When you are checking the oil level of an ATV engine, ensure that the machine is parked on level ground.

Generally speaking there are two methods of checking the oil level on a motorcycle. A window on the side of the engine crankcases (Figure 19-11)

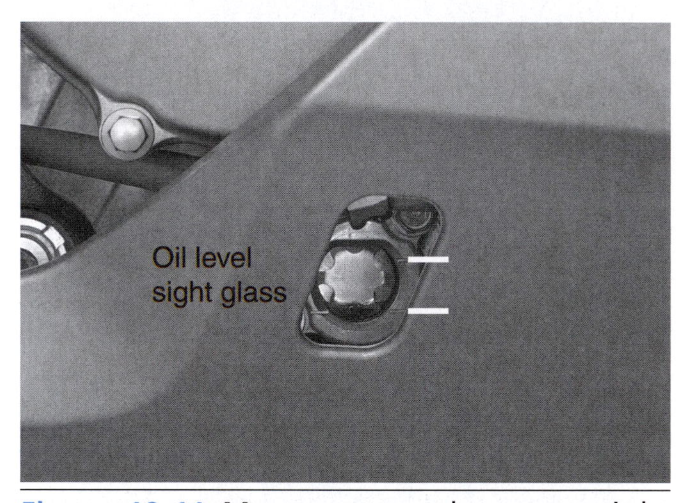

Figure 19-11 Many motorcycles use a sight window to check the oil level. The oil should be between the lines to ensure that the correct amount is in the engine. Copyright by American Honda Motor Co., Inc. and reprinted with permission.

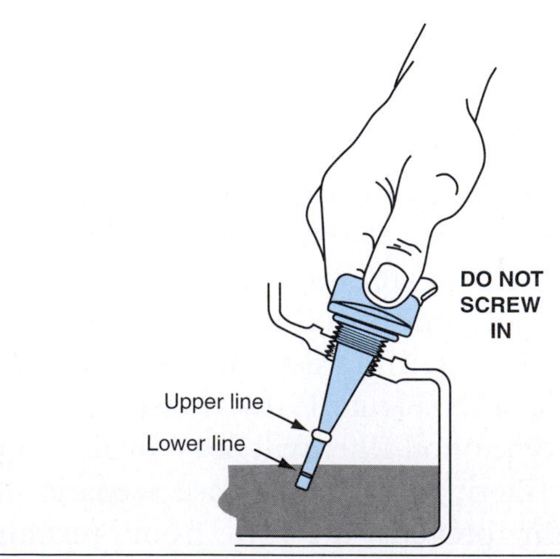

Figure 19-12 The basic way to check the oil level in a motorcycle is with a dipstick as seen here.

allows the inspection of the level by sight. The other and more popular method of checking the oil level is by use of a dipstick (Figure 19-12). This method is similar to an automobile but in most cases the dipstick will have a **screw attachment**. If you were to check the oil level incorrectly with this type of dipstick, the oil level may appear to be too high or too low.

Cooling System Inspection

Liquid-cooled motorcycle engines can be pressure-tested for leaks with a cooling system pressure tester (Figure 19-13) to ensure that the system holds a test pressure for a specified length of time.

Each manufacturer provides specifications for cooling system pressure capabilities. If the system fails the pressure test, you should check hoses, pipe connections, the radiator, and the water pump for leaks. The water pump has a **mechanical seal** to separate the engine oil and the coolant. A mechanical seal is a spring-loaded seal consisting of several parts to seal the rotating impeller in the pump case and prevents water from leaking into and damaging the engine. This seal is a common cause of failures in a liquid-cooled system. The majority of water pumps have a telltale hole on the bottom of the pump. Coolant will leak out of this hole when the mechanical seal has failed, which will require that the pump be replaced. When checking the cooling system, the coolant should be checked with a hydrometer (Figure 19-14) to verify that it has a correct mixture of coolant and distilled water (Figure 19-15).

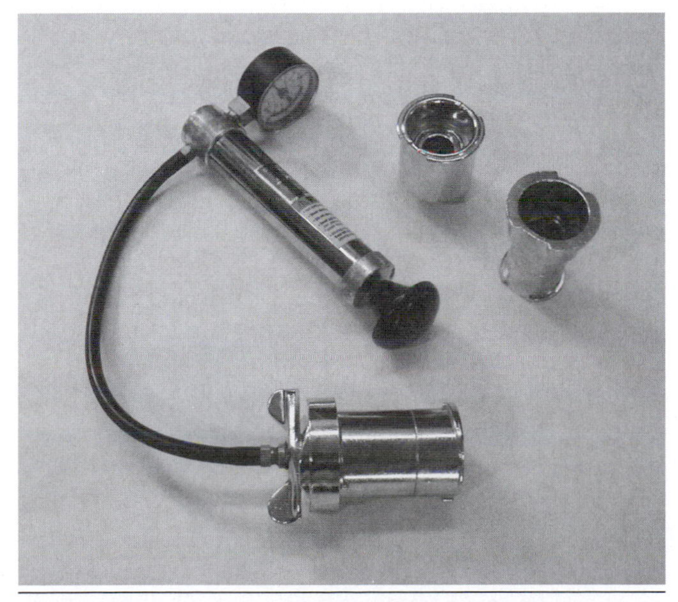

Figure 19-13 A typical cooling system pressure tester is shown here.

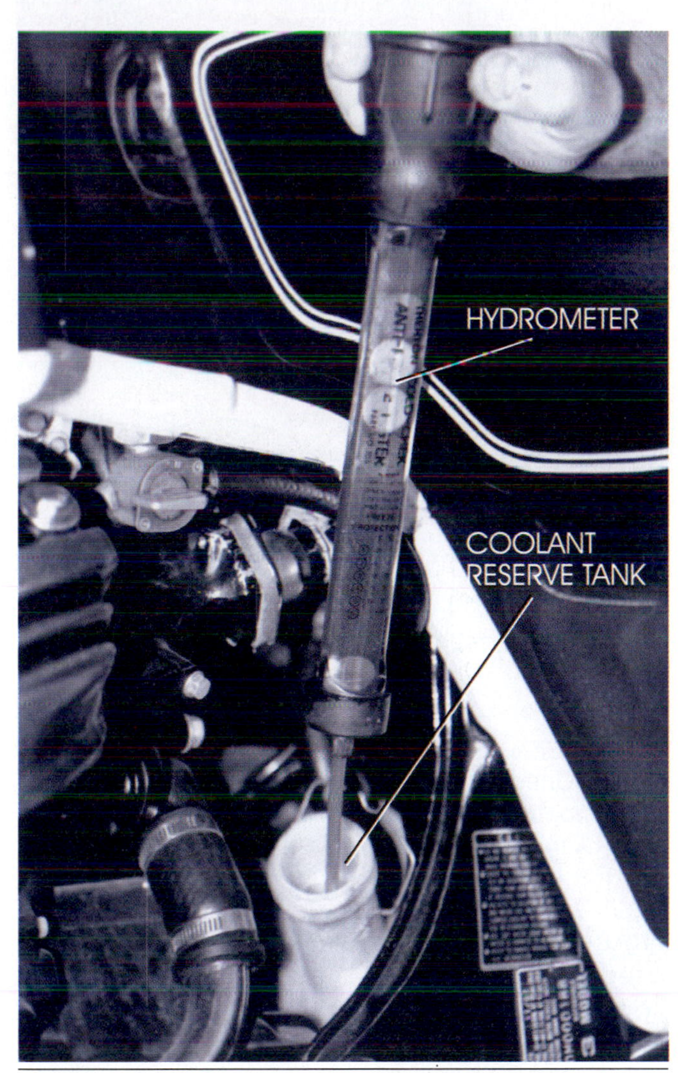

Figure 19-14 A hydrometer checks to ensure that the mixture of coolant and water is correct.

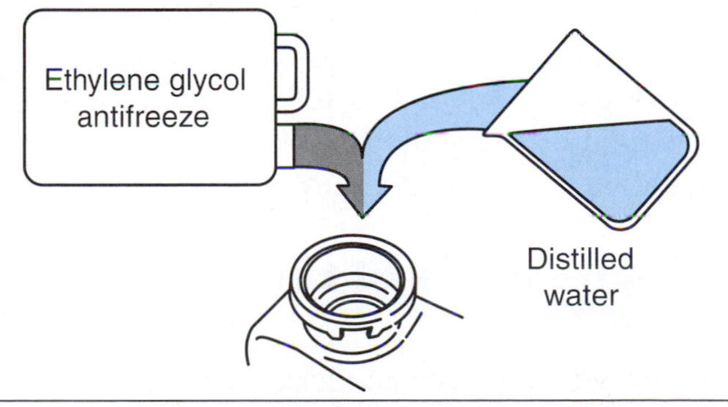

Figure 19-15 A 50/50 mixture of the manufacturer's recommended coolant and distilled water should be used in all motorcycles utilizing liquid cooling.

Figure 19-16 An air-cooled engine must have clean cooling fins accessible to the open air.

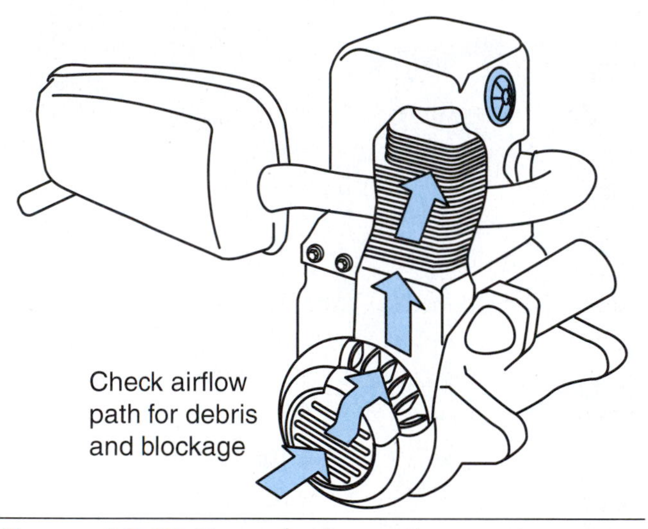

Figure 19-17 Forced air-cooling systems are used on motorcycles that do not have direct contact with open air. They use a fan and engine shrouds to place the air where it is needed. It is important to check for correct airflow on these types of engines.

Air-cooled engines must have clean cooling fins (Figure 19-16). The forced-air cooling systems used on many motor scooters (Figure 19-17) should be inspected after any long-term storage.

Compression and Leak-Down Tests

Compression tests and leak-down tests provide a good indication of the general condition of an engine's piston, rings, and cylinder area. Compression tests are done on both two-stroke and four-stroke engines.

Compression Test

A **compression test** ensures that the engine compression is high enough to heat the fuel-and-air mixture to a combustible level inside the engine's combustion chamber. To do a compression test, remove the spark plug(s). Then, using the compression gauge (Figure 19-18), measure the compression of each cylinder while the engine is being rotated rapidly by the electric starter, or the kick- or pull-starter mechanism. The recommended compression reading will be found in the appropriate manufacturer's service manual. If the compression is below recommendations there may be possible worn parts within. Always remember to hold the throttle control in the wide-open position when checking engine compression to allow the maximum amount of air to be drawn into the engine. If the throttle is not held open, the compression reading will usually read too low.

Unfortunately, the results of a compression test can be deceiving. For example, if the engine is not in stock condition, if the battery does not turn the engine fast enough, or if the testing procedures are not correctly followed, the compression test may indicate that the engine should be disassembled and rebuilt, when the engine is actually in good working condition.

When the compression test has been performed correctly but appears to not be within specification, there are other items you'll need to consider. If the compression readings are all below the service limit, but the readings for all cylinders are relatively close and the engine isn't smoking and is running well, the compression test by itself is seldom a good

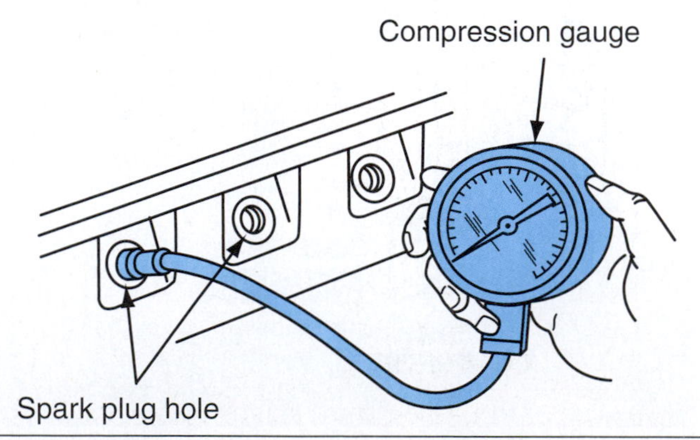

Figure 19-18 A typical compression gauge is shown here attached to a cylinder.

reason to disassemble the engine and do an expensive engine rebuild. If the compression readings for the cylinders of a multi-cylinder engine vary more than 15 percent, there is a good possibility that the engine has a problem that will need extensive repair work.

Leak-Down Test

Leak-down tests are performed on four-stroke motorcycles. A leak-down tester consists of a calibrated pressure gauge that's connected to a pressure regulator, a pressure source, and a flow restrictor (Figure 19-19). As a general rule, a leak-down test provides a better indication of any internal engine problems than a compression test. Leak-down tests are obtained by pressurizing the cylinder with compressed air while the piston is at top-dead center (TDC) on the compression stroke. When this is done, a measurement of the rate at which the air escapes past the rings, piston, and valves is completed. A range of acceptable percentage of air loss is provided by each leak-down tester toolmaker.

A leak-down tester indicates when an engine probably needs repair. It also tells you where the problem is located. By listening for escaping air at the air box, the exhaust system, and engine crankcase filler-cap, you can determine if the problem is being caused by the intake valves, exhaust valves, or the piston and rings. Leak-down testers are available at most quality automotive tool suppliers.

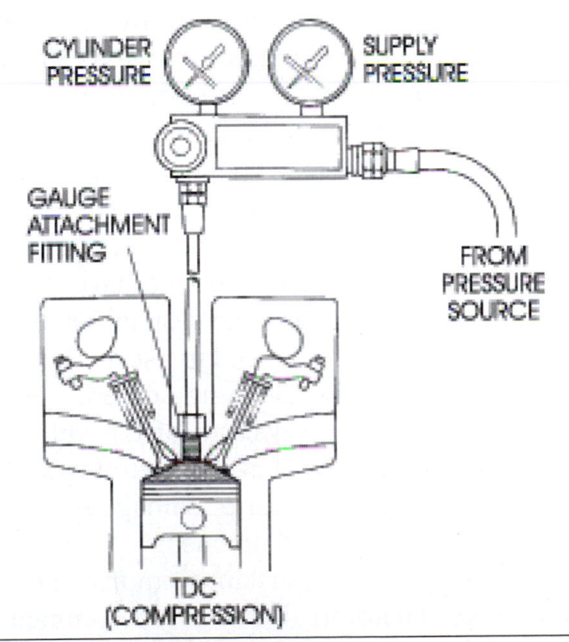

Figure 19-19 Leak-down testers can give the technician a good understanding of any internal engine problems when used correctly.

Valve Adjustment (Four-Stroke Engines)

If the motorcycle has a four-stroke engine, the valves should be inspected for proper adjustment. **Valve clearance** (lash) is necessary to allow for proper valve sealing. When valves aren't properly adjusted, engine performance may be affected. In most cases, the manufacturer will require that the valves be inspected for correct clearance when the engine is at room temperature.

Valve Adjustment Methods

There are different ways manufacturers adjust the valves on a four-stroke motorcycle engine. Here are the most popular types. Note: different engines will require the engine to be in certain positions before adjusting the valves. Be sure to check the appropriate manufacturer's service manual to ensure that the position of the engine is correct before attempting to adjust the valves. The key points for adjusting the valves are as follows:

- The screw-and-locknut (Figure 19-20) valve arrangement uses a screw that can be turned in or out to change the clearances. A locknut is used to hold the screw in place. The screw and locknut may be located on the rocker arm or on a push rod.

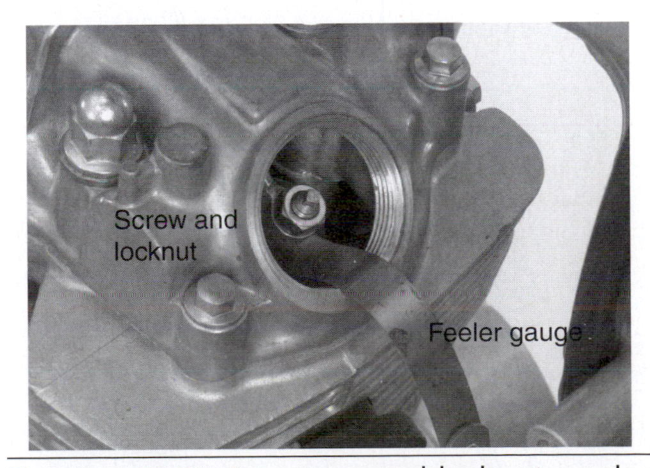

Figure 19-20 A screw-and-locknut valve arrangement is shown here. Copyright by American Honda Motor Co., Inc. and reprinted with permission.

- The shim-and-bucket valve arrangement is used both for valve opening and as a valve adjustment device (Figure 19-21). The shims are used to adjust the valves for proper clearances. Valve clearance is altered by changing the size of the shim. The two popular types of shim-and-bucket adjusters are shim-over-bucket, where the shim rests on top of the bucket, and shim-under-bucket, where the shim rests under the bucket. With the shim-under-bucket design, you must remove the camshaft to replace a shim.

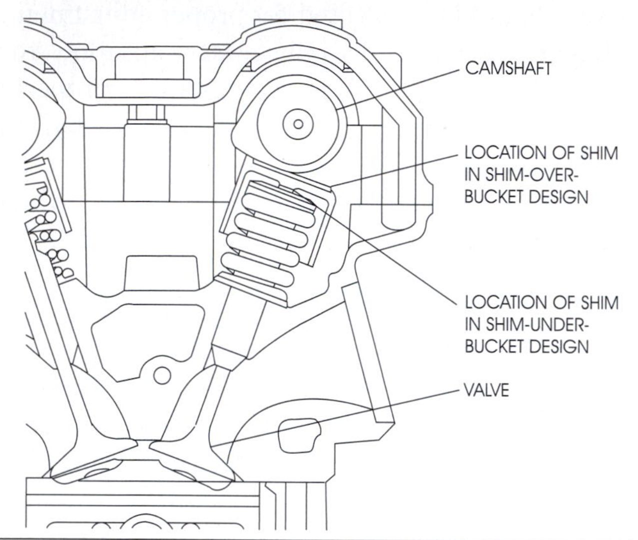

Figure 19-21 A shim-and-bucket valve arrangement is illustrated here. There are two types of shim and bucket systems: shim-over-bucket and the more popular shim-under-bucket. Copyright by American Honda Motor Co., Inc. and reprinted with permission.

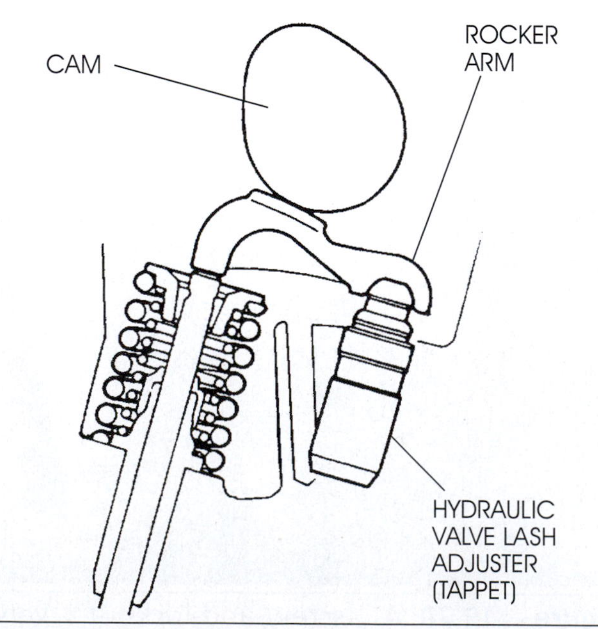

Figure 19-22 Hydraulic valve adjusters require no maintenance. Copyright by American Honda Motor Co., Inc. and reprinted with permission.

- Hydraulic valve lash adjusters (Figure 19-22) automatically adjust for the proper valve clearances by using oil pressure to maintain zero lash at all engine temperatures and rpms. This hydraulic design requires no maintenance.

If required that the valves be adjusted after inspection, always recheck the compression and compare the new reading with the reading obtained before the valves were adjusted. If a valve was too tight before adjustment, the compression will usually increase after the adjustment has been made.

Valve Adjustment Specifications

The service manual for a motorcycle has the manufacturer's recommended specifications for valve adjustments. A good technician doesn't try to remember these specifications, but refers to the manual for this information each time it's needed. There are too many models to remember the specifications of each engine. It would be very easy to confuse the specifications of one model with those of another, causing an incorrect valve adjustment and creating a potential engine performance problem.

Power-Valve Inspection (Two-Stroke Engines)

Two-stroke engines that use **power valves** require maintenance on a regular basis. As we mentioned in an earlier chapter, there are several different types of power valves. Power-valve systems are all designed to do the same thing—they increase the two-stroke engine's power band. They just may do this in different ways due to variations in design.

All power-valve systems need to have the carbon deposits removed from the system. To remove all carbon deposits, you'll need to use a wire brush and a high flash-point cleaning solvent. During cleaning, you should inspect all of the individual components of the power-valve system for wear or any signs of damage. If wear or damage is present, replace the power-valve component.

Manufacturers are constantly improving their power valves. Therefore, it is highly recommended that you follow the procedures given in the appropriate service manual when servicing any type of two-stroke engine power-valve system.

Clutch Adjustment

Clutch systems are either hydraulically operated or are mechanically (cable) linked. The hydraulic style of clutch requires no adjustment. On cable-operated clutch systems, however, you must check the play at the end of the clutch lever (Figure 19-23). If there's excessive play at the lever, the clutch drags and causes the engine to shift with apparent stiffness. A lever with too little play causes a clutch to slip.

If the clutch needs a minor adjustment, you can adjust the play by using the clutch adjuster at the lever end of the clutch cable (Figure 19-24). If more than 10mm (about ½ inch) of adjustment is required, a major cable adjustment is needed. For this, you should screw in the lever-end adjuster and make the major clearance adjustments using the clutch-arm end adjuster of the clutch cable as shown in Figure 19-25. Then, fine-tune the clearance adjustment using the lever-end adjuster of the clutch cable as you did for a minor adjustment.

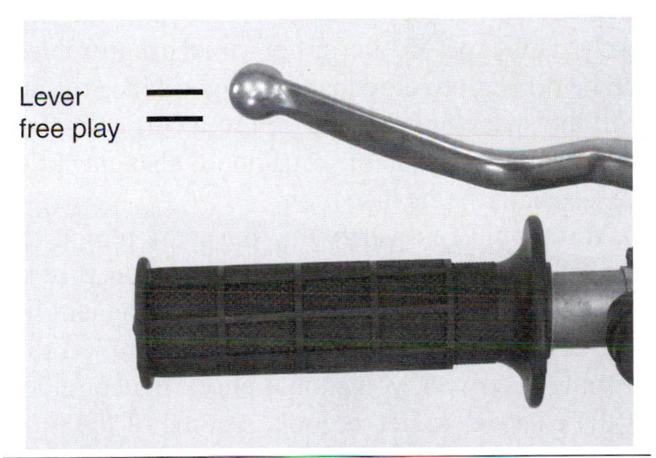

Figure 19-23 Clutch lever free play is indicated here. Copyright by American Honda Motor Co., Inc. and reprinted with permission.

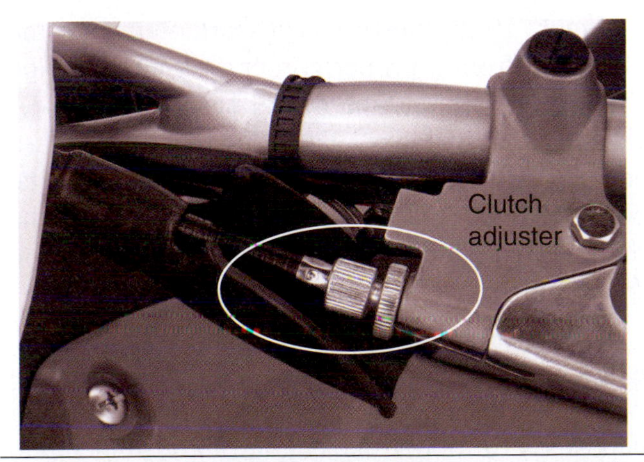

Figure 19-24 Most clutch adjustments can be done by turning the clutch adjuster in or out at the lever end of the clutch cable. Copyright by American Honda Motor Co., Inc. and reprinted with permission.

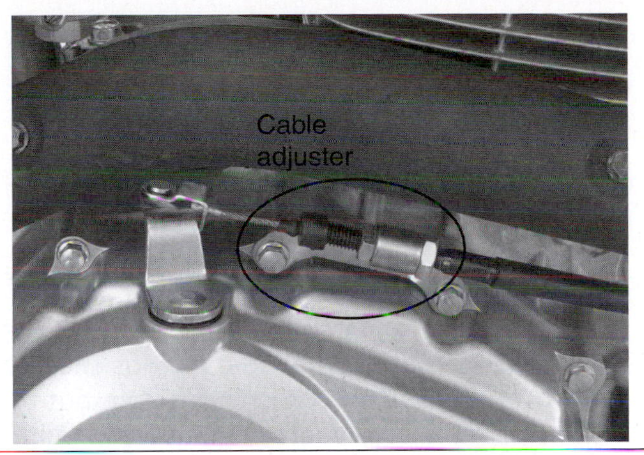

Figure 19-25 If a major clutch adjustment is required use the clutch-arm end adjuster of the clutch cable. Copyright by American Honda Motor Co., Inc. and reprinted with permission.

Spark Plug Inspection and Replacement

On some motorcycles the **spark plug** is one of the easiest components to access during maintenance (Figure 19-26), while on other motorcycles the spark plug may be among the most difficult components to access (Figure 19-27). The spark plug is normally located in the center of the cylinder head combustion chamber.

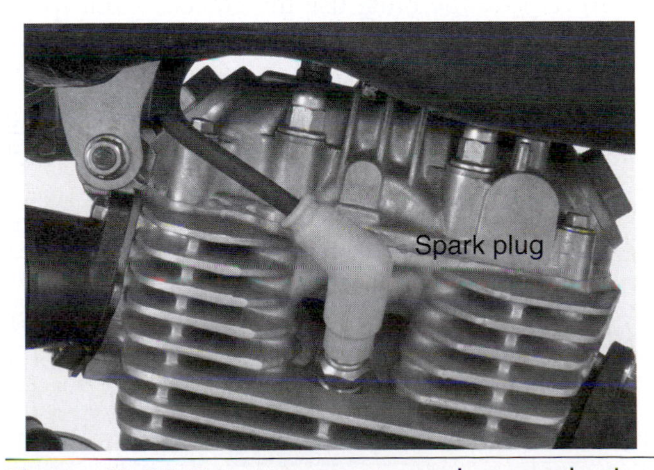

Figure 19-26 For many motorcycles, spark plug access is easy. Copyright by American Honda Motor Co., Inc. and reprinted with permission.

Figure 19-27 To access the spark plugs the fuel tank must be raised, which isn't difficult, but the entire process is involved and time consuming. Courtesy of American Suzuki Motor Corp.

Spark Plug Removal

Be sure to always remove any loose dirt or debris on the cylinder head, near or around the spark plug, before you remove the spark plug. It's very important to prevent any dirt from getting into the engine through the threaded hole in the cylinder head. Remember that the spark plug wire must be disconnected before you remove the spark plug from the engine. If the engine was operated prior to the spark plug removal procedure, you should always allow the engine to cool before attempting to remove the spark plug. Engine heat causes the cylinder head and the spark plug shell to expand. If you try to remove a plug before the engine has cooled, the spark plug may seize, and removing it could damage the cylinder-head threads. When the engine and spark plug have cooled sufficiently, the plug will be much easier to remove and there is less chance of damaging the cylinder head.

To remove the plug, use the correct-sized spark plug socket. We mentioned in an earlier chapter that a spark plug socket is a special socket wrench that's specifically designed for removing and installing spark plugs (Figure 19-28). The spark plug socket has rubber inserts that protect the spark plug's ceramic insulator. The depth of the socket allows it to fit over the top of the spark plug to reach the hexagonal area of the shell. If a spark plug is tightly mounted in the cylinder head, the plug must be carefully removed to prevent it from breaking.

Figure 19-28 A spark plug socket is a special socket wrench that's specifically designed for removing and installing spark plugs.

Spark Plug Inspection

After the spark plug has been removed, inspect it to determine its condition. The condition of a spark plug can tell you much about how an engine is operating. Many experienced motorcycle technicians will begin their troubleshooting procedure process by removing and inspecting the spark plug(s).

When inspecting a spark plug, be sure to check the condition of the ceramic insulator. A damaged insulator can cause a spark plug to fail intermittently. This type of intermittent misfiring problem can be difficult to diagnose. It's a good idea to start with the spark plugs when you're trying to isolate an intermittent problem. You should also check for the issues in the next section.

You should first verify that the spark plug is the correct type for the engine. You can do this by referring to the service manual for the engine you're working on. After you've determined that the plug is correct, you should check the condition of the electrodes. Let us look at some of the most common spark plug conditions.

Normal Spark Plug

Figure 19-29 shows a used spark plug in normal condition from a properly operating engine. Note that the bottom surface of the center electrode is flat and the surfaces of the lower electrode are squared. The electrodes are an ash-gray or light tan color from normal fuel combustion. Note also that there is no buildup of contamination on or around the electrodes.

Oil-Fouled Spark Plug Figure 19-30 shows an example of an oil-fouled spark plug. Oil fouling will cause the plug to be saturated with shiny oil deposits. In a four-stroke engine, an oil-fouled plug may indicate that the piston rings are not sealing the cylinder properly, or oil may be passing through the intake valve stem. A clogged crankcase breather can cause oil-fouled plugs. Remember that a breather is a vent in the crankcase. Thus, a clogged

Figure 19-29 A normal-condition spark plug is shown here. Note that the bottom surface of the center electrode is flat and the surfaces of the lower electrode are squared.

Figure 19-30 Oil fouling will cause the plug to be saturated with shiny oil deposits as can be seen in this photograph.

breather prevents the crankcase from venting properly. Pressure builds up in the crankcase, which can cause oil to be forced up past the piston rings and into the combustion chamber. Any oil in the combustion chamber can foul the spark plug, especially if the compression in the cylinder is below specifications. Oil-fouled spark plugs are more common in two-stroke engines. Remember that in a two-stroke engine, the fuel and oil are premixed in the crankcase. Thus, oil fouling is a potential byproduct of any two-stroke engine operation.

Fuel-Fouled Spark Plug A spark plug that was fouled by excessive fuel can be seen in Figure 19-31. Fuel fouling is indicated by dry, black, fluffy carbon deposits on the spark plug's electrodes. Fuel fouling is most often caused by prolonged operation with a fuel-and-air mixture that's too rich. This is usually caused by a carburetor adjustment problem. A blocked or faulty exhaust valve can also cause fuel fouling. An engine with the choke that is left on for too long an interval can also cause fuel-fouled spark plugs.

Bridged-Gap Spark Plug Both oil fouling and fuel fouling can cause a spark plug condition known as a bridged gap (Figure 19-32). In this

Figure 19-31 A fuel-fouled spark plug is indicated by dry, black, fluffy carbon deposits on the plug's electrodes.

Figure 19-32 A bridged gap is caused by carbon or oil deposit building up in the spark plug's electrode gap until it becomes blocked.

Figure 19-33 This photograph shows a spark plug with an eroded electrode from normal long-term wear.

situation, carbon or oil deposits build up in the spark plug's electrode gap until the gap becomes blocked. A bridged gap will prevent the spark plug from firing properly, resulting in poor engine performance or an engine that won't start. Debris passing through the air filter and into the intake port can also cause a bridged gap.

Electrode-Eroded Spark Plug After many hours of use, spark plug electrodes will begin to erode. This is a normal condition of wear. When erosion occurs the center electrode will appear rounded and the side electrode will have a curve on the inside surface (Figure 19-33). In comparison, a new spark plug has electrodes with flat sharp surfaces.

Spark Plug Gap

The gap between the electrodes of a spark plug must be correct to operate properly. Before you install a spark plug, you should measure the gap between the electrodes. The service or owner's manual for the engine gives the dimension for the proper spark plug gap. The spark plug gap can be checked by using a wire-end spark plug gap tool (Figure 19-34).

Spark Plug Cleaning

Never sand or file the spark plug electrodes and then reinstall the spark plug in an engine. Using sandpaper or a file leaves tiny grooves on the electrodes. These grooves will either burn off or will collect deposits when the engine is operated.

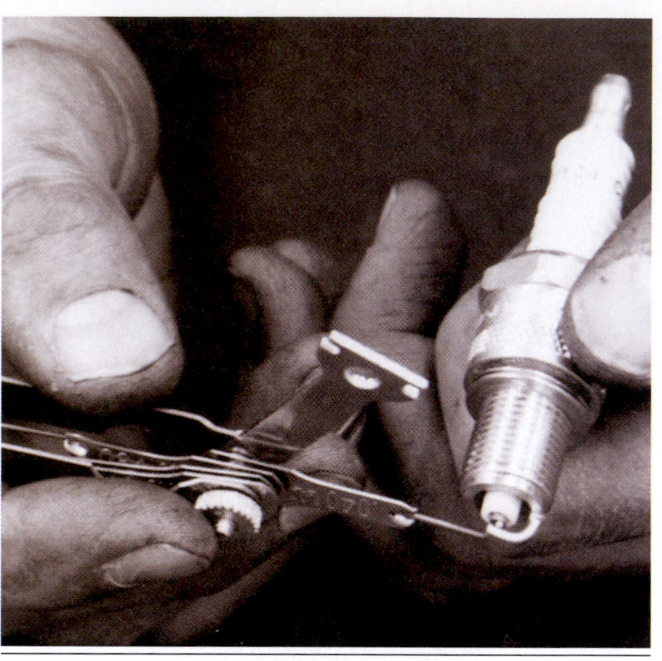

Figure 19-34 The correct method of measuring spark plug end gap is shown here.

Sanding and filing also leaves tiny particles of sand or metal on the electrodes. These particles can get into the engine's cylinder and cause serious damage.

Some spark plug manufacturers have produced small sandblasting machines designed to clean their spark plugs. However, motorcycle and ATV manufacturers strongly recommend against using these sandblasters for the reasons we've just described. If you're ever in doubt about a spark plug's condition, simply replace it.

Spark Plug Installation

To reinstall a spark plug, hold the plug with your fingers and gently screw the plug into the threaded cylinder head opening (Figure 19-35). Do

Figure 19-35 To reinstall a spark plug do not force it into the hole. Hold the plug with your fingers or with the correct spark plug socket and gently screw the plug into the threaded cylinder head opening.

not force the spark plug. The plug should turn at least three full turns into the cylinder head before it shows any signs of resistance. When the resistance point has been reached, use a spark plug socket to tighten the plug into the cylinder head. Be sure to tighten the spark plug with a torque wrench to the manufacturer's specifications.

Battery Inspection

When it comes to batteries you must take extra precautions such as wearing safety glasses and hand protection. **Battery acid** can cause severe burns if it contacts your skin, and will damage clothing. If you accidentally spill any battery acid, the spill should be cleaned up immediately. Use a water and baking soda solution to clean the spill area. This combination helps neutralize the acid.

If the motorcycle has a battery, it should be inspected for cracks in the casing, broken terminals, or other signs of physical damage (Figure 19-36). This includes checking for sulfation or warping on the internal plates if the battery has a transparent outer casing. If any of these conditions are found, the battery should be replaced. You should also ensure that the battery cable connectors make good contact with the battery terminals. If the cable connectors or terminals are corroded or loose, clean and tighten the connections. An application of dielectric grease on the cable connectors and battery terminals helps to prevent corrosion.

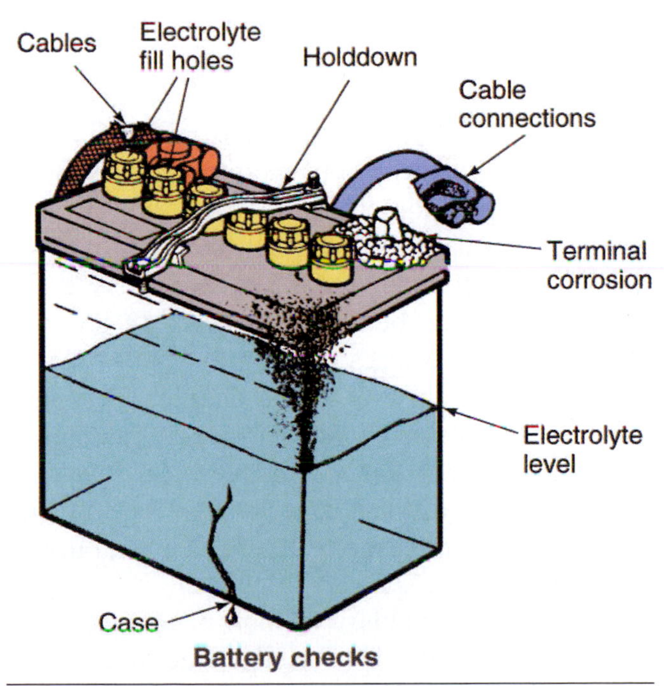

Figure 19-36 When inspecting a battery, look for items such as terminal corrosion and electrolyte level (on conventional batteries). Note that this image has both a maintenance free and conventional battery with filler caps. Maintenance-free batteries do not require electrolyte level inspection.

Conventional Battery Electrolyte Testing

With **conventional batteries**, the condition of the battery is determined by measuring the specific gravity of the electrolyte. The specific gravity is measured with a hydrometer (Figure 19-37). The electrolyte should have a specific gravity of 1.280 to 1.320 (depending on the air temperature). As the battery becomes discharged, this reading decreases.

Conventional Battery Water Level Remember that conventional batteries are filled with a sulfuric acid electrolyte solution. The acid is absorbed into

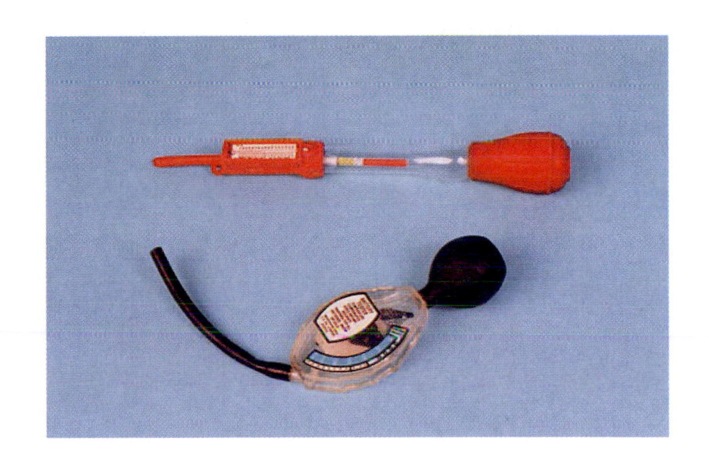

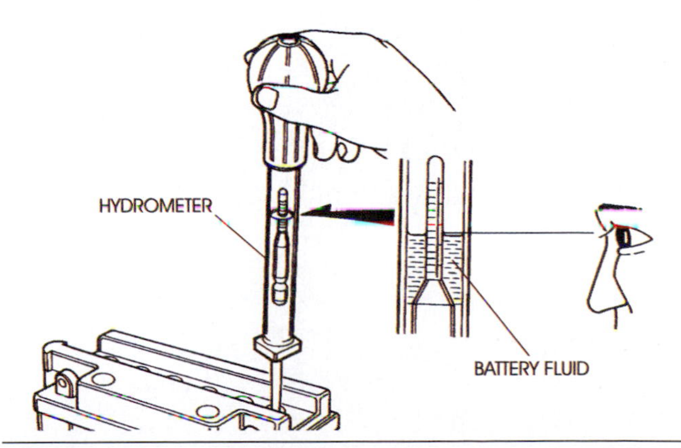

Figure 19-37 Different hydrometers and how one is used are shown here.

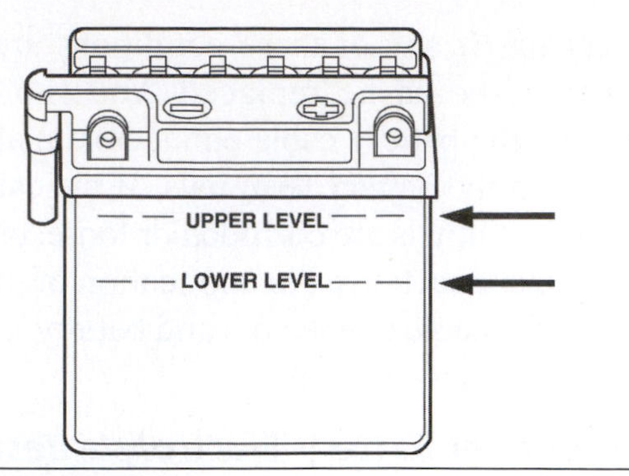

Figure 19-38 If the level is at the lower level or below, replace liquid with distilled water only and fill each cell to the upper level.

the battery plates but the remaining water can evaporate. Look on the outside of the battery for the level on a conventional battery (Figure 19-38). If it is determined that it is necessary to add water to a battery that has a low electrolyte level, only add distilled water (Figure 19-39). Distilled water should be used to prevent minerals and other impurities from contaminating the sulfuric acid and the lead plates in the battery.

Maintenance-Free Batteries

Most all modern motorcycles have **maintenance-free batteries** (Figure 19-40). This type of battery does not require fluid level checks because the battery is sealed. If this type of battery fails to hold a charge, it can be deemed in need of replacement. Before replacing a maintenance-free battery, the charging and regulating circuits should be thoroughly tested.

Figure 19-39 Distilled water can be purchased at any grocery store.

Figure 19-40 A maintenance-free battery has no filler caps and some like the one shown here come from the factory prefilled with acid.

Battery Testing

Due to the high level of quality of today's batteries and the technology that is built into them, in most cases, a battery would not be tested unless a specific request to do so was made or an issue was reported that may be battery-related. In the past and even in some cases today, a battery would be tested for its ability to hold a charge by using a **battery load tester** (Figure 19-41). A load tester places the battery under a heavy electrical load condition (such as using the electric starter) while it is out of the motorcycle to check it's ability to maintain a charge.

The motorcycle industry is now beginning to utilize a tool that has been used in the automotive industry for some time now to measure the health of a battery (Figure 19-42). A **battery conductive analyzer** describes the battery's ability to conduct current. It measures the plate surface available in a battery for chemical reaction. Measuring conductance provides a very reliable indication of the battery's condition and is correlated to battery capacity.

Figure 19-41 A battery load tester places a load on a battery to check its ability to withstand current flow through it.

Figure 19-42 Battery conductance analyzers are becoming more popular in the motorcycle industry due to their ability to accurately tell us the condition of the battery.

This type of tester can be used to detect cell defects, shorts, normal aging, and open circuits in a battery, all of which can cause the battery to fail.

A fully charged battery will have a high conductance reading, up to 110 percent of its internal rating. As a battery ages, the plate surface will sulfate or shed active material, which will lower its capacity and conductance.

The conductance tester will display the service condition of the battery. It will indicate if the battery is good, needs to be recharged and tested again, has failed, or will soon fail. As well as giving a state of charge, this type of tester will also show a state of health (Figure 19-43).

(A) State of health

(B) State of charge

(C) Good or bad

Figure 19-43 A conductance battery tester gives information such as the state of health of the battery (A), the battery's state of charge (B), and also tells if the battery is good or bad (C).

Ignition System Inspection and Adjustment

The ignition of the fuel must occur at the proper time during the compression cycle for a motorcycle engine to develop full power. Because the fuel takes some time to start burning, the spark must occur shortly before the piston starts the power stroke. In virtually every engine, the ignition spark occurs when the piston is still moving upward on the compression stroke. Motorcycle engines have one optimum ignition timing setting that's determined by the manufacturer. This ignition timing setting is listed in the service manual for the engine. If the ignition timing varies from this optimum setting, the engine will potentially lose efficiency and power.

Nearly all modern motorcycles use nonadjustable **electronic ignition systems**. These systems have the correct timing built into the electronic components. It is possible to verify proper ignition operation with the use of a timing light (Figure 19-44). However, in most cases, the ignition system timing is either correct or the system won't function at all. In these cases, component replacement is required to correct the problem. Because most electronic ignition systems use no moving parts, it's a rare occurrence to see a timing problem with them. All older engines were designed so that you can adjust the ignition timing, and most can be adjusted while the engine is running. A **timing light** is used to check the ignition timing.

Figure 19-44 A timing light is used to verify the correct ignition timing of an engine by pointing the light into the inspection hole while the engine is running. Copyright by American Honda Motor Co., Inc. and reprinted with permission.

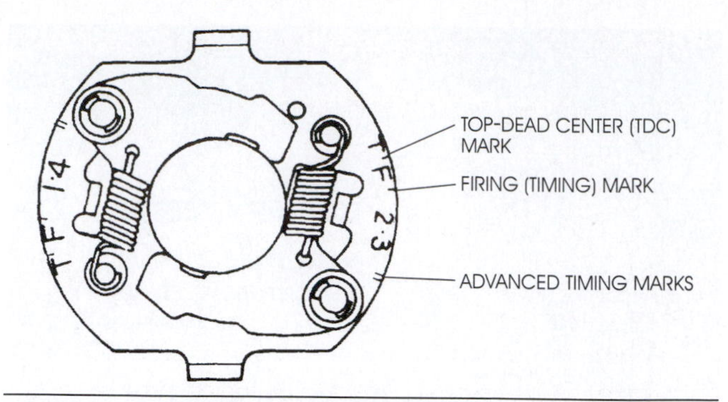

Figure 19-45 An illustration of a centrifugal advance unit is shown here with the three typical timing marks: (T) for top-dead center, (F) for firing mark at idle, and advanced timing marks.

When it is attached to a spark plug, the timing light produces a flash of light each time the spark plug fires. The strobe effect of the timing light freezes the rotating timing marks. This allows you to observe and accurately adjust the timing of the ignition system. Engines that can be timed dynamically usually have centrifugal spark-advance mechanisms. These mechanisms have three timing marks as Figure 19-45 illustrates: a top-dead-center (T) mark, a fire mark (F), and advanced timing marks.

Idle Adjustment

While some motorcycles with fuel injection have an automatic idle speed adjustment integrated within the FI system, most motorcycles have an idle adjustment screw (Figure 19-46) for setting the engine idle speed (engine rpm). The correct idle speed and adjustment screw location is listed in the appropriate manufacturer's service manual.

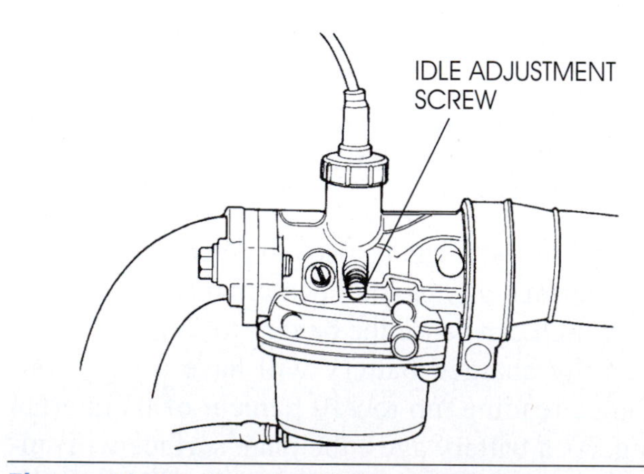

Figure 19-46 A typical idle adjustment screw on a single carburetor is illustrated here.

Carburetor Synchronization

Carburetor synchronization is the process of balancing the output of two or more carburetors so that the amount of fuel-and-air mixture drawn through each one is equal. This is checked by measuring the engine vacuum at each carburetor intake manifold.

When carburetors are used, the operating temperature, smoothness, response, and fuel mileage of a multi-cylinder motorcycle engine depend on proper carburetor synchronization. This is especially critical to the performance of a multi-cylinder engine that has one carburetor per cylinder. The physical linkages and mounting methods vary from model to model, but the basic principles of carburetor synchronization are the same for all multiple-carburetor engines.

To synchronize a multiple-carburetor assembly, you must first install a set of **vacuum gauges** to the intake manifolds of the engine. A vacuum gauge is a tool for measuring pressures below atmospheric pressure (vacuum).This is done by removing the plugs from each cylinder head port and installing the appropriate adapters as illustrated in Figure 19-47, and then installing the vacuum gauges as shown in Figure 19-48.

The procedure to synchronize a set of carburetors is the same on all carburetors, but of course you should always check with the manufacturer's service manual to ensure that the proper procedures are followed. Here is a typical procedure for synchronizing a set of carburetors:

1. With the engine running at its warmed-up operating temperature, adjust the engine idle speed to the manufacturer's specifications.
2. Turn the synchronization screws on each carburetor so that the vacuum between the base carburetor (that's the carburetor with the idle adjustment screw) intake port and each of the other carburetor intake ports is within factory specifications. It's a good habit to adjust the carburetors so that they all show the same vacuum. This results in the smoothest-running engine.
3. Check that the carburetor synchronization adjustment is stable by snapping the throttle control handle several times. This will accelerate the engine and then drop it down to idle speed.
4. When all of the carburetors have been synchronized, check the idle speed and verify that it is still within specification.

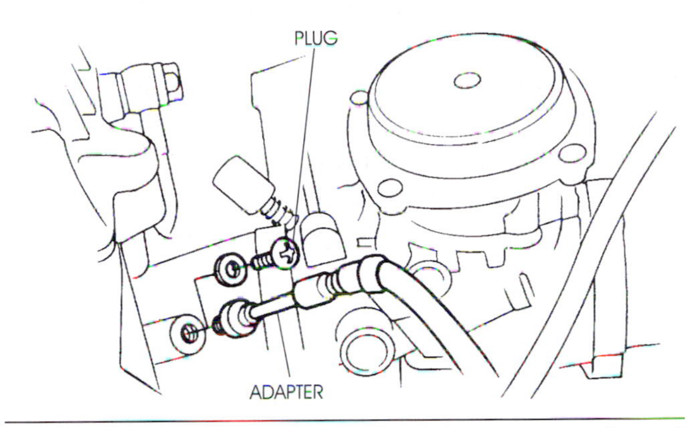

Figure 19-47 Vacuum gauges are placed between the carburetor and intake port to synchronize multi-carbureted engines.

Figure 19-48 A set of vacuum gauges used for carburetor synchronization is pictured here.

Air Filter Inspection

It's very important that you understand and appreciate the importance of motorcycle **air filter** maintenance. The air filter is frequently overlooked during maintenance. The air filter is very important for prolonging the life expectancy of the engine. If dirt and other contaminants are allowed to flow through the intake system, it will damage the engine's internal parts.

Air filters are installed in an air box (Figure 19-49) and are made of various types of materials.

Paper Air Filters

As we discussed in an earlier chapter, the most commonly used motorcycle air filter is the paper filter. This filter consists of laminated paper fibers that are sealed at the ends or sides of the filter. Some paper air filters include supporting inner or outer metal screen shells. The paper used in these air filters is generally molded into a "W" pattern as shown in Figure 19-50. This molded "W" is designed to increase the surface area and to decrease the restriction of air passing through the filter.

The paper air filter must be kept dry and free of oil. If the paper air filter becomes excessively dirty or contaminated with oil, it must be replaced. Don't try to clean a paper air filter with soap and water because this will damage the paper fibers and cause the filter to fail. You should use compressed air to clean a paper air filter as shown in Figure 19-51.

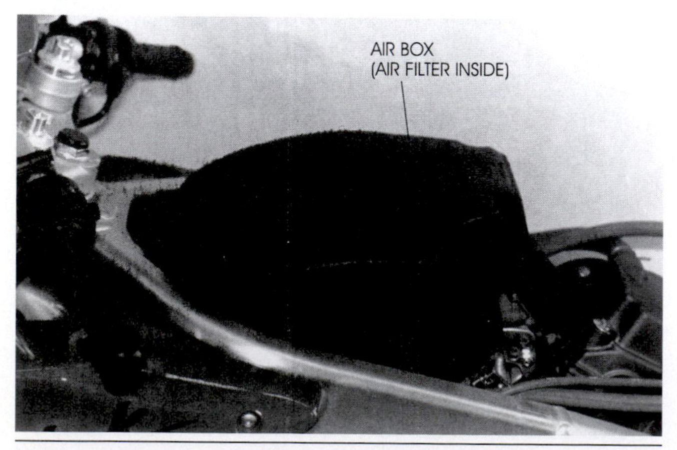

Figure 19-49 Air filters are located in the air box on motorcycles.

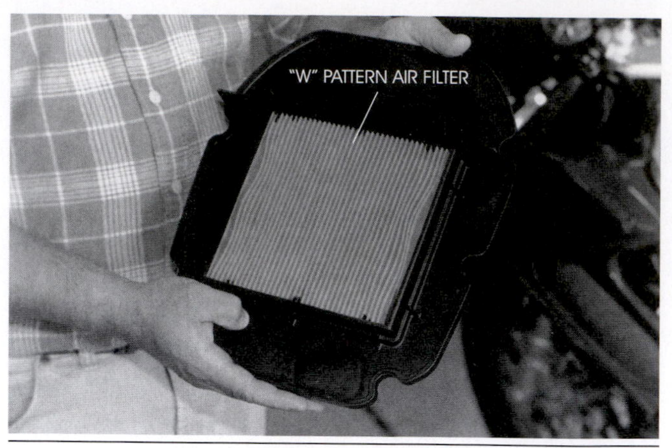

Figure 19-50 A typical paper-type air filter with a pleated "W" pattern is shown here.

Figure 19-51 Cleaning a paper filter with compressed air.

Foam Air Filters

Figure 19-52 shows a foam-type air filter. It uses a special foam and oil to trap dirt and other contaminants. When this type of filter is dirty, you can clean it in a warm, soapy water solution, rinse it, and dry it. When the filter has dried, you must apply special oil specifically made for foam air filters. The excess oil is then squeezed out of the filter before use.

Gauze Air Filters

Figure 19-53 shows a gauze-type air filter. This filter is very similar to the paper air filter but uses a special oil to help trap particles from entering the intake system. Surgical-type gauze is used to trap the dirt as the air passes through the filter. When this type of filter is in need of maintenance, it is cleaned in warm soapy water, rinsed, and dried.

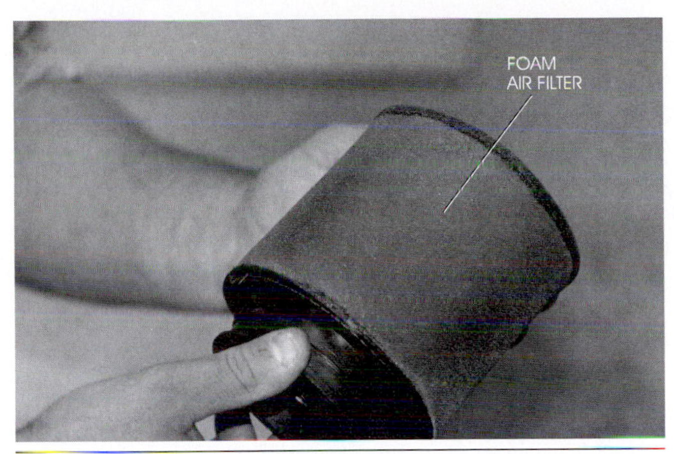

Figure 19-52 Foam air filters are typically found on off-road motorcycles.

Figure 19-53 Gauze-type air filters are similar to paper in design but use a surgical gauze and special oil to keep dirt from entering the intake system.

After drying, you must apply special gauze-filter oil prior to reinstalling.

MOTORCYCLE CHASSIS MAINTENANCE

Motorcycle chassis maintenance includes cleaning, inspecting, and replacing parts that are subject to wear. The technician performing this maintenance adjusts the motorcycle's chassis systems to the settings specified by the manufacturer. Chassis components that require attention during maintenance include the following:

- Cable(s) inspection and adjustment
- Steering-head bearings
- Brakes
- Wheels and tires
- Final drives

Cable Inspection and Adjustment

Cable maintenance consists of inspecting all of the cables for the specified amount of play and ensuring that the cables work freely and don't bind or stick. Cable maintenance also includes checking that the cables are properly lubricated. Figure 19-54 shows a special cable lubricating tool that dispenses lubricant into the cable.

Cable Adjustment

A throttle cable adjustment may need to be performed during cable maintenance. This adjustment consists of creating the correct amount of play between the throttle twist grip or throttle lever and the cable (Figure 19-55). To change the

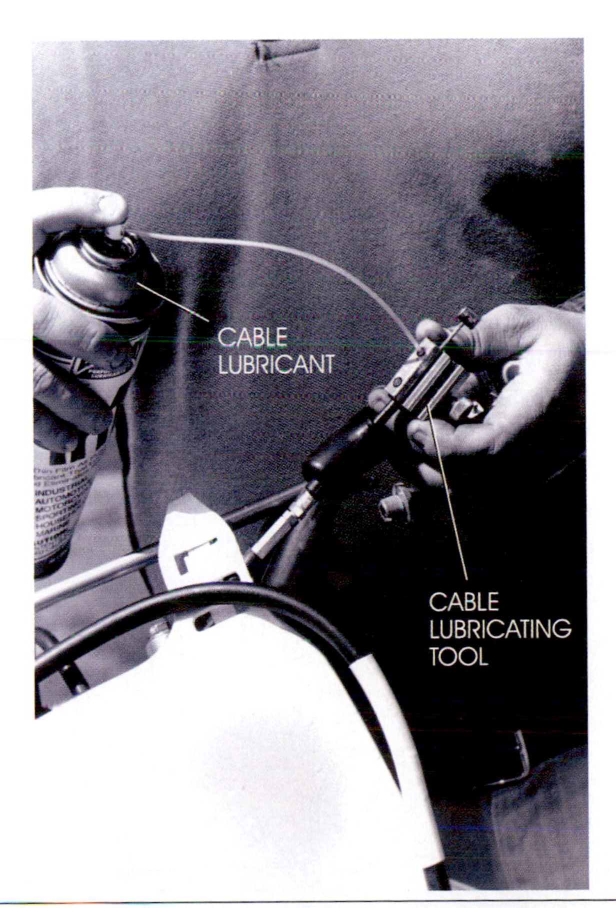

Figure 19-54 A special cable lubrication tool clamps over the desired cable and allows pressurized lubricant to be forced into the cable.

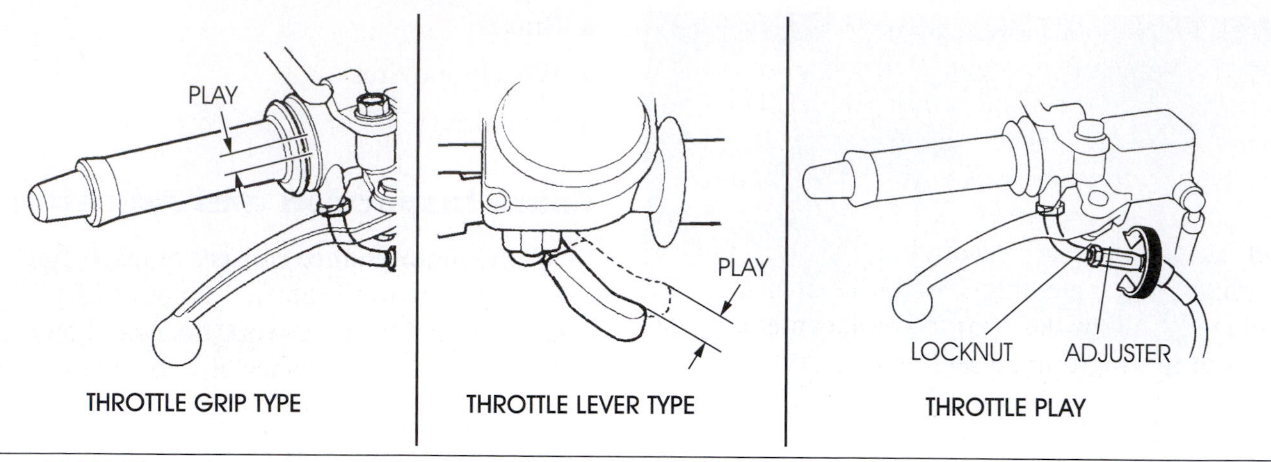

Figure 19-55 Throttle cables require play as stated by the appropriate manufacturer's service manual.

amount of throttle play, use the throttle cable adjuster and locknut.

The clutch and brake cable lever play must also be measured (Figure 19-56). Adjustment to either of these cable types is made at the lever end. Be sure to lubricate the lever pivot bushing as well. When the pivot is left dry, it will wear quickly and will make it difficult to properly adjust the cable free play.

Steering-Head Bearings

A symptom of a loose or worn steering-head bearing assembly would be a loose feeling in the front end while riding the motorcycle (Figure 19-57) or a harsh bump felt through the handlebars when riding over bumps in the road (Figure 19-58).

To check the steering-head bearings on a motorcycle, you must first securely support the machine from beneath the frame, with the front wheel off the ground and free to pivot. Typically, a jack-stand is used to hold the motorcycle in the correct position.

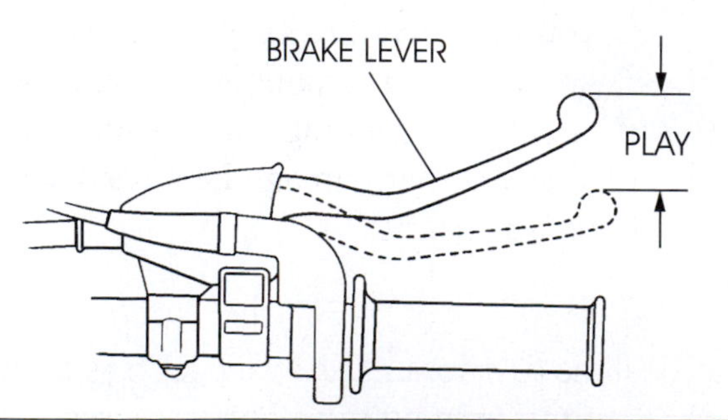

Figure 19-56 Clutch and brake levers that have cable adjustments also require a specified amount of play to function properly.

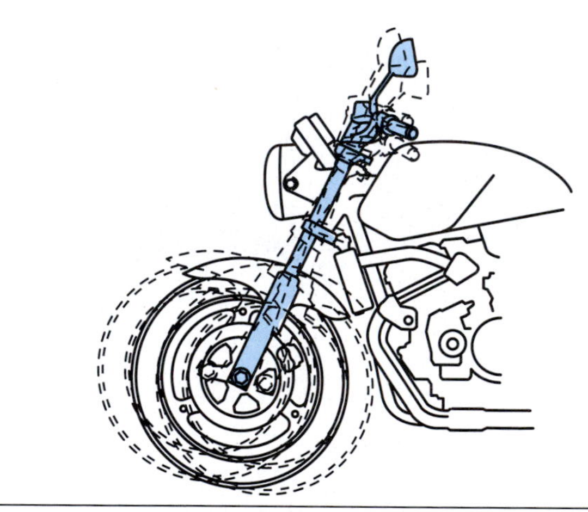

Figure 19-57 A loose steering-head bearing can make the front end feel loose as indicated in this illustration.

Figure 19-58 A harsh up-and-down movement can be felt through the front end with loose steering-head bearings.

Check for smooth movement as you turn the handlebars from left to right. If the operation isn't smooth, binds, or has a heavy feeling when you turn the handlebars, ensure that there isn't any interference from the cables or the wire harness. If the cables and wire harness aren't causing the unusual feel, check for wear or damage to the steering-head bearings. If the steering-head bearings need to be replaced, follow the bearing replacement procedures in the appropriate service manual.

Brake Maintenance

Brake system maintenance consists of inspecting the condition of the brake pads (Figure 19-59A) and disks and/or the brake shoes and drums. You should check for proper cable adjustment on mechanically operated brakes, and check for air in the system of hydraulically operated brakes. To check for air in a hydraulic-brake system, apply pressure on the brake lever (Figure 19-59B) or brake pedal and verify that no air has entered the system. If the lever or pedal feels excessively soft or spongy when operated, this indicates air in the system; the air will need to bleed off. Refer to the appropriate service manual for brake system air-bleeding procedures.

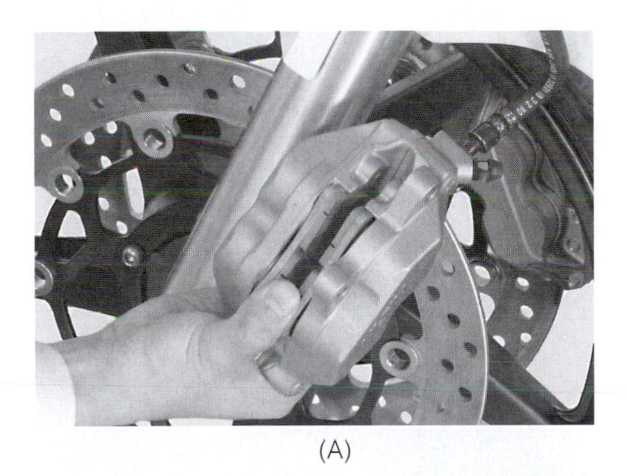

(A)

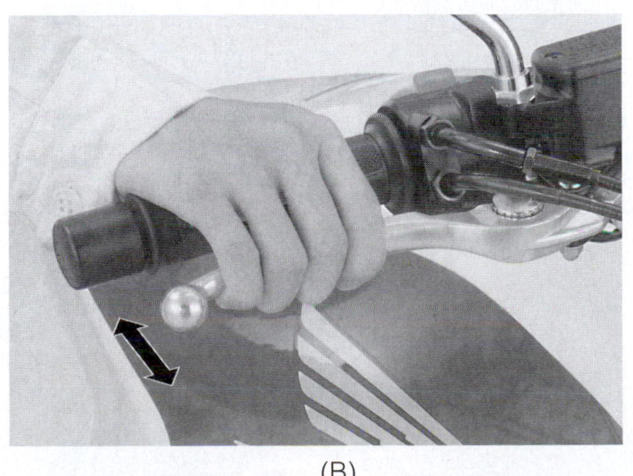

(B)

Figure 19-59 The brake pads are being inspected in (A) and lever feel is being checked in (B). Copyright by American Honda Motor Co., Inc. and reprinted with permission.

Figure 19-60 The brake fluid should be visible through the inspection window and be clear or a light amber color. If the fluid is dark, it should be replaced. Copyright by American Honda Motor Co., Inc. and reprinted with permission.

When inspecting the brake system, also check the condition of the hydraulic fluid in the brake reservoir (Figure 19-60). If the fluid is low or dirty, add fluid or replace it as necessary. Caution should be taken to protect all painted and plastic parts from the brake fluid. Painted and plastic parts are subject to damage if they're exposed to brake fluid.

Wheels and Tires

Inspect the front and rear wheel bearings for wear and damage by checking for excessive play in each wheel with the wheel lifted off the ground and rotated. If you notice areas where the wheel does not rotate smoothly, or produces unusual noises such as rumbling or grinding, the wheel bearings are likely to be worn or damaged and replacement will likely be necessary.

Tire Inspection

Inspect the tires for unusual wear such as **cupping** (Figure 19-61). Cupping (also known as scalloping) is a wear pattern found on tires that start out as flat spots (and eventually turn into depressions) that

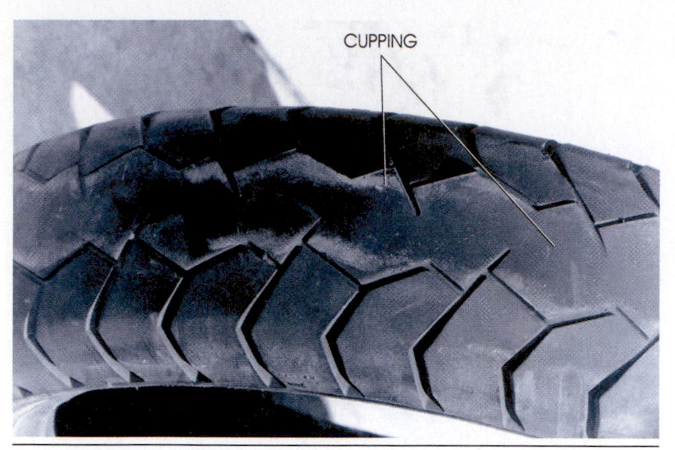

Figure 19-61 Cupping is commonly found on motorcycle tires. It will be accelerated by improper inflation of the tire.

Figure 19-63 Checking for correct tire pressure is one of the most overlooked maintenance procedures on a motorcycle.

develop on the tread of tires (primarily on the front). These flat spots alternate on either side of the center of the tread. If the tires are worn beyond their service limits as determined by inspecting the "wear limit indicators," (Figure 19-62) or have other signs of damage, they should be replaced with new tires.

Tire Pressure Check

One of the most overlooked items on a motorcycle is the **tire air pressure**. It has been reported that a tire that has been run just 2–4 lbs under-inflated for as little as 500 miles will wear prematurely. Improperly inflated tires are subject to uneven wear, and can provide a poor ride. Tire pressure should be checked (Figure 19-63) and set to the factory recommendation. You can find the recommended pressure in the appropriate owner's or service manual, as well as on the tag that can be found on the motorcycle itself (Figure 19-64).

Final Drives

As we learned earlier, there are three types of final drive systems found on motorcycles: chain drive, belt drive, and shaft drive. Each of these systems has specific maintenance requirements.

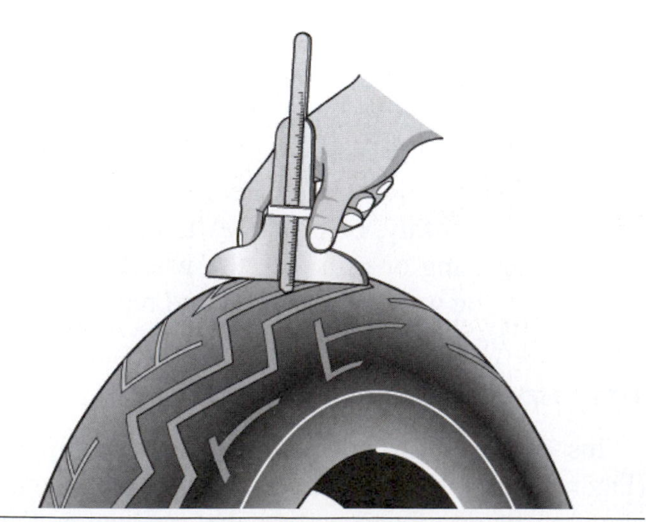

Figure 19-62 Checking the tire tread depth for wear.

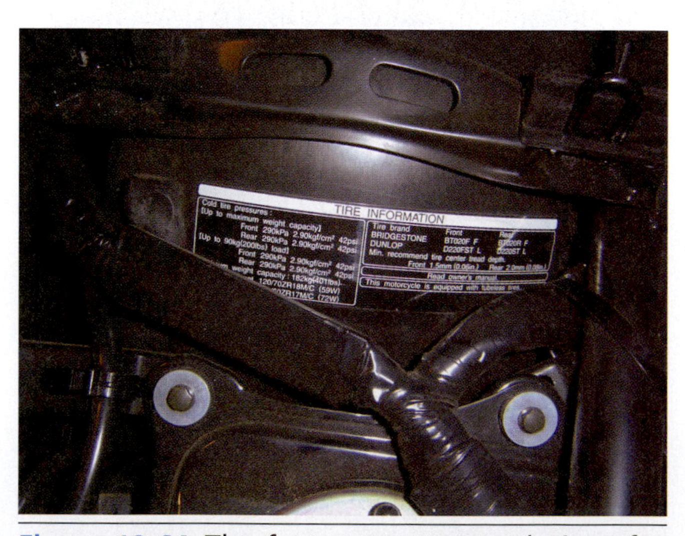

Figure 19-64 The factory recommendations for tire pressure, as well as tire sizes, can be found on a sticker placed on the motorcycle by the manufacturer.

Chain Drive

The **chain-driven final drive** system is the most commonly used motorcycle final drive system. The sprockets and chain of a chain-driven system wear out with use (Figure 19-65 and Figure 19-66).

To obtain the maximum useful life from sprockets and chains, they require consistent inspection and maintenance. The drive chain requires more frequent service than any other final drive system component. The correct tension adjustment (Figure 19-67) and proper lubrication of the drive chain help to extend the useful life of the chain and sprockets.

Nearly all of today's motorcycles that utilize a chain-and-sprocket final drive have a master link to connect the chain. Be sure that the master link is properly installed (Figure 19-68), which will prevent the master link clip from being knocked off due to rotating forces.

Figure 19-65 A sprocket in good condition is shown here.

Figure 19-66 These sprockets are worn out and require replacement.

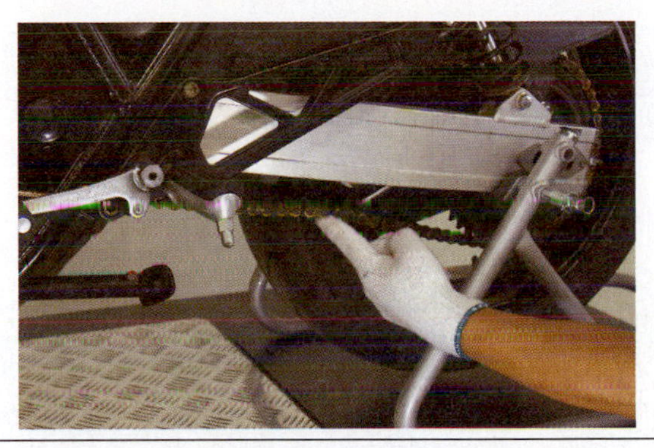

Figure 19-67 This photograph shows the technician checking for proper chain tension.

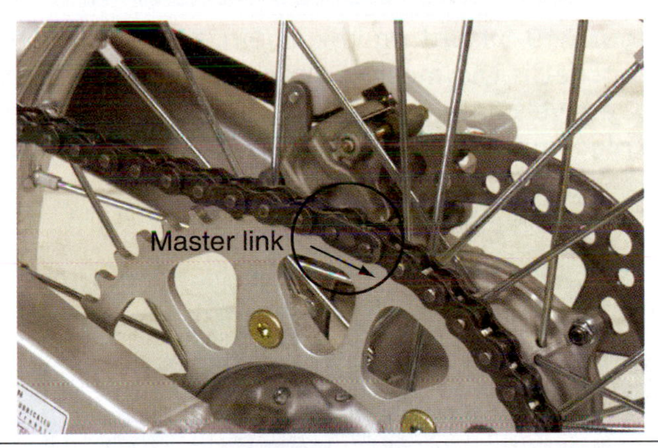

Figure 19-68 This is the correct installation direction of a master link. The clip faces toward the rear of the motorcycle.

The most common way to determine if a chain is worn out while still on the motorcycle is to lift the chain at various points around the rear sprocket. At a point midway between the top and bottom of the sprocket, try to pull the chain away from the sprocket. If you can pull the chain so that one-third of the sprocket tooth shows below the chain (Figure 19-69), the chain should be replaced and the sprocket should be closely inspected for wear as well.

Belt Drive

Belt-driven final drive systems are used on a few select motorcycle models. These systems use a Gilmer-type belt (Figure 19-70) that has teeth molded into it. The belt teeth mesh with a pair of toothed pulleys. The belt requires no lubrication, but must be kept clean and dry. Proper alignment of the belt and pulleys and correct belt tension are extremely critical with this type of final drive system.

Figure 19-69 Lifting the chain away from the sprocket is the most common way to check for excessive chain wear. This chain is in need of replacement.

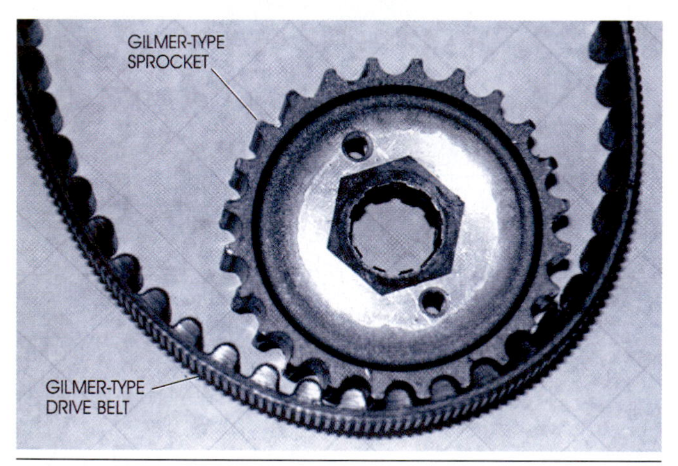

Figure 19-70 A typical Gilmer-type drive belt and sprocket.

Shaft Drive

Shaft drives are the most reliable final drive system. Shaft-driven systems are strong, clean, and require virtually no maintenance. In most cases a shaft-driven final drive system will outlast the machine that it is used on. Shaft-driven system maintenance consists of replacing the gear oil at the appropriate mileage or time intervals, which can be located in the appropriate manufacturer's service manual.

MOTORCYCLE STORAGE PROCEDURES

When a motorcycle will be stored for an extended period, you should take certain steps to reduce the chances of having storage-related problems.

Preparing a Motorcycle for Storage

To prepare a motorcycle or an ATV for storage, you should do the following:

1. Replace the engine oil and filter.
2. If the motorcycle engine is liquid-cooled, be sure that the cooling system is filled with the correct antifreeze solution to prevent the cooling system components from freezing.
3. Fill the fuel tank with fuel and add a fuel stabilizer. The fuel stabilizer will prevent the fuel tank from rusting, and prevent the fuel from deteriorating during storage. Also, turn off the fuel petcock while you're servicing the fuel system. This will prevent fuel seepage into the engine during storage.
4. When used, drain the carburetors. To verify that all of the fuel is out of the carburetor float bowls, start the engine (choke in the ON position) after the bowls have been drained. The engine will run for a brief interval while any remaining fuel is consumed. Fuel-injected motorcycles do not need draining.
5. To prevent rusting in the cylinders, you need to remove the spark plugs from the cylinders, pour a teaspoon of clean engine oil into each cylinder, and cover the spark plug holes with a piece of cloth. Crank the engine several times to disperse the oil in the cylinders. Reinstall the spark plugs.
6. Remove the battery and verify that it's fully charged. Store the battery in an area that's protected from freezing and direct sunlight. You should put the battery on a maintenance charger while it's in storage to prevent the battery from discharging.
7. Wash and dry the motorcycle, then wax all of the painted surfaces. This will protect the vehicle while it is in storage, and will provide a

clean, polished appearance when the vehicle is brought out of storage for the next riding season.

8. Lubricate the drive chain to prevent rusting during storage.
9. Inflate the tires to the factory-recommended air pressure. If possible, store the vehicle with the tires off the ground and protected from direct sunlight.
10. Cover the motorcycle or ATV with a suitable cover and store it in an area that is free of excessive dampness, dust, and chemical fumes.

Removing a Motorcycle or an ATV from Storage

When you remove a motorcycle or an ATV from storage, you should do the following:

1. Uncover and clean the vehicle.
2. Change the engine oil if the vehicle was stored for more than four months.
3. Charge and reinstall the battery as appropriate.
4. Check the following pre-ride inspection items—tire pressure, fluid leaks, chain condition, cable adjustment, smooth throttle movement, brake function, gauge operation, and lights operation.
5. Start the motorcycle and ride the vehicle slowly in a safe riding area to verify that everything is operating correctly.

EMISSION CONTROLS, OPERATION, AND MAINTENANCE

A technician performing engine maintenance on any modern street-legal motorcycle will very likely be working with some type of **emission-control system**. Many off-road motorcycles and ATVs are also now coming equipped with emission controls. It is important to remember that during combustion, the fuel-and-air mixture that burns in the engine produces chemical gases.

- **Carbon monoxide** (CO) is the result of partial combustion. As we mentioned in an earlier chapter, carbon monoxide is a colorless, odorless, poisonous, and potentially deadly gas.
- **Hydrocarbons** (HC) are the unburned (raw or vaporized) fuel.
- **Carbon dioxide** (CO_2) is a gas that results from complete combustion of the fuel.
- **Oxides of nitrogen** (NO_x) are oxidized nitrogen gases resulting from extremely high combustion temperatures.
- **Water** (H_2O) is a byproduct of combustion. Every gallon of burned fuel produces approximately one gallon of water in a vaporized form.

Emission-Control Standards

In the United States, the **Environmental Protection Agency (EPA)** has developed emission standards for all street-legal and some off-road motorcycle manufacturers to follow. Since 1978, motorcycles designed for street use have had to comply with all EPA emission standards. The EPA has established regulations that allow for acceptable noise levels as part of the emission-control standards.

Emission-Control Systems

We'll discuss the basic operation and maintenance of several emission-control devices found on motorcycles. It should be noted that emission-control systems used on street-legal motorcycles do not significantly reduce the power output of motorcycle engines. The following are the emission controls used to reduce street-legal motorcycle exhaust emissions.

- Crankcase-emission controls
- Evaporative-emission controls
- Exhaust-emission controls
- Noise-emission controls

Crankcase-Emission Controls

The engine crankcase has an emission-control system attached to the crankcase breather that prevents the hydrocarbons produced inside of the crankcase from making direct contact with the atmosphere. In most cases, it's done by routing the crankcase breather tube through the air box.

This breather arrangement allows the raw hydrocarbons from the crankcase to be recycled into the combustion chamber and burned during normal combustion. An air-and-oil separator allows condensed crankcase vapors to accumulate. This separator must be checked and emptied periodically. The breather tube has a transparent section on it to indicate when there's oil in the tube. This tube should be closely inspected if the engine is overfilled with oil or if the vehicle is accidentally tipped over. A typical crankcase-emission control system is illustrated in Figure 19-71.

Evaporative-Emission Controls

Motorcycles sold in California that are designed for street use must comply with the **California Air Resources Board (CARB)** requirements for evaporative-emission regulations. Although law in California requires evaporative-emission controls, many manufacturers build motorcycles for use in other states that comply with these requirements as well to help cut down emissions on all of their products.

Evaporative emissions come from the gasoline in the fuel tank. Gasoline vapors are pure hydrocarbons. When the engine isn't running, these vapors are routed into a charcoal canister (EVAP canister); in the canister, the vapors are absorbed and stored (Figure 19-72).

When the engine is running, the vapors are routed through the carburetor and into the engine via a purge-control valve (EVAP purge-control valve). The evaporative-emission carburetor air-vent control valve (EVAP CAV control valve) is opened to allow air to be drawn into the fuel system through the valve. A diagram of a typical evaporative-emission system is illustrated in Figure 19-73.

Maintenance of the evaporative-emission control system includes checking all hoses within the system for cracks and loose connections, and checking for airflow through the purge-control valve (Figure 19-74) by using a vacuum and pressure pump.

The charcoal canister should also be inspected for cracks or other damage. The charcoal canister

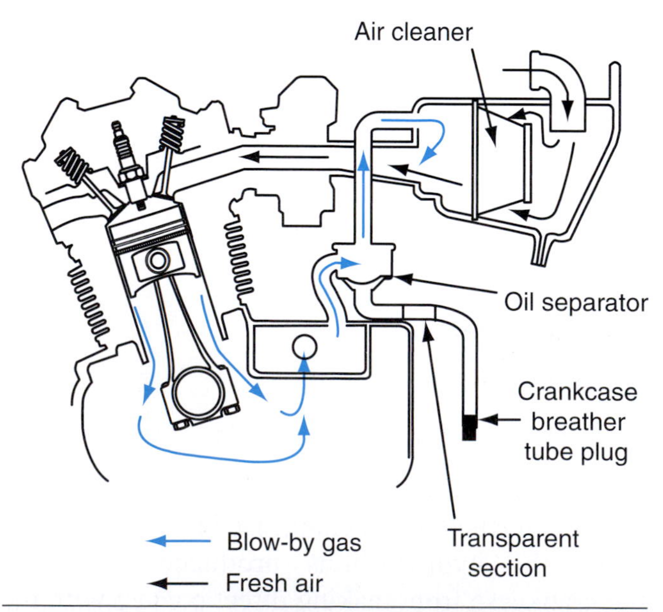

Figure 19-71 Crankcase emission controls prevent blow-by gases from entering the atmosphere by redirecting them back into the air box as illustrated here.

Figure 19-72 A typical charcoal canister is cut in half to show the interior as well as shown whole, as it would be when placed on a motorcycle.

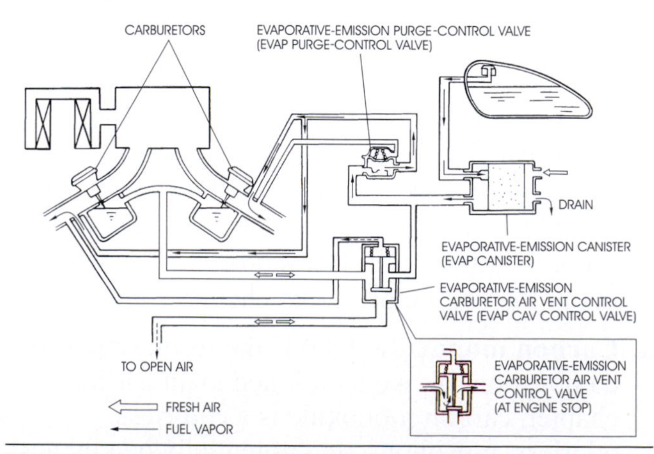

Figure 19-73 The flow of a typical evaporative emission control system is illustrated here.

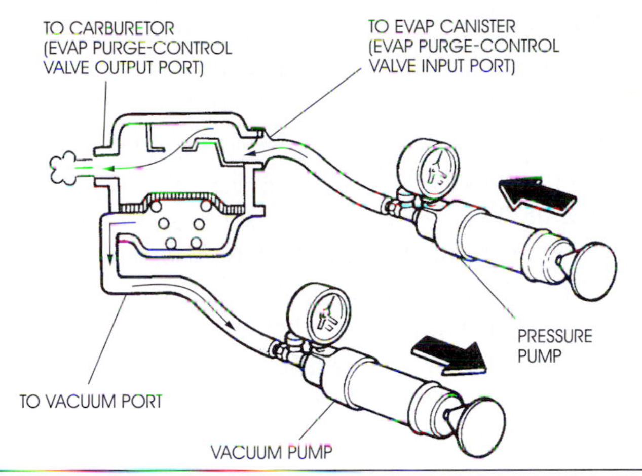

Figure 19-74 Testing a purge-control valve requires a pump for pressure as well as a vacuum pump to verify that it is operating properly.

must be replaced if the motorcycle is ever tipped over and the canister becomes contaminated with raw gasoline.

Exhaust-Emission Controls

Motorcycle exhaust emissions are controlled by three methods. The first method is by creating a lean-burning condition through the fuel system. This creates less pollution because combustion is more complete and efficient. The second method is by using a system that introduces filtered air into the exhaust system via the exhaust port. Figure 19-75 illustrates an example of this type of emission-control system, which allows fresh air to be drawn into the exhaust port whenever there's a negative pressure in the exhaust-port area. The fresh air is used to help complete the burning of any unburned fuel in the exhaust gases. This system also converts the exhaust hydrocarbons and carbon monoxide into carbon dioxide and water. A check valve is used to prevent reverse airflow through the system. Another valve is used to stop the fresh airflow from entering the exhaust system during rapid deceleration, to prevent after burn (backfiring) through the exhaust system. This type of emission control system requires no adjustments, but the components should be inspected on a periodic basis.

The third method may be found on exhaust systems of fuel-injected motorcycles. These motorcycles use a catalytic converter to help reduce pollution. A three-way catalytic converter (Figure 19-76) helps

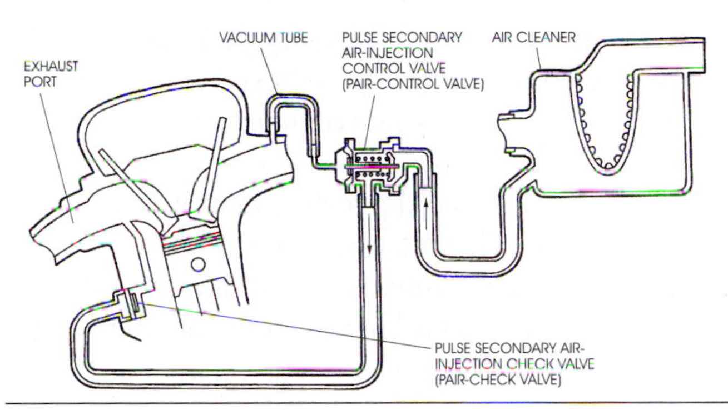

Figure 19-75 Many motorcycles inject fresh air into the exhaust port to assist with completing the burning of fuel in the exhaust gases.

reduce carbon monoxide (CO), hydrocarbons (HC), and oxides of nitrogen (NO_x) molecules.

The catalyst within the converter promotes a chemical reaction that oxidizes the HC and CO that is present in the exhaust gases. The three-way catalytic converter also reduces NO_x by converting NO_x to nitrogen and oxygen (O_2). The models with a three-way catalytic converter are usually equipped with an oxygen (O_2) sensor that tells the control module how much oxygen is in the exhaust. The control module can increase or decrease the amount of oxygen in the exhaust by adjusting the air/fuel ratio. By controlling the air/fuel ratio, any remaining toxic gases in the exhaust are efficiently converted into nontoxic compounds.

Noise-Emission Controls

Noise from a motorcycle does not just come from the exhaust system. A considerable amount of noise is created through the intake system and

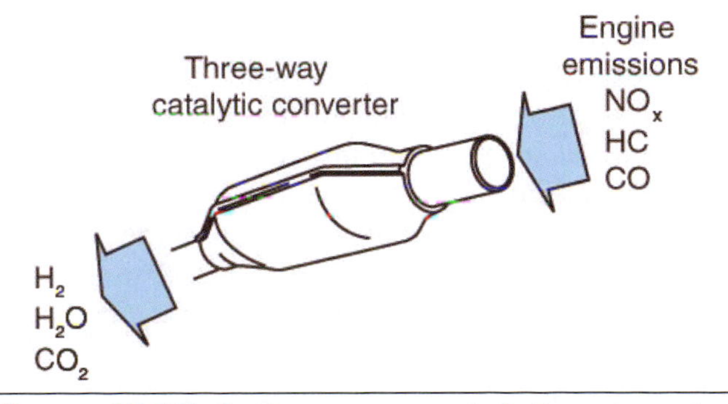

Figure 19-76 A three-way catalytic converter converts the toxic gases of hydrocarbon, carbon monoxide, and oxides of nitrogen into nontoxic hydrogen, water, and carbon dioxide.

even the tires. When a motorcycle is checked for noise emissions a variety of tests are performed including riding by a meter that checks for noise levels at different speeds.

Manufacturers of all street-legal motorcycles use exhaust systems that meet EPA standards. Tampering with these exhaust systems is forbidden by law. The EPA states that federal law prohibits the following:

- The removal or disabling of any noise-control device or element of design except for the purpose of maintenance, repair, or replacement, or
- The use of a vehicle after such device or element of design has been removed/disabled

Among those acts presumed to constitute tampering are the removal or puncturing of the muffler, baffles, header pipes, or any other component that controls exhaust gases.

In addition, removal or puncturing any part of the intake system is considered to be illegal as well as lack of proper emission-control system maintenance. Finally, the law states that for motorcycles equipped with EPA-regulated equipment replacing any moving parts of the vehicle, or parts of the exhaust or intake system with parts other than those specified by the manufacturer, is considered an illegal act.

As you can see, all street-legal motorcycles have strict rules set by the EPA to ensure that noise emissions are followed. These control systems are designed to help keep our atmosphere clean and our neighborhoods quiet, while causing minimal restrictions to the motorcycle's usable power.

Summary

- Understanding the importance of scheduled maintenance intervals for both the engine and chassis is very important to be a successful motorcycle technician. Peak performance from a motorcycle requires that each part be in good working condition and correctly adjusted. Scheduled servicing ensures that marginal parts and out-of-tolerance adjustments are routinely corrected. All manufacturers recommend that motorcycles be serviced at specific mileage or time intervals. The suggested maintenance intervals are listed in the owner's manual to help remind owners and in the service manual to help technicians set up a realistic and appropriate maintenance schedule.
- As a technician, you will perform engine service including cleaning, adjusting, and replacing parts to ensure that parts meet the manufacturer's specifications. Motorcycle chassis maintenance also includes cleaning, inspecting, and replacing parts that are subject to wear. The technician performing this maintenance adjusts the motorcycle's chassis systems to the settings specified by the manufacturer.
- When a motorcycle will be stored for an extended period, there are steps that need to be taken to reduce the chances of having any engine or chassis problems when it comes time to remove the motorcycle from a storage state.
- There are several emission-control devices found on motorcycles. Remember that emission-control systems do not significantly reduce the power output of motorcycle engines but instead create cleaner-burning engines to help keep our air clean and noise at a reasonable level. Federal laws have been put into place to ensure that all manufacturers of motorcycles comply with the standards set by the Environmental Protection Agency (EPA). Emission controls used on motorcycles include the following:
    - Crankcase-emission controls
    - Evaporative-emission controls
    - Exhaust-emission controls
    - Noise-emission controls

Chapter 19 Review Questions

1. If accidentally spill battery acid, what should you use to neutralize it?
   a. A nonflammable household cleaner
   b. A water and baking soda solution
   c. Water
   d. Baking soda
2. A ________ test will provide the best indication of any internal engine problems.
3. Which type of air filter is most commonly used on motorcycles and ATVs?
   a. Foam air filter
   b. Oil bath-type air filter
   c. Gauze-type air filter
   d. Paper-type air filter
4. Motorcycle and ATV manufacturers recommend that their machines be
   a. serviced at specific mileage or time intervals.
   b. serviced every four months.
   c. serviced when they no longer run properly.
   d. tuned-up at least once a year.
5. For every gallon of fuel burned in a motorcycle or ATV engine, approximately how much water is produced?
   a. 1/2 gallon
   b. 1-1/2 gallons
   c. 1 gallon
   d. 2 gallons
6. What is used to replenish a low conventional battery?
   a. Tap water
   b. Baking soda
   c. Sulfuric acid
   d. Distilled water
7. The four-stroke valve adjuster system that requires no maintenance is the___________ arrangement.
   a. hydraulic valve-adjuster
   b. shim-under-bucket
   c. shim-over-bucket
   d. screw-and-locknut
8. All four-stroke motorcycles must have at least one yearly tune-up that includes adjusting the valves. True or False.
9. What is used to test a cooling system for the correct mixture of coolant and water?
   a. Barometer
   b. Manometer
   c. Hydrometer
   d. Thermometer
10. A clutch lever that has too little play will result in a clutch that
    a. grabs.
    b. slips.
    c. creeps.
    d. lurches.
11. If a hydraulic brake lever or brake pedal feels soft or spongy when operated, the system most likely has __________ in it.
    a. dirty fluid
    b. air
    c. water
    d. rust
12. The final drive system that requires the least amount of maintenance is the __________ final drive.
    a. shaft-driven
    b. chain-driven
    c. belt-driven
    d. gear-driven

13. A small hole separates the engine oil and coolant and can be found in a liquid-cooled engine. True or False.
14. The four-stroke valve-adjuster system that requires camshaft removal to make valve clearance adjustments is the
    a. screw and locknut.
    b. shim-over-bucket.
    c. shim-under-bucket.
    d. hydraulic tappet.
15. When used, an exhaust-emission control system directs fresh air into the exhaust port whenever there's a negative pressure in the
    a. carburetor venturi.
    b. exhaust port area
    c. intake port area.
    d. combustion chamber.

CHAPTER

20 Motorcycle Troubleshooting

Learning Objectives

When students have completed the study of this chapter and its laboratory activities they should be able to:

- Understand how to systematically approach problems found when working on a motorcycle
- Understand troubleshooting procedures for the following conditions:
  - Engine problems
  - Fuel system problems
  - Electrical system problems
  - Chassis-related problems
  - Abnormal noise problems

Key Terms

Constant failure

Improper-service failure

Intermittent failure

Symptom

INTRODUCTION

The ability to quickly and correctly troubleshoot motorcycle problems is a major sign of a competent technician. Proper diagnosis of a malfunction makes motorcycle disassembly, repair, and reassembly much easier. If a malfunction has been improperly diagnosed, the repair process will become long and tedious, or even seemingly impossible. In order to be a successful technician, you must possess proficient troubleshooting skills. First and foremost, troubleshooting begins with a thorough knowledge of the following areas:

- Components of a motorcycle
- Understand the job each component performs
- Understand the effect each component has on the overall operation of the motorcycle
- What types of failures/symptoms a bad component will cause

Once you've gained this knowledge, troubleshooting becomes a systematic and controlled approach to solving a problem. You must understand that this knowledge does not always come easily. You must be prepared to make mistakes along the way when diagnosing and repairing a motorcycle. However, make sure you understand what you did incorrectly in those situations and learn from those mistakes.

When troubleshooting, you will need to do the following:

- Gather all available information about the machine's malfunction
- Analyze the symptoms related to the problem
- Pinpoint the most likely cause of the problem

The repair process doesn't begin until after you've gotten a clear picture of what's causing the problem. To achieve this picture, you will have to mentally divide the motorcycle into sections (fuel, engine, electrical, and chassis):

- Picture each component of every section
- Picture each component and its relationship to the other components and decide if each part is functioning properly

For example, suppose a spark plug isn't firing correctly. You must envision the operation of the spark plug in relation to the other motorcycle systems. After doing this, you'll have a variety of possible problems identified, besides the obvious conclusion that the spark plug is bad. The problem could be due to a dirty air filter that's creating an excessively rich fuel mixture. The source of the problem could also be the ignition circuit or the fuel system.

It's imperative that you understand what you're trying to repair before disassembling a machine. Once you begin the disassembly process, the troubleshooting process is over. In our example, cleaning and replacing parts in the fuel system won't solve the problem if the cause is a faulty ignition. Therefore, ensure that you have truly isolated the problem before beginning any repair.

For this chapter, we've compiled a group of tables that cover virtually every aspect of troubleshooting various motorcycle problems. Because it's impossible to cover every type of motorcycle problem, we'll introduce basic diagnostic and troubleshooting techniques and concepts that you can apply to many different situations. Each topic in this chapter contains tables that can be used as a guide when troubleshooting motorcycle problems—thus you will gain a solid foundation for developing your own troubleshooting expertise. Although you'll find there to be many tables here, they by no means contain every possible cause for each problem listed. These tables are intended to aid you in troubleshooting problems by giving you some of the more common causes of the problems listed. To get the most from this chapter, focus your attention on the basic concepts presented herein.

SYSTEMATIC APPROACHES TO SOLVING PROBLEMS

This section covers techniques that will expedite the process of troubleshooting motorcycle problems. Developing a systematic approach to solving problems will help you to perfect your skills as you gain more experience.

TYPES OF PROBLEMS

A **symptom** is an indication of an abnormal condition that you can recognize and identify. An example of a symptom would be a motorcycle that's making a ticking sound when it's idling. The symptom helps you determine the cause of the problem. The following paragraphs cover three types of failures that you may encounter.

Constant Failures

A **constant failure** occurs when a symptom is always present. For instance, a motorcycle is functioning properly and without warning, the engine fails and the rear wheel locks up and will no longer turn. The locked rear wheel is considered a constant failure.

Intermittent Failures

An **intermittent failure** isn't always present. This type of failure increases the difficulty of the troubleshooting process. For example, a motorcycle functions properly with the exception of occasionally blowing a fuse when it hits a bump in the road. The rider replaces the fuse and rides trouble-free until the problem recurs when another large bump is encountered. Chances are, this intermittent problem is caused by multiple factors. In this instance especially, a systematic approach to troubleshooting the problem is required. With any problem (performance, electrical, mechanical, or fuel), a systematic approach allows the problem to be diagnosed in a reasonable amount of time with a high degree of accuracy.

Improper-Service Failures

An **improper-service failure**, as the name implies, is caused by a mistake during the servicing of the machine. Suppose a customer brings an off-road motorcycle to your service department for a new set of tires and the technician servicing the machine fails to properly tighten one of the wheel axle nuts and forgets to install a cotter pin on the nut. After the customer gets the bike back and goes out for a ride, this mistake causes the wheel to wobble and fall off while the customer is riding the motorcycle while out on the trail. Fortunately, most failures caused by improper service aren't this dramatic. It is important not to overlook problems resulting from incorrect service when you are troubleshooting a vehicle.

Beginning the Troubleshooting Process

The proper method of diagnostic troubleshooting consists of four steps that must be followed in the proper sequence. Following these steps will ensure a foolproof approach to the troubleshooting and repair process:

1. Verify the problem.
2. Isolate the problem.
3. Repair the problem.
4. Verify the repair.

When troubleshooting, you must observe the failure and verify that all of the information you've received is accurate and guides you to the troubled area. After you have verified the condition, you're ready to isolate the problem.

Isolating a problem begins with the easiest and most obvious solution to the problem. As the simplest solutions fail to correct the problem, progression to more involved and difficult checks are needed in a step-by-step manner. The most common diagnostic mistake is to overlook the obvious or easiest possible cause of a failure. For example, a motorcycle was functioning properly, then stalled and wouldn't restart. The owner took the motorcycle to a service shop. The technician removed and checked the spark plugs, checked the air filter and performed compression and leak-down tests. When all was said and done, the problem was an empty fuel tank. Unbelievably, this situation is not uncommon and results from poor troubleshooting skills (not starting with the simplest solutions first).

The symptoms of a problem guide you to the specific system you should troubleshoot, provided you have an understanding of how each system works and what it's responsible for. Take the following examples:

- If the battery will not turn the engine over and just clicks when you press the starter button, you can assume that the machine has a discharged battery and possibly a charging system that is failing to provide a proper charge to the battery.

- If gasoline is leaking from the carburetor overflow tube, you can assume that there is an internal carburetor problem that's causing excessive amounts of gasoline to enter the system.

As the severity of problems increases, the knowledge required to isolate problems increases. An example of this is poor engine performance. A performance problem could be caused by an ignition system failure, a mechanical engine problem, or even a fuel-related problem. It is imperative to use all available resources and any information you can gather from your customer to assist you in identifying which system is responsible for the problem.

After you've isolated the problem, you can then repair the problem. In order to repair the problem, you must refer to the specific service manual for the particular machine you are servicing.

When you complete the repair, it is very important to verify the repair. If you cannot verify that the repair was successful and the problem is still present, you must repeat the troubleshooting process, beginning with the verification stage.

Troubleshooting Guides

All manufacturers' service manuals contain troubleshooting checklists and/or tables of possible operating troubles along with their probable causes. These items are designed to aid the technician in troubleshooting and problem solving. All possibilities should be carefully checked because multiple factors may be causing the overall problem. Throughout the tables provided in this chapter, examples of typical problems and possible solutions are provided. These tables have been derived from current service manuals and technical guides to create a generic point of view. However, you should note that the tables provided in this chapter, as well as in the manufacturer's service manuals, are intended only as a guide to diagnosing problems. Always read the detailed information in the specific chapters of the appropriate service manual before performing service work on any system or major component. Remember to adhere to all cautions and warnings.

As you learn more about various motorcycle systems, you may develop a tendency to troubleshoot problems based on your personal experience. This approach can be a gamble that may save you time; but if you guess wrong, it costs you time and money. Do not be afraid to apply your experience to a good troubleshooting routine, but do not underestimate repairs only because the failure looks familiar.

Locating and fixing a problem is very rewarding, provided you use good troubleshooting techniques. Furthermore, the more difficult the problem, the greater the reward when you've solved it. To be successful, the most important barrier to overcome is the lack of self-confidence required to perform the job. The following are some things to keep in mind when you are troubleshooting a problem:

- Always think the problem through.
- Never overlook the obvious.
- Never assume anything.
- Never take shortcuts.
- Never make more than one change or adjustment at a time.
- Always use the appropriate service manual(s) for all removals, replacements, and adjustments.
- Remember to always *verify* the problem, *isolate* the problem, *repair* the problem, and most importantly, *verify the repair*.

TROUBLESHOOTING ENGINE PROBLEMS

There are various conditions that will require the need for troubleshooting an engine condition. These cover a wide spectrum from an engine that does not start to engine overheating. You must have a thorough knowledge of all components of the engine before you can begin troubleshooting. You must know what parts are used, understand how they work, and be aware of their relation to one another.

Refer to the following tables as a guide to the conditions listed:

Engine Does Not Start

Table 20-1: Complaint: Engine Does Not Start/Starting Difficulty

Specific Symptom	Possible Cause
Starter motor not operating	■ Engine stop-switch off ■ Battery voltage low ■ Fuse blown ■ Ignition-switch trouble ■ Neutral-switch faulty ■ Relays not functioning ■ Starter button not contacting ■ Wiring open or short ■ Starter-motor faulty
Starter motor operates but engine doesn't turn over	■ Starter-motor clutch faulty ■ Engine clutch worn
Engine won't turn over	■ Valve seizure ■ Camshaft seizure ■ Rocker arm seizure ■ Cylinder, piston seizure ■ Crankshaft seizure ■ Connecting rod seizure ■ Transmission seizure
Compression low	■ No valve clearance ■ Bent valve ■ Cylinder, piston worn ■ Piston rings worn

Engine Runs Poorly at Low Speed

Table 20-2: Complaint: Runs Poorly at Low Engine Speeds

Specific Symptom	Possible Cause
Spark weak	■ Battery voltage low ■ Spark plug faulty ■ Spark plug cap faulty ■ Ignition coil faulty
Air-and-fuel mixture incorrect	■ Pilot screw improperly adjusted ■ Pilot jet or air passage clogged ■ Air-bleed pipe bleed holes clogged ■ Air filter dirty or missing ■ Choke left on or stuck ■ Fuel level in float bowl incorrect ■ Fuel tank air vent obstructed
Compression low	■ Cylinder head leak ■ Incorrect valve clearance ■ Cylinder, piston worn ■ Piston ring clearance excessive ■ Cylinder head warped ■ Cylinder-head gasket damaged ■ Valve spring broken or weak ■ Valve not seated properly

Engine Runs Poorly at High Speed

Table 20-3: Complaint: Runs Poorly/No Power at High Engine Speeds

Specific Symptom	Possible Cause
Timing/firing incorrect	■ Spark plug fouled ■ Spark plug cap shorted or not in good contact ■ Pickup coil faulty ■ Ignition coil faulty ■ ICM faulty
Air/fuel mixture incorrect	■ Choke stuck ■ Main jet clogged or incorrect size ■ Jet needle or needle jet worn ■ Air jet clogged ■ Fuel level incorrect ■ Air filter dirty or missing ■ Water or foreign matter in fuel ■ Fuel tank air vent obstructed ■ Fuel petcock blocked ■ Fuel line blocked
Compression low	■ Valve clearance incorrect ■ Cylinder, piston worn ■ Piston rings worn ■ Cylinder-head gasket damaged ■ Cylinder head warped ■ Valve spring broken or weak ■ Valve not seated properly
Knocking	■ Carbon buildup in combustion chamber ■ Fuel poor quality or incorrect fuel ■ Spark plug incorrect ■ Ignition timing incorrect
Poor performance	■ Throttle valve won't fully open ■ Clutch slipping ■ Engine overheating ■ Engine oil level too high ■ Engine oil viscosity too high

Engine Overheats

Table 20-4: Complaint: Engine Overheating

Specific Symptom	Possible Cause
Air/fuel mixture incorrect	■ Main jet clogged or wrong size ■ Fuel level in carburetor float bowl too low ■ Carburetor holder loose ■ Air cleaner poorly sealed or missing ■ Air-cleaner duct poorly sealed ■ Air cleaner clogged
Compression high	■ Carbon buildup in combustion chamber
Engine load faulty	■ Clutch slipping ■ Engine oil level too high ■ Engine oil viscosity too high ■ Brake dragging
Lubrication inadequate	■ Engine oil level too low ■ Engine oil poor quality or incorrect
Gauge incorrect	■ Water-temperature gauge broken ■ Water-temperature sensor broken
Coolant incorrect	■ Coolant level too low ■ Coolant deteriorated
Cooling-system component incorrect	■ Radiator clogged ■ Thermostat defective ■ Radiator cap defective ■ Thermostatic fan switch faulty ■ Fan relay faulty ■ Fan motor inoperative ■ Fan blades damaged ■ Water pump faulty

Engine Does Not Reach Operating Temperature

Table 20-5: Complaint: Overcooling	
Specific Symptom	**Possible Cause**
Gauge reads incorrect	▪ Water-temperature gauge faulty ▪ Water-temperature sensor faulty
Cooling-system component incorrect	▪ Thermostatic fan switch trouble ▪ Thermostat trouble

Excessive Exhaust Smoke Troubleshooting

Table 20-6: Complaint: Excessive Exhaust Smoke	
Specific Symptom	**Possible Cause**
White smoke	▪ Piston oil ring worn ▪ Cylinder worn ▪ Valve oil seal damaged ▪ Valve guide worn ▪ Engine oil level too high
Black smoke	▪ Air cleaner dirty ▪ Main jet too large or fallen off ▪ Choke left on or stuck ▪ Fuel level in float bowl too high

Drive Train Troubleshooting

The two most common symptoms found when diagnosing motorcycle transmission problems are the following:

- Transmission is hard to shift
- Transmission jumps out of gear

There are other drive train-related problems as well. The following table provides the most common causes of drive train-related problems:

Table 20-7: Transmission Problems	
Symptom	**Possible Cause**
Hard to shift	▪ Improper clutch adjustment ▪ Incorrect transmission oil ▪ Bent shift forks ▪ Bent shift shaft ▪ Damaged shift drum
Jumps out of gear	▪ Shift fork worn ▪ Gear dogs and/or dog holes worn ▪ Shift drum groove chipped ▪ Gear positioning-lever spring broken ▪ Shift fork pin worn
Doesn't go into gear; shift pedal doesn't return	▪ Clutch not disengaging ▪ Shift fork bent or seized ▪ Gear stuck on the shaft ▪ Shift return spring broken ▪ Shift-mechanism arm broken
Overshifts	▪ Gear positioning-lever spring broken ▪ Shift-mechanism arm spring broken
Clutch Problems	
Symptom	**Possible Cause**
Clutch slipping	▪ Friction plates worn ▪ Steel plates worn ▪ Clutch springs broken or weak ▪ Clutch cable improperly adjusted ▪ Clutch cable sticking ▪ Clutch-release mechanism sticking
Clutch dragging	▪ Clutch cable improperly adjusted ▪ Clutch plates warped ▪ Engine oil viscosity too high ▪ Engine oil level too high ▪ Clutch-release mechanism sticking

FUEL SYSTEM TROUBLESHOOTING

Fuel system troubleshooting is one of the most common motorcycle repair jobs and can be a simple, straightforward, rewarding procedure, or a tedious, complicated, unrewarding chore. The difference between these two extremes lies within your approach to problem solving. You can randomly disassemble and replace components, or you can take a systematic, sensible approach.

When troubleshooting the fuel system-related problem, start with the simple areas such as verifying fuel in the gas tank and fuel flow to the carburetors/injectors.

A plugged gas-tank vent, fuel shut-off valve, or pinched fuel line can be responsible for restricting the fuel supply to the fuel system. Trace through the system in search for blockage. Begin with the vent, then the shut-off valve, and finally the gas line. If any of the components are plugged or restricted, they must be repaired.

Is It Rich or Lean?

Usually, fuel system problems are based on an improper fuel-and-air mixture that's either too rich or too lean. Observe the engine exhaust and check the condition of the spark plug to determine whether the mixture is too rich or too lean. Always keep in mind that a rich or lean mixture can have more than one cause. Too much fuel or not enough air can cause a rich mixture. Too much air or not enough fuel can cause a lean mixture. Either condition can become bad enough that the engine will not start. The following table shows some common carburetor-related symptoms and some likely causes.

Table 20-8: Carburetor-Related Problems

Symptom	Possible Cause
Lean mixture	■ Pilot jet clogged ■ Float level too low ■ Fuel line partially restricted ■ Intake-manifold air leak ■ Fuel pump not working properly ■ Vacuum piston faulty (CV carb)
Rich mixture	■ Choke valve stuck ■ Float level too high ■ Carburetor air jets blocked ■ Air filter element excessively dirty
Engine stalls, hard to start, rough idling	■ Idle speed not properly adjusted ■ Fuel line restricted ■ Fuel mixture incorrect ■ Fuel contaminated/deteriorated ■ Intake-manifold air leak ■ Fuel pump not operating correctly ■ Pilot-circuit blocked ■ Float level incorrect ■ Fuel tank breather clogged
Backfiring or misfiring during deceleration	■ Air cutoff valve faulty ■ Lean mixture in pilot circuit ■ Vacuum line loose or off
Backfiring or misfiring during acceleration	■ Fuel mixture too lean
No fuel flow	■ Fuel petcock blocked ■ Fuel tank air vent obstructed ■ Fuel line blocked ■ Float valve stuck closed
Engine flooded	■ Fuel level in float bowl too high ■ Float valve worn or stuck open ■ Starting technique incorrect

When Is the Problem Apparent?

After determining whether the mixture is too rich or too lean, you must determine in which throttle position the problem occurs to know which circuit needs repair. This means the physical position of the throttle such as one-quarter open or one-half open.

Before starting to work on the fuel system, you should always check some of the external items that can affect carburetion. If the mixture appears rich, check the air filter and the cable to the carburetor choke. If the air cleaner is excessively dirty, the air will have difficulty getting to the engine. If the choke cable is too tight, the choke will be allowing extra fuel into the Venturi or cutting off the air supply, depending on the cold starting device used on the carburetor. If the mixture is too lean, ensure that the fuel is flowing properly from the fuel tank. In addition, inspect the intake manifold for air leaks.

If everything on the external side of the engine is in proper working order, the carburetor will most likely need repairing. The following tables are divided into throttle ranges and provide common causes of rich and lean mixtures. Also included are common repairs for each situation, as well as suggestions to follow if none of the common problems are present.

0—¼ Throttle Opening

Table 20-9: 0–¼ Throttle Opening

Too Rich		Too Lean	
Problem	**Remedy**	**Problem**	**Remedy**
Choke activated	Verify that the choke is in the Off position	Carburetor mounted loosely	Tighten carburetor
Pilot air passage blocked	Blow out passage area with compressed air	Pilot jet plugged	Clean jet with compressed air
Pilot jet loose	Tighten jet	Pilot outlet or bypass ports clogged	Clean with compressed air
Pilot jet air bleed blocked	Clean jet with compressed air	Fuel level too low	Adjust level as per service manual
Fuel level too high	Adjust level as per service manual		
Lean the mixture by turning the adjustment screw one-quarter to one-half turn. (Check service manual to determine if adjustment is CCW or CW.)		Richen the mixture by turning the adjustment screw one-quarter to one-half turn. (Check service manual to determine if adjustment is CCW or CW.)	

¼—½ Throttle Opening

Table 20-10: ¼–½ Throttle Opening

Too Rich		Too Lean	
Problem	**Remedy**	**Problem**	**Remedy**
Pilot jet loose	Tighten	Needle jet blocked	Clean with compressed air
Pilot air passage obstructed	Clean with compressed air	Pilot outlet or bypass ports clogged	Clean with compressed air
Primary air passage blocked	Clean with compressed air	Main jet clogged	Clean with compressed air
Needle jet/jet needle worn	Replace	Fuel level too low	Adjust as needed
Main jet loose	Tighten		
Fuel level too high	Adjust as needed		
Air filter excessively dirty	Clean or replace		

½—¾ Throttle Opening

Table 20-11: ½–¾ Throttle Opening

Too Rich		Too Lean	
Problem	**Remedy**	**Problem**	**Remedy**
Needle jet/jet needle worn	Replace	Main jet clogged	Clean with compressed air
Main jet loose	Tighten	Needle jet blocked	Clean with compressed air
Primary air passage blocked	Clean with compressed air	Fuel level too low	Adjust as needed
Fuel level too high	Adjust as needed		
Air filter excessively dirty	Clean or replace		
If none of the above help the problem, lower the jet needle one position.		If none of the above help the problem, raise the jet needle one position.	

¾—Full Throttle Opening

Table 20-12: ¾–Full Throttle Opening			
Too Rich		Too Lean	
Problem	Remedy	Problem	Remedy
Main jet loose	Tighten	Main jet clogged	Clean
Needle jet/jet needle worn	Replace	Needle jet clogged	Clean
Air filter clogged	Clean or replace	Fuel level low	Adjust as needed
Fuel level too high	Adjust as needed		
If none of the above help the problem, install a smaller main jet.		If none of the above help the problem, install a larger main jet.	

RELATED PROBLEMS FOR OTHER FUEL SYSTEMS

Aside from the carburetor problems previously discussed, other common problems will appear from time to time. The most common of these problems is water in the float bowl. Water is heavier than gasoline and will penetrate the circuits of the carburetor and eventually reach the engine. When this occurs, the engine runs rough or not at all. Draining the float bowl cures the symptom; ultimately, however, the cause needs to be corrected.

Water in the Fuel System

Often cleaning a motorcycle with a high-pressure washer causes water to penetrate the sealing area between the air filter and carburetor or the seal of the fuel cap. If water penetrates either area, it eventually enters the fuel system and causes problems.

Clogged Fuel Tank Vent

Another widespread and potentially baffling problem is a clogged fuel tank vent. Normally, air enters the tank through the vent and replaces the space left by the fuel as it's burned. If the vent is clogged, a vacuum is created in the fuel tank and can restrict the flow of fuel. This restriction can sometimes be enough to cause the engine to stall. Given this scenario, the rider opens the fuel cap to ensure that there's gas in the tank. Opening the fuel tank removes the vacuum, the rider sees fuel, and starts the motorcycle. The machine functions properly for a few more miles until the vacuum is recreated and the fuel flow is slowed or stopped.

When confronted with a baffling fuel system problem, exercising common sense is the only way to rectify the problem. Follow these basic tips when you have reached an impasse:

- Remove yourself from the situation and consider the total process of carburetion.
- Thoroughly think out all options before doing anything drastic.

It is important to implement one change at a time; multiple adjustments made simultaneously amplify the problem.

Electronic Fuel-Injection System Troubleshooting

As mentioned previously, fuel-injected motorcycles are becoming more prevalent every day. Fortunately, there are relatively few problems with fuel-injected motorcycles but when a condition does arise, in most cases the fuel-injection light (FI light), also known in the automotive industry as the malfunction indicator light (MIL) will light up to let the rider know that there is a problem.

When properly activated the FI light will show the technician a code as to what is causing the problem. Along with the FI light, most motorcycle manufacturers are now using electronic diagnostic tools to assist with the troubleshooting of fuel-injected motorcycles due to the high level of technology involved. Unlike the automotive industry, these testers are not regulated as to what information is provided to the technician. Therefore, it is necessary to understand each specific manufacturer's special tool. As the information we are providing here is as generic as possible we will refer you to the specific manufacturer when working with these highly specialized and expensive tools.

Many fuel-injected engine symptoms are different than those found on carbureted engines. As an example, a carbureted engine with low compression tends to run lean. That means that it takes longer to warm up and may not perform as well as a similar machine that has proper compression. Why is this? An engine with lower compression than it should will have less intake port vacuum for the carburetor(s) and less fuel will be drawn into the engine. This makes for a lean running condition. But, fuel-injected engines don't act the same when the compression is low because the fuel is injected into the port regardless of intake vacuum. A fuel-injected engine will generally run rich when it has low compression, just the opposite of a carbureted engine. Remember that recommended compression pressures vary between engines so always check the appropriate service manual specification for the engine you are testing.

The real issue here is that even though the motorcycle will seem to have a fuel-related problem, the fact of the matter is that the problem may be mechanical and no FI light indicator will tell you that.

The real test for the technician with fuel-injected engines is determining the problem when the engine runs poorly and appears to have a fuel-related problem but there is no indication from the FI light. Table 20-13 shows some fuel-injected engine symptoms and some known problems that have been found with these symptoms that will not trigger an FI failure.

Table 20-13: Electronic Fuel Injection Troubleshooting

Symptom	Suggestion	Explanation
Engine will not start; no spark or fuel injection	■ Check for normal "power on" operation. The FI light on the dash and the fuel pump should operate (you can hear the pump working) for about two seconds when the ignition switch is first turned on.	If the FI light stays on or the fuel pump does not operate, inspect the fuses and confirm power and ground circuits to the ECM, including the bank angle sensor if the machine has one.
	■ Check for a failed crankshaft or camshaft position sensor. If these sensors fail, a code may trigger but **only** after the engine has cranked for at least 15 seconds.	Test these components to ensure that no failure has occurred by cranking the engine for 15 to 20 seconds. If the FI light does not turn on, these components are electrically okay.
	■ Inspect the crank and camshaft position sensor rotors for damage.	A bent finger on one of these rotors can cause a no-run condition.

(continued)

Table 20-13: Electronic Fuel Injection Troubleshooting *(continued)*		
Symptom	**Suggestion**	**Explanation**
Engine starts but runs poorly	■ Check that the battery terminals are tight. ■ Check crankshaft and camshaft position sensors for poor contact at their connectors. ■ Check that ground wires are tight at the ground bolt. ■ Check engine compression and cam timing.	
Engine starts, runs rich on all cylinders	■ Disconnect the throttle position sensor and see if there are any changes in the way the engine runs. ■ Measure throttle sensor output voltage with the throttle closed and compare to the manufacturer's specification.	
	■ Inspect fuel pressure.	Verify that the machine has correct pressure by testing it using the specific information in the appropriate service manual.
	■ Inspect for a disconnected MAP sensor hose. ■ Inspect the fuel return hose connecting the fuel pressure regulator and fuel tank for signs of being pinched. ■ Inspect for insufficient or excessive battery voltage.	
Engine starts, runs rich on some but not all cylinders	■ Check the fuel pressure regulator for a leaking diaphragm; may allow fuel into the vacuum hose. ■ Check engine compression. ■ Compare spark plug color between cylinders. ■ Visually inspect the fuel injectors for leaking.	The regulator should hold vacuum when tested.
Engine is hard to start, or has a misfire at mid to high rpm	■ Measure the peak voltage from crankshaft and camshaft position sensors. ■ Inspect the crankshaft and camshaft position rotors for damaged or bent fingers.	Verify the voltage with the appropriate service manual. Generally speaking, it should be above 0.7V at cranking speeds.

ELECTRICAL PROBLEM TROUBLESHOOTING

Of all the problems that come into a motorcycle service department, electrical system problems are usually considered the most difficult to troubleshoot and repair. One of the reasons for this is that many technicians don't fully understand electrical systems, and they can't actually see the electrical system working. They only know the symptoms. For instance, if a charging system stops functioning, you can't see that electricity isn't being produced. All you know is that the battery is dead. Nevertheless, if a tire goes flat, you can see the result of the problem as well as the nail that caused it!

After you've mastered the ability to properly and quickly analyze electrical problems, you'll become a valuable asset to any motorcycle service department. With a complete understanding of how the electrical systems in motorcycles work, you will rarely take more than an hour or so to diagnose any electrical problem. To help you categorize electrical system problems, we will break down this section into four basic areas:

- Charging system troubleshooting
- Ignition system troubleshooting
- DC circuit troubleshooting
- Electric starter-motor troubleshooting

Charging System Troubleshooting

The symptoms found in a charging system that's not operating properly are simple and straightforward. The motorcycle charging system is either undercharging or overcharging.

In the case of a system that's undercharging, the battery will eventually go dead, and the electrical components will no longer function. On some older motorcycles that are used often at night and run at constant low speed with the headlight turned on, the battery may become weak and require charging. This may occur because many older motorcycles had charging systems that didn't function to their full potential until they were running at higher engine rpm. Charging systems in most of today's motorcycles are designed to provide more than adequate electrical output whenever the engine is running. If a battery constantly discharges even though it's been properly maintained and the vehicle has been used frequently, check the charging system before replacing the battery with a new one. Batteries can be quite expensive.

In the case of an electrical charging system that is overcharging the battery, there will undoubtedly be a faulty component in the charging system—most likely, the voltage regulator, which in most cases is integrated into the rectifier as explained in Chapter 15.

Troubleshooting electrical problems isn't difficult. As a matter of fact, it's one of the cleanest jobs you'll be required to do. In most cases, the causes of the problems are as simple as a dirty or loose connection. One manufacturer has let it be known that out of every 150 charging system components that are returned for warranty purposes, only one is actually defective! This tells us that as the technician is diagnosing the problem in the charging system, he or she is fixing the problem without even knowing it. Over 95 percent of all charging system-related problems are connection-related and not actual component problems.

Be sure you know the color codes used for wires before beginning to work on an electrical problem. Every manufacturer uses different colored wires for their electrical circuitry. As you perform each step in the troubleshooting process, check to see if you have corrected the problem.

Use the following tables to supplement the basic charging system troubleshooting procedures we've discussed. The troubleshooting procedures in the tables can be used for any charging system.

Table 20-14: Symptom: Discharging or Weak Charging System		
Step	**If Measurement is Correct**	**If Measurement is Incorrect**
1. Measure the charging voltage at the battery with the engine running at the specified rpm.	a. Check the battery for amperage loss with the key in the Off position. If excessive amperage is being drawn, locate and repair. b. Check the battery with a load tester. Replace battery if necessary.	Go to step 2.
2. Check the voltage between the battery's positive terminal and the ground side of the regulator/rectifier while the engine is running.	The problem is fixed.	a. Check for an open circuit or short in the wire harness. b. Check for poor connections. c. Go to step 3.
3. Check the stator resistance at the point where it connects to the regulator/rectifier with the coupler disconnected.	Go to step 4 (if applicable) or step 5.	a. There's a poor connection at the coupler. b. The charging coil is defective.
4. Check for field coil resistance (if applicable).	Go to step 5.	a. Check for an open circuit. b. The AC generator field coil is defective.
5. Measure the charging voltage at the battery at the specified engine rpm.	The battery is defective.	Go to step 6.
6. Replace the battery with a fully charged battery that's known to be good.	The battery is defective.	The regulator/rectifier is defective.

Table 20-15: Symptom: Overcharging Charging System		
Step	**If Measurement is Correct**	**If Measurement is Incorrect**
1. Check for continuity between the regulator/rectifier ground wire and chassis ground.	Go to step 2 (when applicable) or step 3.	a. Check for proper connections at the regulator/rectifier. b. Check for an open circuit in the wire harness.
2. Check for proper resistance of the field coil wire at the regulator/rectifier coupler (when applicable).	Go to step 3.	a. Check for a short circuit in the field coil. b. Check for a short in the wire harness.
3. Replace the battery with a fully charged battery that's known to be good.	The battery is defective.	Replace the regulator/rectifier.

Ignition System Troubleshooting

Once you've determined that a motorcycle engine's ignition system isn't producing a spark, the next step in the troubleshooting procedure depends on the type of ignition system. If the ignition system uses a breaker-points assembly, the points and condenser are the most likely cause of the problem. To check the points, remove all necessary covers and components. Check the contacts for pitting; check for dirt or moisture between the contacts.

In an electronic ignition system, the problem of no spark may be caused by several different components. Fortunately, all of these components are easy to check. First, check the spark plug to see if it is fouled. Then check to make sure that the engine stop-switch wire is properly connected and functioning correctly. This switch may be a switch that goes to ground or it may be a switch that completes the ignition circuit. Check the service manual wiring schematic to be sure. If that is okay, then check for proper connections at all of the ignition-related components. If these connections appear to be okay, check for proper resistance and AC voltage at the pulse generator and exciter coil (with CDI ignitions). If all of these components are in proper working order, the problem is probably a failure in the ignition control module (ICM). Replace the ICM with a known good component and test the engine. If the engine operates properly, you can assume that the ICM was the problem.

In most motorcycles and ATVs, it's very easy to remove and replace ICMs; but this component is usually quite expensive, so it's important to check all other components before replacing an ICM. Typically, ICMs are very reliable, and the problem is likely to be found in another area of the ignition system.

In a battery-type ignition system, a weak battery can cause ignition failure. Check the battery using a voltmeter to see if the proper voltage (should be at least approximately 12V) is present. Remember that the ignition switch or safety interlock switches can also be the cause of spark failure.

The following tables offer steps to follow with an AC-powered electronic ignition system and a battery-powered electronic ignition system.

Table 20-16: CDI Ignition Troubleshooting

No Spark Condition	
Step	**Action**
1. Disconnect the coupler at the CDI unit.	a. Check for a proper ground connection. b. Measure the resistance of the exciter coil. c. Measure the resistance of the pulser coil. d. Measure the resistance of the ignition-coil primary windings. Note: If any of the above items have an open or short circuit, measure the resistance of the component at the coupler closest to the component.
2. Check for continuity between chassis ground and the ignition stop-switch wire at the ICM.	a. In the Run position, there should be no continuity. b. In the Off position, there should be continuity. Note: If there's continuity when the switch is in the Run position, disconnect the stop switch and check for a spark.
3. Measure the resistance of the ignition-coil secondary winding.	a. If the winding is open, remove the spark-plug cap and retest. b. If the winding is still open after the above test, replace the coil.
4. The exciter coil, pulser coil, ignition coil, and the engine stop switches have all tested good, and all connections have been verified.	Replace the ICM with a known good unit.

Table 20-17: Battery-Powered Electronic Ignition System Troubleshooting	
No Spark Condition	
Step	Action
1. Disconnect the coupler at the ICM.	a. Check for a proper ground connection. b. Measure the resistance of the exciter coil. c. Measure the resistance of the ignition-coil primary windings. Note: If any of the above items have an open or short circuit, measure the resistance of the component at the coupler closest to the component. d. Measure the battery voltage at the ICM with the ignition switch in the On position.
2. Measure the resistance of the ignition-coil secondary winding.	a. If open, remove the spark-plug cap and retest. b. If still open after the above test, replace the coil.
3. The battery has voltage at the ICM, the pulser coil, ignition coil, and the engine stop switches have all tested good, and all connections have been verified.	Replace the ICM with a known good unit.

DC Circuit Troubleshooting

The battery in a motorcycle provides electrical energy to operate the ignition and many other electrical components. We will focus on two components that you'll frequently encounter: lights and switches.

Light Bulbs

Burned-out light bulbs are replaced and not repaired. To check a bulb that has been removed from a circuit, you can use a battery and two wires. One wire is connected to the negative side of the battery and to the ground on the light bulb. The other wire is connected to the positive side of the battery and to the insulated side of the light bulb. If the bulb is good, it will light up.

An ohmmeter can also be used to check light bulbs that have been removed from the circuit. Connecting one lead wire to the ground of the light bulb and the other to the insulated side of the bulb should cause the ohmmeter to show continuity—that is, a complete circuit.

Some light bulbs of different wattage and voltage are the same physical size, so always be sure that the replacement bulb is the same voltage and wattage as the one removed. Check the service manual if you are not certain about what size bulb should be installed.

Light bulbs can go bad due to excessive vibration. Excessive vibration can cause the filament inside the light bulb to break. When this occurs, the bulb must be replaced.

Another problem that you may encounter results from a loose connection in the light bulb socket or circuit. This condition can cause the bulb to get brighter and dimmer, flicker, or not light at all. This problem is corrected by locating the problem and repairing the faulty connection.

Switches

Switches are designed to open and close a circuit. You can check a switch using an ohmmeter. The ohmmeter should indicate continuity when the switch is in the On position and shouldn't indicate continuity when the switch is in the Off position. If a switch is defective, it must be replaced.

Electric Starter-Motor Troubleshooting

Four main troubleshooting problems occur with motorcycle electric starter systems:

- The starter motor turns slowly.
- The starter solenoid makes a clicking sound, but the engine does not turn over.
- The starter motor turns without turning over the engine.
- The starter motor does not turn at all.

Refer to the following table to troubleshoot these starter-motor problems.

Table 20-18: Electric Starter-Motor Troubleshooting

Symptom	Likely Cause	Checks
Starter motor turns slowly	Low charge in the battery	a. Check for a loose battery connection. b. Check for a loose starter-motor cable. c. Check for a faulty starter motor. d. Starter motor turns, but the engine doesn't turn over.
Starter-motor clutch at fault		a. Check for a worn starter clutch. b. Check for a worn starter pinion gear. c. Check for worn or damaged starter-motor idler or reduction gears. d. Check for a broken starter-motor chain.
Starter motor does not turn at all	Faulty fuse or safety device	a. Check all fuses. b. Check all safety devices such as transmission and side-stand lockout devices. c. Check starter solenoid. d. Check for faulty starter motor.
The starter solenoid makes a "click" sound but engine turns over by hand or with kick starter	Faulty solenoid switch or starter motor	a. Connect the starter motor to a battery source that is known to be good. b. If the starter motor turns, the solenoid switch is faulty. c. If the starter motor doesn't turn, the starter motor is defective.

CHASSIS PROBLEM TROUBLESHOOTING

Determining a problem relating to a chassis-related issue from only a verbal complaint can be very difficult. This is because of the many ways in which a chassis can react to different problems. An example is a steering head that shakes at certain speeds. Shake in the steering head could be caused by a problem relating to the front of the motorcycle, or it could be caused by a worn or out-of-balance rear tire. Therefore, you must be careful when attempting to solve a chassis-related problem.

Just as we mentioned in fuel system troubleshooting, whenever you're looking at a chassis problem be sure to make only one adjustment at a time; and also, make only small adjustments each time. Chassis problems can be broken down into the following three categories:

- Handling problems
- Wheel and tire problems
- Brake problems

The following tables give you some of the common symptoms found related to motorcycle chassis systems, as well as some direction that will most likely help to resolve the problem.

Table 20-19: Handling and Chassis Performance

Symptom	Possible Cause
Difficult steering	■ Improper tire pressure(s) ■ Worn tire(s) ■ Worn or excessively tight steering-head bearings ■ Steering-head nut too tight ■ Steering stem bent
Steers off to one side or does not track straight	■ Improperly adjusted fork height ■ Bent axle (front or rear) ■ Bent forks ■ Bent frame ■ Wheels improperly aligned ■ Swing arm bent ■ Worn-out wheel bearings ■ Worn swing-arm bearings
Machine wobbles	■ Bent rim(s) ■ Worn wheel bearings ■ Worn-out tire(s) ■ Tires incorrect for application ■ Tire pressure(s) incorrect
Suspension excessively soft	■ Worn or improper springs ■ Contaminated shock or fork oil ■ Insufficient shock or fork-oil viscosity ■ Suspension air pressure too low (when applicable) ■ Fork-oil level low ■ Incorrect spring adjustment ■ Tire pressure(s) too low ■ Incorrect nitrogen pressure (rear gas shocks) *(continued)*

Table 20-19: Handling and Chassis Performance *(continued)*	
Symptom	**Possible Cause**
Suspension excessively hard	■ Bent fork or shock ■ Fork-oil level too high ■ Fork-oil viscosity too high ■ Suspension air pressure too high (when applicable) ■ Tire pressure(s) too high ■ Incorrect spring adjustment
Handlebars shake excessively	■ Tire(s) worn or out of balance ■ Rim(s) bent ■ Swing-arm pivot worn ■ Wheel bearings worn ■ Steering stem loose

Table 20-20: Wheel and Tire Troubleshooting	
Symptom	**Possible Cause**
Wheel turns hard	■ Improperly adjusted brake ■ Worn wheel bearings
Difficult steering	■ Improper tire pressure(s) ■ Worn tire(s)
Steers off to one side or doesn't track straight	■ Bent axle (front or rear) ■ Wheels improperly aligned ■ Worn wheel bearings
Machine wobbles	■ Bent rim(s) ■ Worn wheel bearings ■ Worn-out tire(s) ■ Tires incorrect for application ■ Tire pressure incorrect
Handlebars shake excessively	■ Tire(s) worn or out of balance ■ Rim(s) bent ■ Worn wheel bearings

Brake Troubleshooting

Table 20-21: Hydraulic Brakes

Symptom	Possible Cause
Brakes feel soft or spongy	■ Air in brake line ■ Worn brake pads or disc ■ Worn or leaking master-cylinder seals ■ Worn or leaking caliper seals ■ Sliding caliper stuck (when applicable) ■ Low or contaminated brake fluid ■ Bent brake lever
Brake lever hard to pull or push	■ Clogged or restricted master-cylinder valve ■ Sticking caliper piston ■ Sliding caliper stuck (when applicable)
Brakes drag	■ Contaminated brake pads ■ Wheel out of alignment ■ Sticking caliper piston ■ Warped brake disc ■ Sliding caliper stuck (when applicable)
Brakes squeak	■ Sticking caliper piston ■ Worn brake disc ■ Contaminated brake pads or disc
Mechanical Brakes	
Symptom	**Possible Cause**
Brakes feel soft or spongy	■ Worn brake pads ■ Worn brake cable
Brake lever hard to pull or push	■ Lever lacking lubrication ■ Brake actuator sticking
Brakes drag	■ Brakes too tight ■ Warped brake drum
Brakes squeak	■ Contaminated brake pads

ABNORMAL NOISE TROUBLESHOOTING

Abnormal noise complaints can be for virtually any part of a motorcycle. The primary areas of concern for abnormal noise are as follows:

- Engine
- Drive train
- Chassis

The following tables indicate classic symptoms and possible problems for abnormal noise.

Table 20-22: Abnormal Engine Noise Troubleshooting

Symptom	Possible Cause
Knocking	■ Carbon buildup in combustion chamber ■ Fuel poor quality or incorrect fuel ■ Spark plug incorrect ■ Overheating ■ Connecting rod big-end clearance excessive ■ Crankshaft bearings worn ■ Connecting rod small-end clearance excessive ■ Balancer bearing worn
Internal slapping noise	■ Cylinder-to-piston clearance excessive ■ Cylinder or piston worn ■ Piston pin or piston holes worn ■ Piston ring worn, broken, or stuck ■ Piston seizure or damage ■ Loose alternater rotor
Ticking noise	■ Valve clearance incorrect ■ Valve spring broken or weak ■ Camshaft bearing worn
Chain noise	■ Camshaft chain tension incorrect ■ Camshaft chain, sprocket or guide worn ■ Balancer gear worn or chipped ■ Starter chain, sprocket or guide worn
External engine noise	■ Primary chain worn ■ Cylinder-head gasket leaking ■ Exhaust pipe leaking at cylinder-head connection

Table 20-23: Abnormal Drive Train Noise Troubleshooting

Symptom	Possible Cause
Clutch noise	■ Weak or damaged damper ■ Clutch-housing/ friction-plate clearance excessive ■ Clutch-housing gear worn
Transmission noise	■ Bearings worn ■ Transmission gears worn or chipped ■ Metal chips jammed in gear teeth ■ Engine oil insufficient
Drive chain noise	■ Drive chain adjusted improperly ■ Chain worn ■ Rear sprocket and/or engine sprocket worn ■ Chain lubrication insufficient ■ Rear wheel misaligned

Table 20-24: Abnormal Chassis Noise Troubleshooting

Symptom	Possible Cause
Front fork noise	■ Oil insufficient or oil worn out ■ Springs weak or broken ■ Springs squeaking
Rear shock noise	■ Shock absorber damaged ■ Springs weak or broken ■ Springs squeaking
Disc-brake noise	■ Pad installed incorrectly ■ Pad surface glazed ■ Disc warped ■ Caliper sticking
Miscellaneous frame noise	■ Brackets, nuts, or bolts not properly mounted or tightened

Chapter 20 Summary

This chapter has covered a wide variety of troubleshooting symptoms as well as possible solutions. The information has been derived from various motorcycle manufacturer service manual suggestions and compiled into one general area.

Use of the tables in this chapter will aid you in finding the problem in a timely manner and ensure that you have repaired the machine correctly the first time.

Chapter 20 Review Questions

1. Name the tool that is used to test a light bulb after it has been removed from its socket.
   a. Voltmeter
   b. Coil tester
   c. Ammeter
   d. Ohmmeter
2. True or False: A motorcycle steering-head vibration will indicate a problem in the front end of the machine. True or False.
3. If the air-and-fuel mixture to a motorcycle engine is too rich, the problem could be caused by
   a. an empty gas tank.
   b. a too-full gas tank.
   c. an air leak between the carburetor and the intake manifold.
   d. leaving the choke on after the engine has warmed up.
4. If you turn on the choke and the engine runs better, this generally indicates a ________ carburetor mixture problem.
5. A possible cause of a transmission that jumps out of gear is
   a. shifting at too low an engine speed.
   b. a damaged shift fork or drum.
   c. excessive play in the clutch lever.
   d. incorrect tire (wrong size) on the rear wheel.
6. An excessive amount of ________ exhaust smoke indicates worn piston rings.
   a. black
   b. brown
   c. blue
   d. white
7. Brake drag can be caused by
   a. a brake lever adjustment that's too loose.
   b. a sticking brake caliper.
   c. air in the brake line.
   d. a bent rim.
8. A four-stroke motorcycle engine with insufficient valve clearance will have
   a. crankshaft seizure.
   b. incorrect ignition timing.
   c. low compression.
   d. extra power at higher speeds.
9. If a charging system is overcharging, the ________ is most likely the faulty component.
10. A hydraulic brake system that has a soft or spongy feel probably has ________ in the brake lines.

Glossary

A

AC: Abbreviation for "alternating current," which is electricity that reverses direction and polarity while flowing through a circuit. Example: 110 volts AC in a household reverses direction and polarity 60 times per second (60 Hz).

Accelerator pump: A small pump that enriches the fuel–air mixture during acceleration.

Acorn nut or cap nut: A nut with a finished or plated surface often used to cover the threaded end of a bolt or stud.

Active combustion: The result of a chain reaction of burning molecules accelerating and the chemical conversion that causes heat to be released very quickly.

Active energy: Energy in use or motion. Also known as kinetic energy.

Additives: Chemicals used in the manufacturing of oil to improve its operating qualities.

Air-Cooled engine design: An engine designed to be cooled by air.

Air filters: Components used to filter the incoming air to the carburetor.

Air mixture screw: A screw that allows an increase or decrease of air into a slow speed circuit of a carburetor.

Air ratchet: A tool to remove nuts and bolts with the assistance of high air pressure.

Allen wrench: A six-sided male-end wrench

All-terrain vehicle (ATV): A separate branch of the motorcycle industry family tree; they are available in a variety of sizes and styles.

Alternating current (AC): The flow of electrons, first in one direction, and then in the opposite direction.

Alternator: An AC "generator" that uses magnetic induction to produce electricity. A revolving magnet and stationary stator windings are used. The current produced is AC.

American National Standards Institute (ANSI): An organization in the United States that sets technical standards.

American Petroleum Institute (API): One of two general automotive agencies created to test, standardize, and classify lubricating oils.

Ammeter: A tool used to measure amperes.

Amp: A measurement of electrical current.

Amperes: Commonly called "amps," which are electrical units of current flow through a circuit (similar to gallons per minute of water through a hose).

Amp hour: Discharge rate of battery in amperes times hours.

Antilock braking: Braking system that prevents the wheels from locking up under excessively hard braking, or when attempting to stop on slippery surfaces such as wet roads.

Armature: A group of rotating conductors that pass through a magnetic field. The current produced is usually DC after passing through a commutator device. Armatures are also used in electric motors.

Atomization: The process of combining air and fuel to create a mixture of liquid droplets suspended in air.

Axial: Side-to-side loads.

Axial tension: The stretching force applied to a bolt when it is tightened into a case or a nut.

B

Babbitt bearings: *See* Plain bearings.

Backlash: The play or loose motion in a gear due to the clearance between two opposing gears.

Ball bearings: The most popular bearing used on motorcycles because they provide the greatest amount of friction reduction and have the ability to handle both axial and radial loads.

Ball peen hammer: A hammer that has two opposing striking surfaces: a flat-faced surface and a rounded surface.

Base carburetor: The carburetor with the idle adjustment screw on a multicarbureted engine.

Base circle: The area of the camshaft that forms a constant radius from the centerline of the journal to the heel.

Basic hand tools: The common tools that are found in almost every workshop toolbox, which includes screwdrivers, hammers, pliers, wrenches, and socket sets.

Battery: An electrochemical device that converts chemical energy to electrical energy. Used to store electrical power to supply uninterrupted energy for an electrical system.

Battery conductive analyzer: A tool to measure a battery's condition.

Beam torque wrench: A wrench to measure torque using a beam bending in response to the torque applied.

Bearings: Friction-reducing device that also reduces free play between shafts to allow for proper spacing and supports different types of loads.

Bench test: Isolated component inspection.

Bleeding: The method used to remove air bubbles from oil lines.

Block diagrams: A more precise schematic of an electrical sub system used to assist a technician with diagnosis of an electrical complaint.

Bolt: A metal rod with external threads on one end and a head on the other. A bolt may also be called a cap screw.

Bolt diameter: The measurement of the outside diameter of a bolt's threads.

Bolt head markings: A way to determine the tensile strength of a bolt identified with lines or slash marks, called grade markings, or by numbers.

Bolt length: A measurement of the bottom of a bolt head to the threaded end of the bolt.

Boost port: Used on two-stroke engines to allow an extra amount of the air-and-fuel mixture to flow into the combustion chamber directly from the intake port area.

Bore: The diameter of an engine's cylinder.

Bottom-dead center (BDC): The point at which a piston is at its lowest position in the cylinder.

Boyle's law: A physical law that states: The product of the pressure and the volume of a given mass of gas is constant if the temperature is not changed.

Brake caliper: The part of a brake system that, when applied by the rider, clamps the brake disc to slow down the motorcycle.

Brake fluid: A type of hydraulic fluid used in brake applications.

Brake horsepower (bhp): The amount of power being delivered to the dynamometer.

Brake lever: A lever mounted to the handlebar (or a pedal on the frame) used for activating the brake.

Brake rotor: Brake system component used to slow down or stop the wheel from rotating.

Braking systems: Systems designed to reduce the machine's kinetic energy by transforming it into heat energy known as friction heat.

Breaker bar: A long bar used to assist with the loosening of nuts and bolts.

Breaker points: Mechanical contacts that are used to stop and start the flow of current through the ignition coil.

Bushing: A component, cylindrical in design, with a lining made of a soft alloy used to allow a shaft to rotate.

C

California Air Resources Board (CARB): The "clean air" agency of the state of California.

Cam ground: Process of machining that makes an oval piston round after it it reaches its operating temperature.

Camshaft: Component used to change rotary motion to reciprocating motion.

Camshaft drive tensioner: Component used to keep the proper tension on the cam chain or cam belt.

Camshaft lift: The distance that the valve actually moves away from the cylinder head.

Capacitor or condenser: A component, which in a discharge state, has a deficiency of electrons and will absorb a small amount of current and hold it until discharged again.

Carbon dioxide (CO_2): The result of complete combustion.

Carbon monoxide (CO): A colorless, odorless, poisonous, and deadly gas that results from partially burned fuel or fuel that is not completely burned during the combustion process.

Carburetor: A device used to mix the proper amounts of air and fuel together.

Carburetor synchronization: The process of balancing the output of two or more carburetors so that the amount of air-and-fuel mixture drawn through each one is equal.

Castle-headed nut: A nut that allows a cotter pin to be installed through a nut and bolt to prevent loosening.

Catalyst: A means to speed up the chemical reaction of something without undergoing any change itself.

Centrifugal clutch: A clutch that uses the engine's rpm to engage and disengage it.

Centrifugal oil filter: A type of oil filter that uses centrifugal force to clean the engine oil.

Chamfered: A process that removes sharp edges of the port to help keep the piston ring from catching as it moves up and down in the cylinder.

Circuit: Composed of three items: a power supply, load, and completed path.

Circuit breaker: Heat-activated switch that interrupts current when overloaded. A circuit breaker can be reset and replaces the function of a fuse.

Clamping force: The force applied by a bolt holding two parts together.

Class A fires: Fires that involve the burning of wood, paper, cardboard, fabric, and other similar fibrous materials.

Class B fires: Fires that involve flammable liquids, gases, and other chemicals.

Class C fires: Fires that involve live electrical equipment such as electrical boxes, panels, circuits, appliances, power tools, machine wiring, junction boxes, wall switches, and wall outlets.

Class D fires: Fires that involve combustible metals such as magnesium, titanium, zirconium, sodium, lithium, and potassium.

Clearance ramps: Parts of the camshaft that take up the valve clearance and open and close the valve.

Click torque wrench: A wrench to measure torque by preloading a "snap" mechanism with a spring to release at the specified torque.

Clutch: A component used to engage and disengage the transmission and the rear wheel from the crankshaft power output.

Clutch drag: A symptom that occurs when the clutch is unable to fully disengage.

Clutch friction discs: Clutch component that transmits the power of the clutch outer to the clutch center through a high-friction material.

Clutch plates: Clutch components used to transfer the power from the clutch friction discs to the clutch center.

Clutch release mechanism: Components used to disengage the power flow from an engine's clutch to the transmission.

Clutch slippage: A symptom that occurs when the clutch does not have the ability to transfer all of an engine's power flow.

Coil: A conductor looped into a coil-type configuration which, when current is passed through, will produce a magnetic field.

Cold start systems: A system used to provide and control a richer-than-normal air-and-fuel mixture necessary to quickly start a cold engine.

Combination bolt: A type of self-tapping bolt that forms female threads when it is screwed into the unthreaded pilot hole.

Combination wrenches: Wrenches with an open end on one side and a closed end on the other.

Combustion: The rapid combining of oxygen molecules with other elements.

Combustion chamber: The small space remaining between the piston and the cylinder head when the piston is at top-dead center.

Combustion chamber volume (CCV): The volume of space when the piston is at top-dead center.

Combustion lag: The time interval between the time that a spark occurs and the energy release of the air–fuel mixture.

Compression: The stage at which the piston rises and compresses the air-and-fuel mixture into the combustion chamber.

Compression check: A test that measures the amount of pressure produced in the combustion chamber of an engine.

Compression gauge: A tool used to measure compression.

Compression ratio: The ratio of the largest cylinder volume to the smallest cylinder volume.

Compression ring: Piston ring closest to the piston crown and is used to seal most of the combustion chamber gases.

Compression stage: When the piston rises and compresses the air-and-fuel mixture trapped inside the combustion chamber.

Compression test: A test to verify that an engine's compression is high enough to heat the air-and-fuel mixture to a combustible level inside the engine's combustion chamber.

Conductor: A wire or material (such as a frame) that allows current to flow through it with very little resistance.

Conductors: A carrier of electricity such as a piece of insulated wire.

Cone-type lock washer: A dished washer made from spring steel.

Constant: A number used in a formula that never changes.

Constant failure: A failure in which a symptom is always present.

Constant velocity carburetor: A carburetor that is venturi-sized and is not controlled by a throttle cable that raises and lowers the throttle slide. Instead, the throttle slide is raised by pressure differences created by the engine as it is operating.

Continuity: Having a continuous electrical path.

Conventional system: One of two common systems of weights and measurements also known as the United States Customary (USC) and Standard system. The conventional system measures in inches.

Counter-balancer: A device that balances the power pulses created by the power strokes in an engine.

Crankcase reed valve: Reed valve induction system that has the intake port located directly on the crankcase.

Crankcases: Parts used to hold all of the engine components together and provide the main engine mounting points.

Crown: Top of the piston and acts as the bottom of the combustion chamber.

Current: The flow of electrons in a circuit.

Cylinder: A circular tube that is closed at one end.

Cylinder head: The component of an engine that seals the upper end of the cylinder.

Cylinder reed valve: An induction design that has the reed valve system in the intake port located on the cylinder.

D

Dampers: Components used to convert the kinetic (bouncing) energy into heat by utilizing friction and oil flow resistance.

DC: Abbreviation for "direct current," which means that the current will only flow in one direction—from positive to negative (conventional theory).

de Carbon shock: A type of damper that uses nitrogen gas in a separate chamber to keep the damper oil from foaming.

Detonation: A condition in which, after the spark plug fires, some of the unburned air–fuel mixture in the combustion chamber explodes spontaneously, set off only by the heat and pressure of the air–fuel mixture that has already been ignited.

Diagnostics: The process of determining what is wrong when something is not working properly.

Dial torque wrench: A wrench to measure torque using a dial gauge to measure the torque applied.

Dielectric: A nonconductor of electricity. Dielectric materials do not allow electricity to flow through them. *See* Insulator.

Diode: A semiconductor often used in a rectifier on motorcycles. A diode has the characteristic of allowing current to pass through in only one direction. Thus it is used to change AC to DC current.

Direct current (DC): The flow of electrons in one direction only.

Direct-drive gear ratio: A gear ratio that is exactly 1:1.

Direct-drive transmissions: A type of transmission that has power flow entering on one shaft and leaves on another shaft of the same axis.

Disc brake: A brake that relies on friction being applied to both sides of a spinning disk by the use of brake pads.

Distilled water: Water that has virtually all of its impurities removed through the process of distillation.

DR-type (deep-recess) bolts: Flange bolts with hex heads and weight reduction recesses in them.

Drum brake: A brake in which the friction is caused by a set of shoes or pads that press against the inner surface of a rotating drum.

Dry lubricants: Used to lubricate without attracting contaminants. Dry lubricants use an evaporating solvent as a carrier.

Dry sump: An engine design that stores its oil supply in a separate oil storage tank.

Duration: Time that the ports are open and is measured in crankshaft degrees.

Dykes piston ring: An L-shaped ring that is only used as a top ring on the piston of a two-stroke engine.

Dynamic: Spinning or rotating in motion; refers to making a test when the component is in use.

Dynamometer (dyno): A measuring instrument that is used to measure power.

E

Elastic range: The range of stretching of a bolt or nut that allows it to return to its original length after use.

Electrical potential: The difference in electrical charge between the two opposing terminals.

Electricity: The flow of electrons through a conductor.

Electric starter: A means to electronically start an engine.

Electrolyte: The sulfuric acid and distilled water solution that batteries are filled with at setup.

Electromagnet: A coil of wire that is wound around a soft iron core, which acts as a magnet when current is passed through it.

Electromotive force (EMF): The pressure of electrons in a circuit. *See* Voltage. Created by difference in potential between positive and negative terminals of power supply. Also called pushing force of electricity.

Electrolysis: The movement of electrons through an electrolyte solution. A battery charges and discharges through electrolysis. Electroplating (chroming) is an example where electrolysis is used to move and deposit metals from one electrode to another. In cooling systems, contaminated (tap water) coolant becomes an electrolyte, allowing electrolysis and the deposition of metal oxide scale on cooling system components.

Electron: The revolving part or moving portion of an atom.

Energy: The ability to do work. Energy itself cannot be seen; however, the results of energy can be seen.

Engine control module (ECM): A main component of an electronic ignition system.

Engine cycle: A complete run through all four stages of operation: intake, compression, power, and exhaust.

Engine displacement: The volume of space that the piston moves as it moves from BDC to TDC in an engine.

Engine seizure: When two or more parts inside an engine are hot enough to melt together.

Environmental Protection Agency (EPA): The government agency that develops emission standards for street-legal vehicles.

Ethanol alcohol: A bio fuel derived from grain and corn that can be used instead of, or as an additive to, gasoline.

Excited-field electromagnetic alternators: An AC-generating system that incorporates a field coil, which is energized into a powerful magnet when DC current is supplied.

Exhaust: The stage at which the piston rises and pushes the exhaust gases out of the cylinder.

Exhaust gas analyzer: A tool that is used to measure the amount of exhaust emissions coming from the engine.

Exhaust port: Used to allow the exhaust gases to escape.

Exhaust stage: When exhaust gases are released from the cylinder.

Exhaust valve: Poppet valve that opens to allow exhaust gases to flow out of the combustion chamber after the air-and-fuel mixture is burned.

F

Field coil: The field coil is an electromagnet. The flux lines may be used for generating electricity, electric motor operation, or operating a solenoid/relay.

Final drive ratio: A numeric comparison between the countershaft of the transmission and the rear wheel.

Fire triangle: The three conditions that must be present for a fire to start: fuel, oxygen, and an ignition source.

Fixed gear: A gear that does not move on the shaft to which it is attached.

Flammable: To ignite easily.

Flat washer: Used to increase the clamping surface under the fastener and prevent the bolt or nut from digging into the component it is attached to.

Flux lines: All the magnetic lines of force from a magnet.

Forced-draft cooling: Air cooling that uses an engine-driven fan, which draws air through ductwork called shrouds that surround the cylinder and the cylinder head.

Fork bushings: Suspension component that prevents lateral movement of the fork during its movement up and down.

Fork oil seal: Seals placed in the top of the outer-fork tube to prevent oil leakage.

Fork springs: Springs that extend the forks and allow movement of the outer fork tubes.

Fractional distillation: The process used to separate a mixture of several liquids, based on their different boiling points.

Frame: The skeleton of the motorcycle in which virtually everything pertaining to the machine is attached.

Free electron: An electron in an atom's outer orbit, which is held only loosely within the atom. Free electrons can move between atoms.

Freewheeling gear: A gear that moves freely on its shaft.

Friction: When resistance to motion is created when two surfaces move against each other, or when a moving surface moves against a stationary one.

Fuel filters: Components used to remove contaminants from the fuel before they reach the fuel system.

Fuel injection: A method of carburetion used on an engine using electronics instead of a carburetor.

Fuel injector: An electronically operated solenoid that turns fuel on and off under pressure.

Fuel mixture screw: A screw that allows an increase or decrease of fuel into a slow speed circuit of a carburetor.

Fuel valves: Also known as fuel petcocks. On/off valves that control the flow of gasoline from the fuel tank to the carburetion system.

Fuse: A short metal strip that is protected by a glass or plastic case, which is designed to melt when current exceeds the rated value.

G

Gaskets: Used to seal the mating surfaces of various parts of the engine.

Gasoline: A volatile, flammable, hydrocarbon, liquid mixture used as a fuel.

Gear ratio: A numerical comparison of the number of revolutions of a drive gear as compared to one revolution of a driven gear.

Gear-type oil pump: An oil pump that consists of a housing and two spur gears; a drive gear attached to the oil pump shaft and a driven gear.

Grease: A lubricant that is suspended in gel.

Ground: A common conductor used to complete electrical circuits (negative side). The ground portion of motorcycle electrical systems is often the frame.

Grounded circuit: A circuit that allows electrical power to flow back to the source after the load, but before the means of control.

H

Hall-effect sensor: A device in which an output voltage is generated in response to the intensity of a magnetic field applied to a wire. Based on the Hall-effect principle, which states: If a current is allowed to flow through a thin conducting material and that material is exposed to a magnetic field, voltage is produced. In essence, a Hall-effect switch is either on or off.

Halon: A chemical used to extinguish fires.

Head size: The distance across the flat or outer sides of a bolt head.

Heli-Coil®: A thread repair device that restores the original thread size if it has been damaged.

Horsepower (hp): The standard unit of measurement to determine the power output of an engine.

Hydrocarbons (HC): Unburned or raw fuel.

Hydrodynamic lubrication: A system of lubrication in which the shape and relative motion of the bearing surfaces causes the formation of a fluid film having sufficient pressure to separate the surfaces.

I

Idle circuit: Carburetor circuit that meters the air-and-fuel mixture at engine idle and up to approximately 1/4-throttle opening.

Ignition: The spark produced by the high-tension coil by which the spark plug "ignites" the air-and-fuel mixture. The contact of a fuel with a spark.

Ignition coil: A component that provides the spark to the engine through the spark plug and is essentially a transformer that consists of two wire windings that are wound around an iron core. The first winding is called the primary winding, and the second winding is called the secondary winding. The secondary winding has many more turns of wire than the primary winding.

Ignition system: A system within an engine with the sole purpose of providing a spark that will ignite the air-and-fuel mixture in the combustion chamber.

Ignition timing: The process of setting the time that a spark will occur in the combustion chamber during the power stroke relative to piston position and crankshaft velocity.

Impact screwdriver: Screwdriver that is used in conjunction with a hammer.

Improper-service failure: A failure caused by a mistake made during the servicing of the machine.

Indirect-drive transmission: A type of transmission that has power flow entering on one shaft and exiting from another shaft on a different axis.

Induction: Method of controlling the intake flow of an engine.

Injector discharge duration: A measurement of time that a fuel injector is in the open position while under pressure.

Insulator: A material that does not conduct electricity and therefore, prevents the passage of electricity. All electrical wires are protected by a plastic or special rubber insulation. *See* dielectric.

Intake stage: The stage at which the piston moves downward and draws an air-and-fuel mixture into the cylinder.

Intake port: Used to allow the gases to enter the engine.

Intake valve: Poppet valve that opens to allow the air-and-fuel mixture to flow into the combustion chamber.

Integrated braking: Braking system that offers the rider the use of both front and rear brakes when stepping on the rear brake pedal, squeezing the front brake lever, or a combination of both.

Intermittent failure: A failure that is not always present.

Internal combustion engine: An engine that uses compressed fuel and air to produce power.

Internal-oil cooling: A method of cooling found on all engines. As the oil is circulated throughout the engine, heat is transferred to the oil from the engine components with which the oil has come in contact.

International Standards Organization (ISO): A worldwide federation of national standards bodies; one each from some 140 countries.

K

Keystone piston ring: A wedge-shaped piston ring.

Kick starter: A means to manually start an engine with a lever that is pushed (kicked) with a person's foot.

L

Labyrinth seal: A component that forms a seal between multiple cylinders when the engine is running on a two-stroke engine by filling with fluid that forms a seal to separate the cylinders.

Law of action and reaction: A physical law that states: For every action there is an equal and opposite reaction.

Law of inertia: A physical law that states: "Anything at rest or in motion tends to remain at rest or in motion until acted upon by an outside force."

Leak-down test: An engine test that allows the measurement of the percentage of air that leaks past the piston rings and valves.

Lines of force: Refers to a magnetic field whose lines run from its north pole to its south pole.

Liquid cooling: An engine design that uses liquid to keep the engine at a desired temperature.

Load: Anything that uses electrical power such as a light bulb, coil, or spark plug.

M

Magnetic induction: When a conductor is moved through a magnetic field, electricity will be "induced" into the conductor when the flux field cuts through the conductor.

Magnetic pulse generator: A generator of single or multiple voltage pulses that uses a magnet to trigger instead of mechanical points.

Magnetism: The characteristic of some (ferrous) metals to align their molecules. The alignment of the object's molecules will cause the object to act as a magnet. Every magnet has both a north and south pole. Like polarities repel, opposites attract. Around every magnet, there is a magnetic field, which contains lines of force.

Main jet circuit: Carburetion circuit that controls fuel between 3/4- and wide-open throttle positions.

Maintenance-free battery: A battery that requires no fluid maintenance.

Maintenance interval: Time frame set by the manufacturer for maintenance of a motorcycle.

Malfunction indicator light (MIL): A fuel injection fault indicator.

Mallet or soft-faced hammer

Master cylinder: The part of a brake system that stores the brake fluid. When pressure is applied to the lever or pedal, the fluid is forced to the brake caliper.

Mechanical seal: A spring-loaded seal consisting of several parts to seal the rotating impeller in the pump case and prevents water from leaking into and damaging the engine.

Mechanical slide carburetor: A carburetor that allows the venturi to be controlled by the rider.

Mechanical work: A force that is applied over a specific distance.

Metric system: One of two common systems of weights and measurements. The metric system is the most widely used system of measurement on motorcycles and ATVs. The metric system is also known as the International System of Units (SI) due to its use around the world. The metric system measures in meters.

Micrometer: A precision measuring tool.

Momentum: The driving force that is the result of motion or movement.

Motorcycle: A two-wheeled automotive vehicle having one or two saddles.

MTBE: Methyl tertiary butyl ether.

Multi meter: A tool used to measure different types of electrical functions.

Multi piece crankshaft: Crankshaft that is cast or forged as multiple parts.

Multi viscosity oil: Oil that is suitable for use under many different climatic and driving conditions.

N

Needle bearings: A variation of the roller bearing.

No-load test: A dynamic test with the component insulated or disconnected from its main system.

NPN: A transistor in which the emitter and collector layers are N-type and the base layer is P-type (negative, positive, negative).

Nut: A fastener that has internal threads and usually is made with a 6-sided outer shape.

O

Octane rating: The measure of a fuel's ability to resist detonation.

Ohm: A measurement of electrical resistance.

Ohm meter: A tool used to measure electrical resistance.

Oil filter bypass valve: A valve that allows oil to bypass a clogged oil filter, providing essential lubrication to critical engine components even if the filter is extremely dirty.

Oil pressure relief valve: A valve used to prevent excessive oil pressure from building up by bleeding excessive oil back into the crankcase.

One-piece crankshaft: Crankshaft that is cast or forged as one part.

Open circuit: A circuit that has an incomplete path for current to flow.

Open-draft cooling: Air cooling that uses the movement of the motorcycle to force air over the fins, removing the excess heat from the engine.

Organic brake pads: Brake pads constructed by mixing non-asbestos fibers, such as glass, rubber, carbon, and Kevlar®, with filler materials and high-temperature resins.

Overall gear ratio: A ratio that compares the number of times a crankshaft turns to the number of turns of the rear wheel.

Over drive gear ratio: A gear ratio that is less than 1:1.
Overhead camshaft: An engine design with the camshaft located on top of the cylinder head.
Oxides of nitrogen (NO_x): Forms of oxidized nitrogen resulting from extremely high combustion temperatures.
Oxygen: A tasteless, odorless, colorless gas that makes up about 20 percent of the air we breathe.
Oxygenated fuels: Gasoline which has been blended with alcohols or ethers that contain oxygen in order to reduce carbon monoxide and other emissions.

P

Parallel circuit: A circuit that has more than one path back to the source of power.
Parallel/series circuit: A circuit that consists of both a series and a parallel circuit.
PASS: An acronym to help remember how to operate a fire extinguisher. Pull (the safety pin), Aim, Squeeze, and Sweep.
Permanent-magnet alternator: The most commonly used type of AC-generating system found on motorcycles that uses permanent magnets incorporated into the rotor.
Permeability: Ability of material to "absorb" magnetic flux (can be temporary or permanent). *See* Reluctance.
Petroleum-based oil: Oil that starts out as crude oil found in the earth.
Piston: A can-shaped metal component that moves up and down inside the cylinder. The main moving part in an engine.
Piston port/crankcase reed: Reed valve induction system that takes the benefits of both the piston port-type of induction to control the lower-speed range of the engine and the crankcase reed induction system for high-speed operation.
Piston ring end gap: The space between the piston ring at its opening when it is compressed in a cylinder.
Piston ring grooves: Grooves cut into the side of a piston to hold the piston rings in place.
Piston ring lands: Support area for the piston rings.
Piston rings: Components that aid in heat transfer from the piston to the cylinder wall, seal in the combustion gases, and prevent excessive oil consumption.
Piston ring side clearance: The space between each piston ring and the inner side of the piston ring groove.
Piston skirt: The load-bearing surface of the piston.
Plain bearings: Precision bearings that are typically made in the shape of a cylindrical sleeve and are designed to withstand extreme loads. Also called Babbitt bearings.
Plastic range: The point at which a bolt is being permanently stretched.
Plastigage: A plastic measuring clearance material that is compressed between bearing surfaces, then compared to a scale to find thickness.
Plunger oil-type pump: An oil pump that consists of a set of check valves, a piston, and a cylinder.
PNP: Transistor in which the emitter and collector layers are P-type and the base layer is N-type (positive, negative, positive).
Polarity: In magnets, polarity is north and south; in electricity, polarity is positive and negative.
Pole: The north "pole" or south "pole" of a magnet. Also refers to the lugs (iron cores) of a stator around which the AC generator's wires are wound.
Poppet valve: Tulip-shaped valve used to control the gases coming into and going out of the engine.
Ports: Holes in a cylinder head (four-stroke) and cylinder (two-stroke) that allow the air-and-fuel mixture to enter the cylinder and exhaust gases to leave the cylinder.
Potential energy: Stored energy.
Power: The rate at which work is accomplished. Power = Work / Time.
Power stage: The stage at which the release of energy pushes the piston back down the cylinder.
Power valves: Valves used on two-stroke engines to widen the range of usable power.
Preignition: The ignition of the fuel/air mixture in an engine before the spark plug actually fires.

Preload: The technical term for the tension caused by tightening the fastener that holds the parts together.

Premixed: The combination of two or more items at a designated ratio.

Press fit: A force fit that is accomplished using a press.

Primary area: The area below the piston crown including the crankcase in a two-stroke engine.

Primary drive: The gear reduction system used for transferring the power from the crankshaft to the clutch.

Primary drive ratio: The gear reduction that is determined from the crankshaft to the clutch of the engine.

Prussian blue: A special blue dye that identifies the contact between the valve face and the valve seat.

Pullers: Tool used to remove gears and press fit components.

Push fit: A force fit that is accomplished manually.

R

Radial: Rotating loads.

Rake: Angle of the steering-stem axis in relation to the ground.

Ratio: A comparison between two values.

Reamer: A long, round cutting tool with cutting edges along its length that operates much like a drill bit.

Reciprocating engines: An engine with a piston that moves alternately up and down inside a cylinder.

Recoil starter: A means to manually start an engine with a pull cord that retracts.

Rectifier: A device that permits electrical current to flow in one direction only.

Regulator: Used to limit the output of a generator or alternator.

Reluctance: Resistance to magnetism. *See* permeability.

Reluctor: Magnetic field interrupter used as a signal generator in ignition systems.

Resistance: The opposition offered to the flow of current in a circuit.

Revolutions per minute (rpm): A measurement of how many complete turns (360°) a crankshaft makes in one minute.

Ring grooves: Points on the piston that allows the piston rings to be installed.

Ring lands: The uncut areas between piston ring grooves.

Rocker arm: Device used to open and close a four-stroke engine valve.

Roller bearings: Similar in design to ball bearings but uses cylindrical-shaped rollers instead of spherical balls.

Rotor: A component that contains a series of magnets and rotates either inside or outside the stationary windings of the stator to create alternating current.

Rotor-type pump: An oil pump that consists of a pair of rotors that squeeze oil through the pump body passages.

RPM: The revolutions per minute at which an engine operates.

S

Schematic: A diagram that shows, by means of graphics and symbols, the electrical connections and functions of a specific circuit arrangement.

Scoring: Deep, vertical scratches in a piston that are normally caused by inadequate lubrication or overheating.

SCR: An abbreviation for silicon-controlled rectifier, which is an electronically controlled switch. *See* Thyristor.

Scraper ring: Used to scrape excessive oil from the cylinder wall on a four-stroke engine.

Screw extractor: A tool used to assist in removing broken bolts.

Scuffing: Wide areas of wear on a piston that usually appear as shiny patches generally caused by inadequate filtering of the air, allowing dirt to be ingested into the cylinder.

Seals: Devices that are designed to prevent leakage between shafts and cases.

Secondary area: The area above the piston crown including the combustion chamber where the air-and-fuel mixture is compressed to prepare for ignition in a two-stroke engine.

Selenium: Similar to silicon materials in characteristics; it is also used as a rectifier on older models.

Self-locking nut: A nut with a spring plate that presses against the thread, making it difficult for the nut to become loose. Self-locking nuts are reusable.

Sequential fuel injection: Fuel injection that allows delivery of fuel to only the cylinder that requires fuel.

Series circuit: A circuit that has only one path back to its source.

Service manuals: Manuals used to assist in the maintenance and repair of a motorcycle.

Shear: The force exerted at 90 degrees to the center line of a bolt.

Short circuit: A circuit that has developed a path to the source of power before it reaches the load in the circuit.

Silicon: A material used in the construction of semiconductors. Because of its characteristics, the material allows current flow only under certain prescribed conditions.

Sine wave: A graphic depiction of the form of alternating current usually taken from an oscilloscope.

Sintered brake pads: Brake pads that have a very high metal content and heat up quickly, which offers more friction earlier.

Sliding gear: A gear that can slide across the axis of the shaft.

Society of Automotive Engineers (SAE): The technical standards board that issues and recommends industry standards.

Solder: Tin/lead alloy with rosin core used to form lower-resistance connections of electrical components or wires.

Solenoid: An electro magnetic coil that mechanically opens and closes a circuit when electric current is run through it.

Spark plug: A component that provides an air gap in which high voltage from the coil causes an arc or spark to provide ignition to the fuel/air mixture.

Spin balancing: A means of balancing a tire with a machine that spins the wheel at a predetermined speed.

Split ring-type lock washer: A washer that is compressed under the pressure of the fastener. Elasticity in the washer is used to prevent loosening.

Sprag: A type of clutch that allows a portion of the clutch to slip when under high-stress situations such as rapid downshifting.

Spring adjusters: Components used on shocks to adjust the set length of the damper spring and therefore, changing the initial spring rate.

Spring rate: The amount of force (lb) required to compress a spring 1 inch.

Squish area/band: Area of the combustion chamber that forces the air-and-fuel mixture into a tight pocket under the spark plug to increase the combustion efficiency.

Stake-type lock nut: A nut that incorporates a metal collar at the top of the nut. A punch is used to stake (bend or indent) the collar of the nut to match a groove in the shaft that it is being used on.

Standard piston ring: Piston ring that is rectangular in shape and is the most popular ring found in the two-stroke engine.

Starter clutch: A one-way clutch mechanism that allows the starter motor to engage only while the starter motor is operating to start the engine. Also known as a sprag clutch.

Static: Stationary. Usually a test made of a stationary component rather than a bench test.

Static balancing: A means of balancing a wheel statically.

Stator: A component that consists of sets of coils, which are used to produce alternating current when used in conjunction with the rotor to power.

Steering stem: A suspension component that supports the fork tubes by clamping around them.

Stellite: An extremely hard metal alloy that resists wear and will not soften at high temperatures. Used on poppet valves.

Stretched bolt: A worn bolt that is identified when a nut threads down the bolt easily and then binds when it reaches the stretched area. At that point it will become hard to turn.

Stroke: The total distance that the piston moves from the top of the cylinder to the bottom of the cylinder.

Stud: A fastener with external threads on each end.

Sump: The lowest portion of the crankcase cavity.

Suspension: A mechanical system of springs or shock absorbers connecting the wheels and axles to the chassis of a motorcycle.

Swing arm: The main component of the rear suspension of most modern motorcycles that is used to hold the rear axle in pace while pivoting vertically to allow the suspension to absorb bumps.

Switch: A device that opens or closes an electrical circuit.

Symptom: An indication of an abnormal condition that can be recognized and identified.

Synthetic-based oil: Man-made oil that lubricates more efficiently and over a larger range of temperatures than standard mineral-based oils.

T

Tapered roller bearings: A bearing with the diameter of one roller end larger than at the other.

Telltale hole: A hole placed in a water pump to allow coolant to leak out when a mechanical seal has failed.

Tensile strength: The amount of pull a fastener can withstand before breaking.

Thermostat: A temperature-sensitive flow valve.

Thermo-switch: A bimetallic switch which, when heated, opens or closes a circuit.

Thread pitch: The coarseness of a thread on a bolt. Thread pitch for metric fasteners is the distance from the top of one thread to the top of the next, which is determined by the number of threads per inch.

Three-wheeler: A type of ATV.

Thyristor: An electronically controlled switch that opens when signaled at the gate and closes after current flow falls. *See* SCR.

Tire cupping: A wear pattern found on tires that start out as flat spots (and eventually turn into depressions), which develop on the tread of tires (primarily on the front). These flat spots alternate on either side of the center of the tread.

Tongued lock plate: A washer that serves the purpose of a washer and locking device.

Top-dead center (TDC): The point at which a piston is at its highest position in the cylinder.

Torque: A measurement of twisting or rotational force.

Torque wrench: A wrench that is used to measure torque.

Torsion: The twisting force applied to the head of a bolt when it is tightened.

Torx bolt: A type of bolt head characterized by a 6-point star-shaped pattern.

Trail: The distance along the ground between an imaginary vertical line through the center of the front axle to the ground and another line through the center of the steering stem to the ground.

Transfer port: Used to transfer the intake gases from the bottom of the cylinder to the combustion chamber through the two-stroke cylinder.

Transmission gear ratio: A numerical comparison of how many turns of the mainshaft cause one revolution of the countershaft.

Transmissions: Components consisting of gears, shafts, and shifting mechanisms that work together to transmit power from the engine to the drive wheel of the machine.

Twin sump: Engine design that separates the oil delivery between the crankshaft piston and valve train from the clutch and transmission.

U

UBS bolts: Bolts with a built in design to resist loosening.

Under-drive gear ratio: A gear ratio that is greater than 1:1.

V

Vacuum gauge: A tool for measuring pressures below atmospheric pressure (vacuum).

Valence electrons: The electrons contained in the outermost electron shell of an atom. Also known as free electrons.

Valve clearance: A measurement between a valve and the valve opening device.

Valve overlap: The time that both intake and exhaust valves are open simultaneously measured in crankshaft degrees.

Valves: Parts that control the air-and-fuel mixture that is drawn into the cylinder and the exhaust gases that are expelled.

Valve seats: Stationary sealing surface that is located in the cylinder head.

Vaporized liquid: A liquid that is converted to a gaseous state through a heating process.

Variable-ratio clutch: A transmission system that provides a variable drive ratio between the engine and the rear wheel.

Vent hoses: Hoses used on fuel tanks and carburetors to permit atmospheric air pressure to enter into certain important areas within the fuel system.

Venturi principle: A principle that states: A gas or liquid that flows through a narrowed-down section (venturi) of a passage will increase in speed and decrease in pressure compared to the speed and pressure in wider sections of the passageway.

Vernier caliper: A commonly used tool that can measure inside and outside diameters as well as depth.

Viscosity: The measurement of a fluid's resistance to flow.

Viscosity index: The number used to indicate the consistency of the oil with changes of temperature. An oil labeled 10W30 is a 10-weight oil at 0 degrees Fahrenheit, but has the viscosity of a 30-weight oil at 210 degrees Fahrenheit.

Volatile: To evaporate easily.

Volt: A measurement of electromotive force.

Voltage: *See* Electromotive force.

Voltage drop: The consumption of the available voltage from the battery as it crosses some form of resistance such as a light bulb or a switch.

Voltage regulator: A device used to maintain constant voltage levels on the electric system.

Voltmeter: A tool used to measure voltage.

W

Water (H_2O): The result of complete combustion.

Water jackets: A series of passageways surrounding the cylinder and combustion chamber in a liquid-cooled engine design.

Water pump: A pump used to circulate liquid in a cooling system.

Watt: The unit of electric power: $W = E \times I$ (Wattage = Voltage × Current).

Well nut: Rubber fasteners with a brass-threaded sleeve installed in the center. When tightened, the insert causes the rubber to expand and hold it in place.

Wet sump: An engine design that stores its oil in the bottom of the engine's crankcase.

Wheelbase: A measurement of a machine from the center lines of the front and rear axles.

Wire gauge: Wire diameter. Usually specified by an AWG (American Wire Gauge) number. The smaller the number, the larger the wire diameter.

Wiring diagram: Similar to a schematic, but less detail. A wiring diagram usually shows components in block form rather than illustrating their internal circuitry.

Wrenches: Used to tighten or loosen nut-and-bolt-type fasteners.

Wrist pin: A cylinder-shaped piston component, used to link the connecting rod to the piston.

Wrist pin boss: Point where the piston attaches to the small end of the connecting rod.

Y

Yield point: The maximum tensile stress that may be impressed upon a bolt without straining it beyond the elastic limit.

Z

Zener diode: Similar to a standard diode, but allows current flow in the reverse direction when the breakdown voltage is reached.

Index

A

Absorbed Glass Mat (AGM), 352
AC. *See* alternating current (AC)
accelerator pump system, 177
AC/DC current, 323–324
AC generators, 332–333, 375–376
acorn nuts, 77
active combustion, 114
active energy, 112
additives, 144–145
adjustable wrenches, 45
advertising specialists, 21
after top dead center (ATDC), 373, 374
air chambers, 412–413
air-cooled engines, 102–103, 159, 448
air cooling, 159
air cutoff valve system, 177–178
air filters, 174–175, 460–461
air mixture screw, 182
air ratchets, 58
air-to-fuel ratios, 169
air tools, 58
aligning punches, 53
Allen wrenches, 47–48
all-terrain vehicles (ATVs), 2, 8, 14–15
alternating current (AC), 32, 323–324, 327, 340
alternator rotor, 286
alternators, 331–332, 340, 346–349
  components, 346–347
  inspecting and testing, 363–366
  permanent-magnet, 347–349
aluminum cylinders, 121
American National Standards Institute (ANSI), 74
American Petroleum Institute (API), 145
ammeters, 326, 328
ampere-hours, 361
amperes (A), 320, 324, 340
amp hour, 340
analog electrical meters, 324–325
angled screwdrivers, 49
antifreeze, 162
antilock braking systems (ABS), 427–428
armature, 340
atomization, 170
atomized liquid, 111
atoms, 315–316
attire, 38
automatic shift transmissions, 221–222
automotive-style automatic transmissions, 222
aviation snips, 51
axel run-out, 433
axial loads, 148
axial tension, 80, 82
axles, 433

B

Babbitt bearing, 297–298
backbone frames, 397
ball bearings, 147–148, 296–297
ball peen hammers, 52
base carburetors, 186
base circle, 118
basic hand tools, 42–56
batteries, 30, 350–353, 375–376
  conventional, 352, 455–456
  definition of, 340
  electricity and, 316–317
  inspecting, 360–362, 455–457
  lead acid, 350–353
  maintenance-free, 352–353, 456
  testing, 456–457
battery acid, 455–456
battery-and-points ignition systems, 385
battery chargers, 361, 363
battery conductive analyzer, 361, 363, 364, 456–457
battery load testers, 456–457
beam-type torque wrench, 62, 83
bearings, 97, 100, 147–150
  ball, 147–148
  inspecting, 248–249, 296–298
  needle, 148–149
  plain, 149
  replacing, 248–249
  roller, 148
  steering-head, 462–463
  tapered roller, 148
  wheel, 433–434
before top dead center (BTDC), 373–374
belt-driven final drive systems, 223–224, 465–466

belt-driven primary drives, 205
belted bias ply tires, 436
bench grinder, 57–58
bench testing, 256, 306, 340
bench vise, 54
bevel gears, 202
bias ply tires, 435–436
bleeding, 153
bleed-type needle jets, 183
block diagrams, 337, 339, 355
BMW, 7
bolt diameter, 72
bolt grades, 74
bolt head markings, 74
bolt length, 72
bolts, 72–73
  stretched, 79–80
  types of, 74–75
boost port, 132
bore, 94
boring cylinders, 232–233
bottom-dead center (BDC), 89, 113
bottom end. *See* lower-end assemblies
box-end wrenches, 44–45
Boyle's law, 111
brake caliper, 423–424, 429
brake fluid, 422–423
brake horsepower (bhp), 92–93
brake levers, 421–422
brake-light circuits, 367–368
brake pads, 424–426
brake rotors, 426
braking systems
  antilock braking systems (ABS), 427–428
  hydraulic disk brakes, 421–427
  integrated braking systems (IBS), 428–430
  introduction to, 419
  maintenance, 463
  mechanical drum brakes, 419–420
  precautions, 426–427
  servicing, 427
  troubleshooting, 493
  types of, 419
breaker bars, 47
breaker points, 380–381
bridged-gap spark plugs, 453–454
British motorcycles, 6–7
brushless-type excited field, 348–349
brush-type excited field, 348
BSA, 6
bushings, 149, 305
butterfly throttle valve, 184–186
bypass ports, 182

C

cable maintenance, 461–462
California Air Resources Board (CARB), 468
caliper assembly, 423–424
cam ground, 121
camshaft drives, 119
camshaft drive systems, 284–285
camshaft drive tensioner, 119
camshaft lift, 118
camshafts, 118–119, 284
  crankshaft rotation and, 286–287
  installing, 285–287
capacitor discharge ignition (CDI), 386–387
capacitors (condensers), 317, 340, 380–381
cap nuts, 77
carbon buildup, 272
carbon dioxide (CO_2), 27, 114, 467
carbon monoxide (CO), 32–33, 114, 469
carbon-monoxide detectors, 29
carburetors, 169–171
  disassembly of multiple, 186
  multiple, 186–187
  problems with, 480
  types of, 180–186
carburetor slides, 182–183
carburetor synchronization, 186–187, 459
carburetor systems, 176–180
career opportunities, 15–21
cartridge fork design, 406
cast-iron cylinders, 120
castle-headed nuts, 76–77
cast wheels, 432–433
catalysts, 112
catalytic converters, 469
C-clamps, 54
center punches, 53
centrifugal advance systems, 374, 458
centrifugal clutch, 208–209, 210
centrifugal oil filters, 158
chain-driven final drive systems, 222–223, 465
chain-driven primary drives, 205
chain noise, 494

chamfered, 128
charging systems, 346–366
 3-phase excited-field electromagnet, 357–358
 3-phase permanent-magnet, 357
 alternators, 346–349
 batteries, 350–353
 full-wave, 356–357
 half-wave, 355–356
 inspecting, 358–366
 introduction to, 346
 operation, 353–355
 rectifiers, 349–350
 troubleshooting, 486–487
 types of, 355–358
 voltage regulators, 350
chassis
 maintenance, 461–466
 troubleshooting, 491–492
chisels, 53–54
choke plate cold start systems, 178
chopper style motorcycles, 400–401
circuit breakers, 340
circuits, 317–319, 340
 brake-light, 367–368
 closed, 317
 DC electrical, 366–368, 489
 electron flow in, 318–319
 hazard relay, 367
 headlight, 367
 horn, 368
 neutral-light, 368
 open, 317–318
 turn signal, 367
 types of, 319
 unwanted conditions, 319
clamping force, 80
clamps, 54
clamp-type mount, 176
Class A fires, 25
Class B fires, 25–26
Class C fires, 26–27
Class D fires, 27
cleaning function, of lubricants, 144
clearance ramps, 118
clearances, 62, 271
click-type torque wrench, 62, 83–84
clunking noises, 221, 293
clutch bearing, 207
clutch center, 207
clutch drag, 212–213
clutches/clutch systems, 100–101, 205–214
 adjustment, 451
 centrifugal clutch, 208–209, 210
 common problems, 212–213
 four-stroke engines, 164
 inspecting, 249, 299–300
 manual clutch, 206–208
 service and adjustment, 213–214
 sprag clutch, 212
 starter, 391
 troubleshooting, 479
 two-stroke engines, 163
 variable-ratio clutch, 209–212
clutch friction discs, 207
clutch noise, 495
clutch outer, 206–207
clutch plates, 207
clutch pressure plates, 207
clutch push rod, 207
clutch release mechanisms, 207
clutch slippage, 212
coils, 340
cold start systems, 178–179
combination (CT) bolts, 75
combination pliers, 50
combination wrenches, 45
combined braking systems (CBS), 428
combustible wastes, 34, 35
combustion, 112
 methods of internal, 113–114
 results of, 114
combustion chamber, 89, 112
combustion chamber volume (CCV), 95
combustion engines
 See also internal-combustion engines
 types of, 112
combustion lag, 113–114
compression check, 238, 267
compression damping, 411–412
compression event, 133
compression gauges, 65
compression ratio, 95–96
compression rings, 122
compression stage, 114
compression stroke, 126, 412
compression tests, 448–449

Computerized Fuel Injection (CFI), 189
condensor, 340, 380–381
conductance, 361
conductance batter tester, 457
conductors, 340
cone-type lock washers, 78
connecting rod bearings, 245, 292
connecting rods, 123, 124, 130, 239, 301
constant failures, 475
constant-mesh transmissions, 216
constants, 94
constant velocity, 181
constant velocity carburetors, 184–186
contact dermatitis, 29–30
contact lens wearers, 36
continuity, 340
conventional batteries, 352, 455–456
conventional system, 71
conventional theory of electricity, 314
coolants, 162
cooling, 144
cooling systems, 159–163, 447–448
counter-balancer, 99
crankcase-emission controls, 467–468
crankcase reed valves, 136
crankcases, 97, 100, 123–124, 130
  inspecting, 246, 295
  separating, 294–295
crankshaft bearings, 245, 292
crankshafts, 96–97, 100, 122–123, 129–130
  inspecting, 253–255, 300–302
  multi-piece, 301–303
  rotation, 286–287
  single-piece, 300–301
crescent wrenches, 45
crimper pliers, 51–52, 359
crosshatching, 121
crown, 121
cubic centimeters (cc), 94
cubic inches (ci), 94
cupping, 463–464
current, 320, 322–323
  alternating (AC), 323–324, 327
  direct (DC), 323, 327–328
  measuring, 327–328
custom cruisers, 11
customer service representatives, 21
cutting pliers, 51, 359
cutting tools, 54–56, 281
CV carburetor slide, 184–186
cylinder bore gauges, 270
cylinder heads, 98–99, 102, 115–118, 127
  inspection of, 230–231, 272–273
cylinder inspection, 231–233
cylinder reed valves, 136
cylinders, 89, 102, 120–121, 127–128
  boring, 232–233
  installing, 285
  measuring, 269–271

D

Daimler, Gottlieb, 3
dampers, 408–414
  function of, 408–409
  operation, 411–413
  pressurized, 413
  types of, 410–411
damper springs, 413–414
DC. *See* direct current (DC)
DC electrical circuits, 366–368, 489
DC motor operating principle, 389
dealerships. *See* motorcycle dealerships
de Carbon shock, 410–411
DeDion-Buton, 3–4
deep sockets, 46
delay valves, 429
Department of Transportation (DOT), 422
detonation, 168, 265, 266
diagnostics
  See also troubleshooting
  four-stroke engines, 258
  two-stroke engines, 227, 228
diagrams
  block, 337, 339, 355
  wiring, 336–339, 341, 342
dial indicators, 61
dial-type torque wrench, 84
diamond frame, 398
dielectric, 340
dies, 55
digital electrical meters, 325
digitally controlled transistorized ignition systems, 387–388
digital/multimeter (DMM), 326–330
digital volt/ohm/amp meter (DVOA), 326–330
diodes, 334–335, 340, 356

direct current (DC), 32, 323, 327–328
direct-drive gear ratio, 203
direct drive system, 3
direct-drive transmissions, 217
direct fuel injection, 188
disc brakes, 419, 421–427
distilled water, 352, 455–456
district parts managers, 21
district sales managers, 21
district service managers, 21
double-cradle frame, 397
dress attire, 38
drill bits, 56
drill press, 57
drills, 56–57
drive chain lubricants, 147
drive chain noise, 495
driven pulley operations, 211–212
drive pulley operation, 209–211
drive train troubleshooting, 479
DR-type (deep-recess) bolts, 74
drum brakes, 419–421
  hydraulic, 420–421
  mechanical, 419–420
dry lubricants, 147
dry-sump lubrication, 155–156
dual-purpose motorcycles, 13
dual-range transmissions, 217
dual rate springs, 414
Ducati, 7
duration, 118, 128
Dykes piston rings, 129
dynamic, 340
dynamometer (dyno), 92

E

earplugs, 37
eczema, 29
elastic range, 81
electrical meters, 324–330
  ammeters, 326
  analog, 324–325
  digital, 325
  multi-meters, 326–330, 360
  ohmmeters, 326
  precautions, 330
  uses of, 325–327
  voltmeters, 325
electrical potential, 321
electrical power, 322
electrical safety, 30–32, 313–314
electrical schematics, 336–339, 341, 353, 354
electrical symbols, 336–339
electrical systems
  charging systems, 346–366
  DC electrical circuits, 366–368
  ignition systems, 372–388
  introduction to, 312–313
  starter systems, 388–391
  troubleshooting, 486–490
electrical terms, 340–342
electrical theories, 314
electric fuel pumps, 173
electricity, 340
  basic principles of, 314–320
  introduction to, 312–313
  properties of, 319–320
  units of, 320–324
electric starters, 97, 222, 333–334, 388–391, 490
electrode-eroded spark plugs, 454
electrodes, 316–317, 379–380
electrolysis, 340
electrolytes, 340, 351, 360
electromagnetism, 331–334
electromagnet excited-field inspection, 365
electromagnets, 331, 340
electromotive force (EMF), 331, 340
electronic components, 334–336
Electronic Control Module (ECM), 193
electronic devices, 334–336
Electronic Fuel Injection (EFI), 189–197
electronic fuel injection systems, 483–485
electronic ignition systems, 458
electronic pointless ignition systems, 386–388
electronic trigger switches, 381–382
electrons, 315–316, 318–319, 340, 341
electron theory, 314
emergency exits, 28
emission-control systems, 467–470
enrichment cold start systems, 178–179
energy, 112
energy-transfer ignition systems, 385
engine bearings, inspecting, 296–298
engine break-in, 242, 308
engine components
  four-stroke engines, 115–124
  two-stroke engines, 127–134
engine configurations, 104–107

Engine Control Module (ECM), 64
engine cooling, 102–104, 159–163
engine cycle, 115
engine design
  four-stroke engines, 96–99
  two-stroke engines, 99–102
engine displacement, 94–95
engine load, 374–375
engine maintenance, 443–455
  compression tests, 448–449
  cooling system inspection, 447–448
  leak-down test, 448, 449
  oil changes, 443–447
  power-valve inspection, 450
  valve adjustments, 449–450
engine oil
  additives, 144–145
  classification, 145–147
  for two-stroke engines, 150–154
  types of, 144–145
engine operation
  basic, 89–91
  four-stroke, 90
  internal-combustion engines, 112–115
  stages of, 131–132
  two-stroke, 91, 131–134
engine overheating, 234
engine ports, 131–133
engine problems, troubleshooting, 483–485
engine ratings, 91–96
engine repairs
  bench testing, 256, 306
  general tips for, 227
  lower-end assemblies, 245–256, 289–309
  procedures, 227, 245, 258–259, 290–291
  top-end assemblies, 227–243, 259–288
engine RPM, 374
engines
  See also four-stroke engines; internal-combustion engines; two-stroke engines
  reciprocating, 89–90
  troubleshooting, 476–479
engine seals
  inspecting, 247–248, 295–296
  leaking, 291–292
  replacing, 247–248
engine seizure, 142
engine sump, 154
entry-level motorcycle technicians, 16
Environmental Protection Agency (EPA), 114, 187, 467, 470
equipment operation, safety issues, 33
Evaporation Emissions Canister (EVAP), 172
evaporative-emission carburetor air-vent control valve (EVAP CAV control valve), 468
evaporative-emission controls, 468
excited-field electromagnet alternators, 347–349
exhaust-emission controls, 469
exhaust event, 134
exhaust gas analyzers, 65–66
exhaust gas safety, 32–33
exhaust port power valves, 130
exhaust ports, 128, 133
exhaust stage, 115
exhaust stroke, 126–127
exhaust systems, 131
exhaust valves, 125
expansion chambers, 131, 134
external engine noise, 494
eye-wash stations, 36

F

failures, 475
fare-nut wrench, 44
fasteners, 72–87
  anatomy, 72–73
  bolts, 72–75
  inspection and cleaning of, 79–80
  nuts, 72–73, 76–77
  preload, 81
  repairing and replacing, 84–87
  stresses on threaded, 80–81
  thread-locking agents and sealers, 78–79
  thread types, 73–74
  tightening and torque, 82–84
  tips for working with, 81–82
  washers, 77–78
feeler gauges, 62–63, 299
fiber oil filters, 158
field coils, 340, 365
files, 55
final drive ratio, 204
final drive systems, 222–224, 464–466
fire classes, 25–27
fire evacuation procedures, 29
fire extinguishers, 27–29

fire safety, 24–29
fire triangle, 25
first gear, 218
fixed gears, 215
fixed Venturi carburetors, 181
flammable, 168
flammable liquids, storage of, 33–34
flange-type mount, 175
flat chisels, 53
flat tip screwdrivers, 49
flat washers, 78
flex handles, 47
float chamber, 176–177
float systems, 176–177
flux lines, 341
foam air filters, 174, 175, 460, 461
foot-pounds (ft-lb), 91
forced-air cooling systems, 448
forced-draft cooling, 159
forced-draft cooling systems, 103
fork bushings, 405
fork oil seals, 406
fort springs, 406
four-stroke cylinders, 120–121
four-stroke engines
  camshafts, 284
  clutch systems, 299–300
  compared with two-stroke engines, 137–139
  components, 115–124
  crankcases, 294–295
  crankshafts, 300–302
  cylinder heads, 272–273
  cylinders, 269–271
  design, 96–99
  diagnostics, 258
  disassembling, 259–261, 294–295
  engine bearings, 296–298
  engine seals, 295–296
  inspecting, 261–284, 295–306
  lower-end assemblies, 289–309
  lubrication, 154–159, 164
  oil changes, 444–447
  oil pumps, 298–299
  operation, 90
  pistons and piston rings, 261–269, 271–272
  reassembling, 285–287, 307–308
  removal, 261
  theory of operation, 124–127
  top-end inspection, 257–288
  transmissions, 302–306
  valve adjustments, 449–450
  valve guides, 277–280
  valve lapping, 283–284
  valves, 273–277
  valve seats, 280–283
fourth gear, 218–219
four-wheelers, 14–15
fractional distillation, 144
fractional distillation process, 168
frames, 395–401
  design, 395–400
  modifications, 400–401
franchised dealerships, 16
free electrons, 341
freewheeling gears, 216
friction, 143
friction-reducing devices, 147–150
front suspension, 402–406
fuel, 168–169
fuel feed systems, 176
fuel filters, 174, 189–190, 191
fuel-fouled spark plugs, 453
fuel induction, 125
fuel injection, 64, 187–197
  direct, 188
  indirect, 188–189
  sequential, 192
  system components, 189–197
fuel injection diagnostic testers, 63–64
fuel injection systems, 483–485
fuel injector failures, 192–193
fuel injectors, 190–193
fuel lines, 173, 190
fuel mixture screw, 182
fuel pressure regulators, 190
fuel pumps, 173–174, 189
fuel systems
  carburetors, 169–171
  carburetors, 176–187
  fuel, 168–169
  fuel delivery system, 171–176
  fuel injection, 187–197
  principles of atomization, 170
  troubleshooting, 480–485
  Venturi Principle, 170–171
fuel tanks, 171–172

fuel tank vents, 483
fuel valves, 172–173
full-wave charging systems, 356–357
fuse, 341

G

garbage disposal, 34
gasket material buildup, 272
gaskets, 97, 101
gasoline, 168–169
gauze air filters, 174, 460–461
gear-driven primary drives, 205
gear ratios, 202–204
gear reduction systems, 204–205
gears, 200–202
  difficulties shifting, 292–293
  inspecting, 251
  jumping out of, 293
  shifting transmission, 217–219
  transmission, 215–216, 251, 302–304
gear-type oil pumps, 156–157
general managers, 20
generator operation, 332–333
generators
  AC, 332–333, 375–376
  magnetic pulse, 381–382
gloves, 38
goggles, 36
grease, 147
ground, 341
grounded circuits, 319

H

hacksaws, 54–55
half-wave charging systems, 355–356
Hall-effect sensor, 382
halon, 27
hammers, 52–54
handling problems, 491–492
hand protection, 38
hand tools, 42–56
  chisels, 53–54
  clamps, 54
  cutting tools, 54–56
  for electrical work, 359–360
  hammers, 52–54
  pliers, 50–52
  punches, 53
  screwdrivers, 48–50
  vises, 54
  wrenches, 43–50
hand tool safety, 38–39
Harley-Davidson Motorcycles, 4–6
Hazard Communication Standard, 30
hazardous chemicals, 29–30
hazardous waste, disposal, 34
hazard relay circuits, 367
headlight circuits, 367
headsets, 37
head size, 72
hearing protection, 37
heavy objects, 34–35
Hedstrom, Carl, 4
helical gears, 200
Heli-Coil, 86
Hendee, Gerge, 4
hex wrenches, 47–48
high-speed movement, 39
high-speed phase, of carburetion, 180
high-tension magneto ignition systems, 383–384
Honda C100 Cub, 7–8
Honda CBR1000RR, 259
Honda CR125R, 229
Honda CRF230F, 259
horizontal crankcases, 294
horizontally opposed multi-cylinder engines, 106–107
horizontally opposed twin-cylinder engines, 105, 106
horn circuits, 368
horsepower (hp), 92–93
hose clamp pliers, 51
hot-rod cruisers, 11–12
housekeeping practices, 33–34
hydraulic brake lines, 422
hydraulic damping units, 406
hydraulic drum brakes, 420–421
hydraulic valve adjusters, 450
hydrocarbons (HC), 114, 168, 467
hydrodynamic lubrication, 149
hydrometer, 163, 360, 447, 455, 469
hydrostatic transmissions, 221–222

I

idle adjustment, 458
idle air bleed passage, 181

idle circuits, 181
idle fuel jet, 182
idle outlet port, 181–182
idler gears, 202
ignition, 341
ignition coils, 376–377
ignition control module (ICM), 488
ignition switches, 376
ignition systems, 372–388
  components, 375–382
  inspection and adjustment, 458
  introduction to, 372
  troubleshooting, 488–489
  types of, 382–388
  workings of, 372–375
ignition timing, 373–375
impact screwdrivers, 49–50
impact wrenches, 58
improper-service failures, 475
incipient fires, 28
Indian Motorcycles, 4
indirect-drive transmissions, 216–219
indirect fuel injection, 188–189
induction, 135
induction systems, two-stroke engines, 135–137
injector discharge duration, 190
in-line multi-cylinder engines, 105–106
inner-fork tubes, 405
insulators, 341
intake event, 133
intake manifolds, 175–176
intake ports, 128
intake stage, 114
intake stroke, 125–126
intake valves, 125
integrated braking systems (IBS), 428–430
intermittent failures, 475
internal combustion, 113–114
internal-combustion engines, 109–139
  See also four-stroke engines; two-stroke engines
  construction of, 112–113
  four-stroke engine components, 115–124
  operation, 112–115
  terminology, 110–112
  theory of operation, 124–127, 131–134
  two-stroke engine components, 127–134
  two-stroke vs. four-stroke engines, 137–139
internal-oil cooling, 159
internal ribs, 396
internal slapping noise, 494
International Standards Organization (ISO), 74
International System of Units (SI), 71
inverted front suspension, 404
iridium, 379

J

Japanese motorcycles, 7–9, 10–11
jet needles, 179–180, 183

K

Kawasaki, 8–9
keystone piston rings, 129
kick starters, 97
kick-starting systems, 222
kilohm (KΩ), 324
kilovolt (KV), 324
knocking noise, 494

L

labyrinth seal, 130
lapping sticks, 283–284
law of action and reaction, 112
law of inertia, 112
laws of motion, 112
lead acid batteries, 350–353
leading link-type suspension, 402
leak-down tests, 267, 448, 449
lifter rod, 207
light bulbs, 489
lights
  test, 65
  timing, 64–65, 458
  warning, 368
lines of force, 341
line wrench, 44
link type suspension, 402
liquid-cooled cylinders, 120
liquid-cooled engines, 103–104
liquid cooling, 159–163
liquid-cooling system testing, 163
load, 341, 374–375
load testers, 361, 363
locking pliers, 50
lot attendants, 18

lower-end assemblies
  bench testing, 256
  common failures, 244–245, 291–293
  disassembling, 294–295
  of four-stroke engine, 96–98, 289–309
  inspecting, 295–306
  reassembling, 307–308
  of two-stroke engines, 99–101, 244–256
low-tension magneto ignition systems, 384–385
lubricants
  dry, 147
  engine oil, 144–147
  specialty, 147
lubrication
  dry-sump, 155–156
  four-stroke engines, 154–159, 164
  hydrodynamic, 149
  introduction to, 142–143
  purposes of, 143–144
  system maintenance, 163–165
  transmission, 153–154
  twin-sump, 156
  two-stroke engines, 150–154, 163
  wet-sump, 156
lung protection, 36–37

M

magnetic coils, 331
magnetic forces, 331–332
magnetic induction, 331–332, 341
magnetic pulse generators, 381–382
magnetism, 330–334, 341
magnetite, 331
magneto ignition systems, 383–385
magnets, types of, 331
main jet, 180
main jet circuit, 183–184
maintenance
  air filters, 460–461
  batteries, 455–457
  brake, 463
  cable, 461–462
  carburetor synchronization, 459
  chassis, 461–466
  clutch adjustment, 451
  compression tests, 448–449
  cooling systems, 447–448
  emission controls and, 467–470
  engine, 443–455
  final drives, 464–466
  idle adjustment, 458
  ignition systems, 458
  intervals, 440–443
  introduction to, 440
  leak-down test, 448, 449
  oil changes, 443–447
  power-valve inspection, 450
  process order, 443
  spark plugs, 451–455
  steering-head bearings, 462–463
  tires, 463–464
  valve adjustments, 449–450
  wheels, 463–464
maintenance-free batteries, 352–353, 361, 456
Malfunction Indicator Lamp (MIL), 193, 197
mallets, 52
manual clutch, 206–208
manual clutch release mechanisms, 207
manual fuel valves, 172
manuals, 67–68
marketing specialists, 21
master cylinder, 422
material handling, safety issues, 34–35
Material Safety Data Sheets (MSDS), 30, 31
matter, 110–111
measurement systems, 71
measuring tools, 59–63
mechanical seals, 447
mechanical slide, 181
mechanical slide carburetors, 181–184, 186
mechanical work, 91–92
megavolt (MV), 324
megohm (MΩ), 324
metatarsal guards, 38
metric system, 43, 71
micrometers, 60–61, 301
mid-range circuits, 183
mid-range phase, of carburation, 179–180
Millet, Alex, 3
milliampere (mA), 324
millivolt (mV), 324
modified frames, 400–401
molybdenum disulfide grease, 305
momentum, 111–112
motor-cross, 14
motorcycle assembler, 18

motorcycle dealerships
  franchised, 16
  opportunities, 16–20
motorcycle industry, opportunities in, 15–21
motorcycle repair instructors, 20
motorcycles
  British, 6–7
  defined, 2
  dual-purpose, 13
  history of, 2–9
  invention of, 2–6
  Japanese, 7–9, 10–11
  off-road, 13–14
  removing from storage, 467
  storage of, 466–467
  street, 10–13
  types of, 10–14
Motorcycle Safety Foundation (MSF), 40
motorcycle technicians, 16, 19
motors
  electromagnetism in, 333–334
  starter, 388–391
motor scooters, 13
multi-cylinder crankshafts, 123, 124, 129–130
multi-cylinder engines, 104–106
multi-meters, 63, 326–330, 360
multi-piece crankshafts, 123, 253–255, 301–302, 303
multi-plate manual clutch, 206–207, 208
multiple carburetors, 186–187
multi-viscosity oil, 146–147
mutual induction, 332
MV Agusta, 7

N

National Electrical Code (NEC), 24
National Fire Protection Association (NFPA), 24
National Safety Council (NSC), 38
needle bearings, 148–149
needle jets, 183
needle nose pliers, 51
neutral gear, 217–218
neutral-light circuits, 368
neutrons, 315
noise-emission controls, 469–470
noises
  transmission, 220–221, 293, 495
  troubleshooting, 494–495
no-load test, 341
nonspoke wheels, 432–433
normally closed solenoid, 333
normally open solenoid, 333
NPN transistors, 336, 341
nuts, 72–73, 76–77

O

Occupational Safety and Health Administration (OSHA), 24, 30
octane rating, 168, 265
off-road motorcycles, 13–14
offset screwdrivers, 49
offset spur gears, 200
ohmmeters, 326, 362, 364
Ohm's law, 321–322
Ohms (Ω), 321–322, 324
oil, 143–145
oil changes, 443–447
oil classification, 145–147
oil control rings, 122
oil filter bypass valve, 158
oil filters, 158, 444–446
oil-fouled spark plugs, 452–453
oil injection systems, 152–153
oil level, checking, 446–447
oil passages, 158–159
oil pressure relief valves, 157–158
oil pumps, 156–157, 298–299
oil residue, 264–265
On-Board Diagnostics (OBD-II), 64
one-piece crankshafts, 122–123
open circuits, 319
open-draft cooling, 159
open-draft cooling systems, 103
open-ended wrenches, 43–44
open stator wire, 364
organic brake pads, 424–425
OSHA. *See* Occupational Safety and Health Administration (OSHA)
outer-fork tubes, 405
overall gear ratio, 204
over-drive gear ratio, 203
overhead camshaft, 99
oxides of nitrogen (NOx), 114, 467, 469
oxygen, 169
oxygenated fuels, 169

P

paper air filters, 174, 460
paper oil filters, 158
parallel circuits, 319
parallel twin-cylinder engines, 105
parts department, 17
parts technician, 17
PASS (Pull, Aim, Squeeze, Sweep), 28
perimeter frame, 398–399
permanent-magnet alternators, 347–349
permanent magnets, 331
permeability, 341
personal protective equipment (PPE), 35–36
petroleum-based oil, 144
Phillips screwdriver tips, 49
pilot air screw, 179
pilot jet air passageway, 179
pin punches, 53
piston assembly, inspection of, 233–239
piston clearance, 271
piston crown, 265
piston port/crankcase reeds, 136
piston port induction, 135
piston ring end gap, 238
piston ring grooves, 236–237, 268–269
piston ring lands, 121
piston rings, 113, 121–122, 129
  inspecting, 237–238, 265–269
  installation, 271–272
  measuring, 237–238, 267–268
  removing, 261–263
piston ring side clearance, 236
pistons, 98, 101–102, 112, 121, 128–129, 233
  defined, 89
  inspecting, 263–265
  installing, 285
  measuring, 234–237, 269
  removing, 261–263
piston skirt, 121
piston-to-cylinder clearance, 238, 270–271
pivotless frame, 399, 400
plain bearings, 149, 297–298
plastic-faced hammers, 52
plastic range, 80, 81
plastigage, 298
plated-aluminum cylinders, 121
platinum electrodes, 379
pliers, 50–52, 359
plunger-type oil pumps, 157
PNP transistors, 336, 341
polarity, 341
poles, 341
poppet valves, 115–116
ports, 127–128, 131–133
post combustion, 114
potential energy, 112
power, 92
power band, 130
power event, 133
power impact wrenches, 58
power port systems, 130
power sources, 375–376
power stage, 115
power stroke, 126
power tools, 56–58
  air tools, 58
  bench grinder, 57–58
  drill press, 57
  drills, 56–57
power tool safety, 39
power-valve inspection, 450
power valves
  maintenance, 241–242
  two-stroke, 239–242
  workings of, 241
Pozidrive tips, 49
precision measuring tools, 59–63
preignition, 265, 266, 378
preload, 81
premixed fuel and oil, 151–152
press fit, 149
pressure differences, 111
pressure relief valves, oil, 157–158
pressure testing, 163
primary area, 131
primary drive ratio, 204
primary drives, 204–205
primary-type needle jets, 183
Prince tips, 49
problems, types of, 475
problem solving. *See* troubleshooting
Programmed Fuel Injection (PGM-FI), 189
progressive rate springs, 414, 415
proportional control valve (PCV), 430
protons, 315–316
Prussian blue, 282

pullers, 59
punches, 53
purge control valves, 468, 469
push fit, 149

Q

quality control specialists, 21

R

race team support technicians, 21
radial clearance, 255, 303
radial loads, 148
radial ply tires, 436
radiator caps, 162
radiator fans, 163
radiators, 161
rake, 400–401
ratios
  compression, 95–96
  defined, 95
reach, spark plug, 378
reamer, 279–280
rear-suspension systems, 407–408, 409
rebound damping, 412
rebuilt engines, starting, 242
rechargeable cordless drills, 57
reciprocating engines, 89–90
recoil starters, 97
rectification, 346
rectifiers, 328, 341, 346, 349–350, 362–363
Reed tips, 49
reed valve induction, 135–136
reed valves, 239
regulators, 341, 362–363
reluctance, 341
remote screwdrivers, 49
repair procedures, 227, 290–291
research and development engineers, 21
resistance, 320–323, 328–329
respirators, 37
respiratory-protection devices, 36–37
retaining-ring pliers, 51
revolutions per minute (rpm), 93–94
rib-joint pliers, 50
riding safety, 39–40
right-to-know laws, 30
ring gears, 202
ring grooves, 121
ring lands, 236
rising rate, 408
road racing, 12
rocker arms, 117
roller bearings, 148
Roper, Sylvester H., 2–3
rotary valve induction, 137
rotors, 341, 346–347, 364–365
rotor-type pumps, 157, 298
RPM (revolutions per minute), 374
rubber hammers, 52
runout, 255

S

SAE (Society of Automotive Engineers), 43
safety glasses, 36
safety issues, 23–40
  attire, 38
  electrical safety, 30–32, 313–314
  equipment operation, 33
  exhaust gas safety, 32–33
  fire safety, 24–29
  hazardous chemicals, 29–30
  hearing protection, 37
  housekeeping practices, 33–34
  lung protection, 36–37
  material handling, 34–35
  personal protective equipment (PPE), 35–36
  riding safety, 39–40
  right-to-know laws, 30
  safety attitude, 24
  safety glasses and goggles, 36
  tool safety, 38–39
sales department, 16–17
scheduled maintenance, 440
  See also maintenance
schematics, 336–339, 341, 353, 354
scientific terms, 110–112
scooters, 13
score marks, 233
scoring, 264
SCR. *See* silicon-controlled rectifier (SCR)
scraper rings, 122
screen oil filters, 158
screw-and-locknut valve arrangement, 449
screwdrivers, 48–50
screw extractors, 55–56, 85–86
scuff marks, 233, 264

sealers, 78–79
sealing function, of lubricants, 144
seals, 97, 100, 149–150
  inspecting, 247–248, 295–296
  leaking, 244, 291–292
  replacing, 247–248
secondary area, 131
secondary compression, 133
secondary master cylinders, 429–430
second gear, 218
sector gears, 202
self-diagnostics, of fuel systems, 193
self-locking nuts, 76
semiconductors, 334
sensors, EFI, 193
sequential fuel injection, 192
series circuits, 319
series/parallel circuits, 319
service department, 17–18
service information library, 67–68
service managers, 19–20
service manuals, 67–68
service technical training instructors, 21
service writers, 19
set-up technician, 18
shaft-driven final drive systems, 224, 466
shear, 80
shears, 51
shift drums, 251, 252, 305, 306
shift forks, 252, 253, 254, 304–305, 306, 307
shifting difficulties, 292–293
shifting mechanisms, 219–220
shim and bucket, 117
shim-and-buckle valve arrangement, 450
shock absorbers, 407–414
short circuits, 319, 364, 365
side clearance, 255
side cutaway, 182–183
silicon, 341
silicon-controlled rectifier (SCR), 335, 349
sine waves, 341
single-cradle frame, 396–397
single-cylinder engines, 104–105
single-piece crankshafts, 300–301
single-plate manual clutch, 207
sintered brake pads, 425
skin injuries, 29–30
sliding gears, 216
sliding t-handles, 47
slot tip screwdrivers, 49
slow-speed phase, of carburation, 179
smoke detectors, 28–29
snips, 51
Society of Automotive Engineers (SAE), 74, 145
socket wrenches, 45–47
soft-faced hammers, 52
solder, 341
soldering gun, 359
solenoids, 333, 391
solid-threaded inserts, 86–87
spark plugs, 377–380
  access, 451
  bridged-gap, 453–454
  cleaning, 454
  electrode-eroded, 454
  fuel-fouled, 453
  gaps, 454
  inspecting, 452
  installing, 454–455
  normal, 452–453
  oil-fouled, 452–453
  removal, 452
special tools, 59–66
specification manuals, 67
specific gravity, 455
specific-gravity test, 163
speed handles, 47
spigot-type mount, 175
spin balancing, 437
split-ring-type lock washers, 78
spoke wheels, 431–432
sport motorcycles, 12
sport-touring motorcycles, 12–13
sprag clutch, 212
spring adjusters, 414–415
spring rate, 414
springs, 408, 409, 413–414
spur gears, 200
squish area, 127
squish band, 127
stake-type lock nuts, 77
standard charge rating (STD), 361, 362
standard damping-rods, 406
standard piston rings, 129
standard street motorcycles, 10–11
standard wrench, 44

starter clutches, 391
starter solenoids, 391
starting mechanisms, 97–98, 101
starting punches, 53
starting systems, 222, 390
  electric, 388–391
  troubleshooting, 490
static, 341
static balancing, 437
stators, 341, 346–347, 365–366
steering-head bearings, 462–463
steering problems, 491–492
steering stems, 404–405
stellite, 277
stop switches, 382
storage, tool, 66–67
storage batteries, 30
storage procedures, 466–467
straightedge, 231
straight rate springs, 414
straight shank punches, 53
street motorcycles, 10–13
stretched bolts, 79–80
stroke, 89, 113
  compression, 126
  exhaust, 126–127
  intake, 125–126
  power, 126
stubby screwdrivers, 49
studs, 75, 76
sulfation, 351, 361
sulfur, 144
sump, 154
suspension systems, 401–408
  dampers, 408–414
  front, 402–406
  rear, 407–408, 409
  swing arms, 406–407
Suzuki, 8–9
swing arms, 406–407
switch devices, 380–382
switches, 341, 367
  stop, 382
  troubleshooting, 489
symbols, electrical, 336–339
symptoms, 227, 258
synthetic-based oils, 145

T

tapered roller bearings, 148
taps, 55
technical advisors, 21
technical illustrators, 20–21
technical writers, 20
telelever-type front suspension, 402–403
telescopic-type front suspension, 403–404
telltale hole, 160–161
tensile strength, 74
tension bolts, 75
terminology, electrical, 340–342
test instruments, 63–66, 360
  compression gauges, 65
  exhaust gas analyzers, 65–66
  fuel injection diagnostic testers, 63–64
  multi-meters, 63
  test lights, 65
  timing lights, 64–65
test lights, 65
thermostats, 161–162
thermo-switch, 341
third gear, 218
threaded fasteners. *See* fasteners
thread inserts, 86–87
thread-locking agents, 78–79
thread pitch, 72
thread types, 73–74
3-phase excited-field electromagnet charging system, 357–358
3-phase permanent-magnet charging system, 357
three-wheelers, 14, 15
throttle body, 193, 197
throttle cables, 461–462
throttle opening, 481–483
thurst face, 234–235
thyristor, 335, 342, 349
ticking noise, 494
tickler systems, 178
timing lights, 64–65, 458
timing marks, 286–287, 373–374
tire air pressure, 464
tire balancing, 436–437
tire construction, 435–436
tires, 434–437, 463–464, 492
tire sizes, 434–435
tongued lock plate washers, 78
toolboxes, 67

tools
  basic hand tools, 42–56
  for electrical work, 359–360
  power tools, 56–58
  precision measuring tools, 59–63
  purchasing, 66
  special tools, 59–66
  storing, 66–67
  test instruments, 63–66
tool safety, 38–39
top-dead center (TDC), 89, 113, 286, 287, 372
top-end assemblies
  cleaning components, 230
  disassembling, 229–230, 259–261
  of four-stroke engine, 98–99, 257–288
  inspecting, 261–284
  reassembling, 285–287
  of two-stroke engines, 101–102, 227–243
torque, 44, 82–84, 93–94
torque wrenches, 61–62, 83–84
torsion, 80
Torx bolts, 74–75
Torx tips, 49
Torx wrenches, 48
touring motorcycles, 12
trail, 400–401
trailing link suspension, 402
transfer event, 134
transfer ports, 128, 133
transistorized ignition systems, 387
transistors, 335–336
transmission gear ratios, 204
transmission gears, 302–304
transmission lubrication, two-stroke engines, 153–154
transmissions, 97, 100–101, 215–222
  assembling, 252–253, 305–306
  automatic shift, 221–222
  clutch systems, 479
  components, 251
  constant-mesh, 216
  direct-drive, 217
  disassembling, 249–250
  dual-range, 217
  electric starters, 490
  four-stroke engines, 164
  gears, 215–216
  gear shifting, 217
  indirect-drive, 216–219
  indirect-drive power flow, 217–219
  inspecting, 249–253, 302–305
  noises, 220–221, 293, 495
  power flow and symptoms, 250
  problems with, 245, 292–293
  problem symptoms, 220–221
  removal, 302
  troubleshooting, 479
  two-stroke engines, 163
transmission shafts, 304
trellis frame, 399–400
triggering switch devices, 380–382
Triumph, 6
Trochoid-type pumps, 298
troubleshooting
  abnormal noises, 494–495
  brakes, 493
  charging systems, 486–487
  chassis, 491–492
  DC circuits, 489
  drive trains, 479
  electrical systems, 486–490
  engine problems, 476–479, 483–485
  fuel injection systems, 483–485
  fuel systems, 480–485
  guides, 476
  ignition systems, 488–489
  introduction to, 474
  process, 475–476
  systematic approaches to, 474–476
  transmissions, 479
  wheels and tires, 492
truing, 432
tubeless tires, 434
tube-type tires, 434
turn signal circuits, 367
twin-cylinder engines, 105
twin-sump lubrication, 156
two-stroke engines
  compared with four-stroke engines, 137–139
  components, 127–134
  design, 99–102
  disassembling top end, 229–230
  engine areas, 131
  engine events, 133–134
  induction systems, 135–137
  lower-end inspection, 244–256

lubrication, 150–154, 163
oil changes, 443–444
operation, 91
power-valve inspection, 450
power valves, 239–242
removal and disassembly, 246
theory of operation, 131–134
top-end inspection, 227–243

U

UBS (uniform bearing stress) bolts, 74
underbone frames, 397–398
Universal Japanese Motorcycle (UJM), 10–11
unloaded. *See* no-load test
upside-down front suspensions, 404

V

vacuum fuel pumps, 173
vacuum gauges, 459
vacuum-operated fuel valves, 172–173
vacuum valve wit electric assist, 173
valence electrons, 342
valve adjustments, 449–450
valve clearance, 117–118, 287, 449
valve-closing devices, 116
valve guides, 116, 277–280
valve lapping, 283–284
valve margin, 276
valve-opening devices, 116
valve operation, 125
valve overlap, 119
valves, 98–99
in four-stroke engines, 273–277
inspecting, 275–277
installing, 285
refacing, 277
valve seats, 116, 280–283
valve springs, 274, 276, 277
valve stems, 276
valve stem seals, 116
valve train, 119–120
vaporized liquid, 111
variable-ratio clutch, 209–212
variable Venturi carburetors, 181
velocity, constant, 181
vent hoses, 173–174
Venturi Principle, 170–171
Vernier caliper, 59–60, 299
vertical crankcases, 294
V-four engines, 106
viscosity, 111, 144
viscosity index, 147
vise grip pliers, 50
vises, 54
volatile, 168
voltage drop, 329
voltage regulators, 346, 350, 356, 357
voltage (V), 320–323
voltmeters, 325
volt/ohmmeters (VOMs), 63, 326–330
volt (V), 320–321, 324
V-twin engines, 105

W

warning lights, 368
washers, 77–78
water (H_2O), 114, 467
water jacket, 103, 120
water pumps, 160–161
watt (W), 342
weight classifications, engine oil, 146
well nuts, 77
wet-sump lubrication, 156
wheel bearings, 433–434
wheel-bearing seals, 433–434
wheels, 430–434, 463–464, 492
wire gauge, 342
wire-mesh oil filters, 158
wire stripper pliers, 51–52, 359
wiring diagrams, 336–339, 341, 342
work, 91–92
work area, cleanliness, 33
worm gears, 202
wrenches, 43–50
adjustable, 45
Allen, 47–48
box-end, 44–45
combination, 45
crescent, 45
flare-nut, 44
line, 44
open-ended, 43–44
socket, 45–47
standard, 44
torque, 83–84
Torx, 48

wrist pin, 239, 271
wrist pin boss, 121

X

X-axis of measurement, 270

Y

Yamaha, 8
Y-axis of measurement, 270
yield point, 80

Z

Zener diodes, 335, 342, 356
zinc, 144

Notes

Notes

Notes

Notes

Notes

Notes